Lecture Notes in Computer Science 2799

Edited by G. Goos, J. Hartmanis, and J. van Leeuwen

Springer
Berlin
Heidelberg
New York
Hong Kong
London
Milan
Paris
Tokyo

Jorge Juan Chico   Enrico Macii (Eds.)

Integrated Circuit and System Design

Power and Timing Modeling, Optimization and Simulation

13th International Workshop, PATMOS 2003
Turin, Italy, September 10-12, 2003
Proceedings

Series Editors

Gerhard Goos, Karlsruhe University, Germany
Juris Hartmanis, Cornell University, NY, USA
Jan van Leeuwen, Utrecht University, The Netherlands

Volume Editors

Jorge Juan Chico
Universidad de Sevilla
Departamento de Tecnologia Electronica
Avenida Reina Mercedes, s/n., 41012 Sevilla, Spain
E-mail: jjchico@dte.us.es

Enrico Macii
Politecnico di Torino
Dipartimento di Automatica e Informatica
Corso Duca degli Abruzzi, 24,10129 Torino, Italy
E-mail: enrico.macii@polito.it

Cataloging-in-Publication Data applied for

A catalog record for this book is available from the Library of Congress

Bibliographic information published by Die Deutsche Bibliothek
Die Deutsche Bibliothek lists this publication in the Deutsche Nationalbibliografie; detailed bibliographic data is available in the Internet at <http://dnb.ddb.de>.

CR Subject Classification (1998): B.7, B.8, C.1, C.4, B.2, B.6, J.6

ISSN 0302-9743
ISBN 3-540-20074-6 Springer-Verlag Berlin Heidelberg New York

This work is subject to copyright. All rights are reserved, whether the whole or part of the material is concerned, specifically the rights of translation, reprinting, re-use of illustrations, recitation, broadcasting, reproduction on microfilms or in any other way, and storage in data banks. Duplication of this publication or parts thereof is permitted only under the provisions of the German Copyright Law of September 9, 1965, in its current version, and permission for use must always be obtained from Springer-Verlag. Violations are liable for prosecution under the German Copyright Law.

Springer-Verlag Berlin Heidelberg New York
a member of BertelsmannSpringer Science+Business Media GmbH

http://www.springer.de

© Springer-Verlag Berlin Heidelberg 2003
Printed in Germany

Typesetting: Camera-ready by author, data conversion PTP-Berlin GmbH
Printed on acid-free paper SPIN 10931929 06/3142 5 4 3 2 1 0

Preface

Welcome to the proceedings of PATMOS 2003. This was the 13th in a series of international workshops held in several locations in Europe. Over the years, PATMOS has gained recognition as one of the major European events devoted to power and timing aspects of integrated circuit and system design. Despite its significant growth and development, PATMOS can still be considered as a very informal forum, featuring high-level scientific presentations together with open discussions and panel sessions in a free and relaxed environment.
This year, PATMOS took place in Turin, Italy, organized by the Politecnico di Torino, with technical co-sponsorship from the IEEE Circuits and Systems Society and the generous support of the European Commission, as well as that of several industrial sponsors, including BullDAST, Cadence, Mentor Graphics, STMicroelectronics, and Synopsys.
The objective of the PATMOS workshop is to provide a forum to discuss and investigate the emerging problems in methodologies and tools for the design of new generations of integrated circuits and systems. A major emphasis of the technical program is on speed and low-power aspects, with particular regard to modeling, characterization, design, and architectures.
A total of 85 contributed papers were received. Many thanks to all the authors that submitted their contributions. In spite of a very dense technical program, we were able to accept only 43 regular papers and 18 posters. Posters differ from regular papers in that they are presented to the audience during three sessions in which authors stand by cardboard displays and answer questions. Authors of regular papers, instead, deliver traditional oral presentations of their contributions during the sessions of the technical program.
Three keynote talks, offered by leading experts from industry and academia, served as starters for the three days of the workshop: Dr. Andrea Cuomo, Corporate VP of STMicroelectronics, Dr. Antun Domic, Senior VP of Synopsys Inc., and Dr. Ricardo Reis, Professor at the Universidad Federal Rio Grande do Sul addressed hot issues regarding architectures for next generation integrated platforms, EDA tools for energy-efficient design, and physical synthesis for high-speed and low-power circuits, respectively.
An industrial session, in which scientists from EDA and IC manufacturing companies illustrated their most recent advances in R&D activities and leading-edge technology development complemented the technical program.
Finally, a one-hour panel session took advantage of the presence of Officers from the European Commission and Coordinators of large EU-funded collaborative projects to provide insights into the new instruments and mechanisms now available in Europe for supporting research and innovation activities.
Last, but not least, a rich social program, paired with an industrial exhibition, guaranteed additional opportunities for establishing new relationships among the workshop participants.

Many thanks to the technical program committee for all their hard work in paper review, paper selection, and session organization. Additional experts also collaborated in the review process, and we acknowledge their contribution. Lists of technical program committee members and of additional reviewers can be found the following pages. Thanks also to the invited speakers for graciously donating their time and to the panelists for an enlightening and entertaining session.
We hope you found the workshop both stimulating and helpful and that you will enjoyed your stay in Turin for the whole duration of the event.

September 2003

Jorge Juan Chico
Enrico Macii

Organization

Organizing Committee

General Chair:	Prof. Enrico Macii, Politecnico di Torino, Italy
Program Chair:	Prof. Jorge Juan Chico, Universidad de Sevilla, Spain
Industrial Chair:	Dr. Roberto Zafalon, STMicroelectronics, Italy
Conference Manager:	Ms. Agnieszka Furman, Politecnico di Torino, Italy

Program Committee

D. Auvergne, U. Montpellier, France
J. Bormans, IMEC, Belgium
J. Figueras, U. Catalunya, Spain
J. A. Carballo, IBM, USA
C. E. Goutis, U. Patras, Greece
A. Guyot, INPG Grenoble, France
R. Hartenstein, U. Kaiserslautern, Germany
S. Jones, U. Loughborough, UK
P. Larsson-Edefors, Chalmers T.U. Sweden
V. Moshnyaga, U. Fukuoka, Japan
W. Nebel, U. Oldenburg, Germany
J.A. Nossek, T.U. Munich, Germany
A. Nunez, Las Palmas, Spain
M. Papaefthymiou, U. Michigan, USA
F. Pessolano, Philips, The Netherlands
H. Pfleiderer, U. Ulm, Germany
C. Piguet, CSEM, Switzerland
R. Reis, U. Porto Alegre, Brazil
M. Robert, U. Montpellier, France
A. Rubio, U. Catalunya, Spain
D. Sciuto, Politecnico di Milano, Italy
D. Soudris, U. Trace, Greece
J. Sparsø, DTU, Denmark
A. Stauffer, EPFL, Switzerland
A. Stempkowsky, Acad. of Sciences, Russia
T. Stouraitis, U. Patras, Greece
A.M. Trullemans-Anckaert, U. Louvain-la-Neuve, Belgium
R. Zafalon, STMicroelectronics, Italy

PATMOS Steering Committee

D. Auvergne, U. Montpellier, France
R. Hartenstein, U. Kaiserslautern, Germany
W. Nebel, U. Oldenburg, Germany
C. Piguet, CSEM, Switzerland
A. Rubio, U. Catalunya, Spain
J. Figueras, U. Catalunya, Spain
B. Ricco, U. Bologna, Italy
D. Soudris, U. Trace, Greece
J. Sparsø, T.U. Denmark
A.M. Trullemans-Anckaert, U. Louvain, Belgium
P. Pirsch, U. Hannover, Germany
B. Hochet, EIVd, Switzerland
A.J. Acosta, U. Sevilla/IMSE-CNM, Spain
J. Juan, U. Sevilla/IMSE-CNM, Spain
E. Macii, Politecnico di Torino, Italy
R. Zafalon, STMicroelectronics, Italy
V. Paliouras, U. Patras, Greece
J. Vounckx , IMEC, Belgium

Executive Sub-committee

President:	Joan Figueras, U. Catalunya, Spain
Vice-president:	Reiner Hartenstein, U. Kaiserslautern, Germany
Secretary:	Wolfgang Nebel, U. Oldenburg, Germany

Reviewers

A. Acosta	J. Luis Güntzel	J. Van Lunteren
D. Auvergne	A. Guyot	A. Macii
N. Azemard	E. Hall	S. Manich
P. Babighian	K. Inoue	K. Masselos
M.J. Bellido	A. Ivaldi	P. Maurine
L. Bisdounis	J. Jachalsky	M. Meijer
A. Bona	R. Jiménez	S. Moch
J.A. Carballo	J. Juan	V. Moshnyaga
J.M. Daga	K. Karagianni	W. Nebel
A. Dehnhardt	C. Kashyap	S. Nikolaidis
M. Donno	H. Klussman	N. Nolte
J. Figueras	O. Koufopavlou	A. Nuñez
L. Friebe	S. Lagudu	M. Papaefthymiou
F. Fummi	C. Lefurgy	F. Pessolano
C. Goutis	J.D. Legat	H.-J. Pfleiderer

C. Piguet
P. Pirsch
C. Psychalinos
R. Reis
J. Rius
F. Salice
A. Schallenberg
D. Sciuto
H. Shafi
C. Silvano
C. Simon-Klar
D. Soudris
J. Sparsø
A. Stempkowsky
G. Theodoridis
A. Tisserand
L. Torres
A.M. Trullemans
M. Valencia
S.-M. Yoo
V. Zaccaria
R. Zafalon
M. Wahle
M. Winter

Sponsors and Industrial Supporters

IEEE Circuits and Systems Society (Technical Co-sponsor)
Politecnico di Torino
European Commission
BullDAST s.r.l.
Cadence Design Systems, Inc.
Mentor Graphics, Italian Site
STMicroelectronics, s.r.l.
Synopsys, Inc.

Table of Contents

Keynote Speech

Gate-Level Modeling and Design

Low Level Modeling and Characterization

Interconnect Modeling and Optimization

Asynchronous Techniques

Keynote Speech

Industrial Session

RTL Power Modeling and Memory Optimisation

High-Level Modeling

Power Efficient Technologies and Designs

Keynote Speech

Communication Modeling and Design

Low Power Issues in Processors and Multimedia

Poster Session 1

Poster Session 2

Poster Session 3

Architectural Challenges for the Next Decade Integrated Platforms

Andrea Cuomo

STMicroelectronics
Via C. Olivetti, 2
Agrate Brianza (Milano), 20041 ITALY
andrea.cuomo@st.com

Abstract. The current trend in microelectronic design indicates the platform-based design paradigm as the solution of choice to simplify the problem of implementing multimillion-gate SoCs with reasonable effort and time. Significant advances in the development of new solutions for platform-based design have come up in the recent past; however, existing platforms are characterized by very strict architectural templates, which limit their usability to very narrow application domains.
Broadening the scope of existing platforms, as well as introducing new architectural concepts targeted towards new types of applications poses significant challenges, both from the research and the industrial perspectives. Purpose of this talk is to provide a look into the future of integrated hardware-software platforms, which will be at the basis of the next geration of electronic products.

J.J. Chico and E. Macii (Eds.): PATMOS 2003, LNCS 2799, p. 1, 2003.
© Springer-Verlag Berlin Heidelberg 2003

Analysis of High-Speed Logic Families

G. Privitera and Francesco Pessolano

Philips Research, Prof. Holstlaan 4, 5656 AA Eindhoven, The Netherlands
francesco.pessolano@philips.com

Abstract. In this paper, we compare the current de-facto standard high-speed family (self-resetting domino logic) with the most promising recently proposed ones: dynamic current mode logic, gate diffusion input logic and race logic architecture. For this purpose, we have used an 8-bit array multiplier in the 0.13 um TSMC CMOS process whose critical path is representative of many modern complex ICs. Each design adopts solely one logic style and it has been optimized with a standard industrial semi-custom flow. As metrics of comparison we have used speed and the energy (E), energy-per-time (ET) and energy-per-time-square (ET2) metrics. The comparison shows that self-resetting domino logic is the best choice when only speed is of concern, while dynamic current mode logic should be the preferred approach for all other cases. The other logic families failed to live up with their original promises of high-performance.

1 Introduction

Low-power design is receiving more and more attention due to the growing number of mobile devices. Traditionally, this problem was approached by purely low-power techniques. However, such an approach is no longer feasible as modern systems also require high-performance along with long battery life. A possible solution being currently explored is the possibility to design a system for its peak performance (i.e. optimized it for speed) and adapt its operating conditions so as to reduce power dissipation. In such a scenario, it is important to identify design styles that allow the achievement of the peak performance instead of styles for low power. Recently two design families have surged as de-facto standard for high-speed: clock-delayed and self-resetting domino logic. However, new proposals have appeared in literature claiming to improve on such digital families.

In this paper, we compare the current de-facto standard high-speed family [1, 2, 3] with the most promising recently proposed logic families, which are dynamic current mode logic (DyCML) [4, 5], gate diffusion input (GDI) logic [6] and race logic architecture (RALA) [7]. We describe the basic principle underlying these logic families. Furthermore, we will describe the result from our comparative analysis based on the design of an 8-bit array multiplier optimized for each approach and designed in the baseline 0.13 um TSMC CMOS process technology. Our comparison is fair, based on industrial requirements and non-biased.

J.J. Chico and E. Macii (Eds.): PATMOS 2003, LNCS 2799, pp. 2–10, 2003.
© Springer-Verlag Berlin Heidelberg 2003

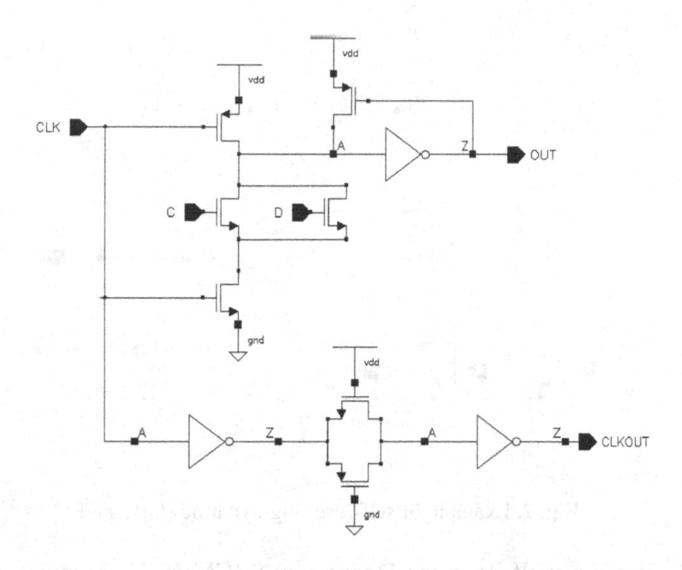

Fig. 1. Example of clock-delayed domino (OR) gate.

This paper is organized as follows. A brief summary on self-resetting and clock-delayed domino logic families, the current de-facto standard for high-speed design, is given in Section II. Section III describes the recently proposed alternative families, while a design comparison is given in Section IV. Some conclusions are drawn in Section V.

2 Current High-Speed Reference Logic Family

Dynamic logic is the underlying concept of all current high-speed logic families. Many flavors of the basic dynamic gates have been presented in literature over the last decade so as to overcome all problems of the original approach. Among these, two families have assumed the position as de-facto standard for high-speed design [1]: clock-delayed domino and self-resetting domino logic. The latter can be also seen as a flavor of the former.

Clock-Delayed (CD) Domino is a single-ended, self-timed logic style, which eliminates the fundamental monotonic signal requirement imposed by standard dynamic domino. It propagates a clock network in parallel to the logic (providing a dedicated clock to each logic stage – Fig.1). Because of the presence of a dedicated clock, CD Domino provides both inverting and non-inverting functions, without the need for differential pairs or cross-coupled latches. The output inverter is no longer essential, because the clock arrives when evaluation has been completed. The delayed clock can be generated within the clock network or each gate can delay the received clock and provide it to the next gate. CD domino's performance and scalability have caused the style to become increasingly popular in high-speed circuits where speeds in the GHz range are required.

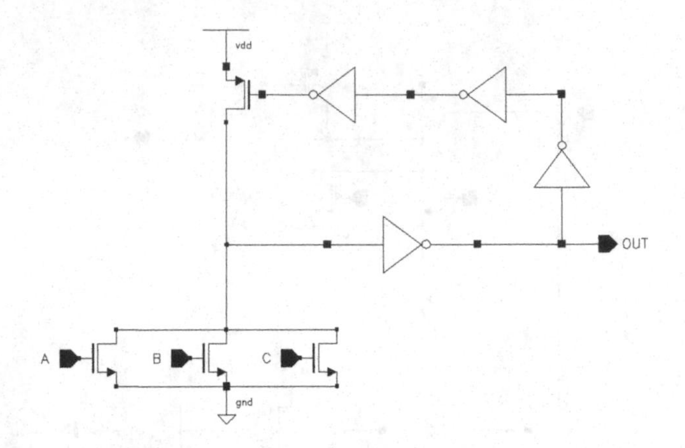

Fig. 2. Example of self-resetting dynamic (OR) gate.

Unlike CD Domino, Self-Resetting Domino or SRCMOS [2, 3] structures generally detect their operating phase and supply their own appropriate clocking (Fig.2). SRCMOS uses the notion of propagating logic as a *wave* or a *bubble*, rather than as a *level.* Unlike the other styles, the output values are available for only a determined duration; the restore wave front chases the logic evaluation wave front through the logic path. The duration of the input (and thus) data wave is also directly related to the power required by this logic style (as it might lead to DC currents as for nMOS logic styles)

SRCMOS designs are extremely fast and dense, but also quite difficult to synchronize: in fact, data waves must transmit so as not to deteriorate while propagating along the cascaded gates. Self-Resetting structures might save power in two ways. When data evaluation does not require dynamic node discharge, the pre-charge device is not active (while it would be clocked in CD domino). In addition, the clock infrastructure is now limited to the latches which launch and receive the path, eliminating clock distribution network. Additional power is spent, however, in the added inverters, which causes DC current.

3 Recent Proposals for High-Speed Design

Recently, many proposals have been described that seem to improve on both above-mentioned logic families. Among these, three logic styles have attracted the most interest as their advantages seemed very clear. Throughout this section, we will give a first description of these logic families that are Dynamic Current Mode Logic (DyCML), Gate-Diffusion Input Logic (GDI) and Race Logic Architecture (RALA). DyCML is the last evolution of low-power high-speed differential logic. GDI is a pass-transistor-based logic aiming at high-speed, low power and compact implementation. Finally, RALA is a completely different high-speed logic concept.

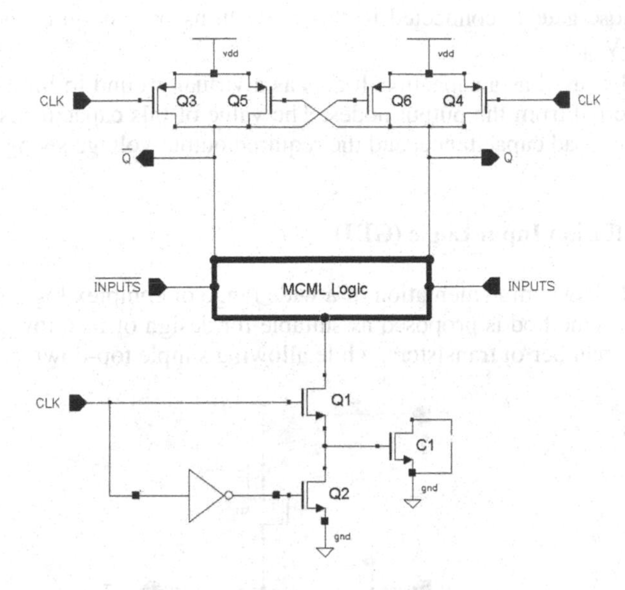

Fig. 3. Basic structure of a DyCML gate.

3.1 Dynamic Current Mode Logic (DyCML)

Dynamic Current Mode Logic (DyCML) [4] is a logic family that adopts the current mode scheme to reduce dynamic power and enhance performance. The main advantages of this kind of logic are reduced output logic swing and high noise immunity. Furthermore, DyCML solves also the short-circuit current problem of standard Current Mode logic so as to yield to large power savings. This is achieved by employing a dynamic current source with a virtual ground. This logic style also adopts active loads, instead of the traditional load resistors to reduce power dissipation and area.

Fig. 3 shows the basic structure of a DyCML logic gate. It consists of a standatrd current-mode logic block (MCML block [5]) for function evaluation, a pre-charge circuit (Q2, Q3, Q4), a dynamic current source (Q1, C1), and a latch to preserve logic value after evaluation (Q5, Q6). The operation of the DyCML is as follows. During the low phase of the clock, the pre-charge transistors Q3, Q4 turn on to charge the output nodes to V_{dd}, while transistor Q2 turns on to discharge capacitor C1 to ground. Meanwhile, transistor Q1 is off, eliminating the dc path from V_{dd} to ground. During the high clock phase, the pre-charge transistors Q2, Q3 and Q4 turn off, while the transistor Q1 switches on creating a current path from the two pre-charged output nodes to the capacitor C1. The latter acts as a virtual ground. These two paths have different impedance depending on the logic function and inputs; therefore, one of the output nodes drops faster than the other node. The cross-connected transistors Q5 and Q6 speed up the evaluation and maintain the logic levels after evaluation. During the evaluation phase, when one of the output nodes drops less than V_{dd}-$|V_{tp}|$, the

transistor whose gate is connected to this node turns on, charging the other output node back to V_{dd}.
Transistor C1 is used as a capacitor. It acts as a virtual ground to limit the amount of charge transferred from the output nodes. The value of this capacitor is dependent on the value of the load capacitance, and the required output voltage swing.

3.2 Gate Diffusion Input Logic (GDI)

GDI logic [6] allows implementation of a wide range of complex logic using only two transistors. This method is proposed as suitable for design of fast, low power circuits, using reduced number of transistors, while allowing simple top-down design.

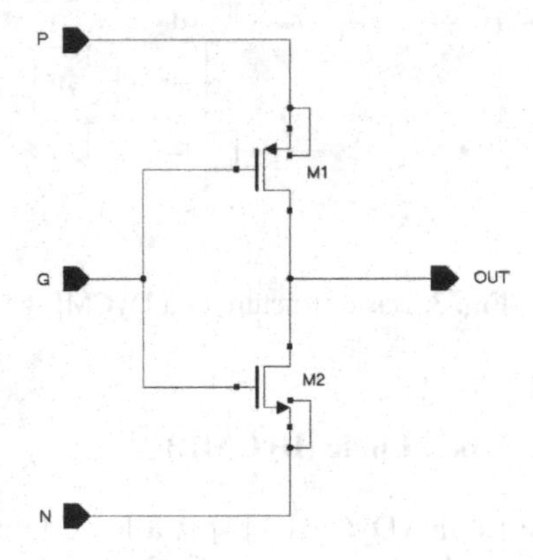

Fig. 4. Example of GDI generic gate.

GDI method is based on the use of a simple cell as shown in Fig.4. At the first glance the cell reminds the standard CMOS inverter, but there are some important differences. GDI cell contains three inputs: input G that is the common gate input of NMOS and PMOS, input P that is the source/drain of PMOS, and input N that is the source/drain of NMOS. The bulks of both NMOS and PMOS are connected to N or P respectively, so that it can be arbitrarily biased. The use of also these two terminals N and P as inputs is what makes GDI different as it allows implementing more complex functionality with a simple inverter-like topology. GDI can be also seen as a pass logic family where both nMOS and pMOS are used at the same time.

3.3 Race Logic Architecture (RALA)

RALA is completely new concept of performing logic operations by means of racing between signals along the interconnection lines, where transistors are used to introduce delays [7]. By detecting the fastest racing signal, any logic operation can be realized.

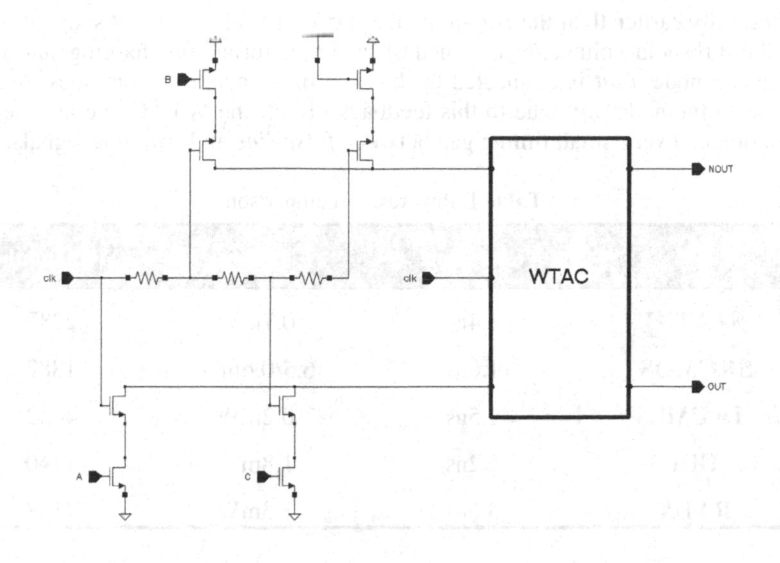

Fig. 5. Basic structure of a RALA gate

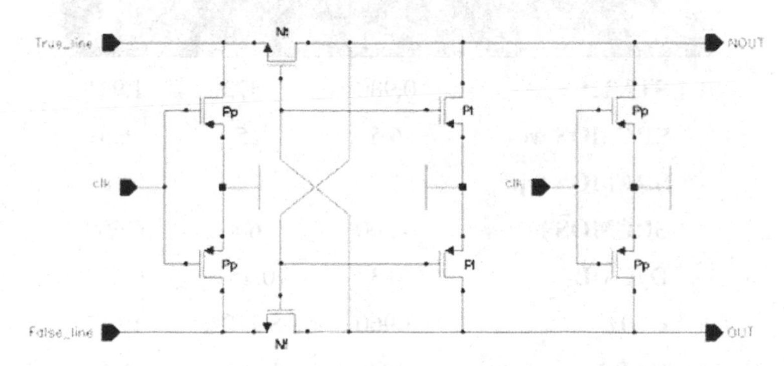

Fig. 6. Typical schematic for the WTAC.

The conceptual diagram of RALA is illustrated in Fig.5 RALA consist of three part: Clock Distribution Line (CDL), Race lines and Winner-Take-All-Circuit (WTAC). The operation of this circuit can be divided into two phases: *pre-charge* and *evaluation*, respectively when clock is low or high. When the clock signal transits from low to high, the clock travels along the transistor array, and switches are triggered sequentially. The termination switch is attached at the end of the CDL; in any case it determines the output when the previous switch did not do it. The race lines consist of a *true-line* and a *false-line*. The WTAC determines which line becomes low earlier. If the true-line becomes first low the output will be high. On the other hand, if the false-line becomes low earlier the output will be low.

The circuit implementation of WTAC is shown in Fig.6. When the clock is low the PMOS transistors Pp pre-charge the four lines (*true-* and *false-line*, *out* and *nout*) to V_{dd}. NMOS transistors N_t and N_f are initially turned on. If the clock rises and the *false-line* falls earlier than the *true-line*, also node *out* falls down. As signal *out* falls below the threshold voltage, N_t is turned off and P_t is turned on charging *nout* to high. Because the node *nout* is connected to the gate of N_f again, it continues to connect *false-line* to the node *out*. Due to this feedback effect, the WTAC operates very fast, and can detect a very small timing gap between *false-line* and *true-line* signals.

Table 1. Raw results comparison

	Cycle time	Power	MOS number
STATIC	1.4ns	0.7mW	2287
SRCMOS	1.0ns	6.5/0.6mW	1887
DyCML	1.5ns	0.2mW	4052
GDI	2.2ns	1.8mW	1740
RALA	3.6ns	1.3mW	7108

Table 2. Power-speed trade-off comparison

	E	ET	ET2
STATIC	0.980	1.372	1.921
SRCMOS wc	6.5	6.5	6.5
SRCMOS typ	1.2	1.2	1.2
SRCMOS bc	0.600	0.600	0.600
DyCML	0.3	0.45	0.675
GDI	3.960	8.712	19.17
RALA	4.68	16.85	60.66

4 Comparison

To have a better comparison among the four families under analysis, an 8-bit array multiplier [10] has been designed in the TSMC CL013G CMOS process. The design has been optimized for each of the above described logic families including static logic (here used as a reference). Each design uses purely one logic style except for the adoption of standard inverter when required. We have selected an 8-bit array multiplier as their critical path consists of 15 combinatorial gates similarly (in number and complexity) to more complex industrial designs (targeting medium-high performance). The multiplier has been designed following the standard industrial flow based on definition of a basic standard cells library. As there is no CAD support for such cells, we have mimicked the semi-custom flow with a full custom one based on such cells library.In our case, such cells were standard Boolean gates, registers, full-adders and multiplexers. Each of these basic cells has been, then, designed as a standard cell. The array has been, thus, composed by plugging these cells together mimicking a semi-automated flow. The final design is, thus, optimized for each logic family so as to lead to a fair comparison.

4.1 Raw Data Comparison

The results of the comparison are summarized in Tab.1. Results have been obtained by means of exhaustive Spice level simulation of the core with estimated parasitic and a proprietary MOS model, which includes also leakage information. The terms of comparison are speed, power consumption and number of transistors used. The power figures have been derived by using signals with period multiple of 4ns and it accounts only for the logic (flip-flops have not been included). This does not invalidate the results since further analysis has shown that our results keep validity also when changing this reference frequency. This value has been selected by looking at the RALA design cycle time plus the flip-flop overhead as it was the slowest design in the lot.

As shown in Tab.1, SRCMOS logic is characterized by the best overall speed results. Delay is 43% smaller when compared to the standard static logic. Power consumption is more than nine times higher than for the static design. However, this represents the worst possible case when nothing is done to prevent static dissipation (as for ratio-ed logic). This means that data is not transmitted as waves, but it is kept constant for the entire clock cycle. Dimensioning the feedback loop and the signal pulse so as to be equal to the delay of the internal inverter chain might reduce SRCMOS power budget to 0.6mW. This value is somehow unrealistic due to tolerances and design constrains, typical power dissipation is about two times larger (1.2mW). The DyCML multiplier shows power consumption about three times lower than for the static design. The delay for the multiplier is roughly the same as for static, while delays increase when cells are used in a standard cell fashion. Area seems to grow double with respect to the static design. However, DyCML also provides the negated output.

GDI logic appears to have the smallest area requirement, but it also needs a triple-well technology. Design is rather simple. Power and speed are 2.5 times and 1.5 times larger respectively. Finally, RALA is the worst family here considered, as it does not show any property of importance.

4.2 Power-Speed Trade-Off Comparison

The values in Tab.1 gives data that is useful when only speed is of interest. As previously mentioned, designing a system now involves selecting the proper design style for achieving the required peak performance. A set of useful metrics has been recently proposed [8, 9] as mean of selecting power in three main application domains: power-driven, high-performance and medium performance applications. The metrics for these domains are energy (E), energy-time-square (ET2) and energy-time (ET) respectively. Tab.2 gives the values derived from Tab.1 for the logic styles under analysis. In this case (Tab.2), we also made explicit the values for SRCMOS worst-case (*wc*), typical case (*typ*) and best-case (*bs*) are discussed in the previous section.

The results show clearly that Dynamic Current Model logic is the choice for low power and medium-performance systems. For high-performance systems, ideal SRCMOS (i.e. with minimum possible data waves) results having the lower ET2 value. However, as this represents the ideal upper limit of SRCMOS it is yet safe to assume that DyCML is a clear choice for all applications domains, even if it requires a higher number of transistors (Tab.1).

5 Conclusions

In this paper, we have described and fairly analyzed the state-of-the art for high-speed digital design and compared them with the latest high-speed logic proposal: self-resetting dynamic logic, gate diffusion input logic, dynamic current mode logic and race logic. A comparison has been performed in terms of power, speed and area. The result of the comparison is that self-resetting logic is definitively the choice for high-speed. Dynamic current mode logic is, however, the best power-speed compromise as showed by the E/ET/ET2 metrics. These results also show that in schemes where peak performance is important and working conditions are modified to reduce power dissipation dynamic current mode logic is the best possible choice. Further work is, thus, required in order to understand how reliable is such logic with respect to voltage scaling and signal integrity.

References

[1] K. Bernstein, K. M. Carrig, C. M. Durham, P. R. Hansen, D. Hogenmiller, E. J. Nowak, N. J. Rohrer, “High Speed CMOS Design Styles”, Kluwer Academic Publishers, 1999.
[2] A. K. Woo, “Static PLA or ROM Circuit with Self-Generated Precharge”, U.S. Patent # 4,728,827, Mar 1, 1988.
[3] W. Hwang, G. D. Gristede, P. Sanda, S. Y. Wang, D. F. Heidel, “Implementation of a Self-Resetting CMOS 64-Bit Parallel Adder with Enhanced Testability”, IEEE Journal of Solid-State Circuits, Vol. 34, No. 8, August 1999.
[4] M. W. Allam, M. I. Elmasry, “Dynamic Current Mode Logic (DyCML): A New Low-Power High-Performance Logic Style”, IEEE Journal of Solid-State Circuits, Vol. 36, No. 3, March 2001.
[5] D. Somasekhar, K. Roy, “Differential Current Switch Logic: A Low Power DCVS Logic Family”, IEEE Journal of Solid-State Circuits, Vol. 31, No. 7, July 1996.
[6] A. Morgenshtein, A. Fish, I. A. Wagner, “Gate-Diffusion Input (GDI) – A Novel Power Efficient Method for Digital Circuits: A Design Methodology”, ASIC/SOC Conference, 2001. Proceedings. 14th Annual IEEE International, 2001.
[7] S. J. Lee, H. J. Yoo, “Race Logic Architecture (RALA): A Novel Logic Concept Using the Race Scheme of Input Variables”, IEEE Journal of Solid-State Circuits, Vol. 37, No. 2, February 2002
[8] A. J. Martin, “Towards an energy complexity of computation”, Information processing Letters, Vol. 77, 2001
[9] R. Gonzalez, B. Gordon, M. Horowitz, “Supply and threshold voltage scaling for low power CMOS”, IEEE Journal of Solid-State Circuits, Vol. 32, No. 8, Aug. 1997
[10] N.H.E. Weste, K. Eshragian, “Principle of CMOS design: a system perspective”, Addison-Wesley publishers, 1993

Low Voltage, Double-Edge-Triggered Flip Flop

Pradeep Varma and Ashutosh Chakraborty

IBM India Research Laboratory, Block 1, Indian Institute of Technology,
Hauz Khas, New Delhi 110016, India
pvarma@in.ibm.com

Abstract. Double-edge-triggered flip flops (DETFFs) are recognized as power-saving flip flops. We study the same from a low voltage perspective [1-1.5V]. We combine a medium-to-high voltage, plain-MOS-style DETFF technique with a clock-skew technique to derive a new DETFF that is suited to low voltages. Speedwise, our result outperforms existing static DETFFs convincingly in the low voltage range. Powerwise, our flip flop beats others for dynamic input in the lower half of the same range. The dynamic counterpart of our static circuit also shows similar power superiority at low voltages.

1 Introduction

Double-edge-triggered flip flops are widely known for their power saving ability [1]-[4], [6]-[9]. By triggering on both clock edges, a DET FF provides the same throughput as a single-edge-triggered flip flop (SETFF) at half the clock frequency. Redundancies such as wasted clock edges (and internal state changes caused by them) are avoided [3], [6], [7]. For a given data throughput, power and energy savings are made not only at the device level, but also at system level due to simpler and less-expensive distribution of a lower-frequency clock [4], [2]. Indeed, the trend towards increasing clock frequencies and pipelining bodes well for power savings via a shift to DET FFs [4]. Low voltage operation is a widespread trend for saving power. In this paper we combine the low voltage and DET trends to provide a comparative study of DETFFs at low voltages [1-1.5V]. The vehicle of our study is a new DETFF derived by combining a clock skew technique [9] with recent work on medium-to-high voltage DETFFs [8]. The resulting flip flop outperforms earlier DET FFs significantly at low voltages.

1.1 Related Work and Comparison Testbed

A qualitative comparison of our flip flop with existing DET FFs (as published in major journals) uncovers costs that our circuit manages to avoid. Static drains are associated with the circuits of [4] due to a V_T drop in internal voltages. Such V_T-reduced logic-high voltages when fed to a CMOS inverter lead to both the inverter transistors being ON (PMOS barely), which leads to a power drain – see [2], and reduced speed. The fix to [4] suggested by [2] has its own power (and speed) costs in terms of a contest/short-based resolution of node voltages

J.J. Chico and E. Macii (Eds.): PATMOS 2003, LNCS 2799, pp. 11–20, 2003.
© Springer-Verlag Berlin Heidelberg 2003

of forward and backward inverters. Compared to [3], we avoid the complexity and cost of clock sharing with doubled area gates. The flip flops of [6] and [1] use many more transistors than us and have fixed internal node voltages that have to be acquired in each clock period independent of the input at D. Such voltages (Q_1, Q_1', Q_2, Q_2' in [6] and N, M, and carryforwards in [1]) introduce unnecessary switching costs.

Quantitative comparison of our work and others is carried out on IBM PowerSpice using MOSIS 0.25 μm N94S parameters. We contain our exercise to static/dynamic flip flops only, thus the (complex) pseudo-static circuit of [10] is left out. The original DET FF [7] is also left out for its uncompetitive complexity overhead (36 transistors). From [1], we simulate the fastest (400 Mhz) circuit. In this, each minimum-area transistor is mapped to a 0.625 μm wide, 0.25 μm long, minimum-area transistor and the larger transistors are scaled accordingly. For [3], simulations with our 0.25 μm parameters require ratioing the weak transistors in the static FF to a more resistive 0.625 μm / 0.8 μm (W/L). All other transistors from [3] and elsewhere are scaled linearly. For example, a length L in a T μm technology becomes L $\times$ (0.25 / T) in our testbed. Though it can be argued that each circuit ought to be scaled as per its unique features, we stick to linear scaling in order to keep the benchmarking exercise practical and also to have one common standard to apply to all circuits.

2 Our Work

Figures 1 and 2 contain our flip flop circuit. Figure 1 shows the latch used in our circuit, and figure 2 shows how two latches are combined to make the flip flop. If the (dashed) feedback branches are removed, then the dynamic counterpart of our (static) flip flop is obtained. Each latch transistor's W/L ratio is shown next to it. Non-latch transistors (fig. 2) are all minimum area (W/L = 0.625 μm / 0.25 μm). The clocked switches in fig. 2 are implemented as CMOS transmission gates. Note that the basic difference in the schema of a DET FF and a SET FF is the use of two latches in parallel in the former and in serial in the latter. Like [4], our DET FF imposes an overhead of two switches and one inverter in combining two latches. We differ in that the switches here precede the latches instead of succeeding them as in [4].

The upper latch in fig. 2 operates in the low phase of the clock and the lower latch operates in the high phase of the clock. The transmission gates feeding the latches operate opposite to them, so that the lower gate is ON only with the upper latch and the upper gate is enabled only with the lower latch. When the clock changes phase, a previously-enabled transmission gate turns off and the latch at its output turns on. The charge at the output of the transmission gate (leftover from just before the gate turned off) gets latched and drives the flip flop output for half the clock period. The remaining clock period is driven by the data latched in the other parallel path. This cycle is then repeated and the flip flop operates as a DET flip flop.

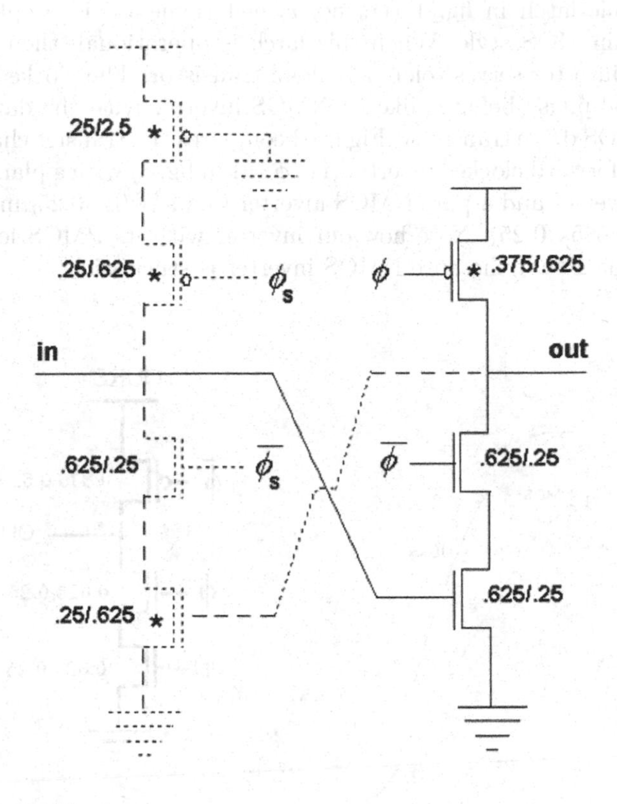

Fig. 1. Latch ($\overline{\phi}$, $\overline{\phi_s}$). Dashed feedback makes latch static. W/L in μm/μm. * denotes weakness

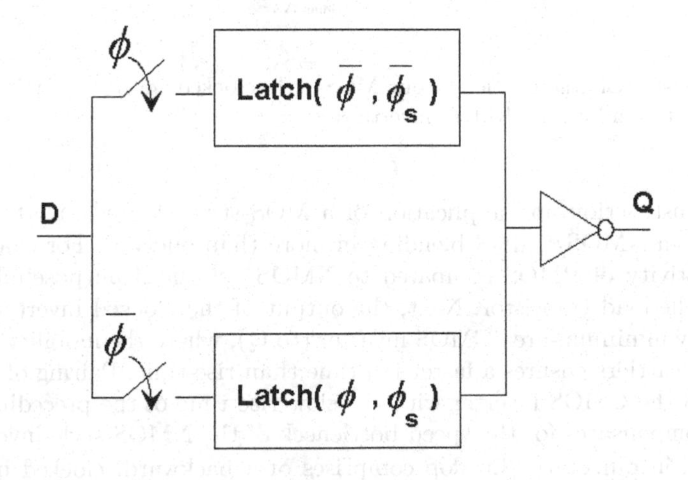

Fig. 2. Our Flip Flop

The dynamic latch in fig. 1 (i.e. no dashed elements) is simply a clocked inverter in plain MOS style. When this latch is operational, then the clocked PMOS in the inverter serves solely as a load transistor. The clocked inverter in this operational phase behaves like an NMOS inverter with the data being fed only to an NMOS drive transistor. Figure 3 compares the transfer characteristics of our enabled forward clocked inverter (i.e. $\phi = 1$ in fig. 3) with a plain minimum-area CMOS inverter and a pure NMOS inverter (load W/L=0.25μm / 0.625μm; drive W/L= 0.625/ 0.25). Note how our inverter with its PMOS load avoids a V_T drop at logic 1 that the pure NMOS inverter is stuck with.

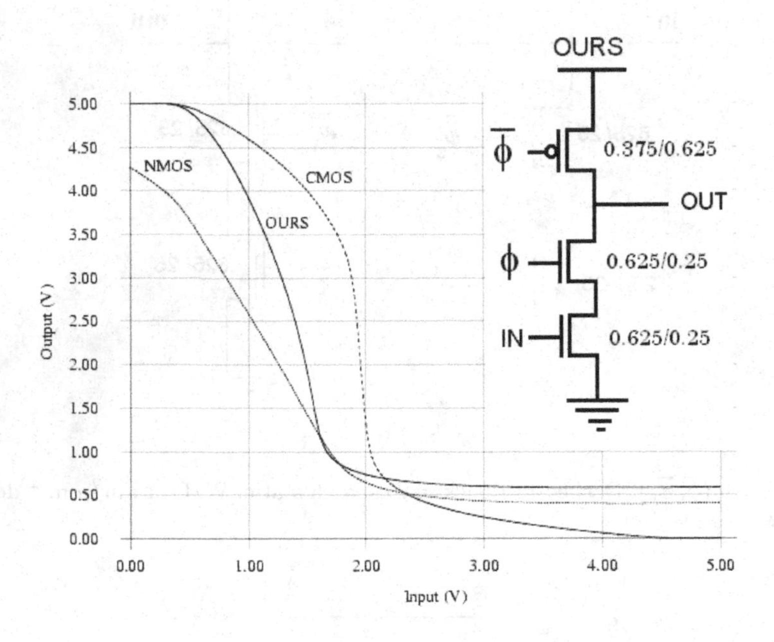

Fig. 3. Transfer Characteristics of our MOS-style clocked inverter, a minimum-area CMOS inverter and a pure NMOS inverter

The construction and application of a MOS-style clocked inverter in fig. 1 benefits from NMOS/CMOS blending in more than one way. For example, the extra resistivity of PMOS compared to NMOS gets used purposefully in constructing the load transistor. Next, the output of the clocked inverter is fed to an ordinary minimum-area CMOS inverter (to Q), where the mobility difference in n and p carriers ensures a faster fall time than rise time. Pairing of the faster fall time of the CMOS inverter with the slow rise time of the preceding clocked inverter compensates for the speed bottleneck of the NMOS-style inverter.

A latch in our static flip flop comprises of a backward, clocked inverter in addition to the forward clocked inverter. The two inverters together make up a back-to-back inverter pair that can hold a node voltage indefinitely even if the clock is stopped at some logic level. The clock signals given to the backward inverter are subscripted with s in figure 1 to indicate that they are delayed/skewed with respect to the ordinary clock signals. A skew of two minimum-area CMOS inverters has been used throughout this work.

In figure 2, when a static latch is enabled, then both the input and output of the latch can race to set each other (if they are at the same logic level). The ordinary method [3], [2], [8] of resolving such races is to make one of the inverters weaker than the other so that it always loses. We eschew this in favor of clock skew [9] to avoid costs including oversized weak inverters.

With clock skew, our latch works as follows: when the latch is enabled, the forward path is enabled before the backward path, which in turn lets the output node be set by the input node before the backward path has had a chance to affect the input node. Thus the fixing of the race between the input and the output is not decided by the *slow* and (power wise) *expensive* way of getting two enabled inverters (one weak, one strong) to short/contest each other.

Unlike [9], we find ourselves having to address the complication of a delayed backward path leaking into an earlier stage. In fig. 2, when the clock changes phase, an enabled latch follows suit, except that its backward inverter has to wait for a total of skew time before turning off. The transmission gate feeding this latch in the meantime turns on and for a brief time (bounded by clock skew), the previously latched data of the flip flop tries to set the D input via the transmission gate. To limit this drain on the preceding stage (D), we propose making the backward inverter weak in order to add a resistive bound to the skew-time bound on the drain. We impose roughly an order of magnitude resistive bound in fig. 1 (compared to the ordinary transmission gates feeding the latches). With the less resistive, path to 0 of the backward inverter decided thus, the more resistive, path to 1 of the inverter is decided by the load/drive ratio of the MOS-style inverter. In a manner similar to [2], we contain the switching capacitance of the clocked load by breaking it into a minimum-area clocked transistor in series with an unswitched, permanently-ON, load transistor. With the backward inverter size decided as above, clock skew enables an independent sizing of the forward clocked inverter. We choose to center this near minimum-area transistors. Thus despite [8]'s transistor savings by use of pass transistors for latch input switches, we still realize an overall 25% gate-area reduction by the use of clock skew in the present work.

3 Performance Comparison

We measured the maximum speed of our circuit by the usual method of taking output Q through a CMOS minimum-area inverter, feeding it back to the input D, and maximizing clock speed until output degrades. An advantage of this method is that it also loads the flip flop in the process of measuring speed as opposed to measuring only within an unloaded context. The output of our static circuit working in this configuration at maximum speed at 1.25 V is shown in Fig. 4. Speed comparison of our static FF with earlier static circuits for [1-1.5V] supply is shown in Fig. 5. The Y-axis of figure 5 is given in clock GHz, and not bit rate, which is twice the clock speed for DETs as usual. The figure shows our circuit (in bold line) outperforming others across the entire range.

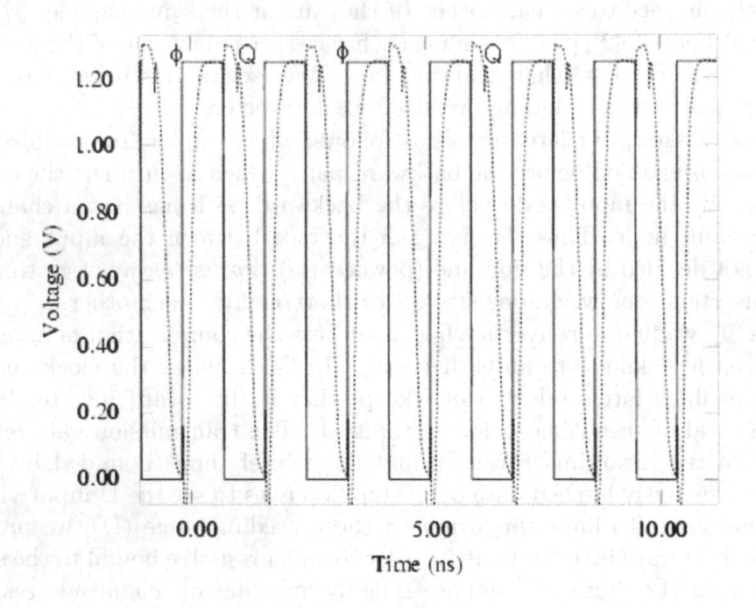

Fig. 4. Maximum frequency toggle response of our static flip flop at 1.25 V

Like [8], our MOS-style clocked inverters concede a power drain in maintaining logic 0 at their output. Although an all 1s input costs less in the present work than [8] (all 1s at 5V and 1.5 GHz in our static circuit cost 1.317mW compared to [8]'s 1.45mW), it is still preferable to invert a bitstream of all 1s before storing them in our flip flop (all 0s cost 0.653mW for our static FF at 5V, 1.5GHz). Figure 6 compares power consumption of our circuit with related work for an all 0s input. Figures 7-9 show power consumption for dynamic vectors.

The x-axis in figures 6-9 shows voltage labels for static FFs only (each prefixed with "s" for static), which increase from left-to-right in the range [1-1.5V]. The figures also show performance of dynamic FFs, including our circuit's dynamic subset (dropping the dashed elements in fig. 1). In order to retain clarity despite reuse of figure space for static and dynamic FF power, we do the following: Voltage (and power) for the dynamic FFs grows right to left (i.e. opposite to static FFs) so that the static and dynamic curves remain distinct. In other words, a voltage label of sK Volts (s for static FF) corresponds to (1.5-K+1) Volts for the dynamic flip flops. All power measurements are made using the method in [5]. Power includes local clock inversion cost (inversion done by min-area CMOS inverters only) in all cases. For our static circuit, local clock inversion covers two

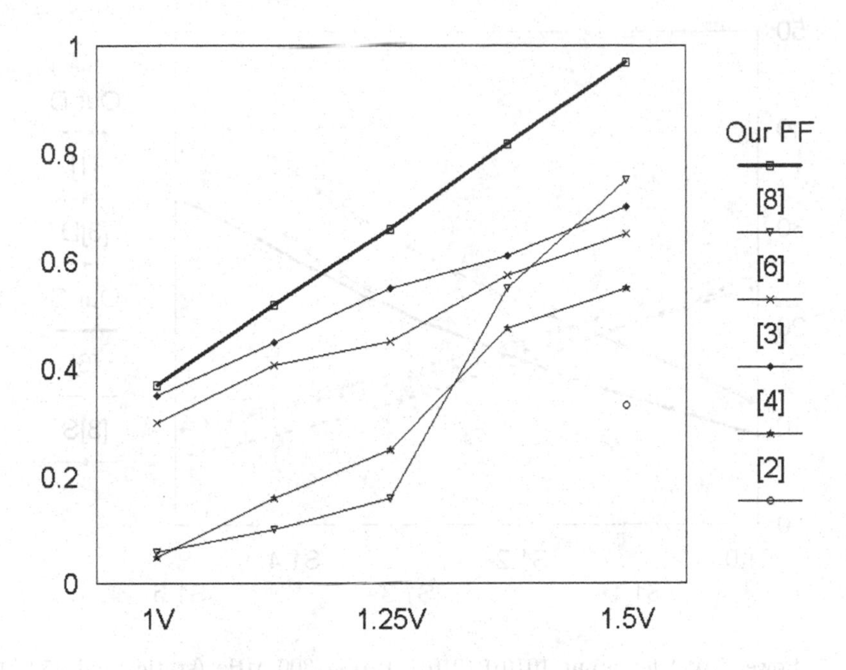

Fig. 5. Maximum speed (ϕ in GHz) comparison

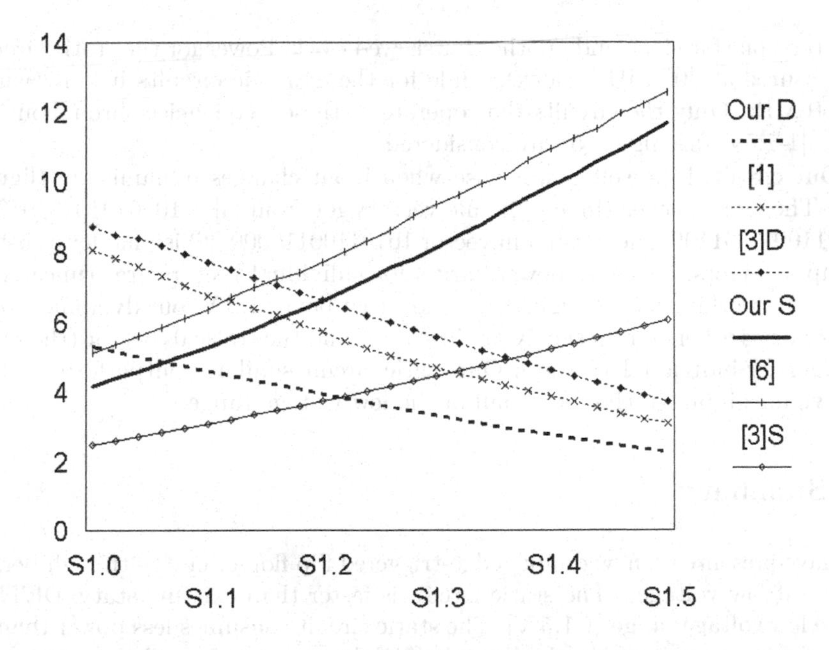

Fig. 6. Power (μW, y-axis) for input 00000000000000 at 300 MHz (static) and 350 MHz (dynamic)

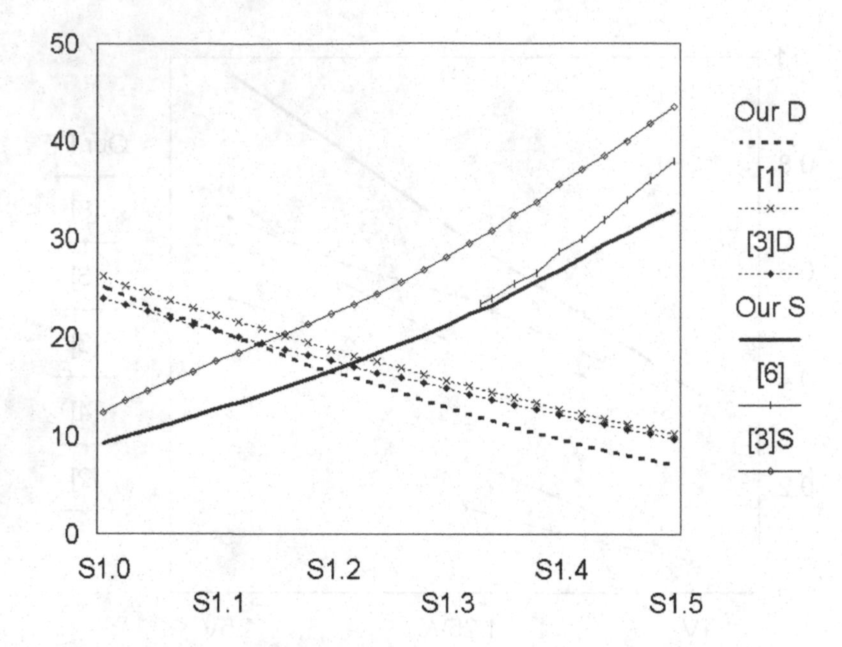

Fig. 7. Power (μW) for input 1010101010101010 at 300 MHz (static) and 350 MHz (dynamic)

inverters, one for clock and another for skewed clock. Power for the static circuits is measured at 300 MHz (clock), while for the dynamic circuits it is measured at 350 MHz. Only the circuits that operate at these frequencies throughout the range [1-1.5V] (see figure 5) are considered.

Our circuit does well power wise when input changes dynamically (figures 7-9). The figures cover three dynamic vectors got from [4] – 1010101010101010, 1100110011001100, and a random vector 10101100110000. [6] is unable to sustain all input vectors, hence its power curves for individual vectors are truncated to cover only working zones. Figures 6-9 show that below 1.2 V, our dynamic circuit outperforms others for all the dynamic vectors and the static 0s vector (the range is larger for individual vectors). Our static circuit similarly outperforms others for dynamic input in the lower half of the low voltage range.

4 Summary

We have presented a new double-edge-triggered flip flop circuit with high performance at low voltages. The static circuit is faster than existing static DETFFs in the low voltage range [1-1.5 V]. The static circuit consumes less power than its peers for dynamic input in the lower half of the above range. The dynamic subset of the static circuit performs similarly, consuming less power in the [1-1.2V] range (at least).

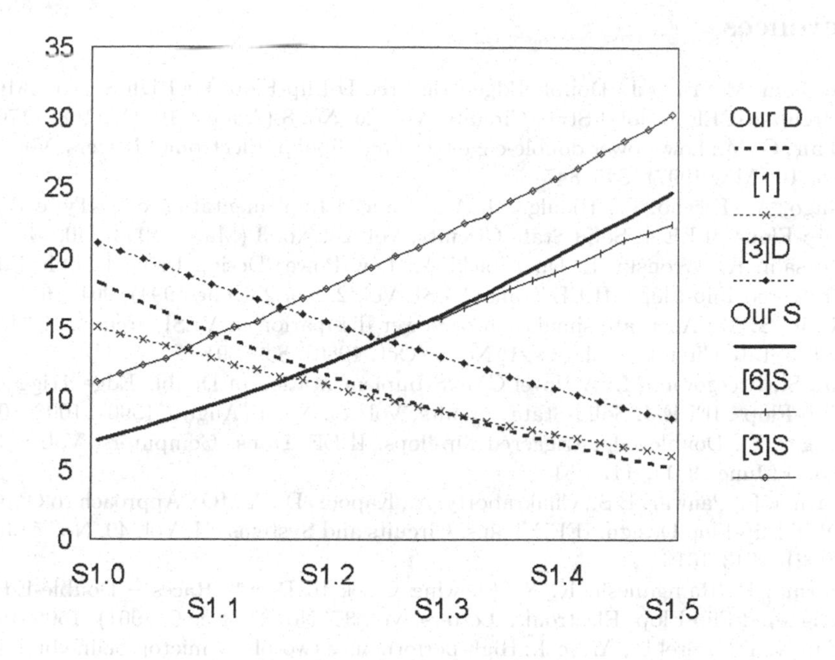

Fig. 8. Power (μW) for input 1100110011001100 at 300 MHz (static) and 350 MHz (dynamic)

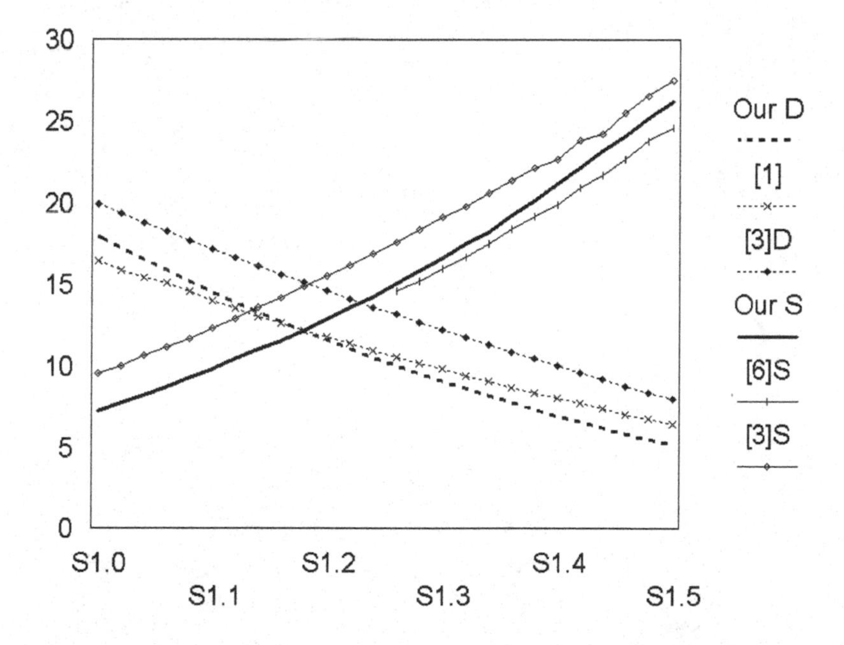

Fig. 9. Power (μW) for random input 10101100110000 at 300 MHz (static) and 350 MHz (dynamic)

References

1. Afghahi, M., Yuan J.: Double Edge-Triggered D-Flip-Flops for High-Speed CMOS Circuits. IEEE J. Solid-State Circuits, Vol. 26, No. 8 (August 1991), 1168–1170
2. Blair, G. M.: Low-power double-edge triggered flipflop. Electronics Letters, Vol. 33, No. 10 (May 1997), 845–847
3. Gago, A., Escano, R., Hidalgo, J. A.: Reduced Implementation of D-Type DET Flip-Flops. IEEE J. Solid-State Circuits, Vol. 28, No. 3 (March 1993), 400–402
4. Hossain, R., Wronski, L. D., Albicki A.: Low Power Design Using Double Edge Triggered Flip-Flops. IEEE Trans. VLSI, Vol. 2, No. 2 (June 1994), 261–265
5. Kang, S. M.: Accurate simulation of power dissipation in VLSI circuits. IEEE J. Solid-State Circuits, Vol. SC-21, No. 5 (Oct. 1986), 889–891
6. Lu, S., Ercegovac, M.: A Novel CMOS Implementation of Double-Edge-Triggered Flip-Flops. IEEE J. Solid-State Circuits, Vol. 25, No. 4 (August 1990), 1008–1010
7. Unger, S.: Double edge-triggered flip-flops. IEEE Trans. Computers, Vol. C-30, No. 6 (June 1981), 447–451
8. Varma, P., Panwar, B.S., Chakraborty, A., Kapoor, D.: A MOS Approach to CMOS DET Flip-Flop Design. IEEE Trans. Circuits and Systems – I, Vol. 49, No. 7 (July 2002), 1013–1016
9. Varma, P., Ramganesh, K. N.: Skewing Clock to Decide Races – Double-Edge-Triggered Flip Flop. Electronics Letters, Vol. 37, No. 25 (Dec 6, 2001), 1506–1507
10. Yun K. Y., Beerel P., Arceo J.: High-performance two-phase micropipeline building blocks: double edge-triggered latches and burst-mode select and toggle circuits. IEE Proc.-Circuits Devices Syst., Vol. 143, No. 5 (Oct 1996), 282–288

A Genetic Bus Encoding Technique for Power Optimization of Embedded Systems

Giuseppe Ascia, Vincenzo Catania, and Maurizio Palesi

University of Catania
Department of Computer Science and Telecommunications Engineering
V.le Andrea Doria, 6 — 95125 Catania, Italy
{gascia,vcatania,mpalesi}@diit.unict.it

Abstract. In this paper we present a genetic approach for the efficient generation of an encoder to minimize switching activity on the high-capacity lines of a communication bus. The approach is a static one in the sense that the encoder is realized ad hoc according to the traffic on the bus. The approach refers to embedded systems in which it is possible to have detailed knowledge of the trace of the patterns transmitted on a bus following execution of a specific application. The approach is compared with the most efficient encoding schemes proposed in the literature on both multiplexed and separate buses. The results obtained demonstrate the validity of the approach, which on average saves up to 50% of the transitions normally required.

1 Introduction

The spread of intelligent, portable electronic systems (digital cameras, cellular phones, PDAs, etc.) has made power consumption one of the main optimization targets in the design of embedded systems. A significant amount of the power dissipated in a system is known to be due to off-chip communication, for example between the microprocessor and the external memory. In the last few years the spread of systems integrated on a single chip, or systems-on-a-chip (SoC), and the use of increasingly advanced technology, has shifted the problem of power consumption on communication buses to the on-chip level as well — for example, in systems interconnecting the cores of a SoC. The ratio between wire and gate capacity has, in fact, gone from 3, in old technologies, to 100 in more recent ones [9] and is continuing to grow, as foreseen in [6]. For this reason, approaches aiming to reduce power consumption, which up to quite recently concentrated on the computing logic, are now tending to focus more and more on the communication system.

As the power dissipated by bus drivers is proportional to the product of the average number of transitions and the capacity of the lines of the bus, it may be a good idea to encode the information transiting on a communication bus so as to minimize the switching activity. Several approach have been proposed and can be grouped into two categories: *static* and *dynamic*.

J.J. Chico and E. Macii (Eds.): PATMOS 2003, LNCS 2799, pp. 21–30, 2003.
© Springer-Verlag Berlin Heidelberg 2003

Static techniques are based on a priori knowledge of the stream of patterns that will travel on the bus to generate an ad hoc encoder which will minimize the switching activity for that stream. The best-known static technique proposed in the literature is the *Beach solution* proposed by Benini *et al.* [1], who analyze the correlation between blocks of bits in consecutive patterns to find an encoding strategy that will reduce transition activity on a bus. In the approach proposed by Henkel *et al.* [4], as the most internal lines of a bus have a greater capacity, the reference trace is used to find a static permutation of bits that will reduce the activity on the lines with the greatest capacity.

Dynamic techniques don't require an advance knowledge of the stream of patterns: encoding decisions are made on the sole basis of past history. In the *bus-invert* technique [7] for data buses whose patterns are typically distributed in a random fashion a word to be transferred is inverted if the Hamming distance between it and the previous word is greater than the size of the bus divided by two. In this way the maximum number of lines that switch will be limited to half the size of the bus. Of course, an additional signaling line is needed to signal the type of encoding applied (inversion or no encoding).

Most of the bus encoding strategies proposed in the literature were envisaged for address buses. They all exploit the strong time correlation between one pattern and the next that is typically observed in an address stream. The high frequency of consecutive addresses, for instance, can be exploited by using *Gray* rather than natural binary encoding [8]. In this way the number of transitions for consecutive patterns will be limited to 1. If additional signaling lines are introduced, even better results can be obtained by using *T0* encoding [2] or one of the several variations on it [3]. The basic idea is to freeze the bus if the address to be sent is consecutive to the one sent previously (the receiver will generate the address locally).

The approach proposed in this paper refers to embedded applications, i.e. ones in which it is possible to know in advance the trace of the patterns transmitted on a communication bus following execution of a specific application. The trace is used to generate a truth table for an encoder that will minimize switching activity on the bus, which can thus be synthesized using any automatic logical synthesis tool. As the problem of searching for the encoder that will minimize switching activity can be viewed as an optimization problem, the exploration strategy adopted uses a heuristic method based on genetic algorithms (GAs).

Furthermore a simple compression algorithm has been proposed to reduce the considerable size of the trace files (used as input in static approaches). Using the compressed memory reference trace file processing times have been drastically reduced with the same the efficiency.

The results obtained on a set of benchmarks confirm the validity of the approach: the saving in terms of transitions is greater than that obtained by the most efficient techniques so far proposed in the literature.

The rest of the paper is organized as follows. In Section 2 we present our GA-based proposal for the generation of an optimal encoder. In Section 3 we present the result obtained. Finally, in Section 4 we draw our conclusions.

2 Our GA-Based Proposal

In this section we will propose a static technique based on GAs for efficient generation of an encoder to minimize switching activity on a bus. The design flows is shown in Figure 1. Henceforward reference will be made to address buses, i.e. buses on which addresses travel (e.g. the addresses of instructions to be executed or general addresses generated by load/store instructions).

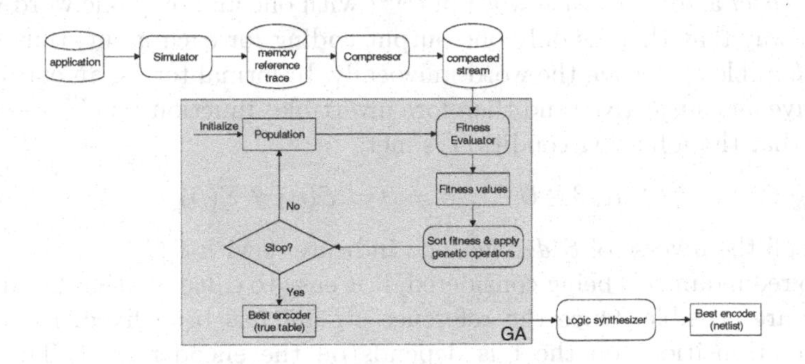

Fig. 1. Encoder design flow.

Starting from the specific application which is executed (simulated, emulated etc.), a memory reference trace file is obtained. In order to facilitate generation of the encoder, these trace file is compressed, as will be explained in the following subsections. The compressed stream is used to generate the optimal encoder which is expressed in the form of a truth table. As encoder generation can be viewed as an optimization problem, we used GAs to explore the design space. Initially a population of encoders is initialized with random encoders and evaluated on the compressed stream. Each encoder in the population has an associated fitness value which represents a measure of its capacity to reduce the number of transitions on the bus. Encoders with higher fitness values are therefore those which determine a lower number of transitions on the bus when stimulated with the compressed stream. The classical genetic operators, suitably redefined for this specific context, are applied to the population and the cycle is repeated until a stop criterion is met. At the end of the process, the individual with the highest fitness value is extracted from the population. This individual will be the optimal encoder being sought. The last step in the flow is therefore logical synthesis of the optimal encoder, which can be done using any automatic logical synthesis tool. To obtain the encoder it will, of course, be sufficient to exchange the encoder input and output columns and perform the synthesis.

The rest of this section is organized as follows. We will first classify encoder generation as a classical optimization problem that can be solved by using design space exploration techniques. We will then solve the problem of the size of the stream of references by means of a simple process of compression that will exponentially increase the computational efficiency of the approach. Then we will deal with the problem of exploring the design space using GAs.

2.1 Formulation of the Problem

Let us consider a binary alphabet to compose words with a fixed length of w bits. Let $U^{(w)}$ be the universe of discourse for words of w bits (i.e. the set of words it is possible to form with w bits). The cardinality of $U^{(w)}$ is therefore 2^W.

An *encoder* associates each word in $U^{(w)}$ with one and only one word in $U^{(w)}$ in such a way that there is only one output coding for each input, thus making the *decoder* able to decode the word univocally. In formal terms, an encoder $\mathcal{E}$ is an injective and surjective (and therefore invertible) function $\mathcal{E} : U^{(w)} \longrightarrow U^{(w)}$, i.e. such that the following condition is met:

$$\forall\, \alpha, \beta \in U^{(w)},\ \alpha \neq \beta \Rightarrow \mathcal{E}(\alpha) \neq \mathcal{E}(\beta) \qquad (1)$$

We will call the inverse of $\mathcal{E}$ *decoder* and indicate it with $\mathcal{E}^{-1}$.

As no redundancy is being considered, it is easy to calculate that $2^w!$ different encoders are possible. Once the reference stream has been fixed, the ensuing number of transitions on the bus depends on the encoder used. The aim is therefore to find the optimal encoder that will minimize the number of transitions on a bus for a specific reference stream.

2.2 Compression of the Reference Stream

Evaluation of an encoder is a computationally onerous task. The memory reference trace file produced following execution of an application typically comprises hundreds of millions of references. It is therefore advisable to compress the stream so as to obtain a stream with an upper bound on the number of patterns. If S is the initial stream and S^* is the compressed one, the optimal encoder obtained using S^* has to be the same as the one that would have been obtained if we had used S as the input to the encoder design flow. The compression is therefore lossless for encoder generation purposes.

Rather than compression, the technique used is based on a different representation of the reference stream. Let us consider a bus with a width of w. A reference stream is a sequence of patterns. Each pattern is an address of w bits. A compressed stream is also a sequence of patterns. A generic pattern is a 3-tuple $\langle r_i, r_j, n_{ij} \rangle$ with $i, j = 0, 1, 2, \ldots, 2^w - 1$ and $i > j$ that specifies the number of occurrences n_{ij} when the references r_i ad r_j are consecutive in S. The meaning of the condition $i > j$ can be explained by observing that, for our purposes, it is only necessary to know what the consecutive addresses are and not their order. If inverted, in fact, the number of transitions does not change. Using this transformation, the maximum number of patterns in S^* wil be:

$$L(w) = 1 + 2 + 3 + 4 + \ldots + (2^w - 1) = \frac{2^w \times (2^w - 1)}{2}$$

whatever the number of patterns in S.

For example, if $S = [12, 3, 12, 12, 5, 7, 12, 5, 7, 5]$ the compressed stream will be made up of the following four patterns $S^* = [\langle 3, 12, 2\rangle, \langle 5, 7, 3\rangle, \langle 5, 12, 2\rangle, \langle 7, 12, 1\rangle]$

2.3 GA-Based Bus Encoding

The design space of an encoder, which includes all the encoders that could possible be realized, grows in a factorial fashion along with the number of words to be encoded, which in turn grows exponentially with the size of the bus.

In general, when the space of configurations is too large to be explored exhaustively, one solution is to use evolutionary techniques. Genetic algorithms have been used in several VLSI design fields [5].

The approach we propose uses genetic algorithms as the optimization tool. Application of GAs to an optimization problem requires definition of the following four attributes: the chromosome, the fitness function, the genetic operators and the stop criterion. Each of these attributes will be defined with specific reference to the case study being investigated in the following subsections.

The Chromosome. The chromosome is a representation of the format of the solution to the problem being investigated. In our case it is a representation of an encoder. The representation we chose consists of encoding the truth table of an encoder. In this way the chromosome will be made up of as many genes as there are rows in the truth table of an encoder. The gene in position i represents encoding of the word i. That is, for an encoder of w bits, we will have 2^w genes. The i-th gene will represent encoding of the binary word that encodes i with w bits.

The chromosome is represented as a table with 2^w rows and 2 columns. Each row corresponds to a gene. Once the generic row i is fixed, the first column represents the encoding of i, while the second gives the position of the gene whose encoding is i.

The Fitness Function. The fitness function measures the fitness of an individual member of the population. In our case the fitness function assigns each encoder a numerical value that measures its capacity to reduce switching activity on a bus. Naturally, the fitness function will depend not only on the encoder but also on the reference stream the encoder is stimulated by.

If E indicates the chromosome that maps the encoder and S^* the compressed reference stream, the number of transitions on the bus caused by S^* when no encoding scheme is applied (binary encoding) is:

$$T_{woe}(S^*) = \sum_{i=1}^{rows(S^*)} H(S^*[i,1], S^*[i,2]) \times S^*[i,3]$$

where H is the *Hamming distance* function and $S^*[i,j]$ is the j-th field of the i-th pattern in the stream S^*. Considering the i-th pattern, $S^*[i,1]$ and $S^*[i,2]$

are the pair of successive references and $S^*[i,3]$ is the number of occurrences of this pair. The number of transitions on the bus due to S^* when it is filtered by the encoder E is:

$$T_{we}(E,S^*) = \sum_{i=1}^{rows(S^*)} H(E[S^*[i,1],1], E[S^*[i,2],1]) \times S^*[i,3]$$

In short, the fitness function is defined as follows:

$$f(E,S^*) = \frac{T_{woe}(S^*) - T_{we}(E,S^*)}{T_{woe}(S^*)} \qquad (2)$$

that is, $f(E,S^*)$ returns the number of transitions saved when the encoder E is used for the stream S^*.

The Genetic Operators. The fitness function defined by Equation (2) is applied to each individual in the population (i.e. to each encoder). The aim of the GA is thus to make the population evolve so as to obtain individuals with increasingly higher fitness values.

The individual with the highest fitness value will always be inserted into the new population (elitism). The encoders making up the population are selected with a probability directly proportional to their fitness value. With a user-defined probability the genetic operators are applied to them and they are inserted into the new population. This selection process is repeated until the new population reaches the size established by the user. Three genetic operators are used: Mutation, Permutation and Cross-over.

They were appropriately redefined so as to guarantee that application to an encoder (in the case of mutation and permutation) or a pair of encoders (in the case of cross-over) always gives rise to an encoder.

Mutation. Mutation is a unary operator that is applied with a certain probability (which we will call *mutation probability*) to an encoder E. Let E be an encoder of w bits. Application of the mutation operator to E consists of varying a single bit of an encoding chosen at random.

More specifically, a random number i between 0 and $2^w - 1$ is extracted and one of the w bits (also chosen at random) of the encoding of i is inverted.

Permutation. Permutation is a unary operator applied with a certain probability (which we will call *permutation probability*) to an encoder E. Let E be an encoder of w bits. Application of the permutation operator to E consists of exchanging the encoding of two randomly chosen words.

Cross-over. Cross-over is a binary operator that is applied to two elements of the population with a certain probability that we will call *cross-over probability.* Let E_1 and E_2 be two encoders of w bits. When applied to the two encoders E_1 and E_2 the cross-over operator swaps the coding of a word chosen at random from the 2^w that are possible. That is, it randomly extracts the word i, and $E_1[i,1]$ are swapped with $E_2[i,1]$. Of course, to guarantee that the condition (1) is met by the new encoders generated, $E_1[i,2]$ and $E_2[i,2]$ have to be updated.

The Stop Criterion. Various stop criteria can be used, for example a user-managed parameter or an automatic criterion (e.g. convergence-based) that stops the process of evolution when no appreciable improvements in the species have been observed for a few generations. The stop criterion we will use is based on a user-defined maximum number of generations.

3 Experimental Results

In this section we will present the results obtained by applying our approach, comparing them with the most effective approaches proposed in the literature.

The application scenario referred to is encoding of the addresses transmitted on a 32-bit bus and generated by a processor during execution of a specific application. Two cases are considered: (i) the bus is multiplexed, (ii) it is a dedicated bus. In the former case the addresses travelling on the bus refer to both fetching instructions and accesses generated by load/store instructions. In the latter case, the bus considered is dedicated address bus for fetching instructions (e.g. the address bus between a processor and an instruction cache).

The 32-bit bus is partitioned with clusters containing the same number of bits and the approach was applied to each cluster separately. It is, in fact, computationally unfeasible to apply the approach to the whole bus, given that the data structure used would require tables of 2^{32} rows to be handled. The cases studied referred to clusters of 4 and 8 bits.

We considered the same reference traces as are used in [1], generated following the execution of specific applications in the field of image processing, automotive control, DSP etc.. More specifically, `dashb` implements a car dashboard controller, `dct` is a discrete cosine transform, `fft` is a fast Fourier transform and `mat_mul` a matrix multiplication.

In all the experiments that will be discussed in the following subsections, we considered a population of 10 individuals, a mutation probability of 50%, a permutation probability of 25%, and a cross-over probability of 25%.

The stop criterion used consists of blocking iterations when no appreciable improvements in the fitness value of the best encoder in the population have been observed for a certain number of generations.

3.1 Address Bus (Fetch + Load/Store)

In this subsection we will comment on the results obtained when encoding is applied to a multiplexed address bus (i.e. one on which addresses generated by both fetch and load/store instructions are travelling).

Table 1 summarizes the results obtained. The first column (*bench*) identifies the benchmark. The second (*trans.*) gives the total number of transitions on the

Table 1. Transitions saving for the address multiplexed bus.

bench	*trans.*	*GEG8*		*GEG4*		*Beach*		*Others*	
		trans.	*saving*	*trans.*	*saving*	*trans.*	*saving*	*trans.*	*saving*
dashb	619680	317622	48.74%	435516	29.71%	443115	28.40%	486200	21.50%
dct	48916	28651	41.40%	33179	32.17%	31472	35.60%	39327	19.60%
fft	138526	67468	51.30%	85405	38.30%	85653	38.10%	100127	27.70%
mat_mul	105950	50552	52.30%	60446	42.90%	60654	42.70%	77384	26.90%
Average saving		48.44%		35.77%		36.20%		23.93%	

bus when no encoding scheme is applied. The remaining columns (in groups of two) give the number of transitions for each approach (*trans.*) and the percent saving in transitions as compared with the case in which no encoding scheme is applied (*saving*). *GEG8* and *GEG4* represent the same implementation of the approach applied to partitioned buses of 4 and 8 bits respectively. *Beach* is the approach proposed in [1]. *Others* indicates the best result obtained by the encoding schemes *Gray* [8], *T0* [2], *Bus-invert* [7], *T0+Bus-invert*, *DualT0* and *DualT0+Bus-invert* [3]. As can be seen, *GEG4* is on average equivalent to *Beach*. Increasing the size of the clusters to 8 bits increases the saving by about 13% as it is possible to exploit the temporal correlation between the references more fully.

3.2 Address Bus (Fetch Only)

When the address bus is not multiplexed the percentage of addresses in sequence increases considerably. Table 2 gives the results obtained on an address bus carrying references to fetch operations alone.

Table 2. Transitions saving for fetch only address bus.

bench	*in-seq*	*trans.*	*GEG8*		*Gray*		*T0*		*GEG8+T0*	
			trans.	*saving*	*trans.*	*saving*	*trans.*	*saving*	*trans.*	*saving*
dashb	55.88%	111258	65694	40.96%	70588	36.55%	41731	62.49%	30182	72.87%
dct	60.31%	11675	6639	43.13%	6885	41.02%	2851	75.58%	1851	84.14%
fft	59.92%	25017	14486	42.09%	15969	36.16%	7021	71.93%	5063	79.76%
mat_mul	63.63%	26814	13802	46.08%	17095	36.24%	7850	70.72%	4345	83.79%
Average savings			43.06%		37.49%		70.18%		80.14%	

In this case *T0* achieves much better savings than the other approaches by exploiting the high percentage of addresses in sequence which do not determine any transitions on the bus. *GEG8* maintains its efficiency with average savings of over 40%. The efficiency of *T0* can be further enhanced by using a hybrid approach *GEG+T0*. Figure 2 shows a scheme of how this can be achieved. The pattern to be transmitted is encoded with both *T0* and *GEG*. If it is in sequence

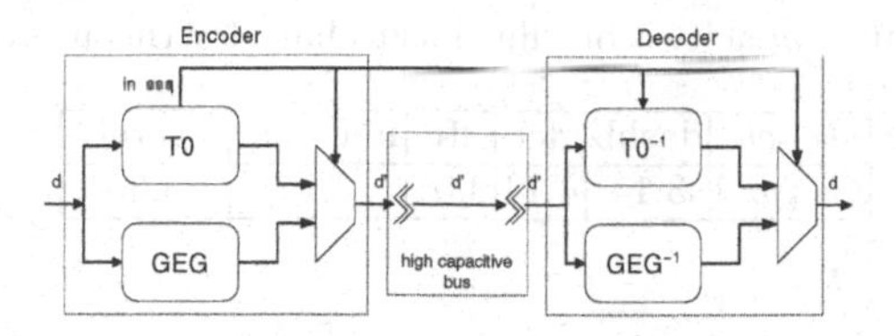

Fig. 2. Block diagram of the *GEG+T0* encoder.

with the previous one, the *T0* encoding is transmitted; otherwise the *GEG* encoding is transmitted. Even though *GEG+T0* is extremely efficient at reducing the amount of power dissipated on the bus, in calculating the saving account has to be taken of the overhead due to power consumption by the encoding/decoding logic. In *GEG+T0*, in fact, this contribution is certainly greater than that of both *T0* and *GEG*, as it contains them both, and both are active at the same time. Another point against *GEG+T0* is that in inherits from *T0* the use of a signaling line that is not present in *GEG*.

3.3 Overal Power Analysis

Table 3 gives the area, delay and power characteristics of the encoders and decoders generated by *GEG* for 8-bit clusters and the benchmarks described previously. The results were obtained using Synopsys DesignCompiler for the synthesis, and Synopsys DesignPower for the power estimation. The circuits were mapped onto a $0.18\mu m$, $1.8V$ gate-library from STMicroelectronics. The clock was set to a conservative frequency of 50 Mhz (i.e. a period of 20 ns). The average delay introduced by the encoder/decoder is, in fact, shorter than 4 ns and so less than 20% of the clock cycle is dedicated to encoding and decoding information.

Table 3. Area, delay and power characteristics of the encoders and decoders.

bench	*Area* (μm^2)		*Delay (ns)*		*Power (mW)*	
	Encoder	*Decoder*	*Encoder*	*Decoder*	*Encoder*	*Decoder*
dashb	25456.64	25477.12	1.93	1.89	0.397	0.511
dct	25989.12	24985.60	1.90	1.97	0.282	0.398
fft	25325.57	25804.80	1.79	2.00	0.352	0.515
mat_mul	21733.38	21729.28	1.87	1.96	0.244	0.462
Average	24626.18	24499.20	1.87	1.96	0.319	0.472

An encoding scheme is advantageous when the power saved on the bus (due to less activity) is greater than the power consumed by the encoding and decoding blocks. Table 4 summarises the minimum capacity a bus line has to have for the approach to be effective for each benchmark.

Table 4. The minimum capacity a bus line has to have for the approach to be effective.

bench	dashb	dct	fft	mat_mul	Average
C_l *(pF)*	3.14	4.14	3.01	3.15	3.36

4 Conclusions

In this paper we have presented a GA-based strategy for designing an encoder that will minimize switching activity on a bus. The strategy, called *Genetic Encoder Generator* (*GEG*) has been compared with the most effective techniques proposed in the literature. The results obtained on a set of specific applications for embedded systems have demonstrated the superiority of our approach, with savings of around 50% on multiplexed address buses (instructions/data) and close to 45% on instruction address buses. In the latter case the *T0* scheme [2] performs better than the approach proposed here, with average savings of 70%. A mixed technique *GEG+T0* (in which a *GEG* and *T0* works concurrently) further enhances the efficiency of *T0*, achieving average savings of 80%.

References

1. Luca Benini, Giovanni De Micheli, Enrico Macii, Massimo Poncino, and Stefano Quer. Power optimization of core-based systems by address bus encoding. *IEEE Transactions on Very Large Scale Integration*, 6(4), December 1998.
2. Luca Benini, Giovanni De Micheli, Enrico Macii, Donatella Sciuto, and Cristina Silvano. Asymptotic zero-transition activity encoding for address busses in low-power microprocessor-based systems. In *Great Lakes Symposium VLSI*, pages 77–82, Urbana, IL, March 1997.
3. Luca Benini, Giovanni De Micheli, Enrico Macii, Donatella Sciuto, and Cristina Silvano. Address bus encoding techniques for system-level power optimization. In *IEEE Design Automation and Test Conference in Europe*, pages 861–866, Paris, France, February 1998.
4. Jörg Henkel, Haris Lekatsas, and Venkata Jakkula. Encoding schemes for address busses in energy efficient SOC design. In *VLSI-SOC 2001 11th International Conference of Very Large Scale Integration*, Montpellier, France, December 2001.
5. Pinaki Mazumder and Elizabeth M. Rudnick. *Genetic Algorithms for VLSI Design, Layout & Test Automation.* Prentice Hall, Inc., 1999.
6. International thechnology roadmap for semiconductors. Semiconductor Industry Association, 1999.
7. Mircea R. Stan and Wayne P. Burleson. Bus invert coding for low power I/O. *IEEE Transactions on VLSI Systems*, 3:49–58, March 1995.
8. C. Su, C. Tsui, and A. Despain. Saving power in the control path of embedded processors. *IEEE Design and Test of computers*, 11(4):24–30, 1994.
9. Neil H. West and Kamran Eshraghian. *Principles of CMOS VLSI Design.* Addison Wesley, 1998.

State Encoding for Low-Power FSMs in FPGA

Luis Mengibar, Luis Entrena, Michael G. Lorenz, and Raúl Sánchez-Reillo

Electronic Technology Department
Universidad Carlos III de Madrid. Spain.
{mengibar, entrena, lorenz, rsreillo}@ing.uc3m.es

Abstract. In this paper, we address the problem of state encoding of FPGA-based Finite State Machines (FSMs) for low power dissipation. Recent work on this topic [1] shows that binary encoding produces best results for small FSMs (up to 8 states) while one-hot encoding produces best results for large FSMs (over 16 states). Departing from these results, we analyze other encoding alternatives that specifically take into account state transition probabilities. More precisely, we consider minimal-bit minimal Hamming distance encoding, zero-one-hot encoding and a partitioned encoding that uses a combination of both minimal-bit encoding and zero-one-hot encoding. Experimental results demonstrate that the proposed encoding techniques usually produce better results than the binary or one-hot encodings. Savings up to 60% can be obtained in the dynamic power dissipation by using the proposed encoding techniques.

1 Introduction

Low-power dissipation is nowadays a major concern in digital system design. Reducing power consumption allows using lower cost packages, cooling systems and power supplies, increases chip reliability and enlarges battery autonomy. Low power is also a key issue in FPGA design because of the increasing capacity and performance of modern FPGAs.

Minimizing power dissipation of Finite State Machines (FSMs) is an important goal, because they are very common in digital systems. Moreover, FSMs often run at full clock speed and contribute significantly to the total power dissipation. An optimal selection of the state encoding is crucial for low-power design of FSMs.

The state encoding problem poses a tradeoff between the code size and the complexity of state encoding and decoding. A minimal-bit encoding reduces the power consumption at the clock lines by using a minimal set of flip-flops. However, it requires more logic for state encoding and decoding, and power dissipation may increase because of the extra capacitance introduced. In the ASIC world, most of the work has been oriented to reduce switching activity by minimizing the Hamming distance of the most probable state transitions. However, in the FPGA case, the impact of the encoding and decoding logic on power consumption is usually higher. Consequently, one-hot encoding is usually recommended by FPGA manufacturers and synthesis tools to reduce power consumption on FSMs [2], [3]. Recent results using Xilinx 4K-series FPGAs [1] show that binary encoding reduces power

J.J. Chico and E. Macii (Eds.): PATMOS 2003, LNCS 2799, pp. 31–40, 2003.
© Springer-Verlag Berlin Heidelberg 2003

consumption for small FSMs (up to 8 states) while one-hot encoding produces best results for large FSMs (over 16 states).

Notwithstanding, reducing the Hamming distance of the most probable state transitions can be useful for low power design of FSMs in FPGAs too. In this work, we address the state encoding problem for LUT-based FPGAs and analyze several approaches to reduce power consumption based on the knowledge of the state transition probabilities. In particular, we consider minimal-bit minimal Hamming distance encoding as an alternative to binary encoding, and zero-one-hot encoding as an alternative to one-hot encoding. Finally, we propose a partitioned encoding that uses a combination of both minimal-bit encoding and zero-one-hot encoding in order to get the best of both approaches. Experimental results demonstrate the effectiveness of the proposed encoding techniques.

The remaining of the paper is as follows. In Section 2, we introduce some basic concepts and review the previous work for low-power state encoding. Section 3 presents the improved state encoding approaches proposed in this work. Section 4 summarizes the experimental results for benchmark FSMs. Finally, Section 5 presents the conclusions of this work.

2 Background

A Finite State Machine (FSM) is defined by a 6-tuple M = (I, O, Q, q0, δ, λ). Where I is a finite set of input symbols, O is a finite set of output symbols, Q is a finite set of states, $q0 \in Q$ is the reset state, $\delta: Q \times I \rightarrow Q$ is the state transition function and $\lambda: Q \times O \rightarrow O$ is the output function.

A finite state machine can be conveniently described by a State Transition Graph (STG) denoted by G(V,E), where the nodes in V represent the states $S_i \in Q$ and the edges in E, labeled with inputs and outputs, represent the transitions between states $S_i, S_j \in Q$.

A FSM with N states requires a state register with al least $\log_2 N$ bits to be properly encoded. The selection of a particular state encoding may have a dramatic impact on the area, delay and power of the circuit that implements the FSM. For low power design, the objective is to find a encoding that minimizes the power dissipation. In a CMOS circuit, dynamic power is the dominant source of power dissipation and can be expressed by the well-known formula

$$P = \sum_n C_n \alpha_n V_{DD}^2 \tag{1}$$

where C_n is the load capacitance at node n, α_n is the switching activity and V_{DD} is the supply voltage. This formula shows that the reduction of dynamic power dissipation must come from a reduction of capacitive loads and switching activity.

In a FSM, the transition from a state S_1 to a state S_2 will produce the switching of as many flip-flops in the state register as the number of bit toggles between the encodings of S_1 and S_2. The number of bit toggles between two states is also known as the *Hamming distance* between them. By reducing the number of bit toggles of the state register in each state transition, the overall switching activity may be possibly reduced [4]. Obviously, most common state transitions would likely have a bigger impact on the power dissipation and should be encoded with the smaller Hamming distance. To this purpose, we can use state transition probabilities. This concept is summarized in the following section.

2.1 State Transition Probabilities

Let $p(S_i)$ denote the *steady state probability* of state S_i, i.e., the probability of the FSM being in state S_i, and $p(S_i, S_j)$ be the *conditional state transition probability* from state S_i to state S_j, i.e., the probability of the FSM being in state S_i and making a transition to S_j.

Conditional state transition probabilities $p(S_i, S_j)$ can be first determined from the static probabilities of the inputs of the FSM and the STG description. Then, steady state probabilities $p(S_i)$ can be determined by modeling the STG as a Markov chain and solving the following system of linear equations [4],[5]:

$$p(S_i) = \sum_j p(S_j)\, p(S_j, S_i)$$
$$\sum_j p(S_i) = 1 \qquad (2)$$

The *total transition probability* $P(S_i, S_j)$ is the probability of the STG making a transition from state S_i to state S_j in a clock cycle and can be obtained from the expression

$$P(S_i, S_j) = p(S_i)\, p(S_i, S_j) \qquad (3)$$

Transition probabilities $P(S_i, S_j)$ can be approximately estimated also by a high-level simulation of the FSM in a context close to the real workload of the design. As the number of clock cycles being simulated increases, the normalized state transition frequency converges to the transition probability [5].

In order to evaluate the switching activity associated to each state, we also define the *total state transition probability* of a state S_i, $P_T(S_i)$, as the probability of the STG making a transition from state S_i to any other state S_j, $i \neq j$, or viceversa

$$P_T(S_i) = \sum_{j, j \neq i} P(S_i, S_j) + \sum_{j, j \neq i} P(S_j, S_i) \qquad (4)$$

$P_T(S_i)$ can be viewed as a measure of state activity. It measures the probability of the STG making a transition to or from a state S_i, excluding self-loops.

2.2 Approaches for Low Power State Encoding

State encoding of FSMs is a classical problem. Traditionally, it has been oriented to minimize area or delay [6],[7],[8]. State encoding for low power has been targeted in the last years. In [4], a state assignment algorithm is proposed that reduces switching activity by minimizing the Hamming distance between the codes of the states with high transition probability, while taking into account the estimated area of the next state and output logic. A probabilistic description of the FSM is used to provide a theoretical framework for low power state encoding. A close approach is presented in [9]. State encoding with non-minimal state register size is focused in [10]. Other recent contributions are provided in [11], [12],[13].

2.3 FPGA State Encoding for Low Power

State encoding for FSMs in LUT-based FPGAs has been specifically addressed in [1]. They measure the power consumption of a set of FSM benchmarks implemented on a FPGA using different encoding approaches, namely binary, one-hot, area optimized using output encoding, and two-hot. They conclude that binary encoding produces the best results for small FSMs (up to 8 states) and one-hot encoding produces the best results for large FSMs (more than 16 states). In addition, they show that power consumption has a strong correlation with the area occupied by the circuit.

3 Improved Techniques for Low Power State Encoding in FPGAs

Previous work on low-power state encoding in FPGAs has concluded that binary encoding is best for small FSMs and one-hot encoding is best for large FSMs. However, reducing the Hamming distance of the most probable state transitions can be useful for low power design of FSMs in FPGAs too. In this section, we introduce three different state encoding techniques that are based on this idea. The first approach reduces the Hamming distance for minimal-bit encoded machines. The second approach focuses on zero-one-hot code. The third approach combines the previous two approaches into an intermediate size encoding in order to get the best of both. Experimental results to validate these approaches will be provided in Section 4.

3.1 Minimal-Bit Minimal Hamming Distance Encoding

According to [1], binary encoding should be used for small FSMs (up to 8 states). The rationale behind this result is that state encoding and decoding in small FSMs, does not take a large amount of combinational logic, and power is benefited from the reduction in the state register size.

However, for a FSM there are many possible state encodings using a minimal code size. Therefore, we can consider a encoding that minimizes the Hamming distance of the encodings weighted by the transition probabilities, that is, a encoding that minimizes the following cost funcion:

$$F = \sum_i \sum_j P(S_i, S_j)\, D(code(S_i), code(S_j)) \tag{5}$$

where $P(S_i, S_j)$ is the transition probability between states S_i and S_j, and $D(code(S_i), code(S_j)$ is the Hamming distance between the encodings of states S_i and S_j.

Algorithms to find a state encoding that minimizes the above cost function are proposed in [4] and [9]. In practice, a minimal-bit minimal Hamming distance code can be exactly determined for small FSMs (up to 8 states) by computing all possible state assigments. For larger FSMs, an approximation can be found by evaluating the cost function over a randomly chosen subset of the possible state encodings.

3.2 Zero-One-Hot Encoding

One-hot encoding uses N bits to encode N states by assigning a 1 in a different position to each of the state codes. This code results in a Hamming distance of 2 for any transition between to different states. However, one of the codes in one-hot encoding is in fact redundant. Zero-one-hot encoding uses just N-1 bits to encode N states by assigning one of the states a code of all 0´s. The Hamming distance in zero-one-hot encoding is 2, except for the transitions involving the all-zero state, that have a distance of 1.

Zero-one-hot encoding can be exploited for low power by assigning the all-zero code to the state with higher total transition probability. This way, the most common state transitions will be benefited from a reduction of the Hamming distance.

3.3 Partitioned Encoding

Minimal-bit encoding uses a minimal set of flip-flops ($\log_2 N$ flip-flops) but requires complex state encoding and decoding logic. On the other hand, one-hot encoding simplifies state encoding and decoding by using additional flip-flops (N flip-flops). We propose a partitioned encoding approach that is intended to exploit the benefits of both minimal-bit and one-hot encodings.

In our partitioned encoding approach, the state set is divided into two subsets according to the total transition probabilities of the states, $P_T(S_i)$. To this purpose we consider a threshold probability PT. The first subset contains the states with $P_T(S_i) \geq PT$ and the second subset contains the states with $P_T(S_i) < PT$.

Then, the state register is partitioned in two fields. Each of the two fields is used to code one of the state subsets and is assigned all zeros for the states in the other subset. For the first subset of states (those having probabilities greater than the threshold probability PT) we use zero-one-hot encoding. For the second subset of states (those having probabilities lesser than the threshold) we use minimal-bit encoding.

Fig. 1 shows one example of partitioned encoding. In this example, the state register has 7 bits. The first 4 bits correspond to the first field, which is used to encode the states with higher activity. Zero-one-hot encoding is used for these states in order to simplify encoding and decoding logic. The last 3 bits correspond to the second field, which is used to encode the states with lower activity. Minimal-bit encoding is preferred for these states in order to reduce the number of flip-flops.

	Field1	Field2
	0000	000
	1000	000
Pi > PT	0100	000
	0010	000
	0001	000
	0000	001
	0000	010
Pi < PT	0000	011
	0000	100
	0000	101
	0000	...

Fig. 1. Example of partitioned encoding

For a FSM with N states and M < N states in the first subset, this encoding approach produces codes of length $M+\log_2(N-M+1)$. The length of the resulting code is an intermediate between the minimal-bit and the zero-one-hot encodings, because it always satisfies the relation $N \geq M+\log_2(N-M+1) \geq \log_2(N)$.

4 Experimental Results

4.1 Experiment Environment

All the FSMs used in the experiments are from the LGSynth93 benchmark set [14]. The original FSMs are described in KISS2 format [15]. In order to minimize the number of states, we use first the tool STAMINA, included in the SIS release [15]. The main features of the original and state minimized FSMs are shown in Table 1.

Since the minimized circuits are described in KISS2 format, we need to translate the benchmark set to a description language that can be read by a commercial tool. To this purpose, we have developed a tool to translate FSMs in KISS2 format to VHDL. At the same time we generate the encodings for the different approaches to be compared, namely binary encoding (BIN), minimal-bit minimal Hamming distance encoding (MHD), one-hot encoding (OH), zero-one-hot encoding (ZOH) and partitioned encoding (PE). For the latter case, a quite conservative threshold probability $P_T = 0.01$ was set. For each benchmark, the VHDL description only differs in the state encoding.

The VHDL despriptions of the benchmarks where synthesized using FPGA Compiler™ from Synopsys. All the circuits were strongly optimized for delay using high effort option. Then Place & Route was run using Foundation 4.1i tools [17] from Xilinx. All the circuits were placed and routed for the same target FPGA, a 2s15tq144-6 (192 Slices, 384 four inputs LUTS and 384 FFs, Spartan II family).

Table 1. Original and minimized benchmark circuits

FSM	Original FSM			Minimized FSM			FSM	Original FSM			Minimized FSM		
	Inputs	Outputs	States	Inputs	Outputs	States		Inputs	Outputs	States	Inputs	Outputs	States
bbara	4	2	10	4	2	7	lion9	2	1	9	2	1	4
bbsse	7	7	16	7	7	13	mark1	5	16	15	5	16	12
bbtas	2	2	6	2	2	6	mc	3	5	4	3	5	4
beecount	3	4	7	3	4	4	opus	5	6	10	5	6	9
cse	7	7	16	7	7	16	planet	7	19	48	7	19	48
dk14	2	5	7	3	5	7	sand	11	9	32	11	9	32
dk15	3	5	4	3	5	4	sse	7	7	16	7	7	13
dk16	2	3	27	2	3	27	styr	9	10	30	9	10	30
dk17	2	3	8	2	3	8	s1	8	6	20	8	6	20
dk27	1	2	7	1	2	7	s1494	8	19	48	8	19	48
dk512	1	3	15	1	3	15	s27	4	1	6	4	1	5
ex1	9	19	20	9	19	18	s386	7	7	13	7	7	13
ex2	2	2	19	2	2	14	s510	19	7	47	19	7	47
ex3	2	2	10	2	2	5	s820	18	19	25	18	19	24
ex4	6	9	14	6	9	14	s832	18	19	25	18	19	24
ex6	5	8	8	5	8	8	tav	4	4	4	4	4	4
ex7	2	2	10	2	2	4	tbk	6	3	32	6	3	16
keyb	7	2	19	7	2	19	train11	2	1	11	2	1	4

Power dissipation was estimated using XPower [18] from Xilinx. To this purpose, a post-layout simulation using a set of random input vectors has been made using Modelsim [16]. A clock frequency of 50 MHz was set for all benchmarks. Power estimation can be considered accurate because we are using the power estimation tool provided by the manufacturer on a completely placed and routed design. On the other hand, Xilinx XPower tool provides estimations for the various sources of power dissipation. This feature allows to isolate dynamic power from other power sources (I/O power, static power, etc...) that are only meaningful in the context of the complete circuit the FSM belongs to. The results reported in the following section report only dynamic power due to internal logic and nets, as given by XPower.

4.2 Benchmark Results

Table 2 shows the results obtained using the workbench described in the previous section for the BIN, MHD, OH and ZOH state encodings. For each example, the area (in terms of LUTs and flip-flops), delay and power measures are given. Partitioned encoding (PE) is only different to zero-one-hot encoding for a subset of the benchmarks. The results obtained with partitioned encoding for these examples are shown in Table 3, along with the OH and ZOH results for comparison purposes.

For almost all the examples, the proposed encoding techniques, MHD, ZOH and PE are the best encodings in terms of dynamic power dissipation. Binary encoding (BIN) gives the best result only in one case (dk15) and one-hot encoding (OH) in another case (s1) out of a total of 38 benchmarks.

For small machines (four states), MHD encoding is the best. Compared with BIN, reductions up to 20.65% can be obtained. For machines with a number of states between five and eigth, either ZOH or MHD are always the best. Power savings up to 55.33% can be obtained (dk17).

Table 2. Area, delay and power for the circuits of the benchmark set

FSM	FSM Descrip.			Area Bin		Area MHD		Area OH		Area ZOH		Delay (ns)				Power (mW)			
	Inputs	Outputs	States	LUTs	FF	LUTs	FF	LUTs	FF	LUTs	FF	BIN	MHD	OH	ZOH	BIN	MHD	OH	ZOH
bbara	4	2	7	21	3	16	3	22	7	18	6	9.3	9.0	10.4	9.5	4.69	4.09	5.21	4.41
bbsse	7	7	13	55	4	50	4	54	13	50	12	13.9	8.1	12.7	13.1	14.01	9.79	13.35	11.82
bbtas	2	2	6	6	3	6	3	9	6	10	5	8.0	8.2	8.1	8.6	1.96	1.84	2.76	2.73
beecount	3	4	4	10	2	10	2	13	4	13	3	9.6	9.6	10.4	10.4	3.60	3.60	4.09	4.09
cse	7	7	16	91	4	90	4	77	16	66	15	13.8	14.4	14.3	13.5	9.91	10.36	13.67	11.26
dk14	3	5	7	44	3	46	3	52	7	50	6	11.4	11.1	13.3	13.1	16.50	15.25	15.75	13.74
dk15	3	5	4	27	2	30	2	38	4	35	3	11.3	11.3	12.9	12.0	10.39	11.97	15.05	13.66
dk16	2	3	27	108	5	98	5	65	27	69	26	16.6	14.4	13.6	12.1	26.84	24.23	15.27	13.57
dk17	2	3	8	23	3	16	3	25	8	22	7	10.2	10.7	14.7	11.9	7.95	3.71	9.04	7.45
dk27	1	2	7	5	3	5	3	11	7	11	6	8.8	8.8	9.2	8.8	2.78	2.26	3.79	3.53
dk512	1	3	15	23	4	24	4	21	15	21	13	10.0	11.3	9.3	12.4	7.02	6.67	5.94	5.53
ex1	9	19	18	108	5	126	5	105	18	85	17	17.2	16.2	15.1	16.2	24.13	23.42	21.87	17.35
ex2	2	2	14	41	4	38	4	31	14	25	10	12.1	11.1	11.2	11.8	14.90	10.68	10.25	9.11
ex3	2	2	5	12	3	13	3	11	5	13	4	9.9	10.7	9.2	9.5	5.12	5.10	4.50	4.49
ex4	6	9	14	28	4	31	4	21	14	18	13	10.6	11.4	12.0	10.3	8.12	6.84	7.60	5.46
ex6	5	8	8	74	3	72	3	51	8	54	7	15.2	15.6	13.7	13.6	21.72	19.51	14.74	15.02
ex7	2	2	4	3	2	3	2	8	4	8	3	8.3	8.3	9.8	9.8	2.31	2.31	3.55	3.55
keyb	7	2	19	109	5	118	5	67	19	69	18	19.2	16.0	14.4	15.3	20.07	20.38	14.20	14.99
lion	2	1	4	3	2	3	2	7	4	7	3	7.7	7.7	8.3	8.2	1.44	1.38	2.44	2.24
lion9	2	1	4	3	2	3	2	5	4	6	3	8.3	8.3	7.7	7.9	1.30	1.19	1.80	1.98
mark1	5	16	12	36	4	36	4	24	12	27	11	10.6	11.1	12.6	11.1	9.61	8.51	17.05	6.84
mc	3	5	4	6	2	6	2	12	4	12	3	9.1	9.6	10.6	10.4	2.72	2.63	3.88	3.94
opus	5	6	9	36	4	39	4	28	9	27	8	12.7	11.9	12.0	12.1	8.33	6.30	7.19	5.97
planet	7	19	48	202	6	207	6	138	48	134	47	18.8	20.5	18.8	16.0	47.69	42.68	33.13	25.71
sand	11	9	32	189	5	223	5	176	32	164	31	16.9	18.1	18.1	16.4	32.37	36.28	35.73	29.58
sse	7	7	13	55	4	50	4	54	13	50	12	13.9	8.1	12.7	13.1	14.01	9.79	13.35	11.82
styr	9	10	30	181	5	214	5	129	30	120	29	16.0	17.8	19.0	16.4	21.17	26.41	22.84	19.56
s1	8	6	20	152	5	148	5	116	20	116	19	16.8	17.5	16.7	16.9	35.07	29.15	23.22	23.59
s1494	8	19	48	261	6	241	6	221	48	208	47	19.7	20.4	19.6	19.0	25.97	23.91	36.13	30.93
s27	4	1	5	19	3	16	3	20	5	20	4	10.7	9.5	9.6	9.9	6.08	5.17	7.16	6.88
s386	7	7	13	54	4	58	4	48	13	42	12	14.6	14.9	13.8	13.7	11.81	9.96	12.53	10.09
s510	19	7	47	99	6	99	6	89	47	80	46	15.2	15.2	15.4	15.6	18.77	18.77	17.29	16.73
s820	18	19	24	160	5	142	5	102	24	106	23	18.0	16.0	16.7	15.2	19.41	21.53	21.44	20.91
s832	18	19	24	148	5	140	5	108	24	101	23	15.0	16.8	14.4	16.4	19.22	21.61	20.65	18.60
tav	4	4	4	9	2	8	2	8	4	8	3	10.2	9.8	9.3	9.3	4.31	3.42	3.44	3.47
tbk	6	3	16	179	4	175	4	135	16	98	15	18.1	20.7	18.5	16.5	42.46	35.83	35.16	23.72
train11	2	1	4	3	2	3	2	5	4	5	3	8.3	8.3	8.1	8.1	1.39	1.39	2.07	2.02
train4	2	1	4	3	2	3	2	5	4	5	3	8.3	8.3	8.1	7.8	1.47	1.29	2.10	2.49

For middle size circuits (nine to sixteen states) the best encoding alternatives are usually ZOH or PE. Compared with OH, dynamic power savings averages 26.66 % with up to 60 % in the case (mark1). In all cases of a set of ten circuits, they always give substantial dynamic power reduction, with a minimun of 6.90 %. PE is slightly better than ZOH. A specific comparison between them is described later on in this section.

For large machines (more than sixteen states), the best encodings are ZOH and PE. Compared with OH, power savings up to 22.40 (ZOH) and 29.90 % (PE) can be achieved.

Table 3. Area, delay and power for comparison between ZOH and PE

FSM	FSM Description			Area OH		Area ZOH		Area PE		Delay (ns)			Power (mW)		
	Inputs	Outputs	States	LUTs	FF	LUTs	FF	LUTs	FF	OH	ZOH	PE	OH	ZOH	PE
bbsse	7	7	13	54	13	50	12	59	8	12.7	13.1	13.8	13.35	11.82	11.27
cse	7	7	16	77	16	66	15	72	9	14.3	13.5	13.6	13.67	11.26	8.44
ex1	9	19	18	105	18	85	17	87	13	15.1	16.2	15.8	21.87	17.35	15.85
keyb	7	2	19	67	19	69	18	88	8	14.4	15.3	16.1	14.20	14.99	13.86
opus	5	6	9	28	9	27	8	29	7	12.0	12.1	12.1	7.19	5.97	5.81
planet	7	19	48	138	48	134	47	147	42	18.8	16.0	17.9	33.13	25.71	28.31
sand	11	9	32	176	32	164	31	195	27	18.1	16.4	19.0	35.73	29.58	34.43
sse	7	7	13	54	13	50	12	59	8	12.7	13.1	13.8	13.35	11.82	11.27
styr	9	10	30	129	30	120	29	152	14	19.0	16.4	16.7	22.84	19.56	16.01
s1494	8	19	48	221	48	208	47	275	12	19.6	19.0	20.9	36.13	30.93	31.97
s386	7	7	13	48	13	42	12	50	8	13.8	13.7	13.3	12.53	10.09	10.00
s820	18	19	24	102	24	106	23	139	9	16.7	15.2	15.5	21.44	20.91	17.39
s832	18	19	24	108	24	101	23	151	9	14.4	16.4	11.1	20.65	18.60	17.64

The data for a specific comparison of ZOH and PE with respect to OH and for a comparison between ZOH and PE are presented in Table 3. From this table we can derive some interesting conclusions. First, ZOH is better than OH in all cases except one (in a set of thirteen benchmarks). The maximum reduction of dissipation obtained is of 22.40 %. PE is better than OH in all cases with power reductions up to 38,26 %. Comparing ZOH with PE, it results that ZOH is better than PE only the very large FSMs, with more than 32 states. It can be concluded that the threshold probability PT was not adequate in these three cases.

Summarizing these results, we can conclude that the proposed state encoding techniques are usually better than either binary encoding or one-hot encoding. MHD is usually the best for small FSMs, while PE is usually the preferred choice for medium and large FSMs.

5 Conclusions

Previous work on state encoding for FSMs implemented with LUT-based FPGAs concluded that binary encoding should be used for small FSMs (up to 8 states), while one-hot encoding produces better results for large FSMS (over 16 states). In this work we have analyzed three alternative approaches for state encoding techniques that take into account the Hamming distance of the encodings in order to minimize the number of bit toggles during state transitions. Experimental results obtained demonstrate that the proposed state encoding techniques produce better results for almost all benchmark examples. In particular, minimal-bit minimal Hamming distance encoding produces best results for small FSMs, reducing the power dissipation up to 53% with respect to binary encoding. On the other hand, zero-one-hot encoding and mainly the proposed partitioned encoding technique produced the best results for large FSMs, reducing the power dissipation up to 60%.

References

[1] G. Sutter, E .Teodorovich, S. López-Buedo, E. Boemo, "Low-Power FSMs in FPGA: Encoding Alternatives", Lecture Notes in Computer Science (LNCS2451): Integrated Circuit Design - Proceedings PATMOS 2002, pp. 459–467, August 2002.

[2] FPGA Compiler II / FPGA Express VHDL Reference Manual v. 1999.05. Synopsys Inc. 1999.

[3] Xilinx Software Manual, Synthesis and Simulation Design Guide: Encoding State. Xilinx Inc., 2000

[4] L. Benini, G. De Micheli, "State Assignment for Low Power Dissipation", IEEE Journal of Solid State Circuits, Vol. 30(3), March 1995.

[5] C-Y Tsui, M. Pedram, A. Despain, "Exact and Approximate Methods for Calculating Signal and Transition Probabilities in FSMs". Proc. Design Automation Conf. pp. 18–23, 1994.

[6] T. Villa, A. Sangiovanni-Vincentelli, "NOVA: State Assignment of Finite State Machines for Optimal Two-Level Logic Optimizations, "IEEE Transactions on computer Aided Design, Vol. 9(9), pp. 905–924, Sept. 1990.

[7] S. Devadas, H. Ma, A .R. Newton, A. Sangiovanni-Vincentelli, "MUSTANG: State Assignment of Finite State MachinesTargeting Multi level Logic Implementations", IEEE Transactions on Computer Aided Design, Vol7 (12). Dec. 1990.

[8] B. Lin, A. R. Newton, "Synthesis of Multiple Level Logic from Simbolic High-Level Description Languages". Proc. of International Conf. VLSI, pp. 187–196, Agosto 1996.

[9] D.-S. Chen, M. Sharrafzadeh, G. Yeap, "State Encoding of Finite State Machines for Low Power Design", IEEE International Symposium on Circuits and Systems ISCAS'95. Vol 3, pp. 2309–2312, 1995.

[10] W. Nöth, R. Kolla, "Spanning Tree Based State Encodin for low power Dissipation" Proc. Design Aut. and Test in Europe DATE'99, pp. 168–174., March 1999.

[11] X. Wu, M. Pedram, L. Wang. "Multi-code state assignment for low power design". IEEE Proc. Circuits, Devices and Systems, vol. 147, no. 5, pp. 271–275. Oct. , 2000

[12] M. Martínez, M. J. Avedillo, J. M. Quintana, J. L. Huertas. "A flexible state assignment algorithm for low power implementations". Proc. Design of Circuits and Integrated Systems Conf., pp. 154–159., Nov. 2001.

[13] M. Koegst, G. Franke, S. T. Rulke, Feske. "A strategy for low power FSM-Design Using multicriteria approach". PATMOS'97, pp. 323–329, 1997

[14] K. McElvain. "LGSynth93 Benchmark Set: Version 4.0". 1993

[15] E. Sentovich, K. Singh, L. Lavagno, C. Moon, R. Murgai, A. Saldanha, P. Stephan, R. Brayton, A. Sangiovanni-Vincentelli. "SIS: A System for Sequential Circuit Synthesis". Tech. Report Mem. No UCB/ERL M92/41, Univ. California Berkeley, 1992.

[16] Modelsim SE/EE Plus 5.4E. User`s manual, V. 5.4. Model Technology Incorporated. Aug. 2000.

[17] Xilinx Foundation 4.1i tools, http://www.xilinx.com/support/library.htm

[18] Xpower, "Xpower Tutorial FPGA Design", Xpower (V1.0) May 11, 2001. Xilinx Inc.

Reduced Leverage of Dual Supply Voltages in Ultra Deep Submicron Technologies

Tim Schoenauer, Joerg Berthold, and Christoph Heer

Infineon Technologies, CL AIP Advanced Macros & Architectures,
Otto-Hahn-Ring 6, 81730 München, Germany
{tim.schoenauer, joerg.berthold, christoph.heer}@infineon.com

Abstract. Dual supply voltage (DSV) is a low-power design technique, which reduces the dynamic power dissipation of a digital circuit [1]. In this paper we will summarize the basic idea of this approach, its benefit and associated costs and outline the dependency of DSV on technology and device parameters. We then evaluate the use of DSV on gate level in the context of the evolving ultra deep submicron (UDSM) technology. Employing DSV exhibits a reduced leverage - especially for leakage sensitive applications – mainly due to a limited reduction of threshold voltages in UDSM technologies and due to the use of multi-threshold devices. Finally we discuss DSV design examples reported in literature and give an outlook on how their benefit is influenced by UDSM technologies.

1 Introduction

For more than three decades the integration density in integrated circuits is growing exponentially following Moore's law. In CMOS-technology this was achieved by downscaling the device geometry leading to gate lengths already well below 100nm. According to the ideal constant-field scaling theory of MOSFETs by Dennard et al. [2], lateral and vertical device dimensions as well as supply and threshold voltages are reduced by a scaling factor s, while doping levels are increased by s. Ignoring second order effects this leads to an improved timing performance of a factor s, a reduced area by s^2, a reduction in power dissipation of s^2 per transistor (assuming the frequency increased by s). Since the number of transistors per area also increases by s^2, the power dissipation per area remains constant. While such a scenario seems extremely attractive, the real evolution of CMOS-technologies has deviated considerably. As will be discussed in more detail in section 3.1, supply voltage and threshold voltage have not scaled as rapidly as device dimensions. A reduced scaling of the supply voltage is motivated by a good timing performance, while the reduction of threshold voltage is limited by leakage current and process variations. One of the consequences is an increase in power dissipation per unit area of high performance ICs [3].

On the other hand there is a strong demand for portable applications where a low power dissipation is a requirement to prolong battery life. Higher power dissipation of increasingly complex circuits causes problems of heat removal and cooling as well.

J.J. Chico and E. Macii (Eds.): PATMOS 2003, LNCS 2799, pp. 41–50, 2003.
© Springer-Verlag Berlin Heidelberg 2003

Hence, for cost-driven consumer electronics a low power dissipation is crucial to allow acceptable packaging costs. For general-purpose microprocessors power dissipation is even becoming a performance limiting factor.

Therefore the demand for design techniques to reduce power dissipation is nowadays higher than ever. Among many low power design techniques, the use of multiple supply voltage is one approach aiming predominantly at the reduction of dynamic power. In section 2 we will summarize the basic idea, benefits and drawbacks of this technique. In section 3 we discuss, what impact UDSM technology has on the efficiency of DSV.

2 Dual Supply Voltages: Overview

In digital CMOS circuits there are three sources of power dissipation [4]. One is the charging and discharging of capacitances due to a changing logic level. A second one is the short-circuit current. It flows from supply to ground while a gate is changing its output, as for a brief interval the p- and n-subnetwork are conducting. Both of these sources of power dissipation are referred to as dynamic power. The third source is static power dissipation due to leakage current.

The use of dual supply voltages primarily aims at reducing the dynamic power. Dynamic power dissipation is typically dominated by charging and discharging of the load capacitance C_L which can be described by

$$P = a \cdot C_L \cdot f \cdot V_{dd}^{\ 2}, \tag{1}$$

where a represents the switching activity, f the clock frequency and V_{dd} the supply voltage. From Eq. 1 it is apparent that lowering the supply voltage is very efficient, since it contributes by the power of two to the dynamic power dissipation. On the other hand lowering the supply voltage reduces the overdrive ($V_{dd} - V_{th}$) of a MOSFET's drain current and thereby compromises timing performance. According to Equ. 2 which is based on the alpha law model of Sakurai et al. [5] the delay t_d of an inverter is given by

$$t_d \propto \frac{C_L \cdot V_{dd}}{(V_{dd} - k \cdot V_{th})^{\alpha}}, \tag{2}$$

where V_{th} is the threshold voltage and α is a fitting constant. A typical value for α modeling short-channel MOSFETs is 1.3. Another fitting constant k has been added, where $k = 1.3$ yields a good match with results from measurements for the recent technology generations. As can be seen from Equ. 1 and Equ. 2, there is a trade-off between power and timing performance, when selecting a supply voltage V_{dd}.

2.1 Basic Idea of Dual Supply Voltages

Employing multiple supply voltages aims at providing timing critical parts of a synchronous circuit with a higher supply voltage V_{ddh} and less timing critical parts with a lower supply voltage V_{ddl}. Thereby timing performance of the overall circuit can be maintained and power dissipation is reduced. This technique may be applied on different levels in a system: on module level, on functional-unit level and on gate level. In this paper we will focus on using dual supply voltages on gate level as proposed by Usami et al. [1].

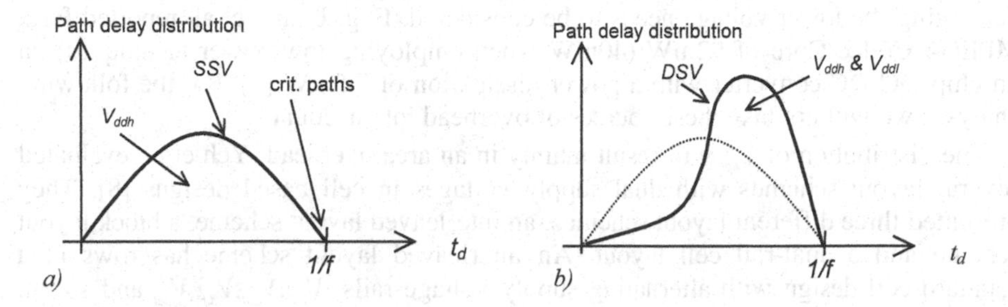

Fig. 1. Path delay distribution of an idealized design with: *a)* single supply voltage (SSV); *b)* dual supply voltage (DSV) with V_{ddl} gates in paths with formerly uncritical paths and V_{ddh} for critical and almost critical paths

For such a case, the idea is to provide gates in timing uncritical paths with V_{ddl} while the rest of the gates are supplied with V_{ddh}. Fig. 1 illustrates this idea within a path delay distribution. Fig. 1a shows a path delay distribution of an idealized circuit. Typically only a small number of paths are critical and limit the maximum clock frequency of the design. In many circuits a large portion of paths may exhibit a path delay of e. g. only half of the clock period or less. Increasing the path delay of such paths as depicted in Fig. 1b by providing all or some of their gates with V_{ddl} will not degrade the timing performance of the overall design even though the paths move towards longer path delays. One can calculate the saving in power dissipation per replaced gate by substituting $V_{dd} = V_{ddh}$ and $V_{dd} = V_{ddl}$ respectively in Equ. 1.

2.2 Overhead of DSV Compared to Single Supply Voltages

While dual supply voltages allow to reduce power dissipation and maintain timing performance this design technique is not for free. Compared to a single supply voltage (SSV), there are three sources of overhead which need to be considered: 1. additional circuitry is required at the interfaces of V_{ddl}-gates to V_{ddh}-gates to convert the voltage levels (level shifters), 2. the second supply voltage V_{ddl} needs to be generated and 3. it needs to be distributed in the chip layout.

The need for level shifter stems from the fact that a V_{ddl}-gate driving a V_{ddh}-gate with logic 1 (electric V_{ddl}) will not completely block the p-MOSFETs of the V_{ddh}-gate and thereby cause an increased static cross current in the V_{ddh}-gate. A level shifter may be realized by six transistors (see e. g. [6] and references therein). Usually level converters are inserted at the output of a flip-flop. Using flip-flops with a built-in level conversion helps reducing the overhead in terms of area, power and delay. In [7] a master-slave flip-flop with a level converter to V_{ddh} in the slave latch (FFLC) is compared to a regular master slave flip-flop using V_{ddh} only. An area increase of 21%, an increased clk-to-q delay of 75% and a power reduction of 37% of a FFLC compared to the regular flip-flop is reported. The power reduction is achieved due to the master latch operating at V_{ddl}.

For system design the overhead due to the generation of the second (lower) supply voltage V_{ddl} might not be negligible. If V_{ddl} is not available in the system anyway, the extra cost of additional circuitry and in any case an extra power dissipation while generating the lower voltage needs to be considered. E. g. Usami et al. reported for a MPEG4 Codec Core of 92mW (40mW when employing low power techniques) an on-chip DC-DC converter with a power dissipation of 7.5mW [7]. For the following analyses we will not take these sources of overhead into account.

The distribution of V_{ddl} will result mainly in an area overhead. Yeh et al. evaluated several layout schemes with dual supply voltages in cell based designs [8]. They presented three different layout schemes: an interleaved layout scheme, a block layout scheme and a dual-rail cell layout. An interleaved layout scheme has rows in a standard-cell design with alternating supply voltage rails: V_{ddh}, V_{ddl}, V_{ddh}, V_{ddl} and so on. The block layout scheme on the other hand allows two sections with each supply voltage per row. A dual-rail cell layout consists of cells which have two separate power rails to carry separate power supply voltages. The results from the study suggest, that an area overhead in the order of 10% (this is a value averaged over several benchmark circuits) needs to be taken into account for a dual supply voltage layout compared to a single supply voltage layout.

Since level shifters introduce additional area, delay and possibly power dissipation, it must be carefully evaluated which gates to supply with V_{ddl}. The number of interfaces between V_{ddl}-gates and V_{ddh}-gates should be kept as low as possible. For this reason it is attractive to replace gates in clusters as introduced by the clustered voltage scaling (CVS) by Usami et al. [1]. Mahnke et al. presented a methodology employing Synopsys Power Compiler for generating a DSV netlist from a SSV netlist by taking advantage of gate sizing feature of Power Compiler [9]. In section 3.4 we discuss results reported in literature from these methodologies in the context of UDSM technology.

Also V_{ddl} must be chosen carefully: if V_{ddl} is too high the power reduction per replaced gate is low, is V_{ddl} too low only very few gates may be replaced and the overall power saving decreases. The optimum V_{ddl} of course depends on the specific circuit. In [9, 6, 10] V_{ddl} was chosen to be in the range of 50% to 80% of V_{ddh}.

3 DSV in UDSM Technologies

In this section we will discuss the potential of DSV in the context of digital CMOS technology generations evolving towards UDSM technology. Two parameters are decisive for the benefit of DSV: the percentage of gates that can be "replaced" (replaced means here supplied by V_{ddl} instead of V_{ddh}) and how much power dissipation is saved per replaced gate. How do these two parameters change when moving towards UDSM technology? They will be mainly influenced by two trends along the downscaling in UDSM technology: voltage downscaling differing for threshold and supply voltage and the use of multi-threshold devices in the same circuit. These trends are discussed in section 3.1 and 3.2 and in section 3.3 respectively.

3.1 Scaling of Threshold vs. Supply Voltage

As already pointed out in the introduction, according to the ideal constant-field scaling theory, supply and threshold voltage should both decrease by a factor of *s* when moving to a new technology generation. As described in the following, this has typically not been the case in the recent CMOS technology generations.

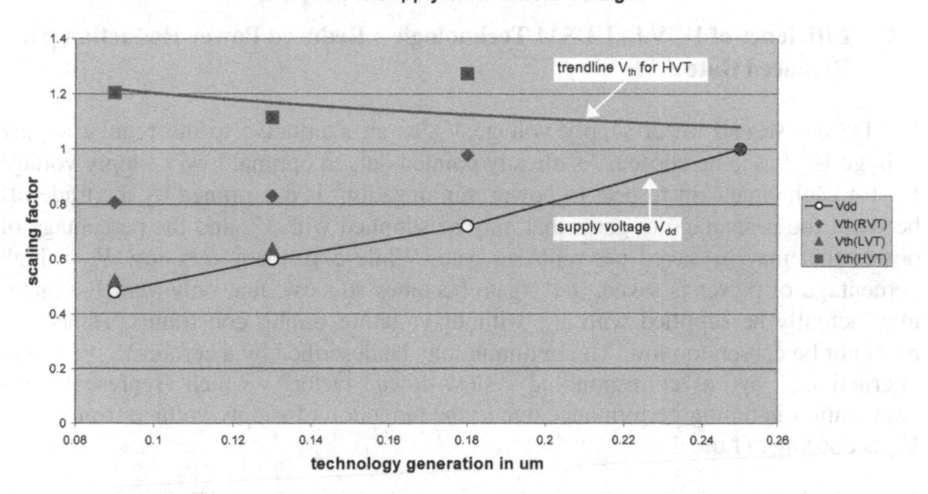

Fig. 2. Scaling of supply and threshold voltage for recent technology generations

Fig. 2 depicts the reduction factor of supply and threshold voltages for the last four technology generations. For the implementation of logic in a 0.25um CMOS technology, devices with only a single threshold voltage were used. In subsequent generations, devices with various threshold voltages evolved: low threshold (LVT) devices for high speed applications on one hand and high threshold voltage (HVT) for low leakage devices on the other hand. LVTs profit from the relatively high overdrive, which results in a high on-current and therefore allows for high speed applications. However the low threshold voltage of these devices induces high leakage currents. Leakage current dominated by subthreshold current of the devices in the off-state grows exponentially with a decreasing threshold voltage. While HVT devices have low leakage currents, they exhibit poor speed performance. Regular threshold (RVT) devices represent a compromise of LVT and HVT devices.

For many power critical applications leakage currents of LVT and RVT devices in 0.13um processes and below are already prohibitively large. This is particularly the case for applications where a circuit remains in standby mode for a considerable amount of time. Such a scenario is relevant for many mobile applications such as mobile phones or PDAs. For these applications HVT devices will be first choice wherever their performance in terms of speed is sufficient. Hence the trendline for the scaling factor of the threshold voltage in Fig. 2 is drawn for HVT devices.

As is depicted in Fig. 2, for these devices the threshold voltage remained about constant while the supply voltage decreased by about half for the recent four technology generations. What is the consequence for the timing performance of transistors? The overdrive decreases which compromises timing performance in general. But also the offset in the overdrive term ($V_{dd} - V_{th}$) given by V_{th} becomes more and more significant in respect to the value of V_{dd}. According to Equ.2 a further drop of the supply voltage (as needed for V_{ddl}) will now degrade timing performance more significantly than before. This will be discussed in more detail in the following section.

3.2 Efficiency of DSV in UDSM Technologies: Reduced Power Reduction per Replaced Gate

For DSV a second lower supply voltage V_{ddl} as an alternative to the regular supply voltage V_{ddh} has to be chosen. As already pointed out, an optimal lower supply voltage V_{ddl} for each circuit (in respect to power minimization) is determined by the trade-off between the percentage of gates that may be supplied with V_{ddl} and the percentage of power that may be saved per replaced gate. While e. g. for a very low V_{ddl} a high percentage of power is saved, a V_{ddl}-gate becomes so slow, that only very few gates may actually be supplied with V_{ddl} without violating timing constraints. Hence V_{ddl} must not be chosen too low. The optimum may be described by a certain V_{ddl} -value or alternatively by a corresponding "slow-down factor", which represents the degradation in timing performance due to the reduction of supply voltage from V_{ddh} to V_{ddl} according to Equ. 2.

Table 1. Voltages for recent technology generations and their different device types (HVT/RVT/LVT: high/regular/low threshold voltage) as well as the corresponding ideal power reduction factor for two slow down factors of 1.5 and 2

Process / um	**0.25**		**0.18**	**0.13**	**0.09**
V_{dd} / V	**2.5**		**1.8**	**1.5**	**1.2**
V_{th} / mV	440	**HVT**	560	490	530
$1-(V_{ddl\ 1.5t} / V_{dd})^2$	0.62		0.43	0.41	0.28
$1-(V_{ddl\ 2t} / V_{dd})^2$	0.76		0.57	0.55	0.39
V_{th} / mV		**RVT**	430	365	355
$1-(V_{ddl\ 1.5t} / V_{dd})^2$			0.52	0.53	0.45
$1-(V_{ddl\ 2t} / V_{dd})^2$			0.67	0.67	0.59
V_{th} / mV		**LVT**		280	230
$1-(V_{ddl\ 1.5t} / V_{dd})^2$				0.60	0.59
$1-(V_{ddl\ 2t} / V_{dd})^2$				0.74	0.74

In the following we will consider two examples: we assume circuits where the power optimum is reached for a slow-down factor of 1.5 and 2. Employing Equ. 2 we are able to calculate a corresponding V_{ddl} for a given V_{ddh} and a given slow-down factor. Subsequently we may calculate the saving in power dissipation per replaced gate by using Equ. 1. According to Equ. 1, replacing a V_{ddh}-gate by a V_{ddl}-gate will reduce power dissipation by a factor of $(V_{ddl} / V_{ddh})^2$. The term $(1 - (V_{ddl} / V_{ddh})^2)$ we will refer to as the ideal power reduction factor in the following. For the different technology generations and their various threshold devices Tab. 1 lists the ideal power reduction factor corresponding to example voltages $V_{ddl\_1.5t}$ and V_{ddl_2t} ($V_{ddl\_1.5t}$ and V_{ddl_2t} cause a transistor delay increase of a factor of 1.5 and 2 respectively).

Fig. 3 illustrates the ideal power reduction factor for different technology generations and their device types. Assuming that leakage current is relevant for power sensitive applications, trendlines for HVT devices are depicted in Fig. 3. The trendlines in Fig. 3 show a decrease of the ideal power reduction factor from approximately 0.6 to 0.3 during the past four technology generations for $V_{ddl\_1.5t}$; for V_{ddl_2t} the ideal power reduction factor decreases from approximately 0.75 to 0.4. In both cases the reduction of dynamic power dissipation per replaced gate decreases by a factor of about 2. This shows a considerable drop in efficiency of the DSV-concept.

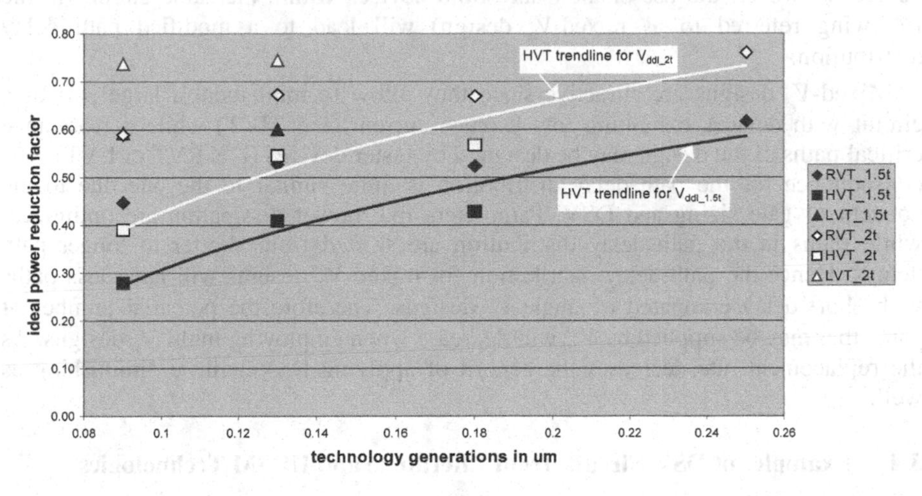

Fig. 3. Ideal power reduction factor for recent technology generations

On the other hand implementing logic with LVT devices, the efficiency of the DSV remains about constant as Fig. 3 shows. However LVT devices in UDSM technologies exhibit a leakage current which is not acceptable for many power sensitive applications. Other low power techniques, such as VTCMOS (variable threshold by reverse substrate biasing) [12] may be applied to overcome subthreshold leakage current for LVT devices. The idea of VTCMOS is to reverse bias the well of a transistor (e. g. during standby-mode), so that the threshold voltage is increased via the body effect. This technique also comes at additional costs such as separate substrate contacts and extra biasing voltages. Moreover, in UDSM technology the body effect decreases reducing the leverage of this technique. Thus LVT devices are not the most likely candidate for low power implementations.

The trend of a decreasing ideal power reduction factor would even become worse when considering the process variations of the threshold voltage. Due to variation in gate length and doping concentration a typical value of about ±70mV needs to be taken into account. This value has practically been constant over the past technology generations and therefore comprises a larger percentage of the nominal value with each new technology generation.

3.3 DSV in Multi-threshold Designs: Reduced Replacement Rate

In the previous section we discussed the power reduction per replaced gate. The second parameter that influences the efficiency of DSV is the achievable replacement rate. A path delay distribution is an indicator of how many paths and therefore gates qualify for replacement. Applying the same library elements and synthesis strategy, the straightforward assumption would be that the path delay distribution for a certain circuit will be more or less the same from one technology to the other. To some degree this assumption may be correct for designs employing only a single threshold device. However, the use of multi-threshold devices within the same circuit (in the following referred to as mixed-V_{th} design) will lead to a modified path delay distribution.

Mixed-V_{th} designs are attractive since they allow to implement a large part of a circuit with devices exhibiting low leakage current (i. e. HVT) while performance critical paths of the design may be designed by faster devices (i. e. RVT or LVT). The consequence for the path delay distribution is quite similar to the one due to the concept of gate sizing and DSV. Parameters like power dissipation are optimized, while paths in the path delay distribution are shifted from shorter to longer path delays. Hence the path delay distribution for mixed-V_{th} designs will have less paths with short delay compared to single-V_{th} designs. Therefore the potential number of gates that may be supplied by V_{ddl} will decrease when employing multi-V_{th} designs. As the replacement rate decreases the benefit of applying DSV will be diminishing as well.

3.4 Examples of DSV Circuits from Literature and UDSM Technologies

Several studies of DSV applications have been reported in literature [1, 11, 7, 6, 9]. Application of DSV comprises decode and control logic blocks of a superscalar microprocessor in a 0.8um technology (V_{ddh} = 5V; V_{ddl} = 3V) [1], random logic modules of a media processor in a 0.3um technology (V_{ddh} = 2.7V; V_{ddl} = 1.9V) [11], a MPEG4 codec core in a 0.3um technology (V_{ddh} = 2.5V; V_{ddl} = 1.75V) [7] and combinational ISCAS'85 benchmarks circuits in [6, 9], where a 0.25um technology (V_{ddh} = 2.5V; V_{ddl} = 1.8V) was chosen in [9]. Hence typical values for V_{ddl} / V_{dd} range between 0.5 and 0.7. Replacement rates range from 51% to 76% for DSV circuit examples in [1, 11, 7], from about 3% to 72% (avg. 26%) in [6] and in average about 20% in [9]. Reported power reduction for combinational logic gates varies strongly with avg. 28% in [11], avg. 30% in [7], avg. about 6% in [9]. Taking into account the power reduction due to FFLCs (see section 2.2) and a V_{ddl}-clock-tree in [11] an avg. power reduction of 47% is reported.

Considering the results from section 3.2, the reported power reduction of combinational logic gates in submicron technologies (0.3um and 0.25um) is likely to be degraded in UDSM technology. Only for applications allowing an implementation of LVT devices the power reduction reported in submicron technologies could continue to be sustained in UDSM technologies, however the drawbacks of LVT devices discussed in section 3.2 need to be kept in mind. If these circuit examples would be implemented using HVT devices e. g. in a 90nm technology, the power reduction would be reduced by about a factor of two thus ranging between average values of 3% to 15%. As discussed in section 3.3, a further decrease in power reduction of combinational logic gates has to be accounted for when a circuit is implemented as a mixed-V_{th} design due to a modified path delay distribution, which leads to a reduced replacement rate.

4 Conclusion

In this paper we examined the benefit of dual supply voltage (DSV) in the context of emerging ultra deep submicron (UDSM) technology. The efficiency of DSV depends on two parameters: 1. the power reduction per cell using the lower supply voltage ("replaced cell") and 2. the overall percentage of replaced cells. The power reduction per replaced cell decreases in UDSM technology for low leakage devices, thereby reducing the leverage of DSV. This is due to a threshold voltage almost remaining constant for low leakage devices from 0.25um to 90nm processes. Moreover, implementing designs with mixed-V_{th} devices will alter the path delay distribution of a circuit in such a way, that there will be more paths with less slack. This will reduce the achievable replacement rate and thereby furthermore diminish the benefit of DSV. Moving from a 0.25um technology towards a 90nm technology, we estimate a reduced leverage to less than half for DSV applications, which require low leakage devices. For high-speed applications, which may accept a high leakage current, the efficiency of DSV is retained. Yet we believe the spectrum of applications for which there is a sufficient leverage justifying the overhead of DSV will be reduced in UDSM technology.

References

[1] K. Usami, M. Horrowitz, "Clustered voltage scaling technique for low-power design," Proceedings of the International Symposium on Low Power Design, April 1995.

[2] R. H. Dennard et al, "Design of Ion-Implanted MOSFETs with Very Small Physical Dimensions," IEEE Journal of Solid-State Circuits, vol. 9, pp. 256–268, Oct. 1974.

[3] B. Razavi, *Design of Analog CMOS Integrated Circuits*, McGraw-Hill, 1999.

[4] K. Roy, S. Prasad, *Low-Power CMOS VLSI Circuit Design*, John Wiley, 2002.

[5] T. Sakurai, R. Newton, "Alpha-Power Law MOSFET Model and its Applications to CMOS Inverter Delay and Other Formulas," IEEE Journal of Solid-State Circuits, April 90.

[6] M. Donno, L. Macchiarulo, A. Macci, E. Macii, M. Poncino, "Enhanced Clustered Voltage Scaling for Low Power," Proceedings of the 12th ACM Great Lakes Symposium on VLSI, GLSVLSI'02, Apr. 2002.

[7] K. Usami et al., "Design Methodology of Ultra Low-power MPEG4 Codec Core Exploiting Voltage Scaling Techniques," DAC98, Proceedings of the 35th Design Automation Conference, 1998.

[8] C. Yeh, Y.-S. Kang, "Lyaout Techniques Supporting the Use of Dual Supply Voltages for Cell-Based Designs," Proceedings of the Design Automation Conference, DAC'99, 1999.

[9] T. Mahnke, S. Panenka, M. Embacher, W. Stechele, W. Hoeld, "Power Optimization Through Dual Supply Voltage Scaling Using Power Compiler," SNUG Europe'02, 2002.
[10] K. Usami, M. Igarashi, "Low-Power Design Methodology and Applications utilizing Dual Supply Voltages," Proceedings of the Asia and South Pacific Design Automation Conference 2000, pp. 123–128, Jan. 2000.
[11] K. Usami et al., "Automated Low-Power Technique Exploiting Multiple Supply Voltages Applied to a Media Processor," IEEE Journal of Solid State Circuits, Vol. 33, No. 3, pp. 463–471, March 1998.
[12] T. Kuroda et al., "A High-Speed Low-Power 0.3um CMOS Gate-Array with Variable Threshold Voltage (VT) Scheme," IEEE CICC'96, pp. 53–56, May 1996.

A Compact Charge-Based Crosstalk Induced Delay Model for Submicronic CMOS Gates

José Luis Rosselló and Jaume Segura

Departament de Física, Universitat Illes Balears, 07071 Palma de Mallorca, Spain

Abstract. In this work we propose a compact analytical model to compute the crosstalk induced delay from a charge-based propagation delay model for submicronic CMOS gates. Crosstalk delay is described as an additional charge to be transferred through the pMOS (nMOS) network of the gate driving the victim node during its rising (falling) output transition. The model accounts for time skew between the victim and aggressor input transitions and includes submicronic effects. It provides an intuitive description of crosstalk delay showing very good agreement with HSPICE simulations for a $0.18\mu m$ technology.

1 Introduction

The constant scaling of feature sizes and supply voltage with the increase in both operating frequency and signal rise/fall times has made digital designs more vulnerable to noise. In modern ICs, interconnect coupling noise (crosstalk) becomes a performance limiting factor that must be analyzed carefully during the design process. If not considered, crosstalk can cause extra delay, logic hazards and logic malfunction.

Fig.1 illustrates the well known crosstalk induced delay effect that appears when two lines (the aggressor and the victim) switch simultaneously. For a victim node falling transition, the gate delay (defined as t_{pHL0}) can be reduced ($t_{pHL,su}$) or increased ($t_{pHL,sd}$) depending on if the aggressor makes a falling or a rising transition respectively. Since large circuits can handle tens of millions of interconnect lines on a single chip, simple and accurate analytical descriptions for crosstalk delay are of high importance for a fast timing verification of the whole chip.

Crosstalk delay has been analyzed in [1] using a waveform iteration strategy. Unfortunately, the time skew between transitions are not considered and no-closed form expression for the worst-case victim delay is provided. More recently crosstalk delay has been modeled analytically in [2]. Analytical expressions are derived that quantify the severity of crosstalk delay and describe qualitatively the dependence on circuit parameters, the rise/fall times of transitions and the skew between transitions. The nMOS (pMOS) network is substituted by a pull-down (pull-up) resistance and short-circuit currents are neglected. The main limitation is that the complex equations obtained for the victim voltage variation cannot be used to obtain a closed-form expression of the propagation delay. Additionally,

J.J. Chico and E. Macii (Eds.): PATMOS 2003, LNCS 2799, pp. 51–59, 2003.
© Springer-Verlag Berlin Heidelberg 2003

other effects as short-circuit currents, with a great impact on propagation delay [3] are not considered in their analysis.

In this work we propose a simple and accurate analytical model to compute the crosstalk delay ($t_{pHL,su}$ and $t_{pHL,sd}$) in submicron CMOS gates. A simple propagation delay model for CMOS inverters [3] is modified to include more complex gates than inverters using the collapsing technique developed in [4]. The model is compared to HSPICE simulations for a $0.18\mu m$ technology showing an excellent agreement.

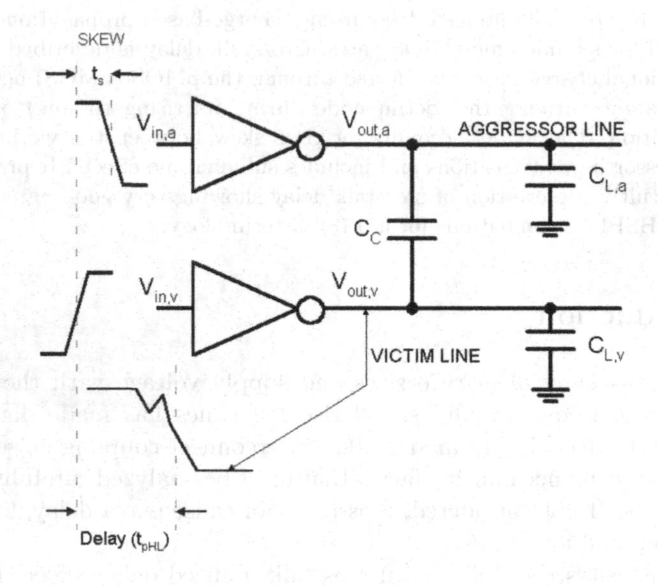

Fig. 1. An additional delay in the victim is induced by the switching transition of $V_{out,a}$ (crosstalk induced delay)

2 Crosstalk Delay Model

Consider the circuit in Fig.1 where the aggressor and the victim gates drive an output capacitance $C_{L,a}$ and $C_{L,v}$ respectively, and there is a coupling capacitance C_c between $V_{out,a}$ and $V_{out,v}$ (that causes the crosstalk between the two lines). The time skew t_s is defined as the time interval from $V_{DD}/2$ at the victim input to $V_{DD}/2$ at the aggressor input. The output interconnect lines are modeled as lumped capacitances. This is a valid approach for medium size interconnects as it has been reported that the relative error of this capacitive model is below 10% for a $2mm$ wire in a $0.18\mu m$ technology [5].

The structure of this paper is as follows, first we present a compact charge-based propagation delay model of CMOS gates that is based on the propagation delay model developed in [3]. Finally, crosstalk effects are included in the model as additional charges to be transferred through the corresponding nMOS/pMOS block.

2.1 Propagation Delay Model for Complex Gates

The propagation delay is usually computed as the time interval between the input and the output crossing $V_{DD}/2$. The crosstalk-free propagation delay (t_{pHL0}) can be described as a function of the charge transferred through the gate using the model in [3] that is based on the nth-power law MOSFET model [6], and on an efficient and accurate transistor collapsing technique for complex gates developed in [4].

For a high to low output transition (for a low to high transition the analysis is equivalent) the propagation delay is given by:

$$t_{pHL0} = \begin{cases} \left[t_n + \left[\frac{Q_f(1+n)}{I_{D0n}^{\langle 1,N_n \rangle}} (t_{in} - t_n)^n \right]^{\frac{1}{1+n}} - \frac{t_{in}}{2} \right] & Q_f < Q_{f_0} \\ \frac{t_{in}}{2} + \frac{Q_f - Q_{f_0}}{I_{D0n}^{\langle 1,N_n \rangle}} & Q_f \geq Q_{f_0} \end{cases} \quad (1)$$

where t_{in} is the input transition time (i.e. the time during which the input is changing), parameter n is the velocity saturation index of the nMOS transistors, $Q_f = C_L V_{DD}/2$ is the charge transferred through the nMOS block until $V_{out} = V_{DD}/2$ (where V_{DD} is the supply voltage), C_L is the output load of the gate, $t_n = (V_{TN}/V_{DD})t_{in}$ is the time when the nMOS block starts to conduct (V_{TN} is the threshold voltage of nMOS), and Q_{f_0} is the charge transferred through the nMOS block when the input transition is finished, given by:

$$Q_{f_0} = \frac{I_{D0n}^{\langle 1,N_n \rangle} (t_{in} - t_n)}{1 + n} \quad (2)$$

Parameter $I_{D0n}^{\langle 1,N_n \rangle}$ is defined as the maximum current that the nMOS block can deliver (modeled as a chain of N_n series-connected transistors) and is obtained from the transistor collapsing technique developed in [4]. For the simple case of a chain with N_n identical transistors, the collapsing technique provides a simple expression for $I_{D0n}^{\langle 1,N_n \rangle}$:

$$I_{D0n}^{\langle 1,N_n \rangle} = \frac{I_{D0n} \left[1 + \frac{\lambda}{2} (V_{D0_n} - V_{DD}) \right]}{1 + (N_n - 1) K_n}$$

where V_{D0_n} is the saturation voltage of nMOS, I_{D0n} is the maximum saturation current of each nMOS (drain current when $V_{GS} = V_{DS} = V_{DD}$) and parameter λ accounts for channel length modulation. The parameter K_n is a technology-dependent parameter given by:

$$K_n = \frac{3 V_{D0_n} n (1 + \gamma_n)}{5 (V_{DD} - V_{TN})} \quad (3)$$

where γ_n is the body effect parameter of nMOS [6].

The transition time at the output t_{out0}, required to evaluate crosstalk delay, can also be expressed as a function of Q_f as [3]:

$$t_{out0} = \begin{cases} \frac{2Q_f}{I_{D0n}^{\langle 1,N_n \rangle}} \left(\frac{I_{D0n}^{\langle 1,N_n \rangle}}{Q_f(1+n)} (t_{in} - t_n) \right)^{\frac{n}{1+n}} & Q_f < Q_{fo} \\ \frac{2Q_f}{I_{D0n}^{\langle 1,N_n \rangle}} & Q_f \geq Q_{fo} \end{cases} \quad (4)$$

Eqs. (1) and (4) are valid for a high to low output transition. Equivalent expressions can be obtained for a low to high output transition.

2.2 Including Short-Circuit Currents

The delay model expressed in eq. (1) is valid for CMOS gates when short-circuit currents are neglected. These currents are included in [3] as an additional charge to be transferred through the pull-down (pull-up) network for an output falling (rising) transition. For a high to low output transition, the charge transferred through the nMOS block is computed as $Q_f = C_L V_{DD}/2 + q_{sc}^f$, where q_{sc}^f is defined as the short-circuit charge transferred during the falling output transition until $V_{out} = V_{DD}/2$. For simplicity we compute this charge as $q_{sc}^f = Q_{sc}^f/2$, where Q_{sc}^f is the total short-circuit charge transferred. For the evaluation of Q_{sc}^f we use a previously developed model described in [7].

2.3 Crosstalk Delay

The impact of crosstalk on the propagation delay is computed accounting for the additional charge injected by the aggressor driver through the coupling capacitance C_c that must be discharged by the victim driver. We consider a falling transition at $V_{out,v}$ slower than a rising transition at $V_{out,a}$ (for other cases the analysis is similar).

Fig.2 shows a high to low transition of the victim node $V_{out,v}$ when the aggressor output switches from low to high for six different skew times represented as $a, b, ..., f$. For case 'a', the aggressor transition (Ag_a) does not impact the victim propagation delay (t_{pHL0}) since the absolute value of the time skew between transitions is too large. The victim output $V_{out,v}$ (transition $V_{out,v\,(a)}$) rises until $V_{out,a} = V_{DD}$ and then falls back to V_{DD} before the rising transition at the victim input is initiated. In this case the crosstalk coupling effect is a glitch that lasts for a time that is defined as t_D (see Fig.3). This characteristic time (t_D) is computed assuming that the ON transistors of the victim gate pMOS block (that discharge the output) are in the linear region and that their drain-source voltage is small. Under these conditions, the current through the pMOS block is:

$$I_p = I_{D0P} \frac{V_{DD} - V_{out}}{V_{D0_p}} \quad (5)$$

where V_{D0_p} is the saturation voltage of pMOS and I_{D0P} is computed from all the ON transistor chains that connect the supply and the output node. Each of

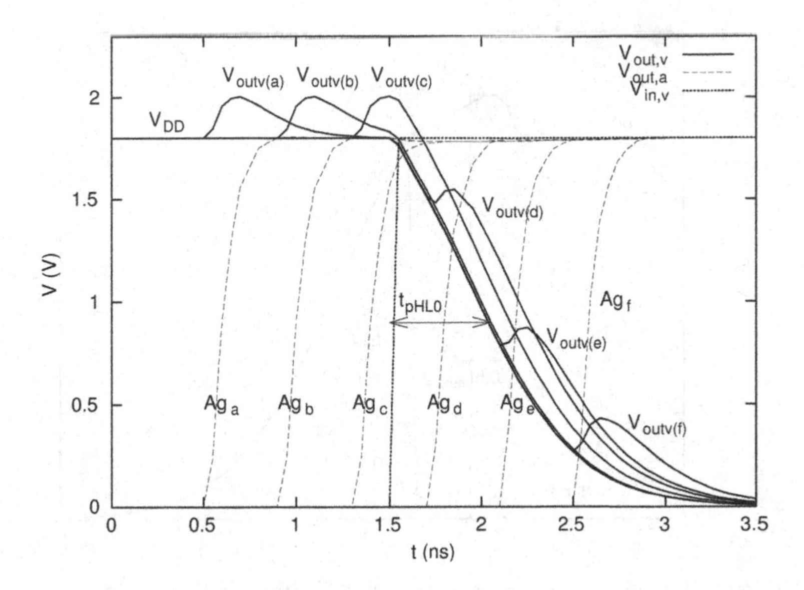

Fig. 2. Crosstalk delay is dependent on the time between transitions $V_{in,a}$ and $V_{in,v}$.

these pMOS chains is collapsed to a single equivalent transistor with maximum saturation current $I_{D0p}^{\langle 1,N_i \rangle}$ (where N_i is the number of transistors in each chain). Finally I_{D0P} is expressed as the sum of these contributions as:

$$I_{D0P} = \sum_{i=1}^{m} I_{D0p}^{\langle 1,N_i \rangle} \tag{6}$$

where m is the number of active chains connecting V_{DD} and $V_{out,v}$. After the time point at which $V_{out,a} = V_{DD}$, the victim line output voltage evolution ($V_{out,v}$) is described as:

$$V_{out,v} = V_{DD} + (V_{\max} - V_{DD})\, e^{-\frac{(t-t_0)I_{D0P}}{C_L V_{D0_p}}} \tag{7}$$

where $C_L = C_{L,v} + C_c$, t_0 is the time at which $V_{out,a} = V_{DD}$ and V_{max} is the maximum voltage at $V_{out,v}$. The characteristic time t_D (relaxation of V_{out} back to V_{DD}) is obtained from (7) leading to:

$$t_D = 2Ln\,(2)\,\frac{V_{D0_p} C_L}{I_{D0P}} \tag{8}$$

It is well known that crosstalk may have an impact on delay only when the aggressor and victim transitions occur within a time window. Otherwise the effect of crosstalk is simply a glitch at the victim line. The time interval during which the coincidence of input transitions may give a crosstalk delay can be defined in terms of a limit time skew value (t_{s_1}). This limit value will depend on the propagation delay of the aggressor driver, the characteristic time t_D, and the aggressor output transition time. Fig.3 illustrates this time relationships

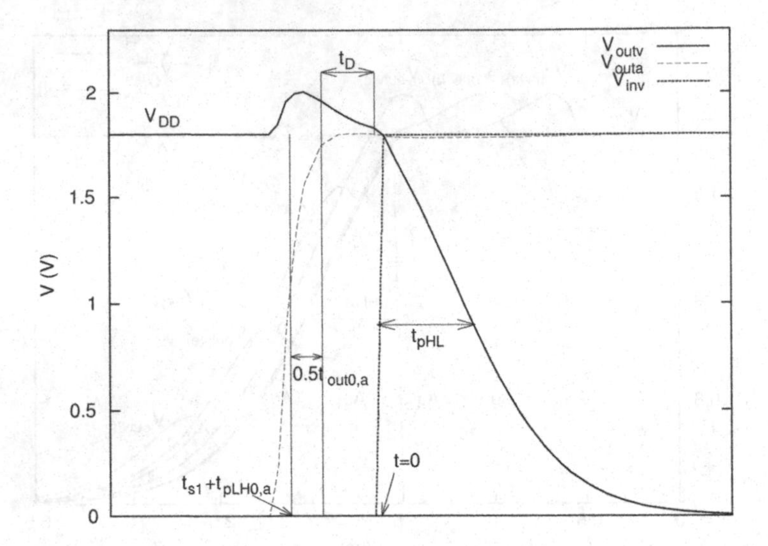

Fig. 3. The aggressor starts to affect the victim voltage variation when $t_s = t_{s1}$.

graphically. The limit time t_{s_1} is obtained equating the time point at which the glitch at $V_{out,v}$ is finished to the beginning of the high to low transition at $V_{out,v}$, i.e.:

$$t_{s_1} + t_{pLH0,a} + \frac{t_{out0,a}}{2} + t_D = t_{pHL0,v} - \frac{t_{out0,v}}{2} \tag{9}$$

where $t_{pHL0,v}$ and $t_{out0,v}$ are the propagation delay and the output transition time of the victim gate respectively that are obtained from (1) and (4) when crosstalk is neglected ($Q_f = C_L V_{DD}/2 + Q_{sc}^f/2$). If $t_s < t_{s_1}$ the effect of crosstalk is a glitch at the victim node and the voltage value is restored by the pMOS devices. When $t_s > t_{s_1}$ the crosstalk will impact delay as an additional charge (defined as Q_c) that must be drained by the nMOS devices during the transition. When $V_{out,a}$ and $V_{out,v}$ start switching at the same time, the value of Q_c is maximum. Defining t_{s_2} as the time skew at this time point, then we have:

$$t_{s_2} + t_{pLH0,a} - \frac{t_{out0,a}}{2} = t_{pHL0,v} - \frac{t_{out0,v}}{2} \tag{10}$$

For $t_s = t_{s_2}$, the coupling charge can be taken as $Q_c = C_c V_{DD}$ since the voltage variation at $V_{out,a}$ occurs during the transition at $V_{out,v}$. For the case at which $t_{s_1} < t_s < t_{s_2}$ we use a linear variation of Q_c with t_s.

$$Q_c = \frac{t_s - t_{s_1}}{t_{s_2} - t_{s_1}} C_c V_{DD} \tag{11}$$

The coupling charge Q_c is equal to $C_c V_{DD}$ for a time skew greater than t_{s_2}. This condition holds if $V_{out,a}$ transition finishes before $V_{out,v}$ reaches $V_{DD}/2$. This limit value for time skew is defined as t_{s_3} and obtained solving:

$$t_{s_3} + t_{pLH0,a} + \frac{t_{out0,a}}{2} = t_{pHL0,v} \tag{12}$$

If $t_{s_2} < t_s < t_{s_3}$ (case d in Fig.2) then Q_c is maximum ($Q_c = C_c V_{DD}$). When $t_s > t_{s_3}$ Q_c starts to decrease until $Q_c = 0$ at $t_s = t_{s_4}$, where t_{s_4} is defined as the time at which the beginning of the transition at $V_{out,a}$ and the time at which $V_{out,v} = V_{DD}/2$ are equal.

$$t_{s_4} + t_{pLH0,a} - \frac{t_{out0,a}}{2} = t_{pHL0,v} \tag{13}$$

In the interval $t_{s_3} < t_s < t_{s_4}$ we use a linear variation of Q_c with t_s as:

$$Q_c = C_c V_{DD} \frac{t_s - t_{s_4}}{t_{s_3} - t_{s_4}} \tag{14}$$

Finally, for $t_s > t_{s_4}$ (transitions e and f in Fig.2) we use $Q_c = 0$. For the case at which $V_{out,a}$ makes a high to low transition, speeding up the propagation delay, a similar model can be obtained changing the sign of Q_c. Once Q_c is obtained, the propagation delay ($t_{pHL,su}$ and $t_{pHL,sd}$) at $V_{out,v}$ is obtained using $Q_f = C_L V_{DD}/2 + Q_{sc}^f/2 + Q_c$ in (1).

3 Results

We compare the model results to HSPICE simulations for a $0.18\mu m$ technology. In Fig.4 we plot the propagation delay limit bounds for a 3-NAND gate for different values of the coupling to load capacitance ratio C_c/C_L. For these cases $t_{s_2} < t_s < t_{s_3}$ and therefore $Q_c = C_c V_{DD}$. The output load value (C_L) is obtained by adding the fan-out of a 3-NAND gate to the coupling capacitance ($C_L = C_{L,v} + C_c$). Therefore, as C_c increases, the delay t_{pHL0} is larger. Fig.4 shows an excellent agreement between HSPICE simulations (using the BSIM3v3 MOSFET model) and the proposed crosstalk delay model.

In Fig.5 we plot the propagation delay of the 3-NAND gate when varying the time skew between $V_{in,v}$ and $V_{in,a}$. Different channel width values of the nMOS transistors of the 3-NAND gate are selected (the pMOS sized appropriately) showing that the proposed model can describe a wide range of design choices.

4 Conclusions and Future Work

A simple description for the evaluation of crosstalk delay has been presented. The model is useful for a fast and accurate signal integrity simulation of large ICs. A very good agreement is achieved between model predictions and HSPICE simulations for a $0.18\mu m$ technology. The description accounts for short-circuit currents and short-channel effects and can be applied to multiple input gates. Experimental measurements for a more faithful validation of the proposed model are under development.

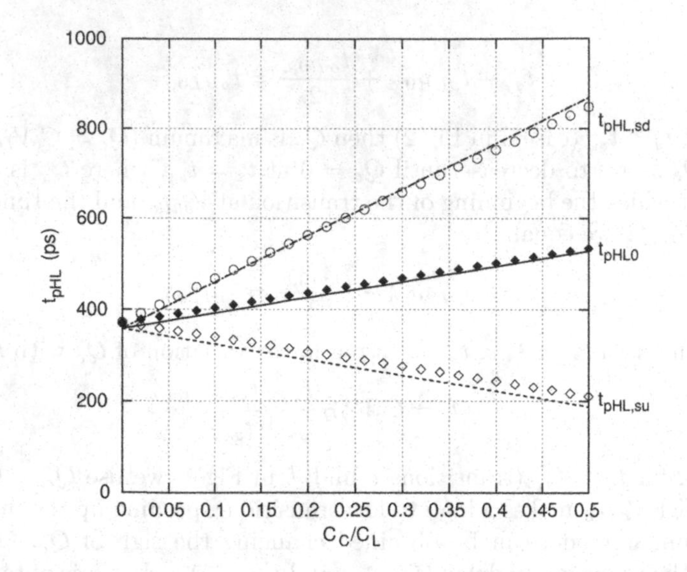

Fig. 4. In this picture we show the propagation delay bounds for a coupling capacitance C_c. Solid lines are model predictions while dots are HSPICE simulations.

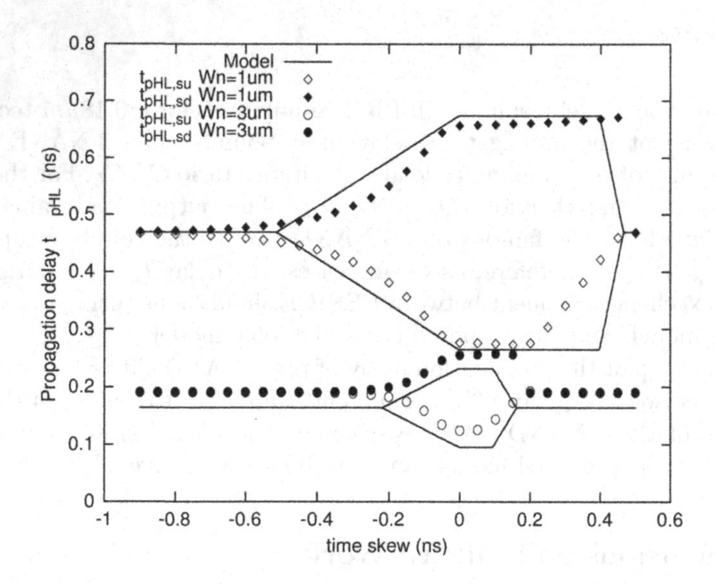

Fig. 5. In this figure we show the influence of the time skew between transition t_s and the propagation delay t_{pHL}. Model predictions are HSPICE simulations while solid line is the proposed model.

Acknowledgment. This work has been supported by the Spanish Government under the project CICYT-TIC02-01238.

References

1. P.D.Gross, R. Arunachalam, K.Rajagopal and L.T.Pileggi, "Determination of worst-case aggressor alignment for delay calculation," *in Proc. Int. Conf. Computer-Aided Design (ICCAD)*, 1998, pp. 212–219
2. W.Y. Chen, S.K. Gupta, and M.A. Breuer, "Analytical Models for Crosstalk Excitation and Propagation in VLSI Circuits," *IEEE Transactions on Computer-Aided Design*, Vol 21, no 10, pp. 1117–1131, October 2002
3. J.L.Rosselló and J. Segura, "A Compact Charge-Based Propagation Delay Model for Submicronic CMOS Buffers" in *Proc. of Power and Timing Modeling, Optim. and Simul. (PATMOS)*, Sevilla, SPAIN, pp. 219–228, Sept. 10–13, 2002
4. J.L.Rosselló and J. Segura, "Power-delay Modeling of Dynamic CMOS Gates for Circuit Optimization" in *Proc. of International Conference on Computer-Aided Design (ICCAD)*, San José CA, USA, pp. 494–499, Nov. 4–8, 2001
5. F. Caignet, S. Delmas-Bendhia and E. Sicard, "The Challege of Signal Integrity in Deep-Submicrometer CMOS Technology," *Proceedings of the IEEE*, vol. 89, no 4, pp.556–573, April 2001
6. T. Sakurai and R. Newton, "A Simple MOSFET Model for Circuit Analysis," *IEEE Transactions on Electron Devices*, vol. 38, pp. 887–894, Apr. 1991
7. J.L. Rosselló and J. Segura, "Charge-Based Analytical Model for the Evaluation of Power Consumption in Submicron CMOS Buffers," *IEEE Transactions on Computer-Aided Design*, Vol 21, no 4, pp. 433–448, April. 2002

CMOS Gate Sizing under Delay Constraint

A. Verle, X. Michel, P. Maurine, N. Azémard, and D. Auvergne

LIRMM, UMR CNRS/Université de Montpellier II, (C5506),
161 rue Ada, 34392 Montpellier, France

Abstract. In this paper we address the problem of delay constraint distribution on a CMOS combinatorial path. We first define a way to determine on any path the reasonable bounds of delay characterizing the structure. Then we define two constraint distribution methods that we compare to the equal delay distribution and to an industrial tool based on the Newton-Raphson like algorithm. Validation is obtained on a 0.25µm process by comparing the different constraint distribution techniques on various benchmarks.

1 Introduction

The goal of gate sizing is to determine optimum sizes for the gates in order that the path delay respect the constraints with the minimum area/power cost. Another parameter to consider is the feasibility of the constraint imposed on a given path. For that, ways must be found to get indicators for exploring the design space, allowing to select one among the available optimization alternatives such as sizing, buffering or technology remapping. The target of this paper is twofold: defining the delay bounds of a given path and determining a way for distributing a delay constraint on this path with the minimum area/power cost.

The problem of transistor sizing has been widely investigated using non linear programming techniques [1] or heuristics based on simple delay models [2]. Recently, in a pedagogical application [3] of the τ model, Sutherland [4], describing the gate delay as the product of electrical and logical efforts, proposed to minimize the delay on a path by imposing an equal effort that is a constant delay on all the elements of the path.

This way to select cell sizes can be proven mathematically exact [3] for a fanout-free path constituted of ideal gates (no parasitic capacitance nor divergence branches). However this evenly budget distribution is far to be the optimal one with respect to delay and area for a real path, on which divergence branches and routing capacitance are not negligible. Starting from the definition of the design space in terms of minimum and maximum delay permissible on a given path, we propose in this paper a design space exploration method allowing an area/power efficient distribution of constraint on a combinatorial path.

The delay bound determination and the constraint distribution method are based on a realistic delay model [5] that is input slope dependent and able to distinguish between falling and rising signals. This model is shortly presented in part 2. In part 3

J.J. Chico and E. Macii (Eds.): PATMOS 2003, LNCS 2799, pp. 60–69, 2003.
© Springer-Verlag Berlin Heidelberg 2003

we give a method for defining delay bounds on a path. Different approaches for distributing a delay constraint are considered in part 4 and compared in part 5 on different benchmarks of increasing complexity. We finally conclude in part 6.

2 Gate Delay Modeling

As previously mentioned sizing at the physical level imposes the use a realistic delay computation that must consider a finite value of the gate input transition time. As developed in [5] we introduce the input slope effect and the related input-to-output coupling in the model as

$$\begin{aligned} t_{HL}(i) &= \frac{v_{TN}}{2}\tau_{INLH}(i-1)+(1+\frac{2C_M}{C_M+C_L})t_{HLstep}(i) \\ t_{LH}(i) &= \frac{v_{TP}}{2}\tau_{INHL}(i-1)+(1+\frac{2C_M}{C_M+C_L})t_{LHstep}(i) \end{aligned} \quad (1)$$

where $v_{TN,P}$ represents the reduced value (V_T/V_{DD}) of the threshold voltage of the N,P transistors, $\tau_{INHL,LH}$ is the duration time of the input signal, taken to be twice the value of the step response of the controlling gate. C_M is the coupling capacitance between the input and output nodes. Indexes (i), (i-1) specify the switching and the controlling gates, respectively.

Following [2] the step response of each edge is defined by the time interval necessary to load (unload) the gate output capacitance under the maximum current available in the structure:

$$\begin{aligned} t_{HLstep} &= \frac{C_L \cdot \Delta V}{I_N} \\ t_{LHstep} &= \frac{C_L \cdot \Delta V}{I_P} \end{aligned} \quad (2)$$

Following the elegant model of [3] the evaluation of this step response on logic gates supplies a general expression given by:

$$\begin{aligned} t_{HLstep} &= \tau \cdot S_{HL} \cdot \frac{C_L}{C_{IN}} \\ t_{LHstep} &= \tau \cdot S_{LH} \cdot \frac{C_L}{C_{IN}} \end{aligned} \quad (3)$$

where τ is a time unit characterizing the process. C_{IN}, the gate input capacitance, is defined in terms of configuration ratio ($k=W_P/W_N$) between the P and N transistors. For simplicity, the S factors (logical effort of [3]) include all the current capability difference between the pull up and pull down equivalent transistors, they are configuration ratio dependent and characterize for each edge, the ratio of current available in an inverter and a gate of identical size.

Then considering an array of gates, the delay path can easily be obtained from (1) and (3) as a technology independent posynomial representation:

$$\frac{t_{HL,LH}}{\tau} = \theta = S_1' \cdot \frac{C_2 + C_{P1}}{C_1} + \cdots + S_{i-1}' \cdot \frac{C_i + C_{Pi-1}}{C_{i-1}} + S_i' \cdot \frac{C_{i+1} + C_{Pi}}{C_i} + \cdots + S_n' \cdot \frac{C_L}{C_n} \quad (4)$$

where the S_i' include the logical effort and the input ramp effect, C_i represents the input capacitance of the gate and C_{pi} the output node total parasitic capacitance, including the interconnect and branching load.

3 Delay Bound Definition

We consider realistic combinatorial paths on which two parameters are known and imposed:

- the output load capacitance of the last gate, that is determined by the input capacitance of the next register,
- the input capacitance of the first gate imposed by the loading conditions of the input register.

In that conditions the path delay is bounded. These bounds can be determined, considering that the delay of a path (4) is a convex function of the gate input capacitance. This is illustrated in Fig.1, where we represent the path delay sensitivity to the transistor sizing of a combinatorial path constituted of 13 gates.

As shown the delay value decreases from a maximum value, obtained when all the transistors have the minimum size, to a minimum value that will be determined below. Note that the maximum delay value is a „reasonable" one, it is very easy to get a much greater value by loading minimum gates with high driving capacity one. This curve illustrates what we define by exploring the design space:

- near the maximum value Θ_{Max} of the delay the path sensitivity to the gate sizing is very important, a small variation of the gate input capacitance results in a large change in delay,
- at the contrary near the minimum Θ_{Min} the sensitivity is becoming very low and in that range any delay improvement is highly area/power expensive.

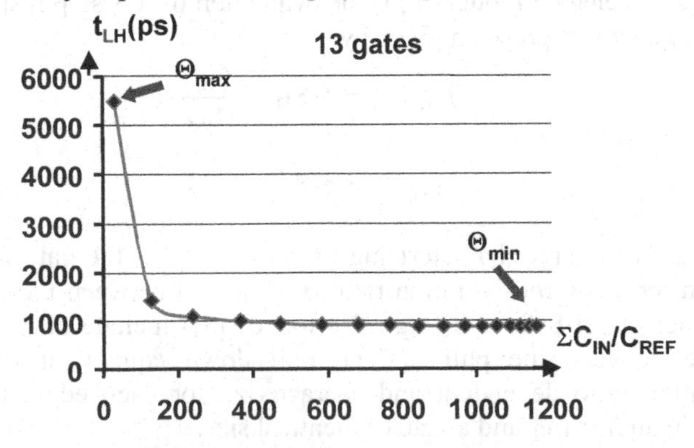

Fig. 1. Illustration of the sensitivity of the path delay to the gate sizing.

Evaluating the feasibility of a delay constraint Θ_c imposes to compare its value to the preceding bounds. If the Θ_c value is closed to the maximum Θ_{Max} the constraint satisfaction will be obtained at reasonable cost by transistor sizing otherwise it would be more profitable to reconfigure the logic or to insert buffers [6]. Let us define these bounds. As previously mentioned we consider for Θ_{Max} the „reasonable" value obtained when all the gates are implemented with transistors of minimum size. For the minimum bounds we just use the posynomial property [7] of (6). Canceling the derivatives of (4) with respect to the gate input capacitances C_i we obtain a set of link equations such as:

$$S'_{i-1} \cdot \frac{C_i}{C_{i-1}} - S'_i \cdot \frac{C_{i+1} + C_{Pi}}{C_i} = 0$$
$$S'_i \cdot \frac{C_{i+1}}{C_i} - S'_{i+1} \cdot \frac{C_{i+2} + C_{Pi+1}}{C_{i+1}} = 0 \ldots\ldots \tag{5}$$

Cell sizes can then be selected to match the minimum delay, by visiting all the gates in a topological order, starting from the output, such as:

$$C_i^2 = \frac{S'_i}{S'_{i-1}} \cdot C_{i-1} \cdot (C_{i+1} + C_{Pi}) \tag{6}$$

This results in a set of n linked equations that can be easily solved by iterations from an initial solution that considers C_{i-1} known and equal to a reference value C_{REF} that can be set to the minimum value available in the library or any other one.

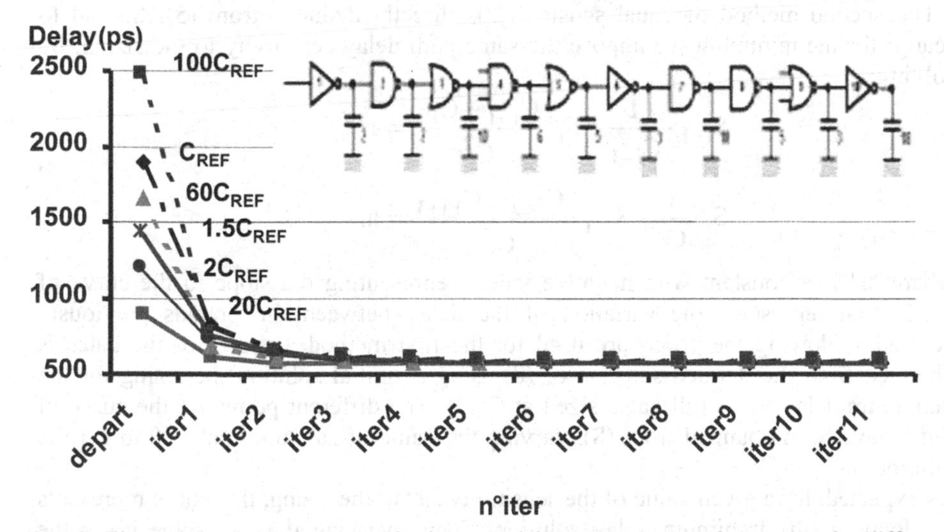

Fig. 2. Illustration of the research of minimum delay on an array of ten gates for different values of the initial reference capacitance; the output load of each gate on the array is given in unit of C_{REF}.

As shown in Fig.2, whatever is the value of the initial C_{i-1} controlling capacitance, only few iterations are required to obtain a fast convergence to the minimum delay value achievable on the array.

4 Constraint Distribution

Determining the possible bounds of delay for a given path topology, the next step is to evaluate the feasibility of a constraint to be imposed on a path. The theory of constant effort or constant delay [4,8] provides an easy way to select the cell size for each stage but for real configuration it is far from the optimum and often results in oversized structures. For that we propose two techniques for the gate size selection in order to satisfy a constraint that we will compare in the next part to the constant delay method.

To define the first method we consider that imposing equal delay to the gates with an important value of the logical effort (S'_i), will result in an important over-sizing of these complex gates. The determination of the lowest delay bound directly provides the corresponding delay distribution on the path, that appears to be the fastest one. So we can use this distribution to define for each gate a weight or gain Θ_i relative to this distribution $\Theta_{Min}=\Sigma\Theta_{Mini}$. In that case we propose to distribute the delay constraint Θ_c on a path using a weight defined on the minimum delay distribution as:

$$\theta_i = \frac{\theta_{Mini} \cdot \theta_c}{\sum_i \theta_{Mini}} \tag{7}$$

Then processing backward from the output of the path, this directly gives, for each gate, the value of Θ_i that determines from (1,3) the size of the corresponding gate.

The second method of equal sensitivity is directly deduced from (5). Instead to search for the minimum we impose the same path delay sensitivity to the sizing, by solving:

$$\begin{aligned} S_{i-1} \cdot \frac{1}{C_{i-1}} - S_i \cdot \frac{C_{i+1} + C_{Pi}}{C_i^2} &= a \\ S_i \cdot \frac{1}{C_i} - S_{i+1} \cdot \frac{C_{i+2} + C_{Pi+1}}{C_{i+1}^2} &= a....... \end{aligned} \tag{8}$$

where "a" is a constant with negative value, representing the slope to the curve of Fig.3, that represents the variation of the delay between the bounds previously defined. Following the procedure used for the first method, the size of the gates is obtained from the iterated solution of (8) using as initial solution the sizing for the maximum delay value (all gates sized at C_{REF}). The different points on the curve of Fig.3 have been obtained from (8), varying the value of „a“, until „a“ = 0 to get the minimum.

As expected, for a given value of the sensitivity „a“ to the sizing, this curve represents the locus of the minimum delay solutions. Thus, varying the „a“ value gives the possibility to explore the design space and to determine the minimum area sizing condition satisfying the delay constraint. As shown in the figure the results are compared with the solutions obtained from an industrial optimization tool (Amps from Synopsys) that gives nearly equivalent exploration.

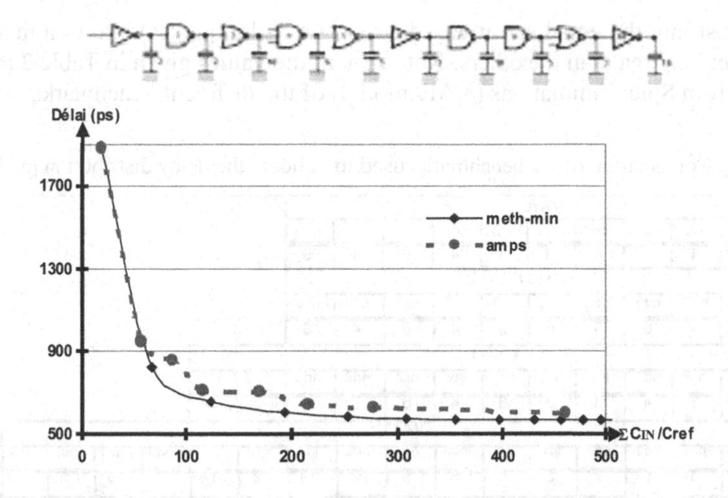

Fig. 3. Exploration of the design space with constant sensitivity constraint.

5 Validation

In order to validate these sizing and constraint distribution techniques we compare on different benchmarks the minimum delay value and the area obtained using the three investigated methods:

- equal distribution of delays (C), ($\Theta_i = \Theta_c/n$), where n is the number of stages,
- weighted distribution (7), (B),
- equal gate sensitivity (8), (A),

and using an industrial tool based on a Newton-Raphson based algorithm (D), [9] (Amps from Synopsys).

These benchmarks are constituted of array of gates with different loading conditions, their structure is shown in Table 1, where the output load of each gate C_P is given in unit of C_{REF} (C_{REF}=3.5fF).

The comparison of the minimum delay values obtained with each technique is given in Table 2 for different paths. The targeted process is the STm 0.25µm with τ = 7.05ps. As shown the lowest minimum value of delay is obtained with both the weighted and the equal sensitivity techniques.

This ascertains the method used to determine the lowest bound of delay on a logical path. Note that around the minimum value of delay the area penalty is very large, in that case this value must be considered more as an indicator for the feasibility of the constraint than as a target to be reached.

The next step is to compare for a given delay constraint the area of implementation obtained with the different distribution techniques. For that we impose on the different benchmarks a delay constraint defined between the bounds previously defined. Then we compare in Table 3 the area corresponding to the gate sizing allowing, with the different techniques, to match the constraint. We can observe that if for a weak constraint value the different techniques appear quite equivalent, for a tighter constraint the equal sensitivity distribution technique (A) allows a match with a much smaller area than the others. Note that all the values given in Table 2 and 3 are obtained from Spice simulations (MM9 model) of the different benchmarks.

Table 1. Composition of the benchmarks used to validate the delay distribution methods.

Ver9																
Type	nd2	nd2	inv	nd2	nr2	nd3	nr2	nd2	nd3							
Cp /Cref	4	6	8	6	6	4	6	4	5							
Ver91																
Type	nd2	nd3	inv	nd2	nr3	nd3	nr2	nd2	nr3							
Cp /Cref	4	6	8	6	6	4	6	4	9							
Ver11																
Type	nr2	nd2	inv	nd2	nr3	inv	nr2	nd2	nd3	nd2	nr3					
Cp /Cref	3	4	3	3	8	4	6	2	8	5	12					
Ver15																
Type	nr3	nd2	nd3	inv	nr3	nd3	nr2	nd2	inv	nr2	inv	nd3	nr3	nd2	nr2	
Cp /Cref	3	5	7	5	8	6	9	2	11	2	14	7	9	5	18	
Ver151																
Type	nr2	nd2	inv	nd3	inv	nd2	nr3	nd2	nr3	inv	nr2	nd3	nr3	inv	nr3	
Cp /Cref	3	7	2	5	8	2	3	7	11	2	8	7	7	3	18	
Ver21																
Type	nr3	nd2	inv	nd2	nr2	nd3	nr2	nd2	nd3	nd2	nr3	nd2				
Cp /Cref	2	6	8	6	8	4	9	2	10	4	9	4				
Type	nr3	nd3	nr2	nd2	nr3	inv	nd2	nd3	inv	nd3	nr2					
Cp /Cref	9	2	9	8	7	7	4	3	11	7	15					
Ver31																
Type	nr2	nd3	nd2	inv	nr2	nd3	nr2	nd2	nd3	nd2	nr3	nd3	nr2	nd2	nr3	inv
Cp /Cref	2	6	9	6	8	4	9	2	10	4	9	2	9	8	7	7
Type	nd2	nd3	inv	nd3	nr2	inv	nr3	inv	nr2	nd2	nd3	nd3	nr3	nd2	nr2	
Cp /Cref	4	3	11	7	8	2	4	10	8	5	9	7	6	4	15	

The weighted distribution (B) still results in a quite equivalent area but the equal delay distribution (C) and Amps (D) may result for quite complex paths in an important increase of area. For some constraint values they may fail to find a solution (xxx).

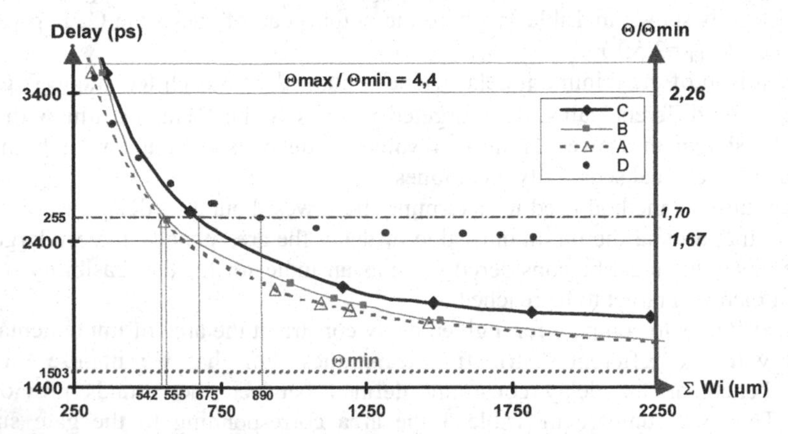

Fig. 4. Illustration of the design space exploration, on the VER31 path, using the different constraint distribution methods.

An illustration of these results is given in Fig.4 where we show for the path VER31 the complete exploration of the design space using the preceding constraint distribution methods. As shown for a delay constraint value smaller than $\Theta_{max}/2$ the gain in area (power) using the equal sensitivity or the weighted distribution method is quite significant.

Table 2. Comparison of the minimum delay values reached with the different sizing techniques: (A) is the equal sensitivity technique (8), (B) the weighted distribution, (C) the equal delay [4], (D) from Amps optimization tool.

Bench	**Gate nb**	**Max. delay value (ps)**	ΣW_i (µm)	Sizing Techn.	Min. delay value (ps)	ΣW_i (µm)
Ver9	9	1399	38	A	544	668
				B	544	668
				C	606	240
				D	602	403
Ver91	9	1874	42	A	620	987
				B	620	987
				C	676	391
				C	633	632
Ver11	11	2085	46	A	698	1448
				B	698	1448
				C	777	440
				D	937	348
Ver15	15	3588	64	A	974	4077
				B	974	4077
				C	1154	800
				D	1224	1047
Ver151	15	3479	3479	A	923	4337
				B	923	4337
				C	1023	1083
				D	960	3067
Ver21	21	4583	94	A	1192	8419
				B	1192	8419
				C	1484	1039
				D	1693	1152
Ver31	31	6560	138	A	1503	21578
				B	1503	21578
				C	1881	2226
				D	2426	1826

6 Conclusion

Based on a simple but realistic delay model for combinatorial gates, we have first determined an easy way to characterize the feasibility of a delay constraint to be imposed on a combinatorial path. We have defined reasonable maximum and real minimum delay bounds. Then we proposed two techniques to match a delay constraint on a path: the equal sensitivity and the weighted method that is a budgeting method. We have applied these methods on different benchmarks with various constraint conditions and compared the area of the resulting implementations with that obtained from an equal delay distribution and with an industrial tool. If for weak constraints the different methods are quite equivalent, for values near the minimum, the proposed methods always find a solution and result in an important area/power saving. Another point to be clarified further is to define at which distance of the minimum delay value it is reasonable to impose a constraint.

Table 3. Comparison of the area (given as the sum of the widths of the transistors) necessary to match the delay constraint in the different distribution techniques.

Bench	Sizing Techn.	$\Theta_{Max}/\Theta_{Min}$	Θ_C/Θ_{Min}	Area (µm)	Θ_C/Θ_{Min}	Area (µm)	Θ_C/Θ_{Min}	Area (µm)	Θ_C/Θ_{Min}	Area (µm)
	A			96		130		159		191
Ver9	B	2.6	1.4	100	1.23	132	1.15	159	1.11	193
	C			103		140		180		240
	D			104		155		245		338
	A			137		197		284		535
Ver91	B	3	1.4	147	1.2	202	1.1	292	1.02	560
	C			161		224		337		xxx
	D			144		202		289		632
	A			66		108		166		310
Ver11	B	3	2.15	80	1.6	134	1.34	195	1.1	330
	C			94		152		228		440
	D			70		127		348		xxx
	A			178		250		367		484
Ver15	B	3.7	1.7	202	1.47	270	1.27	384	1.18	538
	C			229		330		485		800
	D			195		321		789		xxx
	A			302		426		648		1333
Ver151	B	3.8	1.4	310	1.25	432	1.14	636	1.04	1407
	C			410		580		846		xxx
	D			324		486		786		3067
	A			196		406		522		553
Ver21	B	3.8	2.1	214	1.5	421	1.34	529	1.31	558
	C			230		479		651		715
	D			198		441		738		1152
	A			364		490		611		1275
Ver31	B	4.4	2.1	400	1.8	511	1.6	641	1.26	1361
	C			427		591		766		1970
	D			377		670		1826		xxx

References

[1] J. M. Shyu, A. Sangiovanni-Vincentelli, J. Fishburn, A. Dunlop, "Optimization-based transistor sizing" IEEE J. Solid State Circuits, vol.23, n°2, pp.400–409, 1988.

[2] J. Fishburn, A. Dunlop, "TILOS: a posynomial programming approach to transistor sizing" in Proc. Design Automation Conf. 1985,pp.326–328.

[3] C. Mead, M. Rem, "Minimum propagation delays in VLSI", ", IEEE J. Solid State Circuits, vol.SC17, n°4, pp.773–775, 1982.

[4] I. Sutherland, B. Sproull, D. Harris, "Logical Effort: Designing Fast CMOS Circuits", Morgan Kaufmann Publishers, INC., San Francisco, California, 1999.

[5] K. O. Jeppson, "Modeling the influence of the transistor gain ratio and the input-to-output coupling capacitance on the CMOS inverter delay", IEEE J. Solid State Circuits, vol.29, pp.646–654, 1994.

[6] S. Chakraborty, R. Murgai „Lay-out driven timing optimization by generalized DeMorgan transform" IWLS 2001, pp.53–59.

[7] M. Ketkar, K. Kasamsetty, S. Sapatnekar "Convex delay models for transistor sizing" Proc. of the 2000 Design Automation Conf. pp.655–660.

[8] J. Grodstein, E. Lehman, H. Harhness, B. Grundmann, Y. Wanatabe „A delay model for logic synthesis of continuously-sized networks", ICCAD 95, Nov 95.

[9] R. K. Brayton, R. Spence „Sensitivity and Optimization" Elsevier 1980

Process Characterisation for Low VTH and Low Power Design

E. Seebacher, G. Rappitsch, and H. Höller

Austriamicrosystems AG
A 8141 Schloss Premstaetten, Austria

http://www.austriamicrosystems.com

Abstract. We discuss state of the art and new developments for the path to first time right silicon for low VTH and low power analog design. This article touches on a few of the issues that are essential when starting a low VTH or low power design, where the bottom line is a well controlled process technology and the existence of a comprehensive Process Design Kit with accurate SPICE models which include device mismatch parameters and noise parameters. The necessary process characterisation and the requirements for SPICE modelling are described. In this article state of the art MOS transistor modelling especially in the transition region, noise modelling and device mismatch are discussed with regard to low VTH and low power design.

1 Introduction

The „Time to market" is probably the most critical issue in Product development. The path to 'first time right' silicon has been greatly simplified over the last few years in the field of digital chips by using high-level design languages (Verilog or VHDL) in conjunction with synthesis tools and automatic place and route techniques. Analog design and especially low power design has retained a certain mystique with the common notion that it is some kind of artform practised only by a few. Whilst low power analog design is certainly still challenging there are a number of techniques that can be used to reduce the risk.

In particular the key elements for successful mixed-signal designs are accurate compact modelling, the availability of basic library cells and a comprehensive process documentation - preferably all set-up in a quality controlled, standard CAD environment.

In this paper we introduce the necessary process characterisation for low power design suitable for low VTH processes. An overview about the state of the art MOS transistor SPICE modelling challenges is shown, which can effect whether your mixed-signal silicon really works at first time or whether it will take multiple re-spins to achieve your specifications.

J.J. Chico and E. Macii (Eds.): PATMOS 2003, LNCS 2799, pp. 70–79, 2003.
© Springer-Verlag Berlin Heidelberg 2003

2 Process Control and Characterisation

In the last years a lot of low power and low voltage designs have been developed and produced in standard 5V processes with minimum geometries larger or equal than 0.6um and 3.3V process with the minimum geometry down to 0.35um. This processes have typical threshold voltages larger than 0.6V and require high precision SPICE modelling especially in the transition region and for the sub-threshold characteristic of the MOS transistor (see Table 1). For newer process generations as for example 0.25um or 0.18um or special low VTH processes, the threshold voltages of the NMOS transistors have typical values around 0.4V. At the same time the maximum VDS is decreasing significantly, therefore forcing designers more and more to move their operating points towards the transition and sub-threshold region.

Table 1. MOS threshold voltage VTH and gate oxide thickness for different process generations.

Process Generation	VTH [V] NMOS/PMOS	TOX [nm]	VDD[V]
≥0.6um	0.7-0.8	12.5	5
0.35um	0.40-0.5/0.6-0.8	7.5	3.3
0.25um	0.3-0.5/0.4-0.6	6.4	2.5
0.18um	0.3-0.5/0.4-0.6	4.0	1.8
0.13um	0.2-0.3	2.0	1.2
Low VTH processes	0.40-0.5/0.5-0.7	≥12.5	5

If we take the process variation into account VTH values range from 0.30V to 0.50V for NMOS and 0.40V to 0.60V for PMOS transistors (Tab.1). For these process generations and for special low VTH processes with larger minimum geometries it is therefore absolutely necessary to control and characterise the sub-threshold characteristic and the drain–source leakage current of the MOS devices (see Fig.1).
Short and narrow channel effects result in lower threshold voltages for the minimum geometries compared to the large transistor or vice versa (see Fig. 6). To characterise and control such a process all extreme geometries of the MOS transistor have to be measured as Pass/Fail parameters on every single production wafer. Key parameters for such an advanced process monitoring and characterisation are the threshold voltages of large MOS devices e.g. VTH10x10, the threshold voltage for the minimum transistor length $VTH10xL_{min}$, the threshold voltage for the minimum transistor width $VTHW_{min}$ and the threshold voltage of the smallest transistor $VTHW_{min}L_{min}$, as well as the source drain leakage current ILEAK. For low VTH processes it is a must to monitor the ILEAK parameter and define it as a pass/fail parameter.
The threshold voltages VTH and as a consequence the drain-source leakage current must be considered as a function of the temperature to estimate the maximum or worst case leakage current for the design specification.

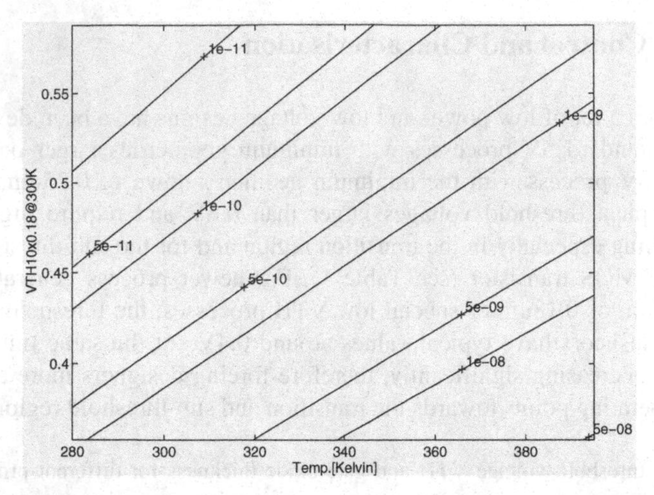

Fig. 1. Contour plot of the source drain leakage current ILEAK at different temperatures and threshold voltage VTH (W/L= 10/0.18) variations in case of a standard 0.18um CMOS Process.

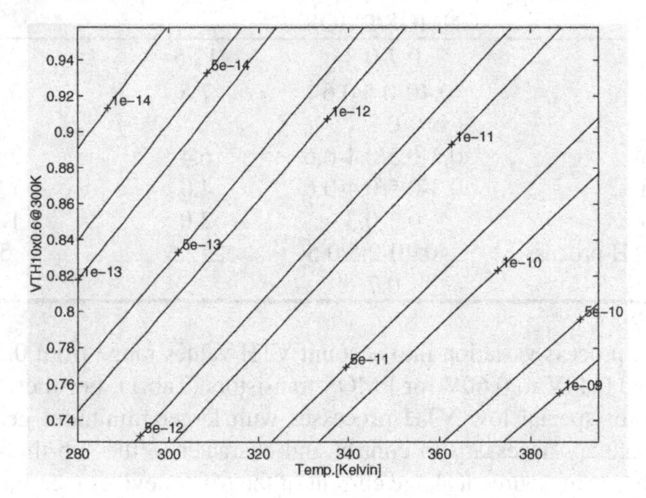

Fig. 2. Contour plot of the source drain leakage current at different temperatures and threshold voltage VTH (W/L= 10/0.6) variations in case for a standard 0.6um CMOS process.

Fig.1 (0.18um process) and Fig.2 (0.60um process) show contour plots of the source drain leakage current as a function of the expected process variation in VTH and the temperature. The lines in the plot specify constant leakage currents. Comparison of the two processes shows a significant increase in the drain-source leakage current in case of the 0.18um process which results from the lower VTH.

3 Compact MOS Transistor Modelling

For analog SPICE simulation of low power circuits it is important to use a physically based model as opposed to purely mathematical models as for example BSIM1 or

BSIM2 [1] or numerically abstract look-up tables. The industry standard BSIM3v3 [2], the Philips LEVEL9 [6] or the EKV [4-5] compact models allow many parameters - such as oxide thickness TOX or long channel threshold voltage VTH - to be set at their actual process values and other parameters to be tuned to achieve a good fit between measured and extracted values.
This process is iterative and must be checked not only for the first order parameters such as the drain current Ids but for the small signal transconductance gm and the small signal output transconductance gds (see Fig.3) also. For analog simulation, with predictable simulator convergence, it is important to use these physical and scaleable models which apply to all geometries and all operating conditions [8]. High precision SPICE modelling especially in the transition region of the MOS transistor for low VTH and sub-micron processes is needed. For low power designers with supply voltages of 1.2V the operating conditions of the MOS transistor is in the transition region between sub-threshold and inversion region, dependent of the geometry.

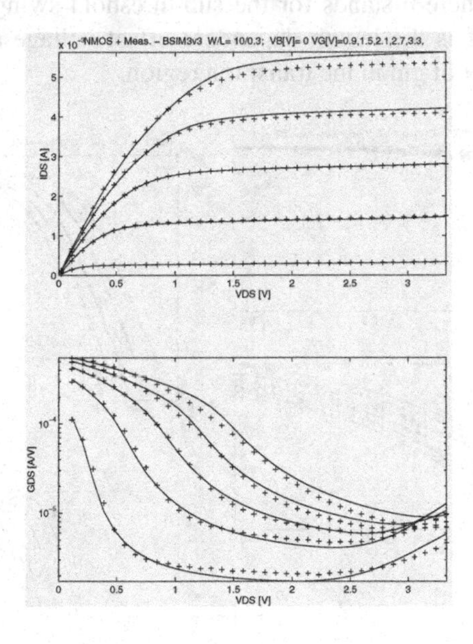

Fig. 3. NMOS output characteristic W/L=10/0.3,VGS=0.9,1.5,2.1,2.7,3.3V; +=measurements, -=BSIM3v3 SPICE model.

State-of-the art piece-wise regional-based compact models like in BSIM3v3 [3] or BSIM4 are developed to describe the behaviour of the transistor in the different operating regions and connect this equations using so called effective voltages to avoid discontinuities (see Eq.1). The old generation of transistor models like the most common SPICE LEVEL2 or LEVEL3 [1] did not care about the transition regions and are therefore not usable for low VTH or low power design. Discontinuities in the first and second order derivative lead to wrong simulation results and mostly to convergence problems. This problems exist in the equations for the currents as well as for the charges and capacitances of the MOS transistor.

$$V_{gsteff} = \frac{n \cdot v_t \cdot \ln\left[1 + e^{\frac{m^* \cdot V_{gst}}{n \cdot v_t}}\right]}{m^* + n \cdot Coxe \sqrt{\frac{2\Phi_s}{q \cdot \varepsilon_{si} N_{dep}}}\, e^{-\frac{(1-m^*)V_{gst} - V_{off}}{n \cdot v_t}}} \tag{1}$$

The new model generation like the BSIM3v3 the Philips Level9 or the advanced BSIM4 [7] model transform the physical voltages like the gate voltage vgs into effective voltages *vgseff* (Eq. 1) to smoothen the transition between the sub-threshold and the inversion region.
As an example equation 1 shows the improved effective gate voltage vgseff of the BSIM4v2 model - where n stands for the sub-threshold swing, vgst is the effective voltage vgs-vth, Voff is the length dependent offset voltage and the parameter m* improves the accuracy of gm in the transition region.

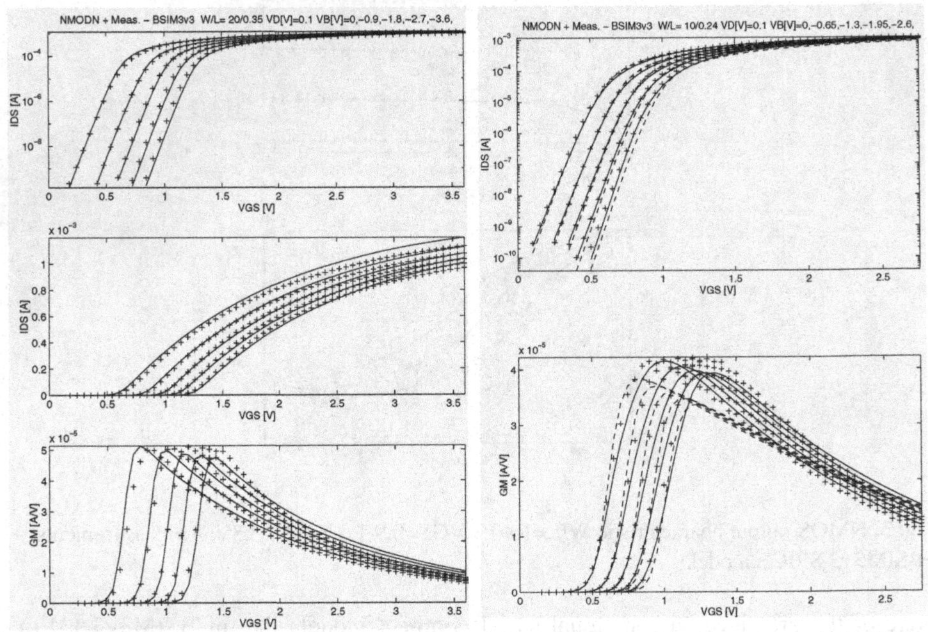

Fig. 4. NMOS transfer characteristic W/L= 10/ 0.3,VBS=0,0.9,1.8,2.7,3.3V; +=measurements, -=BSIM3v3 SPICE model

Fig. 5. NMOS transfer characteristic W/L= 10/0.24,VBS=0,0.65,1.3,1.95,2.6V; +=measurements, --=BSIM3v3 SPICE model, -=BSIM4 SPICE model.

The industry standard model BSIM3v3 which includes a similar effective gate-source voltage shows acceptable results for the sub-threshold characteristic and the transition regions for sub-micron processes larger than 0.25um. The results are shown in Fig. 4 where the transfer characteristic of a MOS transistor with the geometry W/L=10/0.3

is shown. The parameter set has been extracted from 10 different geometries with the use of transfer, output characteristic, gm and gds with local optimisation strategy and high emphasis on low power applications.
Some weaknesses in modelling the transition region has been recognised for the BSIM3v3 model for processes smaller than 0.35um. The BSIM4 model which includes some additional physical effects like the poket implant and additional fitting parameters shows much better results in the transition region compared to the BSIM3v3 model. Fig. 5 shows the transfer characteristic of a W/L=20/0.25um NMOS transistor where the measurements and the models for the BSIM3v3 and the BSIM4 model are compared. Both model-parameter sets are valid for all geometries and operating conditions and have been extracted from 10 different geometries and with the use of the transfer, the output characteristic gm and gds, respectively.

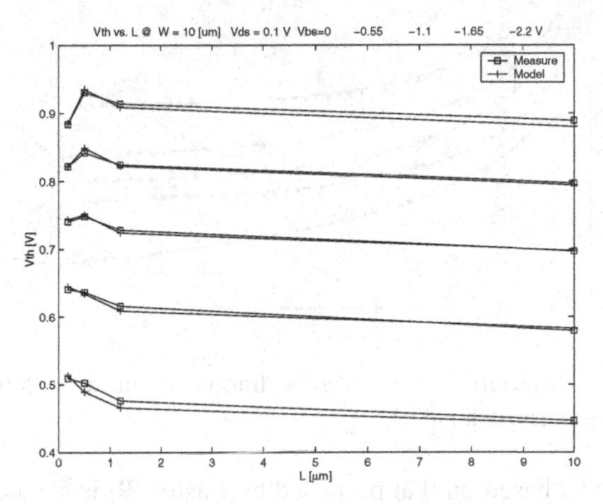

Fig. 6. VTH as function of the transistor length L for different bulk voltages. - BSIM4v2, x measurements

Very important for predicting the leakage current of low VTH processes is the accurate modelling of the geometry dependency of the threshold voltages VTH. Short and narrow channel effects can decrease the threshold voltage and lead to higher leakage currents. In Fig. 6 the threshold voltage VTH as function of different channel lengths and different bulk voltages vbs=0,-0.55,-1.1,-1,65,-2.2V is shown in case of a MOS transistor of a 0.18um process. The comparison of VTH extractions of measurements and the corresponding BSIM4 model is shown.

4 Matching Parameters

In addition to large-scale process variations which may – and will - occur between different wafer lots, silicon performance will also vary on the microscopic scale affecting devices placed on the same die. These local or matching effects are often even

more critical for designs and the availability of statistically accurate matching parameters can be used to significantly reduce the risk of low-yields.
As low power designs operate at voltages close to the threshold voltage, voltage and geometry dependent mismatch parameters for MOS transistors are essentially. The expected MOS transistor current mismatch increases at lower gate-source voltages vgs and the geometry dependency in case of NMOS transistors of a 0.35μm process is shown in Fig.7.

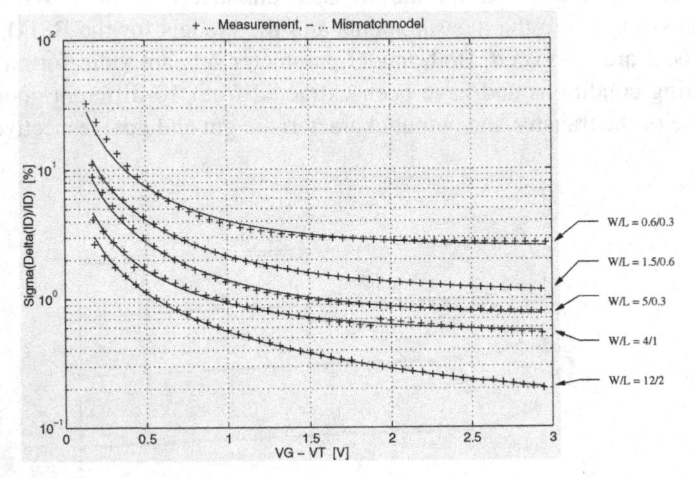

Fig. 7. Mismatch characteristics of NMOS transistors in a 0.35μm CMOS process. +=measurements, —=mismatch model.

A variance model based on that proposed by Bastos [9] is used to fit the measured mismatch data $\sigma(\Delta I_D/I_D)$ that is gained from MOS transistors of different geometries. In this variance model the sensitivity of the fitting parameters (e.g. $\sigma^2(\Delta V_T)$, $\sigma^2(\Delta\kappa)$) ensures that all parameters are equally well determined. This means that the current mismatch performance $\sigma(\Delta I_D/I_D)$ in the operating region around the threshold voltage - the most important region in case of low power design - is dominated by the mismatch of $\sigma^2(\Delta V_T)$. At higher gate voltages $\sigma^2(\Delta\kappa)$ becomes the dominant parameter and in this region additionally the variation of the mobility reduction is important. Another advantage of the variance model approach is that only a single fit is needed, which prevents the mismatch parameters from picking up additionally „fitting noise". Scaling of the mismatch parameters is done by using the well known Pelgrom's law [10] which defines the geometry dependence of the mismatch parameters. This is shown in case of the threshold voltage mismatch $\sigma(\Delta V_T)$ of NMOS transistors of a 0.35μm CMOS process in Fig.8. Statistical SPICE model sets (Monte Carlo models) are provided to enable yield prediction during circuit simulations taking both process variations and mismatch effects into account.

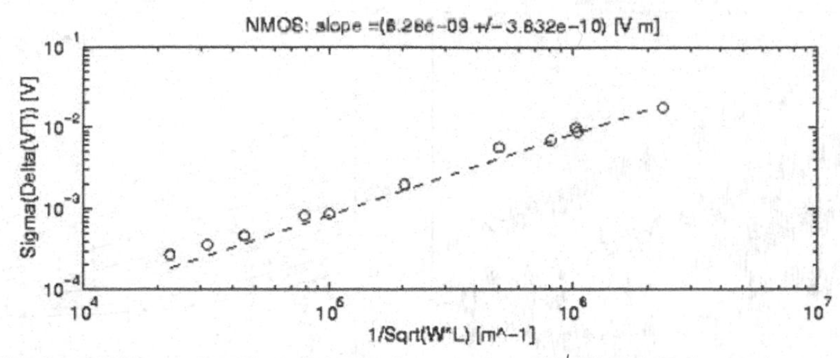

Fig. 8. $\sigma(\Delta V_T)$ as a function of the device geometry $1/\sqrt{(W \cdot L)}$ – Pelgroms law.

5 Noise Modelling

It may occur that an analog design fulfils all of its first order parameters such as the gain or bandwidth, but that other parameters such as the signal to noise ratio are out of spec. Accurate estimation of such effects relies on a process description and SPICE modelling covering not just the standard characteristic curves but also the noise behaviour of each active and passive device (Fig. 9). Low frequency noise (1/f noise) of MOS transistors is of special importance for analog designs (LNA's, A/D converters) since devices are becoming smaller and more compact. Moreover, due to the current limitation in low power designs, the designer is forced to work in weak inversion regions with low gate overdrive. Figure 9 shows the measured low frequency current noise [$A/\sqrt{Hz}$] for a PMOS transistor in a 0.35 um process with fixed bias currents IDS= 1, 5, 20, 100 uA. The IDS= 1 uA bias (lowest curve) has a gate overdrive VGS-VTH=320 mV working in the weak inversion transition region. For the measurement of noise in such regions a special noise measurement hardware unit with a low resolution limit (preferably below 1 $pA/\sqrt{Hz}$]) is needed.

For the modelling of weak inversion noise standard SPICE models (LEVEL 1 or LEVEL 2) are not sufficient for describing noise in the low voltage regions since they do not include an explicit voltage dependence on the gate overdrive. Therefore, more accurate models like the BSIM3V3 noise model [11] have to be used in order to describe all geometric and voltage-dependent effects. In the BSIM3V3 model the carrier densities at the source at the source (N_o) and at the drain (N_l) directly depend on the effective gate voltage overdrive and the noise current spectral density is defined as:

$$S_{i_F} = \frac{1}{C_{OX} \cdot L_{eff}^2} \cdot \frac{I_D}{f^{EF(=1)}} \cdot \mu_{eff} \cdot \\ \cdot f_1 \begin{pmatrix} \mathbf{NOIA, NOIB, NOIC;} \\ N_o(V_{gsteff} - VTH), N_l(V_{gsteff} - VTH) \end{pmatrix} \quad (2)$$

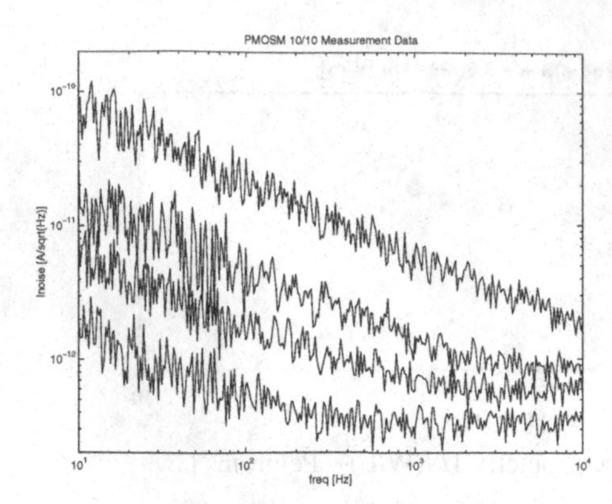

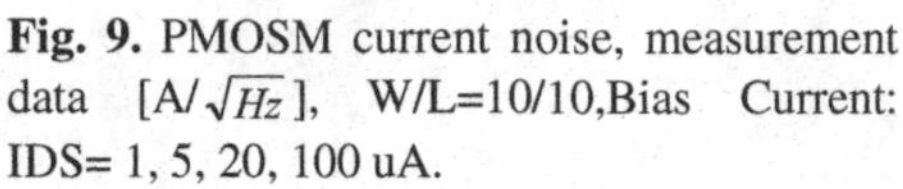
Fig. 9. PMOSM current noise, measurement data [A/$\sqrt{Hz}$], W/L=10/10,Bias Current: IDS= 1, 5, 20, 100 uA.

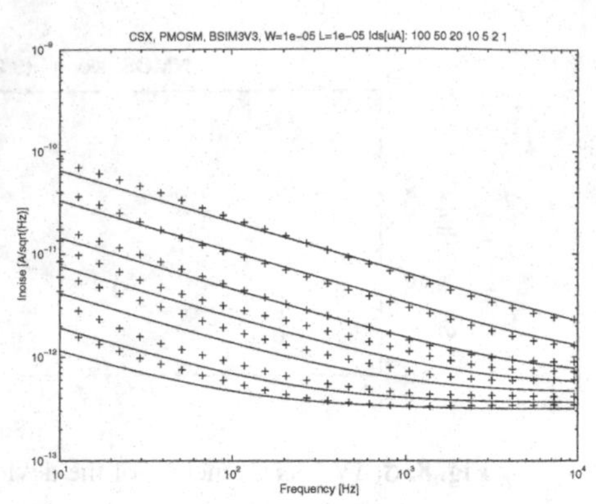

Fig. 10. BSIM3V3 noise parameter extraction for PMOSM measurement (smoothed) vs. extracted [A/$\sqrt{Hz}$], W/L=10/10, Bias Current: IDS= 1, 5, 20, 50, 100 uA, VGS-VTH: 0.32, 0.69, 0.97, 1.39, 2.18, 3.19 V.

Therefore, accurate high accuracy can be achieved in the weak inversion region when parameter extraction is applied to optimise parameters NOIA, NOIB and NOIC. Fig.10 shows the optimisation result of a BSIMV3 noise parameter extraction for a PMOS transistor where accurate results could be achieved for the gate overdrive ranging from 320 mV to 3.19 V from strong to weak inversion. The noise parameters evaluated to NOIA=1.09e+18, NOIB=6.01e+03 and NOIC=1.19e-12. The integration of those noise parameters as part of the MOS transistor model enables the designer the accurate prediction of the 1/f noise corner in the low current region during circuit simulation.

6 Conclusions

In this article important aspects of process control and characterisation of low VTH processes and low power design has been discussed. The threshold dependent drain-source leakage current has been visualised in contour plots for a 0.6um and a 0.18um process. Parameter-sets for state of the art MOS transistor models have been extracted for different process generations. The BSIM4 model shows better results in the transition region of the transistor for sub-half micron processes compared to the industrial standard BSIM3v3. The BSIM3v3 model with a lower number of parameters and shorter simulation time gives acceptable results for processes equal and larger than 0.35 um.

It could be shown that an accurate modelling of the bias dependent mismatch behaviour – especially at low gate voltages - is essential for most low power applications.

To estimate the signal to noise ratio in low power low voltage designs, accurate noise measurements and bias-dependent noise models for MOS transistors are necessary. The development of a special noise measurement hardware and a global optimisation procedure for the BSIM3V3 noise model enables the simulation and prediction of circuit noise in the weak inversion region.

For the development of a low power design with new process generations it is absolutely necessary to take the increased drain-source leakage into account. The usage of highly accurate SPICE models like the BSIM4 model with carefully measured and extracted 1/f noise and matching parameters is absolutely necessary.

References

1. ELDO User's Manual 5.6, Release 2001.2, August 2001, p. 10/258–10/262
2. Cheng. Y., Jeng M.C., Liu Z., Huang J., Chen M., Ko P.K., and Hu C.: and Hu,C.:A physical and scalable I-V model in BSIM3v3 for analog/digital circuit simulation IEEE Electron Devices 1997, 44 (2), pp. 277–287.
3. Department of Electrical Engeneering and Computer Sience: BSIM3v3 Manual, University of California, Berkeley.
4. C.Enz, F. Krummenacher, E. Vittoz, „An analytical MOS transistor model valid in all regions of Operation and dedicated to low-voltage and low current applications", Jornal on Analog Integration Circuits and Signal Processing, Kluwer Academic Pub. 1995, pp 83–114.
5. C.Enz, E. Vittoz, „MOS transistor modeling for low-voltage and low power analog IC design'", Microelectronic Engeneering, vol..39,1997,pp.59–76
6. R.M.D.A. Velghe, D.B.M. Klaassen, F.M. Klaassen, MOS model 9, NL-UR 003/94,1994. Internet: http://www.semiconductors.philips.com/Philips_Models.
7. Weidong Liu, Xiaodong Jin, Kanyu M. Cao, Chenming Hu, Project Director:Professor Chenming Hu: BSIM4.0 MOSFET Model, User's Manual, University of California, Berkeley.
8. B. Ankele, W. Hölzl and P. O'Leary, "Enhanced MOS Parameter Extraction and SPICE Modelling for Mixed Analogue and Digital Circuit Simulation", 1989 IEEE International Conference on Microelectronic Test Structures, Edinburgh 1989
9. J. Bastos, M. Steyaert, A. Pergoot and W. Sansen: *Mismatch-Characterisation-of-Sub micron-MOS-Transistors*, Analog Integrated Circuits and Signal Processing, 12, 95-106(1997), Kluwer Academic Publisher, Boston, 1997.
10. M. Pelgrom, A. Duinmaijer and A.Welbers: *Matching-Properties-of-MOS-Transistors*, IEEE Journal of Solid-State Circuits, Vol.24, October 1989.
11. Kwok K. Hung, Ping K. Ko, Chenming, Hu, Yiu C. Cheng, „A physics based MOSFET Noise Model for Circuit Simulators", IEEE Transactions on Electron Devices, Vol. 37, No. 5, May 1990.

Power and Energy Consumption of CMOS Circuits: Measurement Methods and Experimental Results

Josep Rius, Alejandro Peidro, Salvador Manich, and Rosa Rodriguez

Departament d'Enginyeria Electrònica, Universitat Politècnica de Catalunya
Diagonal 647, 9th floor, 08028 Barcelona, Spain
{rius, peidro, manich, rosa}@eel.upc.es

Abstract. This paper compares a set of measurements of power consumption of CMOS circuits obtained from conventional and non-conventional measurement methods. A description of the advantages and disadvantages of each method is included as well as the precaution measures to prevent measurement errors. Experiments on a 32-bit microprocessor and a standard cell custom circuit prove that by using non-conventional methods it is possible to obtain information unreachable with conventional ammeter measurements.

1 Introduction

Power and energy consumption are critical issues in the design of digital CMOS circuits and systems, and therefore much effort has been devoted to estimate these parameters in the early design phases [1],[2]. Nowadays, a set of useful tools is available, which allow knowing in advance the power consumption of the fabricated circuits.

However, actual measurements are still mandatory since during the manufacture and testing processes, it is necessary to know the real power consumption to check whether the circuit fits the specifications. In addition, at system level or in circuits where power consumption depends to a large extent on the processed data, it is easier and more convenient to perform measurements to know the power/energy consumed by such circuits.

The standard solution for measuring power/energy is to connect an ammeter in series with the power supply pins. This is a simple solution, but the amount of information that can be extracted from this measurement is limited. Thus, other methods have been proposed. In [7], the authors present a summary of several of such techniques. References [3], [4], [8] propose the use of a small resistor in series with the supply pin. In [12], the authors measure the voltage drop in the resistor to determine whether the transient current is abnormal.

Another approach is inspired by the Keating-Meyer technique for I_{DDQ} testing [14] and by the MTCMOS solution to power control [5]. This technique uses a switch and a capacitor in parallel with the supply pins to measure the energy consumed in a given period of time. In [13] the authors present a variation of such a technique that

J.J. Chico and E. Macii (Eds.): PATMOS 2003, LNCS 2799, pp. 80–89, 2003.
© Springer-Verlag Berlin Heidelberg 2003

uses four switches and two identical capacitors to perform the measurement in synchronous digital ICs on a cycle-by-cycle basis. The information published, however, does not deal with the problems related to switch charge injection and capacitor matching, and only relative energy measurements are presented.

The goal of this paper is to test and compare the capabilities and results of each method to show how non-conventional measurement methods allow obtaining information unreachable by conventional methods. Several experiments have been made to collect results on off-chip measurements of the power consumption of two representative CMOS integrated circuits.

The rest of the paper is organized as follows: the next section summarizes the measurement methods. Section 3 describes the results obtained from measurements on a 32-bit microprocessor and a standard cell multiplier. Finally, in section 4 the conclusions of the work are presented.

2 Methods for Measuring the Consumption of CMOS Circuits

By measuring the instantaneous current supplied to the IC, $i_{DD}(t)$, it is possible to calculate the charge, energy and power consumed by an IC. The main problem lies in the proper measurement of $i_{DD}(t)$, which has proved to be difficult in most cases because of the $i_{DD}(t)$ waveform characteristics. That is the reason why in those cases when it is not necessary to obtain an extremely detailed acquisition of the current waveform, other measurement techniques have been proposed.

2.1 Measurement of the Average $i_{DD}(t)$ for Long Periods

In most cases, it is not necessary to know the instantaneous value of $i_{DD}(t)$; an average (I_{DD}) value is sufficient. Average power and energy are easily calculated from I_{DD}.

In this case, the best approach to measure I_{DD} is to connect an ammeter in series with the supply pins. The input pattern is applied to the IC in infinite loops. Thus, the resulting current waveform is periodic. The ammeter averages the power supply current over a window of time, and if the period of the current waveform is much smaller than this window, a stable reading is obtained, showing the average current I_{DD} [3], [9], [10]. This method has many advantages, such as direct display reading, accuracy and small disturbance. However, the drawback of this method is that it does not allow determining the energy consumed in a single transition between two input patterns, or in shorts periods of time. Moreover, in spite of its simplicity, it is necessary to take some care when using an ammeter to measure the average current supplied to an IC [9].

2.2 Non-conventional Measurement Methods

When it is necessary to measure the average current consumed by a circuit in a short period of time (e.g. during the execution of any given sequence of microprocessor instructions or input patterns), the use of an ammeter is not feasible. In addition, the repetition of such a sequence in an infinite loop might not be possible or desirable. Hence, other measurement techniques must be used. There are two methods for solving this problem: (a) measuring the voltage drop on a resistor in series with supply lines (hereafter referred to as R-based measurement); (b) measuring the voltage drops on a capacitor in parallel with the circuit (hereafter referred to as C-based measurement).

R-based Measurements

To obtain the IC current waveform, a resistor R is connected in series with the VDD pins of the circuit, and the average voltage drop ΔV_{avg} on the resistor is measured with an oscilloscope. This voltage drop is intended to be directly proportional to the current drawn by the IC. Thus, the energy consumed during a period T is:

$$E = \frac{V_{DD}^2 T}{R}\left[\frac{\Delta V_{avg}}{V_{DD}} - \left(\frac{\Delta V_{avg}}{V_{DD}}\right)^2\right] = E_0\left[\frac{\Delta V_{avg}}{V_{DD}} - \left(\frac{\Delta V_{avg}}{V_{DD}}\right)^2\right] \tag{1}$$

where $E_0 = V_{DD}^2 T / R$. Expression (1) can be normalized with respect to E_0, and expressions for the average power and current are easily derived from it.

When using this method, special care must be taken for a precise measurement in subjects such as input protection diodes (limited voltage drop on the resistor), decoupling capacitors (undesired filtering) and decrease of the effective supply voltage [7].

C-based Measurements

The method consists in temporarily disconnecting the power supply by means of a switch, keeping the circuit supply pins connected to a capacitor C_{DD}. Then, one or more changes in the inputs are applied and the energy required is supplied by the capacitor, thus decreasing its voltage. By measuring the voltage drop ΔV we can obtain a measure of the energy consumed by the circuit.

The system operates as follows: the switch is closed and the capacitance C_{DD} holds the supply voltage at V_{DD}. Then, the switch is opened and a change in the inputs is applied to the circuit. As a result of the energy consumed by the circuit, the voltage at the supply node decreases ΔV volts. The measuring circuit captures this voltage drop. Finally, the switch is closed. Assuming that the capacitance C_{DD} has a constant value, the energy consumed by the circuit is related to ΔV by means of the following expression:

$$E = \frac{1}{2} C_{DD} V_{DD}^2 - \frac{1}{2} C_{DD}\left(V_{DD} - \Delta V\right)^2 = E_0\left[2\frac{\Delta V}{V_{DD}} - \left(\frac{\Delta V}{V_{DD}}\right)^2\right] \tag{2}$$

where $E_0 = C_{DD} V_{DD}^2 / 2$. Thus, by measuring ΔV it is possible to find E if C_{DD} and V_{DD} are known. From equation (2), it is easy to derive expressions for average power and current in the measuring period of time. Furthermore, equation (2) can be normalized, thus obtaining a figure independent of the exact value of C_{DD}.

As in the case of using a series resistor, several side effects must be taken into account to minimize errors. However, this measurement method has a significant advantage: the feasibility for measuring the energy consumed by a circuit in a single input change. The main drawbacks are the problems derived from EMI sensitivity and the need for a calibration of the capacitor C_{DD} if absolute measurements are required [7].

3 Measurements Using the Described Methods

Several experiments were performed to test the usefulness of each approach, as well as to know the practical limits of each technique. Two circuits have been experimented: a 32-bit ARM7TDMI microprocessor embedded in an AT91R8004 microcontroller, which is mounted on the evaluation board AT91EB1, and a standard cell custom IC. The custom IC implements five instances of an 8×8 bit Guild array multiplier [15], which was designed to measure the relation between logic depth and consumption, as described in [16]. Each circuit block has its own separate clock tree and pipeline registers. Each instance has the same inputs but a different maximum number of cells between successive pipeline registers (this number is defined as *multiplier granularity* β). That is, in the instance with granularity 1, the logic depth between two pipeline registers is one cell; if it is 2, there are two cells between two successive pipeline registers, and so on, [16].

The measuring board (see Fig. 1 left) has four jumpers J1-J4 that allow selecting each one of the different methods to measure the IC consumption. For instance, by connecting J1 to an ammeter, it is possible to measure the average current. Connecting J2, the voltage drop on the resistor R is measured. The final value is $R = 0.737\ \Omega$. To use the C-based method, the J3 and J4 jumpers must be connected. The PMOS transistor and the transmission gate TG work as a single electronically-controlled switch that minimizes the undesirable charge injection produced each time the PMOS is opened and the node VVDD remains floating. The measuring board includes a microcontroller, which manages the proper open/close timing of the "composite" switch. External signals allow the opening and closing operations to be synchronized from an external equipment. In addition, the capacitor C_{DDext} must be connected.

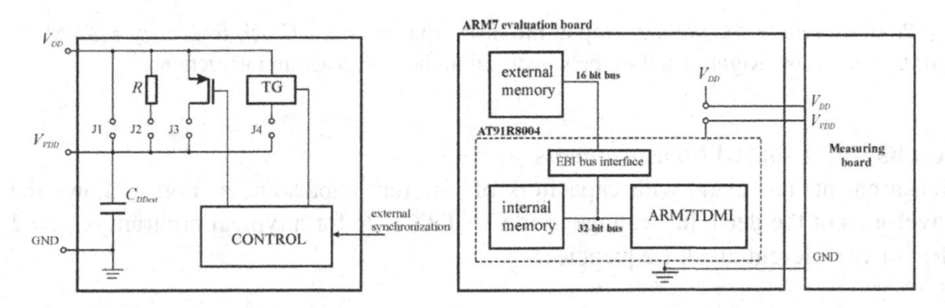

Fig. 1. Left: Measuring board. Right: Block diagram and connection of the ARM board.

3.1 Experiments with the ARM7 Board

The experimental setup is shown in Fig. 1 (right). To perform the measurements, the bridge connecting the V_{DD} voltage to the microcontroller is substituted by the measuring circuit. Small C++ programs are written and the consumption of the ARM7TDMI is measured.

The programs run in both external and internal memory. When they run in internal memory, there is no pad activity, thus measuring the internal consumption of the ARM7 plus the internal static memory. When the programs run in external memory the consumption of the ARM7, the EBI Interface and the pad activity are measured. The bridges in the evaluation board allow the ARM clock frequency to be selected to 4.1, 8.2, 16.4 and 32.8 MHz.

Results for power from ammeter measurements are shown in the second column of Table 1 as a function of frequency for a set of programs running in external memory. These ammeter results are taken as a reference for the other measurement methods.

Results with R-based Measurements

Fig. 2 shows typical waveforms of the voltage drop at the measuring resistor (upper signal). The lowest waveform is a synchronization signal activated at the beginning and at the end of the running program. The third column of Table 1 shows numerical results from expression (1) for power and for the same set of programs as the ammeter measurements.

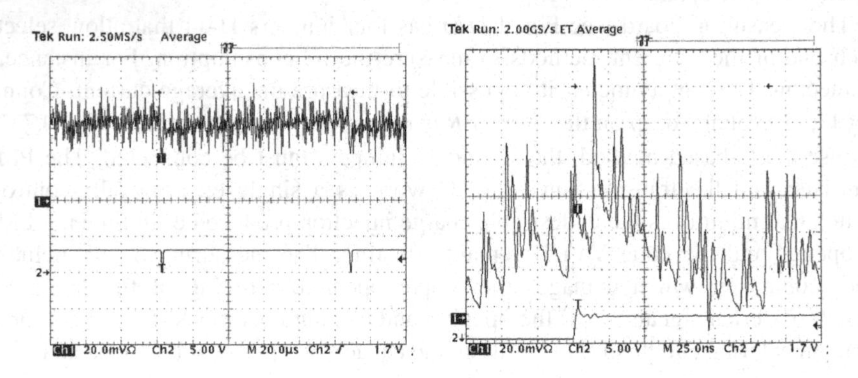

Fig. 2. Waveform of the voltage drop at the measuring resistor. Clock frequency = 32 MHz. Left: General view. Right: detail of the waveform in the first program instructions.

Results with C-based Measurements

Measurements are made with capacitors of different capacitances. Fig. 3 shows the waveforms of the decaying voltage at the VVDD node for a typical program (C_{DD} = 2 µF) at two different clock frequencies.

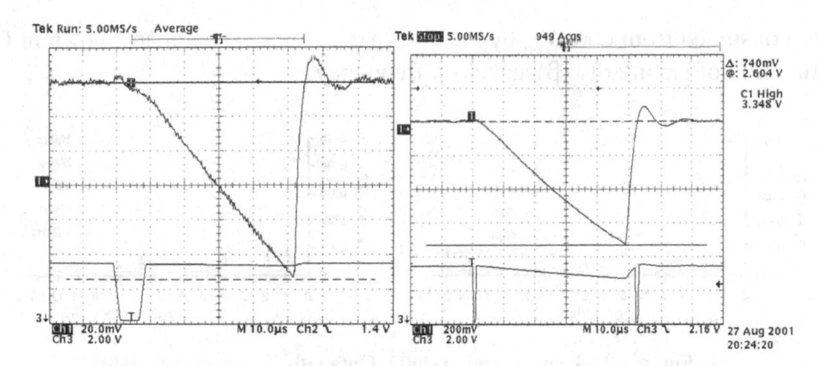

Fig. 3. VVDD waveforms. Clock frequency = 4 MHz (left), 32 MHz (right). The bottom waveform is a synchronization signal.

The results shown in the fifth column of Table 1 were obtained for C_{DD} = 2 µF by using expression (2) and the measured ΔV voltages

Comparison of Methods

To summarize, Table 1 compares the results of the three measurement methods.

Table 1. Results from the different methods.

Frequency [MHz]	P [mW] (ammeter)	P [mW] (R)	difference [%]	P [mW] (C)	difference [%]
32.768	117.68	119.04	1.16	109.78	-6.72
16.384	59.83	61.48	2.76	55.64	-7.00
8.192	29.83	29.14	-2.31	27.63	-7.36
4.096	15.02	14.13	-5.93	16.15	7.55

Taking the ammeter results as reference, the maximum difference is about 6 % for R-based measurements (fourth column of Table 1), and 7.5 % for C-based measurements (sixth column of Table 1). These low percentage differences highlight the usefulness and reliability of such methods for measuring power consumption in short periods of time (a few tens of µs in the examples shown in Figures 2 and 3). Notice that the ammeter measurement requires the program to running in an infinite loop, whereas in the R-based or C-based methods, a single run is sufficient.

3.2 Experiments with the Multiplier

The multiplier has one pin connected to core-V_{DD} and one separated pin connected to pads-V_{DD}, thus allowing an independent measurement of the core or pad consumption. The input patterns were applied from a 16-bit LFSR controlled by an independent clock source, and the multiplier output pins were connected to a Logic Analyzer to test the correctness of the multiplication.

Results Using an Ammeter

The measurements were obtained by applying an infinite loop of 2^{16}-1 pseudo-random input patterns for five clock frequencies and five granularities. The results for the power consumed from core-V_{DD} by the clock signal and datapath are shown in Fig. 4 as a function of granularity (β) and clock frequency.

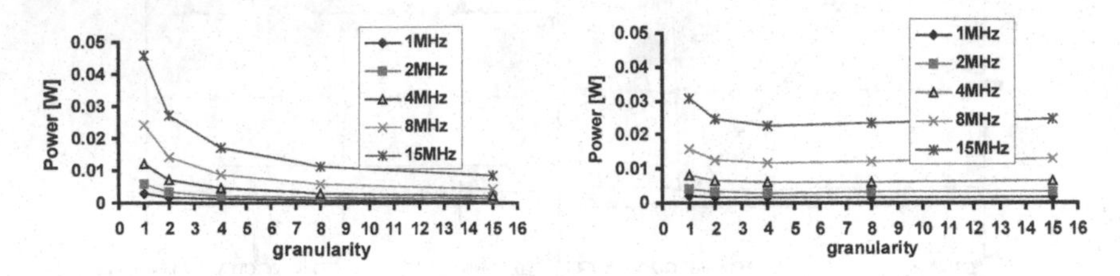

Fig. 4. Clock consumption (left). Datapath consumption (right).

Results with R-based Measurements

The effect of the value of the sensing resistance on the waveform captured by the oscilloscope is clearly visible in the following experiment with the multiplier board. As mentioned above, R-based measurements are based on capturing the voltage drop on the series resistor. However, this resistor, along with the capacitance between VDD and GND and the supply line inductance, forms a series RLC circuit that is excited by the switching activity of the multiplier IC. If the power line inductance is small and the sensing resistance is above a critical value, as is the case of the ARM board, the response of the RLC circuit is overdamped, and the waveform shows mainly the peaks and decaying voltages of the essentially RC equivalent circuit (see Fig. 2). However, if the power line inductance is relatively large and the sensing resistance is small, then the waveform shows an underdamped oscillation, thereby being extremely difficult to extract useful information from the waveform.

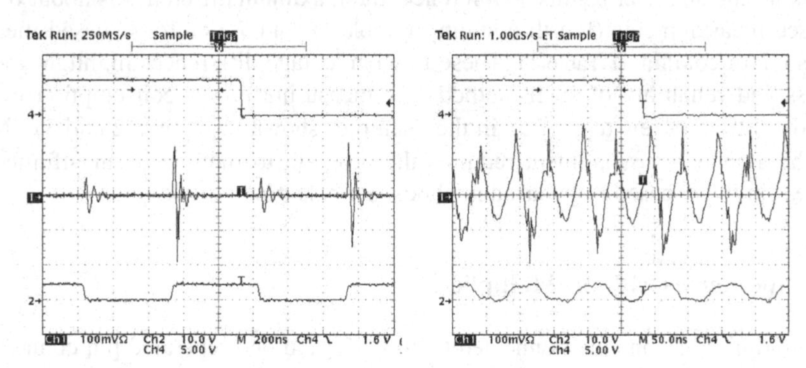

Fig. 5. Bottom signal: clock (1 MHz, left), (10MHz, right). Center signal: resistor waveform.

That is the case of the experiments with the multiplier board. In these experiments, the sensing resistance has the same value (0.737 Ω) as in the ARM experiments (notice that this resistance is on the measuring board), but the power line inductance is higher. Thus, the waveform shows a strong oscillation. This situation is shown in Fig. 5 for a low frequency clock (left), and for a higher frequency clock (right). As can be seen, the portion of signal related to the circuit consumption is very small with respect to the damped oscillation, and therefore a larger resistor would be necessary. This fact must be taken into account if the R-based measurement method is to be used.

Results with C-based Measurements

The external capacitance was C_{DD} = 10 nF and a set of results was obtained. For instance, by disabling the LFSR, the multiplier clock consumption can be measured. As can be seen in Fig. 6 left, there is a superposition of voltage drops that represent such consumption for β equal to 1, 2, 4, 8 and 15.

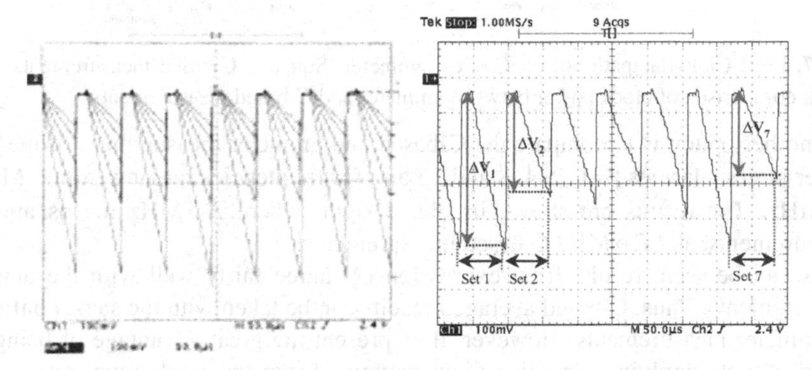

Fig. 6. Left: VVDD waveforms for granularities = 1, 2, 4, 8 and 15. Clock frequency = 200 KHz and C_{DD} = 10 nF. Right: seven successive sets of 10 multiplications. Clock = 200 KHz, β=15.

As can be observed, the consumption due to the different switching capacitance of the clock tree associated to each granularity is clearly visible. In this experiment, the switch is alternatively closed and opened. The period of time in which the switch is open is 51.4 μs. At the end of this period the voltage drop ΔV can be measured and the power and energy easily calculated. As the multiplier clock is a free running signal, power calculated using this method is easily comparable with ammeter measurements.

Differences in the consumption of successive sets of input patterns are also clearly visible, as shown in Fig. 6 right. There, we can see the voltage drops associated to seven consecutive sets of 10 multiplications.

By capturing waveforms like the one shown in Fig. 6 right, it is possible to build a histogram of the IC power consumption for many sets of input patterns. This information is useful, for instance, for testing but also for other purposes, such as the validation of the maximum or RMS power IC specifications.

Comparison of Methods

Only comparison between the ammeter and C-based measurements will be presented. To compare such measurements it is necessary to extract the average consumption of the multiplier for many C-based measurements with random sets of multiplication. This was achieved by averaging the consumption of 50 sets of five multiplications and of 50 sets of twenty-five multiplications. For clock+datapath power with β=15, results are shown in Fig. 7 left.

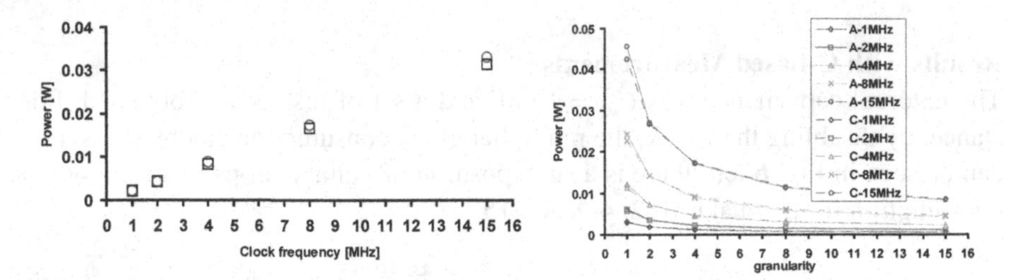

Fig. 7. Left: Clok+datapath power. Circles: ammeter. Squares: C-based measurements. β=15. Right: comparison of clock power between ammeter and C-based measurements.

Another option is to compare the C-based and ammeter measurement of the clock power at granularities β=1, 2, 4, 8 and 15 and for frequencies ranging from 1 MHz to 15 MHz. The results are shown in Fig. 7 right, where A-xMHz means ammeter measurements, and C-xMHz C-based measurements.

As can be seen, results from expression (2) agree fairly well with the ammeter measurements. Thus, C-based averaged results can be taken with the same confidence as ammeter measurements. However, they present the great advantage of being collected without applying repetitive input patterns. From the results, we can conclude that this method has great flexibility to measure power in short periods of time, thus being reliable for power measurements in CMOS circuits.

4 Conclusions

A method that averages i_{DD} current (ammeter) is the simplest method for estimating the average consumption of an IC. Nevertheless, when measuring the power in a transition between two input patterns, or between microprocessor instructions, an average measurement is no longer useful. New methodologies capable of measuring the power consumed in short periods of time must be used. In this paper, two of them have been reported. One measures the power consumed by the IC through the measurement of the voltage drop on a resistor, while the second technique measures the voltage drop on a capacitor. Experiments with representative circuits show that the results obtained from such methods in a single run of input patterns are equivalent to the ones obtained from an ammeter with infinite loops of input patterns. Thus, they have great flexibility and give much more information compared with the ammeter solution. Another interesting feature of these methods is that their results can be normalized, thus obtaining a reading independent of the precise determination of the value of the sensing element.

Acknowledgements. The authors wish to thank E. Bocmo from UAM for providing the Guild multiplier. This work has been partially supported by the CICYT, Projects TIC2001-2246 and TIC2002-03127.

References

1. Farid Najm, "A Survey of Power Estimation Techniques in VLSI Circuits", *IEEE Transactions on VLSI Systems*, pp. 446–455, December 1994.
2. P. Girard, "Survey of Low-Power Testing", *IEEE Design and Test of Computers*, May 2002, pp.82–92.
3. J.T. Russell, M. F. Jacome, "Software Power Estimation and Optimization for High Performance, 32-bit Embedded Microprocessors", *Proceedings of ICCD'98*, pp. 328–333, October 1998.
4. K.A. Jenkins, R.L. Franch, "Measurement of VLSI Power Supply Current by Electron-Beam Probing", *IEEE Journal of Solid-State Circuits*, Vol 27, No 6, pp. 948–950, June 1992
5. S. Mutoh, T. Dousaki, Y. Matsuya, T. Aoki, S. Shigematsu, J. Yamada, "1-V Power Supply High-Speed Digital Circuit Technology with Multithreshold-Voltage CMOS", *IEEE Journal of Solid-State Circuits* , Vol 30, Issue 8, pp. 847–854, August 1995
6. M. Lee, V. Tiwari, S. Malik, M. Fujita, "Power Analysis and Minimization Techniques for Embedded DSP Software", *IEEE Transactions on VLSI Systems*, pp. 123–135, March 1997
7. J. Alcalde, J. Rius, J. Figueras, "Experimental Techniques to Measure Current, Power and Energy in CMOS Integrated Circuits", *Proceedings of DCIS'00, November 2000,* pp. 758–763.
8. T.K. Callaway, E.E. Swartzlander, "Estimating Power Consumption of CMOS Adders", *11th Symposium on Computer Arthmetic Circuits,* pp. 210–216, 1993
9. V. Tiwari, M. Tien-Chien Lee, "Power Analysis of a 32-bit Embedded Microcontroller", *VLSI Design Journal*, Vol. 7, 1998.
10. D. Sarta, D. Trifone, G. Ascia, "A Data Dependent Approach to Instruction Level Power Estimation", *Proc. of IEEE Alessandro Volta Workshop on Low-Power Design*, pp 182–190, March 1999.
11. Hewlett Packard, *"User's Guide HP34401A Multimeter"*, Manual Part Number 34401-90420, October 1992.
12. B. Kruseman, P. Janssen, V. Zieren, "Transient Current Testing of 0.25 μm CMOS Devices", *Proceedings of ITC'99*, pp 47–56, 1999.
13. N. Chang, K. Kim, H.G. Lee, "Cycle-Accurate Energy Consumption Measurement and Analysis: Case Study of ARM7TDMI", *Proceedings of ISLPED'00*, pp. 185–190, July 2000
14. M. Keating, D. Meyer, "A New Approach to Dynamic IDD Testing", *Proceedings of ITC'87*, pp 316–319, October 1987.
15. H. Guild, "Fully Iterative Fast Array for Binay Multiplication and Addition", *Electronic Letters*, pp. 263, Vol 5, No 12, June 1969
16. W. Boemo, S. López-Buedo, C. Santos, J. Jáuregui, J. Meneses, "Logic Depth and Power Consumption: A Comparative Study Between Standard Cells and FPGAs", *Proceedings of DCIS'98*, pp. 402–406, November 1998

Effects of Temperature in Deep-Submicron Global Interconnect Optimization

M.R. Casu, M. Graziano, G. Piccinini, G. Masera, and M. Zamboni

Dipartimento di Elettronica, Politecnico di Torino,
C.so Duca degli Abruzzi 24, I-10129, Torino, Italy

Abstract. The resistance of on-chip interconnects and the current drive of transistors is strongly temperature dependent. As a result, the interconnect performance is affected by the temperature in a sizeable proportion. In this paper we evaluate thermal effects in global *RLC* interconnects and quantify their impact in a standard optimization procedure in which repeaters are used. By evaluating the difference between a simple *RC* and an accurate *RLC* model, we show how the temperature induced increase of resistance may reduce the impact of inductance. We also project the evolution of such effects in future technology nodes, according to the semiconductor roadmap[1].

1 Introduction

Long on-chip interconnects have been usually modeled as *RC* distributed lines and *ad hoc* optimization rules have been consequently developed [1][2]. However, current clock frequencies on the order of and higher than 1 GHz require a suitable *RLC* modelization because the wire inductance is no more negligible. A length-based classification of interconnects that explains when inductance effects have to be taken into account is proposed in [3]. In a recent paper Ismail and Friedman presented the formulation of the interconnect delay as an elegant and compact closed-form function of *RLC* interconnect, driver and receiver parameters [4]. In particular the formula captures the entire range where the interconnect behave as a *RC* or a *RLC* line by combining the effects of the various parameters. In the same paper new design formulas for the computation of the optimal number of repeaters and the length of wire segments between repeaters are also proposed. These are expressed in a form such that when the line behaves as *RC* instead of *RLC*, the classic formulation by Bakoglu still holds [1].

If thermal effects are taken into account, the interconnect resistance must be expressed as a function of the temperature. Usually a linear dependence is accurate within the range of on-chip operating temperatures. Since the amount of wire resistance may change the operating regime of the interconnect from pure *RC* to moderate *RLC* or quasi-*LC* it is important to know the operating temperature for an accurate modelization. The transistor properties like current

[1] This work was founded by the Politecnico di Torino Center of Excellence for Multimedia Radio-Communications (CERCOM).

J.J. Chico and E. Macii (Eds.): PATMOS 2003, LNCS 2799, pp. 90–100, 2003.
© Springer-Verlag Berlin Heidelberg 2003

drive, on-resistance, off-current and threshold voltage are temperature dependent as well. Since the optimal number of repeaters and the optimal wire sizing depend on both wire and device on-resistance it is important to incorporate such effects in the design formulas. Some work in this direction has been done for pure *RC* interconnect taking into account the temperature of the line [5][6] but analysis for global *RLC* lines are still lacking in the literature.

The width of interconnects is expected will continue to scale in future technology nodes as described in the SIA roadmap [7]. The wire resistance will increase and will make interconnects more susceptible to thermal effects. The analysis of this phenomenon and its trend in future technologies is carried out in this work.

In section 2 we introduce the equations of the delay of a *RLC* line and incorporate the temperature dependency in the interconnect and device parameters. Then we evaluate the impact on non-optimized global lines of various lengths in a current VLSI technology. We show the importance of taking inductance into account for accuracy and to avoid timing underestimations. In section 3 the global lines are optimized by a suitable repeater insertion whose optimum number and sizing depends on temperature effects. Then an estimation of the trend in future technologies is proposed in section 4. Finally the conclusion summarizes the achievements of this work.

2 Thermal Effects in *RLC* On-Chip Interconnects

Let's consider a global *RLC* interconnect of length l driven by a CMOS buffer of resistance R_r and charged by the input capacitance C_r of the output buffer. The 50% delay can be computed with good approximation by the following empirical equation developed by Ismail and Friedman[4]

$$T(\zeta) = \frac{1}{\omega_n}(e^{-2.9\zeta^{1.35}} + 1.48\,\zeta) \tag{1}$$

where

$$\omega_n = \frac{1}{\sqrt{Ll(Cl + C_r)}},\ \zeta = \frac{1}{2\sqrt{1 + \frac{C_r}{Cl}}}\left(\frac{Rl}{2Z_0} + \frac{R_r}{Z_0} + \frac{(R_r + Rl)C_r}{\sqrt{LCl^2}}\right),\ Z_0 = \sqrt{\frac{L}{C}}$$

L, R and C are per unit length values ([L]=H/m, [R]=Ω/m, [C]=F/m). For $L \to 0$ (and so $\zeta \to \infty$) eq. (1) converges to the usual delay of a *RC* line [1][2].

It is clear that since the temperature affects all resistive parameters, i.e. line and drivers, its effect on the interconnect optimization should be accounted for. We have taken interconnect and transistor data from the 2001 SIA roadmap [7] and added typical temperature dependence of interconnect resistance and transistor current both fitted in a linear expression:

$$R(T) = \frac{\rho(T_0)}{WH}[1 + \alpha(T - T_0)] \tag{2}$$

$$I_{dsat}(T) = I_{dsat}(T_0)\,[1 + \beta(T - T_0)] \tag{3}$$

where W and H are width and thickness of the line and ρ is the resistivity ([ρ]=Ωm). ρ and α are known being respectively 1.68×10^{-8} Ωm and $0.4°\mathrm{C}^{-1}$

for bulk Copper at 20°C. However, the effective resistivity is higher for the effect of the Cu barrier and is about $2.2\times10^{-8}\ \Omega$m [7]. The temperature dependency changes as well and we derived a higher value of $0.53°\text{C}^{-1}$ from [8]. $I_{dsat}(T_0)$ is one of the roadmap's specifications and is 900 Am^{-1} (i.e. current per device width unit) at $T_0 = 25°$C for high performance devices [7]. This value will be constant in future technology nodes. Using Hspice and the BSIM3v3 MOSFET model we derived an approximate value for β of $-1.1\,\text{Am}^{-1}\text{K}^{-1}$ (negative, because the current decreases as temperature increases). The device resistance is given by

$$R_r(T) = \gamma \cdot V_{dd}/I_{dsat}(T) \qquad (4)$$

where γ is a fitting coefficient [9]. The other LC parameters can be evaluated by means of proper expressions for the typical configuration of a wire embedded between two other wires of the same metal layer and sandwiched between two ground planes [11][12]. Let's consider a line of variable length l implemented in a 130 nm roadmap's technology node [7]. Its parameters are about R=$10^5\ \Omega$/m at 300 K and $1.4{\cdot}10^5\ \Omega$/m at 400 K, C=$2{\cdot}10^{-10}$F/m and L=$2{\cdot}10^{-6}$H/m. In figure 1 is reported the RLC delay of equation (1), for a typical driver-load pair, as a function of length and at various temperatures, together with the RC delay, that is neglecting the inductance L. The percentage difference between the two models is also plotted. Two cases are considered: on the left side a minimum width wire, according to the minimum pitch rule of the 130 nm technology; on the right side a wire with 5 times the minimum width is considered. The minimum width wire behaves as a RC line except for short lengths where the error RC vs. RLC is on the order of 20% at low T and lower than 20% at high T. In the larger line the resistance is reduced and the inductance effect becomes preeminent. The error is much higher and tends to reduce as T and l increase.

The temperature variation leads to a relevant delay change when the interconnect behaves as a RC line, on the order of 50% over the range 300-400 K for minimum width "almost-RC" lines. For the larger "true-RLC" line it is less than 10% over the same range. A proper wire sizing will be then useful not only for the delay reduction, but also for the delay variance minimization if the operating temperature is not known or not precisely controlled.

An accurate estimation of the temperature is needed to improve the correct timing estimation. Since the delays are monotone increasing functions of temperature a worst case approach could also be followed, but this results in an overdesign cost that cannot be acceptable for high performance designs or might impact the global power dissipation.

The RC delay is always smaller than the RLC one regardless of the temperature and the line length. Therefore an accurate RLC modelization is needed to avoid timing underestimations. The RC delay is a stronger function of l at a given temperature. This is not surprising because the RC delay is $\propto l^2$ while the delay of a pure LC lossless transmission line is $\propto l$ (time of flight $\tau_{OF} = l\sqrt{LC}$). The usual technique of placing buffer between RC line segments is very effective because of the square nature of the RC delay with l. In a lossless pure LC line this technique will be ineffective [4]. If we evaluate the optimum number of

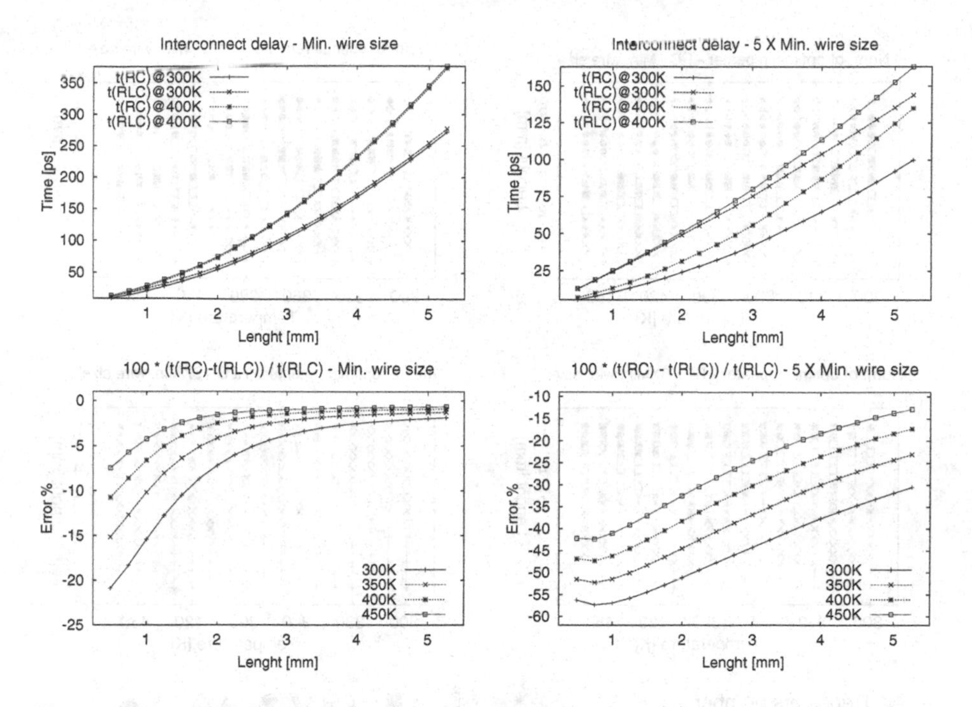

Fig. 1. RC and RLC delays (top) and RC vs. RLC error (bottom) as a function of length and at different temperatures; min. wire width (left) and 5× min. width (right).

repeaters we can expect that a lower number is needed for the *RLC* case with respect to the *RC* one. However, as the temperature increases, the number of repeaters will approach the *RC* optimum number.

3 Interconnect Optimization via Repeaters Insertion

Optimum size and number of repeaters in *RLC* lines are W_{opt} and $N_{opt} = \lceil n_{opt} \rceil$ and are approximated by the following expressions [4]

$$n_{opt} = \frac{\sqrt{\dfrac{CRl^2}{R_{r0}C_{r0}}}}{\left(1 + 0.18\, L^3/(RR_{r0}C_{r0})^3\right)^{0.3}} \qquad W_{opt} = \frac{\sqrt{\dfrac{CR_{r0}}{C_{r0}R}}}{\left(1 + 0.16\, L^3/(RR_{r0}C_{r0})^3\right)^{0.24}} \quad (5)$$

where R_{r0} and C_{r0} are resistance and capacitance of a minimum width driver. The optimum device resistance is given by $R_{ropt} = R_{r0}/W_{opt}$. The above formulas tend to the classic Bakoglu's formulas for *RC* lines when $L \to 0$ [1].

In figure 2 are reported the optimum number of repeaters as a function of length and temperature for the 130 nm global interconnect, assuming that the output load is a fraction of the input capacitance of the repeater itself. At the top of the figure the minimum width has been used and 5× the minimum width at the

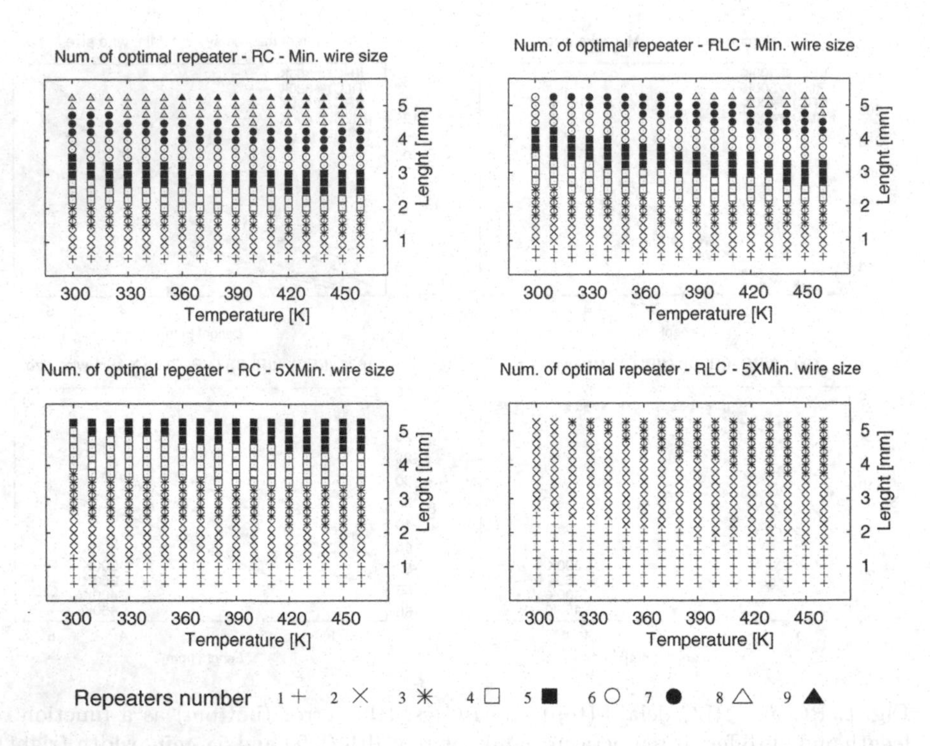

Fig. 2. Optimum number of repeaters for a 130 nm line as a function of length and temperature. On the left the line has been considered *RC* while *RLC* on the right.

bottom. On the left the line has been optimized as if it was a *RC* line (equation (5) with $L \to 0$) while on the right as a *RLC* line (same equation with $L \neq 0$). The minimum width wire requires a high number of repeaters because the line is almost-*RC*. In fact, the top-left graph is very similar to the top-right one. For true *RLC* lines much less repeaters are needed. The erroneous *RC* evaluation of the actual *RLC* delay leads to an excessive number of repeaters. This is on the one hand sub-optimum under the delay point of view but on the other hand detrimental for the power consumption. Moreover the lower sensitivity to temperature of the *RLC* line already observed before is reflected in the optimum number of repeaters that do not vary as for *RC* lines at different T.

In table 1 are reported the values of W_{opt} obtained during the optimization process. The ranges result from the variation of l and T. Once again the minimum width line is confirmed being “almost-RC” with respect to the larger line as shown by the similar values in the first column of table 1. On the contrary for the 5× minimum width lines the inaccurate *RC* model largely overestimates the sizing of the repeaters (second column).

If we plot the delay after optimization we observe that it is now almost linear with length. It is interesting to evaluate the true *RLC* delay of a line optimized

Table 1. W_{opt} size of optimized repeaters for the 130 nm technology node.

	Min wire size W_{opt} $[\mu m]$	5XMin wire size W_{opt} $[\mu m]$
RC model	18.4-21.2	51.1-58.7
RLC model	17-18.6	32.6-38.4

as if it was a *RC* line and to compare it with the delay of the *RLC* optimized line. The corresponding delay is higher because the capacitance of the additional not useful buffers increases it. In figure 3 the two delays are reported as a function of length and at various temperatures. Again the minimum width (left) and the larger wire (right) are considered.

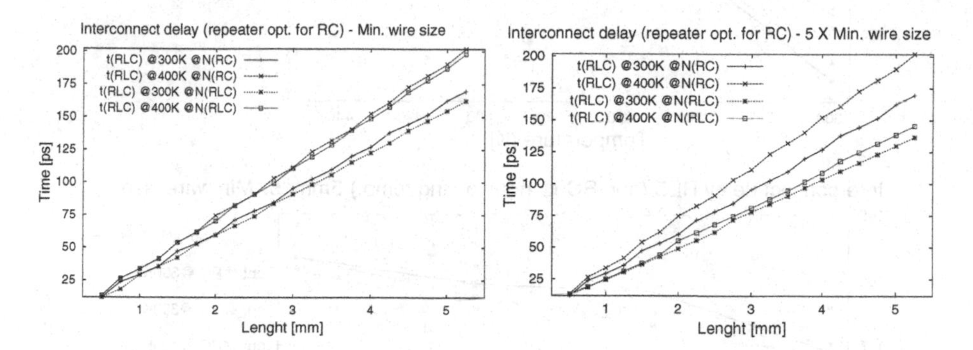

Fig. 3. Optimal RLC delay compared to RLC line optimized like a RC line. Minimum wire width (left) and 5× minimum (right).

The correct estimation of the interconnect temperature as an input for the optimization may be difficult. Moreover thermal gradients are possible such that the temperature is non-uniform along the interconnect length [5][6]. Therefore we have analyzed what happens if the interconnect is optimized at a given temperature and its delay is evaluated at a different temperature. Among the obtained results the most significant are reported in figure 4 for the 5 mm case, minimum width (top graph) and 5× minimum (bottom graph). In the same graphs are also compared the true *RLC* delays of lines correctly optimized with lines optimized as if they were *RC* interconnects. The thicker solid lines are the optimum delays (both *RC* and *RLC*) all over the temperature range while the other curves are optimized only at one temperature and thus are suboptimum over the entire range. Of course, the thick curve crosses the other curves at temperatures where suboptimum lines have been optimized. Sometimes suboptimum is slightly better than optimum because of the integer number of buffers obtained from the $\lceil n_{opt} \rceil$. The true optimum would be obtained by using a non-integer n_{opt} value which is obviously impossible. We observe again that the "almost-*RC*" narrow wire is

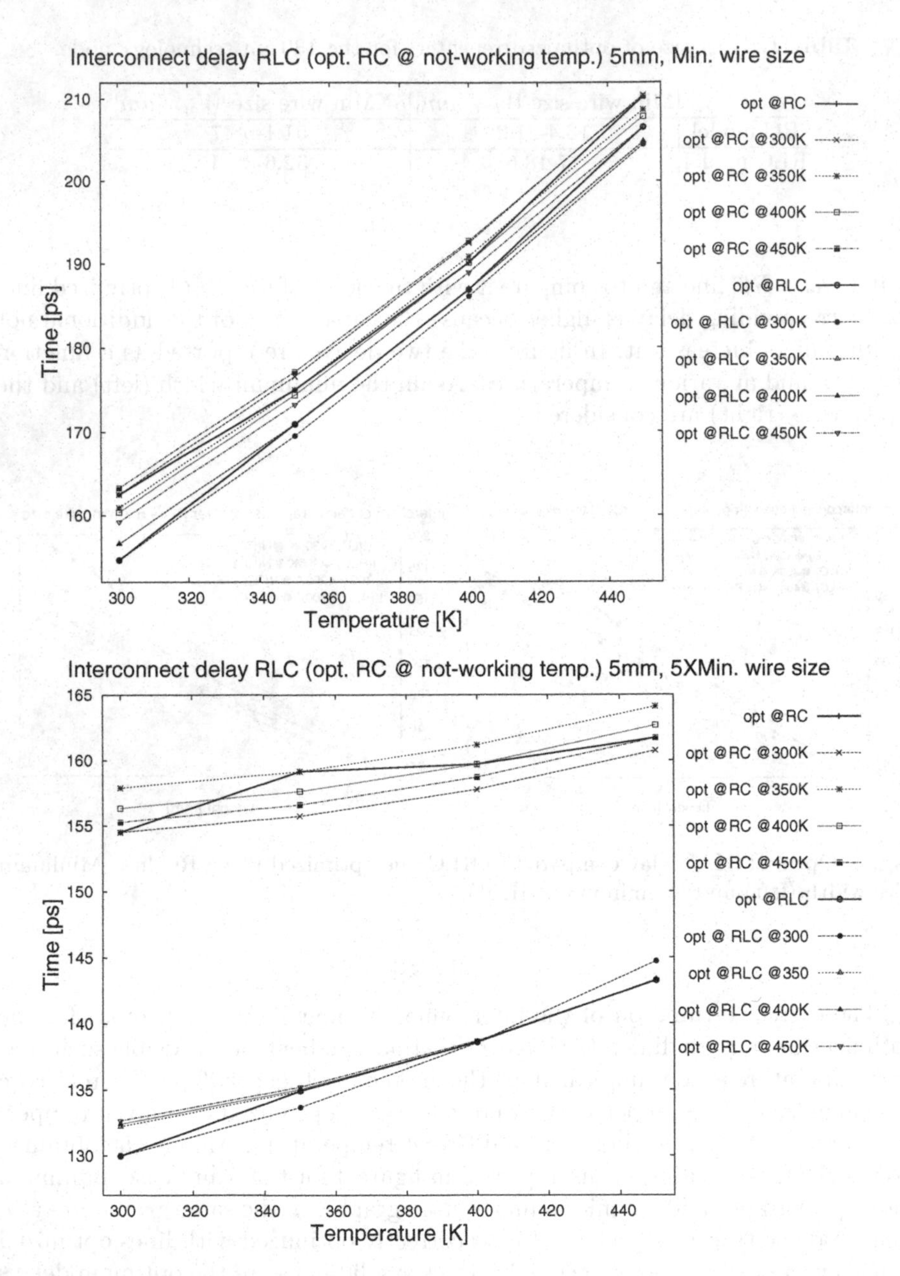

Fig. 4. *RLC* delays for minimum width (top) and 5× minimum (bottom) lines optimized at a given temperature and evaluated at other temperatures within the range.

more sensitive to thermal effects such that a line optimized at 300 K presents a delay increase of about 30% at 400 K ("×" dotted curve in top graph) while the variation for the larger line is 10% ("×" in bottom graph). Moreover the difference between *RC* optimized and *RLC* optimized is slight, less than 5% for minimum wire width (top graph). On the contrary, as already shown in figure 3, the "true-*RLC*" line correctly optimized as *RLC* presents a much smaller delay than the same lines optimized as *RC* (bottom graph in figure 4).

The difference between lines optimized at different temperatures is small meaning that the optimization can be done at a reference temperature. The minimization of the variation over the entire range is obtained by setting the optimization temperature at about the middle of the range. Such approach can be beneficial for the non-critical paths, provided that the maximum variation does not make them critical. On the contrary critical paths have to be treated using a worst case approach. Therefore the designer shall evaluate the maximum temperature over the entire range and optimize the line at that temperature. Since the delay is a monotonically increasing function of temperature this approach will ensure that the delay constraints will be satisfied at all range.

4 Interconnect Optimization in Future Technology Nodes

It is interesting to study the trend of our previous analysis in future roadmap's technology nodes. We evaluated the *RLC* delay of global interconnect lines optimized in the near term technology nodes from 130 nm to 65 nm [7]. The parameters for the 65 nm line are about $R=4.8\cdot 10^5\ \Omega/\mathrm{m}$ at 300 K, $C=1.6\cdot 10^{-10}\mathrm{F/m}$ and $L=2.2\cdot 10^{-6}\mathrm{H/m}$. In figure 5 the *RC* and *RLC* delays as a function of length and at various temperatures are reported for the previous 2001 130 nm and for the 2007 65 nm nodes (top graphs). Both minimum and 5× the minimum width lines are reported. The number of repeaters for the same *RLC* lines are reported in the same figure (bottom graphs) calculated at different temperatures (in the range 300-450 K). The minimum width line in the new technology is much more resistive than the 130 nm line as shown by the complete overlap of *RC* and *RLC* delay curves. As for the larger line the difference is more appreciable but still much lower than for the 130 nm case. We can also observe that at a given length the delay of the 65 nm node is worse than the corresponding 130 nm. One could argue that such comparison is not correct since scaled interconnect lengths should be compared. In this case we should compare the 130 nm delay at length l to the 65 nm delay at length $l \times 65/130 = l/2$ therefore observing a reduction in delay. However the average length of global interconnects is strictly related to the chip size that, according to the SIA roadmap document, does not scale with technology nodes [7]. Thus the comparison at a given length is correct. The overall result is not new because the non-scalability of global wires is a well-known problem [10] but we observe that the impact of both non-scaled length and high temperature make them more *RC* than *RLC* as clear from figures.

For what concerns the optimum number of repeaters, in figure 5 we see that more repeaters are needed at a given length for the 65 nm line. This is also true at scaled lengths ($l \to l/2$). The reason is again that the scaled line is much more resistive than the 130 nm line.

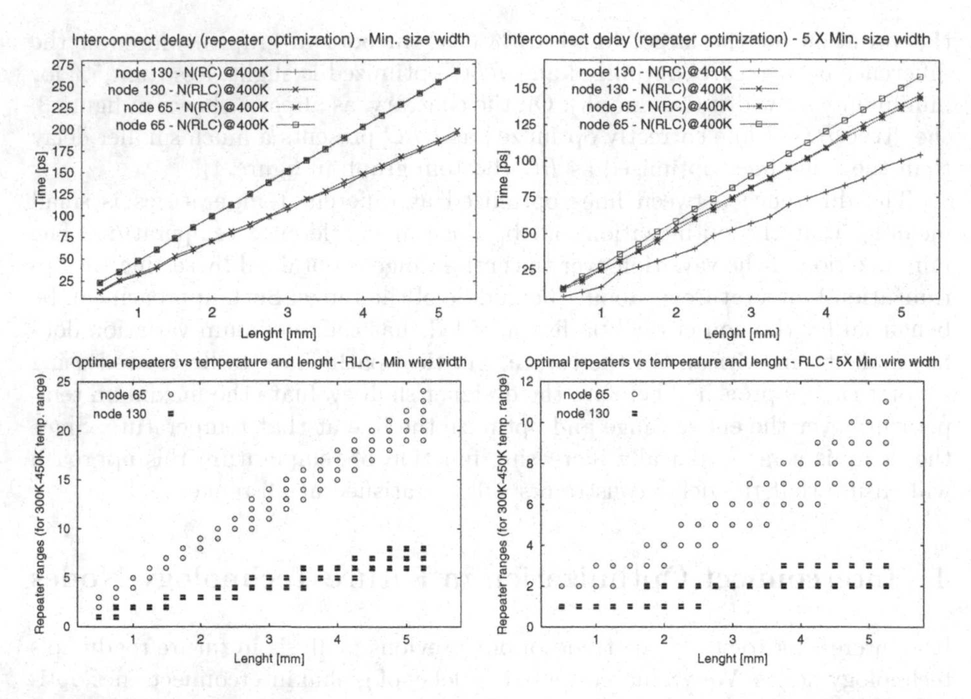

Fig. 5. RLC and RC delays (top) and number of repeaters (bottom) of 130 nm and 65 nm lines: minimum width (left) and 5× minimum width (right) interconnects.

In figures 6 the trend of delay and error *RC* vs. *RLC* is reported for all intermediate nodes from 130 to 65 nm at a length of 5 mm. The delay tends to increase and presents a reduction at 90 nm due to the foreseen introduction of a new low-k material that results in a reduction of capacitance. The difference between the *RC* and *RLC* models tends to diminish for higher temperatures as already shown before and also for scaled technologies approaching the last node. Therefore we conclude that contrarily to the common wisdom, the inductance effect are less dominant than expected in future technology nodes because global wires are more resistive. This phenomenon is exacerbated by the temperature effect that increases the wire resistance.

5 Conclusions

In this paper we have shown how the temperature dependence of interconnect and driver resistance impact the behavior of global wires in scaled VLSI technologies. The effects of temperature are particularly important in *RC* interconnects. We have seen that if inductance effects are important so that a *RLC* modelization is needed, they tend to reduce the impact of temperature. Since the temperature may define if a line behaves as *RC* or as *RLC* by modifying the

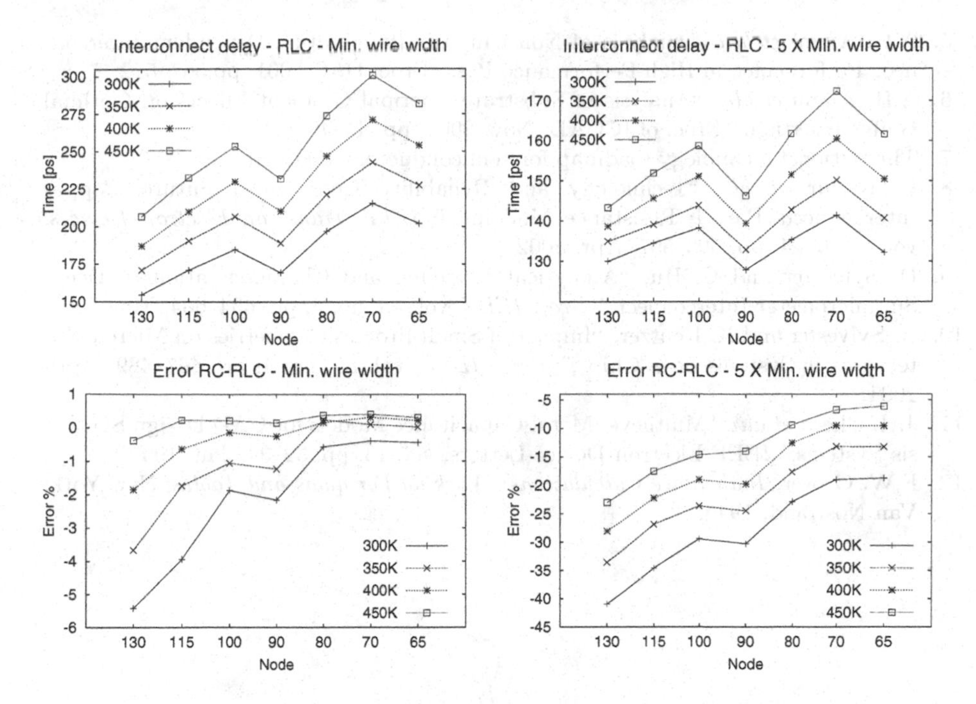

Fig. 6. *RC* and *RLC* delay (top) and difference between *RC* and *RLC* (bottom) of 5 mm global interconnects as a function of the SIA roadmap's technology node.

resistance value, it is very important to incorporate thermal effects into analysis and design. In fact, the optimization of interconnect performance may give rise to strongly different results over the operating range of temperature.

As technology improves and lithography allows to scale wire widths, the resistance per unit length increases and tend to shield inductance effects. As a result we foresee that in future wires the inductance effects will be mitigated and that the effects of temperature will be more and more important.

References

1. H.B. Bakoglu, *Circuits, Interconnections and Packaging for VLSI.* Reading, MA: Addison-Wesley, 1990.
2. T. Sakurai, "Closed Form Expressions for Interconnection Delay, Coupling, and Crosstalk in VLSI's," *IEEE Trans. on Electron Devices,* vol. 40, no. 1, pp. 118–124, Jan. 1993.
3. A. Deutsch *et alii,* "On-chip Wiring Design Challenges for Gigahertz Operation," *Proc. IEEE,* Vol. 89, No. 4, pp. 529–555, April 2001.
4. Y.I. Ismail and E.G.Friedman, "Effects of Inductance on the Propagation Delay and Repeater Insertion in VLSI Circuits," *IEEE Trans. on VLSI,* vol. 8, no. 2, pp. 195–206, April 2000.

5. A.H. Ajami *et alii,* "Analysis of Non-Uniform Temperature-Dependent Interconnect Performance in High Performance ICs," Proc. DAC 2001, pp.567–572.
6. A.H. Ajami *et alii,* "Analysis of Substrate Thermal Gradient Effects on Optimal Buffer Insertion," Proc. of ICCAD, Nov. 2001, pp. 44–48.
7. The national technology roadmap for semiconductors, 2001, SIA.
8. P. Kapur *et al.*, "Technology and Reliability Constrained Future Copper Interconnects–Part I: Resistance Modeling," *IEEE Trans. on Electron Devices*, vol. 49, no. 4, pp. 590–597, Apr. 2002.
9. D. Sylvester and C. Hu, "Analytical Modeling and Characterization of Deep-Submicrometer Interconnect," *Proc. IEEE*, vol. 89, no. 5, pp. 634–664, May 2001.
10. D. Sylvester and K. Keutzer, "Impact of Small Process Geometries on Microarchitectures in Systems on a Chip ," *Proc. IEEE*, vol. 89, no. 4, pp. 467–489, April 2001.
11. J.H. Chern *et alii,* "Multilevel Metal Capacitance Models for CAD Design Synthesis Systems," IEEE Electron Device Letters, vol. 13, pp. 32–34, Jan. 1992.
12. F.W. Grover, *Inductance Calculations: Working Formulas and Tables,* New York: Van Nostrand, 1946.

Interconnect Parasitic Extraction Tool for Radio-Frequency Integrated Circuits

Jérôme Lescot and François J.R. Clément

Cadence Design Systems S.A., ZAC Champfeuillet,
38500 Voiron, France
{lescot, francois}@cadence.com

Abstract. A tool to model interconnect parasitics in radio-frequency (RF) integrated circuits (RFICs) is presented. Accurate modeling is achieved by combining a detailed RLC wire model together with a distributed RC substrate model. Wire geometry is fractured to ensure accurate modeling of wave propagation as well as displacement current due to substrate losses. The wire model includes resistance and coupled capacitance together with self and mutual inductance. The necessity of including a distributed RC model of the substrate is stressed. RLC reduction at the end of parasitic extraction allows for fast simulation of the complete model. Accuracy and performance are demonstrated by comparing the values extracted by the tool against results from existing computer software as well as against silicon measurements.

1 Introduction

As radio spectrum is filling up very quickly, the next generation of wireless applications have to use higher frequency bands. Today, this is typically the case for third generation mobile communication (3G) including UMTS, WCDMA as well as Bluetooth. These applications use the 2.2 GHz or 2.45 GHz bands and include circuitry running at similar frequencies or above. Concurrently, opto-electronic applications using SiGe technology are developed up to 40 GHz.

At such frequencies, interconnect parasitic extraction requires a high-level of detail to provide the accuracy needed by RF designers. The issue is that existing electronic design automation (EDA) solutions do not provide the required level of accuracy. Indeed, commercial solutions focus on the digital and analog/mixed-signal (A/M-S) market for which capacity and performance are more critical than accuracy. Usually such solutions are very fast – several hundreds of thousands of nets per hour – and they extract primarily wire resistance together with lumped and coupling capacitance. The silicon substrate is considered either as a perfect insulator or as a perfect ground plane.

There exist specific features in RF designs that current EDA solutions usually do not support, such as diagonal routing, complex ground planes and wide coplanar lines. Unlike VLSI domain, RF designs are generally sparse and do not have a regular

J.J. Chico and E. Macii (Eds.): PATMOS 2003, LNCS 2799, pp. 101–110, 2003.
© Springer-Verlag Berlin Heidelberg 2003

power-ground distribution network, making current return paths unpredictable. Eventually, the output model must be small enough to afford SPICE-like simulations with a reasonable run time. A full RF block must be completed within a few minutes and wire parasitics for bigger circuits must be extracted at most overnight.

In the past RF designers have been calculating interconnect parasitics by hand to overcome tool limitations. Nevertheless, to reach sufficient accuracy above 2 GHz requires a very fine segmentation of interconnect lines to model slow-wave propagation mode [1]. For such a fine segmentation, parasitic extraction can no longer be calculated by hand in a practical time and requires a dedicated machine computation.

The next section of this article reviews in more detail the performance and accuracy requirements for RF interconnect parasitic extraction. An approach to extract RF interconnect parasitic models is proposed in section 3. Experimental results are then presented in section 4, and the fifth part provides a brief conclusion.

2 Performance and Accuracy Requirements

The requirements for RF parasitic extraction are opposite to digital and A/M-S needs. The number of nets to be modeled is small – typically around hundred – but the level of detail required includes usual resistance (R) and capacitance (C) extraction plus inductance – self (L) and mutual (K) – together with a distributed model of the silicon substrate. As a result the complete model required to capture all RFIC behaviors is a tightly coupled RLCK network.

2.1 RF Signal Propagation Modeling

Most RFICs use lightly-doped wafers with a high-resistivity substrate to limit losses. In heavily doped wafers of CMOS processes, losses in silicon substrate manifest themselves on interconnect behavior as a frequency-dependent series resistance due to magnetically coupled eddy currents in the substrate, which can then be treated as a unique electric node. In this latter case, an implicit model of the substrate is possible [2]. However, in the lightly doped wafers of most RF-focused processes, the substrate can no longer be considered as a unique node. An explicit 3D model is needed to account for electrically coupled displacement currents in the substrate that can be injected by the noisy supply lines through substrate contacts, bulk contacts of transistors, diffused resistors and interconnects. Since displacement currents in a lightly doped substrate dominate eddy currents, a distributed substrate model is needed. As a result of the high resistivity of the substrate material, the current crowding in conductors due to skin and proximity effects, which are regularly ignored in a heavily doped substrate, is a primary loss mechanism in interconnects on lightly doped wafers.

The specific geometries of RF circuits must be supported correctly. Namely, RF interconnect density is low leading to large spacings between wires. RF lines also tend to be wide – typically from a few microns to several tens of microns – and to use forty-five degree angles extensively. Of course, usual characteristics of advanced manufacturing processes must be supported as well. Those include presence of metal fillers – used to maintain the planarity of all metal layers – as well as conformal dielectrics.

2.2 Reduction of Parasitic Circuit Size and Overall Performances

Complying with all these requirements systematically leads to a very large parasitic RLCK network. For small circuits with less than ten nets, the raw extraction result typically includes several thousand self and mutual inductances that render simulation impractical. Consequently, an additional requirement for RF interconnect modeling is to conclude the extraction process with a netlist reduction stage.

2.3 Flow Integration

Flow integration is critical to the ease-of-use of any extraction tool. For digital and A/M-S design flows, parasitic extraction is part of final verification. In contrast, wire parasitics play a critical function in RFICs that must be accounted for in the very early stages of circuit design as well as during full chip final verification. As displayed in Figure 1(a), after a successful design rule check (DRC) or layout versus schematic (LVS) comparison, parasitics can be extracted and back-annotated. The result is then processed with a SPICE-like simulator to check that desired specifications have been met.

3 Extraction Tool to Model RF Interconnects

The diagram in Figure 1(b) depicts the extraction flow implemented in Assura RCX-HF, a new tool dedicated to RF interconnect modeling that implements the technique presented in this paper.

The extraction kernel is integrated in the design flow through a dedicated user interface (UI). The UI allows the selection of all nets or only specific nets. Properties are attached to all selected nets to indicate the maximum frequency of the signal carried by the net together with a choice of three model topologies (RC, RLC and RLCK). After the selection is complete, physical parasitics are extracted and back-annotated before simulation.

Extraction is done in four stages. First, the wire geometry is fractured in simple rectangular shapes with a maximum length equal to one tenth of the wave length. Second, the fractured geometry is sent to a wire processor to compute resistance with skin and proximity effects, inductance – self and mutual – as well as coupling capacitance between wires and from wire to substrate. Next, substrate parasitics are extracted and combined with wire models. Finally, the complete network is reduced to produce a compact netlist using SPICE syntax.

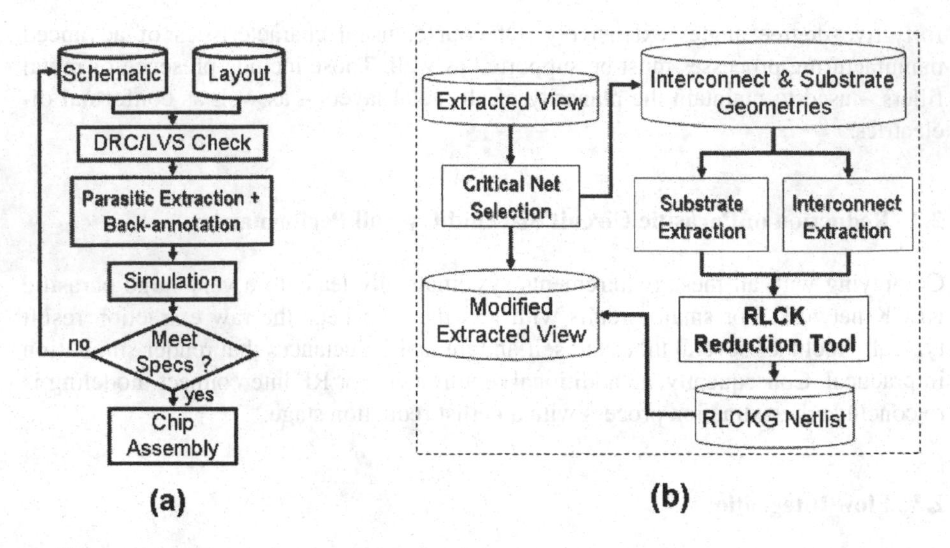

Fig. 1. *(a)* Typical RF design flow. *(b)* Parasitic extraction and back-annotation as implemented in Assura RCX-HF.

3.1 Resistance Model

Assura RCX-HF supports modeling of skin and proximity effects by subdividing the conductor into smaller filaments in the cross-section, each carrying a uniform current. This multi-stranding technique described in [3] is applied in Assura RCX-HF separately on each segment with a non uniform mesh so that the thinner filaments on the surface are at most δ thick where δ is the skin depth of the conductor at the maximum frequency of operation (see Table 1).

Table 1. Wire parameters for TSMC 0.13 µm 8M1P process

Wire Level	Thickness (µm)	Resistivity (Ω.µm)	Skin Depth (µm)	
			@ 5 GHz	@ 10 GHz
metal 2 to 7	0.37	0.021	1.03	0.73
metal 8	0.9	0.018	0.95	0.67

3.2 Capacitance Model

Capacitance extraction in the context of complex geometries with many metal and dielectric layers requires an accurate numerical method. We have selected a boundary element method (BEM) [4] that requires significantly smaller computer resources than the finite element method (FEM), supports the modeling of conformal dielectrics and dummy metals, and is not restricted to Manhattan structures. Since RF wires, like coplanar ground planes, are usually much wider than in an ASIC design, a specific meshing algorithm has been developed to capture the actual charge distribution on wide conductors, which varies most near the edges of the neighboring conductors. Such an enhancement not only increases the accuracy, but the extraction time is even shorter due to the more physical nature of the mesh.

3.3 Inductance Model

Due to the difficulty of obtaining the actual current return paths in an RF design, the Partial Element Equivalent Circuit (PEEC) approach has been proposed by Ruehli [5] to use partial inductance where an infinitely far return path is assumed. As a result, the PEEC approach models magnetic interactions between segments: each line is divided and uses an RLC model for each segment. The model also includes mutual inductance between any two non-perpendicular segments. The partial self and mutual inductances are a function of the geometry of the wire segments and can be computed by using the formulation described in [6]. Since these formulae are exact only when a uniform current flows through the full conductor, additional discretization as described in paragraph 3.1 is needed to capture current crowding effects.

3.4 Distributed Substrate Model

While standard EDA solutions connect metal wires capacitively to a unique ideal ground node for delay computation purposes, the distributed nature of the coupling to substrate in the RF domain requires one separate connection to a substrate mesh for each interconnect segment. An extraction based on the finite-difference method is used to produce an RC model of the substrate [7]. This approach has been preferred to other solutions [8] for its ability to model complex 3-D structures – such as triple wells, deep trenches, etc. – with greater precision. As depicted in Figure 2, each segment of length less than $\lambda/10$ is connected to one port of the substrate model.

To summarize, external ports of the 3D substrate mesh can be classified into three categories: substrate and n-well taps, backside connection when a metallic plating of the die exists, and capacitance to wire connections.

3.5 Netlist Reduction after Combination of Wire and Substrate Models

The main difficulty with partial inductance extraction is the overwhelming size of the problem. Mutual couplings between distant segments could be ignored to make the inductance matrix sparse, however, simply discarding small terms in the partial inductance matrix can render the equivalent circuit unstable.

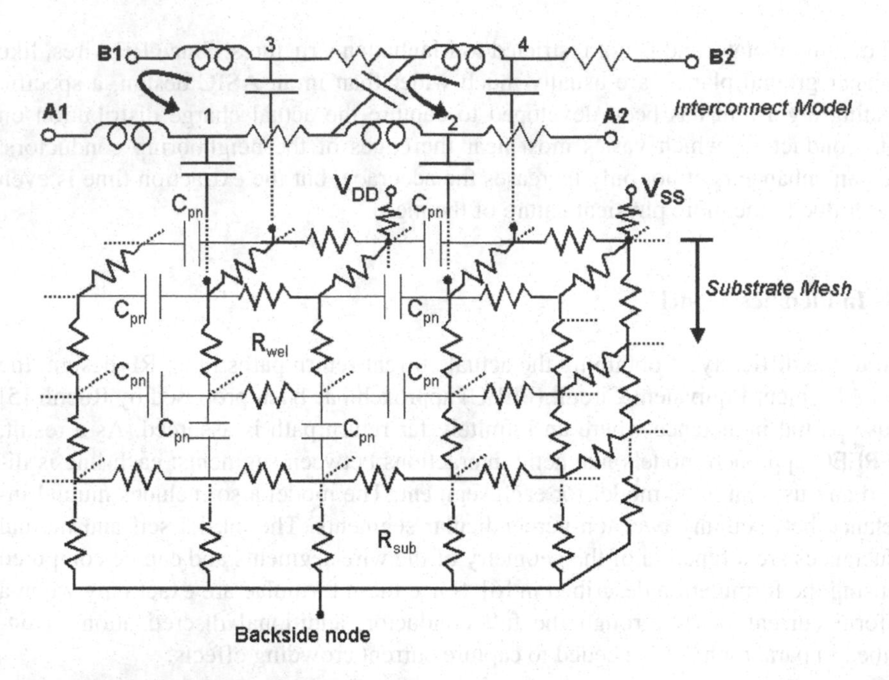

Fig. 2. Equivalent circuit model for two parallel interconnects (*A1* to *A2* and *B1* to *B2*) capacitively coupled to substrate through nodes *1* to *4*. The silicon substrate includes a N-type well modeled by resistors *Rwel* and a P-type bulk represented by resistors *Rsub*. Capacitors *Cpn* model the well-substrate PN junction. In addition, the well and bulk regions are biased through resistive contacts to *Vdd* and *Vss* respectively

A general RLC circuit reduction method for obtaining passive reduced order macromodels is available in Assura RCX-HF as an external module using SPICE exchange format. The description of this Krylov space method is beyond the scope of this paper and an extensive description can be found in [9]. This reduction technique using the PRIMA algorithm -- applied to linear systems using modified nodal analysis (MNA) – is interesting especially when numerous simulation runs are required. To determine the optimal order of approximation, a reliable accuracy measure based on the residual error [10] has been implemented for Assura RCX-HF.

The substrate section of the model is first RC-reduced by using a PACT algorithm [11] that preserves electric properties of the substrate model up to a user-specified frequency. Then, the full model including the interconnect parasitics and the RC-reduced substrate part are RLCK reduced using the PRIMA algorithm.

4 Experimental Results

Accuracy and performance are demonstrated by comparing the tool results with golden reference software as well as with silicon measurements.

4.1 Skin and Proximity Effects

One RF requirement in resistance extraction is the ability of the tool to capture both skin and proximity effects with accuracy on the full frequency of interest. The test vehicle used for validation is an isolated copper conductor of cross-section 4.62 μm by 4.62 μm.

Figure 3 shows the corresponding series resistance per unit length. The results closely agree with the corresponding data presented in [12]. When the same line is in presence of an adjacent parallel line of the same cross-section at a distance of 25 μm, Figure 3 demonstrates the tool's ability to account for proximity effects, whose impact on the equivalent wire series resistance at lower frequencies is more significant than the skin effect.

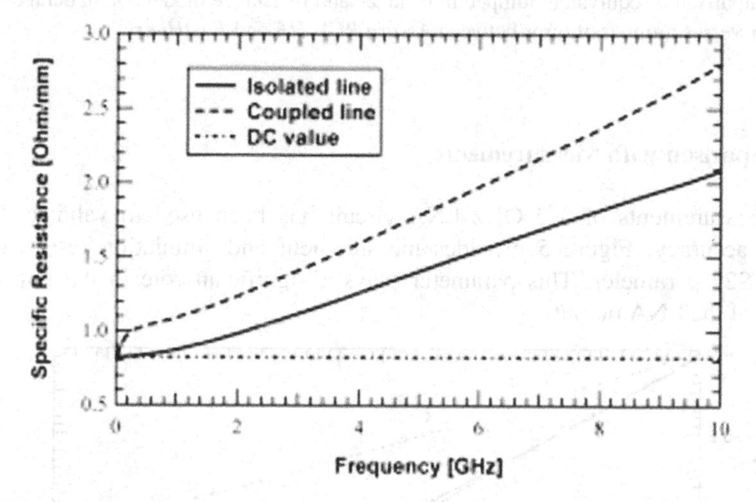

Fig. 3. Wire resistance versus frequency with (*Coupled line*) and without (*Isolated line*) an adjacent parallel conductor. Resistance without skin effect (*DC value*) is provided for reference

4.2 Self- and Mutual Inductance

Sets of validation data are produced using FastHenry [13] to extract the low-frequency lumped inductance of a spiral inductor. The interest of such a structure lies in the fact that its equivalent lumped inductance requires not only the self inductance of each segment but also the positive or negative mutual inductances between all segments. The number of turns of the inductor is varied while other parameters are kept constant: the spacing between coils is 2 μm and the wire width is 10 μm. As shown in Figure 4, good agreement is observed between FastHenry and Assura RCX-HF for both equivalent resistance and inductance of octagonal spiral inductors.

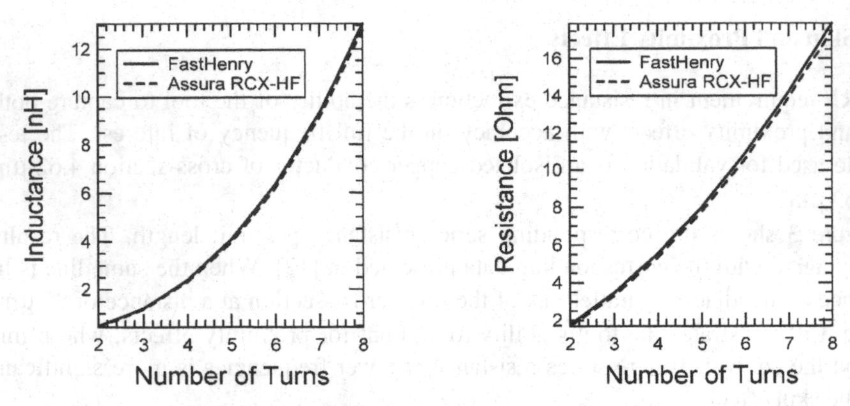

Fig. 4. Comparison of equivalent lumped inductance and resistance models of an octagonal spiral inductor versus number of turns between *Assura RCX-HF* and *FastHenry*

4.3 Comparison with Measurements

Silicon measurements of a 2 GHz LNA circuit has been used to validate Assura RCX-HF accuracy. Figure 5 provides measurement and simulation results of the phase of S22 parameter. This parameter plays a significant role in the impedance matching of the LNA output.

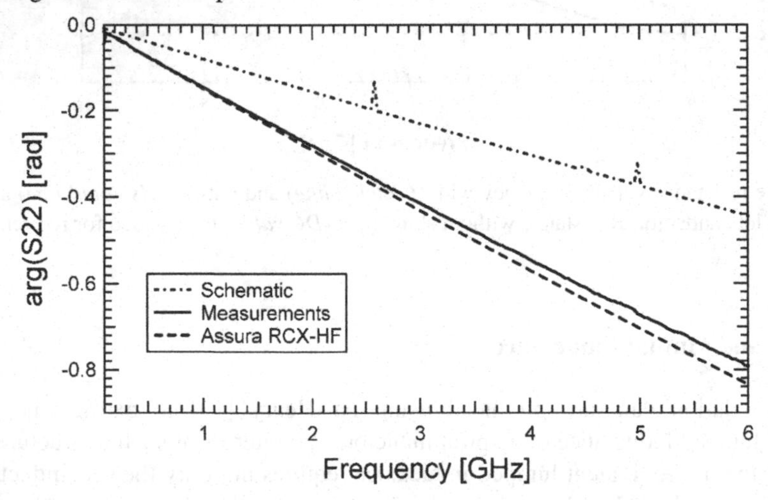

Fig. 5. Phase of S22 parameter – *arg(S22)* – of tested 2 GHz LNA. Silicon measurement (*Measurements*) is compared with simulation of the ideal schematic (*Schematic*) as well as with simulation after back-annotation with Assura RCX-HF interconnect model (*Assura RCX-HF*)

As illustrated in Figure 5, poor agreement was observed between measurements and the initial simulation without parasitics. The starting netlist was extracted using a standard RF design flow. After inserting Assura RCX-HF in the verification flow as described in figure 1(b), a close correlation with measurements was observed. By exploiting the critical net selection capability of the tool, it was possible to demonstrate unexpected coupling between a spiral inductor and RF transmission lines of the input and output stages.

4.4 Performance and Capacity

Performance results for three different test cases are presented below. The circuits include one 4th order Tchebytchev band-pass filter and two LNAs. Extracted netlists model the coupling between integrated spiral inductors. In Table 2, the extracted netlist size before reduction is provided. The number of resistances is far greater than the number of inductances because of the distributed RC substrate model.

Table 2. Summary of extracted model parameters used for performance evaluation

Circuit Name	# of R	# of L	# of C	# of K
Tchebytchev	3,524	206	12,398	15,475
LNA 1	5,207	360	12,642	37,950
LNA 2	10,772	503	54,105	90,039

Table 3 provides extraction and reduction times for the circuits described in Table 2. AC simulation time of Y-parameters are given before and after netlist reduction to exhibit the significant speed-up that is achieved. All extractions have been performed on a SUN Blade 100 with 2 Gbytes of memory and running at 502 MHz.

Table 3. Assura RCX-HF performance

Circuit Name	Extraction Time	Reduction Time	Simulation Time	
			Original Model	RLCK Reduced Model
Tchebytchev	22 min	8 sec	4 min	560 ms
LNA 1	41 min	34 sec	35 min	11 s
LNA 2	80 min	42 sec	32 min	28 s

5 Conclusion

A comprehensive review of RF interconnect model requirements has been presented and a corresponding interconnect extraction tool was introduced that meets all required specifications. We have also demonstrated the ability of the tool to extract instances of passive devices like spiral inductors, micro transformers, as well as regular interconnects. This provides two major improvements over black-box components extracted from measurements or field solver simulation. First, this approach captures the electromagnetic coupling between passive components during the circuit simulation. Second, passive device models are extracted in the context of the existing layout and account for the influence of substrate taps, ground planes or shielding wires in the vicinity of inductors.

Acknowledgements. The authors want to thank Didier Belot and Patrice Garcia from STMicroelectronics for sharing their expertise in RF design as well as for help with silicon measurements.

References

1. H. Hasegawa, M. Furukawa, and H. Yanai, "Properties of Microstrip Line on Si-SiO2 System," *IEEE Transactions on Microwave Theory and Techniques*, vol. 19, pp. 869–881 (1971)
2. D. Sitaram, Y. Zheng, and K. L. Shepard, "Implicit Treatment of Substrate and Power-Ground Losses in Return-Limited Inductance Extraction," presented at International Conference on Computer Aided Design (2002)
3. W. T. Weeks, L. L. Wu, M. F. M. Allister, and A. Singh, "Resistive and Inductive Skin Effect in Rectangular Conductors," *IBM Journal of Research and Development*, vol. 23, pp. 652–660 (1979)
4. M. Bächtold, M. Spasojevic, C. Lage, and P. B. Ljung, "A system for full-chip and critical net parasitic extraction for ULSI interconnects using a fast 3D field solver," *IEEE Transactions on Computer-Aided Design of Integrated Circuits and Systems*, vol. 19, pp. 325–338 (2000)
5. A. E. Ruehli, "Inductance Calculations in a Complex Integrated Circuit Environment," *IBM Journal of Research and Development*, pp. 470–481 (1972)
6. C. Hoer and C. Love, "Exact Inductance Equations for Rectangular Conductors with Approximations to More Complicated Geometries," *IBM Journal of Research of the National Bureau of Standards – Computer Engineering and Instrumentation*, vol. 69C, pp. 127–137 (1965)
7. F. J. R. Clément, "Computer Aided Analysis of Parasitic Substrate Coupling in Mixed Digital-Analog CMOS Integrated Circuits," Ph. D. thesis dissertation, EPFL, Lausanne (1995)
8. E. Carbon, R. Gharpurey, P. Miliozzi, R. G. Meyer, and A. L. Sangiovanni-Vincentelli, *Substrate Noise Analysis and Optimization for IC Design*: Kluwer Academic Publishers (2001)
9. A. Odabasioglu, M. Celik, and L. T. Pileggi, "PRIMA: Passive Reduced-Order Interconnect Macromodeling Algorithm," *IEEE Transactions on Computer-Aided Design of Integrated Circuits and Systems*, vol. 17, pp. 645–654 (1998)
10. A. Odabasioglu, M. Celik, and L. T. Pileggi, "Practical Considerations For Passive Reduction of RLC Circuits," presented at International Conference on Computer Aided Design, San Jose, CA (USA) (1999)
11. K. J. Kerns and A. T. Yang, "Stable and Efficient Reduction of Large, Multiport RC Networks by Pole Analysis via Congruence Transformations," presented at Design Automation Conference, Las Vegas, NV (USA) (1996)
12. M. J. Tsuk and A. J. Kong, "A Hybrid Method for the Calculation of the Resistance and Inductance of Transmission Lines with Arbitrary Cross Sections," *IEEE Transactions on Microwave Theory and Techniques*, vol. 39, pp. 1338–1347 (1991)
13. M. Kamon, L. M. Silveira, C. Smithhisler, and J. White, FastHenry User's Guide, Research Laboratory of Electronics, Department of Electrical Engineering and Computer Science, Massachusetts Institute of Technology

Estimation of Crosstalk Noise for On-Chip Buses

Sampo Tuuna and Jouni Isoaho

Laboratory of Electronics and Communication Systems
Department of Information Technology
University of Turku
FIN-20520 Turku, Finland
{satatu, jisoaho}@utu.fi

Abstract. In this paper an analytical model to estimate crosstalk noise on capacitively and inductively coupled on-chip buses is derived. The analytical nature of the model enables its usage in complex systems where high simulation speed is essential. The model also combines together properties such as inductive coupling, initial conditions, signal rise time, switching time and bit patterns that haven't been included in a single analytical crosstalk model before. The model is compared to three previous crosstalk noise estimation models. The error of the model remains below four percent when compared to HSPICE. It is also demonstrated that for planar buses the five closest neighboring wires constitute up to 95% of the total induced crosstalk noise.

1 Introduction

Global communication has become a bottleneck in deep sub-micron integrated circuits. The speed of communication is not limited by the communicating devices themselves, but the wires connecting them. Wires induce noise onto other lines, which may degrade signal integrity. This noise, crosstalk, is caused by capacitive and inductive coupling between wires. Capacitive coupling is currently dominant, but the importance of inductive coupling increases as signal rise times become faster. Efficient crosstalk models that can be used in computer aided design tools are nowadays necessary since SPICE simulations or model order reduction techniques are too inefficient for today's IC designs consisting of millions of interconnects [1]. Fast models that can be used to verify the amount of noise are therefore needed. Analytical closed-form models are well suited for this task.

Crosstalk noise avoidance is especially important for on-chip buses. In buses several interconnects run parallel to each other for long distances. Over the last decade several models have been proposed for the estimation of noise on coupled interconnects. In [2] a model for up to five coupled lines has been presented. However, the model is based on a single L-segment and doesn't consider inductance. In [3] a Π-model is used, but inductance is neglected. A model for distributed RC lines has been suggested in [4], but signal rise times and inductance are ignored. It has also been assumed that every other wire in the bus carries the same signal. In [5] a model for distributed RLC wires has been presented, but inductive coupling has been ignored. [6] proposes a model that includes mutual

J.J. Chico and E. Macii (Eds.): PATMOS 2003, LNCS 2799, pp. 111–120, 2003.
© Springer-Verlag Berlin Heidelberg 2003

and self inductance, but the different switching times of signals are not included. It has been shown that the effects of inductive coupling can be significant for long interconnects in [7]. The importance of different switching times has been addressed in [8]. None of the mentioned models can take into account the initial states of interconnects. An interconnect that does not reach steady state between cycles may cause a symbol travelling on the line to be corrupted by the initial state caused by an earlier symbol. In this paper an analytical RLC Π-model that includes different switching times, signal rise times, initial conditions and bit patterns is presented. The model considers both capacitive and inductive coupling between interconnects.

2 Crosstalk Noise Model for On-Chip Buses

A set of coupled equations for multiple coupled interconnects can be decoupled using a similarity transform. The diagonalization of the transmission line matrices reduces the problem of solving equations for coupled lines to solving a number of equations for isolated lines. This method has been addressed for instance in [9] and [10]. After the waveforms of the decoupled lines have been obtained, their linear combinations can be used to obtain the waveforms of the original coupled lines. The decoupling of interconnects makes it possible to consider both capacitive and inductive coupling between all wires. It also facilitates the calculations since it makes it possible to consider one interconnect at a time.

The decoupling method can be used for lossless lines by using the similarity transform to define the mode voltage and current matrices V_m and I_m as

$$\begin{aligned} \mathbf{V}(z,t) &= \mathbf{M}_V \mathbf{V}_m(z,t) \\ \mathbf{I}(z,t) &= \mathbf{M}_I \mathbf{I}_m(z,t) \ . \end{aligned} \tag{1}$$

After the decoupling of the transmission lines the driver and receiver ends are still coupled and have to be included in the decoupled model. This is generally done with numerical methods. In [6] a method to analytically include the terminal conditions has been proposed. A homogenous dielectric and lines having the same cross-section and per-unit-length resistance have been assumed. In order to include lossy lines in the decoupling method, the resistance matrix is assumed to be diagonal in the first place. In this case the transformation matrix $\mathbf{M}$ that diagonalizes the capacitance and inductance matrices is the eigenvector matrix of the capacitance or inductance matrix. The driver is modeled as a voltage source with source resistance and the receiver is modeled as a load capacitance. The drivers and receivers of the interconnects are assumed to be similar. The source resistance and load capacitance values stay the same in the coupled and decoupled systems. Under these assumptions a coupled transmission line system including driver and receiver ends can be transformed into a decoupled system with the following formulas. The decoupled system is denoted in the formulas with the superscript '^'.

$$\begin{aligned} \hat{\mathbf{R}} &= \mathbf{R} & \hat{\mathbf{R}}_S &= \mathbf{R} & \hat{\mathbf{L}} &= \mathbf{M}^T \mathbf{L} \mathbf{M} \\ \hat{\mathbf{C}} &= \mathbf{M}^T \mathbf{C} \mathbf{M} & \hat{\mathbf{C}}_L &= \mathbf{C}_L & \hat{\mathbf{V}}_S &= \mathbf{M}^T \mathbf{V}_S \end{aligned} \tag{2}$$

$\mathbf{R}_S$ and $\mathbf{C}_L$ are the source resistance and load capacitance matrices and $\mathbf{V}_S$ is the matrix containing the voltage sources. After the responses of the wires of the decoupled system have been calculated, they have to be combined to obtain the responses of the coupled lines. This can be done with (1). In the following section, the decoupling method is used to obtain an analytical model for a capacitively and inductively coupled bus.

2.1 Model Derivation

Figure 1 demonstrates the need to include initial states in crosstalk modelling. In the figure are represented the input and output voltages of two coupled interconnects. Initially both wires are at quiescent state. In this case the propagation delay of the first wire is 112 ps. However, when the wire switches up the second time, it has not reached steady state, and the delay is increased to 120 ps. Additionally, in the first upwards transition, the noise peak on the other line is 202 mV. Because of the initial state, the second noise peak, at about 2.2 ns, reaches 222 mV. The differences in percentage points are respectively 7.1 % and 9.9 %.

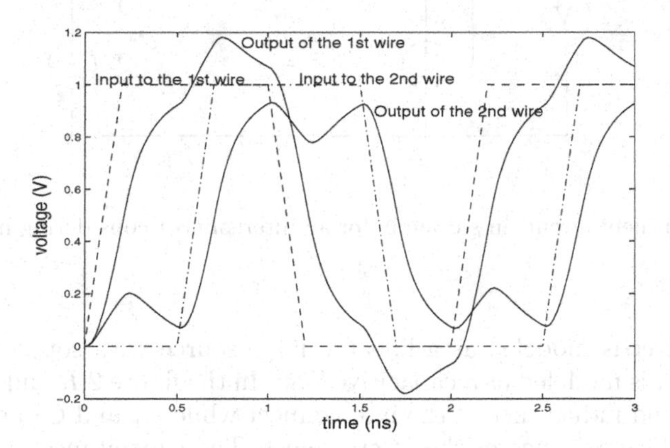

Fig. 1. Two coupled 2 mm long interconnects with initial states

For any lumped linear time-invariant (LTI) circuit its output can be written in the following form [11]

$$Y(s) = \sum_{m=1}^{M} H_{em}(s)X_m(s) + \sum_{n=1}^{N} H_{in}(s)\frac{\lambda_n(0)}{s} \tag{3}$$

where $Y(s)$ is the Laplace transform of the circuit output, $X_m(s)$ is the Laplace transform of the mth independent external voltage or current source, M is the number of external sources, $\frac{\lambda_n(0)}{s}$ is the Laplace transform of the source describing the effect of the value $\lambda_n(0)$ of the nth state variable at $t = 0$, N is the order of the system and $H_{em}(s)$ and $H_{in}(s)$ are functions that relate each external source or initial condition source to the output. If it is assumed that the circuit is initially in quiescent state, the function $H_{em}(s)$ relates the output to the input and is called the transfer function. In the modelling of crosstalk the initial state of the interconnects is generally ignored to enable the usage of transfer function and thus facilitate the calculations.

In the following we will derive an analytical model for a coupled bus including the initial conditions. The bus is assumed to consist of identically sized interconnects that have the same per-unit-length resistance and identical drivers and receivers. The bus is also assumed to be surrounded by a homogenous dielectric. Under these conditions the driver and receiver ends can be included in the model analytically as proposed in [6]. In order to obtain a closed-form time-domain solution the interconnect is modeled using one Π-segment as depicted in figure 2.

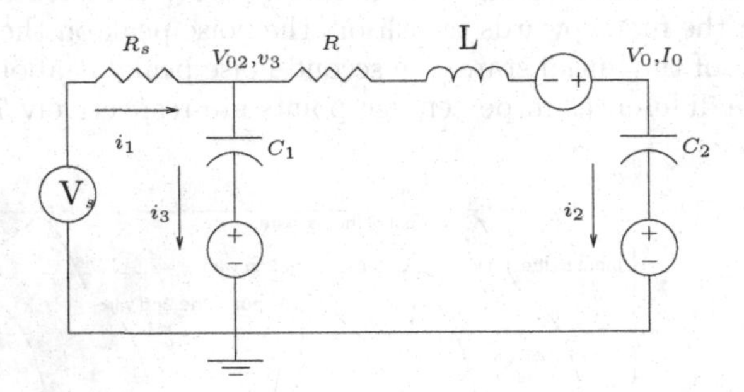

Fig. 2. Equivalent circuit in s-domain for an interconnect considering initial conditions

The driver is modeled as a linear voltage source with source resistance R_s. The receiver is modeled as a capacitive load. In the figure 2 R and L are the total resistance and inductance of the interconnect while C_1 and C_2 are equal to half of the total capacitance of the interconnect. The capacitance of the receiver is added to C_2. The possible initial charges in the capacitors and the initial current in the inductor are included in the model by using their s-domain equivalent circuits. The voltage source v_s is modeled as a superposition of three components v_{s1}, v_{s2} and v_{s3} as shown in figure 3. The components are used to obtain non-simultaneous switching times for interconnects as well as an input signal with a non-zero rise time.

The equations for the components of the voltage source can be written as

$$\begin{aligned} v_{s1}(t) &= \left[\tfrac{v_f - v_i}{t_2 - t_1}(t - t_1) + v_i\right] u(t - t_1) \\ v_{s2}(t) &= \left[-\tfrac{v_f - v_i}{t_2 - t_1}(t - t_1)\right] u(t - t_2) \\ v_{s3}(t) &= v_i \left[u(t) - u(t - t_1)\right] \end{aligned} \tag{4}$$

where $t_2 - t_1$ is the rise time and v_i and v_f are the initial and final values of the input signal. t_1 is the phase of the input signal and $u(t)$ is the unit step function.

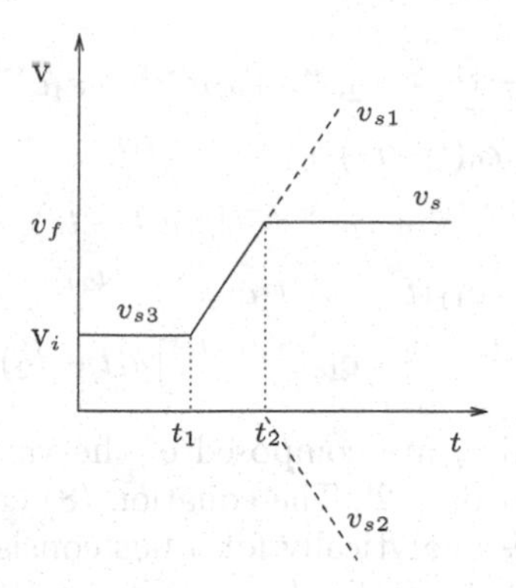

Fig. 3. The voltage source v_s and its components

The following s-domain equations can be derived from the circuit in figure 2. Variables in s-domain have been denoted with capital letters.

$$\begin{aligned} &V_s - I_1 R_s - \left(\tfrac{I_3}{sC_1} + \tfrac{V_{02}}{s}\right) = 0 \\ &\left(\tfrac{I_3}{sC_1} + \tfrac{V_{02}}{s}\right) - I_2 R - (sLI_2 - LI_0) - \left(\tfrac{I_2}{sC_2} + \tfrac{V_0}{s}\right) = 0 \\ &I_1 = I_2 + I_3 \end{aligned} \tag{5}$$

In the equations above V_0 is the initial voltage at the end of the line and V_{02} is the initial voltage of capacitor C_1. I_0 is the initial current flowing through the inductor. The equations can be used to derive an expression for current I_2. By substituting the voltage source components in s-domain into this expression for I_2 and using partial fractions and inverse Laplace transform the time-domain equation for i_2 can be written as

$$\begin{aligned} i_2(t) = &\left[a_1 e^{s_1 t} + a_2 e^{s_2 t} + a_3 e^{s_3 t}\right] u(t) \\ &+ \left[a_4 + a_5 e^{s_1(t-t_1)} + a_6 e^{s_2(t-t_1)} + a_7 e^{s_3(t-t_1)}\right] u(t-t_1) \\ &+ \left[a_8 + a_9 e^{s_1(t-t_2)} + a_{10} e^{s_2(t-t_2)} + a_{11} e^{s_3(t-t_2)}\right] u(t-t_2)\,. \end{aligned} \tag{6}$$

The voltage $v_3(t)$ can be obtained in a similar manner and written as

$$\begin{aligned} v_3(t) = &\, i_2(t) R + v_{out}(t) + \left[b_1 e^{s_1 t} + b_2 e^{s_2 t} + b_3 e^{s_3 t}\right] u(t) \\ &+ \left[b_4 e^{s_1(t-t_1)} + b_5 e^{s_2(t-t_1)} + b_6 e^{s_3(t-t_1)}\right] u(t-t_1) \\ &+ \left[b_7 e^{s_1(t-t_2)} + b_8 e^{s_2(t-t_2)} + b_9 e^{s_3(t-t_2)}\right] u(t-t_2) \end{aligned} \tag{7}$$

where $v_{out}(t)$ is the output voltage of the line. It can be written as

$$\begin{aligned}v_{out}(t) &= \left[c_1 + c_2 e^{s_1 t} + c_3 e^{s_2 t} + c_4 e^{s_3 t}\right] u(t) \\ &+ \left[c_5 + c_6(t - t_1) + c_7 e^{s_1(t-t_1)}\right. \\ &\left. + c_8 e^{s_2(t-t_1)} + c_9 e^{s_3(t-t_1)}\right] u(t - t_1) \\ &+ \left[c_{10} + c_{11}(t - t_2) + c_{12} e^{s_1(t-t_2)}\right. \\ &\left. + c_{13} e^{s_2(t-t_2)} + c_{14} e^{s_3(t-t_2)}\right] u(t - t_2) .\end{aligned} \tag{8}$$

The expressions a_i, b_i and c_i are composed of the variables that are obtained from (5) during the derivation [12]. The equation (8) can be used for decoupled lines to obtain the voltages analytically for a bus consisting of four or less lines.

To model bit patterns, the state of each wire at the end of a clock cycle is passed onto the next calculation by setting the final values of $v_{out}(t)$, $i_2(t)$ and $v_3(t)$ as the initial values V_0, I_0 and V_{02}. Two consequent similar inputs are obtained by setting the same value to the initial voltage v_i and final voltage v_f.

3 Simulation Results

In this section the proposed model is compared to other models and HSPICE. Then the effect of interconnects in a bus on crosstalk noise is studied.

3.1 Model Comparison

The proposed model was compared to other analytical crosstalk models by Dhaou [5], Kahng [13] and Kawaguchi [4]. All models are intended for deep sub-micron design and assume wires running in parallel. Two 4 mm long wires sized at 0.6 μm × 1.2 μm were used in the comparison. The wires were located at 0.6 μm from each other. The electrical characteristics of the wires were obtained with Linpar [14] for 0.13 μm UMC technology. The total capacitance of the wires is 124.1 fF and coupling capacitance is 71.33 fF. The self and mutual inductance are respectively 515.1 pH and 345.6 pH. The resistance of the wires is 23.61 Ω. All values are given per one millimeter. The models were compared to HSPICE simulations with 100 RLC segments. The comparison was performed for three different rise times. The results and the calculated errors for each model are shown in tables 1 and 2. In table 1 are shown the results for the case when one wire is switching and the other remains quiet. Kawaguchi's and Dhaou's models tend to overestimate the noise, since they assume a step input. Kahng's model on the other hand underestimates the noise, because it does not include self or mutual inductance. In table 2 are the results for propagation delay estimation. The delay is defined as the difference between the input signal and output signal at 50 % of the operating voltage. The results demonstrate the need to include both the rise times of signals and inductance in crosstalk models. Models that ignore rise time may clearly overestimate the amount of noise or underestimate the propagation delay. The inclusion of rise times in crosstalk models is needed, since crosstalk may be effectively reduced by using the longest possible rise times.

Table 1. Noise on a quiet wire for different rise times calculated by different models

	New model	Dhaou	Kahng	Kawaguchi	HSPICE
t_r=200ps	0.167V	0.199V	0.151V	0.192V	0.165V
Error	1.2%	20.6%	-8.5%	16.4%	
t_r=100ps	0.192V	0.199V	0.166V	0.192V	0.189V
Error	1.6%	5.3%	-12.2%	1.6%	
t_r=50ps	0.200V	0.199V	0.170V	0.192V	0.195V
Error	2.5%	2.1%	-12.8%	1.5%	

Table 2. Propagation delay when both wires are switching in the same direction

	New model	Kahng	Kawaguchi	HSPICE
t_r=200ps	57 ps	61 ps	-57 ps	57 ps
Error	0%	7.0%	-200.0%	
t_r=100ps	51 ps	60 ps	-7 ps	52 ps
Error	1.9%	15.4%	-113.5%	
t_r=50ps	45 ps	60 ps	43 ps	45 ps
Error	0%	33.3%	4.4%	

The accuracy of the proposed model in different operating speeds and wire lengths was assessed by comparing the calculated results and HSPICE simulations for four coupled copper interconnects. The global interconnects were sized at 0.6 μm $\times$ 1.2 μm. The operating frequency was varied between 100 MHz and 2 GHz while the length of the interconnects was between 0.5 mm and 10 mm. The simulation was run for seven clock cycles with the wires switching in opposite directions so that the effects of crosstalk are at maximum. The rise time of the interconnects was set to 10 percent of the clock cycle, i.e. from 50 ps to 1000 ps. The results are illustrated in figure 4. The error of the proposed model increases with operating speed and wire length. This is due to the limited accuracy of the lumped Π-model. However, the error remains below four percent.

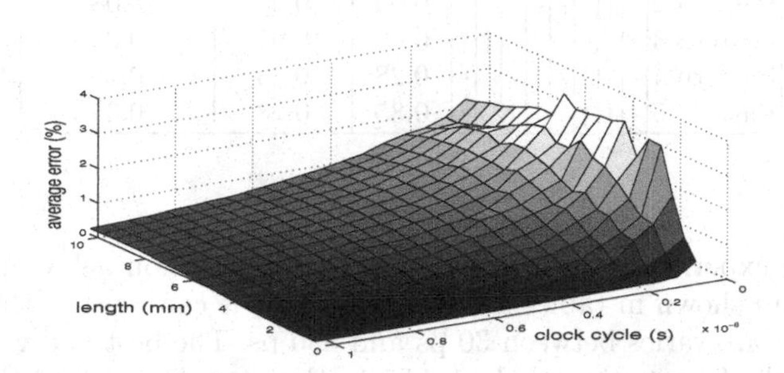

Fig. 4. Difference between the proposed model and HSPICE simulations for four coupled interconnects in 0.13 μm technology

3.2 The Impact of Neighboring Wires on Crosstalk Noise

Given that our model can be used for any number of coupled wires, it is still computationally expensive to diagonalize the matrices for a large number of wires. To reduce the computational cost, we study the impact of each coupled wire on the propagation delay and noise on a victim line. A bus consisting of eight identical interconnects in 0.13 μm technology was used in the investigation. The wires are numbered from one to eight starting from the left. The fourth wire was used as a reference wire in the measurements, since it is most susceptible to crosstalk noise. The simulations were run with HSPICE for three different wire sizes. In the first case, the wires were 3 mm long global wires sized at 0.6 μm and 1.2 μm, and separated by 0.6 μm. In the second case, 1 mm long minimum-sized (0.16 μm × 0.29 μm) wires separated by 0.16 μm were used. Finally, in the last case the wires were also minimum-sized, except for their height that was doubled so that the coupling capacitance was also doubled. In all cases the wires switch simultaneously with a rise time of 100 ps.

The simulation results for crosstalk noise and propagation delay are shown in tables 3 and 4. In these tables, the first eight columns describe the switching status of the wires. The symbols '↑' and '↓' in the tables represent upward and downward transitions while the symbol '-' represents no transition on the wire. From table 3 we see that the maximum amount of noise induced onto the fourth wire is 37 % of the operating voltage. This maximum is reached when all wires except the fourth one are switching. If the wires 2, 3, 5 and 6 are switching, the noise is reduced to 35 %. However, if only the third and fifth wire are switching the noise is reduced to 28 % of the operating voltage. The third and fifth wire thus have the greatest impact on the noise induced onto the fourth wire. Nonetheless, the second and sixth wire still have a considerable effect on noise. As a result, we notice that the wires 2, 3, 5 and 6 constitute up to 95 percent of the total noise. This observation holds also for the two other wire sizes.

Table 3. Noise in an 8-bit bus for different switching patterns and wire properties. The values are normalized to the V_{dd}. The wires are numbered from left to right

	1	2	3	4	5	6	7	8	Global	Minimum	Double height
Pattern 1	↑	↑	↑	-	↑	↑	↑	↑	0.37	0.29	0.35
Pattern 2	↑	↑	-	-	-	↑	↑	↑	0.09	0.06	0.08
Pattern 3	↑	-	-	-	-	-	↑	↑	0.02	0.01	0.02
Pattern 4	-	-	↑	-	↑	-	-	-	0.28	0.23	0.28
Pattern 5	-	↑	↑	-	↑	↑	-	-	0.35	0.28	0.34

The same experiments were conducted for propagation delay measurement. The results are shown in table 4. From this table it can be seen that the delay of the fourth wire varies between 30 ps and 136 ps. The best and worst case are presented in the first section of the table. In the second section of the table, the fourth wire is switching and the wires right next to it are quiet. In case the fourth wire is the only one switching, the delay is 67 ps. If the wires 2,3,5 and 6 are

Table 4. Delay of the fourth wire in an 8 bit bus for different switching patterns and wire properties

	1	2	3	4	5	6	7	8	Global	Minimum	Double height
Pattern 1	↑	↑	↑	↑	↑	↑	↑	↑	30 ps	28 ps	19 ps
Pattern 2	↓	↓	↓	↑	↓	↓	↓	↓	136 ps	102 ps	103 ps
Pattern 3	-	-	-	↑	-	-	-	-	67 ps	58 ps	54 ps
Pattern 4	↑	-	-	↑	-	-	↑	↑	64 ps	56 ps	52 ps
Pattern 5	↓	-	-	↑	-	-	↓	↓	70 ps	60 ps	56 ps
Pattern 6	↑	↑	-	↑	-	↑	↑	↑	54 ps	50 ps	43 ps
Pattern 7	↓	↓	-	↑	-	↓	↓	↓	81 ps	66 ps	64 ps
Pattern 8	↑	↓	↓	↑	↓	↓	↑	↑	136 ps	101 ps	104 ps
Pattern 9	↑	↑	↓	↑	↓	↑	↑	↑	125 ps	90 ps	97 ps

quiet while the others are switching, the delay variation compared to the case when only the fourth wire is switching is less than five percent, i.e. the delay is between 64 ps and 70 ps. However, if only the wires 3 and 5 are quiet, the delay varies by approximately 20 percent, between 54 ps and 81 ps. The activity of the first, seventh and eight wire have little impact on the delay of the fourth wire. This can also be seen in the third section of the table. The experiments for the other wire properties in the two rightmost columns in the table 4 show similar results. Based on the measurements it can be concluded that the two closest wires on both sides of the reference wire cause 95 % of crosstalk noise. When the measurements are repeated using a 10 ps rise time, the proportion is reduced to about 91 % because of the increasing influence of inductive coupling and its long distance effects. These observations can be used to determine a suitable model complexity for evaluating crosstalk noise on a bus.

4 Conclusions

An analytical time-domain model to evaluate crosstalk on capacitively and inductively coupled buses has been presented. The model can be used in the design of high-performance buses since it takes into account the effects of inductance, initial states, and bit patterns. Signal rise times and phases are also included in the model. However, this versatility leads to rather complex equations. It has also been shown that the model achieves good accuracy when compared to previous models and HSPICE. The simulations showed that the error of the model is below four percent for four coupled global interconnects. The impact of initial states was also demonstrated. The initial state of an interconnect affects its propagation delay and the amount of noise induced onto it. It was shown that for planar buses with a signal rise time of 100 ps or more, the two closest wires on both sides of a reference wire cause 95 % of crosstalk noise and delay variation.

References

1. A. Devgan: Efficient Coupled Noise Estimation for On-Chip Interconnects. In IEEE/ACM International Conference on Computer-Aided Design, 1997.
2. G. Servel and D. Deschacht: On-Chip Crosstalk Evaluation Between Adjacent Interconnections. In IEEE International Conference on Electronics, Circuits and Systems, 2000.
3. A. B. Kahng, S. Muddu, N. Pol, and D. Vidhani: Noise Model for Multiple Segmented Coupled RC Interconnects. In International Symposium on Quality in Electronic Design, 2001.
4. H. Kawaguchi and T. Sakurai: Delay and Noise Formulas for Capacitively Coupled Distributed RC Lines. In Proceedings of Asia and South Pacific Design Automation Conference, 1998.
5. I. B. Dhaou, K. Parhi, and H. Tenhunen: Energy Efficient Signaling in Deep Submicron CMOS Technology. Special Issue on Timing Analysis and Optimization for Deep Sub-Micron ICs, VLSI Design Journal, 2002.
6. J. Chen and L. He: A Decoupling Method for Analysis of Coupled RLC Interconnects. In IEEE/ACM International Great Lakes Symposium on VLSI, 2002.
7. S. H. Choi, B. C. Paul and K. Roy: Dynamic Noise Analysis with Capacitive and Inductive Coupling. In IEEE International Conference on VLSI Design, 2002.
8. M. Becer and I. N. Hajj: An Analytical Model for Delay and Crosstalk Estimation in Interconnects under General Switching Conditions. In IEEE International Conference on Electronics, Circuits and Systems, 2000.
9. C. Paul: Analysis of Multiconductor Transmission Lines. John Wiley & Sons, 1994.
10. K. D. Granzow: Digital Transmission Lines, Computer Modelling and Analysis. Oxford University Press, 1998.
11. W. M. Siebert: Circuits, Signals and Systems. The MIT Press, 1986.
12. S. Tuuna: Modelling and Analysis of Crosstalk on High-Performance On-Chip Buses. Master's Thesis, University of Turku, 2002.
13. A. B. Kahng, S. Muddu, and D. Vidhani: Noise and Delay Uncertainty Studies for Coupled RC Interconnects. In IEEE International ASIC/SOC Conference, 1999.
14. A. Djordjevic, M. Bazdar, T. Sarkar, and R. Harrington: LINPAR for Windows: Matrix Parameters for Multiconductor Transmission Lines, Software and User's Manual, Version 2.0. Artech House Publishers, 1999.

A Block-Based Approach for SoC Global Interconnect Electrical Parameters Characterization

M. Addino, M.R. Casu, G. Masera, G. Piccinini, and M. Zamboni

Politecnico di Torino, C.so Duca degli Abruzzi 24, I-10129 Torino, Italy,
{mario.casu,marco.addino}@polito.it

Abstract. A method for SoC global interconnect characterization is presented. Buses are partitioned in blocks whose electrical characterization is done using reduced size primitives and extending the results to the original structure. The accuracy is measured on typical metrics like delays, crosstalk peaks and reabsorbing time. This work is the basis for an automatic evaluator of interconnect metrics to be used in SoC design space explorations and verification.[1]

1 Introduction

In recent years many papers have addressed the problem of on-chip interconnect classification, modeling, order reduction, simulation and parameter extraction [1][2][3]. The reason of this growing interest is that on the one hand interconnect delays were getting preeminent and started dominating gate delays due to the continuous scaling; on the other hand crosstalk effects were starting to affect chip performance, design time and cost spent to reduce their impact to a sustainable level. On top of this, the design methodologies were changing in order to be more adequate to face both the interconnect challenge and the new System-on-Chip (SoC) scenario [4][5]. The twos are strictly related because of the amount of long interconnects needed to connect together the "building blocks" in a SoC.

In order to speed-up the SoC design cycle a stronger interaction between IP, system level and circuit designers would be necessary. On the other hand such interaction should be minimum so as to favor the SoC paradigm based on blocks reuse. This can be achieved through the sharing of a common know-how. IP and system level designers should be aware since from the early design stages of the real performance they can get on silicon. They need proper tools for that. IP designers do not need best possible accuracy and need instead maximum speed for large design space explorations and to characterize the blocks in different conditions and environments. System level architects can trade speed for a better accuracy because they deal with a single design at a time. However the foundation the tools should rely on is the same and consists of a proper modelization of global lines in a SoC environment.

[1] This work was founded by the Politecnico di Torino center of excellence for the development of multimedia radio-communications (CERCOM).

J.J. Chico and E. Macii (Eds.): PATMOS 2003, LNCS 2799, pp. 121–130, 2003.
© Springer-Verlag Berlin Heidelberg 2003

In this work we present the basis for the implementation of a tool for the SoC global interconnect performance estimation. First a classification of interconnect topologies and a block-based approach to tackle their complexity is presented. Then methods for a simplified extraction of parameters in complex and wide blocks are shown together with the accuracy when approximations are used. Finally the conclusions summarize the achievements of this paper.

2 SoC Bus Topologies

SoC global interconnects usually look like buses made of several parallel running wires that connect together the IP's. Long buses suffer from crosstalk and transmission line effects such as multiple reflections, therefore needing proper models and careful designs. The number of possible topologies to route global buses can be reduced from a practical point of view to a finite set. Moreover, using a limited number of pre-characterized topologies can be useful in order to increase the predictability of the interconnect performance. An example is given in figure 1. The identified macro-blocks are *bus* (1), *curves* (2) *cross without contacts* (3), *cross with contacts* (4-6), that may differ in wire number width, spacing and length. However, a single topology may appear many times in a single design or in different similar designs in the same technology. The use of pre-characterized blocks organized in an *interconnect library* can be extremely useful not only in the design phase but also in the verification phase since the parasitic extraction procedure could be easier and faster. Such approach allows the use of accurate field solvers that must handle small blocks instead of a complete layout. The analysis of the performance of global interconnects in a SoC for design *planning* (front-end) and *verification* (back-end) can be done by a suitable partitioning in macro-blocks belonging to the above categories. Finally, the global performance can be retrieved from a suitable combination of blocks performance as it is normally done for logic gates in timing/power analysis.

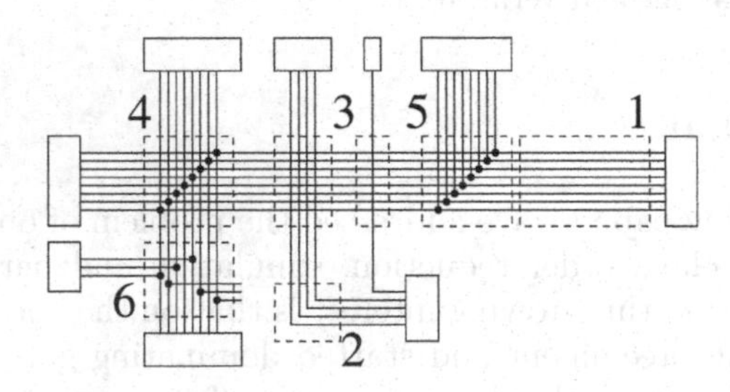

Fig. 1. Example of possible SOC global interconnect topologies

The major simplification of this method is represented by the assumption that the blocks, during the electrical parameters extraction, are considered completely separated and therefore their reciprocal influence is neglected. The accuracy of such approach can be optimized by properly choosing the boundaries of the macro-block. Moreover, the common use of lateral shields (external wires connected to ground or vdd) enforce at the same time the electrical *isolation* of the bus under exam from other interconnect structures and the *predictability* of the electrical properties that may be extracted as if the block was effectively isolated from the rest of the chip. In order to quantify the error introduced by our assumption we characterize, using FASTCAP, a five-by-five orthogonal crossing without contact, like the one sketched in the inset in figure 2. We use 0.13μm feature size technology data for this computation. We evaluate the coupling ca-

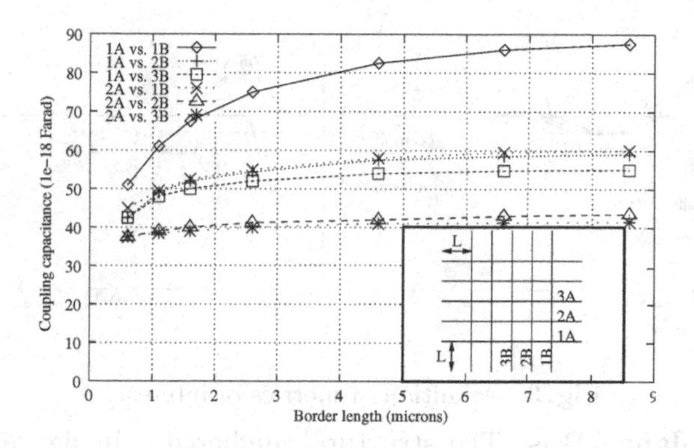

Fig. 2. 5-by-5 cross and coupling capacitance variation due to different border lengths.

pacitance between lines belonging to different layers varying the *Border Length L*. Results in figure 2 show that the variation is relevant only in the case of border lines. For short border length the impact on the total accuracy of the model is minimal for metal lines that do not belong to the border. In addition, if the bus under analysis is provided with external shield wires the influence between macro-blocks during the parameters extraction can be totally neglected also using a short border length.

Another problem is represented by the mutual inductance that is a longer range effect when compared to the coupling capacitance. We distinguish between two types of macro-blocks. Parallel running wires have to be modeled as RLC distributed circuits and can be extracted using 2-D extractors like for instance the 2-D field solver integrated in HSPICE [6]. Crossing with and without contacts can be modeled using lumped RC elements. The latter choice starts from the consideration that in current technologies the side-length of any type of multilayer crossing is well below the bound of 500μm that is under the limit for local interconnect classification [1].

In order to suitably use the macro-blocks we build equivalent circuits to be used in simulators or with analytical models obtained using field-solvers [2][7]. However the complete characterization of relevant size macro-blocks may be very time consuming. Efficient *Reduction Techniques* are therefore needed.

3 Reduction Techniques

We use reduction techniques to simplify the work of field solvers. To validate them we compare reduced and full models on the basis of meaningful metrics such as dynamic delays, noise peaks (Vmax and Vmin), reabsorbing time of noise peaks (tabs, within thresholds) for aggressor and victim lines. Such metrics are reported in figure 3 and are referred to in tabled results in the following. A different approach is used for the various topologies depicted in figure 1.

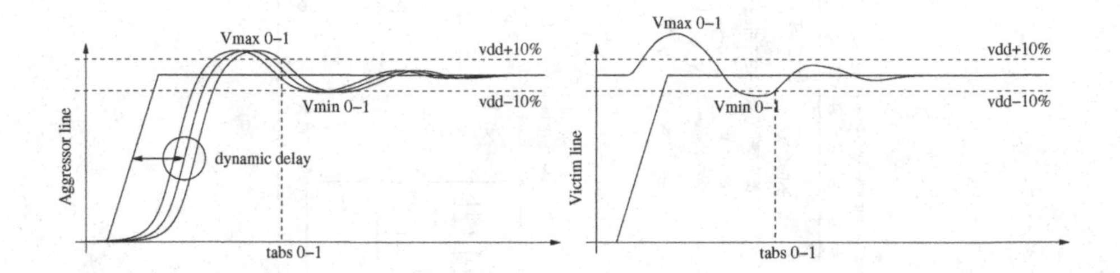

Fig. 3. Definition of metrics of interest.

Uniform Coplanar Bus. The structure numbered 1 in figure 1 has an invariant cross section along the length, thus the parameters extraction can be accomplished by means of a 2-D field solver like the one included in HSPICE [6]. Let's consider a 31-lines uniform coplanar bus. This structure should be modeled using distributed *RLC* equivalent circuits like the HSPICE W-Element [6]. The high number of conductors involved is such that the parameters extraction can require more than an hour on a high-end Sun workstation. In order to derive reduction methods we should exploit the symmetry of the structures, that is reflected in the *RLC* matrices of a coplanar bus. For what concerns the coupling capacitance we observed, as expected, a very fast decay such that after few lines (3-5 for a 0.13μm technology) it becomes negligible. An excellent approximation consists in taking into account only the coupling of the line with the 3-5 lines on each side. On the contrary, the mutual inductance decays very slowly, as shown in figure 4 (left) where the mutual inductance of the first line versus other lines is plotted (Metal 6, minimum spacing, 0.13μm technology). Neglecting the mutual inductance after few lines leads to relevant errors. Our method for the extraction of RLC matrices consists of two steps:

1. Full extraction from a significant structure with geometries equal to the original one (spacing and width), but <u>with less conductors</u>.
2. Matrices previously extracted are <u>extended</u> to the original size.

To define and test some *extension operators* we analyze structure and symmetries of per unit length inductance and capacitance matrices. The diagonal elements, L_{ii} and C_{ii}, represent self inductances and capacitances. C_{ij} and L_{ij} are coupling capacitances and mutual inductances between i-th and j-th lines. Matrices are symmetrical because of the reciprocity theorem and in the special case of the uniformly spaced and wide lines, the structure is also geometrically symmetrical. As a result, matrices are symmetrical with respect to both diago-

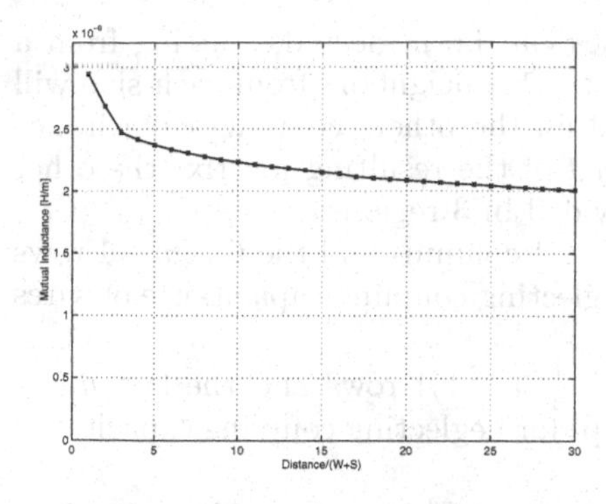

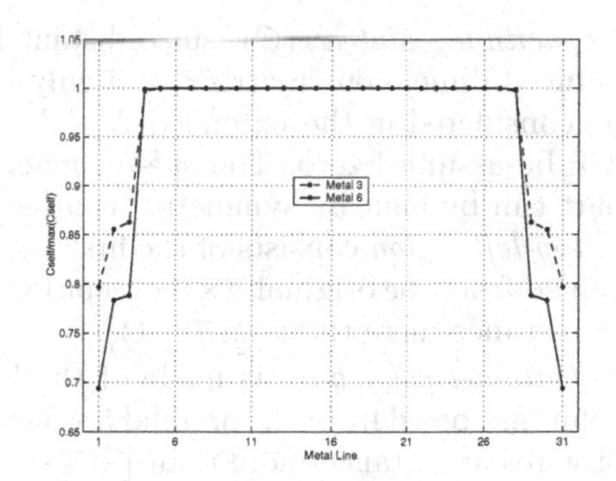

Fig. 4. Bus first line mutual inductance to other lines and normalized self-capacitance of each bus line for various metals.

nals. Moreover, extracted self inductances are all numerically equal and mutual inductances depend only on wire distance and not on position.

For what concerns the capacitance matrix a number of border lines can be found looking at the self and coupling capacitance of each metal lines. In figure 4 (right) we plot the self-capacitances C_{ii} in Metal3 and Metal6 (Lgate=0.13μm technology). The results show that the number of border lines is around 3. Moreover the "border effect" depends on the distance of the ground plane and in presence of grounded external metal lines can be substantially neglected. The approach we follow consists in the extraction of a $\overline{n}$-lines small bus whose parameters are extended to the original N-lines size. The rules we use are the following:
Resistance Matrix (R): only diagonal elements are nonzero and represent resistances per unit length that are all equal in the case of constant width. The extension is straightforward.
Inductance Matrix (L): we use the position invariance of mutual inductance shown before and extend the last element $L_{\overline{n}-1}$ (mutual inductance of two elements whose distance is $\overline{n}-1$) assuming the mutual inductance remains constant after the $\overline{n}$-th metal line. Such assumption, as clear from figure 4 (left), is reasonable and conservative. A possible refinement is a fitting of the curve in figure 4 (left) to be used for the extension. The extended $N \times N$ matrix is:

$$\begin{bmatrix} L & L_1 & \dots & L_{\overline{n}-1} & \dots & \dots & L_{\overline{n}-1} \\ L_1 & L & L_1 & \ddots & \ddots & \vdots & \vdots \\ \vdots & L_1 & L & L_1 & \ddots & L_{\overline{n}-1} & \vdots \\ L_{\overline{n}-1} & \ddots & L_1 & \ddots & \ddots & \ddots & L_{\overline{n}-1} \\ \vdots & L_{\overline{n}-1} & \ddots & \ddots & L & L_1 & \vdots \\ \vdots & \vdots & \ddots & \ddots & L_1 & L & L_1 \\ L_{\overline{n}-1} & \dots & \dots & L_{\overline{n}-1} & \dots & L_1 & L \end{bmatrix}$$

where L is the self inductance and $L_{ij} = L_k$ is the mutual inductance of two wires whose distance is $k = |i-j|$ wires.

Capacitance Matrix (C): since distant lines can be neglected, starting from a reduced dimensions matrix $\overline{n} \times \overline{n}$ only $\lfloor \frac{1}{2}(\overline{n}-1) \rfloor$ neighbors from each side will be considered in the extended $N \times N$ and all the others coupling capacitance will be assumed zero. The *sub-diagonal part* of the resulting matrix (the other part can be built by symmetry) can be divided in 3 regions:
1. *top-left region* consists of the first $\lfloor \frac{1}{2}(\overline{n}-1) \rfloor$ columns and the first $\overline{n}-1$ rows copied from the original $\overline{n} \times \overline{n}$ except for neglecting coupling capacitance of wires at distance more than $\lfloor \frac{1}{2}(\overline{n}-1) \rfloor$;
2. *bottom-right region* is made of the last $\lfloor \frac{1}{2}(\overline{n}-1) \rfloor$ rows and the last $\overline{n}-1$ columns copied from the original $\overline{n} \times \overline{n}$ except for neglecting coupling capacitance of wires at distance more than $\lfloor \frac{1}{2}(\overline{n}-1) \rfloor$;
3. *extension region* includes columns and rows from $\lceil \frac{1}{2}(\overline{n}-1) \rceil$ to $N - \lfloor \frac{1}{2}(\overline{n}-1) \rfloor$. It is created by extending the self-capacitance of the original matrix element $\lceil \frac{1}{2}(\overline{n}-1) \rceil$ and the border elements of top-left and bottom-right regions.

As an example, we report the 10×10 matrix obtained extending a $\overline{n} \times \overline{n} = 7 \times 7$ primitive matrix:

$$\begin{bmatrix} C_{11} & C_{21} & C_{31} & C_{41} & 0 & 0 & 0 & 0 & 0 & 0 \\ C_{21} & C_{22} & C_{32} & C_{42} & C_{52} & 0 & 0 & 0 & 0 & 0 \\ C_{31} & C_{32} & C_{33} & C_{43} & C_{53} & C_{63} & 0 & 0 & 0 & 0 \\ C_{41} & C_{42} & C_{43} & C_{44} & C_{54} & C_{53} & C_{63} & 0 & 0 & 0 \\ 0 & C_{52} & C_{53} & C_{54} & C_{44} & C_{54} & C_{53} & C_{52} & 0 & 0 \\ 0 & 0 & C_{63} & C_{53} & C_{54} & C_{44} & C_{54} & C_{53} & C_{63} & 0 \\ 0 & 0 & 0 & C_{63} & C_{53} & C_{54} & C_{44} & C_{54} & C_{64} & C_{74} \\ 0 & 0 & 0 & 0 & C_{52} & C_{53} & C_{54} & C_{55} & C_{65} & C_{75} \\ 0 & 0 & 0 & 0 & 0 & C_{63} & C_{64} & C_{65} & C_{66} & C_{76} \\ 0 & 0 & 0 & 0 & 0 & 0 & C_{74} & C_{75} & C_{76} & C_{77} \end{bmatrix}$$

The first and second region are shadowed in pale gray while the third one is darker.

In order to find a good trade-off between accuracy and complexity we compared fully characterized 16 and 32 lines buses with extended structures using $\overline{n} = 3, 5, 7$. Here, as typical results, we summarize the metrics extracted from a minimal part of the electrical simulations performed for validation. In table 1 we report the crosstalk on a victim line when N-1=31 aggressor lines are switching (0.13μm technology). We used typical driver resistances and loads (taken from technology data-sheets). Furthermore in table 2 the accuracy of our method for the evaluation of the *dynamic delay* [1] is presented. Three neighbor lines switch at the same in three different configurations: $+++$ means all the lines switch high, $0+0$ means the center line switches high and the others stay

Table 1. Single line crosstalk, bus length=15 mm

Model	**1–0** V_{min}	**0–1** V_{max}	**1–0** t_{abs}	**0–1** t_{abs}	**1–0** V_{min}	**0–1** V_{max}	**1–0** t_{abs}	**0–1** t_{abs}
	Metal 3, center victim				**Metal 3, edge victim**			
Full	0.427 V	1.972 V	8.32 ns	8.268 ns	0.668 V	1.731 V	5.923 ns	5.873 ns
$\overline{n}=7$	+4.8%	+5.0%	−8.6%	−8.8%	+5.0%	+5.1%	−11.4%	−10.5%
$\overline{n}=5$	+4.4%	+4.4%	−13.3%	−13.4%	−12.2%	−12.2%	−8.4%	−8.0%
$\overline{n}=3$	−60%	−60%	20.1%	−20.1%	−37.2%	−37.3%	+13.3%	+13.7%
	Metal 6, center victim				**Metal 6, edge victim**			
Full	0.095 V	2.318 V	2.731 ns	2.742 ns	0.291 V	2.109 V	1.332 ns	1.337 ns
$\overline{n}=7$	−4.3%	−5.2%	+4.0%	−7.0%	−8.1%	−8.6%	+8.0%	+8.0%
$\overline{n}=5$	−4.8%	−5.7%	+2.4%	−7.0%	−15%	−14.6%	+13.4%	+13.5%
$\overline{n}=3$	−66%	−66%	+0.0%	−0.1%	−51%	−23.2%	+108%	+108%

Table 2. Dynamic Delay, bus length=15 mm.

Model	t_d (+++)	t_d (0+0)	t_d(−+−)	t_d (+++)	t_d (0+0)	t_d(−+−)
	Metal 3			Metal 6		
Full	0.656 ns	3.198 ns	5.516 ns	0.327 ns	0.811 ns	1.528 ns
$\overline{n}=7$	−13.9%	−6.1%	−5.5%	+10.75%	−1.6%	+3.2%
$\overline{n}=5$	+0.0%	−7.6%	−4.8%	+11.3%	−6.6%	+0.6%
$\overline{n}=3$	+57.8%	−14%	−21.6%	+121%	−11.2%	−6.5%

low, $-+-$ extern lines switch low and center line switches high. The results show that $\overline{n}=7$ is a good choice. In a couple of minutes using a high-end workstation we can characterize very wide coplanar bus structures with errors that in the typical case are less than 10%.

Curves. We decided to characterize curves (structure 2 in figure 1) using an approach similar to the bus one. We first compute the RC matrices per unit length (we neglect inductance in this case, as for all the cross structures). Then we compute the *effective length* for each wire and the *effective coupling length* to other wires. This heuristic gave similar results in terms of accuracy as for the bus metrics reported in previous tables (not reported here for sake of brevity).

No-Contact Crossing. This block (number 3 in figure 1 consists of $N \geq 1$ lines on layer A and $M \geq 1$ lines on layer B. We have to consider the intra-layer coupling capacitance (A vs. A, B vs. B) and the inter-layer one (A vs. B). The capacitance matrix of this kind of structures can be divided in four blocks:

$$\left[\begin{array}{c|c} A \text{ vs. } A & B \text{ vs. } A \\ \hline A \text{ vs. } B & B \text{ vs. } B \end{array}\right]$$

The whole capacitance matrix and each single block are symmetric. Sub-matrices B vs. A and A vs. B are identical. We chose, as *reduced structure*, a 5×5 crossing because it was a good trade-off between accuracy and complexity.The characterization using FASTCAP requires less than 5 minutes on a Sun ULTRA80.

The extension for intra-layer capacitance follows the same rules used for coplanar buses. FASTCAP, differently from the HSPICE field solver, does not output capacitances per unit length but total capacitances. Therefore a proper scaling must be used to adapt the capacitance extracted from the reduced crossing to a wider one. The sub-matrices B vs. A and A vs. B are extended differently. Let's consider the simple case where N=1, M=32. Simulation results show that the coupling capacitance is almost constant except for the bus border lines. We can extend the sub-matrix A vs. B from 1×5 to 1×32 by replicating the middle element that is almost constant:

$$[\ C_1\ C_2\ C_3\ C_4\ C_5\ ] \Rightarrow [\ C_1\ C_2\ C_3 \cdots C_3\ C_4\ C_5\ ]$$

If $N > 1$ this extension should be applied row-wise and column-wise. Thus the resulting extended sub-matrix for the more general case will be:

$$\begin{bmatrix} C_{11} & C_{12} & C_{13} & \dots & C_{13} & C_{14} & C_{15} \\ C_{21} & C_{22} & C_{23} & \dots & C_{23} & C_{24} & C_{25} \\ C_{31} & C_{32} & C_{33} & \dots & C_{33} & C_{34} & C_{35} \\ \vdots & \vdots & \vdots & \ddots & \vdots & \vdots & \vdots \\ C_{31} & C_{32} & C_{33} & \dots & C_{33} & C_{34} & C_{35} \\ C_{41} & C_{42} & C_{43} & \dots & C_{43} & C_{44} & C_{45} \\ C_{51} & C_{52} & C_{53} & \dots & C_{53} & C_{54} & C_{55} \end{bmatrix}$$

Once obtained the complete capacitance matrix, we can build up the circuit equivalent of the No-Contact Crossing associating to each node a set of capacitance and a series resistance.

In table 3(column 2) are reported the results of the comparison of an extended versus a fully extracted 16×16 crossing (0.13μm Metal 3-4). Also in this case errors on the order of and lower than 10% were obtained with a large CPU time saving (around 100 times faster).

Table 3 (column 3-4) reports a similar comparison for the case 1×32 using a merely capacitive model and a RC one in different configurations of driver resistance and load capacitance. The results are excellent for the RC model and still good for the C one. The time required for the parameters extraction for the reduced structure was only 4% of the time necessary for the full structure.

Table 3. 16×16 and 1×32 crossing no contacts in various load and driver configurations. Extended vs. full model.

Measure	**Ext. RC 16×16**	**Ext. RC 1×32**	**Ext. Purely C 1×32**
X-talk V_{min} 1 − 0	−4.6%	−1.7% ÷ −5.5%	−2.1% ÷ −7.6%
X-talk V_{max} 0 − 1	−4.8%	−1.3% ÷ −6.2%	−2.4% ÷ −8.4%
t_{abs} 1 − 0	−10.7%	−0.0% ÷ −1.7%	−0.0% ÷ −1.7%
t_{abs} 0 − 1	−10.6%	−0.1% ÷ −1.6%	−0.1% ÷ −1.9%
Delay 50% 1 − 0	−5.3%	−0.3% ÷ −2.0%	−0.3% ÷ −3.4%
Delay 50% 0 − 1	−5.9%	−0.2% ÷ −2.3%	−0.2% ÷ −3.3%

Complete crossing with contacts. This N-by-N structure (number 4 in figure 1) is identical to the previous one except from the presence of vias. We first show that the influence of vias in the capacitance extraction process is negligible. If this is true we can reuse the same *reduced structure* of the previous case. For this purpose we compare the following two capacitance matrices: the first one obtained running FASTCAP with the 5×5 cross including vias; the second one is obtained starting from the correspondent no contact crossing (matrix size: 10×10) and manipulating it in the following way:
1. Coupling terms C_{ij} are built as follows:

$$C_{ij} = \frac{C_{AiAj}}{L_5} \cdot L_N + \frac{C_{BiBj}}{L_5} \cdot L_N + C_{AiBj} + C_{AjBi} \tag{1}$$

where C_{AiAj} and C_{BiBj} are the intra-layer coupling capacitance between the i-th and the j-th conductor, C_{AiBj} and C_{AjBi} are the inter-level coupling capacitance and L_5 and L_N are the side length of the crossing respectively in the case of 5 and N lines.

2. Capacitances to ground are retrieved summing the two terms of lines to be connected by vias. By comparing the two matrices we build the error matrix from which results a maximum error of −4.12% and an average error equal to 1.00%. The results show that the impact of vias is negligible for what concerns the capacitance extraction step and that passing from a No-Contact to a Complete cross with contacts does not need an additional extraction. However vias have a significant resistance whose impact on our models have to be evaluated. We compared the crosstalk peaks of two 5-lines M2-M3 crossing connected to four equal coplanar buses with variable length and with resistance taken into account and neglected. As expected, simulation results show that when the via resistance is a small fraction of the total wire resistance (cross edge much lower than bus length) its effect is insignificant. Therefore, depending on the length of the buses connected by the cross, the via resistance will be taken into account or not. If yes the number of terminals of the cross equivalent circuit is doubled because the lines on orthogonal layers are no more at the same potential. In this case the final model we build consists of a capacitive matrix ($2N \times 2N$) obtained by simply taking the correspondent matrix for the no-contact crossing and by adding the resistances. Terms are not summed like in equation (1). Coupling capacitances are slightly overestimated numerically (the contact removes some of the coupling area) but from the electrical point of view it is almost the same (vias short some of the coupling).

Partial Crossing with contacts. This is another N-by-N structure labelled 5 in figure 1. As for the previous case, we try to extrapolate the capacitance matrix starting from the 5-over-5 No-Contact Crossing. Based on our experience, the following techniques give the best results:
1. The coupling capacitance between the i-th and the j-th element (both consisting of a layer A line of fixed length plus a layer B line of variable length depending on its position inside the bus) of an N lines L-Crossing will be evaluated in the following way:

$$C_{ij} = \frac{C_{AiAj}}{L_5} \cdot L_N + \frac{C_{BiBj}}{L_5} \cdot L_c + C_{AiBj} \tag{2}$$

where C_{AiAj} and C_{BiBj} are the intra-layer coupling capacitance between the i-th and the j-th conductor, C_{AiBj} and C_{AjBi} are the inter-level coupling capacitance and L_5 and L_N are the side length of the crossing respectively in the case of 5 and N lines. L_c is the *equivalent coupling length* of B_i and B_j.
2. Again, capacitances to ground are retrieved summing the two terms of lines to be connected by vias.

If we set L_c as the shortest length among B_i and B_j the error matrix shows a maximum error equal to −13.90% and an average error equal to −10.47%. The errors are negative biased because of the choice of L_c. The best approach consists in fitting this value so as to minimize the error. For what concerns the via modelization the same remarks of the complete crossing hold. In table 4 are reported the typical errors comparing the extended structure to the full model and taking vias into account.

Table 4. 16×16 partial crossing: extended vs. full model. Measures taken on the upper bus centerline.

Measure	Ext. Purely C
X-talk ($V_{dd} - V_{min}$) 1–0	−9.2%
X-talk ($V_{max} - V_{dd}$) 0–1	−9.4%
Re-absorbing Time (t_{abs}) 1–0	−13.5%
Re-absorbing Time (t_{abs}) 0–1	−13.7%
Propagation Delay ($t_{d50\%}$) 1–0	−6.9%
Propagation Delay ($t_{d50\%}$) 0–1	−7.1%

Generic crossing. This type of N-by-M crossing (6 in figure 1) is a combination of partial crossings with contacts and no-contact crossings. Based on this we defined simple rules to build the extended RC matrices from a reduced size no-contact N-by-M crossing extracted with FASTCAP. Such rules are combinations of the previously presented and are not reported here for sake of brevity. In any case the errors obtained comparing extended structures to completely extracted crossings are of the same order of previous results.

4 Conclusions and Future Works

In this paper a strategy to tackle the global interconnect complexity in Systems-on-Chip has been presented. We showed that it is possible to find a finite set of bus topologies whose electrical characterization can be done using reduced size primitives with less wires and by extending the results to the original structure. The CPU time saving reaches 95% for wide and complex structures using 3-D field-solvers like FASTCAP and the 2-D field solver integrated in HSPICE. The accuracy is good with errors on the order of and lower than 10% for metrics like delays, crosstalk peaks and reabsorbing time within thresholds.

In future works we will integrate the strategies presented here in a tool for the automated partitioning in macro-blocks of complex global interconnects in Systems-on-Chip. The final goal is that to build an automatic evaluator of global interconnect metrics like delay, power and crosstalk to be used in SoC design space explorations and verification.

References

1. A. Deutsch *et alii*, On-chip Wiring Design Challenges for Gigahertz Operation. *Proc. IEEE*, Vol. 89, No. 4, pp. 529–555, April 2001.
2. D. Sylvester and C. Hu, Analytical Modeling and Characterization of Deep-Submicrometer Interconnect. *Proc. IEEE*, Vol. 89, No. 5, pp. 634–664, May 2001.
3. B. Kleveland *et alii*, High-Frequency Characterization of On-Chip Digital Intecon-nects. *IEEE JSSC*, Vol. 37, No. 6, pp. 716–725, June 2002.
4. J. Cong, An Interconnect-Centric Design Flow for Nanometer Technologies. *Proc. IEEE*, Vol. 89, No. 4, pp. 505–528, April 2001.
5. L. Carloni and A. Sangiovanni-Vincentelli, Coping with Latency in SOC Design. *IEEE Micro*, Vol. 22, No. 5, pp. 24–35, Sept.-Oct. 2002.
6. Avant! Corporation. Avant Star-HSPICE Manual Release 2000.4.
7. J. A. Davis, *A Hierarchy of Interconnects Limits and Opportunities for Gigascale Integration.* PhD thesis, Georgia Institute of Technology, 1999.

Interconnect Driven Low Power High-Level Synthesis*

A. Stammermann, D. Helms, M. Schulte, A. Schulz, and W. Nebel

OFFIS Research Institute
D-26121 Oldenburg, Germany
Stammermann@OFFIS.DE

Abstract. This work is a contribution to high level synthesis for low power systems. While device feature size decreases, interconnect power becomes a dominating factor. Thus it is important that accurate physical information is used during high-level synthesis [1]. We propose a new power optimisation algorithm for RT-level netlists. The optimisation performs simultaneously slicing-tree structure-based floorplanning and functional unit binding and allocation. Since floorplanning, binding and allocation can use the information generated by the other step, the algorithm can greatly optimise the interconnect power. Compared to interconnect unaware power optimised circuits, it shows that interconnect power can be reduced by an average of 42.7 %, while reducing overall power by 21.7 % on an average. The functional unit power remains nearly unchanged. These optimisations are not achieved at the expense of area.

1 Introduction

Recently, several research approaches have been reported taking physical information into account. Most of the proposed algorithms use floorplanning information in high-level synthesis to estimate area and performance more accurately [2, 3]. Similarly, a lot of techniques have already been proposed taking into account power consumption in high-level synthesis [4,5,6,7]. Just a few of these contributions also consider interconnect power [8,9]. For high-level interconnect length estimation the well known Rent's rule is often used [10]. This model requires knowledge of empirical parameters that are computed from actual design instances. This limits the applicability and therefore we do not use Rent's rule.

This work evaluates an approach of simultaneous binding, allocation and floorplanning optimisation. Binding is the task of assigning compatible operations or variables to resources during the high-level synthesis. A low power binding is an assignment in which the power dissipation of the resources is minimal. Binding has a great influence on power dissipation, since different bindings lead to different input streams on the input of resources. Allocation is the choice of the number of resources. Binding and allocation affect the area of the design, the netlist topology (beeing the basis of a floorplan) and the wire activity. In the

* This work was supported by the European Union, IST-2000-30125, POET

J.J. Chico and E. Macii (Eds.): PATMOS 2003, LNCS 2799, pp. 131–140, 2003.
© Springer-Verlag Berlin Heidelberg 2003

following binding will denote the combination of binding and allocation. In order to find a power optimal solution binding and floorplaning must be regarded simultaneously.

A precondition for combining binding and floorplanning is high estimation accuracy of the power consumption of RT-resources and interconnect. In order to determine the power consumption of resources power models describing the power consumption and area of the individual resources at RT level [11] are needed. Interconnect power primarily depends on the wire length of individual wires, the number of vias and the switching activity. We estimate the wire length by generating a slicing tree floorplan. Since a floorplan only affects wires connecting different RT-resources, only the global interconnect is considered. Wires within a resource are encapsulated by the power models.

We use a low power high-level optimisation tool, called ORINOCO® [12, 13], to obtain the RTL circuits and the power consumption of the datapath. ORINOCO®is interconnect unaware. It is amended by our new interconnect power estimation methodes detailed in section 3.

The paper is organized as follows: In section 2 we present a motivation example. In section 3 our proposed optimisation methodology is described. An experimental evaluation is presented in section 4 and conclusions are drawn in section 5.

2 Motivation

Fig. 1 illustrates the effect of different binding solutions on the interconnect length of a register-transfer level (RTL) design. A scheduled data flow graph (SDFG) is given, which contains three generic types of operators: a, b and c and their corresponding functional units are A, B and C. Fig. 2a shows a binding of the SDFG and the corresponding interconnect optimised floorplan. Operators within a grey bar are mapped on the same functional unit. In the floorplan the wires are annotated with their length.

Fig. 2b shows a different binding solution and the updated floorplan. b_3 is re-binded to B_1, b_4 to B_2 and c_1 to C_2. Thereby the total wire lenght decreases by 28 %. This clearly shows the importance of considering interconnect power during the binding step.

3 Interconnect Driven High-Level Synthesis

Regarding the process of high-level synthesis of interconnect shows that we have to think about the wire capacitance and the switching activity. Therefore we discuss the wire capacitance estimation used in our approach, before presenting our new introduced extensions.

3.1 Interconnect Power Estimation

Interconnect power dissipation can be written as $P_{Inter} \approx \sum C_i D_i$, where C_i and D_i are wire capacitance and switching activity for wire i. We derive the wire

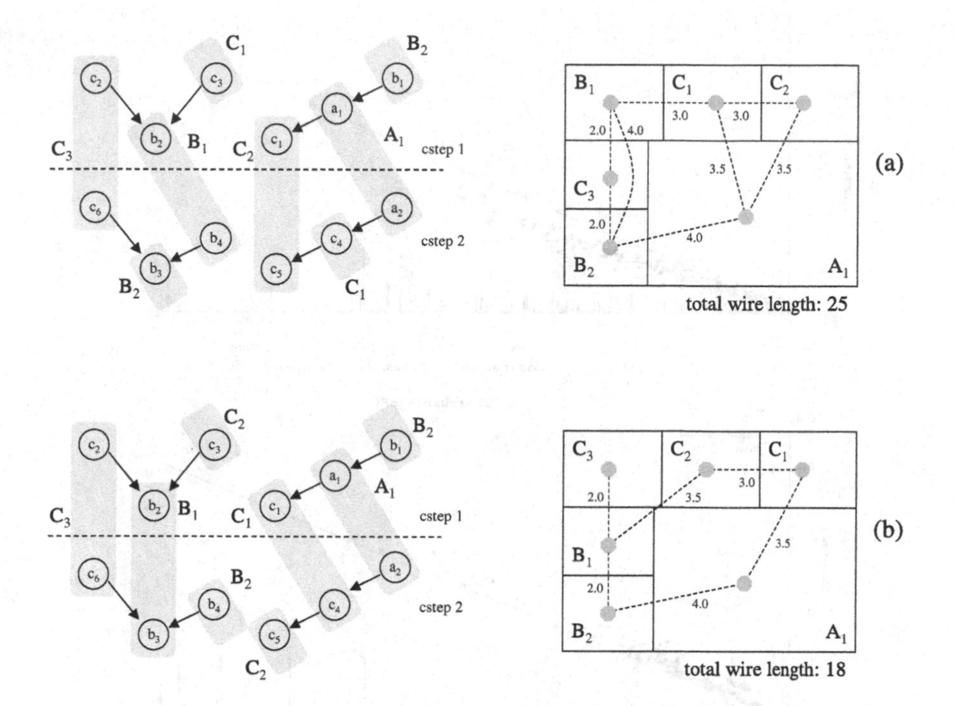

Fig. 1. (a)Original binding (b)New binding and new floorplan

capacitance by using a capacitance model. This model is based on wire length, number of pins and number of branch points. We use a linear regression technique to model the dependencies. *Pins* are the connecting points to RT-resources, e.g. a wire at the input of a multiplier is connected to about 6 gates, that is 6 pins. The number of pins depends on the RT-resource type and can be extracted from the corresponding RT-model. The number of branch points and the wire length is extracted from a floorplan (cf. 3.2).

Our wire length and capacitance estimation is evaluated with a script-based work flow. This flow enables us to impose a floorplan to a commercial tool like Cadence Silicon Ensemble®. Running this flow computes as well the estimated values of the length and capacitance of each wire as the values computed by Silicon Ensemble. For an *fft* the estimated capacitance (length) vs. Silicon Ensemble is plotted in Fig. 2b (Fig. 2a). The curves denote the points where estimated and measured values are identical. The vertical bars show the absolute error. Besides the wire length the number of pins and branch points is a second major contributor for the overall wire capacitance. This impact on the overall capacitance is due to the additional vias for further branches and pins. For the used $0.25\,\mu m$ technology the capacitance (length) estimation has an average std. deviation of 30.2% (29.6%) and an average rel. error o 14.9% (17.5%).

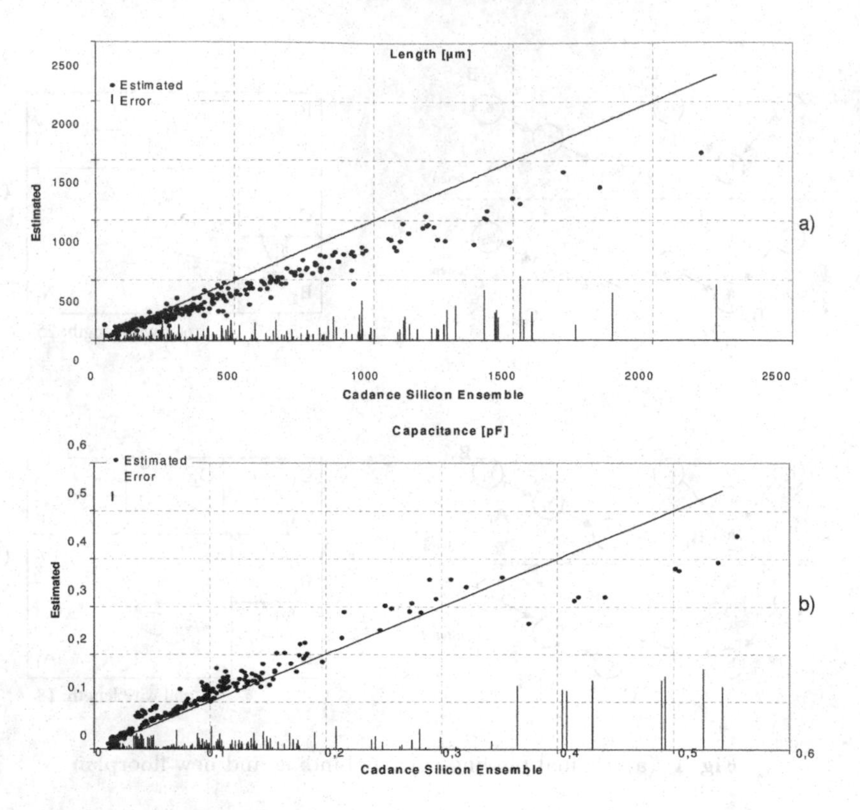

Fig. 2. Estimated (a)length and (b)capacitance vs. Silicon Ensemble for an forced floorplan (fft benchmark)

3.2 Simulated Annealing (SA) Based Floorplanner

For interconnect length estimation an extension of a well known SA based floorplanner by Wong and Liu [14] is used. Simulated annealing is an iterative technique for solving high-dimensional optimisation problems. These techniques switch from one solution (here: floorplan) to another solution in a well-defined way by using 'moves'. This algorithm considers slicing floorplans. A slicing floorplan is a floorplan that can be obtained by recursively cutting a rectangle by either a vertical line or a horizontal line into smaller rectangular regions. A slicing floorplan can be represented by an oriented rooted binary tree. Each internal node of the tree is labeled either $*$ or $+$, corresponding to either a vertical cut or a horizontal cut. Floorplan transformation is achieved by using five types of moves:

1. Swap Leafs F_1
2. Swap Nodes F_2
3. Swap Leaf and Node F_3
4. Shift Leaf or Node F_4
5. Switch direction F_5

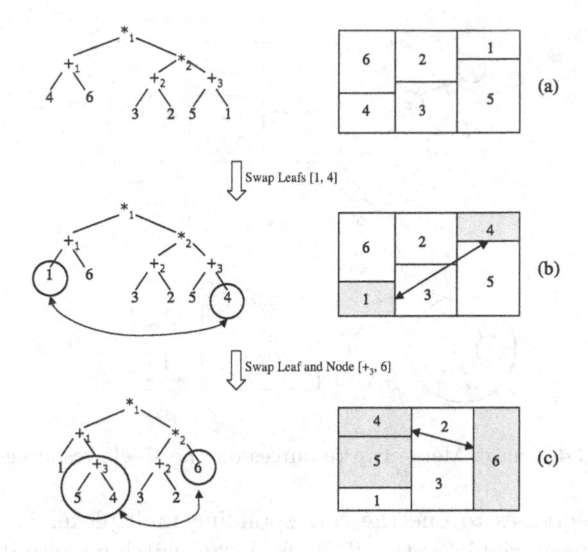

Fig. 3. (a)Initial floorplan (b)Floorplan after F_1 (c)Floorplan after F_3

Fig. 3 illustrates how the moves F_1 and F_3 affects the binary tree (left side) and shows the impact for the corresponding floorplan (right side). Each floorplan considered during SA process is evaluated based on area A and interconnect power P, using a cost function of the form $P + \lambda A$, where λA controls the relative importance of A and P. In [14] the interconnect length is estimated by calculating the Manhattan distance for two pin connections and the minimum spanning tree (MST) for connections with more than two pins. These technique does not suit real wiring because no branch points are considered. Instead, we use Steiner Trees for drawing data transfer wires. To treat the clock distribution network accurately an H-tree (balanced tree) is generated.

3.3 Extended Approach

For our approach we modified the cost function and the SA process. The new cost function is of the form $P_{FU} + P_{wire} + \lambda A$. P_{FU} is the power consumption of the functional units, multiplexer and registers and P_{wire} is the power consumption of the interconnect. λA is the area's contribution to the cost function. The annealing process is amended by three new binding moves. In combination with floorplan moves they allow a variation of the design architecture and corresponding floorplan simultaneously.

1. *Share* B_1
   Share merges two resources res_1 and res_2 to one single resource res_2 (Fig. 4). For such a move to be valid, res_1 and res_2 must be instances of the same type. Moreover, no operation performed by res_1 should have an overlapping lifetime with any operation of res_2. If the number of sources at one input of a resource exceeds one, a new multiplexer is instantiated. If the number

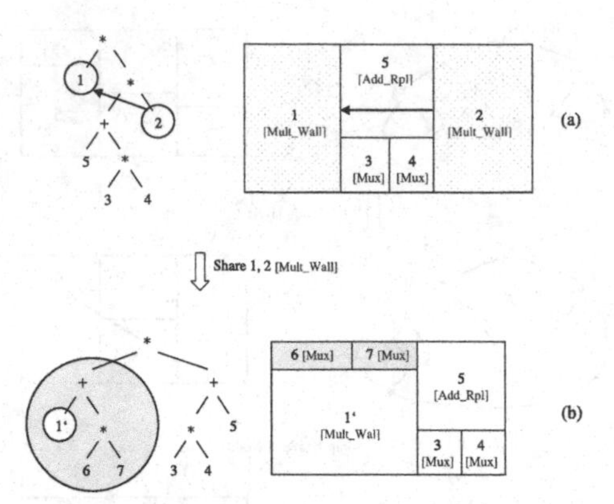

Fig. 4. Share: Merge two resources on one single resource

of sources decreases to one the corresponding multiplexer is dropped out. Sharing resources significantly affects both the switching activity in the data path and the network topology.

2. *Split* B_2
   Split is the reverse of *share*. A single resource is splitted into two resources. Like in move B_1, multiplexers can vanish or appear. Splitting can be done without regarding the lifetime of operations. Apart from potentially reducing switched capacitance, these moves enlarge the avenues for applying other share moves.
3. *Swap* B_3
   Swap interchanges the inputs of commutative operations. Like in move B_1, multiplexers can vanish or appear. This move significantly affects the switching activity in the data path. The influence on the netlist is nearly negligible.

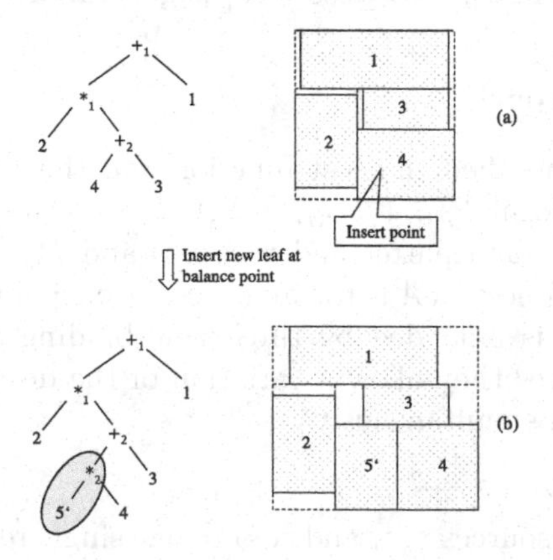

Fig. 5. Inserting new leaf at balance point

$B_1 - B_3$ in combination are able to create every possible binding solution. New components are inserted at their balance point. The balance point is the point, where the new resource would produce the lowest interconnect power. In Fig. 5a this point is inside the left half of leaf 4. Therefore leaf 4 is replaced by a new vertical node with the new leaf 5′ placed on the left side and 4 on the right side. Our floorplanner supports soft macros, which means that leafs are flexible in their aspect-ratio. Therefore inserting or deleting a leaf does not destroy the floorplan. The unused area in Fig. 5b only originates because we limited this ratio. This avoids unrealistic floorplans.

3.4 Optimisation Algorithm

The algorithm itself consists of two nested SA processes (Fig. 6). The inner loop uses floorplan moves ($F_1 - F_5$) optimising the actual floorplan for interconnect power. The outer loop uses the binding moves ($B_1 - B_3$) optimising the actual architecture. By doing so every binding move is followed by a short floorplan annealing process, since binding moves can significantly affect the netlist topology and thus the interconnect power. The effect of a binding move can only be rated after a floorplan update. This can be done rapidly because of the modules'flexibility. Changes in the netlist topology are mended in the actual floorplan. In contrast to previous approaches, a time-consuming floorplan generation from scratch is not necessary.

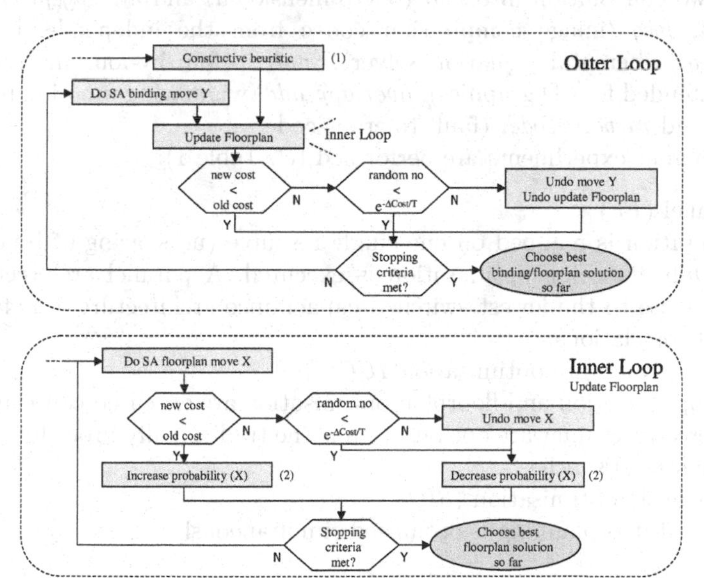

Fig. 6. Inner and outer loop of the optimisation algorithm

In general an annealing move is chosen randomly. If a move leads to a decreased power consumption this move is accepted. If the power is not reduced the move may be acccepted on a probabilistic base. If a generated random number (0 - 1) is smaller than $e^{-\triangle\ Cost/T}$, where $-\triangle\ Cost$ is the power difference and T is the current temperature, the worse solution is accepted. This enables the

SA to escape from local minima. In addition, the probability of choosing a move from the inner loop is decreased or increased depending on the moves acceptance rate (Fig. 6(2)). Iterative algorithms normally start with a constructive heuristic to generate a pre-optimised solution (Fig. 6(1)). In our case we start with the low-power binding heuristic from [7].

4 Experimental Results

Our proposed technique is implemented on top of the low power high-level optimisation tool ORINOCO® [12]. We use the wire capacitance from [15] and an industrial 0.25 μm RTL design library.

Random logic introduces spurious transitions. These transitions cannot be effectively forecasted. Due to this reason we assume that no chaining will be used. This is a sensible assumption for low power design, as otherwise the glitches introduced by the first unit will boost the power of the second [16]. These glitches also contribute significantly to output network power consumption. So far this effect is neglected.

We evaluate eleven algorithmic level benchmarks. An *fdct* (fast discrete cosinus transfomation), an *fft* (fast fourier transform), a one-dimensional *wavelet* transform, two convolution filters *fir* (one-dimensional) and an *img_filter* (two-dimensional), *jpeg* (image compression codec from the independend JPEG group), *diffeq* (differentail equation solver), *matrix* (four-by-four matrix multiplication intended for 3D graphics), *overlapp_add* (windowing function used by ffts), *viterbi* and *turbo_decoder* (fault tolerant codecs).

Three different experiments are performed (cf. Table 1):

1. Full parallel (*FP*)
   Each operation is mapped on one single resource (no sharing of functional units). Only floorplan optimisation is executed. A parallel architecture is typically close to the lowest switched capacitance architecture, due to high temporal correlations.
2. Interconnect unaware optimisation(*IUO*)
   Binding optimisation and floorplan optimisation are executed consecutively. This interconnect unaware optimisation is the traditionally procedure in low power high-level synthesis.
3. Simultaneously optimisation (*SIO*)
   Binding and floorplanning is optimised simultaneously.

To achieve comparable results the total number of moves executed in each experiment are identical. The number of moves is determined depending on the benchmark size. The experiments were performed on a 1.0 GHz Athlon based PC with 256 MB memory. The CPU times vary from 10 seconds for *diffeq* to 218 seconds for *turbo_decoder*.

Table 1 shows the experimental results of the three different experiments and the percentage of energy and area reductions. Since scheduling and thus the timing is fix for each benchmark, energy reduction and power reduction are

Table 1. Experimental results of the different experiments *FP*, *IUO* and *SIO* and the percentage energy and area reductions

		FP [nWs/mm^2]	IUO [nWs/mm^2]	SIO [nWs/mm^2]	IUO vs FP reduction [%]	SIO vs FP reduction [%]	SIO vs IUO reduction [%]
fdct	Data path	508.55	348.40	384.65	31.49	24.36	-10.40
	Interconnect	158.78	125.65	86.11	20.86	45.77	31.47
	Total	667.32	474.05	470.76	28.96	29.46	0.69
	Area	14.211	7.176	7.895	49.50	44.44	-10.02
wavelet	Data path	884.71	843.25	862.76	4.69	2.48	-2.31
	Interconnect	1402.12	1666.55	854.73	-18.86	39.04	48.71
	Total	2286.82	2509.79	1717.49	-9.75	24.90	31.57
	Area	12.116	8.649	8.941	28.62	26.21	-3.38
fir	Data path	626.27	607.19	616.91	3.05	1.50	-1.59
	Interconnect	2526.67	2020.03	954.77	20.05	62.21	52.73
	Total	3152.94	2627.22	1571.68	16.67	50.15	40.18
	Area	11.685	8.68	8.111	25.72	30.59	6.56
fft	Data path	2065.28	2059.79	2068.28	0.27	-0.15	-0.41
	Interconnect	1045.44	1167.87	689.17	-11.71	34.08	40.99
	Total	3110.72	3227.65	2757.45	-3.76	11.36	14.57
	Area	19.82	17.01	11.44	14.18	42.28	32.74
jpeg	Data path	1718.64	1631.85	1636.97	5.05	4.75	-0.31
	Interconnect	1722.54	1626.33	1040.73	5.58	39.58	36.01
	Total	3441.18	3258.17	2677.69	5.32	22.19	17.82
	Area	11.62	6.04	6.48	48.36	44.23	-8.00
viterbi	Data path	172.18	100.78	138.11	41.47	19.79	-37.04
	Interconnect	552.30	647.98	322.97	-17.32	41.52	50.16
	Total	724.48	748.76	461.08	-3.35	36.36	38.42
	Area	9.68	4.11	4.68	57.59	51.65	-14.01
diffeq	Data path	4216.51	3162.04	3187.16	25.01	24.41	-0.79
	Interconnect	755.49	758.36	270.87	-0.38	64.15	64.28
	Total	4971.99	3920.40	3458.03	21.15	30.45	11.79
	Area	3.344	1.832	2.019	45.22	39.62	-10.21
matrix	Data path	2436.95	1700.05	1670.35	30.24	31.46	1.75
	Interconnect	1540.94	848.43	535.44	44.94	65.25	36.89
	Total	3977.88	2548.48	2205.79	35.93	44.55	13.45
	Area	16.01	3.93	3.96	75.45	75.29	-0.64
img_filter	Data path	1914.53	1487.89	1573.02	22.28	17.84	-5.72
	Interconnect	3652.24	3826.40	2527.61	-4.77	30.79	33.94
	Total	5566.77	5314.29	4100.64	4.54	26.34	22.84
	Area	35.08	26.24	28.1	25.19	19.89	-7.08
overlap_add	Data path	787.909	782.498	789.17	0.69	-0.16	-0.85
	Interconnect	403.02	381.694	257.482	5.29	36.11	32.54
	Total	1190.929	1164.192	1046.652	2.25	12.11	10.09
	Area	9.76	7.23	6.57	25.94	32.73	9.16
turbo_decoder	Data path	338.105	254.983	308.036	24.58	8.89	-20.81
	Interconnect	3568.283	3035.708	1767.574	14.93	50.46	41.77
	Total	3906.388	3290.691	2075.61	15.76	46.87	36.92
	Area	11.402	9.529	6.579	16.43	42.30	30.96
average	Data path				17.16	12.29	-7.14
	Interconnect				5.33	46.27	42.68
	Total				10.34	30.43	21.67
	Area				37.47	40.84	2.37

exchangeable. Thus, we will further refer to power as energy. Negative percentage means that the power or area is increased. Please note that the total power of some interconnect unaware optimised benchmarks increases (e.g. *wavelet*), which means that for these benchmarks the traditional optimisation fails. Compared to the traditionally procedure (*IUO*) our proposed technique (*SIO*) reduces the interconnect power for all benchmarks by an average of 42.7 %, while reducing overall power by 21.7 % on an average. The functional unit power just increases sensible for interconnect dominated designs. Compared to *IUO* the area remains nearly unchanged (reduced by an average of 2.4 %).

5 Conclusion

We showed that high-level synthesis has a significant impact on the interconnect power consumption. We proposed a new power optimisation algorithm which simultaneously performs floorplanning and functional unit binding. Experimental results demonstrate the benefit of incorporating interconnect in high-level synthesis for low power and the effectiveness of the proposed technique. In fact, the energy consumption might even increase if the traditional optimisation flow is used. Our technique is implemented on top of the optimisation tool ORINOCO® . Although our technique is general it can be easily incorporated into other high-level synthesis systems.

References

1. L. Scheffer: A roadmap of CAD tool changes for sub-micron interconnect problems. *International Symosium on Physical Design, (ISPD)*, California, United States, 1997
2. J.-P. Weng, A. C. Parker: 3D scheduling: High-level synthesis with floorplanning. *Proc. of Design Automation Conference*, 1992
3. Y.-M. Fang, D. F. Wong: Simultaneous functional-unit binding and floorplanning. *Proc. Int. Conf. Computer-Aided Design*, 1994
4. J. M. Chang, M. Pedram: Register allocation and binding for low power. *Proc. of Design Automation Conference*, 1995
5. R. Mehra, L. M. Guerra, J. M. Rabaey: Low-power architectural synthesis and the impact of exploiting locality *J. VLSI Signal Processing*, 1996
6. A. Raghunathan, N. K. Jha: SCALP: An iterative-improvement-based low power data path synthesis system. *Proc. Int. Conf. Computer-Aided Design*, 1997
7. L. Kruse, E. Schmidt, G. v. Cölln, A. Stammermann, A. Schulz, E. Macii, W. Nebel: Estimation of lower and upper bounds on the power consumption from scheduled data flow graphs. *IEEE Trans. VLSI Systems*, 2001
8. P. Prabhakaran, P. Banerjee: Simultaneous scheduling, binding and floorplanning in high-level synthesis. *Proc. Int. Conf. VLSI Design*, 1998
9. L. Zhong, N.K. Jha: Interconnect-aware High-level Synthesis for Low Power. *ICCAD*, 2002
10. P. Christie, D. Stroobandt: The Interpretation and Application of Rent's Rule. *IEEE Trans. VLSI Systems*, 2000
11. G. Jochens, L. Kruse, E. Schmidt, W. Nebel: A New Paramiterizable Power Macro-Model for Datapath Components. *Proc. of Design, Automation and Test in Europe*, 1999.
12. A. Stammermann, L. Kruse, W. Nebel, A. Pratsch, E. Schmidt, M. Schulte, A. Schulz: System Level Optimization and Design Space Exploration for Low Power. *14th International Symposium on System Synthesis*, Canada, 2001
13. W. Nebel, D. Helms, E. Schmidt, M. Schulte, A. Stammermann: Low Power Design for SoCs. *Information Society in Broadband Europe*, Romania, 2002
14. D. F. Wong and C. L. Liu: A new algorithm for floorplan design. *Proc. of Design Automation Conference*, 1986
15. Semiconductor Industry Association. *National Technology Roadmap for Semiconductors*, San Jose, CA: SIA, 1999
16. J. M. Rabaey: Digital Integrated Circuits. *Prentice Hall*, 1996

Bridging Clock Domains by Synchronizing the Mice in the Mousetrap

Joep Kessels[1], Ad Peeters[1], and Suk-Jin Kim[2]

[1] Philips Research Laboratories, NL-5656 AA Eindhoven, The Netherlands
{joep.kessels,ad.peeters}@philips.com
[2] Kwang-Ju Institute of Science and Technology, Kwang-ju, 500-712 South-Korea
sukjinkim@kjist.ac.kr

Abstract. We present the design of a first-in first-out buffer that can be used to bridge clock domains in GALS (Globally Asynchronous, Locally Synchronous) systems. Both the input and output side of the buffer have an independently clocked interface. The design of these kind of buffers inherently poses the problems of metastability and synchronization failure. In the proposed design the probability of synchronization failure can be decreased exponentially by increasing the buffer size. Consequently, at system level one can trade off between safety and low latency. The design is based on two well-known ideas: pipeline synchronization and mousetrap buffers. We first combine both ideas and then in several steps improve the design.

1 Introduction

Two developments undermine the role of globally clocked VLSI circuits. In the first place, the trend towards system-on-chip designs leads to chips containing several IP modules which all have different cycle times. Secondly, in future technologies it will become increasingly difficult to distribute high-speed low-skew clock signals. Therefore, future chips will contain several locally clocked submodules, which communicate through dedicated glue logic. These heterogeneous systems are called GALS (Globally Asynchronous, Locally Synchronous) systems [2]. Two kinds of GALS systems can be distinguished depending on the way the synchronous submodules communicate.

- In a *clock synchronization* system, the submodules have so-called pausible clocks, which are ring oscillators that can be halted. Safe communication is obtained by synchronizing the clocks [8,9,12,7,5].
- In a *data synchronization* system, the submodules have free-running clocks and the data being communicated from one clock domain to the other is synchronized. A simple solution is the well-known two-register or double-latch synchronizer [8,4]. More elaborate synchronizing schemes are based on first-in first-out buffers. The solution presented in [3] uses (distributed) pointer-based buffers offering both low latency and low power dissipation. However, since reading and writing is done via a bus connecting all cells, it will be

J.J. Chico and E. Macii (Eds.): PATMOS 2003, LNCS 2799, pp. 141–150, 2003.
© Springer-Verlag Berlin Heidelberg 2003

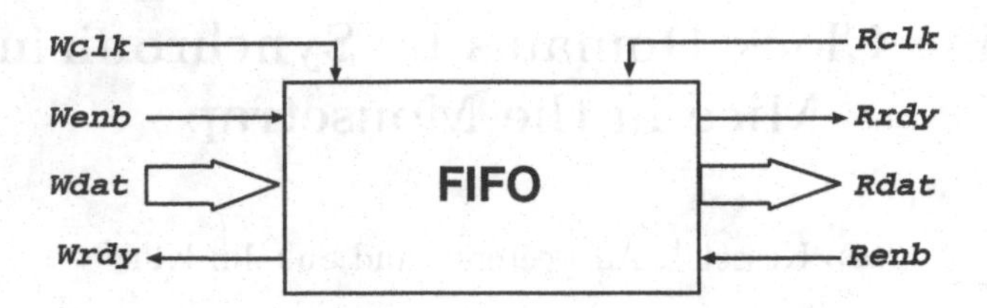

Fig. 1. Interface of synchronizing buffer

difficult to obtain high throughputs. This holds in particular when large distances have to be bridged. In [10] a solution is presented based on ripple buffers. Compared to pointer-based buffers, ripple buffers have longer latencies and dissipate more power, but since they allow a distributed placement with short point-to-point interconnects, they can offer higher throughputs.

Data synchronization systems inherently have to deal with metastable states, which have to be resolved within a given time period. When choosing this period one has to trade off between safety and low latency. Clock synchronization systems are safer in that they wait until the metastable states have been resolved. Moreover, a large subset of clock synchronization systems can be designed without any arbitration [8,9,2,5]. Despite this technical advantage of clock synchronization systems, synchronous designers tend to prefer the more familiar data synchronization systems.

We present a pipeline synchronizer based on the mousetrap buffer [11]. The paper is organized as follows. Section 2 gives the specification of the synchronizing buffer and in section 3 we introduce and analyse the design of the mousetrap buffer. In section 4 we apply pipeline synchronization in the mousetrap buffer. We first combine both ideas and then in several steps improve the design. In section 5 we summarize the differences between the designs. All the designs that we discuss have been implemented with a 16-bit wide data path in a 0.18 μm CMOS technology. They have been simulated using back annotation with estimated wire loads.

2 Specification of the Synchronizing Buffer

We want to design a buffer that can be used to transfer data streams from one clock domain to the other. Fig. 1 shows such a buffer, which has an input side W (for Write) and an output side R (for Read). Each side has a clocked interface with an independent clock (*Wclk* and *Rclk*). All other input/output signals are valid at the rising edge of the corresponding clock signal.

We want the buffer to be able to transfer a data item every clock cycle. Therefore, the clocked protocol at each buffer interface is not a handshake protocol, but a symmetric *rendez-vous* protocol in which both the buffer and the environment indicate their readiness to perform a transfer operation by making a dedicated signal high. The buffer signal is called *rdy* (for ready) and the signal

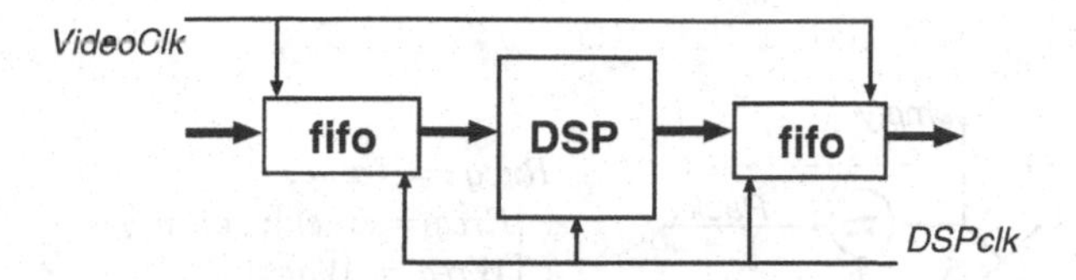

Fig. 2. Synchronizing buffers in video system

from the environment is called *enb* (for enable). A data transfer occurs if both the ready and enable signal are high at a rising clock edge. On the input side, *Wdat* must be valid when *Wenb* is high and on the output side, *Rdat* is valid when *Rrdy* is high. Note that the specification of the interface allows two buffers to be connected directly, provided the two interconnected interfaces share the same clock.

Fig. 2 shows the application of such a buffer in a system in which a DSP is handling a stream of video data. The DSP has a fast clock independent of the video clock. Therefore, two synchronizing buffers are needed: one before and one after the DSP. Since the video streams at both sides cannot be halted, the DSP must be fast enough to keep the ready signals at the video interfaces high.

3 The Mousetrap

The design of the synchronizing buffer is based on the mousetrap buffer [11], which is a ripple buffer using two-phase handshake signalling in the communication between two neighbouring cells. A communication in a *two-phase handshake protocol* consists of only two handshake events: first the sender indicates by means of a transition in a so-called request signal that new data is being offered, then the receiver indicates by means of a transition in an acknowledge signal that the data has been received. Note that each two-phase communication inverts the handshake signal levels, which is in contrast to a four-phase handshake communication which leaves the signal levels invariant.

Fig. 3(a) shows the design of the mousetrap cell (MT). If the two read handshake signals *Rreq* and *Rack* are equal, the cell is empty, otherwise the cell is full. Fig. 3(b) describes the behaviour of the cell for which we use a convention similar to the one introduced in [6]. All cells start by initializing the control latch with outgoing signal *Rreq* to the same value, say false, which implies that initially all handshake signals are equal and, hence, all cells are empty. Subsequently the cell executes an endless loop ('$A; B$' means sequential execution of A and B and 'A^*' means infinite execution of A). In each step of the loop, the cell first waits until the read handshake signals are equal ('$[C]$' means wait until C holds). and when this happens, signal *empty* becomes high and, consequently, the latches in the register become transparent. The cell now waits until signal *Wreq* becomes different from signal *Wack* (the same as signal *Rreq*), which means that its write neighbour is full (offering data). Subsequently the transparent latches take over the input signals ('$A \parallel B$' means concurrent execution of A and B). In

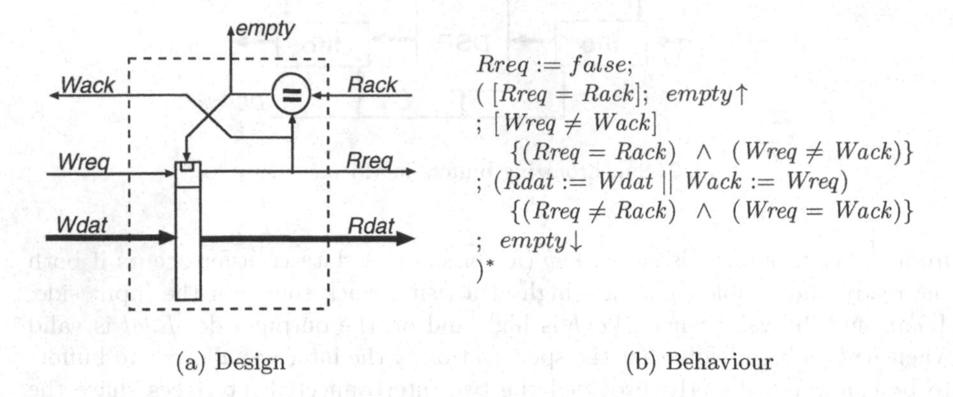

(a) Design (b) Behaviour

Fig. 3. Mousetrap cell (MT)

between curley brackets we give assertions that hold before and after the concurrent assignments. Since the assignment to *Wack* inverts its value (and the value *Rreq*), the read handshake signals become different (cell becomes full) and the write handshake signals become equal (write neighbour becomes empty). Signal *empty* then goes low making the latches opaque and the cell executes a next loop step.

Note that passing a data item (full bucket) is done via the request signals, whereas the acknowledge signals are used to return an empty bucket. Passing a full bucket takes δ(LATCH), [1] whereas passing an empty bucket takes δ(XNOR)+δ(LATCH). Therefore passing empty buckets is the bottleneck when the buffer runs at full speed. We define the *cycle time* of a buffer as the minimum time it takes for one cell to complete a full cycle that is passing a full bucket and receiving and filling an empty one. In our design the total delay adds up to 0.86 nsec leading to a maximum throughput of 1.16 Gsymbols/sec.

We can also observe that returning empty buckets comes with a timing constraint, since a transition in signal *Wack* is already given when the latches are still transparent. The latches will be closed after after a certain delay, called the *extended isochronic fork delay* [1], consisting of the delay of an XNOR-gate plus the hold time of a latch. Therefore a write acknowledge indicates that the write data may be changed only after the extended isochronic fork delay. If the write neighbour is also a mousetrap cell, this requirement is automatically fulfilled (after the handshake signals have become equal, it takes an XNOR-gate delay plus a latch delay before new data can be offered). Note that additional transmission delays between the cells make the design more robust.

The mousetrap buffer has several properties that make it very attractive for clock bridging in GALS systems.

[1] $\delta(A)$ stands for the delay of gate A

- In many GALS systems, clock bridging also implies bridging distances, in which case the transmission delays are important. For this reason a two-phase protocol is much more attractive than a four-phase protocol.
- The mousetrap buffer cell is fast, which means that it allows high clocks rates in the synchronizing buffers. Moreover, for a given clock frequency, a faster buffer has more time to deal with metastable states, which results in a smaller probability of synchronization failure.
- An empty mousetrap buffer has a latency of only one latch delay per cell.
- All cells can be filled, which is exceptional for a buffer based on two-phase handshake signalling.
- Since the design does not contain any special asynchronous elements (such as C-elements), it is more easily understood and accepted by conventional synchronous designers, which are often the designers applying the GALS interfacing circuitry.

4 Pipeline Synchronization

A pipeline-synchronization buffer consists of three sections: a write section, an intermediate section and a read section. The write section synchronizes the input operations with the write clock, while the read section synchronizes the output operations with the read clock. The intermediate section is an asynchronous buffer which serves to decouple the two synchronizing sections. The design of all three sections is based on ripple buffers. The transformation from a ripple buffer into a synchronizing buffer is presented in [10]. This transformation is based on inserting in between two neighbouring cells a component that synchronizes the handshakes with a clock. We define the *maximum clock frequency* of a synchronizing section as the maximum clock frequency at which that section can transfer one data item every clock cycle.

4.1 Synchronization Based on Clock Phases

The synchronizing components in the design presented in [10] are so-called WAIT-components which synchronize handshakes with a clock phase. The basic WAIT-component (called WAIT4), delays the completion of a four-phase handshake until an additional input signal Clk is high. Since signal Clk can go low when the handshake starts, a conflict can occur implying that a so-called arbiter is needed in the design of the component. A *basic arbiter* (also called mutual exclusion element) has two incoming request signals and two outgoing acknowledge signals. An acknowledge signal goes high if the corresponding request signal goes high and it goes low if that request signal goes low. There is one restriction: at most one of the acknowledge signals may be high. Therefore, if one of the acknowledge signals is high and the other request signal goes high, the arbiter should ignore that request until the first request has been served (both request and acknowledge signal are low again). When both request signals go high simultaneously, the arbiter has to decide which acknowledge signal should go high

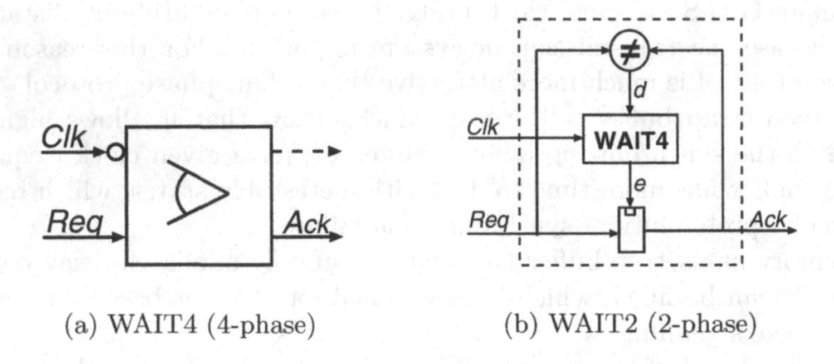

(a) WAIT4 (4-phase) (b) WAIT2 (2-phase)

Fig. 4. Wait components

first. When making the decision the arbiter may reside in a metastable state for an unbounded period of time. Many arbiter designs have been published e.g. in [9,6].

Fig. 4(a) shows the design of the WAIT4-component based on a basic arbiter. Due to the inverter at the clock input, the arbiter is claimed when the clock is low and it is released when the clock is high. Therefore, as long as signal *Clk* is low, the arbiter grants the (virtual) right to proceed to the clock implying that handshakes are paused until *Clk* is high. Signal *Ack* will go high only if both the signals *Req* and *Clk* are high. When the request goes low, the acknowledge signal will also go low .

However, since the mousetrap-buffer uses two-phase handshake signalling, we need a two-phase WAIT-component (called WAIT2). Fig. 4(b) shows the design of such a WAIT2-component based on a WAIT4-component. If the two handshake signals *Req* and *Ack* differ, signal d goes high after which the WAIT4-component waits until signal *Clk* is high before it makes signal e high. Subsequently, the latch becomes transparent and signal *Ack* becomes equal to *Req*. When the handshake signals are equal, the signals d and e become low, and the latch becomes opaque again.

We define the 4TO2-circuit as the additional circuitry needed to convert a WAIT4 into a WAIT2-component. Since this circuit is very similar to the control circuit of a mousetrap cell, it has a similar timing constraint. This design of a WAIT2-component is different from the one presented in [10], which uses two WAIT4-components.

We now give the designs of *phase-synchronizing buffers*, which are buffers that synchronize handshakes with the phases of a clock. For both the write and the read section we present a two-place phase-synchronizing buffer based on WAIT2-components. Fig. 5(a) shows the design of the write synchronizing buffer, which synchronizes the transfer of empty places. Since empty places are transferred by making the acknowledge signal equal to the request signal, the two acknowledge signals are synchronized with the write clock: the left one is delayed until the clock signal is high and the right one until the clock signal is low. Fig. 5(b) shows the design of the read synchronizing buffer, which synchronizes the transfer of

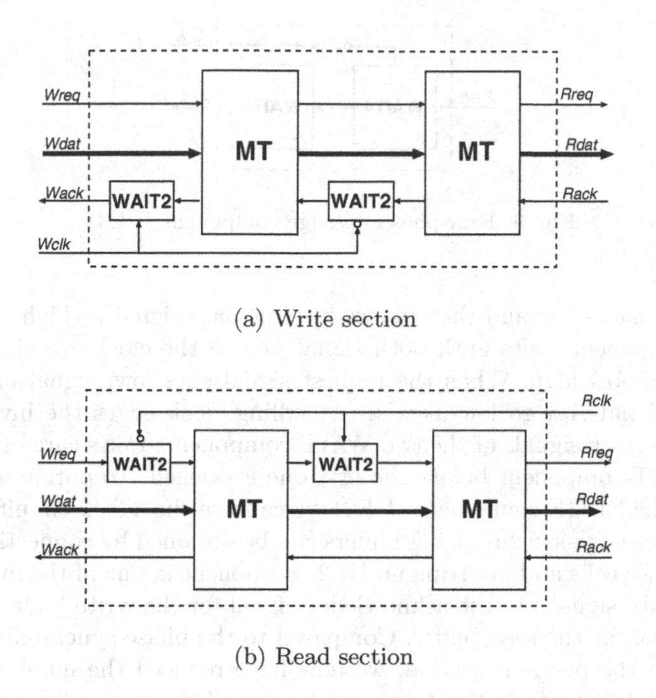

(a) Write section

(b) Read section

Fig. 5. Phase-synchronizing buffers with two mouse-trap cells

data items. Since data items are transferred by changing the request signal, this signal is synchronized with the clock.

We saw that in the free-running buffer the delay in the acknowledge path is δ(XNOR) larger than the delay in the request path (passing empty buckets is the bottleneck). In the design of the write synchronizing buffer we have inserted a WAIT2-component in the acknowledge path, which means that passing empty buckets becomes even slower. The delay difference now becomes δ(XNOR)+δ(WAIT2). Simulations show that full buckets arrive 1.2 ns before the empty ones resulting in a maximum clock frequency of 316 MHz. In the design of the read synchronizing buffer, however, we have inserted a WAIT2 component in the request path leading to a delay difference of $max(\delta(\text{XNOR}),\delta(\text{WAIT2}))$. Since δ(WAIT2) is larger than δ(XNOR), the empty buckets arrive before the full ones. In this case, the time difference is only 0.8 ns resulting in a maximum clock frequency of 403 MHz.

4.2 Synchronization Based on Clock Edges

One can obtain faster synchronizing buffers by applying edge-synchronization instead of phase-synchronization. Fig. 6 shows the UE4-component that synchronizes the upgoing phase of a four-phase handshake with the rising edges of a clock. The component is constructed by connecting two WAIT4-components. The first WAIT4-component waits until both the request signal is high and

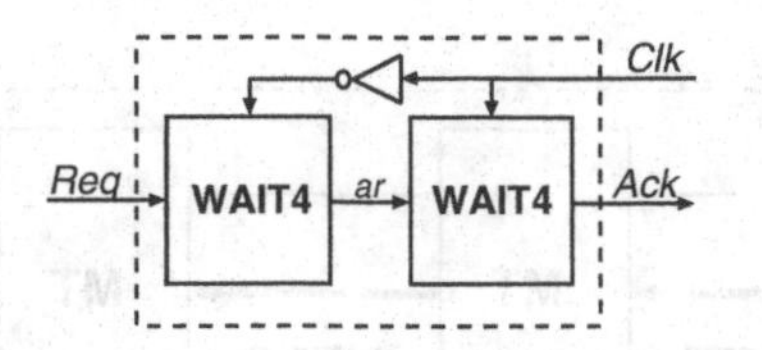

Fig. 6. Four-phase up-edge component (UE4)

the clock signal is low and then makes intermediate signal *ar* high. The second WAIT4-component waits until both signal *ar* and the clock are high and then makes signal *Ack* high. When the request signal goes low, signal *ar* and, consequently, signal *Ack* go low as well. At falling clock edges the inverter delay between the clock signals of the two WAIT-components takes care of closing the second WAIT-component before the first one is opened. To obtain a two-phase version of the EDGE-component (UE2) we can use the 4TO2-circuit again.

Two-place edge-synchronizing buffers can be obtained by connecting two free running buffer cells and inserting an UE2-component in one of the interconnecting handshake signals: in the acknowledge signal for the write buffer and in the request signal for the read buffer. Compared to the phase-synchronizing buffers presented in the previous section, we now have reduced the number of 4TO2-circuits in a two-place buffer from two to one. Therefore, edge-synchronizing buffers are faster than phase-synchronizing ones. The maximum frequency is for the write buffer 478 MHz and for the read buffer 588 MHz.

An edge-synchronizing buffer offers the following advantages when compared to a phase-synchronizing one: it is faster and the timing of the external events in relation to rising clock edges is well-defined (which is very important when we add the circuitry offering the clocked interface).

4.3 Integrated Write Synchronizing Cell

The increase in performance that was obtained by going from clock-phase to clock-edge synchronization came from the fact that for every pair of buffer cells we reduced the number of 4TO2-circuits from two to one. Since there is a strong similarity between the control circuitry of a mousetrap cell and the 4TO2-circuit, it is tempting to find a way to integrate the synchronizing circuit in the mousetrap cell.

Fig. 7(a) shows the design of such an integrated cell in which we use the fact that the latch control signal executes a four-phase handshake. Fig. 7(b) describes the behaviour of the cell. When the cell is empty, it first waits for the next rising clock edge before it makes the latches transparent. Since no circuitry converting a four-phase synchronizer into a two-phase version is needed, the synchronization overhead is again reduced. However, since the cell synchronizes the arrival of empty buckets, it can only be used in the write section.

By including the synchronizing component in the feedback circuit, we have extended the timing requirements with respect to the extended isochronic fork.

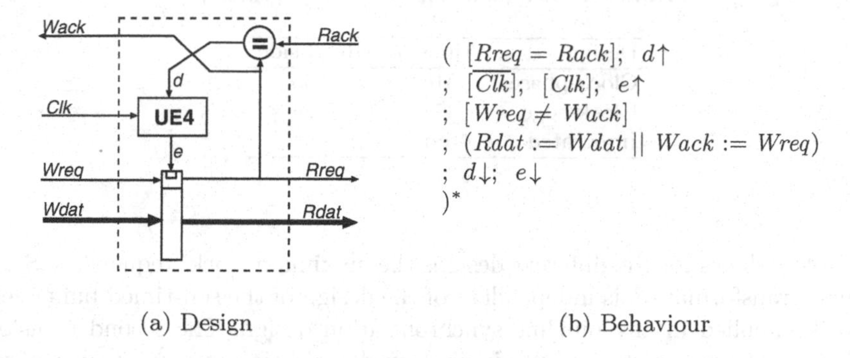

(a) Design (b) Behaviour

Fig. 7. Integrated write synchronizing cell

Since its write neighbour is a free running cell, these extended timing requirements are not automatically fulfilled. One can weaken the timing requirements by inserting an AND-gate with inputs d and e after the UE4-component and thereby speed up the closing of the latch.

The maximum frequency of the write section based on this integrated cell is 581 MHz, which is about 20% faster than the previous design. Note that the new design of the write synchronizing buffer is only marginally slower than the read edge-synchronizing buffer.

5 Concluding Remarks

We presented several designs of a fast buffer that can be used to bridge clock domains in GALS systems. The designs are based on two well-known ideas: pipeline synchronization and mousetrap buffers. We first combined both ideas resulting in a design with *clock-phase* synchronization. In the following two steps we then improved the design.

- In the first transformation we replaced clock-phase synchronization by clock-edge synchronization. This transformation reduces the synchronization overhead. Consequently, the maximum clock frequency is higher (about 50%). Due to an asymmetry in the design of the mousetrap buffer cell, the read section of the synchronizing buffer is about 25% faster than the write section.
- In the second transformation we use the fact that the latch control of the mousetrap buffer cell executes a four phase handshake protocol that starts at the arrival of an empty bucket. Therefore the arrival of empty buckets can be synchronized by incorporating a four-phase synchronizing component in the latch control of the mousetrap cell. Due to this transformation the synchronization overhead is again reduced. However, since this cell only synchronizes the arrival of empty buckets, it can only be used in the write section of the buffer. As a result the new write section offers about the same performance as the read section.

Table 1. Maximum clock frequencies (MHz)

Design	Write side	Read side
Clock-phase	316	403
Clock-edge	478	588
Integrated	581	

Table 1 shows for the different designs the maximum clock frequencies. Since the first transformation is independent of the design of the self-timed buffer cell, it can be applied in any pipeline synchronization design. The second transformation, however, is restricted to pipeline synchronization in mousetrap buffers.

References

1. Kees van Berkel, Ferry Huberts, and Ad Peeters. Stretching quasi delay insensitivity by means of extended isochronic forks. In *Asynchronous Design Methodologies*, pages 99–106. IEEE Computer Society Press, May 1995.
2. Daniel M. Chapiro. *Globally-Asynchronous Locally-Synchronous Systems.* PhD thesis, Stanford University, October 1984.
3. Tiberiu Chelcea and Steven M. Nowick. Robust interfaces for mixed-timing systems with application to latency-insensitive protocols. In *Proc. ACM/IEEE Design Automation Conference*, June 2001.
4. William J. Dally and John W. Poulton. *Digital Systems Engineering.* Cambridge University Press, 1998.
5. Joep Kessels, Ad Peeters, Paul Wielage, and Suk-Jin Kim. Clock synchronization through handshake signalling. In *Proc. International Symposium on Advanced Research in Asynchronous Circuits and Systems*, pages 59–68, April 2002.
6. Alain J. Martin. Programming in VLSI: From communicating processes to delay-insensitive circuits. In C. A. R. Hoare, editor, *Developments in Concurrency and Communication*, UT Year of Programming Series, pages 1–64. Addison-Wesley, 1990.
7. Jens Muttersbach. *Globally-Asynchronous Locally-Synchronous Architectures for VLSI Systems.* PhD thesis, ETH, Zürich, 2001.
8. Miroslav Pečhouček. Anomalous response times of input synchronizers. *IEEE Transactions on Computers*, 25(2):133–139, February 1976.
9. Charles L. Seitz. System timing. In Carver A. Mead and Lynn A. Conway, editors, *Introduction to VLSI Systems*, chapter 7. Addison-Wesley, 1980.
10. Jakov N. Seizovic. Pipeline synchronization. In *Proc. International Symposium on Advanced Research in Asynchronous Circuits and Systems*, pages 87–96, November 1994.
11. Montek Singh and Steven M. Nowick. MOUSETRAP: Ultra-high-speed transition-signaling asynchronous pipelines. In *Proc. International Conf. Computer Design (ICCD)*, pages 9–17, November 2001.
12. Kenneth Y. Yun and A. E. Dooply. Pausible clocking-based heterogeneous systems. *IEEE Transactions on VLSI Systems*, 7(4):482–488, December 1999.

Power-Consumption Reduction in Asynchronous Circuits Using Delay Path Unequalization

Sonia López, Óscar Garnica, Ignacio Hidalgo, Juan Lanchares, and Román Hermida

Departamento de Arquitectura de Computadores y Automática
Universidad Complutense de Madrid, España
{slopezal, ogarnica, hidalgo, julandan, rhermida}@dacya.ucm.es
http://www.dacya.ucm.es

Abstract. The goal of this paper is to present a new technique to reduce power-consumption in circuits with detection of computation completion (asynchronous circuits) by adding delay elements in paths of the circuit. The aim of these new elements is to decrease the number of switchings of those gates placed in the logic cone of the delay element due to the computation completion before those gates receive the new incoming values. We have studied this approach for a set of benchmarks (LGSynth95) and evaluated the trade-off between power-consumption reduction and performance degradation.

1 Introduction

Asynchronous circuits are increasing in popularity among designers because of their advantages. They have no problems associated with clock signal. In addition, circuit performance is the performance of the average case [5], because an asynchronous circuit can start the next computation immediately after the previous computation has completed (completion detection capability). As a result, and particularly in data-dependent computation, asynchronous circuits performance is higher than synchronous [9]. Furthermore, asynchronous circuits may have a better noise and Electromagnetic Compatibility (EMC) properties [3] than synchronous circuits; they are modular and parameter-variation tolerant, and they are insensitive to meta-stability problems.

Finally, asynchronous circuit power consumption is highly dependent on the application. When the clock rate equals to the data rate, little can be gained implementing the circuit as an asynchronous circuit. In any other situation, asynchronous circuits consume less power than synchronous because asynchronous circuits only dissipate energy when and where active. In synchronous circuits, the clock signals cause unnecessary switching activity and the clock signals themselves dissipate a large portion of the total chip power. In [2] there are collected many examples of how asynchronous design helps to increase circuit performance, increase EMC and reduce power-consumption.

J.J. Chico and E. Macii (Eds.): PATMOS 2003, LNCS 2799, pp. 151–160, 2003.
© Springer-Verlag Berlin Heidelberg 2003

However, the totality of these demonstrations and many others are ad-hoc asynchronous implementations. None of them propose a general logic-based technique which uses the key features of asynchronous circuits (essentially, detection of data completion) to reduce the power-consumption.

With these ideas in mind, our main goal is to obtain a method to reduce the power-consumption based in one key decision: the reduction of the switching activity of the circuit slices by delaying the input data of those slices. We delay incoming data exercise all logic gates in the logic cone of the delay element by adding these slices. In the case that computation completion does not require the result from this logic cone then the computation is performed with no switchings in such gates (or at least with a reduction in the number of switchings). This technique obviously increases the delay of some computations, and consequently the average delay of the whole circuit.

The rest of the paper is structured as follows. In Section 2 we present the delay-insensitive circuits and some previous results. In Section 3 we describe this technique, and Section 4 is devoted to present the results we have obtained after building some circuits, from which we draw the conclusions and propose future work.

2 Delay-Insensitive (DI) Circuits

When designing circuits, designers make several assumptions about the wire and gate delays, the number of transitions between emitter and receptor, and the level of integration between control and data. These assumptions define the timing model, the protocol model, and the single-rail or a dual-rail model, respectively.

The delay-insensitive (DI) model [4] assumes unbounded wire and gate delays. In [6], Martin showed that there must be two kinds of data. On the one hand there are calculation data, *d*, which transmit the information involved in the computation. On the other, there are synchronizer data, *s*, which are used to distinguish between two consecutive *d* data. This new kind of data (*s* data) is introduced using dual-rail logic [10]. In the dual-rail approach, the two kinds of data are encoded using two wires, (w_1, w_0). Among the four possible states taken by the dual-rail signals (w_1, w_0), the encoding most used is four-cycle. Hence, the codes $(w_1, w_0) = 00$ or $(w_1, w_0) = 11$ represent the *s* data, $(w_1, w_0) = 01$ represents logic zero and $(w_1, w_0) = 10$ represents logic one.

Finally, DI circuits can be classified according to the communication protocol model. In the four-cycle protocols [8], every computation is executed in two phases (and each phase involves two cycles); a computation phase followed by an synchronism phase. The computation phase consists of the issue of a request and the corresponding issue of an acknowledgment, while the synchronism phase consists of the withdrawal of the request and the corresponding withdrawal of the acknowledgment.

In all experiments we have implemented circuits using the DI paradigm. There are three reasons: first, in this timing model computation completion detection is mandatory, and this feature is also mandatory in order to our technique works properly; second, it is easier, using this timing model, to estimate the power-consumption by counting the number of switchings; and third, it is the most general timing model.

3 Delay Unequalization

In CMOS technologies, the power consumption is highly dependent on the switching activity. In this section we present a new power-saving technique in asynchronous circuits based on switching activity reduction by increasing the delay of some gates of the circuit. We call this technique "*delay unequalization*". But, firstly, let's see why this technique can only be applied to circuits with detection of computation completion.

In circuits without detection of computation completion (i.e. synchronous circuits), output is valid when time gone since inputs were stable goes beyond critical path delay. Once this delay occurs, there is not any new transition in the circuit unless new data are driven in. Thus, in circuits without detection of computation completion, all transitions that could occur in a signal, eventually occur. Those gates whose output values for the current input pattern are identical to the previous values do not switch and do not have hazards. All the remaining gates switch until the final stable state is reached.

On the other hand, in circuits with detection of computation completion (i.e. DI circuits) is not necessary to wait for the critical path delay to be sure that the computation is completed. For this reason, a new computation can start immediately after the previous computation has been completed. The result of the computation can be known even although some signals of the circuit have not reached their final stable value. Once the computation has been completed (end of the computation phase), the synchronism phase is executed and all gates are pre-charged to data s. In this way, all gates that have not switched at this time will not switch for the current inputs at all.

This behavior arises because in digital circuits, for some input patterns, outputs can be solved regardless of the value of some signals (and gates) of the circuit. The technique we propose takes advantage of this fact in order to avoid the switchings in these gates. The key idea is to increase logic transition delay in some gates of the circuit, in such a way that the gates in their logic cones will not receive the new values before the computation has finished.

Figs. 1(a), 1(b) and 1(c) exemplify our technique using a very simple circuit. We are going to study the behavior of this circuit and the number of switchings in two different scenarios. Firstly, let suppose a gate delay of 1ns for all gates in the circuit and an input vector $(a, b, c) = (0, 0, 0)$ (See Fig. 1(b)).

Under this circumstance, output f will be 0 after 2ns regardless of the values of a, b, w_1 and w_2. At time 1ns gates G_1 and G_3 will switch and gates G_2 and G_4 will do it in the following nanosecond. Thus, eventually, all gates will switch (for a total value of 4 transitions), even although it only were necessary the switchings of the gates G_3 and G_4.

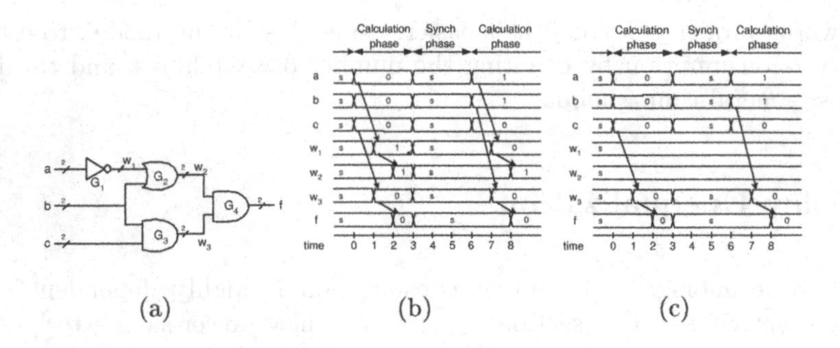

Fig. 1. (a) Example circuit. Number 2 indicates dual-rail logic in all gates and wires of the circuit, (b) Chronogram with non-delayed gates and (c) Chronogram with delayed gates.

Let's now study the second scenario (See Fig. 1(c)). The delay of the gate G_1 is increased up to 3ns and the input vector is as previous. The circuit output is asserted to 0 at 2ns despite the gate delay increment. At this time the circuit driven by signal f detects the value of this signal and resets all gates of the circuit to value s (at time 3ns). As can be observed in Fig. 1(c), gates G_1 and G_2 do not switch and there are switching savings.

Obviously, this approach is only feasible for circuits with detection of computation completion. In circuits without this capability, it is necessary to wait for the critical path delay and then capture the output value.

In the following section we will present the results we have obtained after applying this method to a set of logic benchmarks.

4 Experimental Results

We have measured the switching activity in several well-known logic synthesis benchmarks (LGSynth95[1]) implemented as DI circuits. In order to build the DI implementation of each benchmark we have followed the design flow:

1. Synthesize the circuit using Synopsys and a reduced library which contains gates AND, OR and INV. The delay of each gate is 1ns.
2. Substitute all gates by their DI counterparts. In other words, substitute each gate by a gate with the same functionality but capable of working with dual-rail signals. In order to estimate power-consumption, each DI gate has been described with the capability of counting logic transitions on their outputs.
3. Add the signals which implement the handshake between the circuit and the environment.

The purpose of this transformation is to add the detection of computation completion to the original description of the circuit. So, at this point, each circuit

[1] http://www.cbl.ncsu.edu

has been implemented as a netlist using a reduced library of dual-rail gates (AND2, OR2, INV) capable of counting the total number of switching during the simulation.

Power-consumption has been measured as the number of transitions during the logic simulation. This method can introduce two bias in the measure:

1. [1] states that logic simulations overestimate the power-consumption due to these simulations do not take into account the glitch degradation. This degradation reduces the number of hazards in the circuit. However, in DI circuits there are not glitches so logic simulations do not overestimate the power-consumption due to hazards.
2. The measure of the switching activity has been performed for a reduced set of input patterns. In order to unbiasingly measure the power-consumption, the circuit behavior for these inputs must be representative enough of the general behavior. Switching activity inside a circuit is highly input pattern dependent; hence simulation results are directly related to the specific input patterns used. The two main objectives when selecting a set of input patterns are to reproduce an average behavior of the switching activity and to use a number of patterns small enough in order to limit the cost in computational resources. To select an input pattern set, we have chosen a statistical approach, valid for combinational circuits. In this approach we select a random sample which goodness is given by two parameters: confidence and percentage error. Thus, with a set of 1000 random input patterns we can guarantee that results are within 0.1% of error, with a confidence of 99.9% for every circuit excepting circuit apex2 which error is 0.5% with a confidence of 90% [7]. In order to be sure about the size of the input set we have conducted several simulations with input set of 1000 vector for the circuit apex2. In all cases, the empirical results have a variation coefficient lower than 0.24%, which is in accordance with theoretical prediction. In addition, the same number of input vectors has been used in circuits of equivalent complexity [1].

The following issue to be resolved is to find out the gate for which the maximum saving (if any) is achieved. Fig. 2 illustrates the power-consumption saving when the delay of just one gate of the circuit apex2 is incremented up to 10ns and the delay of the rest of gates remains at 1ns. The saving is measured versus the power-consumption of the original (non-delayed gates) DI circuit.

The range of the power-saving varies from -7.95% to 4.95%. The negative values mean that, for such modification, the power-consumption versus the unmodified implementation of the circuit has increased. This occurs because in the original implementation there are computations (the fastest ones) for which the last gates in the critical path do not switch, but in the modified implementation the delayed gate is placed in such a way that under the same computations all gates in the critical path switch. At current time, we have not found out a heuristic to select such a gate, so we are going to modify logic transition delay in a per gate basis (one gate in each simulation) and select the optimal gate. In

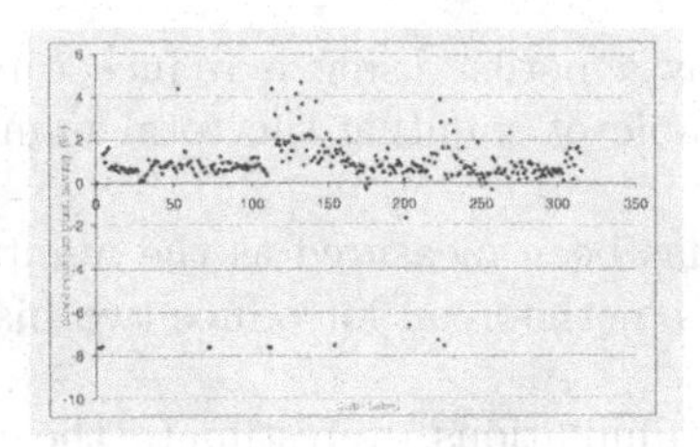

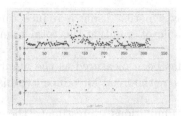

Fig. 2. Power-consumption reduction in circuit apex2. X-axis represents the label of the gate which is modified (from 1 to 315). Y-axis represents the power-consumption saving when unmodified and modified implementations are compared.

usual conditions, gate delay is 1ns. We have studied five different scenarios for the delay of the modified gate: non-modified, gate delayed 10ns, gate delayed 15ns, gate delayed 20ns, and two gates delayed 10ns. Once all gates have been checked, we focus in the gate which provides a lower number of transitions in the whole circuit and analyze the switching activity and the average circuit delay. Fig. 3 illustrates the power-consumption saving using this approach and Fig. 4 presents the average and worst-case delays.

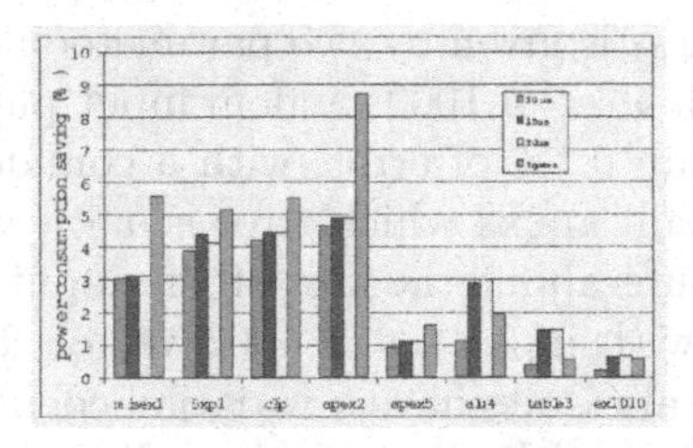

Fig. 3. Power-consumption reduction when modified implementations are compared versus unmodified ones. Circuits are ordered in increasing size.

As per Fig. 3, the power-saving using this technique is highly dependent on circuit topology. Indeed, power-saving ranges from 0.21% for ex1010 to 4.67% for apex2 when just one gate is modified. The reason for this variation is ultimately related to the distribution of the path delays in each circuit. Fig. 5(a) illustrates the distribution of delays for the unmodified and modified implementations of the circuit apex2. This circuit has a high variability in the path delays, and the fastest path is 3 times faster than the slowest one. In addition, 49.5% of the computations require the smallest delay. For these reasons this circuit is very suitable for asynchronous implementations (big differences between the fastest and the slowest computation delay). The same figure illustrates the equivalent distribution when the optimal 10ns-gate has been included in the circuit. This figure shows that only the slowest computations have been affected by this modification, and almost the totality of the computations hold their original delays

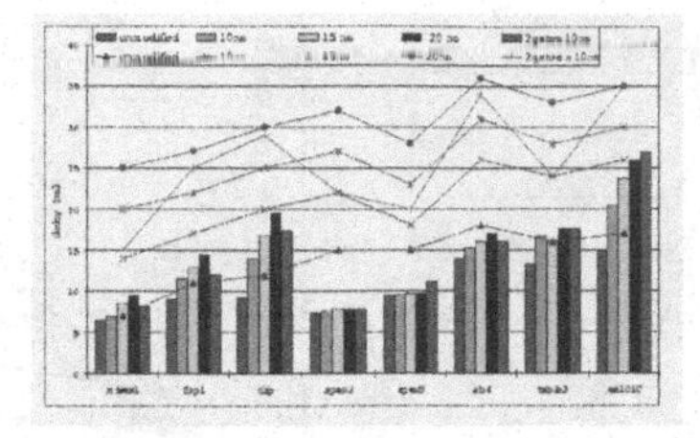

Fig. 4. Average (boxes) and worst-case (lines) delays when modified implementations are compared versus unmodified ones. Circuits are ordered in increasing size.

(97.2% of the computations). However, there are a power-saving in the 73.6% of the computations. The reason is that the gates in those modified paths do not switch because the incoming data do not reach them before the end of the computation. Fig. 6 presents the average power-saving for each delay. The fastest paths are the ones for which the highest power-consumption saving is achieved (around 5.1% of saving in 49.5% of the total computations).

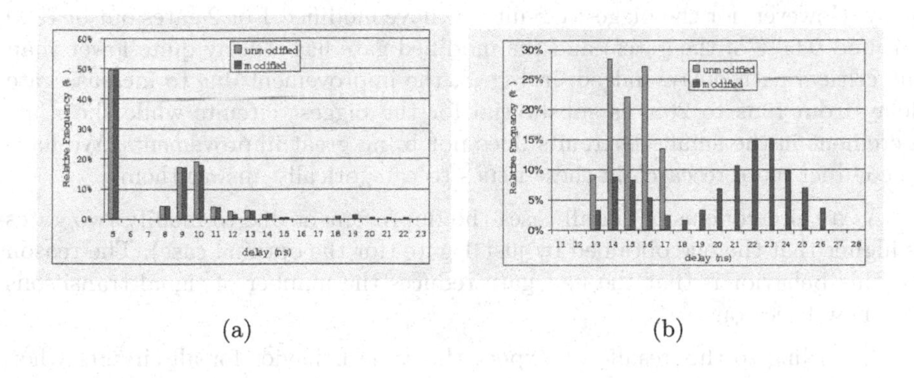

Fig. 5. Delay distribution for unmodified and modified implementations of circuits (a) apex2 and (b) ex1010.

Fig. 5(b) illustrates the distribution of delays for the unmodified and modified implementations of the circuit ex1010. This circuit has a narrow distribution of delays, in such a way that the fastest computation is just 1.44 times faster than the slowest one. In addition, 77% of the computations involve the medium delays (no fastest or slowest ones). For these reasons, this circuit is not very suitable for asynchronous implementations. The same figure illustrates the equivalent distribution when the optimum 10ns-gate has been included in the circuit. Due to the nature of the circuit, all gates have been affected by this modification and the new delay distribution although a bit wider it is not wide enough to achieve higher power savings.

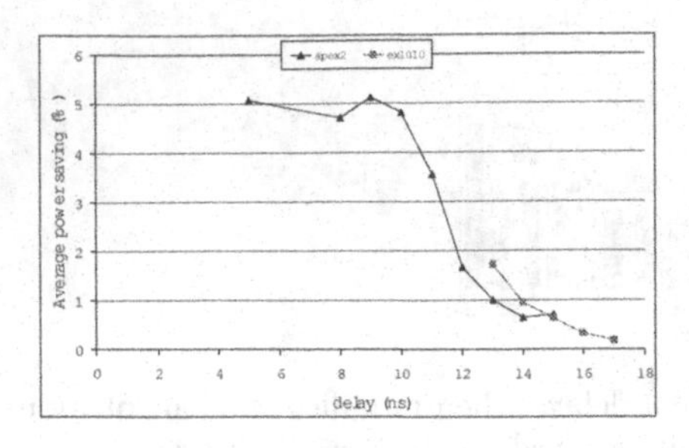

Fig. 6. Average power-saving distribution for each path delay.

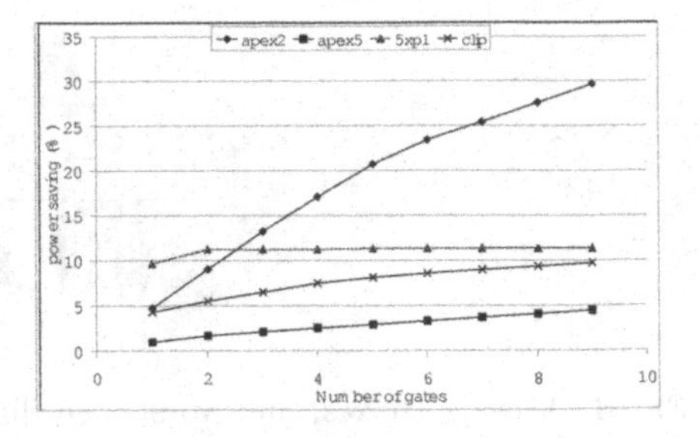

Fig. 7. Power saving for the number of modified gates in the range between 1 and 9.

At this point, we wonder if this feature is typical of big circuits or it is just a matter of the number of gates we have modified and the delay we have used. In the smallest circuits, we have modified 1 or 2 gates out of 300 (around 0.3% of the gates) and the modified gate has a delay in the order of the critical path delay. However, for the biggest circuits we have modified 1 or 2 gates out of 1800 (around 0.06% of the gates) and the modified gate has a delay quite lower than the critical path delay. Indeed, in Fig. 3, the improvement due to increase gate delay from 10ns to 20ns is substantial for the biggest circuits while the same increment in the smallest circuits does not bring great improvements. We have to conduct more research in these issues to categorically answer them.

As a third conclusion, in all cases the improvement due to modify two gates is higher that the one obtained by just 1 gate (for the optimal case). The reason for this behavior is that the new gate reduces the number of signal transitions in a new logic cone.

According to this result, we expect the same behavior for all circuits when the number of modified gates grow. The reason lies in the fact that if all circuit gates would be incremented up to 10ns, the power-consumption of the new circuit would be equals to the one of the unmodified circuit. So the power-consumption saving has a maximum between 0 and maximum number of circuit gates. Following this approach, we have selected for every circuit up to 9 gates using a greedy-like algorithm (see Algorithm 1). Fig. 7 presents the results for circuits misex1, apex5, clip and apex2.

In all cases, the improvement in the power-consumption saving increase when compared to the implementation with just one modified gate. The best circuit is apex2 (around 30% of saving) and the worst is apex5 (around 5% of saving). In addition, except for circuit 5xp1, the circuits have not reached the maximum power-saving. So there is still room for higher power-consumption savings. Circuits alu4, table3 and ex1010 were being simulated at the time of writing this paper. Even although we are using a greedy algorithm to find out the optimal

Algorithm 1 Greedy-like algorithm to search the best gates.

$\{selected\_gates\} = \emptyset$;
$C \leftarrow original\ circuit$;
for $(i = 0; i < 10; i++)$ **do**
  for $(g \in \{circuit\_gates\} - \{selected\_gate_i\})$ **do**
    $D_g \leftarrow increase\_delay(C, g)$;
    $p_g \leftarrow estimate\_power(C_g)$;
  end for
  $selected\_gate_i \leftarrow \{k\ :\ p_k = \max(p)\}$;
  $C \leftarrow D_{selected\_gate_i}$;
  $\{D_g\} \leftarrow \emptyset$;
  $\{p_g\} \leftarrow \emptyset$;
end for

gate, the method requires a lot of computational resources for these circuits (> 1day per circuit running on a Pentium IV at 2.5GHz powered by Linux).

5 Conclusions and Future Work

In this paper we have proposed a new technique (called "*delay unequalization*") to reduce power-consumption in circuits with detection of computation completion. This technique is based in adding delay elements in some gates of the circuit in such a way that gates in the logic cone of the modified gates do not switch for the majority of the computations.

We have applied this technique to a set of benchmarks and evaluated the power-consumption reduction. In all cases, this technique brings power savings with values ranged between 0.21% and 4.67% when just one gated is modified and between 5% and 30% when up to 9 gates are modified. In addition, we have found out that the reason for this variation relies on the distribution of computation delays of the circuits. This approach applies with better results for those circuits in which the new delayed gate modifies only a sub-set of all path delays, and applies with worse results when the topology of the circuit do not hold this premise.

As a future work, we have to better understand the trade-off between power-consumption and performance and attempt to find out non-optimal power-saving solutions without degradation of the worst-case delay. Also, we have to apply this technique to bigger circuits in order to determine its suitability in such cases. In addition, we have to define a heuristic to find the optimal gate without exploring the whole set of gates. Finally, we have to study the applicability of this technique to asynchronous circuits which implement other power-reduction techniques (i.e. PDI circuits).

Acknowledgement. This research has been supported by Spanish Government Grant number TIC 2002/750.

References

1. C. Baena, J. Juan-Chico, M.J. Bellido, P. Ruiz de Clavijo, C.J. Jiménez, and M. Valencia. Measurement of the switching activity of cmos digital circuits at gate level. In Bertrand Hochet, Antonio J. Acosta, and Manuel J. Bellido, editors, *Power and Timing Modeling, Optimization and Simulation (PATMOS)*, pages 353–362, 2002.
2. C.H. (Kees) van Berkel, Mark B. Josephs, and Steven M. Nowick. Scanning the technology: Applications of asynchronous circuits. *Proceedings of the IEEE*, 87 (2): 223–233, February 1999.
3. Kees van Berkel, Ronan Burgess, Joep Kessels, Ad Peeters, Marly Roncken, and Frits Schalij. Asynchronous circuits for low power: A DCC error corrector. *IEEE Design & Test of Computers*, 11(2):22–32, Summer 1994.
4. Scott Hauck. Asynchronous design methodologies: An overview. *Proceedings of the IEEE*, 83(1):69–93, January 1995.
5. D. Kearney. Theoretical limits on the data dependent Performance of asynchronous circuits. In *Proc. International Symposium on Advanced Research in Asynchronous Circuits and Systems*, pages 201–207, April 1999.
6. Alain J. Martin. Asynchronous datapaths and the design of an asynchronous adder. *Formal Methods in System Design*, 1(1):119–137, July 1992.
7. F.N. Najm. A survey of power estimation techniques in vlsi circuits. *IEEE Transactions on VLSI Systems*, 2(4):446–455, December 1994.
8. Takashi Nanya. Challenges to dependable asynchronous processor design. In Tsutomu Sasao, editor, *Logic Synthesis and Optimization*, chapter 9, pages 191–213. Kluwer Academic Publishers, 1993.
9. Steven M. Nowick, Kenneth Y. Yun, and Peter A. Beerel. Speculative completion for the design of high-performance asynchronous dynamic adders. In *Proc. International Symposium on Advanced Research in Asynchronous Circuits and Systems*, pages 210–223. IEEE Computer Society Press, April 1997.
10. Marco Storto and Roberto Saletti. Time-multiplexed dual-rail protocol for low-power delay-insensitive asynchronous communication. In Anne-Marie Trullemans-Anckaert and Jens Sparsø, editors, *Power and Timing Modeling, Optimixation and Simulation (PATMOS)*, pages 127–136, October 1998.

New GALS Technique for Datapath Architectures

Miloš Krstić and Eckhard Grass

IHP, Im Technologiepark 25, 15236 Frankfurt (Oder), Germany
{krstic,grass}@ihp-microelectronics.com

Abstract. A novel Globally Asynchronous Locally Synchronous (GALS) technique applicable to datapath architectures is presented. It is based on a request-driven operation of locally synchronous blocks. Inactivity of the request line is detected with a special time-out circuitry. When time-out occurs, clocking of the locally synchronous block is handed over to a local ring oscillator. Based on this concept, a practical hardware implementation of an asynchronous wrapper is proposed. The simulation results presented demonstrate the potential and performance of the proposed GALS architecture.

1 Introduction

Designing complex wireless communication systems is a very challenging task. Today's industrial trend is focussed on cost reduction by system-on-chip implementation with integration of digital and analog processing parts on a single chip. Digital signal processing algorithms can be implemented with general DSP processors or dedicated datapath oriented hardware. If the system is implemented as a special datapath architecture, it usually contains complex circuit blocks that perform sophisticated arithmetic or trigonometric operations. Often those blocks have point-to-point communication using localised or distributed control. Typically, the communication between those blocks requires high data rates. In many cases, periods of high data throughput are followed by periods of long inactivity, thus causing 'bursty' activity patterns.
The enormous complexity of the modern ASICs for wireless applications creates several problems for system integration and design. One crucial problem is the construction of the global clock tree. This may lead to a substantial slow-down of the design process. The synchronous global clock also generates increased electromagnetic interference (EMI), which can lead to integration problems between analog and digital circuits. Furthermore, power demands of such complex systems are usually high and hence, power consumption must be minimised. This can partly be achieved by using known methods for minimisation and localisation of switching power like clock gating or asynchronous design. However, clock gating makes the design of the clock tree even harder. An additional problem is the integration of prelayouted hardware IP-cores from different vendors for specific process technologies. Those blocks are individually tested but embedding them into the remaining circuitry is often not trivial.

J.J. Chico and E. Macii (Eds.): PATMOS 2003, LNCS 2799, pp. 161–170, 2003.
© Springer-Verlag Berlin Heidelberg 2003

Recently, Globally Asynchronous Locally Synchronous (GALS) architectures are proposed as an elegant solution for many problems described above [1, 2, 3, 4]. However, for datapath organised circuits in the area of mobile communications, the known implementations of GALS do not fully utilise the potential of this technique. In this paper we will introduce a novel GALS concept and architecture, optimised for ASIC datapath structures which are widely used in wireless communication systems.
The paper is structured as follows: in Section 2 the current state-of-the-art is given. In Section 3 and 4 we introduce the principle mechanism of our approach, and explain some details of it's implementation. Finally, in Section 5 some results are presented and conclusions are drawn in Section 6.

2 Discussion of Existing Techniques

Several existing approaches allow the integration of different hardware blocks on one chip without global synchronicity. In GALS – Globally Asynchronous Locally Synchronous systems local blocks are functioning fully synchronously. Those synchronous circuit blocks are driven by local clocks generated by stoppable ring oscillators. Data transfer between different blocks normally entails stopping of the local clocks in order to avoid metastability problems. Many years after the first GALS proposal [5], this idea is reactivated and a working GALS architecture is described in [1]. However, this solution results in poor performance for data-transfer intensive multi-channel systems. Further improvements were made in [2]. For the first time the term 'asynchronous wrapper' is used for an asynchronous interface surrounding locally synchronous modules. Such an asynchronous wrapper consists of input and output ports and local clock generation circuitry. Input and output ports can be implemented either as demand or poll ports. The locally synchronous part can be designed in standard fashion. This concept is further elaborated in [3], and better arbitration of concurrent requests is achieved. In [4] another serious GALS problem is tackled – the generation of stable local clocks with a determined frequency. It is suggested to self-calibrate local clocks by introducing an additional low frequency stable clock as a reference.
All solutions described so far are oriented towards a very general application. Consequently, they are not optimised for specific systems and environments. Power saving mechanisms cannot be implemented easily. The demand port of the wrapper, which performs the clock-gating function, is controlled by the synchronous part of the circuit. This means, a standard synchronous design cannot be easily deployed in such GALS environment. Additional functions have to be incorporated to facilitate interaction with the asynchronous wrapper. The synchronous designer has to identify and signal a state, which indicates the lack of input data and the demand for new input tokens. This can be a difficult task in some cases. Furthermore, circuits are generally targeted for relatively low data rate applications. Additionally, in some cases it is necessary to insert an additional FIFO between GALS blocks to prevent performance deterioration due to local clock stretching [6].
Moreover, normally ring oscillators constructed from standard cells are deployed. They cannot generate a stable clock signal at a defined frequency. However, fre-

quency fluctuations complicate data-transfer with data producers or consumers such as ADCs and DACs operating at a predefined and stabilised clock. In communication systems it is nearly always necessary to communicate with blocks such as ADC or DAC. One way to resolve this problem is to over-constrain locally synchronous blocks to work at a higher local frequency. Another solution could be a mechanism to calibrate the local clock. However, this would further complicate the design and cause additional cost.
We believe that the technique presented in this paper will, for datapath architectures, result in better performance than known implementations.

3 Novel GALS Architecture

Our work is mainly motivated by the aim to avoid unnecessary transitions and the attendant waste of energy during data transfer between GALS blocks. A secondary issue is to achieve high data throughput with low latency.
It appears plausible that for a constant input data stream the GALS blocks should operate in quasi-synchronous mode. This way no unnecessary transitions are generated. A possible solution to achieve a quasi-synchronous operation is to operate GALS blocks in a request-driven mode, i.e. to synchronise the local clock generators with the request input signal. However, for bursty signals, when there is no input activity, the data stored inside the locally synchronous pipeline has to be flushed out. This can be achieved by switching to a mode of operations in which the local clock generator drives the GALS block independently. To control the transition from request driven operation to flushing out data, a time-out function is proposed. The time-out function is triggered when no request has appeared at the input of a GALS block, but data is still stored in internal pipeline stages. It is also conceivable to switch from a request driven (push) operation to an acknowledge driven (pull) operation in this situation. However this idea was rejected since it would be difficult to deal with isolated tokens inside the synchronous pipeline. In particular, if the pull operation doesn't result in a token being propagated to the primary output, an internal oscillator would still be required. A much more sophisticated control would be required to facilitate this approach, which is in principle feasible.
Conceptually, in our approach the locally synchronous circuit is designed to be driven both by the incoming request as well as the local clock signal. The request input signal is generated from the asynchronous wrapper of the previous block. It is synchronised with the transferred data, and can be considered as a token carrier. When there is no incoming request signal for a certain period of time, the circuit enters a new state where it can internally generate clocks using a local ring oscillator. The number of internally generated clocks is set to match the number of cycles needed to empty the locally synchronous pipeline. When there is no valid token in the synchronous block, the local clock will stall and the circuit remains inactive until the next request transition, indicating a fresh data token occurs at the input.
More complex and demanding is the scenario when after time-out and starting of the local-clock generation, a new request appears before the synchronous pipeline is emptied. In this case it is necessary to complete the present local clock cycle to pre-

vent metastability at data inputs. Subsequently it is possible to safely hand over clock generation from the local ring oscillator to the input request line. To deal with this situation it is necessary to implement additional circuitry to prevent metastability and hazards in the system.
The proposed architecture has several potential advantages. As with all GALS systems, no global clock-tree is required. The clock signal is generated by 'multiplexing' the local clock and the input requests. Due to the request driven operation the local clock usually does not need to precisely match the frequency of the global clock or the datarate. Hence, there are little constraints for the design of the ring oscillators. Locally synchronous blocks do not have to be 'overconstrained' and can be designed for clock frequency f_c as with standard synchronous designs. Using the proposed concept the circuit immediately responds to input requests and hence unnecessary delays are avoided. A token-flow approach, often deployed in a synchronous environment anyway, seems to be a more natural style to design synchronous blocks for GALS application than the design rules proposed for GALS architectures in [3]. The suggested architecture offers an efficient power-saving mechanism, similar to clock gating. A particular synchronous block is clocked only when there is new data at the input, or there is a need to flush data from internal pipeline stages. In all other cases there is no activity within this block. In a complex system, designed with the proposed GALS technique, locally synchronous modules are driven by independent clocks. These clocks will have arbitrary phases due to varying delays in different asynchronous wrappers. During local clock generation they have even different frequencies. In this way EMI and the attendant generation of substrate and supply noise can be reduced.
The target applications of the proposed GALS technique are datapath architectures, often deployed for baseband processing in communication systems. For designing 'joins' and 'forks' between asynchronous wrappers, standard asynchronous methods can be employed. From a top-level view our architecture is equivalent to an asynchronous design with coarse granularity. In contrast to 'normal' asynchronous circuits, a stage is not restricted to a single register with combinational logic, but can be a big and deeply pipelined synchronous block. Standard GALS implementations as advocated in [3] do not well support communication with ADCs and DACs running at a fixed stable clock f_c. The local clock signals within those GALS blocks would have to be tuned to a higher frequency than f_c, or large interface FIFOs must be used. With bursty data transfer it can be shown that stretching of the local clock at every single data transmission will significantly diminish performance. We think that our approach is better suited for the particular systems described above.

4 Implementation of Novel GALS Architecture

A detailed block diagram of a proposed GALS block is shown in Fig. 1. The asynchronous wrapper consists of input and output port(s), pausable clock generator, time-out generator, and a clock control circuit. From Fig. 1 it can be seen that the *INT_CLK* signal is generated by an OR operation of signals *REQ_INT* and *LCLKM*.

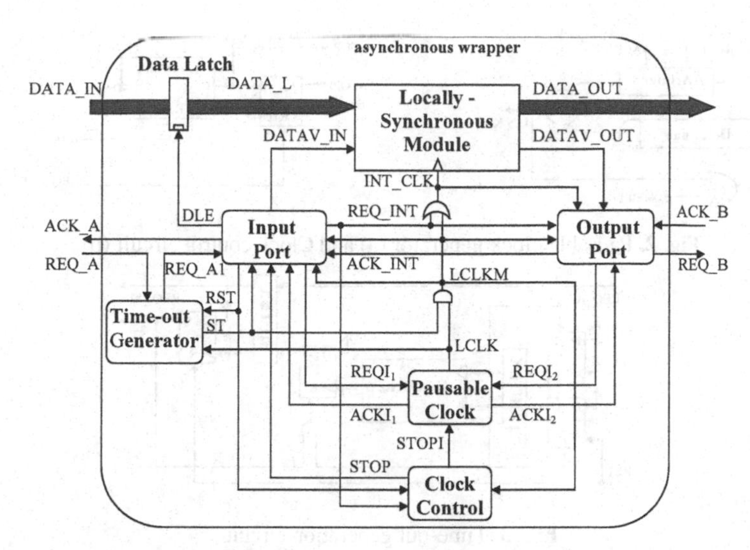

Fig. 1. Block diagram of proposed Asynchronous wrapper

Signal *REQ_INT* is generated by the input port. *LCLKM* is the output *LCLK* from the pausable clock generator, gated with signal *ST* as shown in Fig. 1. *REQ_INT* and *LCLKM* are mutually exclusive, so hazard-free operation of *INT_CLK* is preserved. Input data are buffered in a transparent latch. This is needed to prevent metastability at the input of the locally synchronous block. The operation of this data latch is controlled by signal *DLE* with the latch being transparent when *DLE* is asserted. Signal *DLE* is asserted after a transition of the local clock, when previously latched data is already written into the register stage of the locally synchronous module. The input of the locally synchronous block is assumed to have no additional registers. In the following all blocks shown in Fig. 1 will be described in detail.

Pausable clock: This local clock generator is implemented as a ring oscillator, and is shown in detail in Fig. 2a. Different to the implementation in [3] it is possible to stop the clock with an additional signal *STOPI* which is activated in two cases: Firstly, immediately after reset, to prevent the activation of the clock oscillator before the first request signal arrives at the local block. Secondly, after time-out, i.e. when the number of local clock cycles is equal to the number of cycles necessary to output all valid data tokens stored inside the pipeline. In this situation, the local clock is stopped to avoid unnecessary waste of energy.

Clock control: Pausable clock operation is controlled by a block named *Clock control* as shown in Fig. 2b. This block generates two output signals: *STOPI* and *STOP*. Signal *STOP* is a control signal for the asynchronous finite state machine (AFSM) in the input port. When *STOP* is asserted, the local clock will be halted. This signal is activated when the counter, clocked with the local clock, reaches the number equal to the depth of the synchronous pipeline. Signal *STOPI* is directly used as a control signal for the ring oscillator. The D-flip-flop is used to hold this signal in asserted state until a new request arrives.

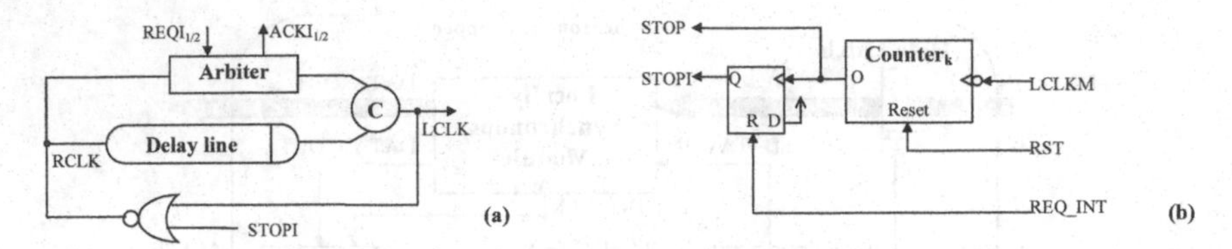

Fig. 2. Pausable clock generator (a) and Clock control circuit (b)

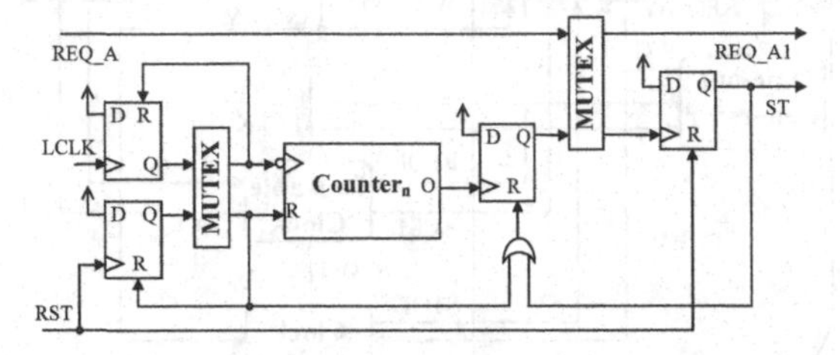

Fig. 3. Time-out generation circuit

Time-out generator: The time-out generation unit is implemented with a small number of hardware components as shown in Fig. 3. Nevertheless, it has to support a relatively complex operating scenario. Generally it consists of one counter which counts the number of negative edges of the local clock. This counter is designed as a standard synchronous counter. When reaching its final value it generates a time-out signal. The counter's resets signal *RST* is activated once during every input port handshake. *RST* and the clock signal are not a priori mutually exclusive, potentially giving rise to metastable behavior of this counter. To avoid metastability, one mutual exclusion (mutex) element and two flip-flops must be inserted. This resolves the simultaneous appearance of the rising clock edge with the falling edge of the reset signal. Another situation which must be dealt with is the appearance of an external request signal simultaneously with the time-out signal (*REQ_A* and *ST*, respectively in Fig. 3). This condition may violate the assumed burst mode operation and cause erroneous operation of the input port AFSM. In order to resolve this potential problem, an additional mutex and two flip-flops are added. To make the line *REQ_A1* available for most of the time, the time-out signal entering the mutex circuit should be active for only a short period of time. This behavior is achieved using two flip-flops as shown in Fig. 3.

Input port: The input port mainly consists of an input controller plus some supporting circuitry shown in Fig. 4a. This input controller must guarantee safe data transfer and is implemented as an AFSM working in burst mode. The specifications of the input controller is given in Fig. 5a. In the normal mode of operation it just reacts to the input requests and initiates 'clock cycles' (signal *REQ_INT*) for every incoming request. If there is no activity on input lines for a certain period of the time, signal *ST* is activated (time-out). Now the circuit waits for two possible events. The first one is the completion of the expected number internal clock cycles (indicated by signal *STOP*). This would bring the AFSM back into its initial state. The second possible event is the appearance of an input request in the middle of the pipeline flushing

process. The already started local cycle must be safely completed and clock generation control must be handed over to the input requests.
The additional circuitry in Fig. 4a is to disable, during local clock generation, the acknowledge signal *ACK_INT* generated by the output port. *ACK_INT* is enabled once again after transition from local clock generation to the request-driven operation by activation of signal *ACKEN*. Signal *ACKC*, which is fed to the input port AFSM as shown in Fig. 4a, is also used for generation of the control signal *DLE* for the input data latch.
Output port: A block diagram of the output port is shown in Fig. 4b. It mainly consists of an AFSM output controller and two D-type flip-flops. The flip-flops are used to condition the signals indicating 'data output valid' *(DOV)* and 'not valid' *(DONV)* for use in the AFSM. Since the AFSM is event driven and not level driven, the level based signal *DATAV_OUT* is transformed into the two event based signals *DOV* and *DONV*, synchronised with the internal clock *INT_CLK*. The AFSM, with the specifications given in Fig. 5b, can be excited from two mutually exclusive sources: from the internal request *REQ_INT* and from the local clock *LCLKM*. If there is no valid data at the output, any request indicated either by activation of *REQ_INT* or *LCLKM* is immediately acknowledged by activating signal *ACK_INT*. When output data is to be transferred to the next GALS block, an output handshake (signals *REQ_B* and *ACK_B*) must be performed. In this case the internal handshake (signals *REQ_INT* and *ACK_INT*) must be coupled with the output handshake. When both *DOV* is activated and signal *LCLKM* is asserted by the local clock, the local clock must be stretched using signals $REQI_2$ and $ACKI_2$, until an output handshake is performed. This will prevent a new clock cycle before completion of the output data transfer. The implemented circuit is based on worst-case assumptions. It allows that in the middle of flushing the pipeline, a new request at the input port may appear. This complicates the structure of the system considerably and adds additional control and arbitration circuitry. Thus, for certain applications the asynchronous wrapper can be simplified.
Our proposed GALS technique has several features so far not reported in the available literature. A new concept of using the asynchronous handshake as clock initiation for the local synchronous blocks is proposed and implemented. A new time-out function for detection of input inactivity is described in detail. We have developed a specific circuit for defining the time-out period. This circuit is hazard- and metastabillity free and is used both, for generation of the pausable clock as well as the time-out signal.

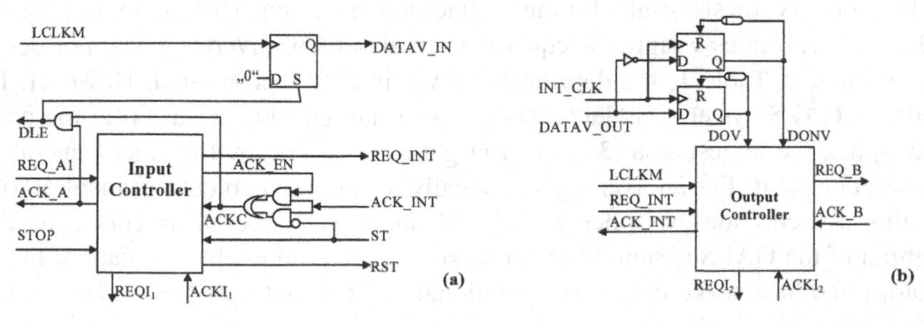

Fig. 4. Input (a) and Output (b) port structure

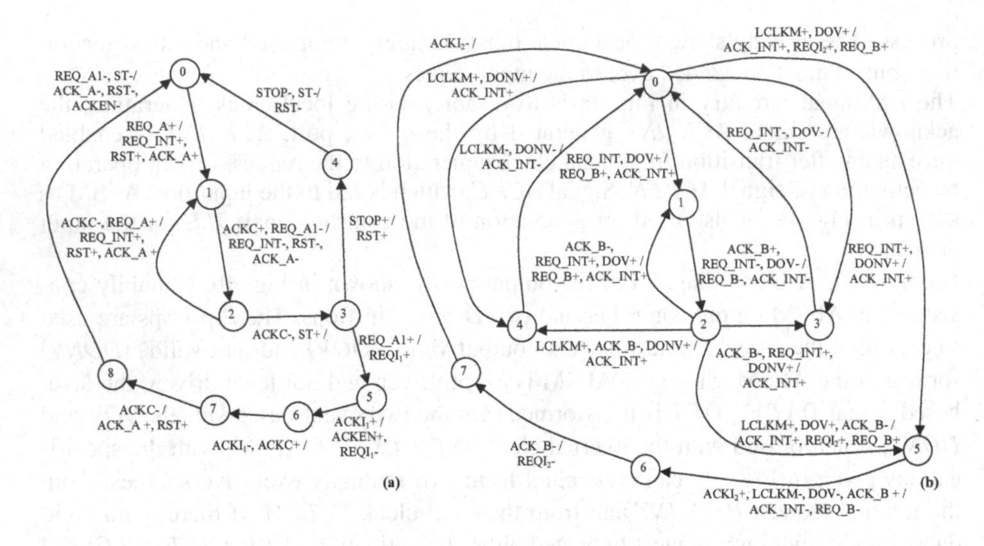

Fig. 5. Specification of the input (a) and output (b) controller AFSM

5 Results

Our design flow for developing this GALS system was a combination of a standard synchronous design-flow with specific asynchronous synthesis tools. All synchronous circuits are designed in VHDL and synthesised for our in-house 0.25 μm SiGe:C BiCMOS standard cell library using Synopsys. All asynchronous controllers are modeled as asynchronous finite-state machines and subsequently synthesised using the 3D tool [7]. 3D guarantees hazard-free synthesis of extended and normal burst-mode circuits. The logic equations generated by 3D are manually translated into structural VHDL and subsequently synthesised. All circuits are fitted with a reset logic, necessary for initialisation. Behavioral and postsynthesis simulations are performed using a standard VHDL-Verilog simulator. The local clock generator is designed for a frequency of 20 MHz. For higher frequencies it will occupy less silicon area.

In Table 1 some synthesis results for our technology are given. The circuit area for the complete asynchronous wrapper is equivalent to about 530 inverter gates. For deriving the figures in Table 1, the data latch shown in Fig. 1 is omitted. However, for designing a GALS system this latch has to taken into consideration. From the complete wrapper, the largest area (340 inverter gates) is needed by the clock generation and time-out circuit. For an average size locally synchronous block of about 50,000 gates, this asynchronous wrapper would add about 1 % overall silicon area. The throughput of the GALS system after synthesis was determined by simulation. In our technology the handshake circuit is operational up to about 100 MSps. This is fast enough for moderate speed applications.

Table 1. Asynchronous wrapper circuit area

Hardware block	Cell area (μm²)	Number of inverter gates
Input Controller	5951	98
Output Controller	6073	100
Local Clock Generation and Time-out Detection	20495	336
Asynchronous Wrapper (Total)	32519	534

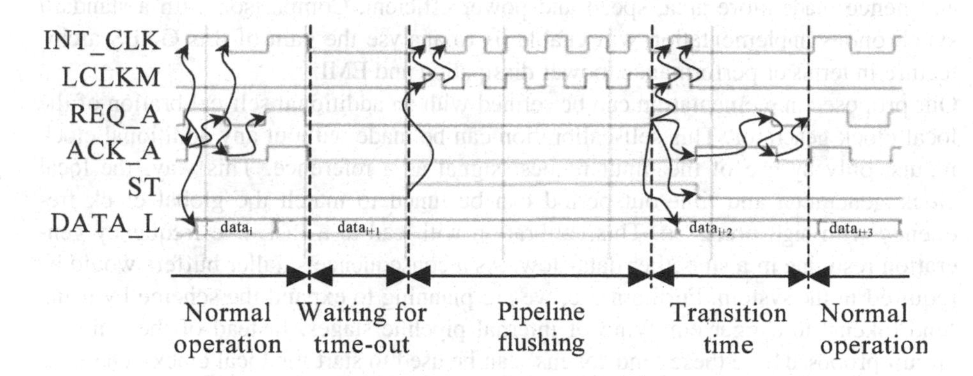

Fig. 6. Internal operation of the asynchronous wrapper

In Fig. 6 a simulation trace for different modes of operation is shown. This simulation is based on a GALS system consisting of a synthesised asynchronous wrapper fitted with a behavioral model of a 'dummy' synchronous block. The 'dummy' synchronous block is a 21-stage FIFO. The simulated system consists of three cascaded GALS blocks which are fitted with an ideal token producer and a consumer at input and output respectively. Fig. 6 shows the signals within one asynchronous GALS wrapper. In this figure we can see that *INT_CLK* is generated from signal *LCLKM* and *REQ_A*. Four different modes of operation of the asynchronous wrapper are shown. First the 'normal' mode of operation can be observed, where a standard handshake is performed on the lines *REQ_A* and *ACK_A*. Every request signal is interpreted as a new clock cycle. When *REQ_A* is kept at '0', the circuit enters a second state: waiting for time-out. During this period the internal clock signal is disabled. The time-out event is indicated by activation of signal *ST*. Subsequently the transition into a third mode occurs. The local clock signal *LCLKM* will be activated and in turn drive the *INT_CLK* signal. Further, the *REQ_A* line indicates the arrival of new data before deactivation of *LCLKM*. Now the system enters a transition mode. In this mode the initiated local clock cycle must be completed and subsequently control of the internal clock signal is handed over to the request line. In the last period of the simulation trace the circuit has resumed 'normal' request-driven operation.

6 Conclusions and Further Work

In this paper a novel technique for asynchronous communication between synchronous blocks is presented. The novel GALS technique reported here, allows easy integration of synchronous blocks into a complex system. It makes use of a new asynchronous wrapper which is particularly suited for datapath architectures. This wrapper implies natural and simple rules for synchronous block design without hardware overhead. Our proposed method also offers explicit and effective power saving mechanisms. However, for specific applications, the circuits can be further optimised and hence made more area, speed and power efficient. Comparison with a standard synchronous implementation will enable us to analyse the gain of this GALS architecture in terms of performance, power dissipation and EMI.
Our proposed implementation can be refined with an additional self-calibration of the local clock generator. This self-calibration can be made without any additional clock inputs, only by use of the input request signal as a reference. This way, the local clock generation and time-out period can be tuned to match the global clock frequency with high precision. This calibration will lead to a PLL-like frequency generation resulting in a smoother data-flow. As a consequence smaller buffers would be required in the system. Furthermore, we are planning to expand the scheme by using 'end-tokens' to trigger emptying of internal pipeline stages. Instead of the time-out circuit proposed here these 'end-tokens' can be used to start the local clock generator. Work is being carried out to adapt the scheme for GALS blocks requiring a higher local clock frequency than their input data rate.
A subjective comparison of the GALS implementation shows that the design process is easier, faster and less error prone when compared with the equivalent synchronous design. This gives us hope that the scheme can be successfully deployed to enhance the design of datapath architectures.

References

1. K. Y. Yun, R. P. Donohue, Pausible Clocking: A first step toward heterogeneous systems, In Proc. International Conference on Computer Design (ICCD), October 1996.
2. D. S. Bormann, P. Y. K. Cheoung, Asynchronous Wrapper for Heterogeneous Systems, In Proc. International Conf. Computer Design (ICCD), October 1997.
3. J. Muttersbach, Globally-Asynchronous Locally-Synchronous Architectures for VLSI Systems, Doctor of Technical Sciences Dissertation, ETH Zurich, Switzerland, 2001.
4. S. Moore, et al., Self Calibrating Clocks for Globally Asynchronous Locally Synchronous System, In Proc. International Conference on Computer Design (ICCD), September 2000.
5. D. M. Chapiro, Globally-Asynchronous Locally-Synchronous Systems, PhD thesis, Stanford University, October 1984.
6. S. Moore, et al., Point to Point GALS interconnect, In Proc. International Symposium on Asynchronous Circuits and Systems, pp. 69–75, April 2002.
7. K. Y. Yun, D. Dill, Automatic synthesis of extended burst-mode circuits: Part I and II, IEEE Transactions on Computer-Aided Design, 18(2), pp. 101–132, February 1999.

Power/Area Tradeoffs in 1-of-M Parallel-Prefix Asynchronous Adders

João Leonardo Fragoso, Gilles Sicard, and Marc Renaudin

TIMA Laboratory / CIS Group
46, avenue Félix Viallet
38031, Grenoble Cedex, FRANCE
{joao.fragoso, gilles.sicard, marc.renaudin}@imag.fr

Abstract. This work describes generalized structures to design 1-of-M QDI (Quasi Delay-Insensitive) asynchronous adders. Those structures allow to design from simple ripple-carry adders to faster parallel-prefix adders. The proposed method is fully automated and integrated in TAST (TIMA Asynchronous Synthesis Tool) tools suite. This paper also demonstrates that the most widely used dual-rail encoding (binary representation in QDI circuits) is not the best solution for numbers' representation in asynchronous circuits. In fact, according to the domain of values to be represented increasing the radix leads to parallel-prefix adders with lower area, delay and power consumption. Hence, this work enables the designer to optimize his/her design by choosing the appropriate 1-of-M number representation.

1 Introduction

Properties of asynchronous or self-timed circuits present the ability to reduce the design complexity and to improve the reliability of VLSI circuits [1]. In addition, when compared to synchronous circuits, asynchronous design leads to high-speed, low-power, reduced electromagnetic interference circuits [2-5]. Asynchronous circuits are potentially faster since the latency can be computed by averaging all computation delays, while clocked systems use the worse case delay to define the operation frequency. As well, if an asynchronous circuit has no data to process, it will not operate. On the contrary, the clock signal causes activity in a synchronous circuit and it consumes power even if this circuit has no data to process. Finally, clocked systems concentrate the EMI spectrums on clock frequency harmonics, while clockless circuits exhibit a smooth EMI spectrum.

These attributes increase the interest in asynchronous circuits design. However, the asynchronous CAD tools are still immature. In addition, these tools must deal with logic and time constraints and it carries out an even more complex problem than custom synchronous design. On the other hand, in last years, the asynchronous community has done a great effort to offer methodologies and tools to support asynchronous design.

As synchronous design, asynchronous methodologies can improve their results using dedicated compilers to generate specific modules, as memories and arithmetic

J.J. Chico and E. Macii (Eds.): PATMOS 2003, LNCS 2799, pp. 171–180, 2003.
© Springer-Verlag Berlin Heidelberg 2003

operators. These compilers explore the regularity and architecture knowledge to optimize the circuit generated. This paper presents generalized adders that are used in TAST [6]. TAST (acronym for TIMA Asynchronous Synthesis Tools) is an open framework for the design of asynchronous circuits. The input language is an enhanced version of CHP. The adders discussed herein are automatically generated, and their generalization allows to directly produce circuits processing 1-of-M encoded data. So, no data-converting interface is needed to embed a generated operator as part of a circuit described in CHP.

2 1-of-M Data Encoding

The communication between asynchronous blocks is made by a set of wires and a communication protocol (a channel), as shown in Fig. 1.a. A data scheme using one-wire per bit is called bundled-data. In bundled-data (Fig. 1.b) or single-rail encoding, a request line is always associated with data lines to specify when data are valid. Alternatively, each data bit can be encoded using two wires, commonly called dual-rail encoding. Obviously, dual-rail extensions are possible when a data value is coded using a set of wires with only one active. These encodings are denoted 1-of-M (as known as one-hot encoding) where M is the number of wires used (Fig. 1.c). 1-of-M encoding allows recognizing when data is valid and a request line is no longer needed. An invalid state, when all wires are set down to zero, is used to separate two valid data. So, an integer value can be encoded in any scheme presented. Nevertheless, we concentrate the discussion on 1-of-M encoding and first formalize the numbers' representation.

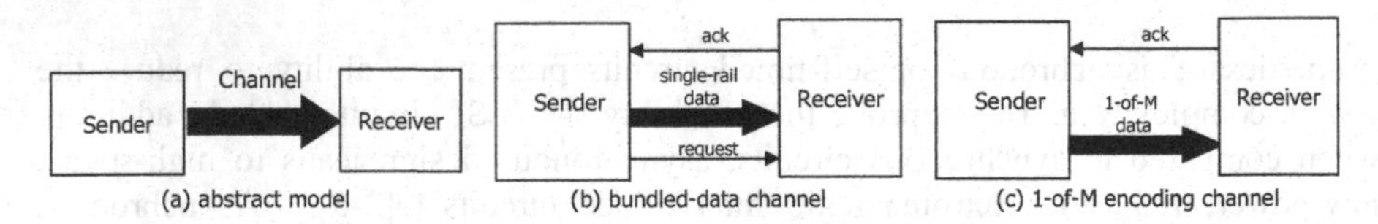

Fig. 1. Asynchronous communication channels

The number of digits d required to encode an integer value v in a radix M is expressed by $d = \lceil \log_M(v) \rceil$. In a QDI circuit, each digit is encoded using M wires and the total number of wires is defined by $d \times M$. Consequently, a data encoding defines the radix representation of a circuit and this representation contains only one active wire per digit. Table 1.a shows an example of unsigned values that can be represented using 3 digits in dual-rail and 2 digits in 1-of-3 rails. In Table 1.a dots are used to separate digits. Clearly, signed values can also be represented and, as in binary representation, radix complement is used to encode signed values. However, the signed values coded with d digits are radix dependent and their range is different for even and odd radixes.

Table 1.b shows signed values that can be encoded using 3 dual-rail digits and 2 digits in 1-of-3 rails. At this point, it is easy to realize that designing an interface to convert a number in radix M_1 to a number in radix M_2 may not be trivial, and this interface may be complex. Therefore, it is interesting to be able to design operators that use the same data encoding of the circuit where this operator is embedded.

Table 1. (a) Unsigned and (b) signed integer values coded in dual-rail and 1-of-3 rail

(a)

Unsigned Value	Dual-rail	1-of-3
0	01.01.01	001.001
1	01.01.10	001.010
2	01.10.01	001.100
3	01.10.10	010.001
4	10.01.01	010.010
5	10.01.10	010.100
6	10.10.01	100.001
7	10.10.10	100.010
8		100.100

(b)

Signed Value	Dual-rail	Signed Value	1-of-3
0	01.01.01	0	001.001
1	**01.01.10**	**1**	**001.010**
2	01.10.01	2	001.100
3	01.10.10	3	010.001
-4	10.01.01	4	010.010
-3	10.01.10	-4	010.100
-2	**10.10.01**	-3	100.001
-1	10.10.10	**-2**	**100.010**
		-1	100.100

Moreover, *M-1* (radix minus one) complement of 1-of-M encoded values does not require any inverter but only wire permutation. Table 1.b bold values show dual-rail 1-complement and 1-of-3 rails 2 complement. Hence, a 1-of-M encoding adder can be changed into a subtractor by just exchanging the wires of each digit of an operand.

3 TAST Design Flow

TAST (acronym for TIMA Asynchronous Synthesis Tools) is a methodology to synthesize a whole asynchronous circuit (control and data-path) [6-7]. The flow starts from high-level modelling using the CHP language (Communicating Hardware Processes). The CHP language, initially proposed by Martin [8], is naturally adopted in this work because: (1) it includes non-deterministic choice structures required to model arbitration [9][10], and (2) it is very well suited to model and synthesize delay-insensitive circuits [8].

In [6] and [7], some new features are added to improve the CHP language. The 1-of-M encoding is one of the novelties we brought to CHP. The MR[M][d] (multi-rails) and SMR[M][d] (signed multi rails) types allow declaring unsigned/signed variables and channels built of d digits of 1-of-M rails each.

The initial CHP program is compiled into an intermediate format (Petri nets and Data Flow Graphs). From this format, delay-insensitive gate-level implementations are automatically derived and their schematics obtained following a formal procedure that is beyond the scope of this paper. All gate-level implementations of CHP programs are described using VHDL gate netlists. When the synthesizer finds an arithmetic operation, the arithmetic operator compiler is invoked to generate the required operator. The synthesizer informs the compiler about the operation, length and base parameters. The compiler generates a VHDL description for the new operator and this operator is instantiated into the produced netlist.

4 Generalize 1-of-M Addition

The binary addition is a well-know arithmetic operation:

$$s = (a+b+ci)\bmod(2) \qquad co = (a+b+ci)/2 \tag{1}$$

Obviously, those equations could be generalized for any radix replacing the value 2 in equation (1) by M, where M is the radix. However, the carry result (co) is always less or equal to one independently of the radix chosen to encode a, b and s. Consequently, the carry signal can always be encoded in dual-rail. To explain the generalization of the s and co computations, 1-of-3 rails code are used in all the following examples. Lowercase indexed variables are used to indicate digits and uppercase indexed variables are used to indicate wires within a digit.

The sum and carry-out results can be expressed by a matrix computation, where columns and lines are the indexes of active wires in a and b inputs. Each matrix cell contains the indexes of active s and co output wires. Table 2.a shows the matrix computation for 1-of-3 rails with carry-in equal to zero. After the computation of all minterms, it is easy to see that the inverse diagonal of the matrix always affects the same output wire of the sum. Hence, each wire of the sum is the logic addition of all the terms on the inverse diagonal. When carry-in equals to one, the rule is the same, but the diagonals are left-shifted, as shown in Table 2.b.

Table 2. Result for S and Co for 1-of-3 rails encoding (a) Ci equal zero (b) Ci equal one

ci=0		a		
co / s		0	1	2
	0	0/0	0/1	0/2
b	1	0/1	0/2	1/0
	2	0/2	1/0	1/1

ci=1		a		
co / s		0	1	2
	0	0/1	0/2	1/0
b	1	0/2	1/0	1/1
	2	1/0	1/1	1/2

So, the carry-in signal can be used as a multiplexer selection to define if the sum result should be left-shifted. In this way, sum computation is split in two stages: (1) A and B addition and (2) carry-in addition. Furthermore, the first wire of carry-out (carry-out equals to 0) is defined by the logic addition of all the elements above the secondary diagonal. The second wire (carry-out equals to 1) is the logic addition of the elements under the secondary diagonal, free of the carry-in value. The carry-in defines where the secondary diagonal is included either into the first wire or into the second wire function of the carry-out. It means that the cells above the secondary diagonal kill the value of carry-out (co=0), the cells on the secondary diagonal propagate the value of carry-in to carry-out and the cells under the secondary diagonal generate the value of carry out (co=1).

Table 3. Result for kill (K), propagate (P) and generate (G) functions for 1-of-3

Kpg		a		
		0	1	2
	0	K_0	K_1	P
b	1	K_1	P	G_0
	2	P	G_0	G_1

Table 2.a can be rewritten to express this function as shown in Table 3. In Table 3, the K and G values are identified by the inverse diagonal indexes. It is important to capture this information because the K and G values belonging to a given inverse diagonal are involved in the computation of the sum's wires. This implementation allows using the same sub-function to perform both the carry-out and sum computations. The following equations define the generation for all K, P and G, where M is the radix value:

$$K_j = \sum_{i=0}^{j} \left(A_i \cdot B_{j-i}\right) \forall j < M-1 \Rightarrow K = \sum_{j=0}^{M-2} K_j \tag{2}$$

$$P = \sum_{i=0}^{M-1} \left(A_i \cdot B_{M-i-1}\right)$$

$$G_j = \sum_{i=j+1}^{M-1} \left(A_i \cdot B_{M+j-i}\right) \forall j < M-1 \Rightarrow G = \sum_{j=0}^{M-2} G_j$$

in this way, all inverse diagonals are defined as:

$$T_j = K_j + G_j, \forall j < M-1$$
$$T_{M-1} = P \tag{3}$$

and, finally, the sum can be defined as:

$$S_j = Ci_0 \cdot T_j + Ci_1 \cdot T_{(j-1)\bmod(M)}, \forall j = \{0,\ldots, M-1\} \tag{4}$$

and the carry-out as:

$$Co_1 = G + Ci_1 \cdot T_{M-1} \qquad Co_0 = K + Ci_0 \cdot T_{M-1} \tag{5}$$

To define a half-adder (without the input carry-in), the sum and carry-out outputs can directly be calculated from the equations K, P and G. Thus, the full-adder generation is generalized for 1-of-M rails encoding. Fig. 2 shows the circuit resulting of the full-adder generation using 1-of-3 rail operands.

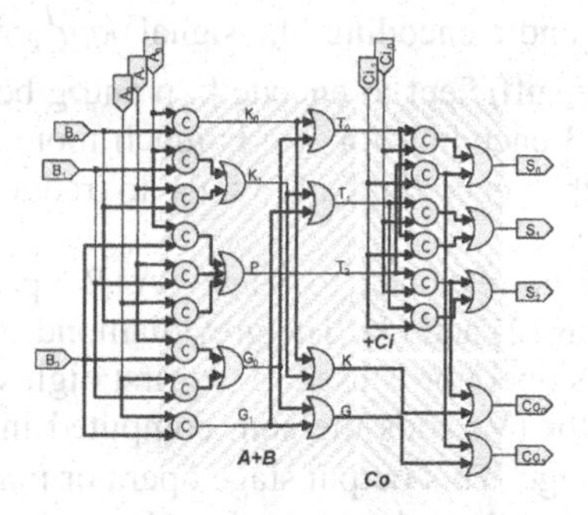

Fig. 2. Generation result for a Full-Adder in 1-of-3 rails encoding

In Fig. 2, it is possible to distinguish the two computation stages. The first one, labeled "A+B", calculates the matrix of Table 3 (equation 2), signals T (equation 3) and signals K and G needed to compute carry out. The second one, labeled "+Ci" and "Co" computes sum and carry-out using the carry-in signal. This generalized full-adder can be replicated d-times to design ripple-carry adders processing d digits that are 1-of-M encoded.

4.1 1-of-M Parallel-Prefix Adders

Faster adders using 1-of-M encoding can also be generalized. The aim is to allow the carry-out resolution in parallel. We must then separate the sum from carry-out computation. Here, we will explore the carry lookahead structure proposed by Sklansky in [11].

Fig. 3.a shows the basic DAG (directed acyclic graph) structure of *d*-digits parallel-prefix adder. This structure is composed by three stages: (1) an input-stage where the sum of a_i and b_i is added, (2) the prefix computation stage (or carry tree) and (3) an output stage where the sum computed in the input stage is added to the carry. Fig. 3.b shows the prefix computation tree proposed by Sklansky.

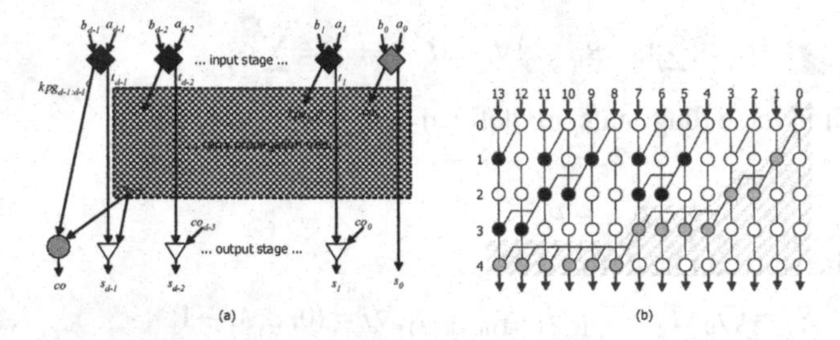

Fig. 3. Basic parallel-prefix structure

Looking at the generalized full-adder, those three Sklansky adder stages can be effortlessly identified. Therefore, we will define five 1-of-M logic operators shown in Fig. 4 that compose the Sklansky graph.

The black diamond operator (◆) is used in input stages to evaluate t_i (sum result) and the prefix $kpg^0_{i:i}$ (kill, propagate, generate). The a_i, b_i and t_i signals are 1-of-M encoded. Despite the a_i, b_i and t_i encoding, the signal $kpg^l_{i:k}$ is always encoded using 1-of-3 rails. Three rails are sufficient to encode k, p and g because they are mutually exclusive. Note that, 1-of-3 encoding for *kpg* is much more efficient than using two dual-rail digits. In fact, this encoding allows us to reduce the number of wires required in computation trees.

The diamond operator corresponds to stage "A+B" presented in Fig. 2. This operator computes equations (2) and (3). The grey diamond operator (◆) corresponds to a simple half-adder. This operator is used in the first digit when there is no carry-in signal. The triangle operator (∇) adds the sum computed in the first stage with the carry-out of the previous stage. This output stage operator matches up the stage "Cin" shown in Fig. 2 and this operand resolves equation (4).

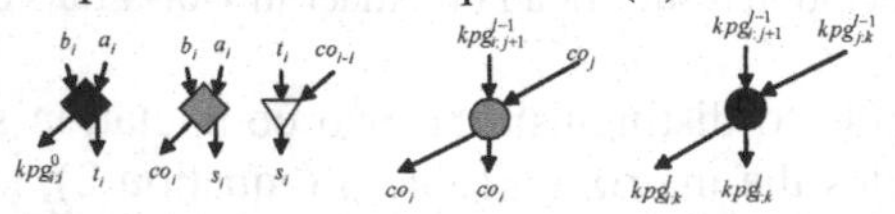

Fig. 4. 1-of-M encoding parallel-prefix operators

In the computation tree, white nodes (○) just pass the input signal unchanged to the next tree line. In order to ease the comprehension, an empty first line is added to the Sklansky tree shown in Fig. 3.b.

The $kpg^{l}_{i:k}$ 3-rails encoded signal means the prefix group of digits *i*,..., *k* at line l. To compute a new prefix group, the • operator can be applied according to:

$$kpg^{l}_{i:k} = (K^{l}_{i:k}, P^{l}_{i:k}, G^{l}_{i:k}) = kpg^{l-1}_{i:j+1} \bullet kpg^{l-1}_{j:k} = (K^{l-1}_{i:j+1} + P^{l-1}_{i:j+1} K^{l-1}_{j:k}, P^{l-1}_{i:j+1} P^{l-1}_{j:k}, G^{l-1}_{i:j+1} + P^{l-1}_{i:j+1} G^{l-1}_{j:k}) \quad (6)$$

Fig. 5.a shows the implementation for the dot operator. This operator corresponds to the black nodes in the computation tree.

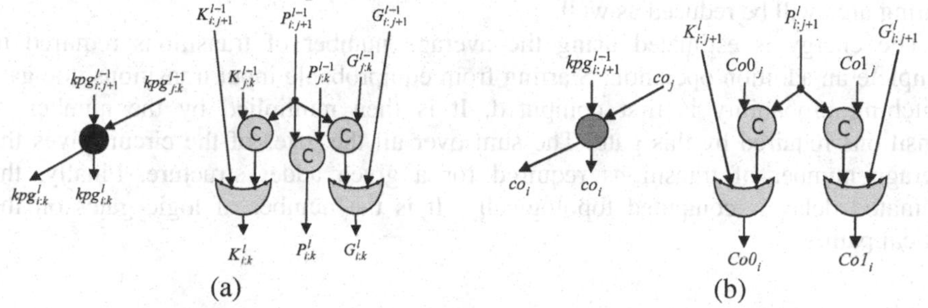

Fig. 5. (a) The dot operator circuit for 1-of-M QDI circuits and (b) The simplified dot operator for 1-of-M QDI circuits

At level m with all prefix groups from *i* down to 0 calculated, the signal *P* is always equal to zero since a carry-out was either generated or killed at this digit. This means that $co_i = (Co0_i, Co1_i) = (K^{m}_{i:0}, G^{m}_{i:0})$. So, $co_i = kpg^{m}_{i:0}$ and the carry-out signal can again be encoded in dual-rail (two wires). The computation tree grey nodes correspond to operator that computes co_i. It is easy to demonstrate that the right operator of a grey node is always a precedent grey node output. In this way, the dot operator can be simplified at the grey nodes in order to reduce the number of wires needed. The grey node operation can be expressed by:

$$co_i = kpg^{l-1}_{i:j+1} \bullet co_j = (K^{l-1}_{i:j+1} + P^{l-1}_{i:j+1} Co0_j, G^{l-1}_{i:j+1} + P^{l-1}_{i:j+1} Co1_j) = (Co0_i, Co1_i) \quad (7)$$

Fig. 5.b shows the implementation of the simplified dot operator and this circuit binds the "Co" labeled region of Fig. 2. So, the right operand input and the output of grey nodes are 2-rails encoded and the tree highlighted region in Fig. 3.b corresponds to the region using 2-rails encoding. This simplification implies a significant reduction in complexity of the computation tree.

In addition, a grey node is used in the output stage to compute the last carry-out signal, provided that all outputs must be completed before acknowledging the adder inputs. This grey node replaces the last computation tree column as shown in Figure 3.b. Therefore, it is not necessary to speed up the last carry-out computation. When a carry-in signal is used, to keep the separation between sum and carry computation, the carry-in signal is directly included into the computation tree and the first grey diamond node is replaced by a black one. This implies right shifting the computation tree. The generalization presented, also allows designing the Brent-Kung adders of ʃ [12] and the Kogge-Stone adders of ʃ [13]

5 Results

The VHDL description of the adders presented are fully automatically generated for any 1-of-M data encodings and any number of digits. For comparison, the circuits' areas were estimated based on standard-cell transistors number. The complexity of a C-element is eight transistors and an OR2-element is six transistors. Obviously, the estimated area does not take routing into account. However, the comparisons remain fair because the number of wires required to encode the operands of the compared adder structures is kept constant. Consequently, if the transistor count is reduced, the routing area will be reduced as well.

The energy is estimated using the average number of transitions required to complete an addition operation. Starting from equiprobable input transitions, the gate switching probability is first computed. It is then multiplied by the number of transitions required by this gate. The sum over all the gates of the circuit gives the average number of transitions required for a given adder structure. Finally, the estimated delay is computed topologically. It is the number of logic gates on the critical path.

5.1 1-of-M Ripple-Carry Adders

Table 4 shows the estimated area and energy for various ripple carry adders compared to an equivalent dual-rail implementation.

Table 4. Ripple-carry results generation compared to an equivalent dual-rail implementation

Encoding	1-of-3	1-of-4	1-of-5
Area	+22.4%	+50.0%	+83.5%
Energy	-27.3%	-40.6%	-48.1%

As guessed, the adder complexity increases with the radix (data encoding) due to the size of the computation matrix (M^2 elements). The full-adder generation is based on minterms synthesis [14] applied to a two stage structure of the full adder. Other asynchronous circuit design methodologies are presented in [15] [16]. These methodologies optimize the circuit when not all the minterms are needed. However, for an adder, all the minterms must be computed and these methodologies lead to an area overhead. The dual-rail version of the generalized full-adder presented in this paper has a complexity of 100 transistors whereas the optimized version presented in [15] has a higher complexity of 118 transistors. Moreover, the implementation presented in [16] is not easily generalized. In addition, the power consumption reduces as the radix increases, since the number of digits and, consequently, the number of active wires is reduced given the same representation number set. For the same reason, the critical path is reduced due to the reduced number of digits involved in the computation.

5.2 Sklansky Adders

The same conclusion applies to Parallel-prefix adders in terms of power consumption and delay, and for the same reasons than the ripple-carry adders. On the other hand, the complexity in Sklansky adders has not a similar behavior. This is because the computation tree is only a function of the number of digits and is independent of the input data encoding.

Increasing the radix implies increasing input and output stages complexity. However, because the number of digits needed to encode the same set of values is lower, the complexity of the computation tree is smaller. So, beyond a number of digits, increasing the radix and reducing the number of digits results in a simpler adder. This is because the tree reduction compensates the increasing of the input and output stage complexities. Additionally, reducing number of digits and computation tree also implies reducing the critical-path length. Consequently, increasing the radix for large operators implies reducing the area, the power consumption and the delay of the generated adders.

Table 5 shows a comparison between two adders for the same domain of values (a 128-bit data-path). It shows two Sklansky adders without carry-in signal. The first one uses 128 dual-rail digits and the second one uses 64 digits using the 1-of-4 encoding. Raising the radix from 2 to 4 leads to a 5.53% area reduction and a 50.94% average power reduction. These results show that dual rail encoding is not the optimal representation to design arithmetic operator with large set of input values.

Table 5. Comparaison between dual-rail and 1-of-4 rails

domain	encoding	digits	#wires	area	energy
2^{128}	dual-rail	128	256	23,332	1,169.25
	1-of-4	64		22,042	573.63
% reduction				**5.53**	**50.94**

6 Conclusions

A generalized method to design 1-of-M QDI adders has been presented. This fully automated method allows generating adders and subtractors using any radix. The generated operators can be embedded in 1-of-M QDI circuits and no data-converting interface is required. The adders obtained have a reduced area when compared to other optimized implementations. Moreover, the use of an asynchronous library and complex cells could significantly improve the results.

In addition, the presented work shows that dual-rail encoding is not always the best representation for all domains of values when fast adders are required. Increasing the radix can lead to reduce area/power and delay characteristics. The proposed method is fully automated and integrated in the TAST tools suite. The adder generator is also available as an independent tool to generate and evaluate the proposed adder structures. Future works will be focused on multiplier generation following a similar approach.

References

1. J. T. Udding. A formal model for defining and classifying delay-insensitive circuits and systems. Distributed Computing, 1:197–204, 1986.
2. K. Stevens, et.al. An asynchronous instruction length decoder. IEEE JSSC, v36(2), pp 217–228, 2001.
3. W. Coates, et.al. FLETzero : an asynchronous switching experiment. Proceedings of ASYNC, pp 173–182, 2001.
4. J. Woods, P. Day, S. Furber, J. Garside, N. Paver, S. Temple. AMULET1: an asynchronous ARM microprocessor. IEEE Transaction on Computers, v46(4), pp 385–398, 1997.
5. M. Renaudin. Asynchronous circuits and systems: a promising design alternative. Journal of microelectronic engineering, 54: 133–149, 2000.
6. A.V. Dinh Duc, et.al. TAST CAD Tools. In: 2nd Asynchronous Circuit Design Workshop. Munich-Germany, 28–29 January, 2002.
7. M. Renaudin, P. Vivet, F. Robin. A Design Framework for Asynchronous/Synchronous Circuits Based on CHP to HDL Translation. In: International Symposium on Advanced Research in Asynchronous Circuits and Systems (ASYNC), Barcelona, Spain, April 19–21, pp 135–144, 1999.
8. A.J. Martin. Synthesis of Asynchronous VLSI Circuits. Internal Report, Caltech-CS_TR-93–28, California Institute of Technology, Pasadena, 1993.
9. J-B. Rigaud, M. Renaudin. Modeling and design/synthesis of arbitration problems. AINT'2000, Proceedings of the Asynchronous Interfaces: Tools, Techniques and Implementations Work-shop, TU Delft, The Netherlands, July 19–20th 2000, pp. 121–128.
10. J-B. Rigaud, J. Quartana, L. Fesquet, M. Renaudin, "High-Level Modeling and Design of Asynchronous Arbiters for On-Chip Communication Systems", Design Automation and Test Conference (DATE), 4–7 March, 2002, Paris, France.
11. J. Sklansky. Conditional-sum addition logic, IRE Trans. Electronic Computers, 9(2):226–231, 1960.
12. R. Brent, H. Kung. A Regular Layout for Parallel Adders. IEEE Transactions on Computers, vol. C-31, no. 3: 260–264. March, 1982.
13. P. Kogge, H. Stone. A Parallel Algorithm for the Efficient Solution of General Class of Recurrence Equation. IEEE Trans. on Computers, vol(22), no. 8: 783–791. Aug, 1973.
14. J. Sparsø, J. Staunstrup, M. Dantzer-Sørensen. Design of delay insensitive circuits using multi-ring structures. In: Proceedings of EURO–DAC, Germany, 1992. pp. 15–20.
15. I. David, R.Ginosar, M. Yoeli. An efficient implementation of Boolean functions as self-timed circuits. IEEE Transactions on Computers, vol. 41, no. 1:2–11. January, 1992.
16. A. Kondratyev, K. Lwin. Design of asynchronous circuits using synchronous CAD tools. IEEE Design and Test of Computers, vol. 19, no. 4:107–117. July, 2002.

Static Implementation of QDI Asynchronous Primitives

P. Maurine, J.B. Rigaud, F. Bouesse, G. Sicard, and M. Renaudin

LIRMM 161 rue Ada 34392 Montpellier Cedex 5 France
TIMA Laboratory 46, avenue Félix Viallet 38031 Grenoble Cedex France

Abstract. To fairly compare the performance of an asynchronous ASIC to its homologous synchronous one requires the availability of a dedicated asynchronous library. In this paper we present TAL_130nm a standard cell library dedicated to the design of QDI asynchronous circuits. Cell selection and sizing rules applied to develop TAL_130nm are detailed. It is shown that significant area and power savings as well as speed improvements can be obtained.

1 Introduction

If asynchronous circuits can outperform synchronous ICs in many application domains such as security and automotive [11], the design of integrated circuits still remains essentially limited to the realization of synchronous chips. One reason can explain this fact: no CAD suite has been proposed by the EDA industry to provide a useful and tractable design framework. However, some academic tools have been or are under development [1,2,3].

Among them TAST [3] is dedicated to the design of micropipeline (μP) and Quasi Delay Insensitive (QDI) circuits [11]. Its main characteristic is to target a standard cell approach. Unfortunately, it is uncommon to find in typical libraries (dedicated to synchronous circuit design) basic asynchronous primitives such as C-elements. Consequently, the designer of QDI asynchronous IC, adopting a standard cell approach, must implement the required boolean functions on the basis of AO222 gate [1,9]. It results in sub optimal physical implementations as illustrated on figure (1) that gives evidence of the power and area savings that can be obtained from the development of a library dedicated to the design of asynchronous circuits. Within this context, we developed TAL_130nm (TIMA Asynchronous Library), a standard cell library dedicated to the design of QDI asynchronous circuits.

This paper aims to introduce the methods we used and the choice we made to design TAL. It is organized as follows. In section II, the structural specificities of QDI gates are introduced. This section also describes two sizing criteria, deduced from a first order delay model, allowing reducing area cost while maintaining the throughput. In section IV, we deduce from the first order delay models of both static and ratioed CMOS structures two sizing criteria allowing reducing the area cost of any QDI gate while maintaining its throughput. Finally, section IV reports the performance of the gates designed following our sizing strategy and compare them to gates implemented using basic AO222 gates borrowed from a standard synchronous library.

NB : the meaning of the different notations used throughout the paper is given in table 1.

J.J. Chico and E. Macii (Eds.): PATMOS 2003, LNCS 2799, pp. 181–191, 2003.
© Springer-Verlag Berlin Heidelberg 2003

2 QDI Element Specificities and Library Sizing Strategy

2.1 QDI Element Specificities

Depending on the desired robustness to process, voltage and temperature variations, handshake technology offers a large variety of asynchronous circuit styles and a large number of communication protocols. Our aim is not here to give an exhaustive list of all the possible alternatives, but to introduce the main specificities of the primitives required to design 4-phase QDI circuits.
For such circuits, the data transfer through a channel starts by the emission of a request signal encoded into the data, and finishes by the emission of an acknowledge signal. During this time interval, which is a priori unknown, the incoming data must be hold in order to guarantee the quasi-delay-insensitivity property. This implies the intensive use of logical gate including a state holding element (usually a latch) or a feedback loop.

As we target a CMOS implementation, it results from the preceding consideration that most of the required primitive are composite or complex positive gates. Indeed they can be decomposed in one or more simple dynamic logic gates and a state holding element. In fig.1 we give possible decompositions of a 3-input Muller gate and a COR222 gate, both widely used to implement basic logic such as „And", „Or", „Xor" in multi-rail design style.

2.2 Library Sizing Strategy

Due to their composite structure, different sizing strategies can be applied to the library. The one we adopted is based on the five following design rules:
①: balance at first order the amplitudes of the currents flowing through the N and P arrays in order to balance the active and RTZ phases.
②: designing at least the drives X0, X1, X2, X4 for each functionality in order to accommodate a large range of loads. (Many gates have been designed in drives 0,1,2,4,8,12)
③: design each drive in order to ensure that, independently of the logic function, its output driver has the same current capability that the equivalent inverter. As an example, the last stage of the logic decomposition of the 3-input Muller gate (M3) of drive Xj is sized in order to deliver the same switching current than the inverter of drive Xj.
④: minimize the area by designing each cell in order to accommodate weak and important loads in two functional stages. This means that only the two last stages of the COR222 decomposition will be sized in order to accommodate the output load; the preceding stage being designed for a minimum area cost. This is equivalent to targeting implementations with low input capacitance values. Such strategy may allow the most frequent possible use of weak drives without compromising too much the speed performances.
⑤: avoid whenever possible logic decompositions in which the state holding element drives the output node. In figure (1f), the placement of the output inverter and the

latch can be interchanged, but it is preferable as suggested in [10] to place the latch first and to let the inverter drive the output node in order to minimize the cell area according to ③.

3 First Order Models and Sizing Criteria

In order to achieve high speed performance and to ensure the correct behaviour of the state holding element, we need first order delay models of both static and ratioed CMOS structures. This section's aim is to briefly present the models we adopted and the gate sizing strategy we deduced from them. We first introduce the first order model of the drain to source current they are based on.

3.1 Drain Source Current Model

The drain source current model adopted is the α-power law model proposed in [4] considering for simplicity that α=1. It allows us to model the behaviour of the transistors for which the current saturation occurs by carrier velocity saturation phenomenon. Thus, the expressions of the drain source current considered afterward are:

$$I_{DS,N/P} = \begin{cases} 0 \\ \dfrac{\mu_{0,N/P} \cdot C_{OX}}{L_{GEO}} \cdot W_{N/P} \left(V_{GS,N/P} - V_{TN/P}\right) \cdot V_{DS,N/P} \\ K_{N/P} \cdot W_{N/P} \cdot \left(V_{GS,N/P} - V_{TN/P}\right) \cdot V_{DS,N/P} \end{cases} \quad (1)$$

3.2 Step Response of a Static CMOS Structure

As a first order delay model for all the static structures, we use the generalization of the inverter step response proposed by Mead [11]:

$$t_{HLS(HLS)} = \frac{C_L \cdot \Delta V}{I_{N(P)}} = \tau \cdot \frac{C_L}{C_{N(P)}} \quad (2)$$

The inverter step response can be generalized to all the static gates by reducing each gate to an equivalent inverter [3, 5, 6]. To do so, one can estimate the ratio $\Delta W_{N,P}$ between the current that a transistor is likely to deliver and the current that the associated serial array of transistor can deliver. Then following (2), the step responses of a logical gate can be expressed as:

$$t_{HLS} = \tau \cdot \Delta W_N \cdot \frac{C_L}{C_N} = \tau \cdot \Delta W_N \cdot F_O \cdot \frac{(1+k)}{2}$$
$$t_{LHS} = \tau \cdot R \cdot \Delta W_P \cdot F_O \cdot \frac{(1+k)}{2 \cdot k} \quad (3)$$

From (3), we get, defining adequately $SD_{N,P}$ the following expressions:

$$t_{HLS(LHS)} = \tau \cdot SD_{N(P)} \cdot F_O \quad (4)$$

3.3 Step Response of a Ratioed Structure

Let us consider a ratioed structure loaded by an infinite load as represented on figure 2. It corresponds to the worst-case configuration, as the output driver is not able to discharge significantly the output node Z before the node Z_{INT} stabilized.

The step response of a MOS gate being defined as the time necessary for the structure to charge or discharge its output voltage up to or down to $V_{DD}/2$, we solved the differential equation:

$$I_P^L(t) - I_N^H(t) = -C_L \cdot \frac{dV_{OUT}}{dt} \tag{5}$$

to obtain the output voltage evolution:

$$V_{OUT}(t) = \frac{\alpha_P \cdot m_n \cdot W_P^L}{\alpha_N^{'} \cdot \Delta W_P \cdot W_N^H} \cdot \left(1 - e^{-\frac{\alpha_N^{'} \cdot W_N^H}{m_n \cdot C_L} \cdot t}\right) \tag{6}$$

This expression increases quickly to finally reach an asymptotic value. Let us note ΔV_{LH} the corresponding voltage variation of node Z_{INT}:

$$\Delta V_{LH} = \lim_{t \to \infty} V_{OUT}(t) = \frac{\alpha_P \cdot m_n \cdot W_P^L}{\alpha_N^{'} \cdot \Delta W_P \cdot W_N^H} \tag{7}$$

To ensure a correct behaviour of the latch, the limit must be at least equal to the inversion voltage V_{INV} of the output driver. However, in order to design high speed latches, and as this limit is reached asymptotically, it is required to satisfy:

$$\Delta V_{LH} > V_{INV} \tag{8}$$

To ensure a maximal security of operation as well as a good switching speed, we set $\Delta V_{LH} = \Delta V_{HL} = V_{DD}$ as our standard. This standard leads, while designing the latch, to respect the following sizing ratios,

$$K_{P^L(N^L) \to N^H(P^H)} = \frac{W_{P(N)}^L}{W_{N(P)}^H} = \frac{\alpha_{N(P)}^{'}}{m_{n(p)}} \cdot \frac{\Delta W_{P(N)}}{\alpha_{P(N)}} \cdot \Delta V_{LH(HL)} \tag{9}$$

reported in table 2 for the considered 130nm process. The analysis of these results clearly shows that a single N^L transistor delivers enough current to control the latch. Therefore, we set m_p to one. On the contrary, it appears necessary to reduce the current capabilities of the N^H transistor array in order to avoid an area expensive oversizing of the P^L transistor array. This explains why we set m_p value to 2.

Knowing how to size the latches in order to ensure a correct behaviour, we can evaluate from (6) and from its dual expression, the step responses $t_{HLS,LHS}^R$ of the ratioed structure represented in fig.2.

$$t_{HLS}^R = -\frac{m_p \cdot C_L}{\alpha_P^{'} \cdot W_P^H} \cdot \ln\left(1 - \frac{V_{DD} \cdot \Delta W_N \cdot \alpha_P^{'} \cdot W_P^H}{2 \cdot m_p \cdot \alpha_N \cdot W_N^L}\right) \tag{10a}$$

$$t_{LHS}^R = -\frac{m_n \cdot C_L}{\alpha_N^{'} \cdot W_N^H} \cdot \ln\left(1 - \frac{V_{DD} \cdot \Delta W_P \cdot \alpha_N^{'} \cdot W_N^H}{2 \cdot m_n \cdot \alpha_P \cdot W_P^L}\right) \tag{10b}$$

With m_n and m_p set respectively to two and one, expressions 10a and 10b) become:

$$t_{HLS}^{R} = t_{HLS}\left(1+\frac{V_{DD}}{4}\cdot\frac{\Delta W_N\cdot\alpha_P^{'}\cdot W_P^H}{m_p\cdot\alpha_N\cdot W_N^L}\right) = t_{HLS}\cdot\beta_{HLS}$$
$$t_{LHS}^{R} = t_{LHS}\left(1+\frac{V_{DD}}{4}\cdot\frac{\Delta W_P\cdot\alpha_N^{'}\cdot W_N^H}{m_n\cdot\alpha_P\cdot W_P^L}\right) = t_{LHS}\cdot\beta_{LHS} \qquad (11a,b)$$

where t_{LHS} (t_{LHS}) is the step response of the pull-up (pull down) associated to the structure of fig.2 in the absence of the transistor N^H (P^H), and the term β_{LHS} (β_{HLS}) corresponds to the slowing down factor induced by transistor N^H (P^H). Thus, considering expression (11), it seems that the ratioed structure represented on fig.2 behaves as a static structure for which the transistors have the following widths: W_P^L/β_{LHS} and W_N^L/β_{HLS}.

3.4 Gate Sizing

As explain in section II-b, where the library sizing strategy is defined, we want to take advantage of the composite structure of the QDI primitives in order to minimize the area. The application of this strategy results in only sizing the two last stages so that the preceding stages are sized at the minimal area cost. This leads to consider the structure of fig.3, in order to determine the tapering factor [7,8] to be applied to the logic decomposition to minimize the propagation delays. In order to respect rule ⑤, all functionalities are decomposed in such a way that the state holding element do not control the output driver. However, it appears to be too area-expensive for Muller gates. Consequently, two cases have been studied.

Case 1 : the last stages is a static CMOS structure (usually an inverter)

Let us express at first order, the propagation delays ($\Theta\approx\Theta_{HL}\approx\Theta_{LH}$) of the three last stages of the generic structure of fig.3. Using the generalized step response and considering that the internal configuration ratio is equal to R (see rule ①), we get :

$$\frac{\Theta}{\tau} = SD_{N(q-2)}\cdot F_{O(q-2)} + SD_{P(q-1)}\cdot F_{O(q-1)} + SD_{N(q)}\cdot F_{O(q)} \qquad (12)$$

with:

$$F_{0(k)} = {C_{k+1}}/{C_k} \qquad (13)$$

Evaluating the optimal value ($d\theta/dC_{(q-1)}=0$) of the input capacitance of stage (q-1), C_{q-1}^{opt}, we get:

$$C_{q-1}^{opt} = \sqrt{{SD_{P(q-1)}}/{SD_{N(q-2)}}\cdot C_q\cdot C_{q-2}} \qquad (14)$$

To estimate the quality of this sizing criterion, we applied a derating factor η to $C_{i\text{-}opt}$ ($C_{q-1}= C_{q-1}^{opt}\cdot\eta$). Then, for some implementations, we simulated the propagation delay value of the structures. As an example, fig.4 illustrates the variation of the propagation delays with respect to η in the case of a Muller2 (for 2 different loading conditions: Foo = 2 and 5). As shown, the structure sized according to (14) is closed to the optimal one.

***Case 2:* the last stage is the state holding element (Muller gate only):**
Let us again express the first order propagation delay (Θ) of the three last stages of the generic structures of fig.3. We get:

$$\frac{\Theta}{\tau} = SD_{N(q-2)} \cdot F_{O(q-2)} + \beta_{LHS(q-1)} \cdot SD_{P(q-1)} \cdot F_{O(q-1)} + SD_{N(q)} \cdot F_{O(q)} \quad (15)$$

It can be rewritten as:

$$\frac{\Theta}{\tau} = SD_{N(q-2)} \cdot F_{O(q-2)} + \left(1 + \frac{\varphi}{C_{q-1}}\right) \cdot SD_{P(q-1)} \cdot F_{O(q-1)} + SD_{N(q)} \cdot F_{O(q)} \quad (16)$$

The evaluation of the C_{q-1}^{opt} value minimizing equ.16 leads solving the Ferro-Cardan equation:

$$C_{q-1}^{3} - A^{-1} \cdot C_{i} - 2\varphi \cdot A^{-1} = 0 \quad (17)$$

with: $$A = \frac{SD_{N(q-2)}}{SD_{P(q-1)}} \cdot \frac{1}{C_{q-2} \cdot C_{q}} \qquad \varphi = \frac{V_{DD}}{4} \cdot \frac{\Delta W_{P(q-1)}}{m_{n}} \cdot \frac{\alpha'_{N}}{\alpha_{P}} \cdot \frac{1+k}{k} \cdot C_{OX} \cdot L_{GEO} \quad (18)$$

Two sub-cases have to be considered.

Case 2.a: β_{LHS} and β_{HLS} are close to 1

It corresponds to the case of C_{i+1}'s high values or equivalently to the case of strong drives. In that case, the solution of eq.17 is eq.19.

Case 2.b: β_{LHS} and β_{HLS} are greater than 1

It corresponds to the case of weak drives. The resolution leads then to the solution:

$$C_{q-1}^{opt} = \sqrt[3]{\frac{\varphi}{A} \cdot \left(1 + \sqrt{1 - \frac{1}{27 \cdot A \cdot \varphi^{2}}}\right)} + \sqrt[3]{\frac{\varphi}{A} \cdot \left(1 - \sqrt{1 - \frac{1}{27 \cdot A \cdot \varphi^{2}}}\right)} \quad (19)$$

4 Results

We designed thirty functionalities that are very frequently used, using the 130nm process from STMicroelectronics and an industrial automatic layout generator. Table 3 reports typical values of the area reduction factor when compared to AO222 based implementations. As shown, the average area reduction factor obtained for all gates is 1.9.
As it was difficult to detail herein speed and power performances for all the gates with respect to their AO222 based implementations, the results obtained for three representative gates are detailed : Muller2_X2, Muller3_X2 and COR222_X2.
These gates are representative of many others as the electrical paths involved in the switching process are the same or practically the same than various other implemented gates such as: Muller4, COR211, COR221, COR22, COR21, COR222_Ackin_Set ….
The simulation protocol used to compare the proposed implementations to the AO222 based implementations is described in fig.3. With such a protocol, the Foi and Foo values variation enable to analyze the effect of the input ramp and of the load on the performances.

Fig.5 reports the speed improvement and the power saving obtained with respect to AO222 based implementations. As shown, in the typical design range (Foi and Foo ranging from 1 to 10), we can conclude that for almost identical speed performances (speed improvement between –15% and 15%), the cells designed using our strategy are significantly smaller and consumes less power, except for the Muller2 gate.
This exception can be explained by one main reason. The output driver of the Muller gates is a latch. Such a structure burns a large amount of power while driving a large capacitance. However, for Muller3 and 4, this extra power consumption is easily balanced by the fact that we only need one latch to implement them while the corresponding AO222 based implementations require respectively two and three feedback loops.

5 Conclusion

Taking advantage of the composite structure of QDI gates, we designed a complete library. Using a generalized step delay model, we have been able to obtain gates that are two times smaller, while maintaining their power and speed performances (compared to AO222 based implementations). These results clearly demonstrate that to obtain a fair comparison between asynchronous and synchronous ASICs, one need to develop dedicated libraries. Indeed, the area reduction of the cells strongly impacts the routing, and thus the global performances of a given circuit, both in terms of speed and power. Future works will be focused on quantifying the gain brought by TAL in terms of speed and power by the design of significant asynchronous prototypes chips.

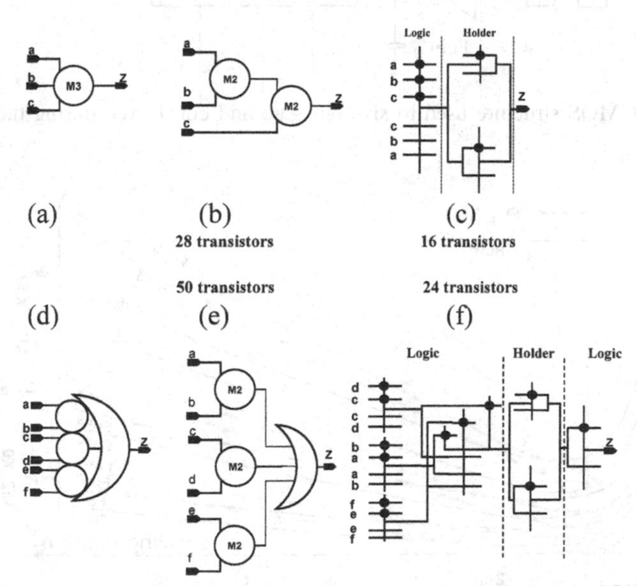

Fig. 1. (a,d) Symbols of Muller3 and COR222 gates, (b,d) Muller3 and COR222 decompositions in AO222 based „design style", (c,f) Muller3 and COR222 schematics requiring a minimal number of transistors.

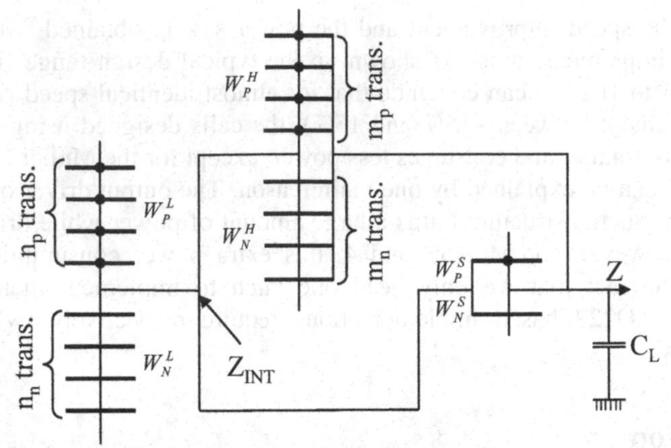

Fig. 2. Generic CMOS ratioed structure

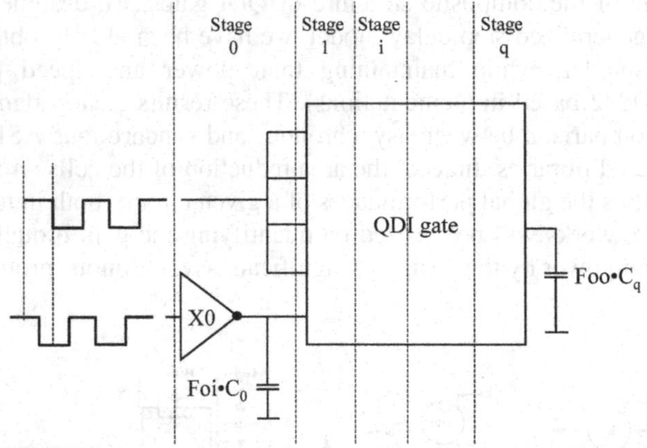

Fig. 3. Generic CMOS structure used to size our gate and considered during the comparison protocol

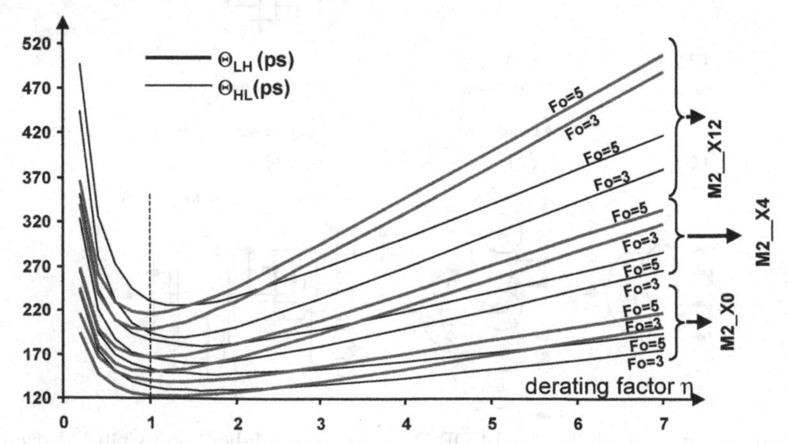

Fig. 4. Quality evaluation of the sizing criterion (14); case of a Muller implementation. ($Fo=C_L/Ci+1$)

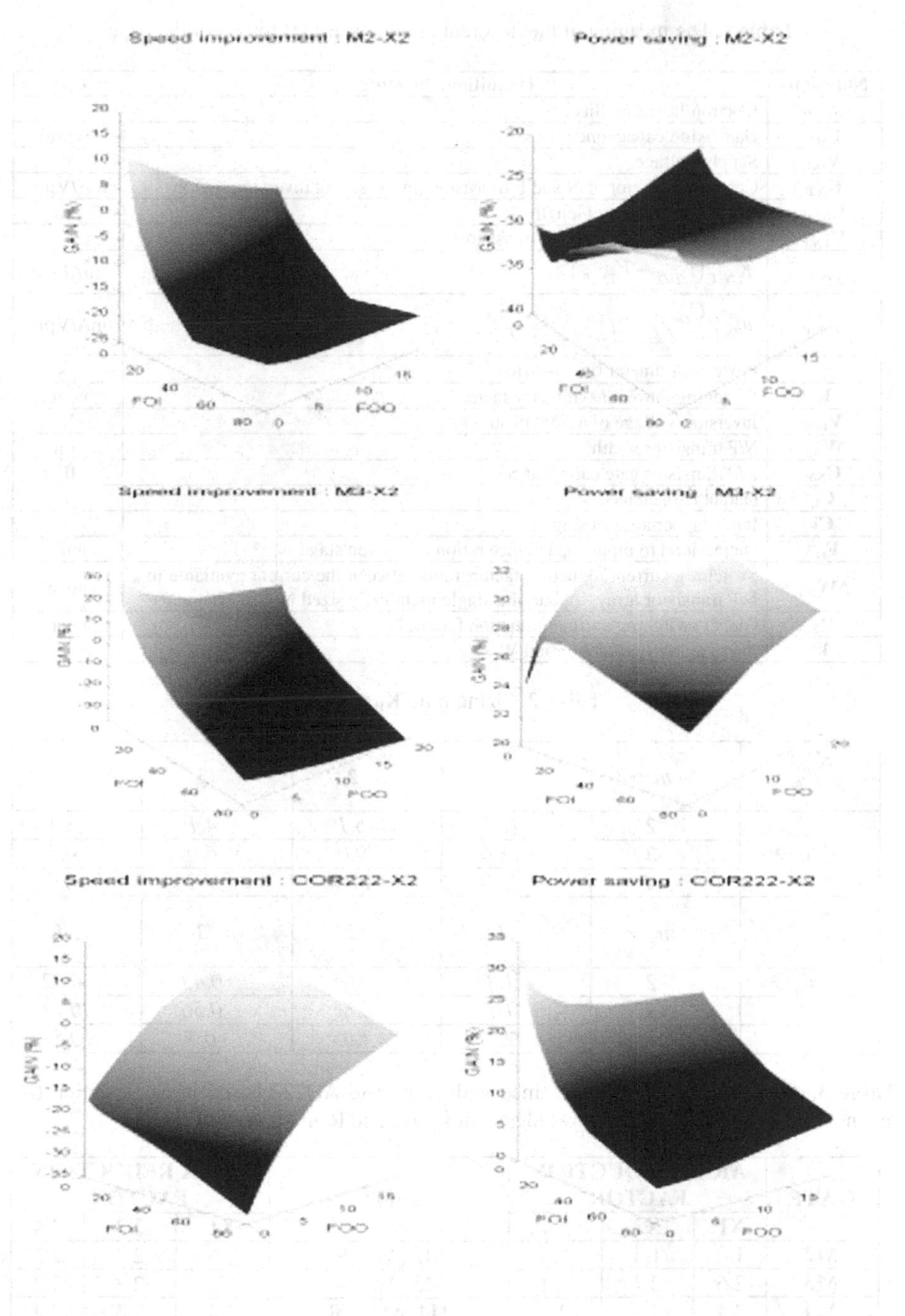

Fig. 5. Speed improvement and the power saving obtained with respect to AO222 based implementations for various loading and controlling design conditions.

Table 1. The meaning of the different notations used throughout the paper

Notations	Definition, meaning	Unit
$\mu_{0N,P}$	Electron/hole mobility	
C_{OX}	Gate oxide capacitance	$fF/\mu m^2$
V_{DD}	Supply voltage	V
$K_{N,P}$	Conduction factor of N and P transistors in the strong inversion region	mA/Vμm
L_{GEO}	Geometrical channel length	μm
$V_{TN,P}$	Threshold voltages of N/P transistors	V
$\alpha_{N/P}$	$K_{N,P}(V_{DD}-V_{TN,P})$	mA/μm
$\alpha'_{N/P}$	$\mu_{0N,P}\frac{C_{OX}}{L_{GEO}}(V_{DD}-V_{TN,P})$	mA/Vμm
τ	Process parameter time metric	ps
R	Switching current asymmetry factor	none
V_{INV}	Inversion voltage of a CMOS structure	V
$W_{N,P}$	N/P transistor width	μm
$C_{N/P}$	N/P transistor gate capacitance	fF
C_L	Output capacitance	fF
Ci	Input capacitance of stage i.	
F_O^L	Output load to input capacitance ration of a given stage	none
$\Delta W_{N,P}$	Switching current reduction factor: ratio between the current available in a N/P transistor array to that of a single identically sized N/P transistor.	none
$SD_{N,P}$	Global switching current reduction factor.	none
k	Internal configuration ration W_P/W_N	none

Table 2. Sizing ratio $K_{P,N^L \rightarrow N,P^H}$

$K_{P^L \rightarrow N^H}$	$m \rightarrow$	*1*	2	3	4
	2	*12.2*	*6.1*	*4.1*	*3.1*
$n \rightarrow$	3	*18.0*	*9.0*	*6.0*	*4.5*
	4	*23.8*	*11.9*	*7.9*	*6.0*
$K_{N^L \rightarrow P^H}$	$m_p \rightarrow$	*1*	2	3	4
$n_n \rightarrow$	2	*1.3*	*0.65*	*0.43*	*0.32*
	3	*1.7*	*0.85*	*0.56*	*0.42*
	4	*2.1*	*1.05*	*0.7*	*0.53*

Table 3. Area reduction factors obtained with respect to AO222 based implementation (Ack means that there is an input acknowledgement signal, and R an input reset signal)

GATE	AREA REDUCTION FACTOR			GATE	AREA REDUCTION FACTOR		
	X1	X2	X4		X1	X2	X4
M2	1.0	1.4	1.5	M2_Ack_R	2.5	2.3	2.3
M3	2.6	2.7	2.4	M3_Ack_R	2.3	2.3	2.2
M4	3.4	3.5	2.3	M4_ACK_R	2.4	2.4	2.4
COR222	1.8	1.8	2.0	COR222_Ack_R	2.5	2.5	2.1
COR221	1.3	1.3	1.5	COR221_Ack_R	2.2	2.2	2.1
COR211	0.9	0.9	1.0	COR211_Ack_R	1.7	1.5	1.9

References

[1] T. Chelcea, A. Bardsley, D. A. Edwards, S. M. Nowick" A Burst-Mode Oriented Back-End for the Balsa Synthesis System " Proceedings of DATE '02, pp. 330–337 Paris, March 2002
[2] J. Cortadella, M. Kishinevsky, A. Kondratyev, L. Lavagno and A. Yakovlev "Petrify: a tool for manipulating concurrent specifications and synthesis of asynchronous controllers" IEICE Transactions on Information and Systems, Vol. E80-D, No. 3, March 1997, pages 315–325.
[3] M. Renaudin et al, "TAST", Tutorial given at the 8th international Symposium on Advanced Research in Asynchronous Circuits and Systems, Manchester, UK, Apr. 8–11, 2002
[4] T. Sakurai and A.R. Newton, "Alpha-power model, and its application to CMOS inverter delay and other formulas", J. of Solid State Circuits vol.25, pp.584–594, April 1990.
[5] P. Maurine, M. Rezzoug, N. Azemard, D. Auvergne, "Transition Time modelling in Deep Submicron CMOS" IEEE trans. on CAD of IC and Systems, vol. 21, n°11, November 2002
[6] A. Chatzigeorgiou , S. Nikolaidis, I. Tsoukalas," A modelling technique for CMOS gates" IEEE Trans on CAD of Integrated Circuits and Systems, Vol.18, n°5, May 1999
[7] B. S. Cherkauer, E. G. Friedman " A Unified Design Methodology for CMOS Tapered Buffers" IEEE Transactions on VLSI Systems, vol.3,n°1,pp.99–111, March 1995
[8] N. Hedenstierna, K. O. Jeppson "CMOS Circuit Speed and Buffer Optimization" IEEE transactions on CAD, vol. CAD-6, n°2, pp. 270-281, March 1987
[9] C. Piguet, J. Zhand "Electrical Design of Dynamic and Static Speed Independent CMOS Circuits from Signal Transistion Graphs" PATMOS '98, pp. 357–366,1998.
[10] I. Sutherland, B. Sproull, D. Harris, "Logical Effort: Designing Fast CMOS Circuits", Morgan Kaufmann Publishers, INC., San Francisco, California, 1999.
[11] C. Mead N. Conway „Introduction to VLSI systems“ Reading MA: Addison Wesley 1980
[12] M. Renaudin, "Asynchronous Circuits and Systems : a promising design alternative", in "Microelectronics for Telecommunications : managing high complexity and mobility" (MIGAS 2000), special issue of the Microelectronics-Engineering Journal, Elsevier Science, Guest Editors : P. Senn, M. Renaudin, J. Boussey, Vol. 54, N° 1–2, December 2000, pp. 133–149.

The Emergence of Design for Energy Efficiency: An EDA Perspective

Antun Domic

Synopsys Inc.
Mountain View, CA 94043 USA
antun.domic@synopsys.com

Abstract. As feature size is getting smaller, gate counts and clock speed are increasing, and so does power consumption. Designing for low power consuption has thus become mandatory for a large fraction of the ICs that will hit the market in the near future.
This presentation offers the view of the EDA world to the problem of low power design, highligthing the current status of design methodologies and tools, as well as providing a forecast of how the EDA vendors intend to face the power problem for the new generations of electronic technologies.

J.J. Chico and E. Macii (Eds.): PATMOS 2003, LNCS 2799, p. 192, 2003.
© Springer-Verlag Berlin Heidelberg 2003

The Most Complete Mixed-Signal Simulation Solution with ADVance MS

Jean Oudinot

Mentor Graphics
13/15, rue Jeanne Braconnier
Meudon La Foret, 92360 France
jean_oudinot@mentorg.com

Abstract. The Mentor Graphics ADVance MS (ADMS) simulator gives designers the most comprehensive environment they need to develop complex analog and mixed-signal (AMS) circuit and System-on-Chip (SoC) designs. The single-kernel ADMS architecture combines the advantages of several high-performance simulation engines: Eldo for analog, ModelSim for digital, Mach for fast, transistor-level and Eldo RF for radio frequency simulation. It uses a single netlist hierarchy and is language-neutral, allowing designers to freely mix VHDL, Verilog, VHDL-AMS, Verilog-A, SPICE and C anywhere in their designs. This makes ADMS the most efficient solution for top-down design and bottom-up verification for the AMS SoC development.

J.J. Chico and E. Macii (Eds.): PATMOS 2003, LNCS 2799, p. 193, 2003.
© Springer-Verlag Berlin Heidelberg 2003

Signal Integrity and Power Supply Network Analysis of Deep SubMicron Chips

Louis K. Scheffer

Cadence
555 River Oaks Parkway
San Jose, CA 95134 USA
lou@cadence.com

Abstract. In the recent past, chip designers were mainly worried about logical correctness, adherence to design rules, and chip timing. Neither the power supply network, nor the signal nets, needed to be analyzed in detail for analog effects or possible reliability problems.
However, in modern DSM chip technologies, these simplifications are no longer good enough. The power supply network, for example, must be designed from the beginning as opposed to added as an afterthought, and must take into account many other aspects of the chip, from the package in which it will be placed, to the detailed needs of the logic that composes the blocks.
Furthermore, reliability issues are even more worrying, for these cannot= be found by testing when the chip is new - they develop over time and may cause a chip to fail in the field that worked perfectly well on the tester. Electromigration, wire self heat, and hot-electron effects fall into this category, affecting the power supply wires, the signal wires, and the gates respectively.
A new generation of tools has been developed to address these problems, both for power supply networks and signal nets. This paper and presentation will discuss the motivation, technology, and operation of these tools. The power supply analysis tools combine static and dynamic analysis to ensure that the quality of the delivered power is sufficiently high and that the supply network will be reliable over the life of the chip. In addition, since power supply voltage has a big effect on both timing and=
signal integrity, the power supply analysis tools must relay their findings to these other tools, and these other tools must correctly account for these variations.
The signal net analysis tools look at different analog effects. They look at changes in timing induced by changes in power supplies and the behavior of nearby nets. Furthermore, they examine the gates and wires for any potential reliability problems.
Finally, we will discuss some of the strategies employed by tools such as placement, routing, and synthesis, and how they attempt to avoid these problems in the first place.
Such a flow, combining both avoidance and accurate analysis, is necessary to allow both high productivity and adherence to the more complex constraints of DSM design.

J.J. Chico and E. Macii (Eds.): PATMOS 2003, LNCS 2799, p. 194, 2003.
© Springer-Verlag Berlin Heidelberg 2003

Power Management in Synopsys Galaxy Design Platform

Synopsys Inc.
Mountain View,CA 94043 USA

Abstract. Designers continue to be challenged with the need to manage power together with timing and signal integrity throughout the design flow. Synopsys Low Power solution optimizes and verifies your low power designs within the Galaxy^{TM} Design Platform. Learn how to optimize for dynamic and leakage power using Power Compiler^{TM} within Synopsys industry-leading synthesis flow. Discover how PrimePower precisely debugs your design through accurate peak power verification.

J.J. Chico and E. Macii (Eds.): PATMOS 2003, LNCS 2799, p. 195, 2003.
© Springer-Verlag Berlin Heidelberg 2003

Open Multimedia Platform for Next-Generation Mobile Devices

STMicroelectronics
Via C. Olivetti, 2
Agrate Brianza (Milano), 20041 ITALY

Abstract. STMicroelectronics is at the forefront of the digital multimedia revolution, enabling the consumer electronics industry with highly integrated platformbased solutions. As a leading supplier of integrated circuits (ICs) for mobile phones, ST also aims to make compelling multimedia a practical reality for handheld devices. An ambitious development effort has led to a remarkable application processor that enables smart phones, wireless PDAs, Internet appliances and car entertainment systems to playback media content, record pictures and video clips, and perform audiovisual communicate with other systems.

J.J. Chico and E. Macii (Eds.): PATMOS 2003, LNCS 2799, p. 196, 2003.
© Springer-Verlag Berlin Heidelberg 2003

Statistical Power Estimation of Behavioral Descriptions*

B. Arts[1], A. Bellu[2], L. Benini[3], N. van der Eng[1], M. Heijligers[1],
E. Macii[2], A. Milia[2], R. Maro[2**], H. Munk[1], and F. Theeuwen[1]

[1] Philips Research Labs, ED&T/Synthesis, Eindhoven, NL
[2] Politecnico di Torino, DAUIN, Torino, I
[3] Università di Bologna, DEIS, Bologna, I

Abstract. Power estimation of behavioral descriptions is a difficult task, as it entails inferring the hardware architecture on which the behavioral specification will be mapped through synthesis *before* the synthesis is actually performed. To cope with the uncertainties related to handling behavioral descriptions, we introduce the concept of statistical estimation, and we sketch how a prototype statistical power estimator has been implemented within a high-level design exploration framework (the AspeCts environment).

1 Introduction

The addition of the power dimension to the already large area/speed design space dramatically expands the number of available design alternatives. Therefore, in a power-conscious synthesis flow, it is of increasing importance the ability of estimating the impact of the various choices made by the designers or the effects of automatic optimizations on the final power budget.

1.1 Previous Work

Most of the research on power estimation has initially focused on gate and transistor levels, and only recently it has extended to higher levels of abstraction (i.e, RTL and behavioral).

RTL power estimation is at the transition point between research and industrial applications. RTL power estimators are commercially available [1,2,3] and are now gaining widespread acceptance in the design practice.

Our work focuses on an even higher level of abstraction, namely, the behavioral level. The main challenge in behavioral power estimation is the filling of the abstraction gap that does exist between specification and implementation. A behavioral specification leaves a significant amount of freedom for hardware implementation: Scheduling of operations, type of functional units, assignment. In other words, a single behavioral specification can be mapped to a wide variety of hardware architectures, characterized by largely different power consumption.

* This work was supported by Philips Electronics B.V. under grant n. 200/2001.
** Now at Motorola, Torino Labs.

J.J. Chico and E. Macii (Eds.): PATMOS 2003, LNCS 2799, pp. 197–207, 2003.
© Springer-Verlag Berlin Heidelberg 2003

Most behavioral power estimation approaches [4] adopt a highly simplified power consumption model, where an energy cost is associated to the operations performed in the behavioral specification, and total energy is computed by summing the cost of all operations. The main shortcoming of this model is that it does not account for two critical factors, namely: (i) The presence of hardware which is required for correctly performing a computation, but it is not explicitly represented as operations in the behavioral description. (ii) The switching activity dependency of power consumption. Neglecting these factors can lead to inaccurate estimates, therefore adequate countermeasures have to be taken.

The approach adopted to overcome the accuracy limitation of the cost-per-operation model follows the direction of specifying more details on how the mapping of behavior onto implementation is performed in the target design flow. This is similar to the "fast synthesis" step performed by advanced RTL power estimators, where a simplified gate mapping algorithm is applied to logic blocks for which a macro-model does not exist. In behavioral power estimation, information on operation scheduling and on the number and type of functional units can be taken into account to reduce uncertainty on the final hardware implementation. However, given the high-level of abstraction of this information, a significant level of uncertainty remains on the final hardware implementation. Hence, several authors have remarked that power estimation at the behavioral level should not produce a single-value estimate, but it should identify ranges representing uncertainty intervals. A few techniques have been proposed to compute these intervals. A deterministic lower-bound vs. upper bound solution was proposed in [5]. In alternative, probabilistic approaches, modeling the uncertainty on power estimation as the sample space of a random variable have been explored in [6]. The techniques of this paper follow and extend the latter approach.

1.2 Contribution

We introduce the concept of statistical power estimation for hardware components that are not explicitly instantiated in the behavioral description being considered (e.g., steering logic, data-path registers, wiring and control logic). We call these components *implicit*, because they are required in the RTL implementation, and yet they are not explicitly represented in the design database created from the input description (C code, in our case). To estimate the power consumed by these components, we perform Monte-Carlo sampling of multiple instantiations, that is, we consider architectures for which different assumptions for the implicit components are made. Then, we obtain a distribution of average power values, on which we perform statistical analysis.

Besides models for functional units, we also propose statistical models for steering logic, clock, wiring, controller and memory ; this with the purpose of enabling the implementation of a complete power estimation prototype tool.

The estimator was developed as an add-on library to the AspeCts behavioral exploration tool by Philips [7], and accuracy assessment was made against data determined on synthesized RTL descriptions (derived from some benchmarks) by means of state-of-the-art tools (namely, BullDAST PowerChecker [3]).

2 The AspeCts Environment

The design of electronic systems can roughly be divided into two phases. In the first phase, the functionality of the system is specified, typically defined by a collection of communicating processes. Each of these processes is specified by means of a behavioral C algorithm. The second phase concerns the actual implementation of the system by defining a specific hardware architecture consisting of a collection of connected components and mapping the processes onto these components.

A system-level simulator offers the capability of investigating user-defined mappings, by visualizing the performance of the system as a whole, so that the designer can make design trade-offs.

The input of a system-level simulator consists of a system-level specification, a hardware structure made of a collection of connected components, and a mapping of processes to components. Because the whole system is under investigation, and because a specification tends to change even after the implementation phase has started, this might result in many different mappings. If an algorithm is mapped to a dedicated, but not yet designed, hardware component, it is not realistic to describe each possible mapping by a dedicated implementation or specific production code.

Therefore, system-level simulation tools provide the possibility of characterizing a mapping by means of a performance model. The simulation of the behavioral system-level description is driven by the performance models provided. In practice, a designer is confronted with algorithms not developed by himself. Besides that, the designer has to cope with some predefined hardware constraints, relate this to the algorithms, and keep track of the overall implementation consequences. This makes it very difficult to generate realistic performance models, without detailed and time-consuming analysis of the code. Automating these tasks is thus desirable.

The AspeCts environment fulfills these goals. It consists of two main parts: The first one is in charge of executing a C algorithm, scheduling all activities that take place, and storing such activities in a database. In the second part, the database is used to perform memory analysis and optimization based on the scheduled activities.

2.1 Execution Tracing, Data-Flow Analysis, and Scheduling

To be able to investigate the performance of a C algorithm with respect to its typical usage, a simulation-based approach was chosen.

During program simulation, the execution trace is determined, i.e., the sequence of operations which are executed when running the program. Timing information to the arguments of these operations is then added; scheduling and data-dependency analysis is thus implicitly performed.

AspeCts supports different scheduling strategies, including sequential, data-flow based and control-flow based. To consider implementation hints provided by the designer, the scheduler in AspeCts also accepts various kinds of constraints, including timing constraints and functional resource constraints.

2.2 Memory Analysis

Memory becomes more and more important in the design of electronic systems, as area, delay and power dissipation, heavily depend on the chosen memory architectures. A designer can be supported in making design decision by good estimations of the required memory capacity and bandwidth. Therefore, analysis is required to prove the feasibility of a chosen memory architecture for a given algorithmic specification. Because most scalars are stored in data-path registers and because compilers and high-level synthesis tools can handle scalars quite well, these steps are mainly intended for multi-dimensional arrays. In C terminology, this means that we assume that memory is closely related to the concept of C arrays. To provide unlimited memory bandwidth and size for all arrays, and hence enable a maximum parallel schedule where all array accesses can be scheduled without restrictions, the method assumes that each array is "mapped to" a separate memory. To offer the designer a way to define specific memory structures, the following tasks can be performed: Memory allocation, binding, compaction, sizing, bandwidth analysis and correctness checks. The final feedback provided to the designer includes, for each memory array, information about its size, the number of productions and consumptions, and the maximum number of values stored in the array simultaneously.

3 Power Modeling in AspeCts

3.1 Functional Units

One of the major sources of error in power estimation originates from input pattern dependency. Our functional unit models are sensitive to the average switching activity and signal probability at their inputs. Estimates of these quantities are produced by sampling the values propagating through the various operations during the execution of an AspeCts model. The activity-driven macro-models are look-up tables (LUT), which have proved to be very reliable and robust over a wide range of input statistics.

LUT models make use of a tabular representation of the functional relation between boundary statistics and power. The number of parameters used to represent boundary statistics has a great impact on model size, since each independent variable adds a dimension to the parameter space. We used three parameters (the average input signal probability P_{in}, the average input activity D_{in} and the average output activity D_{out}), leading to a 3-dimensional LUT.

Discretization is applied to the parameters in order to obtain a finite number of configurations to be associated to LUT entries. Characterization consists of performing low-level power simulation experiments for all configurations of the discretized boundary parameters, in order to characterize the corresponding LUT entry by storing the power value obtained in it.

The key advantage of LUT models over regression equations is their capabilities of representing non-linear dependencies. On the other hand, the proper use of LUT models involves a critical trade-off between model size (the number of entries in a LUT is $L * N$, where L is the number of discretization levels and N the number of parameters), characterization effort and accuracy.

3.2 Steering Logic

The assumption on the structure of the steering logic is that it is implemented by means of a multiplexer network. The basic brick of the steering logic net is thus the 2-to-1 MUX. An activity-sensitive power model for this element is characterized as any other functional component for the chosen technology library. The size and shape of the steering logic network is determined by the number of inputs such a network is supposed to convey to the inputs of the shared functional units and shared registers. This approach, which implicitly assumes a quasi-balanced tree network of 2-to-1 MUXes, offers a very simple mean for estimating the steering logic power. In fact, the process does not require the introduction of scaling factors for extending the characterized model to components with wider bit-width, nor it requires additional runs of the model characterization procedure. The drawback of this solution, however, stands in the fact that, when real hardware is synthesized, steering logic trees are usually unbalanced and, most important, other components than 2-to-1 MUXes are instantiated. As a consequence, power estimates achieved with this model tend to be pessimistic.

3.3 Clock and Wires

For what concerns the estimation of the power consumed by the clock tree network and by the interconnect wires, the models we have developed are somehow reminiscent of the models used by BullDAST PowerChecker [3] in the context of RTL power analysis. They are based on the observation that the power consumed by the clock tree (interconnect wires) is in strict relationship with the complexity of the clock network (interconnect network); this, in turn, is very much related to the number of sequential elements that the clock signal has to drive (number of estimated cells in the gate-level netlist of the design). The models we have adopted are analytical and depend, primarily, on the number of estimated registers (estimated cells) of the input description.

3.4 Controller

The logic that controls steering of MUXes and handling of loops and conditionals is not explicitly represented in any AspeCts internal data structure. Yet, it is a known fact that control logic can consume a non-negligible fraction of the total power, therefore it should be taken into account to improve the accuracy of power estimation. The structure of the control logic, a finite-state-machine (FSM), is in essence determined by the scheduling. Once the schedule is known, the controller has to follow the sequence of control steps to activate the required components with control signals that drive the steering logic.

For a pure data-flow specification, control flow is straight-line (i.e., no conditionals and loops) and the FSM is very simple. It is a chain of states with an added transition from the last state to the first (which represents the re-start of a computation) and optionally a self-loop in the first state (which represents the reset state, if present).

Let us assume, for the time being, a simple single-chain structure for the FSM (this assumption will be relaxed later). Controller power estimation is based on a simple characterization procedure. In a preliminary pre-processing step, a data set is created by instantiating several *chain* FSMs, with varying number of states and outputs, synthesizing them and estimating their power consumption via gate-level simulation. Then, the data are collected in a look-up table and an interpolation/extrapolation rule is provided. The final results of the characterization procedure can be viewed as the construction of a power model $P(N_{states}, N_{outputs})$, which returns an expected power consumption for a chain FSM with number of states N_{states} and number of outputs N_{ouputs}. Notice that this model is not input pattern sensitive because the FSM is autonomous. In presence of a reset-state self-loop, we can add another sensitivity parameter in the model, namely the percentage of time in which the reset signal is asserted.

If the specification is not a pure data-flow, but it contains significant control-flow elements, the model of the FSM power is more complex, but it is obtained via a similar characterization process. In this case, the modeling process is based on the assumption that a FSM for a general specification, with non-trivial control flow is mostly chain-based, with multiple-exit states, corresponding to conditionals, and feedback edges, corresponding to loops. In this case, the FSM logic will be more complex to account for the additional edges, hence the power model contains a third parameter, i.e., the number of *extra edges* w.r.t. a single-chain FSM (where the number of edges is the same as the number of states). Thus, the general power model becomes $P(N_{states}, N_{outputs}, N_{extraedges})$. Characterization of the model is performed with a similar process as for single-chain FSMs, with the only difference that a larger sample FSM set has to be generated, in order to sample also the third parameter $N_{extraedges}$.

It is clear that, in this case, the path followed by the FSM is data-dependent; therefore, we expect input pattern dependence, to some degree, of the FSM power consumption. However, this effect is quite minor, especially if a flat FSM with logarithmic state encoding is generated, because all the next state logic is shared and most of the output logic has, as input, all the state registers.

3.5 Memories

Power modeling for memories (i.e., the pre-characterization of cost-per-access constants) does not leverage the characterization flow assumed for functional units, because memories are either instantiated as hard macros or they are created directly at the layout level by memory generators. If hard memory macros are available, their energy-per-access constants are usually specified in their datasheets. Similarly, memory generators usually provide power views to designers. In case of incompletely specified memory macros, where the power view is absent, we provide a parametric technology-based model, which determines cost-per-access estimates for memory macros given a few basic technology-dependent parameters, such as wire capacitance, drain and gate capacitances, power supply voltage. This default power model is based on CACTI, a memory performance and power estimation tool widely used in the research literature [8].

4 Statistical Power Estimation

The main purpose of the statistical estimator, called in the sequel Monte-Carlo AspeCts (MC-AspeCts), is to provide a quantitative assessment of the uncertainty associated with power estimation at a high level of abstraction.

4.1 The Concept

It is not realistic assuming that the accuracy of behavioral (C-level) power estimation can be comparable to that of RTL estimation: Design choices associated to these degrees of freedom are not fully specified, and a margin of uncertainty still remains on the consumption of the final hardware. Providing a single-value estimate automatically implies an arbitrary selection of a hardware configuration, which might be far away from the final implementation. It is therefore important giving the designer a feeling for the uncertainty while, at the same time, providing him/her with a meaningful power estimate.

To be more specific, let us focus on a concrete example: Consider a behavioral description where 4 operations have to be mapped onto two functional modules (e.g., four additions mapped onto two adders), as depicted in the data-flow of Figure 1. In AspeCts, the user specifies a resource constraint (i.e., two adders), but he/she does not explicitly binds operations to functional modules. Internally, AspeCts will map operations onto functional modules (e.g., OP1, OP2, executed by adder ADD1, and OP3, OP4 executed by ADD2), but this assignment is not guaranteed to be the same as in the final implementation. Indeed, many other assignments, which are compatible with the given scheduling and allocation constraints, are possible. These assignments do not violate neither user constraints nor scheduling decisions, but they can have a significant impact on power consumption. This is because the sharing pattern of a functional module determines the switching activity at its inputs and outputs, which in turns determines its switching activity and its internal power dissipation.

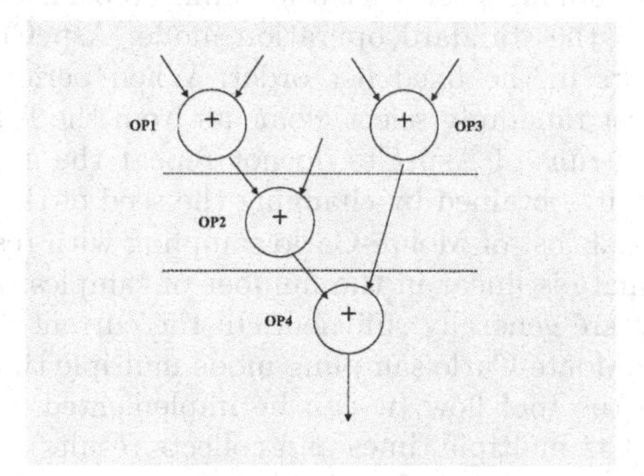

Fig. 1. Example of Data-Flow Graph.

To better explain this concept, let us consider the data-flow graph of Figure 1 with a resource constraint of 2 adders. The assignment of OP1 and OP2 to ADD1 and OP3, OP4 to ADD2 are compatible with the schedule and with the resource constraint, but so is the assignment of OP1, OP2, OP4 to ADD1 and OP3 to ADD2. Clearly, the switching activity and, consequently the power consumption of ADD1 and ADD2 can change significantly in the two cases.

In addition, we note that the two bindings produce completely different steering logic (MUX trees) at the inputs of the two adders: Four 2-input MUXes in the first case, and two 3-input MUXes in the second (at the inputs of ADD1 only). To address the unavoidable uncertainty on switching activity and on steering logic structure caused by the high level of abstraction at which we perform power estimation, we view the power consumption as a random variable P. The sample space of such a variable is the space of all operation bindings compatible with scheduling and resource constraints, and the values of the variable are the power dissipation estimates for a given binding. In order to estimate power for a design we should not limit ourselves to providing a single value of the random variable; instead, we should characterize its distribution in terms of mean value and variance. This would allow a designer to perform design space exploration without going through the complete design cycle for different architectural options.

The statistical distribution of the random variable P is unknown a priori. To estimate its mean value and variance, we resort to Monte-Carlo sampling. In other words, we generate a number of bindings, compatible with resource constraints and scheduling, and we estimate the power consumption for each one of them. Exploration of the sample space stops when a Monte-Carlo convergence test confirms that a stable average value and variance estimate has been obtained. Thanks to the central limit theorem, we know that Monte-Carlo convergence is very rapid, and in practice a few tens of sample points are sufficient to obtain reliable estimates.

4.2 Implementation in AspeCts

Monte-Carlo sampling of the entire design space of legal bindings is made quite easy in the AspeCts framework. At each scheduling step a list of "free" functional units is available. In the standard operation mode, AspeCts scans the list in order, assigning units in the fixed list order. When performing Monte-Carlo sampling, we can just randomly select elements from the list. We only have to ensure that different runs of AspeCts do not repeat the same pseudo-random sequence. This is easily obtained by changing the seed of the RAND function.

The computational cost of Monte-Carlo sampling with respect to the cost of a single power estimate is linear in the number of samples. As stated before, a few tens of samples are generally sufficient. In the current implementation, we just run AspeCts in Monte-Carlo sampling mode multiple times. This choice has minimal impact on the tool flow (it can be implemented with a simple script that calls MC-AspeCts multiple times and collects results) and it can be easily parallelized on multiple machines for speed-up purposes.

The final output of MC-AspeCts is an estimate of the mean value $\mathtt{mean}(P)$ of random variable P and its variance $\mathtt{var}(P)$. It should be clear that $\mathtt{mean}(P)$ and $\mathtt{var}(P)$ give information on what is the expected power consumption of an implementation of the design, and how much uncertainty we have on this estimate because of the degrees of freedom left unspecified at the high level of abstraction. In addition to the estimate regarding the whole design, the breakdown of power consumption over the different components of the design is optionally provided.

5 Experimental Results

We have performed extensive benchmarking of the MC-AspeCts tool. In the sequel, we first report some results concerning the analysis of the Monte-Carlo convergence of the estimation. Then, we show some results about the accuracy we have achieved in the estimation with respect to synthesized RTL descriptions obtained by imposing resource and throughput constraints similar to those used for AspeCts estimation. RTL estimates have been determined using BullDAST PowerChecker; model characterization for both MC-AspeCts and PowerChecker was done using a Philips 0.18um CMOS technology library.

5.1 Results on Monte-Carlo Convergence

The example we consider is the simple loop shown below:

```
for (I=0; I<n; I++) {
  A = vect[I] * x; B = vect[I} * y;
  C = vect[I] * z; D = vect[I] * w;
  F = A + B; G = C + D;
  Out = F + G;
}
```

This example enables, for a given set of resource constraints, different assignments of data to operators and registers, and thus well exercises the calculation of mean value and variance of the power random variable. The results obtained for a given test-bench (that is, a vector of input samples, whose length was fixed to 1000 integer numbers) are reported in Figure 2.

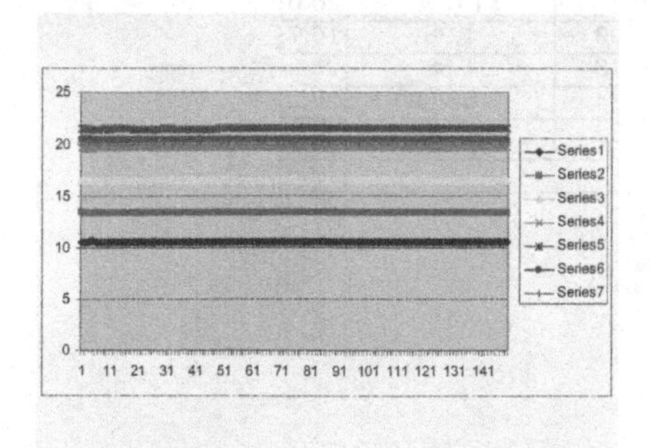

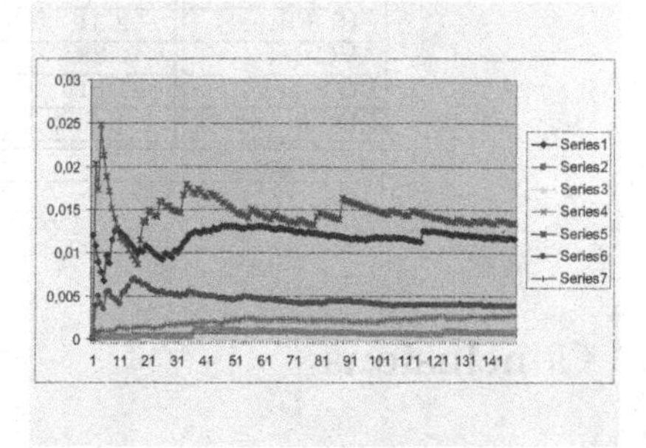

Fig. 2. Mean Value (left) and Variance (right) for Simple Example.

The curves in the diagrams, identified as *Series*, refer to different configurations of the resource constraints. Depending on the degrees of freedom in the allocation of data to resources, the variance of power is larger and requires a longer time to settle (the configuration in which there is no freedom, i.e., 4 multipliers and 2 adders, has null variance and it is not shown in the diagram).

The results confirm that Monte-Carlo convergence is achieved very quickly (no more than 50 simulation are needed to reach stability in the variance). Also, the mean value is very stable and provides a good indication of the *average case* power consumption of the generic architecture. Last, variance analysis allows to early determining *risky* constraint configurations, that is, configurations for which power consumption may vary after the architecture is synthesized.

5.2 Power Estimation Results

Eight examples, specified in C, constitute the benchmark suite for assessing the estimation accuracy of MC-AspeCts. For each example, one or more streams of input patterns characterized by different correlations were defined; this with the goal of testing the tool under different operating conditions of the benchmarks. Specific ad-hoc files (e.g., image files) were also used for some benchmarks.

Table 1 collects the data (power values are in mW). The estimation error is around 22%, on average, with values of the variance well below 0.40, indicating that the power random variable has different stability depending on the benchmark. By looking at the breakdown of power estimates over the various components (not reported here for space reasons), we observe that the major sources of inaccuracy come from MUX, wiring and clock power estimation, while results are satisfactory for what concerns functional units, control logic and memories.

Table 1. Statistical Estimation Results.

Benchmark	*MC-AspeCts*		*PowerChecker*	Δ
	Mean Val	*Variance*		[%]
GCD	1.95	0.18	1.71	14.0
BSORT	178.17	0.31	151.53	17.6
SSORT	127.53	0.12	107.82	18.3
Compl-Mult	89.46	0.37	74.17	20.6
FIR Filter	2.31	0.09	1.98	16.7
AVG-Int	2.38	0.13	1.88	26.6
PCX-Dec	218.19	0.34	168.89	29.2
EDGE-Detect	18.75	0.22	13.72	36.7
Average				22.6

6 Conclusions

We have proposed a solution to the problem of estimating power consumption of a behavioral descvription based on the concept of statistical power modeling. This concept has been implemented into a prototype tool built on top of an industry-strength design exploration environment. The estimation results obtained on a set of benchmarks have been validated against RTL estimation results determined using a commercial power estimator.

References

1. Synopsys, Inc., *Synopsys PowerCompiler*, www.synopsys.com.
2. Sequence Design, Inc., *PowerTheater*, www.sequencedesign.com.
3. BullDAST s.r.l., *PowerChecker*, www.bulldast.com.
4. R. San Martin, J. P. Knight, "Power-Profiler: Optimizing ASICs Power Consumption at the Behavioral Level," *DAC-32*, pp. 42–47, 1995.
5. L. Kruse, *et al.*, "Estimation of Lower and Upper Bounds on the Power Consumption from Scheduled Data Flow Graphs," *IEEE TVLSI*, Vol. 9, pp. 3–14, 2001.
6. D. Bruni, A. Bogliolo, L. Benini, "Statistical Design Space Exploration for Application-Specific Unit Synthesis," *DAC-38*, pp. 641–646, 2001.
7. M. J. M. Heijligers, A. Hogenhuis, "Analyzing Architectural AspeCts of Behavioral Descriptions," *Conf. on Embedded System Design*, pp. 239–250, 2000
8. S. J. E. Wilton, N. P. Jouppi, ' 'CACTI: An Enhanced Cache Access and Cycle Time Model," *IEEE JSSC*, Vol. 31, pp. 677–688, 1996.

A Statistical Power Model for Non-synthetic RTL Operators*

Maurizio Bruno[1], Alberto Macii[2], and Massimo Poncino[3]

[1] BullDast s.r.l., Torino, ITALY
[2] Politecnico di Torino, Torino, ITALY
[3] Università di Verona, Verona, ITALY

Abstract. Power estimation at the Register-Transfer level is usually narrowed down to the problem of building accurate power models for the RTL (synthetic) operators. In this work we show that, when RTL power estimation is integrated into a realistic design flow, other types of primitives need to be accurately modeled. In particular, we show that most of the RTL functionality is realized by sparse logic elements.
We thus propose statistical power models for these primitives, that we have validated on a set of industrial benchmarks.

1 Introduction

The need of pre-synthesis power estimation tools has long been considered as crucial in a design flow, and particularly at the register-transfer (RT) level, which can be considered the standard entry level for most designers. Interestingly, from the research point of view, RTL power estimation is deemed today as a solved problem, as witnessed by the significant reduction of the literature on topic in the last few years.
Research on RTL power estimation has in fact defined an estimation paradigm that has shown to be superior to any other, in terms of both accuracy, robustness, and flexibility: The *macro-modeling* approach. This paradigm relies on a view of the RTL description as a set of instantiated "modules" (often called *soft macros* or *synthetic operators*), corresponding to typical HDL operators (such as adders, multipliers, or comparators), or more complex blocks (such as counters, or shift registers). Besides these modules, a RTL description includes also a finite-state machine that represents the controller. This model fits nicely into the Finite State Machine with Datapath (FSMD) model of [1], and is common to all the estimation approaches based on macromodeling. Under this assumption, RTL power estimation amounts to the construction of accurate *power models* for the synthetic operators, and possibly for the controller; the total power consumption of the design will simply consist of summing all the contributions given by the individual power models, plus the contribution of the controller.
Although this view may differ from what is the designers' perception of RTL (i.e., an HDL description containing clocked processes), it is somehow consistent

* This work is supported, in part, by the EC under contract IST-2001-30125 "POET"

J.J. Chico and E. Macii (Eds.): PATMOS 2003, LNCS 2799, pp. 208–218, 2003.
© Springer-Verlag Berlin Heidelberg 2003

with the internal format kept by conventional RTL synthesis tools as a result of the translation of the HDL description. This process is called, in the VHDL terminology, *elaboration*. Unfortunately, this consistency is only apparent: Elaboration builds a structure based on automatic inferencing rules that have limited semantic power. Therefore, the view of RTL structure as a FSMD is too ideal. The result of elaboration consists basically of the instantiation of four basic types of primitives: *gates*, *macros*, *selectors* (i.e., multiplexers), and *memory elements* (i.e., flip-flops).
Apparently, the difference with the FSMD model is limited to how the controller is specified: Since macros represent RTL operators, gates, selectors and memory elements should represent the building blocks that constitute the controller. In practice, however, the degree of inferencing is quite limited, and only the basic HDL operators are inferred ("+","-", and "*"). For example, registers of the datapath are not instantiated as a macro, and are translated as a set of memory elements. In other terms, the granularity of RTL designs is quite small, and only a few macros are typically exposed in the RTL internal database. As an example, Table 1 shows a breakdown of the internal RTL database for a set of industrial-strength designs, in terms of instances of each type.

Table 1. Breakdown of the Basic Primitive Blocks.

Benchmarks	*Gates* [%]	*Selectors* [%]	*Mem. Elem.* [%]	*Macros* [%]
Bench1	36.9	14.4	46.8	1.9
Bench2	81.5	4.3	13.8	0.4
Bench3	37.5	11.2	50.8	0.5
Bench4	71.8	13.3	14.6	0.3
Bench5	60.0	7.9	29.9	2.2
Bench6	55.4	5.1	34.3	5.2

Although the distribution of the various block types varies significantly across different benchmarks, the percentage of macros is relatively low (1.7% on average). Even if the weights of the blocks are not uniform (one single macro is equivalent to several gates and memory elements), the relative importance of other primitives with respect to macros is much higher. This analysis suggests that, in order to achieve accurate power estimates, accurate power models for the other primitives are even more essential than macro power models.
Estimation of power consumption for gates, multiplexers and flip-flops (called *non-synthetic operators* hereafter) at the RTL is quite different from the case of macros. In principle, their power models are quite well understood (e.g., a NAND gate); the difficulty here is due to two main facts:

- The RTL netlist is quite far from the actual netlist that will be produced by synthesis, in terms of both the total number of primitives and their distribution. For instance, the synthesis step will optimize the design under given constraints;
- The RTL netlist is expressed in terms of *technology-independent* primitives; synthesis, on the contrary, will map primitives onto instances of a library cell.

These two observations leads us the main issue behind estimation of these types of blocks at the RTL: More than a true power modeling problem, it is a problem of estimating *the impact of RTL synthesis*.
This paper presents a solution to this problem, that has been implemented into the PowerChecker RTL power estimation tool by BullDAST [3]. Our solution is based on a statistical characterization of these non-synthetic operators, and, thanks to a novel pre-characterization scheme, it allows to estimate the impact of RTL synthesis on these primitives within a 20% of the gate-level estimates. In addition, our methodology also accounts for the power consumption due to the wires contained in the design. Preliminary results on a set of industrial-strength benchmarks shows the effectiveness of the proposed estimation scheme.

2 RTL Power Estimation Flow

2.1 RTL vs. Gate Power Estimation

Figure 1 summarizes the typical estimation flow at the RTL and gate-level.

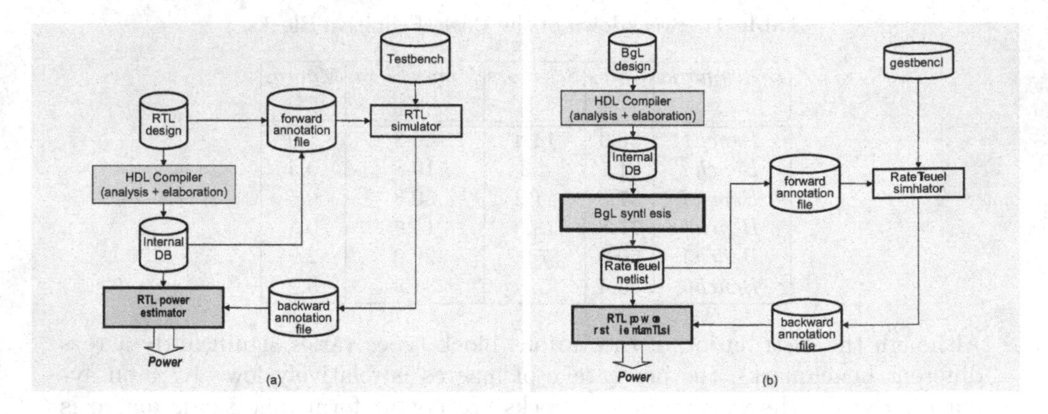

Fig. 1. RTL (a) and Gate-Level (b) Power Estimation Flows.

The starting point of the two flows is a design written in some HDL. After analysis and elaboration by the HDL compiler, the design is translated into a technology-independent internal format, that contain the four types of components mentioned in Section 1: Macros, gates, memory elements, and selectors. The two flows start differentiating at this point. In true RTL estimation (Figure 1-(a)), a *forward annotation file* is produced, that contains the list of nets to be monitored during (RTL) simulation. RTL simulation takes this file, the HDL description and a testbench, to produce a *backward annotation file*, consisting of all the nets specified in the forward annotation file, this time annotated with switching activity and static probability values. An RTL power estimator takes the internal database produced in the first step and this activity information, and calculates a power estimate. This estimate is basically obtained by exercising

specific *power models* for the objects of the internal database with the activity values derived from simulation.
In gate-level estimation (Figure 1-(b)), we notice two main differences. First, *RTL synthesis* comes into play. This operation translates and optimizes the internal database into a *gate-level netlist.* The synthesis is carried out under some constraints, and the elements of the gate-level netlist are instances of a fully-characterized technology library. The second difference is relative to the granularity of the simulation. The simulator that produces activity information now operate on a forward annotation file built from the gate-level netlist, and is therefore more "precise".
The comparison of the two flows exposes the sources of inaccuracy of RTL vs. gate-level estimation. First, the use of the internal database as a working description. This description is structural, but has little to do with the synthesized gate-level netlist (but the functionality, of course). Second, the granularity of simulation. RTL simulation will annotate only a subset of the nets annotated in the gate-level flow. These two differences are also the sources of the speed-up of RTL versus gate-level estimation.

2.2 PowerChecker Estimation Flow

Before getting into the details of the power modeling solution proposed in this work, we first need to stick our discussion to a specific power estimation tool. In the following, we will refer to PowerChecker, a RT-level power estimator by BullDAST [3]. PowerChecker, thanks to its power modeling capabilities, allows to handle HDL operators, memories and control as RTL design objects. Power models can be either pre-characterized, or built on-line, and an effective caching mechanism enables easy model re-use to improve estimation efficiency.
PowerChecker maps quite closely to the estimation flow of Figure 1-(a): The internal database is called PowerChecker *Internal Database* (PID), and its elements are called *design objects.* Concerning the interface to the RTL simulators, PowerChecker manages various annotation file formats, such as Synopsys SAIF, as well as VCD. This allows to support various RTL simulators; currently, PowerChecker supports Synopsys VSS [4], Model Technology's ModelSim [5], and Cadence's NC-SIM [6]. For further details on the modeling capabilities of PowerChecker, the reader is referred to [7].
The PID internal database consists of five types of design objects, summarized in Table 2: Currently, PowerChecker only supports memories as HARD MACRO

Table 2. Types of PowerChecker Design Objects.

Design Object	*Description*
GATE	logic functions such as AND,OR,NOT
MUX	2-to-1 selectors
REG	memory elements or registers
SOFT MACRO	synthesizable, parametric RTL modules such as adders, multipliers, etc.
HARD MACRO	legacy blocks, such as cores, memories, etc.

modules. In its first release, PowerChecker used accurate models for both HARD and SOFT MACROs (i.e., the synthetic operators), and estimated the power due to the non-synthetic operators using low-effort synthesis: All non-MACRO objects were grouped into a single entity, that was synthesized onto the target technology library. This paradigm provides quite high accuracy; however, its effectiveness in terms of estimation speed is based on the assumption that non-MACROs objects represent the minority of the objects, and that their contribution to the total power is marginal. Table 1 shows that this assumption is not very correct; using low-effort synthesis may imply synthesizing most of the RTL design.
This means that, besides models for MACROs, we must devise power models for non-MACROs object as well. The next section describes how this model is derived.

3 Power Models

Building a power model for non-MACRO objects that is not based on synthesis is equivalent to estimate the impact of RTL synthesis on that object. In this sense, it is clear such a model can only be based on a statistical characterization of a set of synthesis experiments.
The important point to understand is that, to achieve high estimation accuracy, we cannot be completely independent of the synthesis tool that will be used in the actual design flow. Therefore, before discussing possible power models, we must define the synthesis flow in which the RTL power estimator will be embedded. In our experiment, we refer to Synopsys DesignCompiler [2], that can be considered as the de-facto standard for RTL synthesis. However, the procedure could be applied to any synthesis flow.

3.1 Model Parameters

The choice of the parameters of the model is dictated by two constraints:

1. They should be quantities available at the RT level, and that can be extracted from either the PID (*complexity* parameters), or from RTL simulation results (*activity* parameters);
2. They should be as much insensitive as possible to the effects of synthesis.

Although the second requirements seems very restrictive, at the RTL there exist the so-called *synthesis invariants*, design objects that are preserved through the synthesis process. Typical synthesis invariant elements are the I/O pins of the top entity, the pins of the blocks instantiated within a design, and the registers. In our context, REG objects are therefore good candidates for being parameters of the model. The number of REG, in particular, can be used as a complexity parameters, and the activity (as provided by RTL simulation) of REG objects can be used as an activity parameter. In order to increase the robustness, however, we have also included into the model both complexity and activity factors for all other non-synthetic design objects, namely GATE and MUX objects.

Therefore, the complexity parameters extracted from the PID are: N_R, the number of REG objects, N_M, the number of MUX objects, and N_G, the number of GATE objects.
The activity parameters extracted from the results of RTL simulation are: A_R, the average toggle rate of the pins of REG objects, A_M, the average toggle rate of the pins of MUX objects; and A_G, the average toggle rate of the pins of GATE objects.

3.2 Model Exploration and Design

Having defined the parameters, we defined then the following model template:

$$P = k_0 + k_1 \cdot T_1 + k_2 \cdot T_2 + ... + k_i \cdot T_i$$

where k_i are the weights of the model, and $T_i = X_1 \cdot X_2 \cdot ... \cdot X_j$ represents the generic, j-th order term of the model. The X_j denotes the generic parameter, that is, in our case $X_j \in (N_R, N_M, N_G, A_R, A_M, A_G)$; i denotes the number of terms of the model, and j the maximum order of the model.
Since the model is empirical, any a priori choice of i and j would be arbitrary. Therefore, we have set up an exploration procedure that allowed us to choice the best values of i and j. Since the cardinality of the set (i, j) exponentially increases the exploration time, we have limited the choice to $i \in (1, 2, 3, 4)$ and $j \in (1, 2)$. Notice that for a given (i, j) pair, there are several possible models, characterized by the absence or presence of some of the parameters.
The exploration engine works as follows. Given a (i, j) pair, for each feasible model, it runs a characterization procedure over a set of 15 RTL benchmarks of various complexities. The characterization consists of (i) the isolation of the non-MACRO objects, (ii) their grouping into a single block, and (iii) its synthesis onto a given technology library. The power consumption of this implementation, under a given testbench is stored as a sample point of the exploration space. Notice that *only the power consumption due to the cells* is recorded; power due to wires is not considered here. After characterization is completed, least squares regression is run on the set of measured point to achieve an interpolating equation, that represents the model.
Table 3 reports the results (in terms of estimation errors with respect to gate-level estimates of the non-synthetic portion of the design). Each entry shows the results for the best model of that cardinality. Notice that the benchmarks shown in the table *are not* those used for characterization.
The model that yielded the least estimation error is the (4,2) model:

$$P = 2.472 + 505,254 \cdot N_G \cdot A_G - 599,630 \cdot N_M \cdot A_M + 221,711 \cdot N_R \cdot A_R \quad (1)$$

From Table 3 we can observe that the accuracy of the models increases as the values of (i, j) increase. Moreover, the models with $j = 2$ are better than models with $j = 1$. This fact shows that it is essential to have both structural and simulation parameters, and neither type of parameters can be neglected. Furthermore, the parameters related to REG objects, as expected, showed to be an

Table 3. Estimation Error Results.

Design	*(i,j)*							
	$(1,1)$ [%]	$(1,2)$ [%]	$(2,1)$ [%]	$(2,2)$ [%]	$(3,1)$ [%]	$(3,2)$ [%]	$(4,1)$ [%]	$(4,2)$ [%]
Bench1	29.4	11.3	6.7	4.9	19.3	1.9	16.4	3.3
Bench2	23.6	16.1	81.8	12.8	66.9	7.9	62.1	12.3
Bench3	25.5	23.3	23.6	23.1	18.6	18.0	15.6	11.1
Bench4	57.2	11.3	43.6	3.4	1.9	12.7	1.1	1.5
Bench5	48.1	15.4	29.5	11.8	34.9	4.4	22.9	11.9
Bench6	26.4	8.0	16.0	5.5	10.2	6.5	9.1	0.5
Avg.	35.0	14.2	33.5	10.3	25.3	8.6	21.2	6.7

essential parameter of the model. In fact, parameters A_R and N_R are included in every model that gives minimum estimation error.
To summarize, the model exploration has been run off-line to determine the best equation (Equation 1) that minimizes the estimation error (of the non-synthetic design objects of the design) over a set of RTL benchmarks. This equation is used in PowerChecker to estimate the power due to these objects; the values of the parameters are available after HDL compilation (those related to complexity) and RTL simulation (those related to activity).

3.3 Technology Independence

The models described in the previous section are technology dependent, that is, the characterization is refers to a given technology library. In fact, different libraries (e.g., with different cell types and complexities, or different features size) would result in different synthesized implementation, and thus different models.
In order to obtain a model that is independent of the technology library, we have devised an original solution that works as follows. Let us call L_{ref} the reference library used for the characterization phase, when building the model, and let L_{new} be the actual technology library to be used in the actual synthesis flow.
To determine how the base model applies to the new applies, the model is *scaled* according to a *scaling factor* K. In practice, if $P_{L_{ref}}$ is the original power model, the model for the new library L_{new} will simply be $P_{L_{ref}} = K \cdot P_{L_{ref}}$. The issue is thus how to determine K.
Since the target technology library is specified up-front in the synthesis flow, the computation of K is based on scaling a single "golden" instance G of the library. More precisely, when PowerChecker is first installed, a sample HDL description (for instance, a `a AND b`) statement) is synthesized onto L_{new}, under random inputs, to get a value $P_{L_{new}}(G)$. The tool internally stores the value of the power consumption of the golden description for L_{ref}, namely $P_{L_{ref}}(G)$. The scaling factor is then simply obtained as $K = P_{L_{new}}(G)/P_{L_{ref}}(G)$.

It is important that the golden description G is simple; that is, when synthesized, it should map onto a few non-MACRO design objects. In the current implementation, the golden description is that of an logic AND operator.

4 Wires Power Model

As explicitly mentioned in the previous section, the issue of the power due to wires has been completely disregarded so far. Needless to say, the contribution of wires to the total power cannot be neglected. This is especially true for the clock net: The clock nets fans out to all the REG objects, thus loading a very large capacitance that has to be charged and discharged. Furthermore, the clock net is usually the net with the higher toggle rate. Because of this unbalanced distribution, PowerChecker uses two separates models: One for the clock net (P_{clk}), and one for the other nets (P_{net}). Hence, the total power dissipated by all the nets can be expressed as the sum of the two components.
To illustrate the basic principle of the wire power models, consider a single wire. Calculation of its power consumption requires the toggle rate α of the wire, and its capacitance C. While the toggle rate can be easily extracted from RTL simulation, the capacitance C is not directly available at the RT level. In fact, C consists of two components: C_{int}, the internal wire capacitance, and C_{load}, the contribution due to the load connected to the wire. The former strictly depends on the technology; the latter cannot be estimated in terms of a net's fanout, because in the PID each cell has many redundant connections that will disappear during synthesis.
Therefore, the only wires whose capacitance can be modeled are those that are inputs of synthesis invariant components, that is, REG objects. Once we have extracted the power consumed by such wires, we compute the total net power by means of an empirical model.
The model construction for REG input wires is based on the assumption that each REG will map onto a flip-flop of the target technology library. The model construction flow proceeds then as follows. First, a register is synthesized onto the target library in order to extract its area ($Area_R$). Second, the number of registers (N_R) of the design is extracted from the PID.
Based on these two quantities, a model of the *total area* of the design is computed empirically. For a given set of benchmarks, the total area has been measured after synthesis, and related to N_R and $Area_R$. The extrapolation of the measured values yielded the following model

$$Area = 4.19 \cdot N_R \cdot Area_R$$

Given the value of area, the wire load model of the library gives the correct range of loads to be used, and the relative capacitances. For instance, in the Synopsys DesignCompiler flow, this range is defined by the so-called *Wire Loading Model Selection Group*. This range yields the value of C_{int}.
Concerning the contribution due to the load, since we only consider REG objects, we can extract C_{load} directly from the gate-level report, as a by-product

of the area extraction phase. These two values allow us to compute the total capacitance C_{tot}, referred to a net connected to a REG object.
Moving from these considerations, we can now compute the power due to the clock net (P_{clk}), with the usual formula $P_{clk} = 0.5 \cdot \alpha_{clk} \cdot C_{tot} \cdot N_R^{clk} \cdot V_{dd}^2$, where N_R^{clk} is the number of registers connected to the clock net. In the case of design with multiple clocks, this formula can be applied to each clock net in the design.
For what concerns the P_{net}, the first step consists of computing the power of all wires connected to REG objects (P_{wire_reg}). Next an empirical model, which accounts for all the other nets, will be generated. In formula:

$$P_{wire\_reg} = \sum_{i=1}^{n} (0.5 \cdot \alpha_i \cdot C_{tot} \cdot N_R^i \cdot V_{dd}^2)$$

Where n is the number of nets connected to one register at least. Finally, P_{net} can be obtained computed with the following empirical model:

$$P_{net} = 3.97 \cdot P_{wire\_reg} + 3070$$

Also in this case, the numeric values reported in the model are extracted from a regression process performed on P_{net} and P_{wire_reg}.

5 Experimental Results

The above modeling framework has been implemented in PowerChecker, to replace the estimation of non-MACRO objects based on fast synthesis. The model has been applied to a set of industrial designs.
The first experiments concerns the accuracy of PowerChecker estimates, compared to those at gate-level (namely, Synopsys PowerCompiler).

Table 4. Estimation Results for Non-Synthetic Objects.

Design	*Technology A*			*Technology B*		
	Gate Level (μW)	PowerChecker (μW)	*Rel. Err.* (%)	*Gate Level* (μW)	PowerChecker (μW)	*Rel. Err.* (%)
Bench1	279.2	268.9	3.7	502.3	611.1	21.6
Bench2	489.8	405.9	17.1	973.0	922.4	5.2
Bench3	436.0	510.3	17.0	836.1	1159.7	38.7
Bench4	1227.0	1185.8	3.3	2220.3	2694.9	21.3
Bench5	1195.3	1265.7	5.9	2413.2	2876.5	19.2
Bench6	3762.1	3760.8	0.1	7238.4	8547.2	18.1

Table 4 reports the power contribution due to the non-MACRO portion of each design, in order to evaluate the accuracy of the models. In particular, only the power due to the cells (without wire contribution is reported). The table contains two sets of columns. The first set, labeled *Technology A* is relative to results obtained using as technology the same used to characterize the power models. The

second set of columns *Technology B* reports the results calculated with another technology, and by applying the scaling procedure described in Section 3.3. We notice that, as expected, in the case of *Technology A* the accuracy of the power model is better than the case of *Technology B*. However, the accuracy achieved by the scaling procedure is acceptable, with the main advantage that we do not have to re-synthesize the design if we change the target technology library. In fact, if the technology changes, PowerChecker computes the new technology scaling factor that must be applied during the estimation phase without changing the model.
The second experiment concerns the validation of the wire model. Table 5 shows the same results of Table 4, this time where only the contribution due to the wires has been reported. In this case, however, we have reported only the power figures computed with *Technology A*, since, as mentioned in Section 4, we always need to synthesize a single register onto the target library to extract its area value.

Table 5. Results for the Wires.

Design	*Gate Level* (μW)	PowerChecker (μW)	*Rel. Err.* (%)
Bench1	18.8	21.5	14.6
Bench2	46.6	32.0	31.2
Bench3	8.1	10.9	34.1
Bench4	72.7	60.3	17.0
Bench5	121.4	138.6	14.2
Bench6	219.5	251.0	14.3

Also for the wires, the relative estimation error is limited (below 35%, and 20.9% on average). Notice that these values are referred to all the wires except the clock net, which is treated separately because of its relative importance. Estimates for the clock net are even more accurate (estimation error below 5%), since the clock net is one of the synthesis invariants; hence, we can know the exact value of its load capacitance.
Table 6 shows which is the impact, in terms of power consumption, of each type of design objects. The major contributors are the non-synthetic objects, except for Benchmarks 4 and 6 where the clock net is dominant.
Finally Table 7 shows what speedup we can expect by using PowerChecker for RTL estimation with respect to gate-level estimation. On average, we have measured a 15x speedup.

6 Conclusions

Power models for non-synthetic operators with small granularity are essential for accurate estimates at the RTL. This is particularly true when RTL estimation is coupled to a realistic design flow based on commercial synthesis tools.
We have proposed a statistical power model for those objects, that (i) provides good accuracy, (ii) can be made technology-independent through a pre-characterization phase, and (iii) includes an accurate, statistical power model

Table 6. Power Estimation Breakdown.

Benchmarks	*Macros* [%]	*Non Synthetic* [%]	*Wires* [%]	*Clock* [%]
Bench1	5.7	79.3	5.3	9.7
Bench2	14.6	68.1	6.4	10.9
Bench3	0.7	82.4	1.5	15.4
Bench4	0.1	4.0	0.2	95.7
Bench5	25.1	58.7	5.9	10.3
Bench6	1.4	6.2	0.4	92.0

Table 7. Estimation Time.

Design	*Gate Level* (minutes)	PowerChecker (minutes)	*Speedup* (%)
Bench1	2	1	2x
Bench2	44	3	14x
Bench3	166	9	18x
Bench4	33	16	2x
Bench5	103	6	16x
Bench6	38	1	38x

for the wires connecting these objects. Results on a set of industrial benchmarks has shown very good accuracy for both cell (8% on average) and wire power (20% on average), as well as significant speedup (up to 38x) with respect to gate-level estimation.

References

1. D.D. Gajski, N.D. Dutt, A.C-H. Wu, S.Y-L. Lin, "High-Level Synthesis Introduction to Chip and System Design," *Kluwer Academic Publishers*, 1992.
2. Synopsys DesignCompiler, *www.synopsys.com/products/logic/design_compiler.html*
3. BullDast PowerChecker, *www.bulldast.com/powerchecker.html*
4. Synopsys VSS, *www.synopsys.com/products/simulation/vss_cs.html*
5. ModelTechnology ModelSim, *http://www.model.com/products/default.asp*
6. Cadence NS-SIM, *http://www.cadence.com/datasheets/affirma_nc_sim.html*
7. A. Bogliolo, I. Colonescu, R. Corgnati, E. Macii, M. Poncino, "An RTL Power Estimation Tool with On-Line Model Building Capabilities", *PATMOS'2001*, Yverdon-les-bains, Switzerland, September 2001.

Energy Efficient Register Renaming[1]

Gurhan Kucuk, Oguz Ergin, Dmitry Ponomarev, and Kanad Ghose

Department of Computer Science,
State University of New York, Binghamton, NY 13902–6000
{gurhan,oguz,dima,ghose}@cs.binghamton.edu
http://www.cs.binghamton.edu/~lowpower

Abstract. Modern microprocessor designs implement register renaming using register alias tables (RATs), which maintain the mapping between architectural and physical registers. Because of the non–trivial power that is dissipated in a disproportionately small area, the power density in the RAT is significantly higher than in some other datapath components. In this paper, we propose mechanisms to reduce the RAT power and the power density by exploiting the fundamental observation that most of the generated register values are used by the instructions in close proximity to the instruction producing a value. Our first technique disables the RAT lookup for a source register if that register is a destination of an earlier instruction dispatched in the same cycle. The second technique eliminates some of the remaining RAT read accesses even if the source register value is produced by an instruction dispatched in an earlier cycle. This is done by buffering a small number of recent register address translations in a set of external latches and satisfying some RAT lookup requests from these latches. The net result of applying both techniques is a 30% reduction in the RAT energy with no performance penalty, little additional complexity and no cycle time degradation.

1 Introduction

Dynamically scheduled superscalar processors use aggressive out–of–order execution mechanisms to maximize performance by harvesting available parallelism in the sequential programs. Each successive generation of superscalars increases the number of instructions that are issued in a cycle and also uses larger instruction windows in order to consider more instructions for scheduling. The inevitable consequence of such an approach is a dramatic increase in the overall datapath complexity, power consumption and also power density, especially in high–frequency implementations. While in the past power was a consideration mainly in the domain of embedded systems, today it is an important design constraint for high–performance microprocessors. Unless power dissipation is controlled through technology–independent techniques, the areal power density will soon become comparable to that of nuclear reactors [11] leading to intermittent and permanent failures on the die and also creating serious challenges for the cooling facilities. Furthermore, the areal power density distribution across a typical chip is highly skewed, being lower over the on–chip caches and significantly higher elsewhere, resulting in the presence of the localized hot spots on the chip. The non–uniform thermal stresses that result are problematic.

One on–chip structure with a high power density is the Register Alias Table (RAT). RAT maintains the register address translations needed for handling the true data

[1] supported in part by DARPA through contract number FC 306020020525 under the PAC–C program, the NSF through award no. MIP 9504767 & EIA 9911099 and IEEC at SUNY Binghamton.

J.J. Chico and E. Macii (Eds.): PATMOS 2003, LNCS 2799, pp. 219–228, 2003.
© Springer-Verlag Berlin Heidelberg 2003

dependencies. True data dependencies are handled by assigning a new physical register for every new result that is produced into a register. The RAT maintains information to locate the most recent instance of an architectural register. Register renaming is a technique used in all modern superscalar processors to cope with false data dependencies by assigning a new physical register to each produced result. The mappings between the logical and the physical registers are maintained in the RAT, so that each instruction can identify its source *physical* registers by performing the RAT lookups indexed by the addresses of the source *logical* (architectural) registers. The read and write accesses to the RAT, as well as the actions needed for checkpointing and the state restoration, result in a significant amount of power dissipated in the RAT. For example, about 4% of the overall processor's power is dissipated in the rename unit of the Pentium Pro [8]. Even higher percentage of the overall power – 14% – is attributed to the RAT in the global power analysis performed in [4]. When coupled with the relatively small area occupied by the RAT, this creates a hot spot, where the power density is significantly higher than in some other datapath components, such as the on–chip caches.

In this paper, we introduce mechanisms to reduce the RAT power and the power density by exploiting the fundamental observation that most of the generated register values are used by the instructions that are in close proximity to the instruction producing a value. Specifically, we propose two methods to reduce the RAT power dissipation. Our first technique disables the RAT lookup for a source register if that register is a destination of an earlier instruction dispatched in the same cycle. The second technique eliminates some of the remaining RAT read accesses even if the source register value is produced by an instruction dispatched in an earlier cycle.

The rest of the paper is organized as follows. To motivate our work, we describe the RAT complexities and the sources of associated energy dissipations in Section 2. We present the details of our energy reduction techniques in Sections 3 and 4. Our simulation methodology is described in Section 5, and we present and discuss the simulation results in Section 6. Section 7 reviews the related work and we offer our concluding remarks in Section 8.

2 The RAT Complexity

For this study, we used a RISC–type ISA, where the instructions may have at most two source registers and one destination register. We further assumed that the RAT is implemented as a multi–ported register file, where the number of entries is equal to the number of architectural general–purpose registers in the ISA. The RAT is indexed by the architectural register address to permit a direct lookup. The width of each RAT entry is equal to the number of bits in a physical register address. An alternative design is to have the number of entries in the RAT equal to the number of physical registers, such that each RAT entry stores the logical register corresponding to a given physical register and a single bit to indicate if the entry corresponds to the most recent instance of the architectural register. In this scheme, as implemented in the Alpha 21264 [7], the RAT lookup is performed by doing the associative search using the logical register address as the key. We did not consider this variation of the RAT in this paper, because it is inherently less energy–efficient than the direct–lookup implementation, due to the large dissipations that occur during frequent associative lookups. One way to address this problem is to use a recently proposed dissipate–on–match comparator [3] in the associative logic within the RAT, but such an evaluation is beyond the scope of this paper.

In a W–way superscalar machine, up to W instructions may undergo renaming in the same cycle. Thus, 2*W register address translations may have to be performed in a cycle to obtain the physical addresses of the source registers. In addition, up to W new physical registers may have to be allocated to hold the new results. The number of ports needed on the RAT in a W–way superscalar machine is quite significant. Specifically, 2*W read ports are needed to translate the source register addresses and W write ports are needed to update W RAT entries for the destinations of the co–dispatched instructions. In addition, before the destination register mapping is updated in the RAT, the old value has to be checkpointed in the reorder buffer for possible branch misprediction recovery. If the instruction that overwrites the entry is later discovered to be on the mispredicted path, the old mapping, saved within the reorder buffer, is used to restore the state of the RAT. W read ports, needed for such checkpointing, bring the total port requirements on the RAT to 3*W read ports and W write ports.

The energy dissipations take place in the RAT in the course of the following events:
1) Obtaining physical register addresses of the source operands.
2) Checkpointing the old mapping of the destination register.
3) Writing to the RAT for establishing the new mapping for the destination register.

3 Exploiting the Intra-group Dependencies

In a superscalar machine, there may be sequential dependencies among the group of instructions that are co–dispatched within a cycle. To take care of such dependencies, register renaming must logically create the same effect as renaming the instructions individually and sequentially in program order. A sequential implementation of register renaming will be too expensive and will dictate the use of a slower clock. To avoid this, the accesses to the RAT and dependencies are handled as follows:

Step 1. The following substeps are performed in parallel:

(a) RAT reads for the sources of each of the co–dispatched instructions are performed in parallel, assuming that no dependencies exist among the instructions.

(b) New physical registers are allocated for the destination registers of all of the co–dispatched instructions.

(c) Data dependencies among the instructions are noted, using a set of comparators. The address of each destination register in a group of instructions is compared against the sources of all following instructions in the group and if a match occurs, the dependency is detected.

Step 2. If a data dependency is detected among a pair of instructions, the source physical register for the dependent instruction as read out from the RAT is replaced with the allocated destination register address of the instruction producing the source to preserve the true dependencies. After resolving dependencies as described, the RAT is updated with the addresses of the new destinations.

The above steps, including the concurrent substeps within Step 1, avoid a bottleneck that would otherwise result from performing the RAT lookup and dependency handling in a strictly sequential manner among the co–dispatched instructions. The price paid in this approach is in the form of redundant accesses to the RAT for the mappings of source registers that are renamed because of dependencies within the group of co–dispatched instructions. If a dependency is not detected, the source mapping obtained from the RAT is used, otherwise it is tossed out.

A considerable amount of energy is expended within the multi–ported register file that implements the RAT because of the concurrent accesses to it within a single cycle. It is thus useful to consider techniques for reducing the energy dissipation within the RAT. Our first proposed solution disables parts of the RAT read accesses if the intra–group data dependency is noted, as described by Step 1(c). The outputs of the comparators, corresponding to a given source, are NOR–ed and the output of the NOR gate is used as the enabling signal for the sensing logic on the bitlines used for reading the physical address mapping of this source from the RAT. For example, consider three instructions – I1, I2 and I3 (in program order) that are renamed in the same cycle. A source for the instruction I3 is compared for possible intra–group dependencies against the destination of I1 and the destination of I2. If one of these comparators indicates a match (the output of the comparator stays precharged at the logical "1"), the output of the NOR–gate becomes zero and the sense amps used for reading the bitlines of a source of the instruction I3 are not activated, thus avoiding the energy dissipation in the course of sensing. Measurable power savings can be realized, because sense amps contribute to a large fraction of the overall read energy. In the rest of the paper we abbreviate this technique as the *CSense* (Conditional Sensing).

Figure 1 shows the percentage of the RAT lookups for the source registers that can be aborted if the *CSense* is used for a 4–way and a 6–way processor. Results are presented for the simulated execution of a subset of the SPEC 2000 benchmarks, including both integer and floating point codes. Detailed processor configurations are described in Section 5. On the average across all of the simulated benchmarks, around 31% of the RAT read accesses can be aborted for a 4–way processor and around 41% for a 6–way processor. The latter number is higher, because more instructions are dispatched in the same cycle, thus increasing the possibility that the most recent definition of a source register is within the same instruction group. Notice, finally, that our technique does not prolong the cycle time, because the output of the NOR gate is driven to the sense amp before the wordline driver completes the driving of the word line. Our simulations of the actual RAT layouts in a 0.18 micron 6–metal layer TSMC process (Section 5) show that the decoder delay is about 150 ps, the delay of the wordline driver is about 100 ps, the bitline delay to create the small voltage difference across the bitline pair is 60 ps and at that point the sense amp is activated. The comparator delay is about 120 ps and the delay of the three–input NOR–gate is 60 ps. Consequently, the signal that controls the activation of the sense amp is available 180 ps after the beginning of the cycle, while the sense amp is normally activated after 310 ps are elapsed since the beginning of the cycle. The sense amp control signal is thus available well in advance of when it needs to be used, leaving enough time to route the signal to the sense amps, if need be. These delays were obtained using highly optimized hand–crafted layouts of the RAT assuming 32 architectural registers and 4–way wide dispatch/renaming. Therefore, the *CSense* can be applied without any increase in the cycle time.

4 Buffering Recent Address Translations

Our simulations of the SPEC 2000 benchmarks show that the dependent instructions are usually very close in proximity to each other. If the register needed as a source is not defined by an earlier instruction dispatched in the same cycle, then it is likely defined by an instruction dispatched one or at most a few cycles earlier. We exploit this behavior by caching recently updated RAT entries in a small number of associatively–addressed

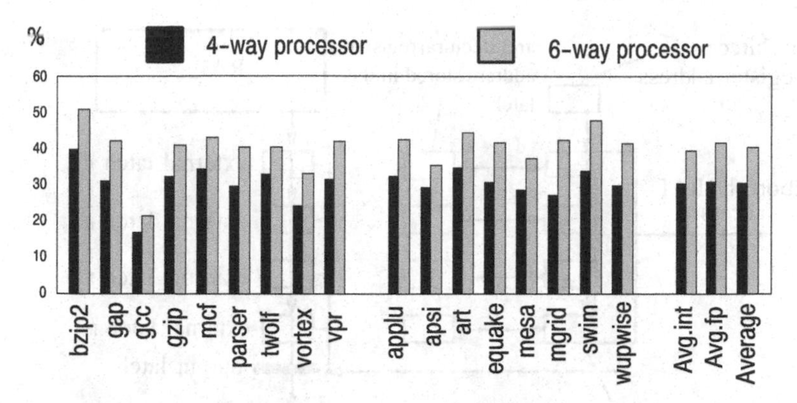

Fig. 1. The percentage of the source operands that are produced by the instructions codispatched in the same cycle

latches (ELs), external to the RAT. The basic idea of avoiding a RAT access using ELs was inspired by the work of [5], where multiple line buffers were used to reduce the overall cache energy dissipations. The RAT access for a source register now proceeds as follows:

(a) Start accessing the RAT and at the same time address the ELs to see if the desired entry is located in one of the ELs.

(b) If a matching entry is found, discontinue the access from the RAT.

As long as the overhead of accessing the ELs is less than the energy spent in accessing the RAT before the RAT accessing is aborted, this technique will result in an overall energy saving.

Figure 2 depicts a RAT with multiple ELs (4 in this case). The hardware augmentations to the basic register renaming structure are as follows. First, we need four latches to hold the addresses of each of the four most recently accessed architectural registers. Second, four comparators are used to compare the address of the architectural register, whose lookup in the RAT is being performed against the register addresses located in the four ELs.

Assuming a two–phase clock, the access steps for a RAT with multiple external latches are as follows:

Phase 1:

(a) Precharge the RAT for a read access.

(b) Start the decoding of the register address.

(c) Simultaneously, compare the register address with the register addresses stored in the latches.

Phase 2:

If a match occurs to an EL (a latch hit), abort the readout from the RAT. Otherwise (on a latch miss), proceed with the regular RAT access.

Notice that on a read miss on the associatively–addressed ELs, the data is not brought from the RAT array into the latches. This is so for the following reason. A large percentage of the register values are consumed by just one instruction; this was observed in [2] and also noticed in our simulations. If this is the case, then bringing the data that was once read from the RAT into the ELs will only result in polluting the latches with

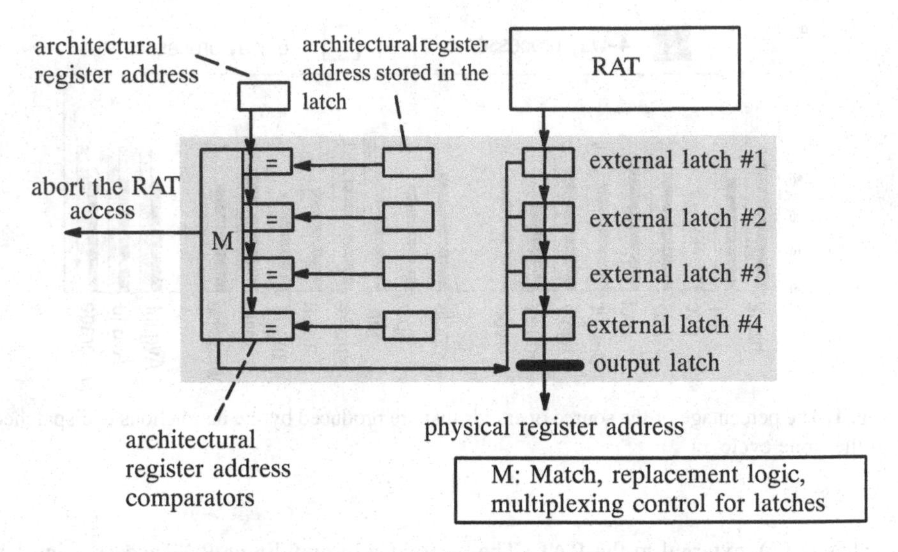

Fig. 2. Renaming Logic with Four External Latches (ELs)

unusable data in most situations. In addition, extra energy dissipation occurs if such data movement is needed. Because of this, we only record the translation in the ELs when the physical register is allocated. In other words, the update of the RAT and the update of the ELs proceed in parallel. A victim EL is selected randomly for setting up the new entry. Notice, that as a consequence of this policy, there is no need to write the translation information back to the RAT once an entry is evicted from the ELs.

It is critical to limit the energy spent in associatively addressing the ELs. To accomplish this, we make use of comparators that dissipate energy on a match (for example, as introduced in [3]), instead of traditional CAM cells or comparators that dissipate energy on a mismatch. The use of these comparators actually helps in two ways. First, as at most one of the external latches will have the matching entry, energy is dissipated within at most one comparator during the associative lookup. Second, the comparators of [3] actually have a faster response time than the traditional comparators or CAM cell. (The difference in timing between the comparator of [3] and the traditional comparator is, however, very small so even the traditional pull–down comparator can be used in our scheme, albeit with higher power dissipated in the latches.) For the same reason as in the *CSense* scheme, the detection of a match in the external latches is completed well in advance of the normal sense amp activation. Therefore, the energy dissipation in the sense amps can be avoided without compromising the cycle time. These savings exceed that spent in locating a matching entry within a latch array consisting of four external latches, as shown in the result section. Also, the use of dissipate–on–match comparators within the intra–group dependency checking logic is not an attractive solution from the energy standpoint because of a higher percentage of match situations. Our detailed analysis, using microarchitectural data about the bit patterns of the comparands indicate that the use of traditional comparators is a more energy–efficient approach for the use in the intra–block dependency checking logic.

5 Simulation Methodology

To evaluate the energy impact of the proposed techniques, we designed and used the AccuPower toolsuite [10]. The widely–used Simplescalar simulator [1] was significantly modified to implement *true hardware level, cycle–by–cycle* simulation models for realistic superscalar processor. The main difference from the original Simplescalar code is that we split the Register Update Unit into the issue queue, the reorder buffer and the physical register file. It is important, because in real processors the number of entries in all these structures, as well as the number of ports to these are quite disparate. The configuration of a simulated 4–way superscalar processor is shown in Table 1. For simulating a 6–way machine, we increased the window size and the cache dimensions proportionately.

We simulated the execution of 9 integer (*bzip2*, *gap*, *gcc*, *gzip*, *mcf*, *parser*, *twolf*, *vortex* and *vpr*) and 8 floating point (*applu*, *apsi*, *art*, *equake*, *mesa*, *mgrid*, *swim* and *wupwise*) benchmarks from SPEC 2000 suite. Benchmarks were compiled using the Simplescalar gcc compiler that generates code in the portable ISA (PISA) format. Reference inputs were used for all the simulated benchmarks. The results from the simulation of the first 2 billion instructions were discarded and the results from the execution of the following 200 million instructions were used for all of the benchmarks.

For estimating the energy/power dissipations for the key datapath components, the event counts gleaned from the simulator were used, along with the energy dissipations measured from the actual VLSI layouts using SPICE. Hand–crafted CMOS layouts for the RAT in a 0.18 micron 6 metal layer CMOS process (TSMC) were used to get an accurate idea of the energy dissipations for each type of transition. A 2 GHz clock and a V_{dd} of 1.8 volts were assumed for all the measurements.

Table 1. Architectural configuration of a simulated 4–way superscalar processor

Parameter	Configuration
Machine width	4–wide fetch, 4–wide issue, 4–wide commit
Window size	32 entry issue queue, 96 entry ROB (integrating physical register file), 32 entry load/store queue
Function Units and Latency (total/issue)	4 Int Add (1/1), 1 Int Mult (3/1) / Div (20/19), 2 Load/Store (2/1), 4 FP Add (2), 1FP Mult (4/1) / Div (12/12) / Sqrt (24/24)
L1 I–cache	32 KB, 2–way set–associative, 32 byte line, 2 cycles hit time
L1 D–cache	32 KB, 4–way set–associative, 32 byte line, 2 cycles hit time
L2 Cache combined	512 KB, 4–way set–associative, 128 byte line, 4 cycles hit time
BTB	4096 entry, 4–way set–associative
Branch Predictor	Combined with 1K entry Gshare, 10 bit global history, 4K entry bimodal, 1K entry selector
Memory	128 bit wide, 60 cycles first chunk, 2 cycles interchunk
TLB	64 entry (I), 128 entry (D), fully associative, 30 cycles miss latency

6 Results and Discussions

Figure 3 shows the hit ratio to the ELs – that is, the percentage of the RAT accesses that can be satisfied from the ELs. Separate results are presented for integer and floating point registers, with the use of 4 and 8 ELs. With the use of 4 ELs, the average hit ratio is about 44% for the integer ELs, and about 30% for the floating–point ELs. Adding four more entries to the ELs increases the hit ratios only slightly (about 53% for integer, and about 38% for floating–point ELs) because, again, the use of most register values is very close in proximity to the definitions of those registers. On the other hand, a higher number of ELs increases the complexity and power.

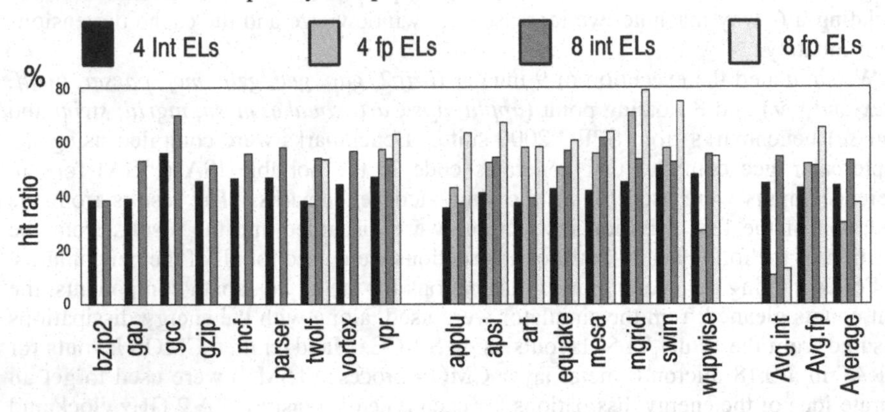

Fig. 3. The hit ratio to integer and floating–point External Latches (ELs)

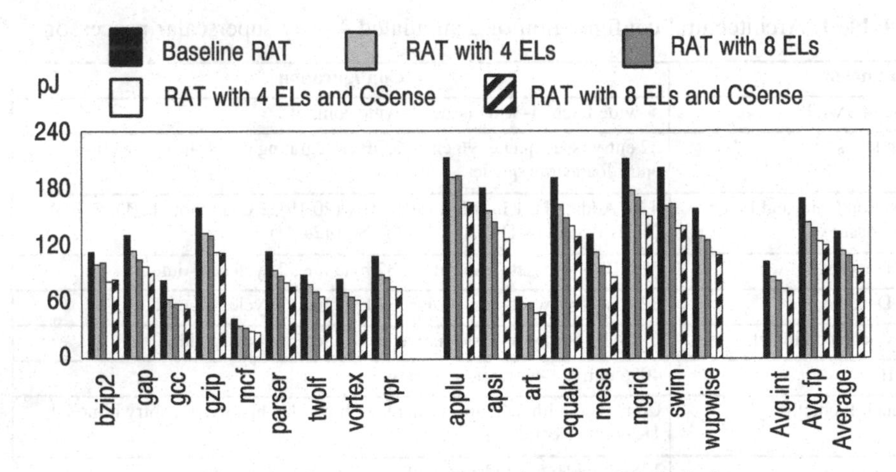

Fig. 4. Energy of the baseline and proposed RAT designs

The percentages shown in Figure 3 do not account for the matches within the co–dispatched group, as those matches are detected by the intra–group dependency checking logic. Combined, the percentages shown in Figure 1 and Figure 3 represent the total percentage of cases when the sense amps can be disabled if the two techniques are used in conjunction.

Figure 4 shows the energy reduction achievable by applying the proposed techniques. The first bar shows the energy dissipation of the baseline RAT. The second and third bars show the energy impact of adding four and eight ELs, respectively. The last two bars show the energy reduction in the RAT if ELs are used in conjunction with the *CSense*. The average energy savings with the use of four and eight ELs are around 15% and 19%, respectively. The combination of both techniques results in about 27% energy savings on the average for four ELs and 30% energy reduction for eight ELs. Our analysis also indicate that the use of eight ELs is the optimal configuration, because the overall energy increases if the number of ELs goes beyond eight, due to the additional complexity and power of managing the ELs. At the same time, the percentage of EL hits stays nearly unchanged, as described earlier.

7 Related Work

Moshovos proposed to reduce the power consumption of the register alias table in [9]. The proposed optimizations reduce power of the renaming unit in two ways. First, the number of read and write ports needed on the register alias table is reduced. This is done by exploiting the fact that most instructions do not use the maximum number of source and destination register operands. Additionally, the intra–block dependence detection logic is used to avoid accessing the register alias table for those operands that have a RAW or a WAW dependence with a preceding, simultaneously decoded instruction. The technique of [8] only steers the sources that actually need to be translated through the RAT lookup to the RAT ports. The source addresses that are not used (for example, some instructions have only one source) or those produced by the instruction in the same block do not have to be translated through the RAT array. All sources that need to be translated using the RAT access are first identified and then they are steered to the available RAT ports. If a port is not available, the renaming blocks. This incurs an inherent performance penalty in terms of IPCs and, in addition, stretches the cycle time, because the filtering of source addresses and the RAT accesses are done serially. Simulation results show that for an aggressive 8–way superscalar machine it is possible to reduce the number of read ports from 24 to 12 and the number of write ports from 8 to 6 with a performance penalty of only 0.5% on the average across the SPEC 2000 benchmarks. The second optimization reduces the number of checkpoints that are needed to implement aggressive control speculation and rapid recovery from the branch mispredictions. This is done by allowing out–of–order control flow resolution as an alternative to conventional in–order resolution, where the checkpoint corresponding to a branch can not be discarded till this branch itself as well as all preceding branches are resolved.

In [6], Liu and Lu suggested using the hierarchical RAT. A small, first–level RAT is used to hold the mappings of the most recent renamed registers. Instructions access this small RAT first and only on a miss access the large full–blown second–level RAT. Because the process is serialized, the performance degradation is unavoidable as at least one extra cycle is needed in the front–end of the pipeline, thus increasing the branch misprediction penalty.

In contrast to these techniques, our proposed mechanisms do not have any performance penalty, nor do they increase the cycle time of the processor.

8 Concluding Remarks

We proposed two complementary techniques to reduce the energy dissipation within the register alias tables of modern superscalar microprocessors. The first technique uses the intra–group dependency checking logic already in place to disable the activation of the sense amps within the RAT when the register address to be read is redefined by an earlier co–dispatched instruction. The second technique extends this approach one step further by placing a small number of associatively–addressed latches in front of the RAT to cache a few most recent translations. Again, if the register translation is found in these latches, the activation of the sense amps within the RAT is aborted. Combining the two proposed techniques results in a 30% reduction in the power dissipation of the RAT. The power savings comes with no performance penalty, little additional complexity and no increase in the processor's cycle time.

References

1. Burger, D. and Austin, T. M., "The SimpleScalar tool set: Version 2.0", Tech. Report, Dept. of CS, Univ. of Wisconsin–Madison, June 1997 and documentation for all Simplescalar releases.
2. Cruz, J–L., Gonzalez, A., Valero, M. et. al., "Multiple–Banked Register File Architecture", *in Proceedings 27th International Symposium on Computer Architecture*, 2000, pp. 316–325.
3. Ergin, O., et.al., "A Circuit–Level Implementation of Fast, Energy–Efficient CMOS Comparators for High–Performance Microprocessors", *in Proceedings of ICCD*, 2002.
4. Folegnani, D., Gonzalez, A., "Energy–Effective Issue Logic", *in Proceedings of International Symposium on Computer Architecture*, July 2001.
5. Ghose, K. and Kamble, M., "Reducing Power in Superscalar Processor Caches Using Subbanking, Multiple Line Buffers and Bit–Line Segmentation", *in Proceedings of International Symposium on Low Power Electronics and Design (ISLPED'99)*, August 1999, pp.70–75.
6. Liu, T., Lu, S., "Performance Improvement with Circuit Level Speculation", *in Proceedings of the 33rd International Symposium on Microarchitecture*, 2000.
7. Kessler, R.E., "The Alpha 21264 Microprocessor", IEEE Micro, 19(2) (March 1999), pp. 24–36.
8. Manne, S., Klauser, A., Grunwald, D., "Pipeline Gating: Speculation Control for Energy Reduction", *in Proceedings of the 25th International Symposium on Computer Architecture (ISCA)*, 1998, pp. 132–141.
9. Moshovos, A., "Power–Aware Register Renaming", Technical Report, University of Toronto, August, 2002.
10. Ponomarev, D., Kucuk, G., and Ghose, K., "AccuPower: an Accurate Power Estimation Tool for Superscalar Microprocessors", *in Proceedings of 5th Design, Automation and Test in Europe Conference (DATE–02)*, March, 2002.
11. Pollack, F., "New Microarchitecture Challenges in the Coming Generations of CMOS Process Technologies", Keynote Presentation, *32nd International Symposium on Microarchitecture*, November 1999.

Stand-by Power Reduction for Storage Circuits

S. Cserveny, J.-M. Masgonty, and C. Piguet

CSEM SA, Neuchâtel, CH
stefan.cserveny@csem.ch

Abstract. Stand-by power reduction for storage circuits, which have to retain data, is obtained through limited locally switched source-body biasing. The stand-by leakage current is reduced by using a source-body bias not exceeding the value that guaranties safe data retention and less leaking non-minimum length transistors. This bias is short-circuited in active mode to improve the speed and the noise margin, especially for low supply voltages; however, this is made for a fraction of the circuit containing the activated part, allowing a trade-off between switching power and leakage. For a SRAM in a 0.18 µm process the leakage is reduced more than 25 times without speed or noise margin loss.

1 Introduction

In processor-based systems-on-chip the memories limit most of the time the speed and are the main part of the power consumption. A lot of work has been done to improve their performances, however, new approaches are required to take into account the trend in scaled down deep sub-micron technologies toward an increased contribution of the static consumption in the total power consumption [1-6]. The main reason for this increase is the reduction of the transistor threshold voltages.

In a previous paper [7], proposals have been made for low-power SRAM and ROM memories working in a large range of supply voltages. However, with all bit lines kept precharged at the supply value, the proposed techniques are not favorable for the static leakage. If a ROM can be switched off in the stand-by mode, or precharging is done only for the selected bit lines before a read (with the corresponding speed penalty), for the 6 transistor SRAM cell enough supply voltage has to be present all the time in order to keep the information stored in the cell.

Negative body biasing increases the NMOS transistor threshold voltage and therefore reduces the main leakage component, the cut-off transistor subthreshold current. A positive source-body bias has the same effect and can be applied to the devices that are processed without a separate well, however it reduces the available voltage swing and degrades the noise margin of the SRAM cell. Another important feature to be considered is the speed reduction resulting from the increased threshold voltage, which can be very severe when a lower than nominal supply voltage is considered.

This paper presents an approach based on the source-body biasing method for the reduction of the subthreshold leakage, with the aim of limiting the normally associated speed and noise margin degradation by switching it locally. In the same time, this bias is limited at a value guaranteeing enough noise margins for the stored data.

In this paper the case of a SRAM is considered, however the same approach can be applied to any blocks containing storage circuits (flip-flops, registers, etc).

J.J. Chico and E. Macii (Eds.): PATMOS 2003, LNCS 2799, pp. 229–238, 2003.
© Springer-Verlag Berlin Heidelberg 2003

2 Leakage Reduction

In the deep sub-micron processes, as they scale down, there is a tendency toward a static leakage strongly dominated by the subthreshold current of the turned-off MOS transistors. In even deeper nanometric processes there is an important tunneling gate current leakage [6], however this can be neglected in the above 100 nm processes.

As the subthreshold leakage is much higher for the NMOS than that of the PMOS, the leakage reduction will be applied here only for the NMOS transistors, however, if necessary, the proposed methods can be applied as well for both. Notice also that the leaky NMOS, usually, have no separate well in a standard digital process.

In order to allow source-body biasing in the 6 transistors SRAM cell, the common source of the cross-coupled inverter NMOS (SN in Figure 1) is not connected to the body. Body pick-ups can be provided in each cell or for a group of cells and they are connected to the VSS ground.

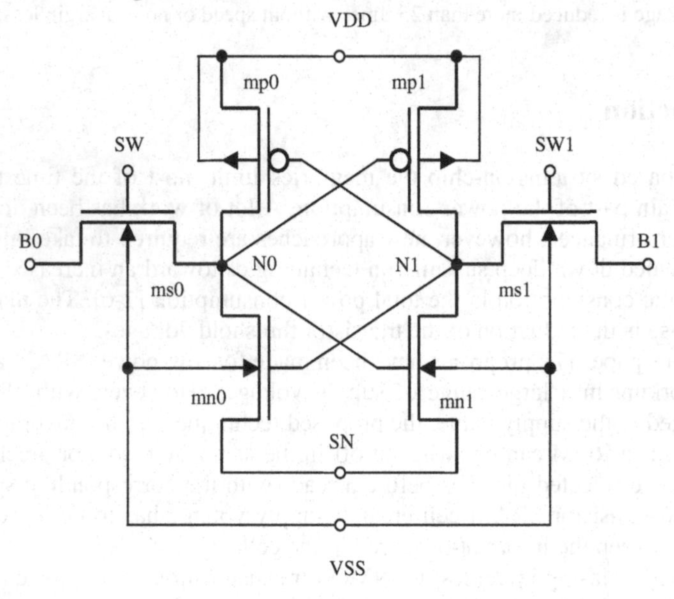

Fig. 1. SRAM cell with separate NMOS source connection SN

In this figure 1 the possibility for separate select gate signals SW and SW1 has been shown, as for the asymmetrical cell described in [7], in which read is performed only on the bit line B0 when only SW goes high while both are activated for write; however, a symmetrical cell, which is selected for read and write with the same select word signal SW ≡ SW1, can also be considered.

A positive VSB bias between VSN and VSS will reduce the subthreshold leakage of a non-selected cell (SW and SW1 at VSS) as shown in the Figure 2.

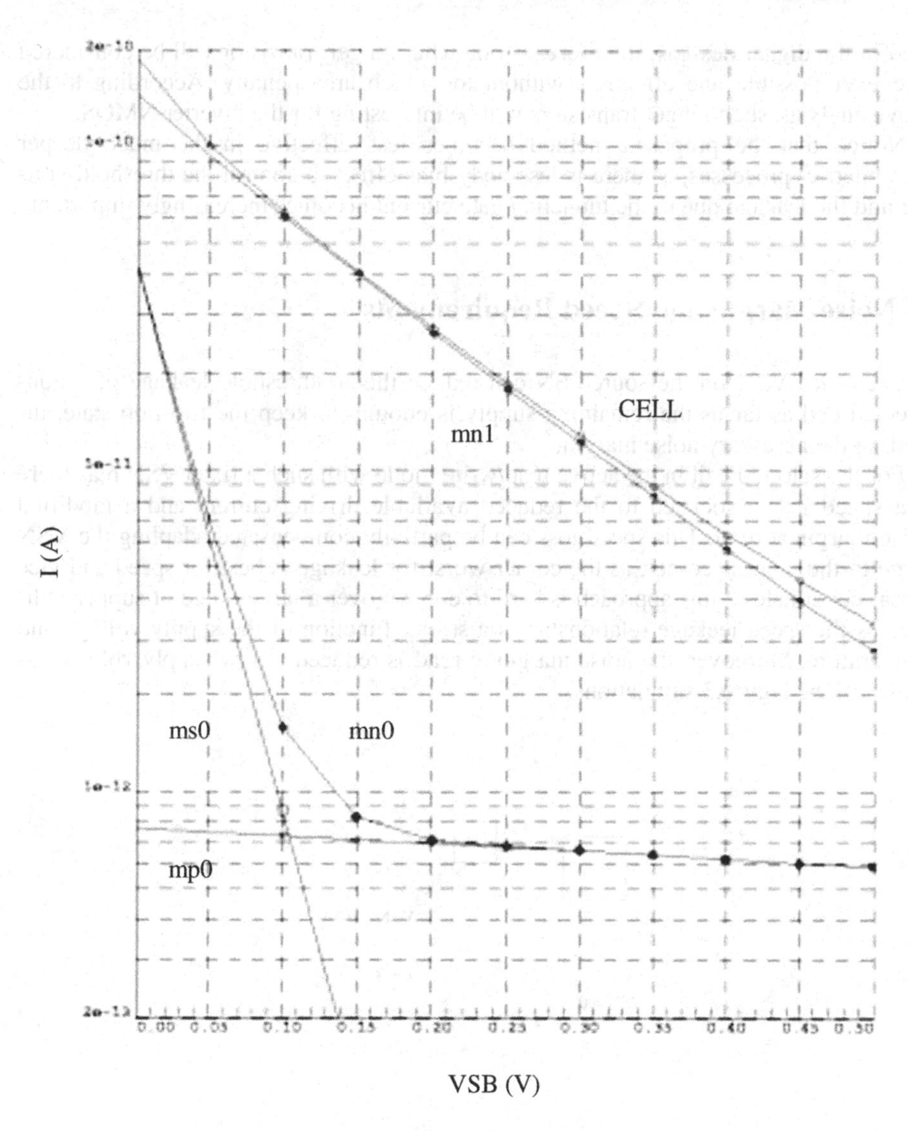

Fig. 2. The simulated cell current (logarithmic scale in A) and its decomposition into the contributions of its transistors as a function of the source-body bias (linear scale in V)

The result in this Figure 2 is for a minimum size symmetrical cell in a 0.18 μm process (mn0, mn1, mp0 and mp1 have W = 0.22 μm and L = 0.18 μm) for the state with the nodes N0 low (VSN) and N1 high (VDD). The cell leakage is dominated by the transistor mn1, i.e. the inverter NMOS which is cut off; the VGS = 0 leakage of this transistor is controlled by its $V_{TH}(V_{SB})$. The other branch contribution due to mp0 (a PMOS) and ms0 (a longer NMOS select transistor whose VGS becomes negative) is much less.

Furthermore, the ratio between the on (at VD = VG) and off (at VGS = 0) currents in these deep sub-micron processes is worst at the minimum channel length, normally used in the digital designs; therefore, a somewhat longer transistor will be considered whenever possible and effective, without too much area penalty. According to the above analysis, such longer transistors will be interesting for the inverter NMOS.

Notice that the proposed method becomes less effective in the much deeper (nanometric) processes, as there is less body bias effect to control the threshold voltage and the leakage due to the tunneling gate current becomes increasingly important.

3 Noise Margin and Speed Requirements

A fixed bias VSN on the source SN can reduce the subthreshold leakage of a non-selected cell as far as the remaining supply is enough to keep the flip-flop state, including the necessary noise margin.

For the selected cell in the active read/write mode with such a fixed VSN bias there is a speed loss associated to the reduced available driving current and a modified noise margin at read. This speed loss can be partially compensated adapting the VSN value to the process corner, as the corner worst for leakage is best for speed and vice versa; nevertheless, this approach is hard to control over a large range of supply voltages as the speed/leakage relationship is a strong function of the supply voltage and temperature. Moreover, the noise margin at read is reduced at low supply voltages as shown in the Figure.3 simulations.

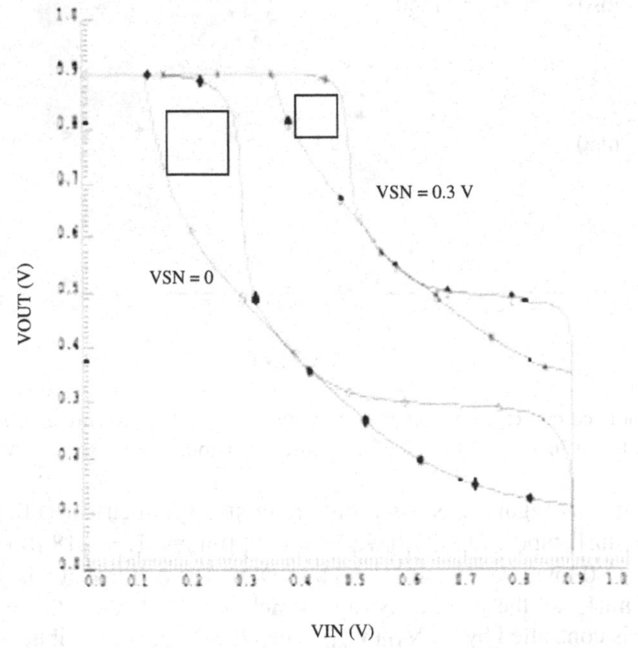

Fig. 3. Worst case (Nfast Pslow 125 °C) read access static noise margin analysis for the minimum size symmetrical cell at VSN = 0 and VSN = 0.3 V for VDD = 0.9 V

Such a low VDD is important for large supply range and portable applications.

The noise margin, represented [8,9] by the maximum size square that can be nested into the cross-coupled voltage transfer characteristics, is visibly reduced by the source-body bias at this low value of VDD. Notice also that the crossing of the transfer characteristics changes from 3 to 5 points.

For the non-selected cell, important for the stand-by leakage, only the inverter transfer characteristics (without select MOS loading) come into account, and even if the VSN increase reduces the noise margin, it remains reasonably high if the VSN does not increase too much, as shown in the Figure 4 for the same voltages as above.

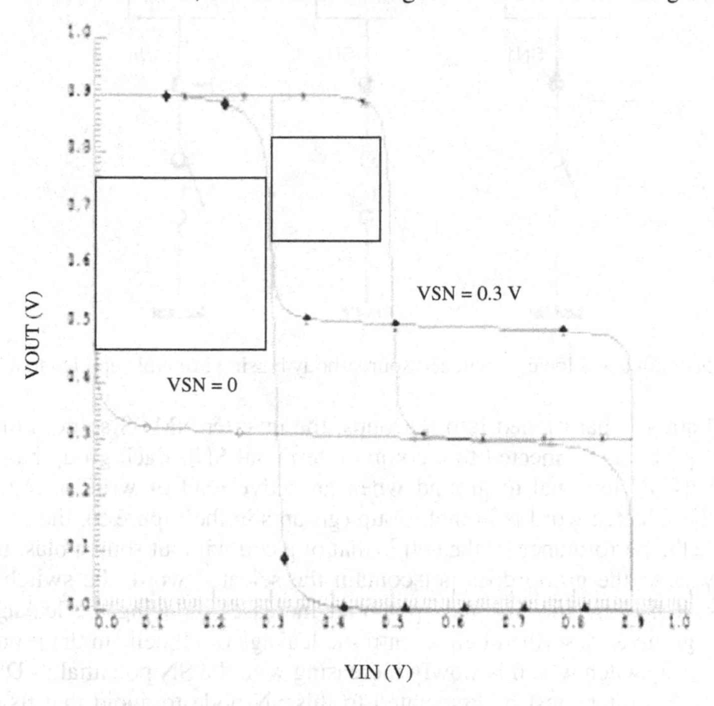

Fig. 4. Worst case (Nfast Pslow 125 °C) stand-by static noise margin analysis for the minimum size symmetrical cell at VSN = 0 and VSN = 0.3 V for VDD = 0.9 V

4 Locally Switched Source-Body Bias

The VSN bias, useful for static leakage reduction, is acceptable in stand-by if its value does not exceed the limit at which the noise margin of the stored information becomes too small, but it degrades the speed and the noise margin at read. Therefore, it will be interesting to switch it off in the active read mode; however, the relatively high capacitance associated to this SN node, about 6 to8 times larger than the bit line charge of the same cell, is a challenge for such a switching.

It is proposed here to do the switching of the VSN voltage between the active and stand-by modes locally assigning a switch to a group of cells that have their SN sources connected together, as shown in the Figure 5.

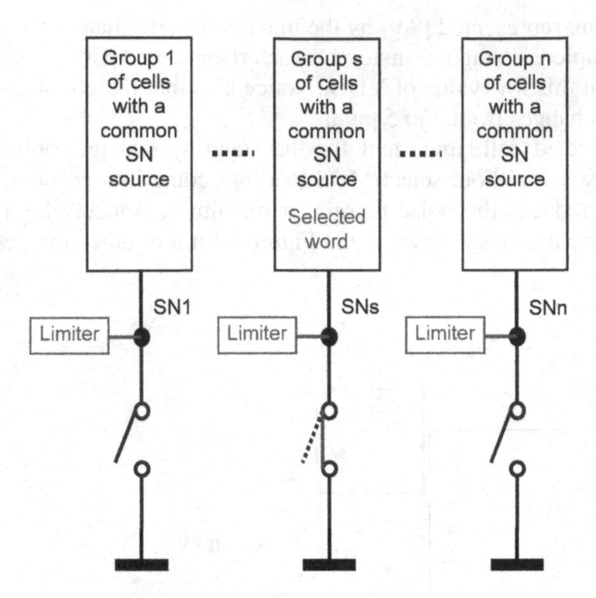

Fig. 5. The limited and locally switched source-body biasing principle applied to a SRAM

The cell array is partitioned into n groups, the inverter NMOS sources of all cells from a group i being connected to a common terminal SNi. Each group has a switch connecting its SN terminal to ground when an active read or write operation takes place and the selected word is in that group (group s in the Figure 5), therefore in the active mode the performance of the cell is that of a cell without source bias. However, in stand-by, or if the group does not contain the selected word, the switch is open. With the switch open, the SN node potential increases reducing the leakage of the cells in that group as described before until the leakage of all cells in the group equals that of the open switch which is slowly increasing with the SN potential (VDS effect); nevertheless, a limiter must be associated to this SN node to avoid that its potential becomes too high, guaranteeing enough margin for the stored state.

The group size and the switch design are optimized compromising the equilibrium between the leakages of the cells in the group and the switch with the voltage drop in the activated switch and the SNs node switching power loss. The switch is a NMOS that has to be strong enough compared to the read current of one word, the selected word, i.e. strong compared to the driving capability of the cells (select and inverter NMOS in series), and in the same time weak enough to leak without source-body bias as little as the desired leakage for all words in the group with source-body bias at the acceptable VSN potential. On the other hand, the number of words in the group cannot be increased too much in order to limit the total capacitive load on the SN node at a value keeping the SN switching power loss much less than the functional dynamic power consumption. In particular, this last requirement shows why the local switching is needed contrary to a global SN switching for the whole memory.

The Figure 6 shows the most flexible approach to implement the described principle: groups are built with nrs rows in a one-word column and the switch size is adapted to the number of bits in a word.

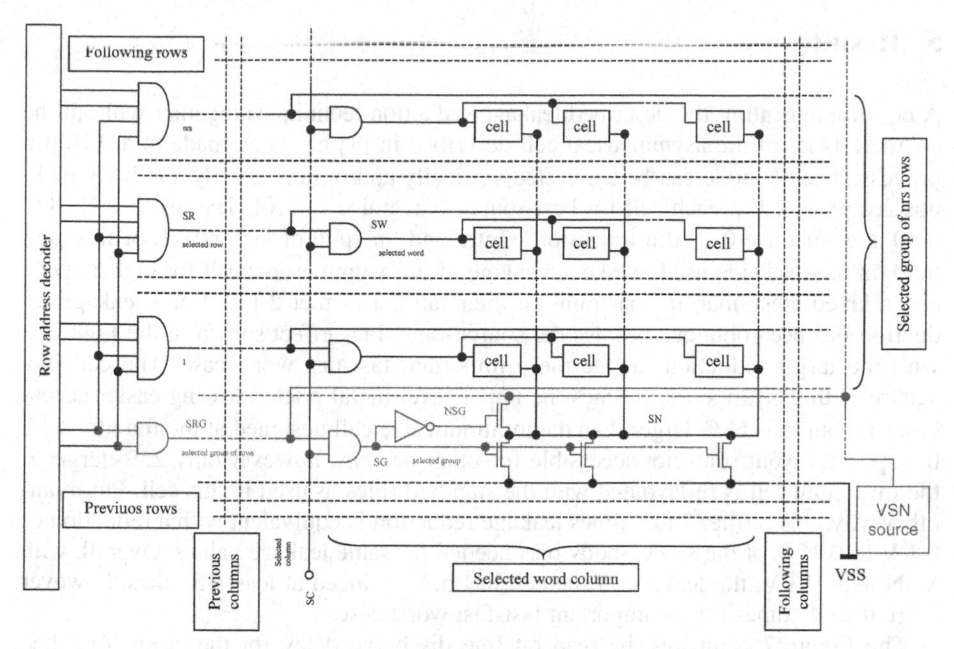

Fig. 6. Example of a SRAM organization implementing the locally switched and limited source-body biasing

In this implementation, with nrs an integer power of 2 (for example 16 or 32), the selection of the group of rows is made by a simple partial row decoding, which, combined with the one-word column selection signal SC, generates the group activation signal SG closing the NMOS switch for that group. The limitation is obtained by connecting the SN node of the group to a VSN source with a NMOS pass transistor activated by the complement of the SG signal, therefore connecting all groups except the activated one. Contrary to the fixed bias approach in which the whole active current goes through the biasing source, here the VSN source has to deliver or absorb only a very small current, the difference between the small leakages of the cells and the switches which are essentially equilibrated by design.

Other implementations were considered depending on the size and specific design constraints of the memory. When possible, it is quite interesting to make a one full row group. In some situations another interesting approach is to activate the group only by address decoding, reducing the number of group switching at the expense of a larger leakage in the group that remains selected. It is important to notice that the SN switching power loss depends on the group size and its relevance to the total power dissipation depends on the number of groups in the memory. Even more important is to notice that the limiter should only limit the SN potential increase; the Figure 6 shows only one of the different implementations that have been considered to fulfill this function. Diode limiters, as considered in [10] for the typical case, are very interesting, however should be carefully designed and checked over all parameter corners.

5 Results

A cell implementing the described leakage reduction techniques together with all the characteristics of the asymmetrical cell described in [7] has been made in a 0.18 μm process. The SN node can be connected vertically and/or horizontally and body pick-ups are provided in each cell for best source bias noise control. The inverter NMOS mn0 and mn1 use 0.28 μm transistor length, and, in spite of larger W/L of ms0 and mn0 on the read side used to take advantage of the asymmetrical cell for higher speed and relaxed noise margin constraint on their ratio, a further 2 to 3 times leakage reduction has been obtained besides the source-body bias effect shown in the Figure 2, with the larger reduction for the more important fast-fast worst case. The cell is a square with 2.8 μm sides, 5 times the upper level metal pitch allowing easier access blocs layout; it is 68% larger than the minimum size cell designed at the foundry with the special layout rules not accessible for other designs, however only 25% larger if the minimum cell is re-layouted with the standard rules as used in this cell. Put in another way, this further 2 to 3 times leakage reduction is equivalent with a reduction of 0.1 V to 0.15 V of the source-body bias needed for same leakage values. Overall, with VSN near 0.3 V, the leakage of the cell has been reduced at least 25 times, however more than 40 times for the important fast-fast worst case.

The Figure 7 compares the read bit line discharge delay for the group switched source-body bias with the fixed bias approach.

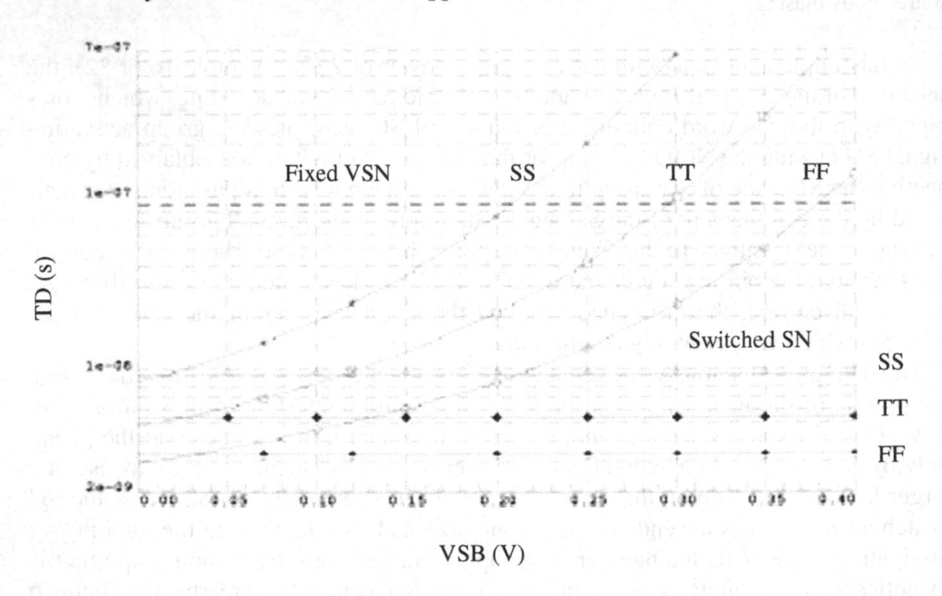

Fig. 7. Simulated bit line discharge delay (s) as a function of the source-body bias (V) for 256 rows at 25 °C considering the new cell and group switch design and all bits in the word at 1, compared to the fixed VSN bias result for the same cell and conditions

As expected, the equivalent VSN bias for the same delay, shown by the crossings in the Figure 7, correspond to the voltage drop in the switch. For the worst case considered here (all bits in the read word at 1, i.e. maximum current in the switch), this drop varies between 19 mV for SS (worst case for delay) and 28 mV for FF.

When the group switching is compared to the fixed VSN = 0 with the same cell, the about 10% maximum speed loss and the noise margin reduction related to these few mV are really not important.

The only area increase, compared to the previous design in [7], is due to the group switches. If the switches are implemented as in the Figure 6, the switches for two groups need an area about that of one row; therefore the area penalty for 16 word groups is 3%; if the groups are organized in the rows, the area increase is even less.

Following the strong leakage reduction in the cell array, it was important to design the remaining blocks of the memory in order to keep the total stand-by current of the memory still dominated by the cells. The stand-by leakage of the other blocks has been limited at well under 10% of the total stand-by current of the memory adapting some control signals for better low leakage behavior of the blocks and by using for the most leaky elements, identified according to their function, similar approaches such as source biasing, however without the state retention constraints, and/or longer than minimum length transistors. This did not require any extra area, some more switches used for the bit line pull down transistors could be placed in the very large input signals buffer area already defined by the length of the select row and the height of the read/write blocks

The tape out of the first integration of a complex portable wireless sensor network circuit that uses several sizes of this SRAM memory is expected in august 2003, therefore silicon measurements are not yet available. However, the different size memories are automatically generated and their performances, as needed for the high level synthesis of the application, are characterized by simulation. The foundry BSIM3v3 files for this 0.18 μm process have been used in all these simulations, as in the few simulation results shown in this paper.

References

[1] S. Kosonocky, M. Immediato, P. Cottrell, T. Hook, R. Mann, J. Brown: Enhanced Multi-Threshold (MTCMOS) Circuits Using Variable Well Bias, Proceedings ISLPED'01, pp. 165–169

[2] S. Narendra, S. Borkar, V. De, D. Antoniadis, A. Chandrakasan: Scaling of Stack Effect and its Application for Leakage Reduction, Proceedings ISLPED'01, pp. 195–200

[3] C. Kim, K. Roy: Dynamic Vt SRAM: A Leakage Tolerant Cache Memory for Low Voltage Microprocessors, Proceedings ISLPED'02, pp. 251–254

[4] N. Azizi, A. Moshovos, F. Najm: Low-Leakage Asymmetric-Cell SRAM, Proceedings ISLPED'02, pp. 48–51

[5] C. Piguet, S. Cserveny, J-F. Perotto, J-M. Masgonty: Techniques de circuits et méthodes de conception pour réduire la consommation statique dans les technologies profondément submicroniques, Proceedings FTFC'03, pp. 21–29

[6] F. Hamzaoglu, M. Stan: Circuit-Level Techniques to Control Gate Leakage for sub-100nm CMOS, Proceedings ISLPED'02, pp. 60–63

[7] J. -M. Masgonty, S. Cserveny, C. Piguet: Low Power SRAM and ROM Memories, Proceedings PATMOS 2001, paper 7.4

[8] E. Seevinck, F. J. List, J. Lohstroh: Static-Noise Margin Analysis of MOS SRAM Cells, IEEE J. Solid-State Circuits, vol. 22, pp748–754, Oct. 1987
[9] A. J. Bhavnagarwala, X. Tang, J. D. Meindl: The Impact of Intrinsic Device Fluctuations on CMOS SRAM Cell Stability, IEEE J. Solid-State Circuits, vol. 36, pp658–665, April 2001
[10] T. Enomoto, Y. Oka, H. Shikano, T. Harada: A Self-Controllable-Voltage-Level (SVL) Circuit for Low-Power High-Speed CMOS Circuits, Proceedings ESSCIRC 2002, pp 411–414

A Unified Framework for Power-Aware Design of Embedded Systems*

José L. Ayala and Marisa López-Vallejo

Departamento de Ingeniería Electrónica
Universidad Politécnica de Madrid (Spain)
{jayala,marisa}@die.upm.es

Abstract. This paper presents a new tool for power estimation of caches built inside a unified framework for the design of embedded systems. The estimator, which works in parallel with the functional simulator of the system, has been designed to deal with different target architectures, providing high flexibility. The estimation is based on an improved analytical power model that provides high accuracy on the estimation. The proposed framework has been verified with benchmarks from the MiBench suite, obtaining good results in terms of accuracy and execution time.

1 Introduction

Power consumption has become a critical design issue in processor based systems. This effect is even more appreciable in mobile embedded systems, where the limited battery capacity, the portability of the devices and the packaging materials require low power consumption and energy dissipation. When designing complex high-performance low-power embedded systems, the designers have to analyze multiple hardware and software options by experimenting with different architectural and algorithmic tradeoffs. In most cases the information about power consumption is not available until the whole system has been specified and the prototyping results can be collected. Due to this limitation it is necessary to perform different iterations through the design flow until the power and performance objectives have been satisfied, increasing a key parameter, the *time-to market*. In order to optimize the design of power efficient embedded systems, big efforts have to be carried out to provide information of energy dissipation from the very early design phases.

When dealing with the processor of an embedded system, there are several parameters that have a strong relation with the performance-power tradeoff. Mainly, the target processor and the memory cache hierarchy [9]. In this paper we present the design and use of an accurate estimation tool for cache power consumption. The tool has been designed to be highly integrated in a retargetable design tool chain. With this work the most important source of power dissipation in the system is completely characterized from the very early design phases,

* This work was supported by the Spanish Ministry of Science and Technology under contract TIC2000-0583-C02-02.

J.J. Chico and E. Macii (Eds.): PATMOS 2003, LNCS 2799, pp. 239–248, 2003.
© Springer-Verlag Berlin Heidelberg 2003

and the subsequent design takes into account the power dissipation as a constraint. The proposed methodology is extensible to new architectures, providing a helpful framework for the design of power-constrained embedded systems.

The power estimation of the cache is based on an analytical model that has been improved by including the effects of the switching activity. Simulation results show how this fact can drastically impact in the final estimated energy.

The paper is structured as follows. Next, related work is reviewed. Section 3 presents the main objectives of this work while section 4 describes the power estimation tool and its basic enhancements. Finally estimation results are presented and some conclusions are drawn.

2 Related Work

In the last few years special emphasis has been put on power estimation and high-level optimization tools attending to the designers' demands. Some of these works are based on analytical power models that can predict with high accuracy the power dissipation in certain processor modules (cache, system clock, data path, etc). The tool presented in this paper uses the analytical power model for caches developed by Kamble and Ghose [7]. This analytical model can be very accurate[1] but can also exhibit big deviations due to the statistical distribution of the switching activity value of the bus, that is assumed homogeneous. This is due to the lack of exact simulation results (the exact value of the switching activity can only be known when an application runs on the target architecture and the transferred data and addresses are tracked). This limitation has been solved with the present work.

Current research on power estimation has moved to higher level and very early design phases. The aim of this shift is to accelerate the complex design flow of embedded systems and lead the designer decisions with power consumption constraints. Important research has been carried out in the area of high-level power estimation by macro-model characterization [8], instruction characterization [10] or VHDL description analysis [1]. The problems with these approaches are, respectively, the extremely high dependence with the target architecture, the lack of simulation information (therefore, a statistical homogeneous distribution is still assumed) and long simulation and design time due to the low-level description. Moreover, power simulators like Wattch [2] or SimplePower [13] achieve accurate power estimations in a short time, but they are still limited to simple and unrealistic target architectures.

Finally, compiler [6] and source code [11] estimation tools have been proposed, which partially could solve the target dependence problem. However, they are still narrowly coated to a macro-model target characterization where the module dependencies are not properly managed and the static analysis does not allow explicit calculation of statistical parameters. Summarizing, it can be said that there is no a single retargetable framework to design embedded systems with both performance and power constraints.

[1] 3% of estimation error.

3 Objectives

Current tools for designing embedded systems do not include power estimation as a design constraint. Available utilities to estimate power dissipation in the datapath or the cache are not fully integrated in the design flow, and do not consider realistic processors as target architectures. Furthermore, existing power estimators for caches do not reflect the inherent data correlation due to the lack of simulation information at instruction level; this usually conducts to estimation errors that make difficult the design and analysis of low power policies.

The goal of our research work is to provide an accurate cache power consumption estimation tool highly integrated in a retargetable design tool chain. Such unified framework integrates the processor and cache functional simulators with a power estimation tool that can provide accurate estimations for many different real target architectures. Moreover, the overflow of this power estimation must be negligible in the design time or the design phases, but it must extend the simulator output with the power results. The proposed methodology is also applicable to new architectures, providing a fast and useful way to analyze power consumption on the devised systems (i.e. *fully extensible*).

In this way, the resulting unified framework provides the following advantages for the designers of embedded systems: automatic generation of cross-design tools, parallel output of functional simulation and accurate cache power estimation, negligible learning time for the user and flexibility on the selection of the target processor. This is a very helpful environment for a designer, who can use these tools for many different applications, as it is the study of the influence of compiler optimizations. Such application will be described in section 5.2.

4 Design of the Tool

The design of this tool is based on CGEN ("Cpu tool GENerator") [4], a unified framework and toolkit for writing programs like assemblers, disassemblers and simulators. CGEN is centered around application independent descriptions of different CPUs[2] (ARM, MIPS32, PowerPC...), many of them present in current embedded systems. Our work is built over the CGEN core to generate embedded system design tools with explicit information of power dissipation in caches. As CGEN successfully does with cross-design tools, the cache power estimator can be automatically generated for the whole set of available target processors in order to free the designer from this annoying task.

4.1 Brief Description of the Analytical Model

As it was previously said, in order to get an accurate estimation of cache power dissipation, the exact memory access pattern must be known to feed the analytical model of power consumption [7]. This model provides mathematical expressions for the several energy dissipation sources in the cache architecture and

[2] Plus environment, like ABI.

takes as arguments parameters of different kinds: architectural (cache size, way configuration, word line size, etc), technological (line and device input/output capacitors, etc) and statistical (switching bus activity, cache accesses, etc). Such analytical model uses expressions like the ones presented in equations 1 and 2 for the energy dissipated per read and write access, where `swf` is the switching factor parameter, `T` (tag size), S_t (status bits per block frame), `L` (line size) and `m` (number of ways in a set associative cache) are architectural parameters, C_x are technology terms, and the rest are the statistical arguments that describe the access. V_{dd} represents the source voltage and V_s is the voltage swing on the bit lines during sensing.

$$E_{bit\_r} = swf \cdot V_{dd} \cdot V_s \cdot C_{bit\_rw} \cdot N_{array\_accs} \cdot 2 \cdot m \cdot (T + S_t + 8 \cdot L) \quad (1)$$

$$E_{bit\_w} = swf \cdot V_{dd} \left[(\frac{1}{2} \cdot V_{dd} - V_s) \cdot N_{bit\_w} + (\frac{1}{2} \cdot V_{dd} + V_s) \cdot N'_{bit\_w} \right] \cdot C_{bit\_rw} \quad (2)$$

These expressions can be deduced from the general equation for the power dissipation $E = swf \cdot C \cdot V_{dd}^2$ taking into account the particular cache architecture and the statistics for the access pattern.

While the technology and architectural parameters are known from the architecture design phase, the access pattern depends on the particular data that are in use. Furthermore, to calculate the exact value of the switching activity, perfect knowledge of the transferred data and addresses is needed. Such simulation detail can only be achieved at instruction level, when the memory and register contents are also available.

The switching activity factor reflects the number of bit changes at the data or address bus divided by the bus width (bit change factor). The usual approach to deal with this unavailable term is supposing half of the bits change their value, what is far away from being true when strong data or address correlation appears in the application and a reduced number of bits switches. Therefore, over and underestimation on cache energy consumption can be explained by this limitation in the analytical model. The approach presented here overcomes this past limitation by explicitly calculating the switching activity term thanks to the information provided by the functional simulator.

4.2 Description of the Power Estimation Tool

To have access information, the running code must generate a detailed memory trace (memory address accessed, data transferred, type of the access) per memory access. To get a machine independent implementation of this trace facility, we have supplied the target machine with four new virtual registers (*trace registers*) storing the trace information: `tr0` stores the memory address accessed, `tr1` is a counter of the memory accesses, `tr2` is the bit flag that indicates a read or write access, and `tr3` stores the data read/written from/to memory. The counter `tr1` is needed to track the memory access instructions and generate the required trace. These trace registers are identically defined in every target processor. In

```
(dnh h-tr "trace registers"
() ; attributes
(register WI (4))
(keyword "" ((tr0 0) (tr1 1) (tr2 2) (tr3 3)))
() ()
)

(dni (.sym ld suffix) (.str "ld" suffix)
((PIPE O))
(.str "ld" suffix " $dr,@$sr")
(+ OP1_2 op2-op dr sr)
(sequence ()
(set (reg h-tr 3) (ext-op WI (mem mode sr)))
;data transfered
(set dr (ext-op WI (mem mode sr)))
;execution
(set (reg h-tr 1) (add (reg h-tr 1) (const 1)))
;access counter
(set (reg h-tr 0) (ext-op WI sr))
;memory address
(set (reg h-tr 2) (const 0)))
;memory read
((m32r/d (unit u-load))
(m32rx (unit u-load))))
```

Fig. 1. RTL tracing code

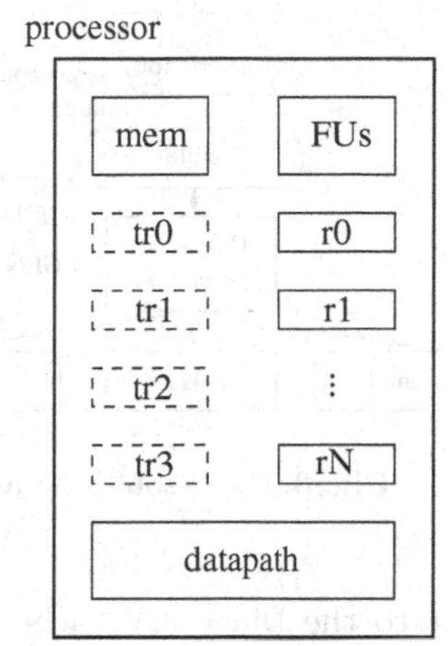

Fig. 2. Processor architecture

the same way, an identical approach is followed by the read and write operations in such trace registers. The specification of these four registers has been done in the CGEN's RTL language[3] and also the simple modifications to the processor ISA, both of them automatically accomplished by a text parser. Figure 1 shows the trace registers specification and a traced `load` instruction for a MIPS32 processor, where it can be observed how the description of the trace registers and their read and write operations are independent of the target processor (on condition that they are included in every memory access instruction). Figure 2 shows the schematic view of the processor architecture, where the virtual trace registers, used only for simulation purposes, have been represented.

The information stored in these extra registers can be read by the simulator as a normal hardware register. The simulator code in charge of reading this trace information and its analysis is also common for all the different target architectures, and has been coded in the simulator kernel. The trace information stored in these registers allows us to create a trace file which feeds the Dinero IV [3] cache simulator tool. Dinero IV is a cache simulator for memory reference traces which simulates a memory hierarchy consisting of various caches connected as one or more trees, with reference sources (the processors) at the leaves and a memory at each root, and which works in parallel with the processor simulator.

The flow of the code that performs memory tracing works as follows. For every new instruction, the trace registers are read to retrieve their information. If a new memory access happens, the trace information is stored in a file and

[3] A subset of the Scheme language.

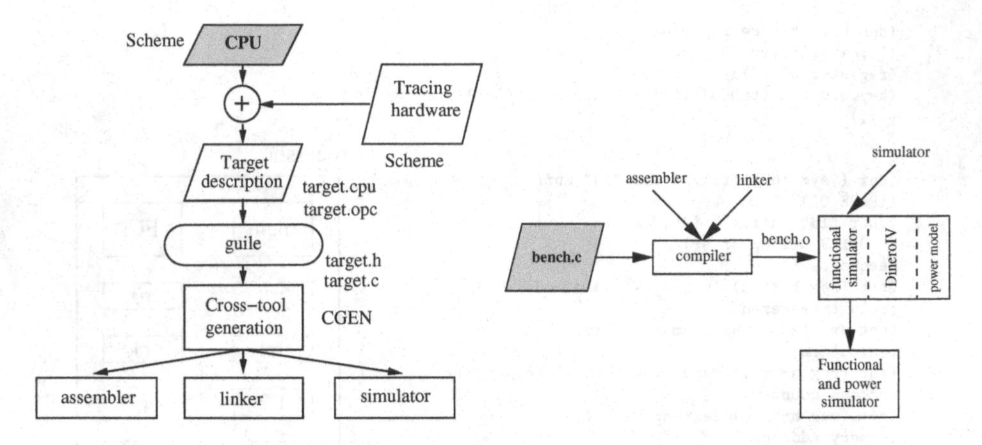

Fig. 3. Cross-tool generation

Fig. 4. Extended design flow

sent to the Dinero IV cache simulator when the size of this package of traces reaches a minimum value.

The Dinero IV functionality has been extended to perform the estimation of cache power consumption. The analytical power model has been coded in the simulator kernel and calculates the following energy expressions: energy consumption associated with the bit line precharge, energy consumption associated with the data read and write, energy dissipated in the data and address buses and energy consumed by the combinational logic (sense amplifiers, latches, address decoders...). The architectural parameters used in these expressions are based on a 0.8 μm^4 cache implementation [12]. Moreover, the statistical parameter *switching activity* has been explicitly calculated, what allows a more accurate estimation. The calculus of the switching activity parameter is obviously enabled by the detailed tracing output and allows to obtain the temporal evolution of the energy consumption with respect to the data and address correlation.

Figure 3 shows the proposed extended cross-tool generation for the design of embedded systems with explicit cache power estimation. As it can be noticed, the designer must only provide the hardware description of the new target architecture which will be the core processor of the embedded system. This hardware description is automatically extended with the tracing registers and, from there on, the rest of tools are generated.

Figure 4 shows the design flow corresponding to a typical cross-design environment (host and target machine are not the same) with the particularity of concurrent functional and power simulation. In this case, the designer should only provide the source code for the benchmark, as in the common case. This source code is cross-compiled and linked and, after that, simulated for the target processor with the new tool. In this way, the simulation output has been upgraded with power consumption information.

[4] Technology available when writing this paper. It can be easily updated with new info.

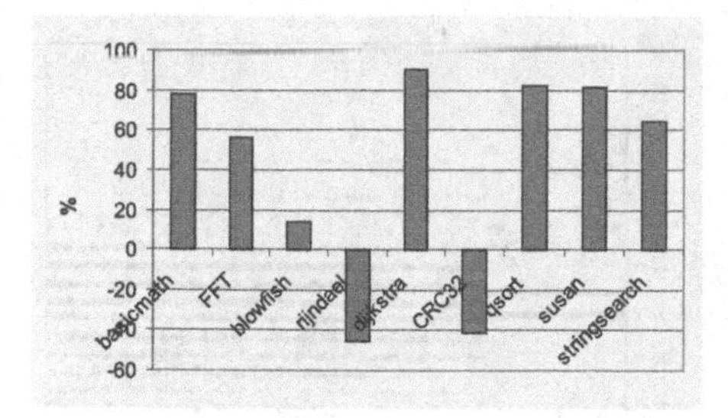

Fig. 5. Difference between average values (power consumption per write access)

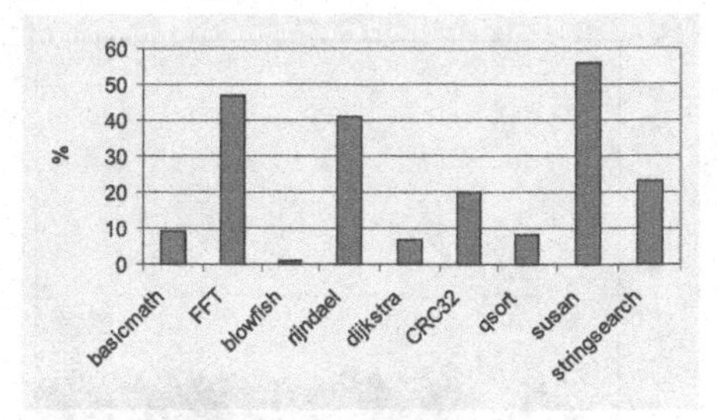

Fig. 6. Difference between standard deviations

5 Experimental Results

The proposed methodology and tools have been applied to different real benchmarks from the MiBench suite [5], compiled on an x86 machine for a MIPS32 target processor. However, any processor among MCore, i960, ARM, PowerPC or others could have been selected with similar results. The simulation time has also been analyzed resulting on a low overhead for the power estimation with respect to the functional simulation. This estimation overhead is directly related to the desired granularity (power estimation per number of simulated instructions) and to the quality of the implementation. We have obtained a simulation time with power estimation information in the same order of the functional simulation time. For instance, when simulating a medium size benchmark (*stringsearch*), the simulation time with the power estimation goes from 1.011 s to 1.400 s[5]

5.1 Switching Activity

For the whole set of benchmarks, the improvement on accuracy related to the exact calculation of the switching activity has been first analyzed.

Figure 5 shows the percentual difference between the average values of energy consumption on this data cache for the write operation. Positive values mean overestimation of the energy consumption by the homogenous model, while negative values mean underestimation of the energy consumption. It can be observed how the simplified model overestimates the power consumption in most cases because it is not able to take into account data correlations which appreciably reduce the switching factor. On the other hand, there are two cases (*rijndael* and *CRC32*) where data presented extremely low correlation and, therefore, the real switching factor is bigger than the homogeneous assumption. This underestimation gives rise to bigger power consumption than the predicted by the simplified model. This situation is specially undesirable in the design of embedded systems with power constraints, since it can cause malfunctioning of the system.

[5] Simulation time can be easily improved with parallel programming techniques.

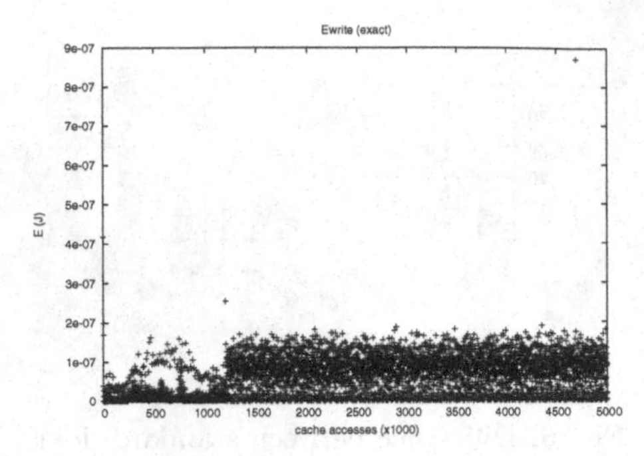

Fig. 7. Susan: exact switching factor. Write operation.

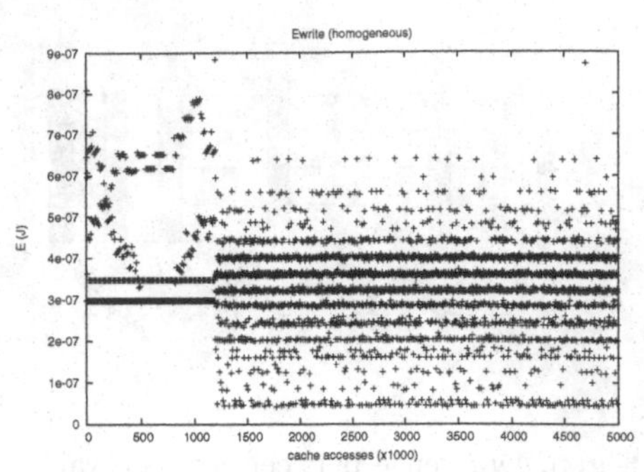

Fig. 8. Susan: homogeneous switching factor. Write operation.

Figure 6 shows the percentual difference between the standard deviation of both data distributions (exact model and homogeneously distributed model) measured for several benchmarks. This metric has been selected to analyze how spread out the power distribution is. As it can be noticed, the big errors exhibited by some of the benchmarks are due to the fact that the real power distribution is far away from a homogeneous distribution.

From the whole set of experiments carried out, we have selected two benchmarks to analyze in depth. One of them is *stringsearch*, a test that searches given words in phrases, and the other one is *susan*, an image recognition package.

Now we will analyze the effect of the explicit calculation of the switching activity on energy consumption. Figures 7 and 8 show the energy bitline consumption per write access in data cache when using the calculated value of the switching activity or the homogeneous distribution, respectively, in the *susan* benchmark. The X axis represents every 1000 memory accesses, while the Y axis represents the value of the energy consumption in Joules. Both tests have been compiled without optimizations and simulated for 5 millions of memory accesses.

As it can be noticed, the exact switching factor not only reflects an appreciable difference in the energy consumption per write access ($1 \cdot 10^{-7} J$ on average for the exact version, and $4 \cdot 10^{-7} J$ on average for the homogeneous version), but also a more accurate profile of the energy consumption evolution (increasing low-energy values for the earlier memory accesses, when the image file is being read from disk and written into memory). Similar behavior can be found for the *stringsearch* benchmark showing, in this case, $7 \cdot 10^{-8} J$ on average for the exact version and $3 \cdot 10^{-7} J$ on average for the homogeneous version.

The effect of the switching factor in the energy consumed per read operation is less influencing. The reduced correlation on the read accesses exhibited by the simulated applications, the linear product of the switching factor in the mathematical expressions, and the small values involved (energy per read access is much smaller than energy per write access) explain such difference.

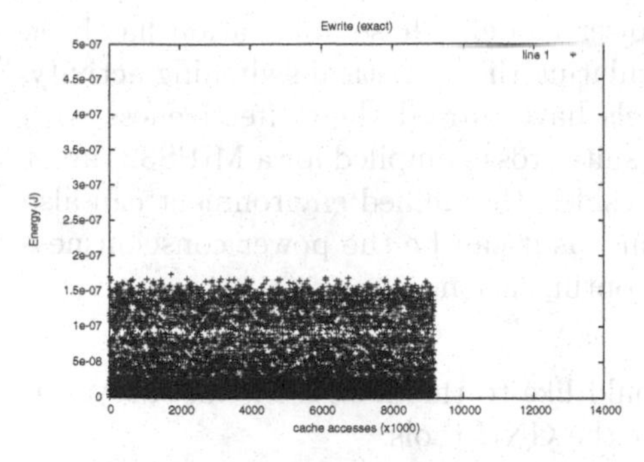

Fig. 9. Stringsearch: with compiler optimizations

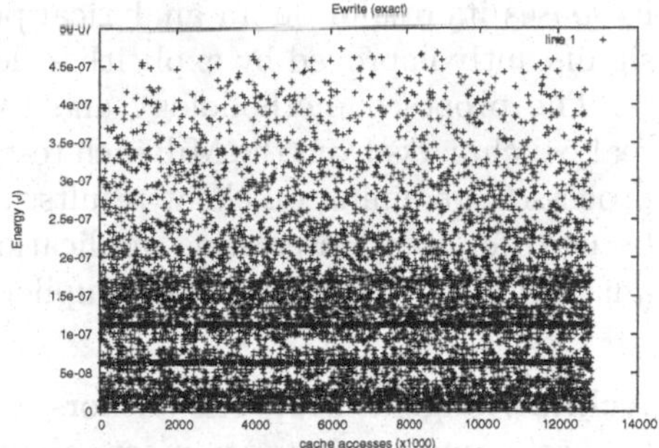

Fig. 10. Stringsearch: without compiler optimizations

5.2 Compiler Optimizations

The proposed methodology and tools can be used to evaluate energy aware compiler optimizations, next step in our research. Current performance-oriented compiler optimizations can also be characterized in terms of power awareness. For such objective, two versions of the *basicmath* benchmark have been compared, one without compiler optimizations (`-O0` option) and another one with compiler optimizations (`-O3` option). As figures 9 and 10 show, the effect of this set of compiler optimizations in power consumption per write access can result in lower power dissipation on average per instruction and, additionally, reduced number of memory accesses[6]. Both benchmarks have been simulated up to completion. We can conclude that there is a reduction on the energy consumption when compiler optimizations are applied. It is partially due to the decrease of the number of instructions and memory accesses, and it is also caused by a lower energy per instruction produced by the increased data correlation.

6 Conclusions

We have presented a unified framework of cross-design tools, which supplies a helpful environment for the efficient design of power-aware embedded systems, providing power dissipation information from the very early phases of the design flow. Furthermore, the tools, which are automatically generated, allow free selection of the main power consumers in embedded systems, i.e. target architecture and cache hierarchy configuration. Only by providing energy information in parallel with the performance simulation, the designers can lead their decisions based on both constraints without large computing overhead. Moreover, the design flow presents high flexibility since the power estimation for the cache hierarchy can be performed for different target processors.

The design of the cache power estimation tool has been carried out using the unified environment of GNU cross-design tools. The power estimation util-

[6] In this way, the total energy consumption is also minimized.

ity bases its results on an analytical power model, whose application has been significantly improved by explicitly calculating the statistical switching activity.

The proposed methodology and tools have proved their effectiveness with real benchmarks from the MiBench test suite cross-compiled for a MIPS32 target processor, obtaining excellent results. Besides, the unified environment can also be used for many interesting applications, as it can be the power consciousness analysis of current and future compiler optimizations.

Acknowledgements. The authors would like to thank José Manuel Moya for his useful and precious comments about the GNU tools.

References

1. J. M. Alcântara, A. C. C. Vieira, F Gálvez-Durand, and V. Castro. A methodology for dynamic power consumption estimation using VHDL descriptions. In *Symposium on Integrated Circuits and Systems Design*, 2002.
2. David Brooks, Vivek Tiwari, and Margaret Martonosi. Wattch: a framework for architectural-level power analysis and optimizations. In *International Symposium on Computer Architecture*, 2000.
3. Jan Edler and Mark D. Hill. Dinero IV, 2002. http://www.cs.wisc.edu/~markhill/DineroIV/.
4. GNU. CGEN, 2003. http://sources.redhat.com/cgen/.
5. Matthew R. Guthaus, Jeffrey S. Ringenberg, Dan Ernst, Todd M. Austin, Trevor Mudge, and Richard B. Brown. MiBench: A free, commercially representative embedded benchmark suite. In *Annual Workshop on Workload Characterization*, 2001.
6. I. Kadayif, M. Kandemir, N. Vijaykrishnan, M. J. Irwin, and A. Sivasubramaniam. EAC: A compiler framework for high-level energy estimation and optimization. In *Design and Test in Europe*, 2002.
7. M. Kamble and K. Ghose. Analytical energy dissipation models for low power caches. In *International Symposium on Low Power Electronics*, 1997.
8. Johann Laurent, Eric Senn, Nathalie Julien, and Eric Martin. High-level energy estimation for DSP systems. In *International Workshop on Power and Timing Modeling, Optimization and Simulation*, 2001.
9. B. Moyer. Low-power design for embedded processors. *Proceedings of the IEEE*, 89(11), November 2001.
10. S. Nikolaidis and Th. Laopoulos. Instruction-level power consumption estimation embedded processors low-power applications. In *International Workshop on Intelligent Data Acquisition Computing Systems: Technology and Applications*, 2001.
11. E. Senn, N. Julien, and E. Martin. Power consumption estimation of a C program for data-intensive applications. In *International Workshop on Power and Timing Modeling, Optimization and Simulation*, 2002.
12. S. E. Wilton and N. Jouppi. An enhanced access and cycle time model for on chip caches. Technical Report 93/5, DEC WRL, 1994.
13. W. Ye, N. Vijaykrishnan, M. Kandemir, and M. J. Irwin. The design and use of SimplePower: A cycle-accurate energy estimation tool. In *Design Automation Conference*, 2000.

A Flexible Framework for Fast Multi-objective Design Space Exploration of Embedded Systems

Gianluca Palermo[1], Cristina Silvano[1], and Vittorio Zaccaria[2]

[1] DEI, Politecnico di Milano, Milano Italy
{gpalermo, silvano}@elet.polimi.it
[2] STMicroelectronics, Agrate Brianza, Italy
{vittorio.zaccaria}@st.com

Abstract. The evaluation of the best system-level architecture in terms of energy and performance is of mainly importance for a broad range of embedded SOC platforms. In this paper, we address the problem of the efficient exploration of the architectural design space for parameterized microprocessor-based systems. The architectural design space is multi-objective, so our aim is to find all the Pareto-optimal configurations representing the best power-performance design trade-offs by varying the architectural parameters of the target system. In particular, the paper presents a Design Space Exploration (DSE) framework tuned to efficiently derive Pareto-optimal curves. The main characteristics of the proposed framework consist of its flexibility and modularity, mainly in terms of target architecture, related system-level executable models, exploration algorithms and system-level metrics. The analysis of the proposed framework has been carried out for a parameterized superscalar architecture executing a selected set of benchmarks. The reported results have shown a reduction of the simulation time of up to three orders of magnitude with respect to the full search strategy, while maintaining a good level of accuracy (under 4% on average).

1 Introduction

Decreasing energy consumption without a relevant impact on performance is a 'must' during the design of a broad range of embedded applications. Evaluation of energy-delay metrics at the system-level is of fundamental importance for embedded applications characterized by low-power and high-performance requirements. The growing diffusion of SOC embedded applications based on the platform-based design approach requires a flexible tuning framework to assist the phase of *Design Space Exploration (DSE)*. In general, different applications could impose different energy and performance requirements. The overall goal of the DSE phase is to optimally configure the parameterized SOC platform in terms of both energy and performance requirements depending on the given application. In general, parameterized embedded System-On-Chip architectures must be optimally tuned to find the best energy-delay trade-offs for the given classes of applications. The value assignment to each one of the system-level parameters

J.J. Chico and E. Macii (Eds.): PATMOS 2003, LNCS 2799, pp. 249–258, 2003.
© Springer-Verlag Berlin Heidelberg 2003

can significantly impact the overall performance and power consumption of the given embedded architecture. To explore the large design space for the target architecture, an approach based on the full search of the optimal architectural parameters at the system-level with respect to the energy-delay cost function can be computationally very costly due to the long simulation time required to explore the wide space of parameters.

The problem addressed in this paper consists of defining a flexible Design Space Exploration (DSE) framework to efficiently explore the multi-objective design space in order to find a good approximation of Pareto-optimal curves representing the best compromise between the interesting design objectives, mainly energy and delay. The proposed DSE framework is flexible and modular in terms of: target architecture and related system-level executable models, exploration algorithms, and system-level metrics. The target SOC platform consists of a parameterized superscalar architecture. The set of tunable parameters is mainly related to the target microprocessor, the memory hierarchy and the system-level interconnection buses. Each parameterized component of the target architecture is provided with the corresponding system-level executable model to dynamically profile the given application to derive the information related to power and performance metrics.

The main goal of the selected set of exploration algorithms is its efficiency with respect to the full search exploration algorithm, while preserving a good level of accuracy. Up to now, the exploration algorithms plugged in the DSE framework are the Random Search Pareto (RSP) technique, the Pareto Simulated Annealing (PSA), and the Pareto Reactive Tabu Search (PRTS). The algorithms have been tuned to efficiently derive a good approximation of Pareto-optimal curves. The analysis of the proposed framework has been carried out for a parameterized superscalar architecture executing a set of multimedia applications. The reported results have shown a reduction of the simulation time of up to three orders of magnitude with respect to full search strategy, while maintaining a good level of accuracy (under 4% on average).

The rest of the paper is organized as follows. A review of the most significant works appeared in literature concerning the DSE problem is reported in Section 2. The DSE problem is stated in Section 3 along with the problem of the approximation of Parte curves. The proposed DSE framework is described in Section 4, while Section 5 discusses the experimental results carried out to evaluate the efficiency of the proposed framework for a superscalar configurable target architecture. Finally some concluding remarks have been reported in Section 6.

2 Background

Several system-level estimation and exploration methods have been recently proposed in literature targeting power-performance tradeoffs from the system-level standpoint [1], [2], [3], [4], [5], [6], [7].

The SimpleScalar toolset [5] is based on a set of MIPS-based architectural simulators focusing on different abstraction levels to evaluate the effects of some high-level algorithmic, architectural and compilation trade-offs. The SimpleScalar framework provides the basic simulation-based infrastructure to explore both processor architectures and memory subsystems. However, SimpleScalar does not support power analysis. Based on the SimpleScalar simulators, SimplePower [8] can be considered one of the first efforts to evaluate the different contributions to the energy budget at the system-level. The SimplePower energy estimation environment consists of a compilation framework and an energy simulator that captures the cycle-accurated energy consumed by the SimpleScalar architecture, the memory system and the buses. More recently, the Wattch architectural-level framework has been proposed in [6] to analyze power with respect to performance tradeoffs with a good level of accuracy with respect to lower-level estimation approaches. Wattch represents an extension of the SimpleScalar simulators to support power analysis at the architectural level. Wattch provides a framework to explore different system configurations and optimization strategies to save power, in particular focusing on processor and memory subsystems.

The *Avalanche* framework [2] evaluates simultaneously the energy-performance tradeoffs for software, memory and hardware for embedded systems. The Avalanche framework mainly focuses on the processor and memory subsystems. The work in [4] proposes a system-level technique to find low-power high-performance superscalar processors tailored to specific user applications. Low-power design optimization techniques for high-performance processors have been investigated in [7] from the architectural and compiler standpoints. A trade-off analysis of power performance effects of SOC (System-On-Chip) architectures has been recently presented in [9], where the authors propose a simulation-based approach to configure the parameters related to the caches and the buses.

Among DSE framework appeared in literature so far, very few approaches have been introduced recently to approximate Pareto-curve construction for computer architecture design [9] [10].

In general, the most trivial approach to determine the Pareto-optimal configurations into a large design space with respect to a multi-objective design optimization criteria consists of the comprehensive exploration of the configuration space. This *brute force* approach can be feasible only if the number of parameters in the configuration space is very limited. On the contrary, it is quite common to find a design space composed of tens of parameters, leading to an exponential analysis time. Thus, traditional heuristics must be used. When the design space is too large to be exhaustively explored, heuristic methods must be adopted to find acceptable near-optimal solutions. The problem of the efficient construction of Pareto curves has been often addressed by using domain-specific algorithms. For example, the high-level synthesis scheduling problem (minimization of latency with area constraints) implicitly needs to consider an approximation of the Pareto curve of the area/latency design evaluation space. For this problem, specific approaches have been proposed in the past [11]. However, due to the high generality of the design space of a platform design, domain-specific algorithms are very difficult to find, thus one should resort to traditional heuristics.

Platune [9] is an optimization framework that exploits the concept of parameter independence to individuate approximate Pareto curves without performing the exhaustive search over the whole design space. The authors define two parameters as interdependent if changing the value of one of them impacts the optimal parameter value of the other. In this case, all the combinations of these two parameters must be analyzed to find an optimal configuration. However, if the parameters are independent, the two subspaces can be analyzed separately, leading to a reduced simulation time. The main drawback of this approach is that parameter independence must be specified by the user by means of a dependency graph since no automatic methods are proposed for such task. Platune is not a modular framework since it allows only the exploration of a MIPS based system and its goodness has not been compared with simpler approaches such as random search or full parameter space exploration.

More recently, Palesi et al. [10] extended Platune by applying genetic algorithms to optimize dependent parameters, resorting to the default Platune policy when independent parameters are specified by the user. However, their approach is always based on an a-priori parameter dependency graph to be given by the user and it is compared only with the default Platune policy.

3 Design Space Exploration

To meet the time-to-market constraints, modern design techniques oriented to the reuse of intellectual proprieties are increasing their importance. For example, the use of a customizable System-On-Chip (SOC) platforms [12] where a stable microprocessor-based architecture can be easily extended and customized for a range of applications, enable a quick deployment and low cost high level design flow. For example, even considering a simple embedded microprocessor-based architecture composed of the CPU and the memory hierarchy, the identification of the optimal system configuration by trading off power and performance still leads to the analysis of too many alternatives. The overall goal of this work aims at overcoming such problems by providing a methodology and a design framework to drive the designer towards optimal solutions in a cost-effective manner.

3.1 Parameterized Design Space Definition

We define the *design space* as the set of all the feasible architectural implementations of a platform. A possibile configuration of the target architecture is mapped to a generic point in the design space as the vector $a \in \mathcal{A}$ where $\mathcal{A}$ is the architectural space defined as:

$$\mathcal{A} = S_{p_1} \times \ldots S_{p_l} \ldots \times S_{p_n}$$

where S_{p_l} is the ordered set of possible configurations for parameter p_l and "$\times$" is the cartesian product. Associated with each point a of the design space, there are a set of evaluation functions (or *metrics*). The *design evaluation space* is the multi-dimensional space spanned by these evaluation functions.

The problem afforded in this paper consists of platform optimization, that is searching for the best design, i.e., an implementation that optimizes all the objectives within the design evaluation space. However, the optimization problem involves the minimization (maximization) of *multiple objectives* making the definition of optimality not unique. To address this problem, let us introduce the definition of *Pareto point* for a minimization problem [13]. A Pareto point is a point of the design space, for which there is no other point with at least an inferior objective, all others being inferior or equal. Obviously, a Pareto point is a global optimum in the case of a monodimensional design space, while in the case of multi-dimensional design evaluation space, the Pareto points form a trade-off curve or surface called Pareto curve.

In general, Pareto points are solutions to *multi-objective* or *constrained* optimization problems. For example, we can be interested in minimizing the power consumption under a delay constraint or viceversa. The solution of this problem is straightforward if the Pareto curve is available. However, a Pareto curve for a specific platform is available only when all the points in the design space have been characterized in terms of objective functions. This is often unfeasible due to the cardinality of the design space and to the long simulation time needed for computing the evaluation functions.

Our target problem consists of finding a good approximation of the Pareto curves by trading off the accuracy of the approximations and the time needed for their construction. In a system-level description, the designer is required to specify the boundaries of the design space for each parameter (for example, the maximum size to be considered for the cache size) and other possible freezing constraints on some parameters.

4 Proposed Design Space Exploration Framework

The overall goal of this work aims at providing a methodology and a retargetable tool to drive the designer towards near-optimal solutions, with the given multiple constraints, in a cost-effective fashion. The final product of the framework is a Pareto curve of configurations within the design evaluation space of the given application. To meet our goal we implemented a skeleton for an extendible and easy to use framework for multi-objective exploration. The proposed DSE framework is flexible and modular in terms of:

- target architecture and related system-level executable models;
- exploration algorithms;
- system-level metrics.

Given a new target architecture, the user can plug in the new architecture module in the framework, while the othermodules can be considered as ready-to-use black boxes. Similarly, given a new exploration alghoritm or a new executable module, the user can plug-in the new module in the framework.

The proposed framework is shown in Figure 1. It is mainly composed in two modules: System Description Module (SDM) and Design Space Exploration Module (DSEM). The DSEM receives as input the description of the possible design configurations (i.e. the target *design space*) and the application for which the system has been designed and for which the optimal configuration must be found (SDM). The framework explores the design space by using an iterative optimization technique with respect to different metrics.

The *Optimizer* module is responsible for choosing, from the design space, a set of candidate optimal points to be evaluated. Once selected, each point is mapped into a specific instance of the target architecture by the *"Architecture Mapping"* module. Providing a 'mapping' between the high-level description interface used by the Optimizer and the actual specification of the target architecture, this module enables the evaluation of each point by means of simulation based on the executable model of the target system. In the presented framework, the model of the target architecture can be described at different abstraction levels, being available the corresponding simulator.

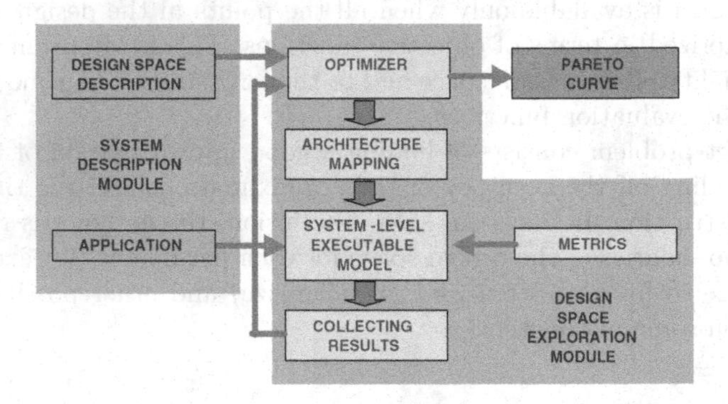

Fig. 1. Proposed design space exploration framework

Up to now, the approximation of the Pareto curve could be performed by a set of already plugged in exploration algorithms:

- Random Search Pareto(RSP) [14]. RSP algorithm is derived from *Monte Carlo* methods. In general, the main characteristic of *Monte Carlo* methods is the use of random sampling techniques to come up with a solution of the target problem. The random sampling technique has been proved to be one of the best techniques to avoid falling into local minima.
- Pareto Simulated Annealing (PSA) [15]. Simulated annealing is a Monte Carlo approach for minimizing such multivariate functions. The term simulated annealing derives from the analogy with the physical process of heating and then slowly cooling a substance to obtain a strong crystalline structure. In the Simulated Annealing algorithm a new configuration is constructed by imposing a random displacement. If the cost function of this new state is less

than the previous one, the change is accepted unconditionally and the system is updated. If the cost function is greater, the new configuration is accepted probabilistically; the acceptance possibility decreases with the temperature. This procedure allows the system to move consistently towards lower cost function states, thus 'jumping' out of local minima due to the probabilistic acceptance of some upward moves. The PSA is an evolution of SA for multi-objective optimization. At each step of PSA, the starting point is not a single configuration but a set of configurations (*Partial Pareto Set*).
- Pareto Reactive Tabu Search (PRTS) [16]. The Tabu Search (TS) is an iterative algorithm that explores the design space by means of 'moves'. The key concept behind the algorithm is the tabu list, i.e., a list containing prohibited moves that, usually, consist of the most recently visited points. The reason of the tabu list is to avoid to stuck into local minima. Recent studies [16] have demonstrated that the length of this list is a determining factor to reduce the possibility to stuck into local minima and a careful tuning of the list length is of fundamental importance for the success of the algorithm. The Reactive Tabu Search is an evolution of the Tabu Search algorithm that exploits an *adaptative prohibition period*, paired with an *escape mechanism*, to afford the tuning problem. In RTS, the prohibition period of a specific solution increases with the frequency of the visits to that solution. Moreover, to avoid the possibility of a cyclic exploration, an escape mechanism is used in order to escape from local minimum. The escape is usually implemented by generating a random walk.

Our effort has been devoted to the tuning phase of the parameters requested in the described set of algorithms.

5 Experimental Results

In this section, we present the experimental results obtained by applying the proposed DSE framework to optimize a superscalar microprocessor-based system. The first subsection of this paragraph describes the target architecture and the related design evaluation space; the second subsection discusses the application of the framework to two case studies.

5.1 Target System Architecture

In general, a superscalar architecture is composed of many parameters, so that the design space to explore is quite large. Our analysis has been focused on those design parameters significantly impacting the performance and the energy consumption. Each instance of the virtual architecture has been described in terms of the following parameters:

- $\mathbf{S}_{\mathbf{s}_i}, \mathbf{S}_{\mathbf{s}_d}, \mathbf{S}_{\mathbf{s}_{u2}}$ are the ordered sets of the possible sizes of the I/D L1 caches (from 2 KByte to 16 KByte) and unified L2 cache (from 16 KByte to 128 KByte).

- $\mathbf{S_{b_i}}, \mathbf{S_{b_d}}, \mathbf{S_{b_{u2}}}$ are the ordered sets of the possible block sizes of the I/D L1 caches (from 16 Byte to 32 Byte) and unified L2 cache (from 32 Byte to 64 Byte).
- $\mathbf{S_{a_i}}, \mathbf{S_{a_d}}, \mathbf{S_{a_{u2}}}$ are the ordered sets of the possible associativity values of the I/D L1 caches (from 1 way to 2 ways for the I-cache and from 2 ways to 4 ways for the D-cache) and unified L2 cache (from 4 ways to 8 ways).
- $\mathbf{S_{ia}}, \mathbf{S_{im}}$ are the ordered sets of the possible number of integer ALUs and multipliers (from 1 to 2).
- $\mathbf{S_{fpa}}, \mathbf{S_{fpm}}$ are the ordered sets of the possible number of floating point ALUs and multipliers (from 1 to 2).
- $\mathbf{S_{iw}}$ are the ordered sets of the possible issue width sizes (from 2 to 8).

Wattch simulator [6] has been used as our target architectural simulator providing a dynamic profiling of energy and delay.

5.2 Application of the Methodology

In this subsection, we report the results in terms of the efficiency and accuracy of the application of our DSE methodology to a set of benchmarks composed of a set of DCT transforms and FIR filters as well as other numerical algorithms written in C language.

To validate our exploration methodology we carried out two parallel exploration flows: the first is based on the exhaustive search, while the second is based on the given algorithm. Each benchmark has been optimized independently with the two flows and the resulting Pareto curves have been compared. In the full search case, the optimizer analyzes the global design space, so the number of simulations to be executed is 196608. This corresponds to approximately 370 hours of simulations for the selected set of benchmarks. Table 1 shows, in the first row, the number of visited configurations (‰) with respect to the total number of configurations. In the second row, Table 1 shows the average percentage error of the approximate Pareto curve obtained with the PSA algorithm with respect to the exhaustive search. To evaluate the accuracy of the approximate curve we used the error defined in *[omitted for blind review]* Table 1 shows a fast convergence for the PSA algorithm due to an accurate tuning of its parameters: we found an average error under the 4% reducing up to three orders of magnitude the simulation time.

Figure 2 presents the results obtained by applying the proposed framework to two benchmarks: the FDCT and FIR1. In this case, the optimization module used to find the Pareto curves is the PSA algorithm. Figure 2 shows the scatter plot of all the configurations generated in the energy-delay space by the PSA algorithm (light gray) and the corresponding Pareto configurations (dark gray). Once the Pareto curve has been found, a constrained design space exploration problem could select its solution among the Pareto points. For our two cases studies based on the PSA algorithm, we found the approximated Pareto curve in approximately one hour of simulation with respect to a week requested by the exhaustive search.

Table 1. Efficiency and accuracy of the PSA-based exploration vs. full search

Visited Configuration [‰]	0.5	1	5	10	50
Average Error [%]	6	4	2	1	0.5

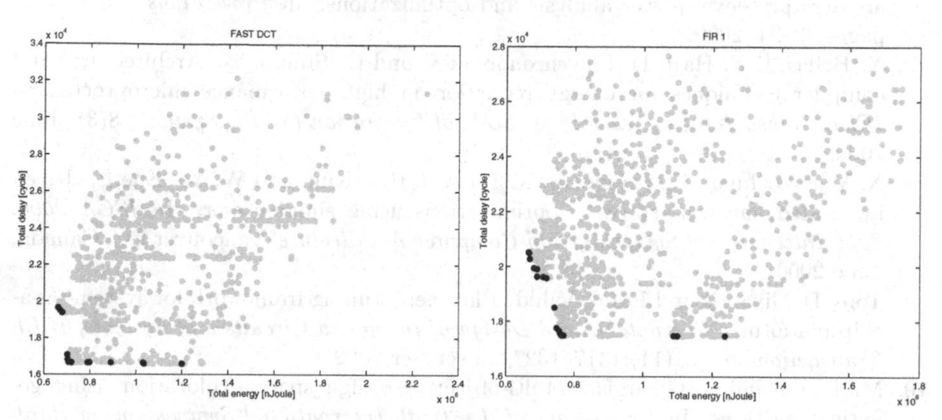

Fig. 2. Energy-delay PSA-based exploration results for FAST DCT and FIR1

6 Conclusions

In this paper, a flexible design space exploration framework has been proposed to efficiently derive Pareto-optimal curves. The framework is completely configurable in terms of target architecture, metrics and exploration algorithm. The paper discusses also an application of the proposed exploration technique based on the Pareto Simulated Annealing algorithm, comparing the results with the full search exploration. For the selected set of benchmarks, the PSA techniques is up to three orders of magnitude faster than the full search, while maintaining its accuracy within 4% on average.

References

1. C. L. Su and A. M. Despain. Cache design trade-offs for power and performance optimization: A case study. In *ISLPED-95: ACM/IEEE Int. Symposium on Low Power Electronics and Design*, 1995.
2. Y. Li and J. Henkel. A framework for estimating and minimizing energy dissipation of embedded hw/sw systems. In *DAC-35: ACM/IEEE Design Automation Conference*, June 1998.
3. J. K. Kin, M. Gupta, and W. H. Mangione-Smith. Filtering Memory References to Increase Energy Efficiency. *IEEE Trans. on Computers*, 49(1), Jan. 2000.
4. T. M. Conte, K. N. Menezes, S. W. Sathaye, and M. C. Toburen. System-level power consumption modeling and tradeoff analysis techniques for superscalar processor design. *IEEE Trans. on Very Large Scale Integration (VLSI) Systems*, 8(2):129–137, Apr. 2000.

5. Doug Burger, Todd M. Austin, and Steve Bennett. Evaluating future microprocessors: The simplescalar tool set. Technical Report CS-TR-1996-1308, University of Wisconsin, 1996.
6. David Brooks, Vivek Tiwari, and Margaret Martonosi. Wattch: a framework for architectural-level power analysis and optimizations. In *Proceedings ISCA 2000*, pages 83–94, 2000.
7. N. Bellas, I. N. Hajj, D. Polychronopoulos, and G. Stamoulis. Architectural and compiler techniques for energy reduction in high-performance microprocessors. *IEEE Transactions on Very Large Scale of Integration (VLSI) Systems*, 8(3), June 2000.
8. N. Vijaykrishnan, M. Kandemir, M.J. Irwin, H.S. Kim, and W. Ye. Energy-driven integrated hardware-software optimizations using simplepower. In *ISCA 2000: 2000 International Symposium on Computer Architecture*, Vancouver BC, Canada, June 2000.
9. Tony D. Givargis and Frank Vahid. Platune: a tuning framework for system-on-a-chip platforms. *Computer-Aided Design of Integrated Circuits and Systems, IEEE Transactions on*, 21(11):1317–1327, November 2002.
10. M. Palesi and T Givargis. Multi-objective design space exploration using genetic algorithms. In *Proceedings of the Tenth International Symposium on Hardware/Software Codesign, 2002. CODES 2002*, May 6–8 2002.
11. D. Gajski, N. Dutt, A. Wu, and S. Lin. *High-Level Synthesis, Introduction to Chip and System Design.* Kluwer Academic Publishers, 1994.
12. K. Keutzer, S. Malik, A. R. Newton, J. Rabaey, and A. Sangiovanni-Vincentelli. System level design: Orthogonolization of concerns and platform-based design. *IEEE Transactions on Computer-Aided Design of Integrated Circuits and Systems*, 19(12):1523–1543, December 2000.
13. A. Aho, J. Hopcroft, and J. Ullman. *Data Structures and Algorithms.* Addison-Wesley, Reading, MA, USA, 1983.
14. Anatoly A. Zhigljavsky. *Theory of global random search*, volume 65. Kluwer Academic Publishers Group, Dordrecht, 1991.
15. Jaszkiewicz A. Czyak P. Pareto simulated annealing - a metaheuristic technique for multiple-objective combinatorial optimisation. *Journal of Multi-Criteria Decision Analysis*, (7):34–47, April 1998.
16. R. Battiti and G. Tecchiolli. The reactive tabu search. *ORSA Journal on Computing*, 6(2):126–140, 1994.

High-Level Area and Current Estimation*

Fei Li[1], Lei He[1], Joe Basile[2], Rakesh J. Patel[2], and Hema Ramamurthy[2]

[1] EE Department, University of California, Los Angeles, CA
[2] Intel Corporation, San Jose, CA

Abstract. Reducing the ever-growing leakage current is critical to high performance and power efficient designs. We present an in-depth study of high-level leakage modeling and reduction in the context of a full custom design environment. We propose a methodology to estimate the circuit area, minimum and maximum leakage current, and maximum power-up current, introduced by leakage reduction using sleep transistor insertion, for any given logic function. We build novel estimation metrics based on logic synthesis and gate level analysis using only a small number of typical circuits, but no further logic synthesis and gate level analysis are needed during our estimation. Compared to time-consuming logic synthesis and gate level analysis, the average errors for circuits from a leading industrial design project are 23.59% for area, 21.44% for maximum power-up current. In contrast, estimation based on quick synthesis leads to 11x area difference in gate count for an 8bit adder.

1 Introduction

As VLSI technology advances, leakage power becomes an ever growing power component. Dynamic power management via power gating at system and circuit levels is effective to reduce both leakage and dynamic power. Figure 1 (a) shows a system with a multi-channel voltage regulation module (VRM). The VRM channels can be configured to supply power *independently* for individual modules. Therefore, modules can be turned on or off at appropriate times for power reduction. Power gating at the circuit level is also called MTCMOS (see Figure 1 (b)). A PMOS sleep transistor with a high threshold voltage connects the power supply to the virtual Vdd. The sleep transistor is turned on when the function block is needed, and is turned off otherwise.[1] We use MTCMOS to study power gating in this paper and the idea can be extended to VRM.

Key questions in applying power gating include: (i) How to estimate the leakage reduction by power gating and how to decide the area overhead of power

* This research was partially supported by the NSF CAREER Award 0093273, SRC grant 2002-HJ-1008 and a grant from Intel Design Science and Technology Committee. Address comments to lhe@ee.ucla.edu.

[1] Instead of the PMOS sleep transistor, an NMOS sleep transistor can be inserted between the ground and virtual ground and I_p to be presented later becomes the discharging current in this case. For simplicity of presentation, we assume PMOS sleep transistors in this paper.

J.J. Chico and E. Macii (Eds.): PATMOS 2003, LNCS 2799, pp. 259–268, 2003.
© Springer-Verlag Berlin Heidelberg 2003

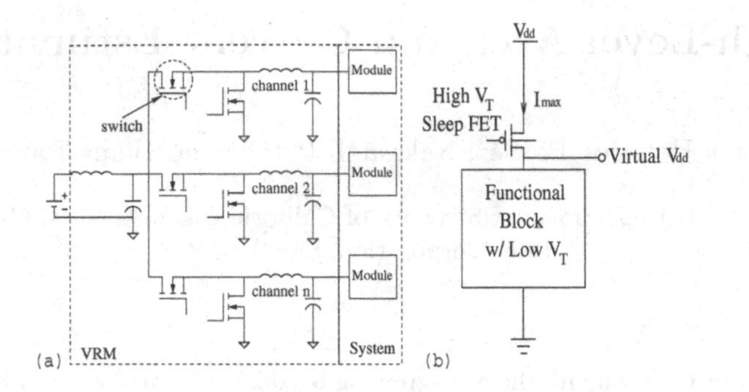

Fig. 1. Power gating at (a) system level and (b) circuit level.

gating? The answer determines whether power gating is worthwhile for a given design, and (ii) How to answer the above question at an early design stage without performing time-consuming logic synthesis and gate level analysis. Early decision making is needed to deal with time-to-market pressure. This paper presents an in-depth study of high-level leakage modeling and reduction by using commercial synthesis tools such as Design Compiler.

In Section 2, we propose a method to estimate the gate count for a given logic function without performing logic synthesis. We show that the quick synthesis leads to 11x difference for a simple adder, and further validate and improve an area estimation technique that was originally developed for a library with limited number of cells [1]. The improved estimation method has an average error of 23.59%. As shown in [2], all nodes in a power-gated module are at logic "0" state, and must be brought to valid logic states by power-up current (I_p) before useful computation can begin. Further, I_p depends on the input vector. Its maximum value must be known to design reliable sleep transistors and VRM. In Section 3, we propose a high-level metric to estimate the maximum I_p without performing logic synthesis and gate-level I_p analysis. We verify this metric by a newly developed gate-level analysis for accurate I_p. In all sections, we use the design environment of a leading industrial high-performance CPU design project. There are hundreds of cells with various sizes (1x to 65x) in the library. All experiments are carried out on a number of typical circuits. The circuits are specified in Verilog and synthesized by Design Compiler to verify our high-level estimations. Due to the need of IP protection, we report normalized current value in this paper.

2 Area Estimation

2.1 Overview

Table 1 presents synthesis results for adders where *synthesis 1* uses logic functions with intermediate variables, and *synthesis 2* uses equivalent logic functions without intermediate variables. 11x difference in gate count is observed for an

Table 1. Area count based on quick synthesis

Circuit	Synthesis 1	Synthesis 2
1bit adder	3	3
4bit adder	20	16
8bit adder	42	490

8bit adder. It shows that quick synthesis using Verilog specified at a higher abstraction level does not necessarily lead to a good estimation. Instead of using quick synthesis, we apply and improve the high-level area estimation in [1].

We summarize the estimation flow from [1] in Figure 2. It contains a one-time pre-characterization, where gate-count $\mathcal{A}$ is pre-characterized as a function $\mathcal{F}$ of the linear measure $\mathcal{L}$ and output entropy $\mathcal{H}$. Then, a *multi-output function* (MOF) is transformed into a *single output function* (SOF) by adding a *m-to-1* MUX, where m is the number of outputs in the original MOF. $\mathcal{L}$ and $\mathcal{H}$ are calculated for the SOF to look up the pre-characterized table and obtain gate count. Removing MUX from this gate count leads to $\mathcal{A}$ for the original MOF.

We improve the original estimation method in two ways. First, it is claimed in [1] that SOFs with the same output entropy $\mathcal{H}$ and same linear measure $\mathcal{L}$ have the same $\mathcal{A}$. However, we find that it may not be true for VLSI functions implemented with a rich cell library. Functions with smaller output probability of logic '1' have fewer gate count under the same linear measure. Therefore, we have pre-characterized $\mathcal{A}$ as a function $\mathcal{F}(\mathcal{L}, \mathcal{P})$, where $\mathcal{P}$ is the output probability. Since complementary probabilities lead to the same entropy, our pre-characterization is more detailed compared to that in [1]. Further, we have developed an output clustering algorithm to partition the original MOF into sub-functions (called sub-MOFs) with minimum support set overlap, and have improved the efficiency and accuracy of the high-level estimation. We summarize our estimation flow with the difference highlighted in Figure 2, and describe each step and our implementation details in the following sections.

2.2 Linear Measure

Linear measure $\mathcal{L}$ is determined by on and off-sets of an SOF as $\mathcal{L} = \mathcal{L}_1 + \mathcal{L}_0$, where $\mathcal{L}_1$ and $\mathcal{L}_0$ are the linear measure for the on-set and off-set, respectively. $\mathcal{L}_1$ is further defined as $\mathcal{L}_1(f) = \sum_{i=1}^{N} c_i p_i$ ($\mathcal{L}_0$ can be defined similarly). N is the number of different sizes of all the prime implicants in a minimal cover of function f. The size of a prime implicant is the number of literals in it. c_i is one distinct prime implicant size. p_i is a weight of prime implicants with size c_i and can be computed in the following way. Suppose all the input vectors to the logic function can occur with the same probability. Let $c_1, c_2, \ldots, c_N$ be sorted in a decreasing order, and weight p_i be the probability that one random input vector matches all the prime implicants with size c_i but not by the prime implicants with size from c_1 to c_{i-1}, $1 < i \leq N$. For $i = 1$, p_1 is just the probability that one random input vector matches prime implicants with size c_1. Here "a matching" means

Estimation of gate count $\mathcal{A}$ in [1]:
Pre-characterization
1. Pre-characterize $\mathcal{A}$ as $\mathcal{F}(\mathcal{L}, \mathcal{H})$ using randomly generated SOFs. $\mathcal{H}$ is the output entropy.
Mapping
1. Partition the MOF into sub-MOFs randomly;
2. Transform each sub-MOF into an SOF by adding MUX;
3. Compute $\mathcal{L}$ and $\mathcal{H}$;
4. Obtain gate count $\mathcal{A}'$ of transformed SOF from $\mathcal{F}(\mathcal{L}, \mathcal{H})$;
5. Remove MUX from $\mathcal{A}'$ to get gate count of original MOF;
6. Obtain final estimate by adding up gate-count for all the sub-MOFs.

Estimation of gate count $\mathcal{A}$ in this paper:
Pre-characterization
1. Pre-characterize $\mathcal{A}$ as $\mathcal{F}(\mathcal{L}, \mathcal{P})$ using randomly generated SOFs. $\mathcal{P}$ is the **output probability**.
Mapping
1. Partition the MOF to **minimize support-set overlap**;
2. Transform each sub-MOF into an SOF by adding MUX;
3. Compute **output probability** $\mathcal{P}$ and linear measure $\mathcal{L}$;
4. Obtain gate count $\mathcal{A}'$ of transformed SOF from $\boldsymbol{\mathcal{F}(\mathcal{L}, \mathcal{P})}$;
5. Remove MUX from $\mathcal{A}'$ to get gate count of original MOF.
6. Obtain final estimate by adding up gate-count for all the sub-MOFs.

Fig. 2. Estimation of gate count $\mathcal{A}$

that the intersection operation between the vector and the prime implicant is consistent. Note that p_i satisfies the equation $\sum_{i=1}^{N} p_i = \mathcal{P}(f)$, where $\mathcal{P}(f)$ is the probability to satisfy function f.

The minimum cover of an SOF can be obtained by two-level logic minimization [3]. To compute the weight p_i, a straightforward approach is to make the minimum cover disjoint and compute the probability exactly. However, in practice, this exact approach turns out to be very expensive. In our experiments, when the number of inputs is larger than 10, the program using the exact approach does not finish within reasonable time. But with p_i defined as the probability, $\mathcal{L}_1(f)$ can be viewed as a random variable $\mathcal{L}_1'(f)$ with certain probability distribution. For each random input vector, the variable $\mathcal{L}_1'(f)$ takes a certain value 'randomly'. With probability of $1 - P(f)$, $\mathcal{L}_1(f)$ takes the value '0'. Then, $\mathcal{L}_1(f)$ becomes the *mean* of the random variable $\mathcal{L}_1'(f)$. By assuming that the variable $\mathcal{L}_1'(f)$ takes a Gaussian distribution, we use Monte Carlo simulation technique to estimate the mean value efficiently.

2.3 Output Probability and Gate-Count Recovery

The output probability can be obtained as a by-product of Monte Carlo simulation. Since weight p_i satisfies $\sum_{i=1}^{N} p_i = P(f)$, we can keep record of all the p_i during the Monte Carlo simulation. When simulation process satisfies the stopping criteria, the output probability can be obtained easily. To recover the gate count of the original MOF, the estimated gate count for the transformed SOF is subtracted by $\alpha\mathcal{A}_{mux}$. $\mathcal{A}_{mux}$ is the gate count of the complete multiplexer we have inserted, and α is the coefficient to get the *reduced* multiplexor gate count due to the logic optimization.

2.4 Output Clustering

As the number of primary outputs increases, the time to calculate the minimum cover of a function increases non-linearly. To make the two-level optimization more efficient, one may partition the original MOF into sub-MOFs by output clustering, and then estimate for each sub-MOF individually. The gate-count of the original MOF is the sum of gate-counts for all the sub-MOFs. However, estimation errors may be introduced due to the overlap of the support sets of the sub-MOFs. We propose to partition the outputs with minimum support set overlap (see Figure 3). A PO-graph is constructed with vertices representing the Primary Outputs (POs). If two POs have support set overlap, there is an edge connecting the two corresponding vertices. The edge weight is the size of the common support set. The vertex weight is the sum of the weights of all edges connected to this vertex. There are two loops in the algorithm. In each iteration of the inner loop, the vertex with the minimum weight is deleted and the weights are updated for edges and vertices that connect the deleted vertex. It continues until the number of remaining vertices is less or equal to the pre-specified cluster size. The PO-graph is then re-constructed with all the POs that have not been clustered. The algorithm continues until all the outputs are clustered and the PO-graph becomes empty.

```
Output Clustering Algorithm:
Construct the PO-graph;
While (PO-graph is not empty)
begin
        While (# of remaining vertices > CLUSTER_SIZE)
        begin
                Delete the vertex v with the minimum weight;
                Update weights for edges and vertices connecting v;
        end
        Obtain one cluster of POs using remaining vertices;
        Re-construct PO-graph excluding POs already in clusters;
end
```

Fig. 3. Output clustering algorithm

2.5 Experimental Results

We compare area estimation methods in Figure 4, where x-axis is the circuit ID number and y-axis is the gate count. During the Monte Carlo simulation to calculate the linear measure, we choose the parameters of confidence and error as 96% and 3%, respectively. The actual gate count is obtained by the synthesis using Design Compiler. The method with random output clustering has an average absolute error of 39.36%. By applying our output clustering algorithm to minimize support set overlap, we reduce the average absolute error to 23.59%. Such estimation errors are much smaller compared to the 11x gate-count difference in Table 1. Note that different descriptions of a given logic function do not change the $\mathcal{L}$ and $\mathcal{P}$, and therefore do not affect the estimation results by our approach. High-level estimation costs over 100x less runtime compared to logic synthesis.

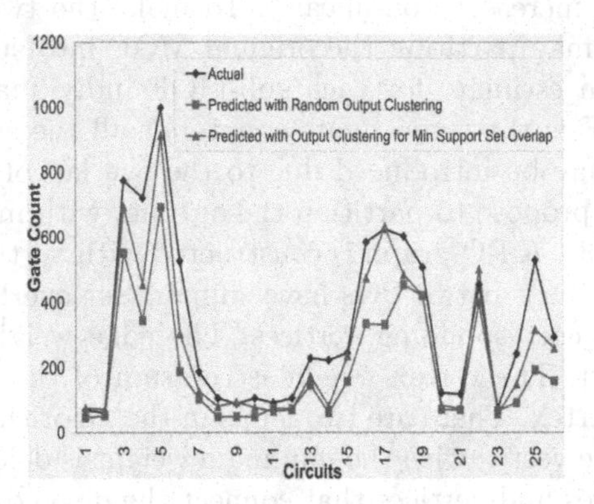

Fig. 4. Comparison between actual and predicted gate-count.

3 Power-Up Current Estimation

Given the Boolean function f of a combinational logic block and the target cell library, our high-level estimation finds the maximum power-up current $I_p(f)$ when the logic block is implemented with the given cell library for power gating. A Boolean function can be implemented under different constraints, but we assume the min-area implementation in this paper. We propose the following high-level metric $\mathcal{M}_p$ for $I_p(f)$:

$$I_p(f) \propto \mathcal{M}_p(f) = \mathcal{I}_{avg} \cdot \mathcal{A} \tag{1}$$

where $\mathcal{A}$ is the gate count estimated using the method in Section 2, and $\mathcal{I}_{avg}$ is the weighted average I_p to be discussed in Section 3.2. Because an accurate gate-level estimator is required for the calculation of $\mathcal{I}_{avg}$ and verification of $\mathcal{M}(f)$, we introduce our gate-level estimation in the next section.

3.1 Gate-Level Estimation

Background Knowledge. Power-up current (I_p) is different from the normal switching current (I_s). I_s depends on two successive circuit states S_1 and S_2, which are determined by two successive input vectors V_1 and V_2 for combinational circuits. As discussed in Section 1, I_p can be viewed as a special case of I_s where the state S_1 before power-up is logic "0" for all the nodes. Because no input vector leads to a circuit state with all nodes at logic "0" for non-trivial circuits, the maximum I_p is in general different from the maximum I_s. Moreover, the I_p of a circuit is *solely* decided by the circuit state S_2, and therefore decided by a single input vector when the circuit is powered up. To illustrate that I_p depends on the input vectors, we present the I_p obtained by SPICE simulation for an 8-bit adder under two different input vectors in Table 2. The difference of the maximum I_p is about 24%. It is obvious that I_p is greatly affected by the input vector when the circuit is powered up. We refer I_p element to be the power-up current generated by an individual gate, and give the following observation related to timing:

Observation 1 *If a set of gates are controlled by one single sleep transistor, all these gates are powered up simultaneously. I.e., all the I_p elements for these gates have the same starting time.*

Table 2. Maximum I_p of an 8bit adder.

Circuit	vector1	vector2	difference
Adder8	1830	2260	23.50%

ATPG-based algorithms have been proposed in [2]. It is assumed that the power-up current is proportional to the total charge in the circuit after power-up, and the charge for one single gate with output value "1" is proportional to its fanout number. Therefore, the gate fanout number is used as the figure of merit of the power-up current (I_p) for the gate with output value "1". ATPG algorithm is performed to find the logic vector that maximizes the figure of merit. However, this algorithm does not take the current waveform in the time domain into account. The vector obtained by ATPG algorithm has to be further used in SPICE simulation to obtain the I_p value.

To achieve more accurate estimation and obtain I_p value directly, we need a current model that can capture the current waveform. We apply the piece-wise linear (PWL) function to model the I_p element. SPICE simulation is used to get the power-up current waveform and the waveform is linearized at different regions to build the PWL model for each cell in the library. Our PWL model considers the following four dimensions: gate type, input pin number, gate size, and post-powerup output logic value. Note that a much simplified PWL model, the right-triangle current model has been successfully used in [4] for maximum switching current estimation.

Genetic Algorithm. Since exhaustive search for the input vector that generates the maximum I_p is infeasible, we apply Genetic Algorithm (GA) in our gate-level estimation. We encode the solution, input vector, into a string so that the length of the string is equal to the number of primary inputs. Each bit in the string is either '1' or '0'. The initial population is randomly generated. The population size is proportional to the number of primary inputs. The fitness value is chosen as the maximum I_p value under the input vector represented by the string. The I_p value is obtained by waveform simulation with PWL current model.

Tournament selection is used in our selection process. From the current generation, we randomly pick two strings and select the one with the higher fitness value. After that, the two strings are removed from the current generation. We repeat this procedure until the current generation becomes empty. By doing this, we divide the original strings into inferior and superior groups. We keep record of the strings in the superior group and put these two groups together to carry out tournament selection again. The two superior groups generated in the two tournaments are combined to go through crossover and mutation, and produce the new generation. The string with the highest fitness will be selected twice so that the best solution so far will stay in the next generation. Since strings with lower fitness have higher probability of being dropped, the average fitness tends to increase by each generation.

The crossover scheme we use is the one-point crossover algorithm. One bit position is randomly chosen for two parent-strings and they are crossed at that position to get the two child-strings. After crossover, we further use a simple mutation scheme that flip each bit in the string with equal probability. The new generation is produced after crossover and mutation, and is ready to go through a new iteration of natural selection. The algorithm stops after the number of generations exceeds a pre-defined number. We summarize the algorithm in Figure 5.

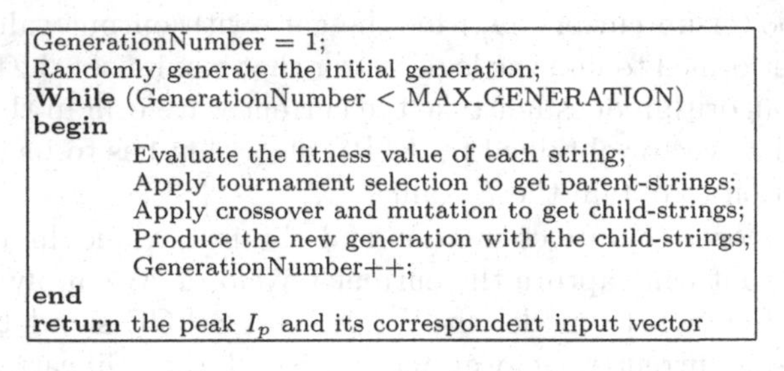

Fig. 5. Generic algorithm for gate-level estimation.

We carry out experiments and compare the results of Genetic Algorithm to that of simulations with 5000 random vectors. Under the same PWL current model, GA achieves up to 27% estimation improvement to approach the upper bound of power-up current, The average improvement for all the circuits is 6%.

3.2 Calculation of $\mathcal{I}_{avg}$ and Experimental Results

$\mathcal{I}_{avg}$ is not simply the average I_p element for all cells in a library. The frequency of cells used in logic synthesis should be taken into account. We assume that the logic synthesis results for a few typical circuits (or random logic functions) are available. We calculate $\mathcal{I}_{avg}$ in a regression-based way as follows: We compute the average maximum I_p per gate for n typical circuits by applying the gate-level estimation. We then increase n until the resulting value becomes a "constant". We treat this constant value as $\mathcal{I}_{avg}$. In Figure 6 (a), we plot $\mathcal{I}_{avg}$ with respect to the number of circuits used to calculate $\mathcal{I}_{avg}$. The figure shows that the change of the $\mathcal{I}_{avg}$ value is relatively large when the number of circuits is small (less than 10 in the figure). After the number of circuits increases to 20, the value of $\mathcal{I}_{avg}$ becomes very stable and can be used as our high-level metric $\mathcal{M}_p$.

To validate our regression-based $\mathcal{I}_{avg}$, we use the computed value of $\mathcal{I}_{avg}$ under PWL model and the accurate gate count to get the high-level metric $\mathcal{M}_p$. We compare the gate-level estimation $I_p(ckt)$ by Genetic Algorithm to the metric $\mathcal{M}_p$ in Figure 6 (b). The average absolute error between $I_p(ckt)$ and $\mathcal{M}_p$ is 12.02%. Note that the circuits in Figure 6 (b) are different from those used to compute $\mathcal{I}_{avg}$ for the purpose of the verification of metric $\mathcal{M}_p$.

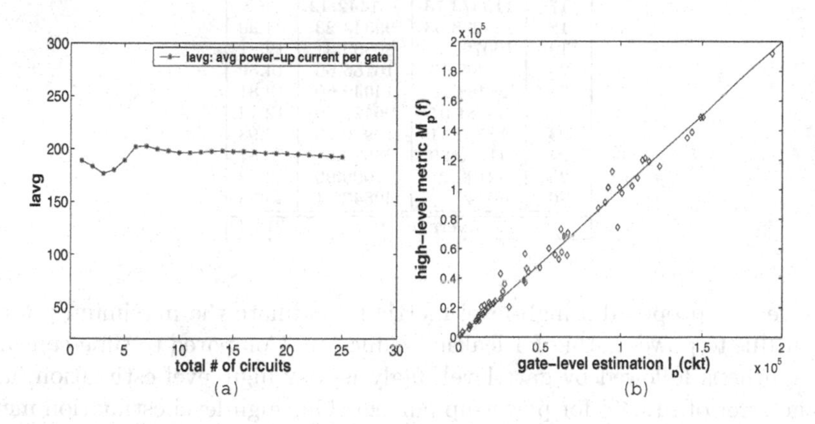

Fig. 6. (a) $\mathcal{I}_{avg}(PWL)$ w.r.t. number of circuits. (b) Comparison between gate-level estimation $I_p(ckt)$ and high-level metric $\mathcal{M}_p$.

Furthermore, we compare the maximum I_p using estimated $\mathcal{I}_{avg}$ and $\mathcal{A}$ to the maximum I_p obtained via logic synthesis followed by gate-level analysis. As shown in Table 3, shows that the average estimation error is 21.44%. We measure gate-level analysis runtime as the time for logic synthesis and Genetic Algorithm, and measure the high-level estimation runtime as the time for area estimation and application of the formula $\mathcal{M}_p(f) = \mathcal{I}_{avg} \cdot \mathcal{A}$ (pre-characterization only has one-time cost and is ignored in the runtime comparison). Our high-level estimation achieves more than 200x run-time speedup for large test circuits.

4 Conclusions and Discussions

Using design examples and design environment of a leading industrial CPU project, we have presented an improved high-level area estimation method. The estimation has an average error of 23.59% for designs using a rich cell library.

Table 3. Results of high-level I_p estimation.

PWL current model			
circuit id	gate-level est. $I_p(ckt)$	high-level est. $I_p(ckt)$	Abs. Err(%)
1	12234.00	14626.98	19.56
2	11756.12	13857.14	17.87
3	150437.28	133374.96	11.34
4	143767.61	87569.42	39.09
5	193818.20	175523.76	9.44
6	92024.79	58700.38	36.21
7	28170.35	24442.45	13.23
8	13975.08	15589.28	11.55
9	16349.61	18283.73	11.83
10	30964.78	15011.90	51.52
11	16687.65	12702.38	23.88
12	18577.41	13664.68	26.44
13	42565.20	33488.08	21.33
14	25710.94	15781.74	38.62
15	49502.92	45420.62	8.25
16	94697.20	91418.62	3.46
17	113573.73	121442.42	6.93
18	124298.63	93343.23	24.90
19	100798.74	80833.31	19.81
20	18907.54	16166.66	14.50
21	21486.25	14049.60	19.81
22	85981.01	96422.59	12.14
23	12599.41	10970.23	12.93
24	41482.89	28676.58	30.87
25	92199.21	61009.90	33.83
26	40449.54	49847.21	23.23
Average			21.44

We have also proposed a high-level metric to estimate the maximum power-up current due to power gating for leakage reduction. Compared to time-consuming logic synthesis followed by gate-level analysis, our high-level estimation has an average error of 21.44% for power-up current. Our high-level estimation method can be readily applied to estimating the area overhead due to the sleep transistor insertion in power gating. There is reliability constraint for sleep transistors, i.e., avoidance of damaging the sleep transistor by a large transient current. We can obtain the maximum transient current as the bigger one between the maximum power-up current and maximum switching current, and size the sleep transistor to satisfy the reliability constraint.

References

1. M. Nemani and N. Najm, "High-level area and power estimation for VLSI circuits," *IEEE Trans. on Computer-Aided Design of Integrated Circuits and Systems*, vol. 18, pp. 697–713, June 1999.
2. F. Li and L. He, "Maximum current estimation considering power gating," in *Proc. Intl. Symp. on Physical Design*, pp. 106–111, 2001.
3. G. D. Micheli, *Synthesis and Optimization of Digital Circuits.* New York: McGraw Hill, 1994.
4. A. Krstic and K.-T. Cheng, "Vector generation for maximum instantaneous current through supply lines for CMOS circuits," in *Proc. Design Automation Conf.*, pp. 383–388, June 1997.

Switching Activity Estimation in Non-linear Architectures

Alberto García-Ortiz, Lukusa Kabulepa, and Manfred Glesner

Institute of Microelectronic Systems, Darmstadt University of Technology
Karlstrasse 15, D-64283 Darmstadt (Germany)
agarcia@mes.tu-darmstadt.de

Abstract. Energy estimation in the early stages of the design flow has a paramount importance for the successful implementation of modern low power systems. With this goal, the present work proposes a technique for switching activity estimation in non-linear architectures, where the classical techniques based on Gaussian models cannot be applied. Using the characterization of the probability density function with a projection in an orthogonal polynomial base, and a symbolic propagation mechanism, a technique is presented to calculate the switching activity from the moments of the signal. The approach has been validated with practical circuits from the wireless communication arena. Comparisons with reference bit level simulations and previous works are reported to assess the accuracy of the technique.

1 Introduction

The increasing demand of battery-powered devices, and the necessity of integrating high performance (and thus power hungry) digital systems in a single chip , make indispensable the integration of low power estimation and optimization techniques in the ICs design flow. Since the impact of these techniques is especially significant at higher levels of abstraction, tools working at the architectural and system level are highly desired.

In order to estimate the power consumption in DSP architectures, several techniques have been proposed [6,1,8,7] to obtain the switching activity using only word-level parameters of the signal, which can be efficiently estimated at high levels of abstraction. In these techniques, the word-level statistics (typically the standard deviation and temporal correlation) are propagated through the architecture, and used to perform bit level estimations of the switching activity.

The main drawback of those approaches lies in the underlying premise assuming that the signals own a Gaussian distribution. This assumption reduces the usability of the estimation procedures in practical design scenarios, as it is the case of non-linear architectures. An estimation procedure without that restriction has been explicitly addressed in [4], where a very accurate formulation is presented for uncorrelated signals. Nevertheless, since this technique requires the knowledge of the complete probability density function (PDF), its use is restricted for practical applications.

In this work we present a technique that alleviates the aforementioned limitation by requiring only the moments of the signal. Since these moments can be more easily

J.J. Chico and E. Macii (Eds.): PATMOS 2003, LNCS 2799, pp. 269–278, 2003.
© Springer-Verlag Berlin Heidelberg 2003

propagated through the design, the approach has a higher practical significance. An alternative methodology to our technique is described in [2], but since it requires expensive numerical integration procedures, we do not consider it further.

The paper is organized as follows: next section introduces the general estimation methodology and the different orthogonal bases considered in the work. Afterwards, a technique for propagating the moments is introduced in sec. 3. With this background, the estimation of the switching activity in an OFDM synchronization unit is addressed in sec. 4. The paper finishes with some concluding remarks.

2 Estimation Approach

The goal of the paper is to model the bit-level switching activity (t_i) in non-linear architectures, using just the moments of the signals, and additionally to provide a mechanism to propagate those moments through the design. As in [4,3], we consider only signals which are temporally uncorrelated; hence, t_i can be obtained from the bit probability (p_i) with the expression $t_i = 2p_i(1-p_i)$. Certainly, the assumption of uncorrelation restricts the scope of the present work, but it represents a necessary first step towards the more general case of correlated distributions. Additionally, it has an important significance in practical applications, as we show in sec. 4.

The efficient characterization of the signal PDF represents a major obstacle to estimate p_i. Different parametric and non-parametric approaches have been proposed for that characterization (see [9] for a detailed exposition). In this paper, we employed a linear combination orthogonal polynomial functions as we reported in [3], where the reader is referred for further details.

Basically, the PDF of the signal, $f(x)$, is approximated as:

$$f(x) \approx \hat{f}(x; c_0, c_1, \cdots, c_n) = \sum_{i=0}^{N-1} c_i q_i(x) \phi(x) \tag{1}$$

where c_i are the fitting coefficients, $\phi(x)$ a positive function that defines a scalar product, and $q_i(x)$ a set of polynomials that are orthogonal for the given norm. The approximation order N represents the number of linear terms in the approximation. Given a weighted function $w(x) = \phi(x)^{-1}$, the estimation error $\int_{-\infty}^{+\infty} [f(x) - \hat{f}(x)]^2 w(x)\, dx$ can be minimized by choosing the fitting parameters as:

$$c_k = \frac{\int_{-\infty}^{+\infty} f(x) q_k(x)\, dx}{\int_{-\infty}^{+\infty} q_k^2(x) \phi(x)\, dx} = \frac{\sum_j a_{k,j} \gamma_j}{\int_{-\infty}^{+\infty} q_k^2(x) \phi(x)\, dx} \tag{2}$$

where $a_{k,j}$ denotes the jth coefficient of the polynomial $q_k(x)$, and γ_j the jth moment of $f(x)$, formally defined by:

$$\gamma_j = \mathbf{E}[x^j] = \int_{-\infty}^{+\infty} x^j f(x)\, dx \tag{3}$$

In this work we have considered three families of orthogonal polynomials, Legendre, Laguerre, and Hermite.

Legendre polynomials. This family uses a constant weight and norm functions inside the interval $[-1,1]$. Thus, the approximation error gets uniform distributed inside that interval. In fact, the expansion in Legendre polynomials corresponds to the minimum square error polynomial fitting of the function $f(x)$. If the original signal is not in the range $[-1,1]$, a linear transformation can be used prior to the estimation in order to move the working interval to that range.

Laguerre polynomials. The Laguerre polynomials, $L_n(x)$, are defined in the range $[0,+\infty[$ to be orthogonal with the norm function $\phi(x) = exp(-x)$. Since this function is not constant, there is a weight factor in the definition of the error. The larger values of x are more strongly pondered, and thus, the approximation trends to be more accurate in the tales of $f(x)$. The best approximation results used to be achieved when the function $f(x)$ is scaled in such a way that the mean value is equals to one.

Hermite polynomials. The Gaussian kernel $\phi(x) = \frac{1}{\sqrt{2\pi}} e^{-x^2/2}$ is used by the Hermite polynomials as norm function. Similar to the Laguerre polynomials, the weight function strengths the accuracy for larger values of x; but in this case, the weight is even stronger. Therefore, the approximation becomes poor if $f(x)$ differs significantly with respect to a Gaussian distribution. As we show in sec. 2.1, this fact makes the Hermite base un-robust for our estimation approach. The expansion of a PDF with a Gaussian kernel and Hermite polynomials has been very often used in statistics with the name of Gram-Charlier expansion. The best approximation results used to be achieved when the function $f(x)$ is linearly transformed to have zero mean, and unit variance.

2.1 Analysis of Accuracy

A key issue in the technique is the analysis of the accuracy provided by the bases previously presented. As a typical scenario, we estimate the transition activity and bit probability in signals with a Chi-square distribution of different degrees of freedom ($\chi^2(n)$). Varying this parameter, a family of continuous PDF with very different characteristics can be produced. Since the χ^2 is always positive, unsigned binary code is assumed for all the signals.

The PDF is approximated for each signal, polynomial family, using Eq. (1) and Eq. (2). The moments required for the estimation are calculated analytically to reduce numerical precision related errors. The so estimated bit probability is compared with the reference value obtained by direct numerical integration of $f(x)$. The experiment was conducted for different orders of approximation. Fig. 1 depicts the average error for the different distributions and its standard deviation. The results show that the Laguerre base is the one which overall performs better in terms of mean square error. Except for the simulation concerning the χ^2 with one degree of freedom, errors smaller than 0.02 can be obtained, using just the first five moments of the signal. For the $\chi^2(1)$, the convergence is degraded because in this distribution the PDF is not bound when $x = 0$.

For both, the Legendre and Laguerre bases, the quality of the approximation increases with the order of the approximation (see Fig. 1), showing that they are quite robust and stable. The opposite tendency is observed for the Hermite polynomials, that are very sensitive to the order of the approximation. If more than four moments are taken, and $f(x)$ differs significantly from a Gaussian distribution, the approximation gets notably deteriorated. It is worth noting that the accuracy of the Hermite base gets improved with the number of degrees of freedom. The rationale behind is that the χ^2 distributions tends to a Gaussian in this case. After this preliminary analysis, we concentrate our focus on the Laguerre base that seems to be the most promising one.

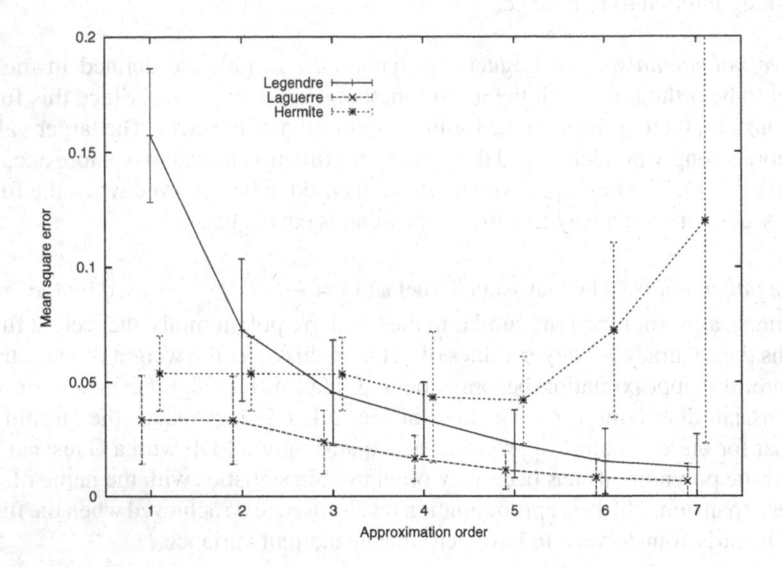

Fig. 1. Mean squared error in the approximation of the switching-activity.

2.2 Particularization for a Laguerre Base

Once we have an approximation of $f(x)$, the technique developed in [4] can be used to get an estimation of p_i. Nevertheless, we can get profit of the linear form of Eq. (1) to produce faster estimations. The idea is to pre-compute the bit probability of each one of the functions $L_n(x)\exp(-x)$ involved in the expansion of $f(x)$. Then, the total bit probability can be obtained by weighting with c_k the contribution of each of term. Since in practice, the order of the approximation is small, this approach is computationally very effective.

The technique that we employ for estimating the characteristic bit probability of each Laguerre polynomial, is again an expansion with a series of Laguerre polynomials. The main advantages of the method are its simplicity, accuracy and computational efficiency. Let $p_i\{L_n\}$ denote the probability of the *normalized* ith bit of the nth Laguerre polynomial. (See [4] for the definition of normalized bit). Then we have:

$$p_i\{L_n\} = \sum_{k=0}^{M-1} a_k\, L_k(2^{-i}) \tag{4}$$

With a_k calculated by minimum squares. Experimentally we have found that it is sufficient to consider $M = n+2$ terms in the approximation to get reasonable results. For $n = 5$ and $M = 7$, the larger discrepancy appears for L_5, with a value of 0.006.

3 Symbolic Propagation of Moments

In order to obtain the statistical moments required by the current technique, two main approaches can be employed: either by using word level simulators such as MATLAB and SystemC, or by means of symbolic techniques. Simulation approaches are conceptually simpler and can be integrated in current industrial design flows; however they require a complete executable implementation of the system, and longer runtime executions than symbolic approaches. The main idea of the symbolic techniques is to propagate the statistical properties of the signals through the design. As follows, we describe a propagation approach for the statistical moments focusing in different non-linear arithmetic operators. The technique here provided represents one of the main advantages of the moment-based approach with respect to the PDF-based methodology.

Let $\gamma_{x,j}$ and $\gamma_{y,j}$ denote the jth raw moments of the inputs of a block, and $\gamma_{z,j}$ at its output. Then:

Square: Since $z = x^2$, it is straightforward that:

$$\gamma_{z,j} = \mathbf{E}[x^2] = \gamma_{x,(2j)}$$

Power: For a positive integer power n, the output moments can be easily calculated. Since $z = x^n$, we have:

$$\gamma_{z,j} = \mathbf{E}[x^n] = \gamma_{x,(n\cdot j)}$$

Multiplication: The raw moments at the output of the multiplier can be calculated using the input moments and the spatial correlation factor between the inputs. In the particular case of uncorrelated inputs, the expression gets reduced to:

$$\gamma_{z,j} = \mathbf{E}[(x \cdot y)^j] = \gamma_{x,j}\gamma_{y,j}$$

Absolute value: For a module calculating the absolute value of a signal, the moments at the output cannot be obtained using a finite number of moments at the input. The exact value can only be obtained for even orders as follows:

$$\gamma_{z,(2j)} = \mathbf{E}[|x|^{2j}] = \mathbf{E}[x^{2j}] = \gamma_{x,(2j)}$$

The key point is that for an even order, the absolute value can be removed from inside the expectation operator; however, for odd orders that is not the case, and hence, the output moments cannot be found in closed form. Nevertheless, for smooth distributions, the odd moments can be approximated by using interpolation between two consecutive even values. It is worth noting that a direct polynomial interpolation of the raw moments cannot be used. The rationale behind is the different measuring units that each moment has. For example, $\gamma_{z,2}$ depends on the second power of z, while $\gamma_{z,4}$ does it on the fourth; hence, they cannot be combined. This problem can be overcome by normalizing the moments with the nth root function according to the moment degree. Thus,

$$Ez^{2j+1} \approx \left(\frac{\sqrt[2j]{\mathbf{E}[z^{2j}]} + \sqrt[2j+2]{\mathbf{E}[z^{2j+2}]}}{2} \right)^{2j+1} \tag{5}$$

As an example of the accuracy of the technique, Fig. 2 compares the exact and estimated raw moments of the absolute value of a Gaussian signal. The results show an excellent approximation, which gets only slightly degraded for the moment of order one, where extrapolation is used. In this case, the exact value is $\sqrt{2/\pi} \approx 0.80$, whereas the approximation equals 0.84.

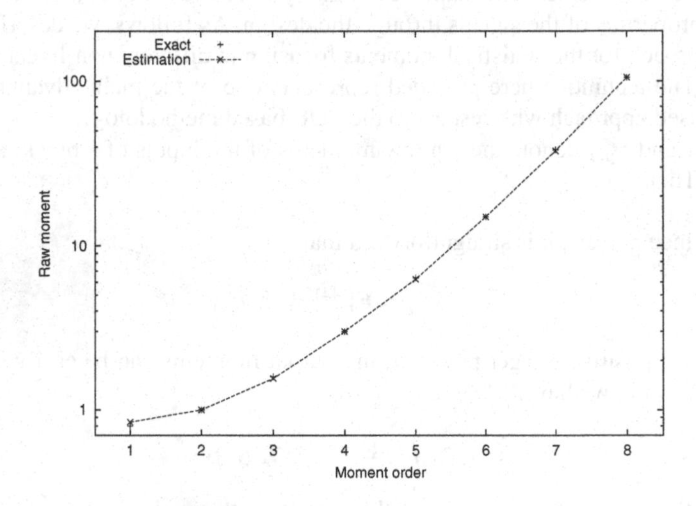

Fig. 2. Exact (points) and estimated (lines) raw moments at the output of an abs. operator.

4 Experimental Results

The previously presented moment-based approach has been validated with an extensive set of bit-level simulations. In a first experiment, the transition activity has been estimated for six different distributions: a Rayleigh, three χ^2 with orders 1 to 3, a sinus-shaped PDF, and the absolute value of a Gaussian. Five moments and the five terms for the approximation of the bit-level probability using Laguerre polynomials (see Eq. (4)) are employed. Although the accuracy is slightly smaller that those of the PDF-based approach [4], the simulation results show a very good estimation of t_i in all the scenarios. The distributions that are close to an exponential dependency (e.g, χ^2 of orders two and three, and the absolute value of a Gaussian) the maximum error per bit is smaller than 0.05, which represents an excellent accuracy. For the rest of the distributions the error is slightly larger, with a maximum value around 0.11 for the Rayleigh distribution. Especially interesting is the χ^2 distribution of order 1. Since it is not bound for

$x = 0$, a Laguerre approximation cannot completely describe its shape. However, five moments are enough to provide a fair approximation with maximum errors around 0.02. Furthermore, since that error is positive for some bits, while negative for the others, the approximation of the total transition activity is very accurate.

In a second experiment, the switching activity in the synchronization unit of an OFDM (Orthogonal Frequency Division Multiplex) receiver is analyzed. The goal is to show the validity and accuracy of the proposed technique in real applications.

4.1 Synchronization Algorithm and Hardware Architecture

For emerging high data rate wireless LANs such as ETSI HiperLAN/2 and IEEE 802.11a, OFDM is an promising modulation scheme because of its high spectral efficiency and its robustness against channel dispersions in a frequency-selective environment. Despite the excellent properties of this modulation scheme, the performance of an OFDM system can be substantially reduced by the occurrence of synchronization errors. In contrast to continuous transmission such as broadcasting, packet-oriented OFDM applications require burst synchronization schemes that can provide rapidly a reliable estimation of the synchronization parameters. Burst OFDM synchronization in a frequency-selective fading environment relies mostly on the detection of repeated preambles via correlation of the received samples. If a preamble sequence $\{d_n;\ n = 0, ..., N_s - 1\}$ is transmitted over a frequency-selective slow-fading channel, the received samples can be expressed as

$$r_n = w_n + e^{j2\pi n \xi_f} \sum_{k=0}^{L-1} h_k \, d_{n-k} \tag{6}$$

where ξ_f is the frequency offset normalized to the inter-carrier spacing, N is the OFDM symbol length, w_n is Gaussian noise and L represents the number of channel response samples h_k. Different algorithms (see [5] and references therein) have been proposed for the estimation of the frame start position and the parameter ξ_f. The basic idea lies in the evaluation of the signal autocorrelation function, which presents a maximum at the beginning of a data burst. In multi-path environments the spectral power of the signal is not constant, and therefore the autocorrelation does not suffice to achieve the synchronization. In order to account for this effect, more sophisticated algorithms (see Fig. 3) construct a timing metric which employs the complex correlation (S_n) and the spectral power (P_n) inside a window of length N_s. Thus, they require the calculation of the following parameters:

$$S_n = \sum_{m=0}^{N_s-1} r^*_{n+m} \, r_{n+m+N_s} \tag{7}$$

$$P_n = \sum_{m=0}^{N_s-1} |r_{n+m}|^2 \tag{8}$$

that can be implemented in hardware with a non-linear circuit. For example, Fig. 3 portrays the circuits for calculating S_n. It is observed that the real and imaginary parts of the input signals (I and Q channels respectively) are combined via multipliers and hence they are non-linear.

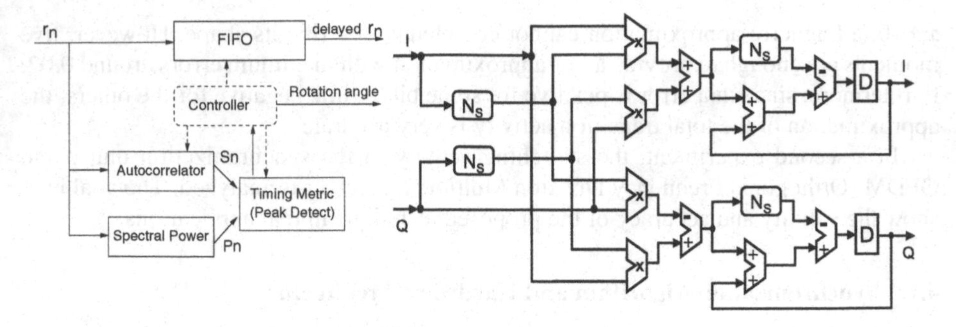

Fig. 3. (Left) Top level architecture of the coarse grain synchronization unit and (right) schematic of a hardware implementation of the autocorrelation function.

4.2 Energy Estimation

The switching activity at the different nodes of the architecture has been measured and compared with our estimation approach. The excitation used for the bit-level simulations refers to an Hiperlan/2 scenario, with a 64-QAM transmission mode, and the standardized type-A channel model. The signal is quantized with 8 bits, and has a standard deviation of approximately 32. Since the correlation factor is very small, (it equals 0.23) our current framework can be employed.

Fig. 4 represents the estimated and measured values for some representative nodes of the circuit used to calculate P_n. Those nodes correspond to the output of the module which squares the input (it is the worst case estimation), and the output of the next adder. In order to make the figure more legible, the values concerning the second node have been shifted two bits to the right.

The approximation was carried out by propagating symbolically the moments of the incoming signal as described in sec. 3 and using a Laguerre base. Only the first five moments where used for the analysis. As it can be observed, the proposed technique exhibits an excellent accuracy, despite of the small correlation of the input signal. The total cumulative error in the worst case is about 6%, and just 3% for the typical cases.

In oder to compare the accuracy of the current formulation with previous approaches, the technique proposed by Lundberg and Roy in [7] has been also implemented. Although the approximation is quite accurate for one of the nodes, for the other one the estimation is notably degraded. In fact, this degradation increases when the PDF differs from the absolute value of a Gaussian signal, since this is the only scenario considered in that work. The experimental results show that this method of Roy can lead to errors higher than 16% in a non-Gaussian scenario.

The switching activity in the nodes of the circuit computing S_n has been also analyzed. In principle, since the autocorrelation can be positive or negative, the switching activity using two's complement representation approximately equals 0.5 (observe that the input signal is poorly correlated). However, as the multipliers are implemented in sign magnitude to decrease the power consumption, the switching activity at the input and output of the multiplier differs from that value. Fig. 5 depicts the measured and estimated values for these nodes. The analysis has been performed, both, when the mo-

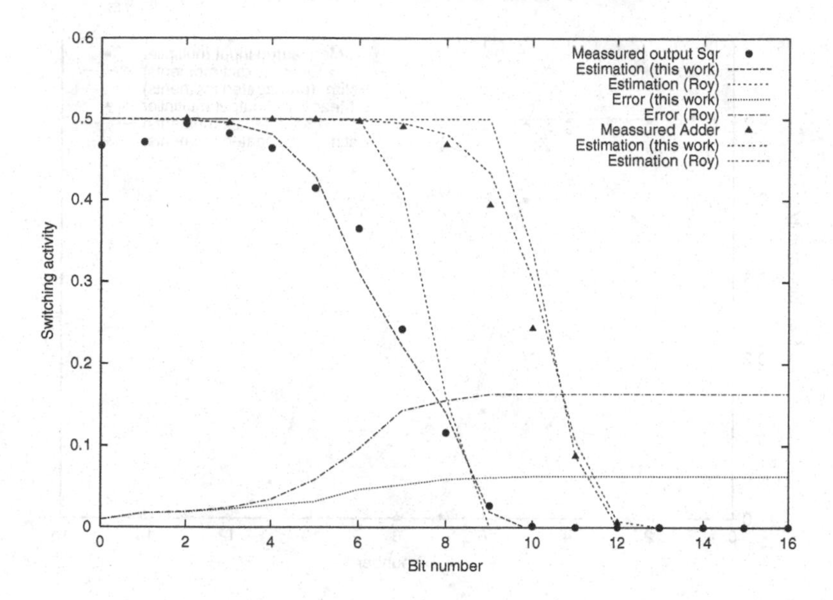

Fig. 4. Estimated and measured switching activity for the circuit calculating P_n.

ments are symbolically approximated, and when they are exactly calculated. The results show that the symbolic propagation induces a small increase in the approximation, that nevertheless is accurate enough for the purposes of early energy evaluation.

5 Conclusion

A *moment-based* estimation approach for the switching activity is presented in this paper. It allows energy estimation in signals with general uncorrelated distributions, and thus, in non-linear architectures. The approach is not restricted to Gaussian distributions, and it is simpler that the previous PDF-based techniques. The reported analysis of the accuracy of different polynomial bases shows that the Laguerre family represents a good candidate for the estimation, especially in unsigned scenarios. The paper also presents an efficient particularization of the general template for this base, and a technique for propagating the raw moments of the input signals through the design. For this propagation, the main difficulty is induced by modules, such as the absolute value operator, where approximated interpolation procedure are required.

Finally, extensive experimental results concerning the synchronization unit of an OFDM transceiver have been reported to assess the accuracy and applicability of the proposed approach. The comparison with previous techniques shows a significant reduction of the estimation error from 16% to a typical value of 3%.

The presented results are used as a foundation for addressing the estimation of general correlated distributions.

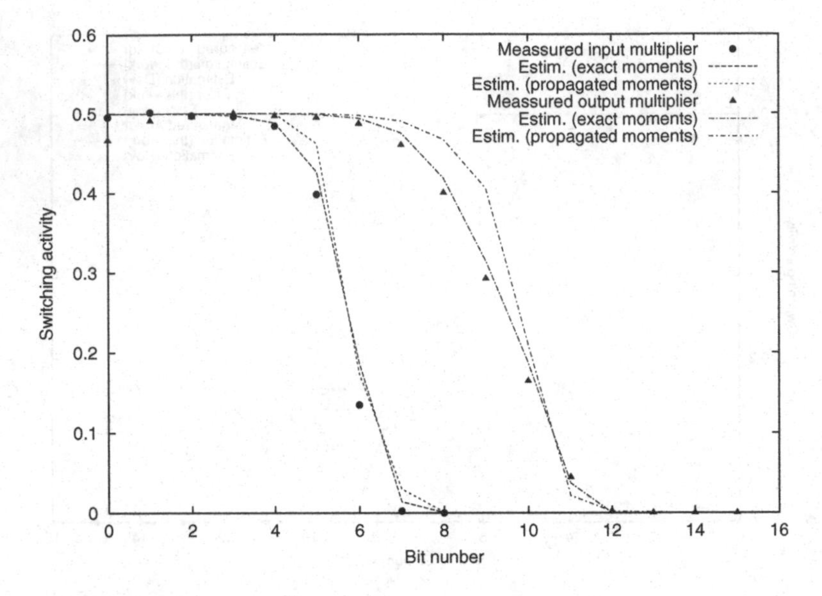

Fig. 5. Estimated and measured switching activity for the circuit calculating S_n.

References

1. S. Bobba, I. N. Hajj, and N. R. Shanbhag. Analytical expressions for average bit statistics of signal lines in DSP architectures. In *ISCAS*, pages 33–36, 1998.
2. Jui-Ming Chang and Massoud Pedram. *Power optimization and synthesis at behavioural and system levels using formal methods*. Kluwer Academic Publishers, 1999.
3. Alberto García-Ortiz and Manfred Glesner. Moment-based estimation of switching activity in non-linear architectures. In *Int. Symp. on Signal, Circuit and Systems*, July 2003.
4. Alberto García-Ortiz, Lukusa D. Kabulepa, and Manfred Glesner. Transition activity estimation for generic data distributions. In *Int. Symp. on Circuits and Systems (ISCAS)*, pages 468–471, May 2002.
5. Lukusa D. Kabulepa, Alberto García-Ortiz, and Manfred Glesner. Design of an efficient OFDM burst synchronization scheme. In *Int. Symp. on Circuits and Systems (ISCAS)*, pages 449–452, May 2002.
6. P.E. Landman and J.M. Rabaey. Architectural power analysis: the dual bit type method. *IEEE Trans. on VLSI Systems*, 3:173–187, June 1995.
7. M. Lundberg, K. Muhammad, K. Roy, and S. K. Wilson. A novel approach to high-level switching activity modeling with applications to low-power DSP system synthesis. *IEEE Trans. on Signal Processing*, 49:3157–3167, December 2001.
8. J. H. Satyanarayana and K. K. Parhi. Theoretical analysis of word-level switching activity in the presence of glitching and correlation. *IEEE Trans. on VLSI Systems*, pages 148–159, April 2000.
9. Bernard W. Silverman. *Density estimation for statistics and data analysis*. Monographs on Statistics aand appplied probability. Chapman & Hall, 1986.

Instruction Level Energy Modeling for Pipelined Processors*

S. Nikolaidis[1], N. Kavvadias[1], T. Laopoulos[1], L. Bisdounis[2], and S. Blionas[2]

[1]Department of Physics,
Aristotle University of Thessaloniki, 54124 Thessaloniki, Greece
snikolaid@physics.auth.gr
[2]INTRACOM S.A., Peania, Greece

Abstract. A new method for creating instruction level energy models for pipelined processors is introduced. This method is based on measuring the instantaneous current drawn by the processor during the execution of the instructions. An appropriate instrumentation set up was established for this purpose. According to the proposed method the energy costs (base and inter-instruction costs) are modeled in relation to a reference instruction (e.g. NOP). These costs incorporate inter-cycle energy components, which cancel each other when they are summed to produce the energy consumption of a program resulting in estimates with high accuracy. This is confirmed by the results. Also the dependencies of the energy consumption on the instruction parameters (e.g. operands, addresses) are studied and modeled in an efficient way.

1 Introduction

Embedded computer systems are characterized by the presence of dedicated processors which execute application specific software. A large number of embedded computing applications are power or energy critical, that is power constraints form an important part of the design specification [1]. Recently, significant research in low power design and power estimation and analysis has been developed. The determination of methods for the accurate calculation of the energy consumption in processors is required for system level energy consumption evaluation and optimization.

Power analysis techniques for embedded processors that employ physical measurements were firstly suggested in mid 90's. Significant effort on software optimization for minimizing power dissipation is found in [1]-[4] where a technique based on physical measurements is developed. Power characterization is done with the extraction of cost factors for the average current drawn by the processor as it repeatedly executes short instruction sequences. The base cost for an instruction is determined by constructing a loop with several instances of the same instruction. Inter-instruction effect induced when executing different adjacent instructions, is measured by replacing the one-instruction loops used in the measurement of base costs, with loops consisting of appropriate instruction pairs. The sum of the power base cost of the instructions

* This work was supported by EASY project, IST-2000-30093, funded by the European Union

J.J. Chico and E. Macii (Eds.): PATMOS 2003, LNCS 2799, pp. 279–288, 2003.
© Springer-Verlag Berlin Heidelberg 2003

executed in a program, refined by the power cost of the inter-instruction effects, are considered to provide the power cost of the program. This method has been validated for commercial targets based on embedded core processors.

The majority of work published on the field of measurement-based techniques, refers to the Tiwari method [1] as a base point. By Tiwari method only average power estimates can be utilized for modeling task, since the measurements are taken with a standard digital ammeter. Direct application of the Tiwari technique is found in [5] where an extensive study of the ARM7TDMI processor core is reported. In [6], physical measurements for the processor current are also obtained by a precise amperemeter. However, power modeling effort is more sophisticated, as architectural-level model parameters are introduced and integrated within the power model. These consist of the weight of instruction fields or data words, the Hamming-distance between adjacent ones, and basic costs for accessing the CPU, external memory and activating/deactivating functional units.

Instantaneous current is firstly measured in [7], where a digitizing oscilloscope is used for reading the voltage difference over a precision resistor that is inserted between the power supply and the core supply pin of the processor. Instantaneous power is then calculated directly from the voltage waveform from which average figures are extracted to guide instruction power modeling. Resistor-based methodologies suffer from supply voltage fluctuations over the processor which reduces the accuracy of the method.

All the above techniques acquire the current drawn by the processor on instruction execution. A complex circuit topology for cycle-accurate energy measurement is proposed in [8,9], which is based on instrumenting charge transfer using switched capacitors. The switches repeat on/off actions alternately. A switch pair is charged with the power supply voltage during a clock cycle and is discharged during the next cycle powering the processor. The change in the voltage level across the capacitors is used for the calculation of the consumed energy in a clock cycle. However, this method can not provide detail information for the shape of the current waveform, which may be significantly useful in some applications and also in case high quality power models including the architectural characteristics of the processor are required. In order to measure the energy variations, various (*ref*, *test*) instruction pairs are formed, where *ref* notes a reference instruction of choice and *test* the instruction to be characterized. This setup combined with the above modeling concept are then utilized to obtain an energy consumption model for the ARM7 processor.

2 Instruction-Level Energy Modeling for Pipelined Processors

The proposed method is based on the measurement of the instantaneous current drawn by the processor during the execution of the instructions. The instrumentation set-up for measuring the instantaneous current has been presented in [10]. The current sensing circuit is a high performance current mirror, which is inserted in the supply line and copies the drawn current on a resistor. In this way the supply voltage fluctuation at the processor is minimized increasing the accuracy of the measurements. A high-speed digitized oscilloscope is used for monitoring the voltage drop on the resistor.

The method is developed for pipelined processors like the ARM7 (three-stage pipeline). However, its application for non-pipelined processors is straightforward.

2.1 Energy Consumed in Pipelined Processors

The current drawn by the processor in a clock cycle is the result of the concurrent process of many instructions due to the pipeline structure of the processor. So, it is rather impossible to isolate the current drawn from the pipeline stage, which processes the test instruction. Instead, we have to find a method for calculating the energy consumed by an instruction from the current drawn in the clock cycles needed for its execution.

In each clock cycle the instantaneous current, $i(t)$ is monitored and the energy consumed in that cycle, E_{cycle}, is calculated as:

$$E_{cycle} = V_{DD} \int_0^T i(t)dt \tag{1}$$

where V_{DD} is the value of the supply voltage and T is the clock period. The energy which is consumed as an instruction passes through a n-stage pipeline, $E\_M(Instr)$, is distributed in n clock cycles and it is found by summing the energy components of the n cycles:

$$E\_M(Instr) = E_{cycle\_1} + E_{cycle\_2} + \cdots + E_{cycle\_n} \,. \tag{2}$$

In n clock cycles n instructions are executed. Suppose a set of instructions with only one instance of the *test* instruction, and with the others being the same, let us say a reference instruction (probably the NOP), running through the pipeline. Then in n clock cycles one *test* instruction and n-1 reference instructions are executed, and the energy corresponding to the test instruction is calculated as:

$$E(Instr) = E\_M(Instr) - (n-1)E(ref) \,. \tag{3}$$

Since the energy budget of a program corresponds to the sum of the energy of each instruction, such a model seems to lead to an overestimation because instead of modeling the circuit state changes from one instruction to another, we model every instruction as a circuit state change (change from test to NOP instruction), which results in counting the number of circuit state changes twice. To overcome this problem the inter-instruction effect is also modeled in a similar way, and in this case the value of this effect is expected to be rather (at least in most cases) negative compensating this overhead.

2.2 Base Instruction Cost

In the proposed method as base instruction cost we mean the amount of energy which is consumed for the execution of an instruction after the execution of a reference instruction (NOP). In order to calculate the base instruction costs, loops including an instance of the test instruction and a number of instances of reference instruction are executed on the processor. Figure 1 shows the pipeline states during the clock cycles required for the execution of the instruction on a three-stage pipelined processor. The instruction is executed in three cycles. In each clock cycle the current is monitored and the energy is calculated.

Pipeline Stages	3-stage pipeline operation				
IF	NOP	NOP	Instr	NOP	NOP
ID	NOP	NOP	NOP	Instr	NOP
EX	NOP	NOP	NOP	NOP	Instr
Clock Cycles	n-1	n	n+1	n+2	n+3

Fig. 1. Pipeline states during the execution of instructions

The energy consumed in these three stages is calculated as:

$$E\_M(Instr) = E_{cycle\_n+1} + E_{cycle\_n+2} + E_{cycle\_n+3} \tag{4}$$

In these three cycles one test instruction and two reference instructions are executed. This amount of energy contains inter-cycle effects due to the change of the pipeline states and it can be expressed as

$$E\_M(Instr) = E_{Instr} + 2E_{NOP} + E_{NOP,Instr} + E_{Instr,NOP} + E_{NOP,NOP} \tag{5}$$

where E_{inst} corresponds to the real energy of the instruction, E_{NOP} to the energy of the NOP instruction and the others to the inter-cycle effects between the cycles n,n+1 ($E_{NOP,Instr1}$), n+1,n+2 ($E_{Instr1,NOP}$) and n+2,n+3 ($E_{NOP,NOP}$).
The energy of the instruction is calculated as:

$$E(Instr) = E\_M(Instr) - 2E_{NOP} \tag{6}$$

and corresponds to

$$E(Instr) = E_{Instr} + E_{NOP,Instr} + E_{Instr,NOP} + E_{NOP,NOP} \ . \tag{7}$$

The energy of the NOP instruction is:

$$E_{NOP} = E_{cycle\_n} \ . \tag{8}$$

2.3 Inter-instruction Effect Cost

The instruction sequence for measuring the inter-instruction effect is shown in Figure 2. Appropriate loops will be executed on the processor. In this case, in four clock cycles one *Instr1*, one *Instr2* and two reference instructions are executed.

Pipeline Stages	3-stage pipeline operation				
IF	NOP	Instr1	Instr2	NOP	NOP
ID	NOP	NOP	Instr1	Instr2	NOP
EX	NOP	NOP	NOP	Instr1	Instr2
Clock Cycles	n	n+1	n+2	n+3	n+4

Fig. 2. Pipeline states during the execution of instructions for measuring inter-instruction costs

Consequently, it is:

$$E\_M(Instr1, Instr2) = E_{cycle\_n+1} + E_{cycle\_n+2} + E_{cycle\_n+3} + E_{cycle\_n+4} \tag{9}$$

which corresponds to:

$$E\_M(Instr1, Instr2) = E_{Instr1} + E_{Instr2} + 2E_{NOP} + E_{NOP,Instr1} + E_{Instr1,Instr2} + E_{Instr2,NOP} + E_{NOP,NOP} \tag{10}$$

while the inter-instruction effect is figured as:

$$E(Instr1, Instr2) = E\_M(Instr1, Instr2) - E(Instr1) - E(Instr2) - 2E_{NOP} \tag{11}$$

and corresponds to:

$$E(Instr1, Instr2) = E_{Instr1,Instr2} - E_{Instr1,NOP} - E_{NOP,Instr2} - E_{NOP,NOP} \tag{12}$$

2.4 Modeling the Energy of a Program

The total energy consumed when a program is executed can be estimated by summing the base costs of the instructions and the inter-instruction costs. Doing that, it is observed that the inter-cycle costs included in the expressions (7) and (12) cancel each other and finally a sum including only the actual energy components of each instruction and the corresponding inter-instruction effects appears. Indeed, only three additional components appear, one inter-cycle component of the first instruction of the program and one of the last instruction and an inter-instruction effect between NOP instructions. These three components will determine the maximum theoretic error of the method. An example is given in Figure 3 where the application of the method for a program consisting of three instructions ADD, MOV and CMP is analyzed. As it is shown, all the inter-cycle costs included in equations (7) and (12) are eliminated except the three ones mentioned before. In this way the actual energy components of the instructions are isolated from secondary effects and high accurate estimates for the processor can be created.

$$E(ADD) = E_{ADD} + E_{NOP,ADD} + E_{ADD,NOP} + E_{NOP,NOP}$$
$$E(MOV) = E_{MOV} + E_{NOP,MOV} + E_{MOV,NOP} + E_{NOP,NOP}$$
$$E(CMP) = E_{CMP} + E_{NOP,CMP} + E_{CMP,NOP} + E_{NOP,NOP}$$
$$E(ADD,MOV) = E_{ADD,MOV} - E_{ADD,NOP} - E_{NOP,MOV} - E_{NOP,NOP}$$
$$+ \quad E(MOV,CMP) = E_{MOV,CMP} - E_{MOV,NOP} - E_{NOP,CMP} - E_{NOP,NOP}$$

$$E(\mathrm{Pr}\,ogram) = E_{ADD} + E_{MOV} + E_{CMP} + E_{ADD,MOV} + E_{MOV,CMP} + E_{NOP,ADD} + E_{CMP,NOP} + E_{NOP,NOP}$$

Fig. 3. Estimation of the energy of a program consisting by three instructions

It has to be mentioned that a hypothesis was used so that the inter-cycle costs can cancel each other. The inter-cycle effect as shown in the expressions of the base cost and the inter-instruction cost is not the same since in these two cases the pipeline stages do not necessary correspond to the same state. However, according to our results this approximation does not influence significantly the accuracy of the method.

3 Accurate Instruction-Level Energy Modeling

As it was mentioned above, the energy consumed during the execution of instructions can be distinguished in two amounts. The base cost, energy amount needed for the execution of the operations which are imposed by the instructions, and the inter-instruction cost which corresponds to an energy overhead due to the changes in the state of the processor provoked by the successive execution of different instructions. Measurements for determining these two energy amounts for each instruction of the ARM7TDMI processor were taken and presented in [11]. However the base costs in [11] were for specific operand and address values (zero operand and immediate values and specific address values to minimize the effect of 1s). This base cost is called *pure* base cost.

We have observed in our measurements that there is a dependency of the energy consumption of the instructions on the values of their parameters (operand values, addresses). To create accurate models this dependency has to be determined. Consequently the dependency of the energy of the instructions on the register numbers, register values, immediate values, operand values, operand addresses and fetch addresses was studied. Additional measurements were taken to satisfy this necessity. It was observed by the measurements that there is a close to linear dependency of the energy on these parameters, versus the number of 1s in their word space. In this way the effect of any of the above *energy-sensitive factors* was efficiently modeled by a coefficient. Incorporating these effects in our models the proposed method keeps its promised accuracy while it becomes very attractive since it can be easily implemented in software as an estimation tool.

Making some appropriate experiments we observed that the effect of each energy-sensitive factor on the energy cost of the instruction is independent of the effect of the other factors. The effects on the energy of these factors are uncorrelated as can be observed in Table 1. The distortion of our results from this conclusion is, most of the time, less than 2-3% and only in some marginal cases becomes more than 7%. According to this conclusion, the effect of the energy-sensitive factors can simply be added to give the total energy amount.

Table 1. Comparison of measured to calculated instruction energy costs (nJoules) due to additional dependencies

	Measured		Calculated		
Instruction formation	$E_{base}+E_{oprd}$	$E_{base}+E_{oprd}+E_{regn}$	$E_{base}+E_{oprd}$	$E_{base}+E_{oprd}+E_{regn}$	% diff
ADD Rd,Rn,Rs,ASR Rm	2.58	2.61	2.57	2.61	-0.04
ADDRd,Rn,Rs,ASR imm	1.57	1.60	1.55	1.60	-0.27
ADD Rd,Rn,Rs	1.51	1.59	1.51	1.56	1.71
ADD Rd,Rn,Rs,RRX	1.51	1.63	1.52	1.64	-1.19
LDR Rd, [Rn,Rs]	3.07	3.27	2.97	3.29	-0.63
STR Rd, [Rn,Rs]	2.28	2.48	2.23	2.43	2.01

Other sources of energy consumption are conditions of the processor which lead to an overhead in clock cycles because of the appearance of idle cycles. This is the case of the appearance of pipeline stalls, which was measured and modeled.

According to the above, the energy, E_i, consumed during the execution of the i instruction can be modeled as:

$$E_i = b_i + \sum_j a_{i,j} N_{i,j} \tag{13}$$

where b_i is the pure base cost of the i instruction, $a_{i,j}$ and $N_{i,j}$ is the coefficient and the number of 1s of the j energy-sensitive factor of the i instruction, respectively.

Each instruction may contain some of these energy-sensitive factors. The effect of all the factors included in an instruction have to be taken into account to create the energy budget of the instruction.

Having modeled the energy cost of the instructions, the energy consumed for running a program of n instructions can be estimated as:

$$PE = \sum_1^n E_i + \sum_1^{n-1} O_{i,i+1} + \sum \varepsilon \tag{14}$$

where $O_{i,j}$ is the inter-instruction cost of the instructions i and j, and ε is the cost of a pipeline stall.

4 Pure Base Cost and Inter-instruction Cost Models – Results

The complete models for the instruction-level energy consumption of the ARM7TDMI created according to the proposed methodology can be found in [12]. Pure base costs of all the instructions and for all the addressing modes are given. Since the number of the possible instruction pairs (taking into account the addressing modes) is enormous, groups of instructions and groups of addressing modes according to the resources they utilize, have been formed and inter-instruction costs have been given only for representatives from these groups. In this way we keep the size of the required model values reasonable without significant degradation of the accuracy (error less than 5% in the inter-instruction cost by using representative instructions).

Most of the values of the inter-instruction costs have negative sign as it was expected. The contribution of the inter-instruction costs, as they are calculated according to the proposed method, remains small. As it can be observed by our models most of the inter-instruction costs are less than 5% of the corresponding pure base costs while almost all the cases are covered by a 15% percentage. Also, there is no symmetry in the inter-instruction cost for a pair of instructions. For example, the execution of the LDR after the ADD present a cost of 0.064nJoule while the execution of ADD after LDR present an cost of –0.122nJoule. This seems more reasonable from the case of symmetric inter-instruction costs as Tiwary method supposes.

To determine the accuracy of the method a number of programs with various instructions have been created. In these instructions the operand values were kept zero and thus the effect of energy sensitive factors wasn't taken into account. The energy consumed during the execution of each program was calculated directly from measurements of the instantaneous drawn current and also calculated by using the derived instruction-level energy models. The error was found to be up to 1.5%.

5 Models of the Energy Sensitive Factors Effect – Results

The dependency on the energy of the energy sensitive factors was also studied. Energy depends on the number of 1s in the word structures of these entities. The Hamming distance between the corresponding word fields of successive instructions is not considered here since according to the followed modeling methodology, where NOP is used as a reference instruction, Hamming distance equals to the number of 1s of the corresponding fields of the instructions.

The effect of each factor was studied separately from the others since, as it can be verified by the results, the correlation among the effect of these factors is insignificant. This energy dependency can be approximated with sufficient accuracy by linear functions. Coefficients were derived for each instruction for any energy sensitive factor. However, appropriate grouping of the instructions is used to keep reasonable the number of required coefficients to increase the applicability of the method without significant loss in the accuracy.

The grouping of the instructions for the derivation of the coefficients and the corresponding measurements is presented in [12]. Some results are given here. In Figure 4 the effect of the register number for data-processing instructions in register addressing mode is presented. Actual physical measurements were compared with the estimated energy values, as they are produced using the selected coefficients, for various instructions. The error in most cases was less than 3%. The energy consumption versus the number of 1s in operand values for the STR instruction is shown in Figure 5. The linear dependency is obvious in both figures.

To evaluate the absolute accuracy of our modeling approach, real kernels were used as benchmarks. The corresponding assembly list have been extracted from C programs by utilizing the facilities of the *armcc* tool, shipped with the ARM ADS software distribution. The energy consumption during the cycles of the program was calculated by direct measurement of the drawn current and by using the derived models and equation (14). According to our results the error of our approach in real life programs was found to be less than 5%.

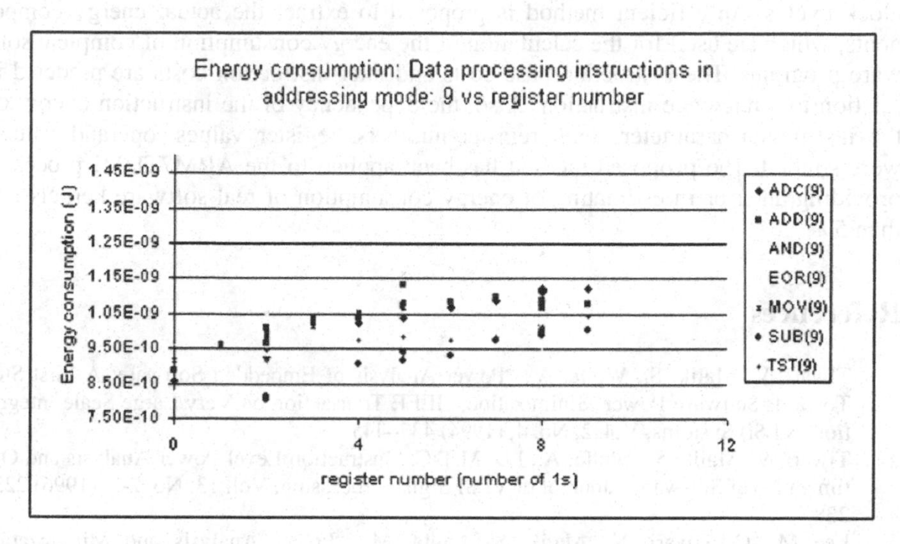

Fig. 4. The effect of register number for data-processing instructions in register addressing mode

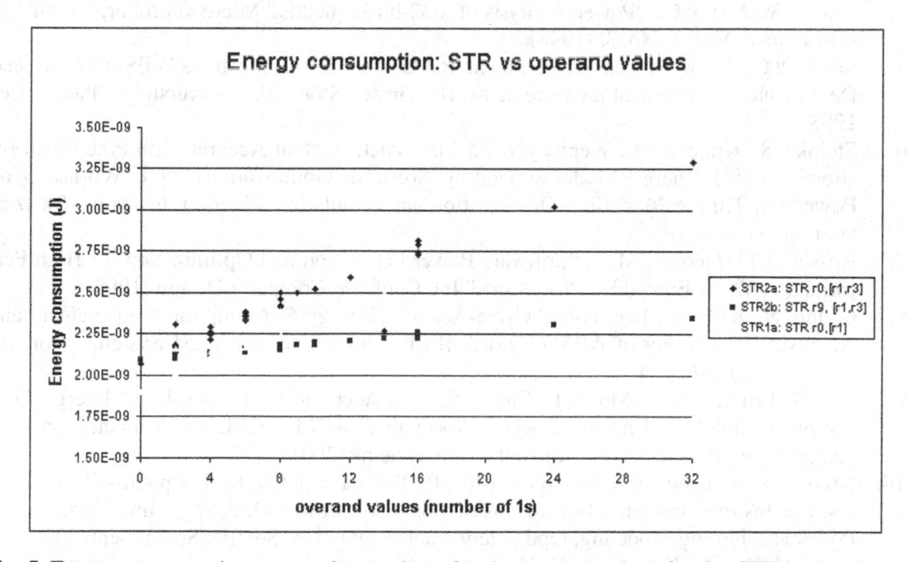

Fig. 5. Energy consumption versus the number of 1s in operand values for the STR instruction

6 Conclusions

High quality instruction level energy models can be derived for pipelined processors by monitoring the instantaneous current drawn by the processor at each clock cycle.

Knowing the current waveform the corresponding energy consumed at a clock cycle can be calculated. However, energy components of an instruction exist in different clock cycles. An efficient method is proposed to extract the actual energy components, which are used for the calculation of the energy consumption of complete software programs. The instruction base costs and inter-instruction costs are modeled in relation to a reference instruction. Also, the dependency of the instruction energy on the instruction parameters (e.g. register numbers, register values, operand values) were studied. The proposed method has been applied to the ARM7TDMI processor providing an error in estimating of energy consumption of real software kernels less than 5%.

References

1. Tiwari, V., Malik, S., Wolfe, A., "Power Analysis of Embedded Software: A First Step Towards Software Power Minimization," IEEE Transaction on Very Large Scale Integration (VLSI) Systems, Vol. 2, No. 4, (1994) 437–445
2. Tiwari, V., Malik, S., Wolfe, A., Lee M.T-C. "Instruction Level Power Analysis and Optimization of Software," Journal of VLSI Signal Processing, Vol. 13, No. 2-3, (1996) 223–238
3. Lee M.T-C, Tiwari, V., Malik, S., Fujita, M., "Power Analysis and Minimization Tehniques for Embedded DSP Software," IEEE Transactions on Very Large Scale Integration (VLSI) Systems, (1997) 123–135
4. Tiwari, V., Lee T.C., "Power Analysis of a 32-bit Embedded Microcontroller," VLSI Design Journal, Vol. 7, No. 3, (1998)
5. SOFLOPO, "Low Power Development for Embedded Applications", ESPRIT project: Deliverable 2.2: Physical measurements. By Thanos Stouraitis, University of Patras, Dec 1998
6. Steinke, S., Knauer, M., Wehmeyer, L., Marwedel, P., "An Accurate and Fine Grain Instruction-Level Energy Model supporting Software Optimizations," Int. Workshop on Power and Timing Modeling, Optimization and Simulation, Yverdon-les-bains, Switzerland (2001)
7. Russell, J.T., Jacome, M.F., "Software Power Estimation and Optimization for High Performance, 32-bit Embedded Processors," Int. Conf. On Computer Design (1998)
8. Chang, N., Kim, K., Lee, H-G. "Cycle-Accurate Energy Consumption Measurement and Analysis: Case Study of ARM7TDMI," IEEE Transactions on VLSI Systems, Vol. 10, No. 2, (2002) 146–154
9. Lee, S., Ermedahl, A., Min S-L, Chang N., "An Accurate Instruction-Level Energy Consumption Model for Embedded RISC Processors," ACM SIGPLAN Workshop on Languages, Compilers and Tools for Embedded Systems (2001)
10. Nikolaidis, S., Kavvadias, N., Neofotistos, P., Kosmatopoulos, K., Laopoulos, T., Bisdounis, L., "Instrumentation set-up for Instruction Level Power Modeling," Int. Workshop on Power and Timing Modeling, Optimization and Simulation, Seville, Spain, Sept. 2002
11. S. Nikolaidis, N. Kavvadias, P. Neofotistos, "Instruction level power measurements and analysis", IST-2000-30093/EASY Project, Deliverable 15, Sept 2002. http://easy.intranet.gr
12. S. Nikolaidis, N. Kavvadias, P. Neofotistos, "Instruction level power models for embedded processors", IST-2000-30093/EASY Project, Deliv. 21, Dec 2002. http://easy.intranet.gr

Power Estimation Approach of Dynamic Data Storage on a Hardware Software Boundary Level*

Marc Leeman[1], David Atienza[2,3], Francky Catthoor[3], V. De Florio[1], G. Deconinck[1], J.M. Mendias[2], and R. Lauwereins[3]

[1] ESAT/K.U.LEUVEN, Kasteelpark Arenberg 10, 3001 – Leuven, Belgium;
[2] DACYA/U.C.M., Avenida Complutense s/n, 28040 – Madrid, Spain;
[3] IMEC vzw, Kapeldreef 75, 3000 Leuven, Belgium.

Abstract. In current multimedia applications like 3D graphical processing or games, the run-time memory management support has to allow real-time memory de/allocation, retrieving and data processing. The implementations of these algorithms for embedded platforms require high speed, low power and large data storage capacity. Due to the large hardware/software co-design space, high-level implementation cost estimates are required to avoid expensive design modifications late in the implementation. In this paper, we present an approach designed to do that. Based on memory accesses, normalised memory usage[1] and power estimates, the algorithm code is refined. Furthermore, optimal implementations for the dynamic data types involved can be selected with a considerable power contribution reduction.

1 Introduction

The fast growth in the variety, complexity and functionality of multimedia applications and platforms impose a high demand of memory and performance. This results in high cost and power consumption systems while current markets demand low power consumption ones. In addition, most of the new multimedia algorithms rely heavily on the use of dynamic memory due to the unpredictability of the input data at compile-time. This, combined with the memory hungry nature of certain parts of the algorithms, makes the dynamic memory subsystem one of the main sources of power consumption.

With the aforementioned characteristics, classical hardware (HW) design improvements, like voltage or technology scaling can only partially compensate for the growing HW/software (SW) gap [10]. In the last years, the boundary of what is considered to be critical design improvements for very large scale integration systems has been shifting consistently towards the SW side.

Even though a lot of current SW dominated transformations [3] result in platform independent improvements, they are critical for embedded devices due to the more constrained HW specification and especially of the memory hierarchy. Next to performance

* This work is partially supported by the Spanish Government Research Grant TIC2002/0750, the Fund for Scientific Research - Flanders (Belgium, F.W.O.) through project G.0036.99 and a Postdoctoral Fellowship for Geert Deconinck.

[1] The sum of the memory used at a time slice, multiplied by the time. This amount is then divided by one run of the algorithm

J.J. Chico and E. Macii (Eds.): PATMOS 2003, LNCS 2799, pp. 289–298, 2003.
© Springer-Verlag Berlin Heidelberg 2003

related techniques, power consumption is of paramount importance for hand-held devices. Even in devices that are not dependent on batteries, energy has become an issue due to the circuit reliability and packaging costs [15]. As a result, optimisation for embedded systems has three optimisation goals that cannot be seen independently from each other: memory usage, power consumptions and performance.

In order to optimise embedded system designs, detailed power consumption profiling must be available at an early stage of the design flow. Unfortunately, they do not exist presently at this level for the dynamically (de)allocated data type (further called Dynamic Data Type or DDT) implementations. In order to evaluate this accurately today, simulations would be necessary at a much closer level to the final implementation on a certain platform, e.g. at instruction (ISA) or cycle accurate HW level. Since each implementation of a DDT defines how the memory is accessed and allocated, they form an important factor for power consumption.

In this paper we explain how a high-level (i.e. from C++ code) profiling approach is able to analyse and extract the necessary information for a power-aware refinement of the DDT implementations involved in multimedia applications at run-time. To this end, the three important factors that influence the power consumption and overall performance of the memory subsystem are studied, i.e. the memory usage pattern over time, the amount of memory accesses and the data access mechanisms. This power analysis approach is a crucial enabling step to allow subsequent optimisations and refinements. In the latter stages, the DDT implementations can be optimised based on relative power contribution estimates early on during system integration, enabling large savings in the design-time of the system.

The remainder of this paper is organised as follows. In Section 2, we describe some related work. In Section 3, we explain the high-level profiling phase and further refinements that it allows. In Section 4, we describe our drivers and present the experimental results. Finally, in Section 5, we state our conclusions.

2 Related Work

A large body of SW power estimation techniques have been proposed at lower abstraction levels, starting from code that is already executable on the final platform. One of the first papers is [14] and many contributions have added to that. But none of these has explicitly modelled the contribution of the dynamically allocated data types in the memory hierarchy of the platform. Work to obtain accurate figures on a higher level is more recent (e.g. [2,15]).

Although the level of such estimations has been extended to the assembly code and also C code, they are based on an analysis and design space without run-time analysis. This is not sufficient to deal with dynamic memory applications. In fact, in algorithms governed by dynamic memory accesses and storage (such as multimedia applications) the control flow and accesses to the DDTs are unknown at compile-time, and the aforementioned run-time analysis becomes necessary.

Most of the power estimation systems focus on obtaining accurate absolute values for HW-SW systems. In the framework of DDT optimisations and refinements, this is an overkill and we are mainly interested in the *power contribution* of the dynamic memory sub-system. As such, accurate relative figures are more important.

Several analytical and abstract power estimation models at the architecture-level have received more attention recently [4] since they are needed for high-level power analysis in very large scale integration systems. However, they do not focus on the dynamic memory hierarchy of the systems and they are not able to analyse the power consumption from the DDTs at the SW level for dynamic data-dominated applications like multimedia applications.

To solve this gap in the power analysis context with respect to global dynamic memory profiling, the approach we propose is inspired partially from [17], but it is clearly different in a number of important parts. Basically, this reference handles applications from the network routing domain, where only one simple DDT is used and the main focus is in the multiple tasks running on the system. In fact, once the system has been initialised, the usage of the DDTs in that application domain does not vary much and averages around the same values. A snapshot of the memory footprint at any time during execution gives a reasonably good image of the memory behaviour. Therefore, in [17], a detailed profiling is not performed at run-time for DDTs and the memory footprint is used to determine the memory contribution in the power estimations.

In addition, according to the characteristics of certain parts of multimedia applications, several transformations for DDTs [16] and design methodologies [3] have made significant headway for static data profiling and optimisations taking into account static memory access patterns to physical memories. Also, the access to the data at a SW level has to take power consumption into account and research has been started to propose suitable power-aware data structure transformations at this level for embedded systems [5].

3 Description of the Approach

3.1 General Framework

The main objective of our high-level profiling is to provide the necessary run-time information of the DDTs used by an algorithm in a very early stage of the development flow. To do this, the algorithm that needs to be ported and optimised must go through a number of phases. First of all, the source code is analysed (including its structure in classes) and the profiling code to extract accurate information of the DDTs at run-time is inserted. Secondly, it is executed and the profiling information is stored. Finally, this information is processed to get the necessary power estimations and memory accesses reports in a post-processing phase.

After the analysis, the detailed power and timing representation of the DDTs is used to analyse and optimise their interactions in the global source code flow (i.e. intermediate variable elimination transformation [9]). Finally, the refinement of the DDTs can be performed with an exploration that uses the same profiling framework to evaluate possible trade-offs between power consumption, memory footprint and performance. Figure 1 gives an schematic overview of the overall optimisation approach that we propose, which is not the topic of this paper but it is summarised here to show the context of our power analysis approach. For a more in-depth discussion, see [8].

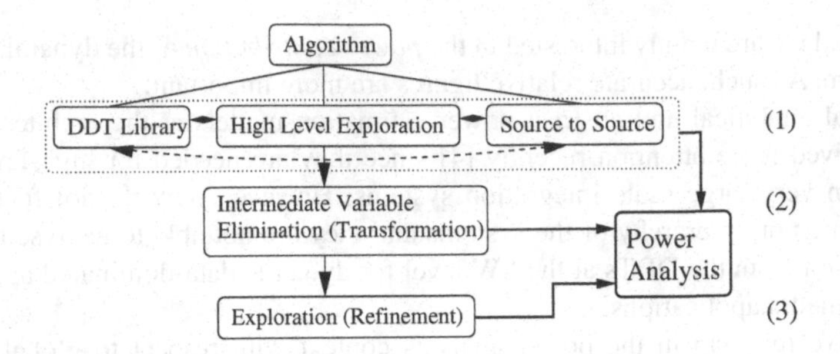

Fig. 1. Overview of the used system-level refinement approach

One of the most important characteristics of this approach is that it is based on a phase-wise exploration and refinement. Every phase is ideally self-contained and once an algorithm is optimised at that level, it can be handed down to the next phase in the design flow, i.e. a more HW oriented optimisation. As a result, the development team still has sufficient freedom to make (significant) changes without expensive re-iterations through the entire development flow.

The main features of our power analysis approach to support the optimisation and refinement steps are outlined in the following:

multiple and complex dynamic data types: All the considered multimedia applications employ a number of complex dynamic data types with very different behaviour.

automatic instrumentation and insertion of profile objects: Instead of tedious manual code transformations, an automatic tool has been developed to support our approach. It analyses the dynamic memory accesses in the application and modify the custom DDTs sources from developers to include all the information required for profiling.

structured reports generation: In applications which multiple DDTs, an analysis of the hierarchy of classes and structure of the source code must be done. Therefore, a profile framework has been developed to collect run-time profile information analysing where memory is used in the different parts of the DDTs, i.e. allocations blocks, intermediate layers, etc.

detailed power and timing information acquisition: In multimedia applications, during an algorithm run, data sets can be dominant and non-existent in other parts. As such, memory can be re-used and dynamic memory usage can vary a lot. Normalised memory usage is used to give a better representation of a DDTs impact. Profile runs gather detailed memory access patterns and memory usage for power consumption estimates.

layered DDT library: The implementation of complex DDTs can be considered as layered implementations of basic DDTs[2] [13]. A library provides standardised interfaces to the most commonly used N-layered implementations.

source to source transformations are possible: Using structured profiling reports, a global source code optimisation is able to eliminate temporary buffers that introduce memory movements between intra-algorithmic phases without any useful processing of data [9].

[2] All DDTs are a combination of a list, array and tree

3.2 Profiling Phase

As pointed out previously, it is mandatory to obtain run-time information about the DDTs to optimise the system. The developer has several choices to modify and explore the DDT search space. When the developer wants to evaluate the internal data structures provided with the algorithm or has his own library in C++. In that case, the *automatic insertion of profile collectors* modifies and instruments the sources including the profile framework. A second option is linking the algorithm sources with the provided library of multi-layered DDTs. The library provides standardised interfaces that need to be integrated in the algorithm sources. Finally, a third option is to explore DDTs not yet included in his custom sources or in the DDT library implementations. In that case, a modular approach for composing multi-layered DDTs is provided, based on mixins [12].

The search space for DDTs is then explored with an heuristic or exhaustive fashion, depending on the complexity of the program. The profile runs enable the framework to extract detailed run-time information. Finally, an automated post processing extracts timing information and power estimates of each DDT combination used in the exploration run. In Figure 2, the timing visualisation obtained for the multimedia drivers are shown.

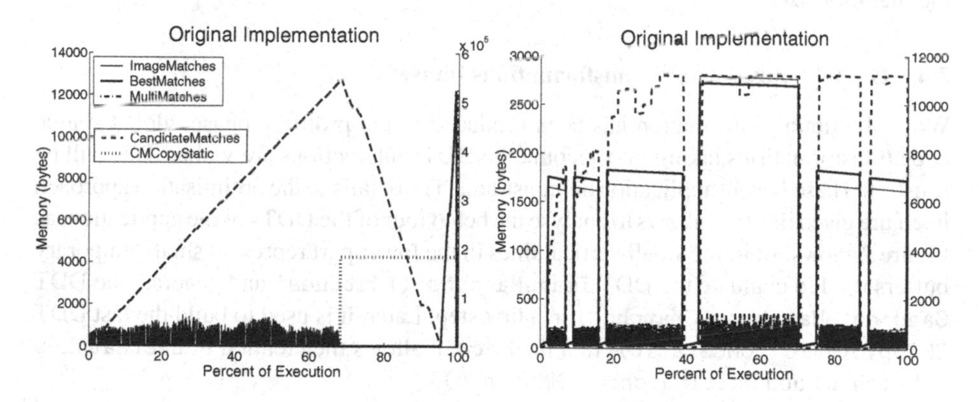

Fig. 2. On the left, memory behaviour of the matching algorithm between 2 frames (only *CandidateMatches* and *CMCopyStatic* plotted on the right axis). On the right, a similar plot for processing 6 frames in the game engine. The dashed line should be plotted on the right axis.

3.3 Memory Power Models

As it has been explained, for the profiling phase a realistic model for the dynamic memory subsystem is necessary. We have used initially the CACTI model [7], which is a complete energy/delay/area model for embedded SRAMs. It has two main advantages. First, a clear hierarchy in the modelling of the different memory components at four different levels. The first level includes modelling of transistors and interconnect wires. The second level is where these devices are combined to form the memory circuits, i.e. address decoder, SRAM cell, etc. For the delay, the Horowitz approximation [6] is used, while the energy consumption depends only on the equivalent circuit capacitance, supply voltage and swing voltage. The last level consists of an exploration phase that returns (among other results) the least power consumption values for an optimal partitioning scheme for the specific memory. The second main advantage of CACTI is the fact that it is scalable to different technology nodes.

With the aforementioned model, we have represented the different sizes of memories required and compared with real data-sheet values from Trimedia for the on-chip memories caches with a size of 32 KB and an SRAM of 1 MB at 166 MHz with .18 μm technology. We also compared to a very recent model for large SRAMS [1].

From this, it became clear that the main drawbacks of this CACTI model are the outdated circuits and the old technology parameters to scale under .18 μm technology. It was built originally developed considering the .8 μm technology node. Consequently, we are also using a variation of it where we try to make more accurate the energy and delay contribution of the sense amplifiers and the address decoder using a certain percentage of the total sub-bank energy and delay (instead of a constant as in the original CACTI model). Therefore, the sense amplifiers contribute 20% in memory and the address decoder contributes 30% in memory energy consumption and 20% in delay. This way the results are more accurate since the delay and energy of these components depend on the memory size.

3.4 Global Source Code Transformations Phase

When the timing information has been produced in the profiling phase, global source code transformations taking into account the DDTs interactions are viable. This will be illustrated based on an application demonstrator. The details of the optimisation approach itself are given in [9]. It allows to analyse the behaviour of the DDTs of the application, as Figure 2 shows. In it, the small vertical lines in the lower part represent small temporary buffers used to evaluate the DDT `ImageMatches (IMatches)` and generate the DDT `CandidateMatches (CMatches)` in a first step. Later, it is used to build the fast DDT `CMCopyStatic (CMCStatic)`. In a final step, it allows the creation of `BestMatches (BMatches)` and `MultiMatches (MMatches)`.

In fact, most of the multimedia applications are written in this previously explained *data production and consumption* fashion. The algorithm can be subdivided in smaller components (often functions or method calls) with internal and specifically designed DDTs, which receive an input buffer, do some processing and produce the output.

As such, the dynamic data structures passed between functionality components become critical bottlenecks (comparable to physical memory bandwidth congestions). When these intermediate DDTs are not used for any other purpose and there is an injective relation in the data-flow, they can be removed [9]. Figure 3 shows the results obtained after these transformations for our multimedia drivers.

3.5 Dynamic Data Type Refinement Using High-Level Profiling

In the final phase of the approach proposed, alternative complex DDT implementations from our library are evaluated to refine the original implementations of the DDTs. For each implementation in the library, the same high-level profile framework explained in

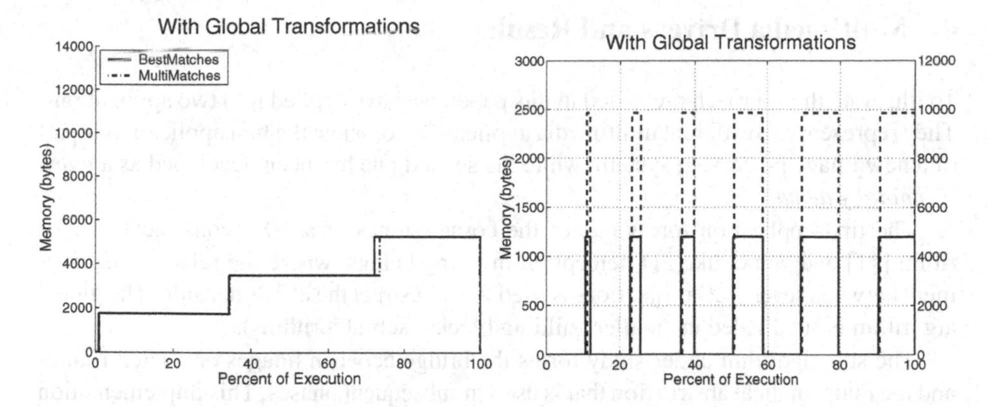

Fig. 3. Global source code transformations allowed the removal of 3 DDTs in the matching algorithm (left) and 2 DDTs in the game engine (right), saving memory footprint and power. The final code runs up to 10× faster.

Section 3.2 is employed. In this refinement, optimal solutions are determined by a combination of the objectives pursued (i.e. power, memory accesses and normalised memory usage). As a result, a number of *Pareto optimal* [3] points, which represent different DDT implementations, are obtained. Then, the developer decides according to his constraints and requirements. An example of the Pareto points with the power models used for one of our multimedia drivers is shown in Figure 4.

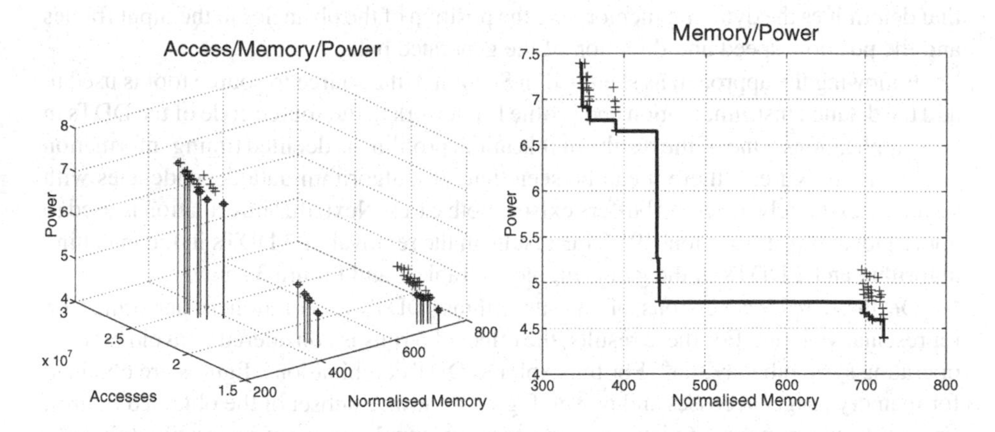

Fig. 4. The left figure shows the combination of pareto optimal solutions in the 3D game. The global pareto points are projected in the Memory/Accesses plain. They form a pareto curve. The left shows a projection of this 3D space in the Power/Memory plane. These figures are obtained with a Cacti based model for .18 μm technology.

[3] A point is called Pareto optimal, when it is impossible to improve one objective without worsening any other objective.

4 Multimedia Drivers and Results

To illustrate the approach presented in this paper, we have applied it to two applications. They represent two different multimedia application domains: the first application is part of a new image processing system, while the second one has been developed as a *game technology demo*.

The first application forms one of the corner-stones of a 3D reconstruction algorithm [11] and works like 3D perception in living beings, where the relative displacement between several 2D projections is used to reconstruct the 3rd dimension. The global algorithm is subdivided in smaller building blocks (sub-algorithms).

The sub-algorithm under study forms the bridge between images or related frames and the mathematical abstraction that is used in subsequent phases. This implementation matches corners [11] detected in 2 subsequent frames (images) and the operations on images are particularly memory intensive (a 640 $\times$ 480 image consumes over 1 MB). This algorithm uses internally several DDTs that, due to the partial image-dependency related data, do not fit in internal memory of current embedded processors. The size of these DDTs is fixed by a number of factors (e.g. structure and textures in the images) determined outside the algorithm and are uncertain at compile-time. In this phase, the accesses to the images are randomised and classic image access optimisations like row dominated accesses versus column wise accesses are not relevant.

The second application where the methodology has been applied is a 3D simulation game driven by a frame grabbing device. In a frame, obstacles are detected in the scene. In the free-space area, balls are rendered. These balls can move according to 3 degrees of freedom (up/down, left/right, front/back). When the ball reaches a wall, it either bounces off the obstacle or gets stuck to the obstacle. In this case study, the uncertainty that determines the dynamic memory are the position of the obstacles in the input frames and the position, speed and direction of the generated balls.

Following the approach as sketched in Section 3, the source to source tool is used to add the detailed instrumentation and profile framework to the source code of the DDTs in both applications. One of the results of this initial profiling is detailed timing information as shown in Figure 2, there it can be seen that intra-algorithm data dependencies with small but extremely accessed buffers exist in both cases. Next, this information is used to apply global transformations [9]. This results in the removal of 3 DDTs in the matching algorithm and 2 DDTs in the game engine, as Table 1 and Figure 3 show.

On these *refined* versions of the algorithms, DDTs exploration is performed on representative input. For these results, the effect of 5 runs is considered to avoid random operating system behaviour[4]. For the explored DDT combinations, figures are obtained for memory usage, accesses and power. Figure 4 shows a subset of the obtained figures. From a designers point of view, only the pareto optimal points are interesting (circled). For both applications, the results for the least power consumption are shown in Table 1.

Even though the figures of the memory model change, the relative values are similar and let us select the same "optimal" DDTs. For `BMatches` and for `MMatches`, the final DDT implementation consist of a 2-layered array structure, with an external dynamic array of 10 positions, then each position consist of another array of 146 basic positions

[4] The results of the profiling runs was very similar, only minor variations were observed

of 3 floats each. For the game engine also the DDTs are 2-layered array structures, the DDTs that contain the walls are now implemented as a dynamic array in the first level of 10 positions. Then, another one of 56 basic elements for the vertical walls and 26 for the horizontal ones. In both cases, the basic elements consist of 6 floats. Finally, the highly accessed balls are implemented as a first dynamic array of 10 positions where dynamic arrays of 179 basic elements of 1 float each are stored.

In the end, as Table 1 shows, an improvement of normalised memory footprint up to 99.97% and power consumption up to 99.99% compared to the original implementation of the corner matching algorithm. Similarly, 26.6% and 45.5% (or 79.9% depending on the technology used) for the 3D simulation game. In addition, there was a final speedup (when the DDTs were refined in the last phase of the approach proposed) of almost 2 orders of magnitude for the matching algorithm and 1 order of magnitude for the 3D simulation game.

Table 1. Refinement results of the DDTs for both driver applications. Between parenthesis the percentage saved in power consumption and memory footprint (fprint) are given. The initial DDTs removed in the final version are marked as RM.

DDTs	orig. mem. fprint (B)	orig. power .18μm (μJ)	orig. power .13μm (μJ)	final mem. fprint (B)	final power .18μm (μJ)	final power .13μm (μJ)
IMatches	5.14×10^2	0.30×10^3	0.18×10^3	RM	RM	RM
CMatches	2.75×10^5	3.03×10^3	3.03×10^3	RM	RM	RM
CMCStatic	1.08×10^5	3.92×10^5	4.48×10^4	RM	RM	RM
MMatches	3.62×10^2	0.03×10^2	0.02×10^1	3.81×10^3	0.02×10^2	0.02×10^1
BMatches	3.07×10^2	0.04×10^2	0.02×10^1	3.81×10^3	0.03×10^2	0.02×10^1
Total: matching	3.85×10^5	3.95×10^5	4.80×10^4	7.63×10^3 (99.97%)	0.05×10^2 (99.99%)	0.04×10^1 (99.99%)
VerWalls	1.72×10^3	2.96×10^3	2.94×10^3	8.30×10^2	2.01×10^3	1.52×10^2
VWallsBump	5.53×10^1	0.16×10^3	0.04×10^2	RM	RM	RM
HorWalls	1.64×10^3	0.15×10^3	0.28×10^3	6.99×10^2	1.77×10^3	1.28×10^2
HWallsBump	6.23×10^1	1.82×10^3	0.04×10^2	RM	RM	RM
Balls	8.42×10^3	9.33×10^3	1.91×10^3	7.20×10^3	4.18×10^3	7.54×10^2
Total: 3D game	1.19×10^4	1.46×10^4	5.14×10^3	8.73×10^3 (26.6%)	7.96×10^3 (45.5%)	1.03×10^3 (79.9%)

5 Conclusions

Power and energy estimations at an early phase of system implementation has become an increasingly important concern. Power and timing estimations at a very high-level, i.e. SW level, early in the system design process for the DDTs present in modern multimedia applications are not available yet. At the same time, accurate estimations from lower-level models, e.g. RTL-level or gate-level, suffer from unacceptable long computing times and capture information that is not (yet) relevant. Moreover, they come too late in the global design flow of the final system, which implies a very costly set of iterations through the design flow for any change in the first phases. In this paper, a fast and consistent system-level profiling approach that can be used to overcome the previous limitations is presented. It allows the designers to profile, analyse and refine the implementations and effects of the dynamic data types from their applications in a very early stage of the design flow taking into account power consumption, memory footprint and performance. Its effectiveness is illustrated using two different and complex multimedia examples. Finally, we have also explained how different power models can be used to obtain accurate results for a specific final platform if it is needed.

References

1. B. S. Amrutur et al. Speed and Power Scaling of SRAM's. *IEEE Trans. on Solid-State Circuits*, 35(2) (2000)
2. L. Benin et al. A power modeling and estimation framework for vliw-based embedded systems. In *Proc. of PATMOS* , Yverdon Les Bains, Switzerland (2001) 2.1.1–2.1.10
3. F. Catthoor et al. *Custom Memory Management Methodology – Exploration of Memory Organisation for Embedded Multimedia System Design.* Kluwer Academic Publishers, Boston, USA (1998)
4. R. Y. Chen et al. Speed and Power Scaling of SRAM's. *ACM Trans. on Design Automation of Electronic Systems*, 6(1) (2001)
5. E. G. Daylight et al. Incorporating energy efficient data structures into modular software implementations for internet-based embedded systems. In *Proc. of wrkshp on Software Performance* (2002)
6. M. A. Horowitz. Timing models for mos circuits. Technical report, Technical Report SEL83-003, Integrated Circuits Lab. Stanford Univ. (1983)
7. N. Jouppi. Western research laboratory, cacti, (2002) `http://research.compaq.com/wrl/people/jouppi/CACTI.html`.
8. M. Leeman et al. Methodology for refinement and optimisation of DM management for embedded systems in multimedia applications. In *Proc. of SiPS*, Seoul, Korea (2003)
9. M. Leeman et al. Intermediate variable elimination in a global context for a 3d multimedia application. In *Proc. of ICME* , Baltimore, MD (2003)
10. T. Mudge. Power: A first class architectural design constraint. *IEEE Computer*, 34(4):52–58, (2001)
11. M. Pollefeys et al. Metric 3D surface reconstruction from uncalibrated image sequences. In *Lecture Notes in Computer Science*, volume 1506, Proc. SMILE Wrkshp (post-ECCV'98), Springer-Verlag (1998) 139–153.
12. Y. Smaragdakis et al. Implementing layered designs with mixin layers. *Lecture Notes in Computer Science*, 1445:550 (1998)
13. B. Stroustrup. *The C++ Programming Language*. Addison-Wesley Publishing Company, Inc., Harlow, England (1997)
14. V. Tiwari et al. Power analysis of embedded software: A first step towards software power minimization. In *Proc. of ICCAD*, San Jose, California, USA (1994)
15. N. Vijaykrishnan et al. Evaluating integrated hardware-software optimizations using a unified energy estimation framework. *IEEE Transactions on Computers* (2003) 52(1):59–75
16. S. Wuytack et al. Global communication and memory optimizing transformations for low power systems. In *IEEE Intnl. wrkshp on Low Power Design*, Napa CA, (1994) 203–208.
17. C. Ykman et al. Dynamic Memory Management Methodology Applied to Embedded Telecom Network Systems. *IEEE Transactions on VLSI Systems* (2002)

An Adiabatic Charge Pump Based Charge Recycling Design Style

Vineela Manne and Akhilesh Tyagi

Department of Electrical and Computer Engineering,
Iowa State University, Ames, Iowa, U.S.A.
{vineelam,tyagi}@iastate.edu

Abstract. Adiabatic CMOS circuits reduce the energy consumption by supplying the charge at a rate significantly lower than the inherent RC delay of a gate. However, such a technique inherently trades energy for delay. In this work, we propose to incorporate an adiabatic charge pump to recycle charge for energy reduction. The adiabatic component, charge pump, is placed away from the critical paths. The adiabatic delays are overlapped with the computing path logic delays resulting in no change in the computation speed. One implementation scheme that taps the ground-bound charge in a capacitor (virtual ground) and then uses an adiabatic charge-pump to feed internal virtual *Vdd* nodes is described. The method has been implemented in DSP computations such as FIR filter, DCT/IDCT filters and FFT filters. Complete design details and performance analysis are presented. Simulation results in SPICE indicate that the proposed scheme reduces energy consumption in these DSP circuits by as much as 15% with no loss in performance.

1 Introduction

The rapid growth in the deployment of highly complex digital circuits in communication, signal processing and wireless applications has created a large market for low-cost, high throughput and low-power devices. Considerable research efforts over the last decade have been focused on building digital circuits that can operate at high clock frequency [1,2]. However, many applications require high performance, and yet need to consume low energy. Hence, the challenge is not just low power & energy design, but it is to design with lowest possible energy for a given performance.

Power optimization and low power design techniques can be implemented at various levels of abstraction: technology, circuit, architecture and algorithm [3]. In this paper, a new method of conserving power by tapping the charge bound for the ground terminal (ground-bound charge/current) which is recycled to virtual internal V_{dd} terminals is proposed. This reduces energy wastage resulting in lower energy consumption.

The main sources of power consumption in CMOS circuits can be classified as: static power dissipation occurring due to leakage current arising from substrate

J.J. Chico and E. Macii (Eds.): PATMOS 2003, LNCS 2799, pp. 299–308, 2003.
© Springer-Verlag Berlin Heidelberg 2003

injection and sub-threshold effects, and dynamic power dissipation arising due to short circuit dissipation and switching dissipation caused by charging and discharging of the parasitic load capacitances. The aim of this work is to tap the ground-bound charge which is lost by any combination of these mechanisms and recycle it by generating internal power supplies (as will be explained in the later sections). This scheme achieves overall energy savings of up to 15% with no performance degradation for a variety of DSP computations.

2 Block Level Description

The conceptual description of the proposed scheme is given in Fig.1. The main strategy is to collect the ground-bound charge from one module (a logic block) into a capacitor, which acts as virtual ground. This charge is used to pump up the voltage of other capacitors (virtual V_{dd}s) to a higher level using an adiabatic charge pump circuit. These virtual V_{dd}s are then used to power other suitable sub-modules (logic blocks).

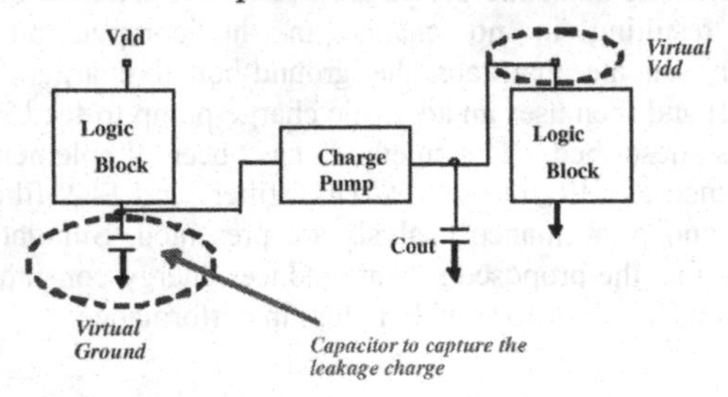

Fig. 1. Conceptual Description of the Proposed Architecture

This scheme is most applicable in systems with considerable switching activity that contain a large number of transistors. Although not a necessary condition, it suits pipelined logic blocks well. The earlier pipeline stages can be driven by clean V_{dd} terminals (and virtual ground terminals). The charge collected and recycled from the earlier stages can drive the later pipeline stages through virtual V_{dd} nodes. In other words, the charge recycling mechanism forms its own parallel pipeline dovetailed with the logic pipeline. This charge recycling pipelining is inserted explicitly as a design technique. Note that the charge recycling pipeline need not necessarily be synchronous variety in our schema. In fact, the scheme adopted by us deploys self-timed (asynchronous) control for charge recycling. The following discussion and the specific DSP filter implementations described later illustrate the concept further.

Such systems are first divided into conceptual logic blocks with specific functionality. Each of these logic blocks is then associated with either a virtual ground node or a virtual supply node. Depending on the transition activity in the blocks, the block with large number of transistors and large ground-bound charge is chosen as a 'source block'. The ground node of the 'source block' is connected to a capacitor that collects all the leakage charge from that particular logic block and serves as "virtual ground". The voltage on the "virtual ground capacitor" is continuously monitored to ensure that it still serves as logic '0' (a predetermined voltage should not be exceeded) and does not hinder the performance of the circuit. Once the capacitor reaches the desirable voltage threshold, it is disconnected from the logic block. Any further ground-bound charge from the 'source block' is either collected in another capacitor or dumped to ground. The charged capacitor is then connected to a charge pump circuit that amplifies the DC voltage to a higher level so as to be able to act as 'virtual supply'.

Having generated a higher voltage on the output of the charge pump, a suitable sub-block in the system is chosen to act as the "target block". The supply for this "target block" is provided by the charge pump output. The choice of this "target block" is dependent on certain design issues. Firstly, the delay between the operation of the "source block" and the "target block" should match the delay introduced by the charge pump circuit. In other words, it is necessary that the "target block" needs to function certain time duration after the "source block" so that the charge dumped by "source block" could be used to generate the required supply. This is also the main rationale behind the suitability of this scheme to systems that can be divided into sub-blocks that operate sequentially as indicated earlier.

Secondly, the energy available at the output of the charge pump should exceed the energy required by the "target block" to undergo a certain number of transitions. This places an indirect constraint on the complexity, or in other terms the 'transition activity' of the "target block". Since each logic transition is associated with charging and discharging of certain load capacitance, for almost all simple systems, the energy required by the block can be computed considering an average input case. Based on the energy requirement, an appropriate load block is then chosen as "target block".

As the complexity of the system under consideration increases, more than one block can be chosen as "source block" and more than one block as "target block". The virtual ground consists of a sea of capacitors that are dynamically connected to various "source blocks" as they dump charge. This not only ensures the availability of a charged capacitor for use by the charge pump at any instant of time, but also increases the efficiency of the circuit by making sure that almost all the dumped charge is collected for recycling. The charge pump can then be continuously operated by time multiplexing the various charged capacitors. In summary, the source charge can be either spatially distributed, or temporally distributed, or a combination of both. Note that this scheme also provides a limited degree of voltage scaling naturally. The logic blocks with a virtual V_{dd} or ground operate with a lower voltage swing.

3 Proposed Architecture

Charge pumps are widely used as DC-DC converting circuits that generate voltages higher than the available supply voltage. Most charge pumps are based on the Dickson charge pump circuit [4]. Although the traditional charge pump well serves the purpose of boosting the voltage, a slight modification to the traditional scheme has been adopted in this work to achieve low power operation.

3.1 Modification to Standard Charge Pump

As explained in Section 2, the conceptual idea of the proposed architecture is to collect the dumped charge from a logic block into a capacitor and then use a charge pump circuit to boost the virtual supply voltage to a higher level so as to act as virtual supply for a "target block". Although the additional area overhead due to charge pump is not an important issue, the energy consumed by the charge pump itself, during the process of voltage boosting is critical. It needs to be ensured that the additional energy spent in the charge pump operation does not constitute a significant portion of the total energy saved by the scheme, thereby rendering the scheme inapplicable. If we consider the operation of traditional charge pump in our application, with the "virtual ground capacitor" as the input source of energy, every time the charge is shared between the "virtual ground capacitor" and capacitor C_P of the charge pump, half of the energy is dissipated in the switches. In other words, if the charging of the capacitor C_P occurs abruptly, then the well known ½ CV^2 formula expresses the dissipated energy where V is the voltage difference between the input capacitor and C_P. This loss is inevitable regardless of the network design parameters. However, according to the "adiabatic charging principle", the dissipation for charging a given node capacitance to a particular voltage, can to a first approximation be asymptotically reduced to zero, if the charging time tends to infinity [6, 7]. In other words, since the dissipation during charge distribution is directly proportional to square of the voltage difference, if the voltage difference between the two nodes is reduced, the corresponding dissipation also reduces.

With this perspective, it is better to charge capacitor C_P in incremental steps of ΔV rather than in one step. This then requires the availability of multiple input capacitors to the charge pump, with input voltages V_1, V_2, V_3..... V_N, such that

$$V_1 = \Delta V; \quad V_2 = 2\Delta V; \quad V_3 = 3\Delta V; \quad ......V_N = N\Delta V \tag{1}$$

where $\Delta V = Vmax/N$.

The energy dissipated during the charging process is then given by

$$E_{Dissipated} = \frac{C_L V_{\max}^{2}}{N} \tag{2}$$

where N is the total number of stages of input. Thus increasing the number of input stages reduces the amount of dissipated energy.

However, as the number of stages is increased, not only does the complexity of the circuit increase, but the issues such as additional energy dissipation in control logic to monitor the input capacitor voltages, leakage effects from the multiple input capacitors, speed of operation, area overhead *etc* also become critical. Hence a trade-off must be reached between energy dissipation and circuit complexity. In this work, the number of input stages to the charge pump was therefore, restricted to three. The circuit diagram of the three-stage input model for a charge pump is given in Fig 2.

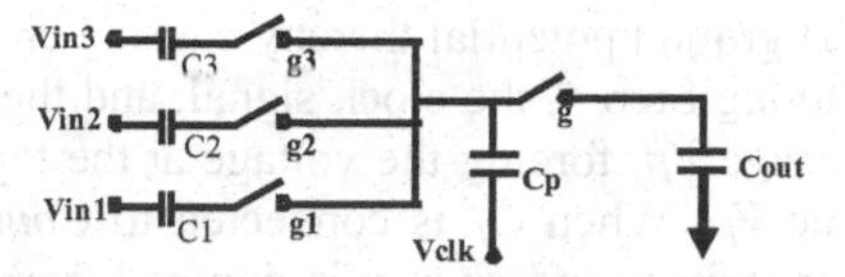

Fig. 2. Three-stage input model for a Charge Pump

3.2 Circuit Operation

The inputs to the charge pump are three different virtual ground capacitors with the capacitor voltages (*Vin1<Vin2<Vin3*) differing by a pre-determined constant value. These voltage values on the input capacitors are achieved by monitoring the charging of the capacitors and disconnecting them from the "source block" once the required voltage is achieved. This task of correctly observing the voltage and breaking the connection between the "virtual ground capacitor" and the "source block" and then connecting the "virtual ground capacitor" to the appropriate input of the charge pump depending on the voltage level is controlled by a voltage comparator.

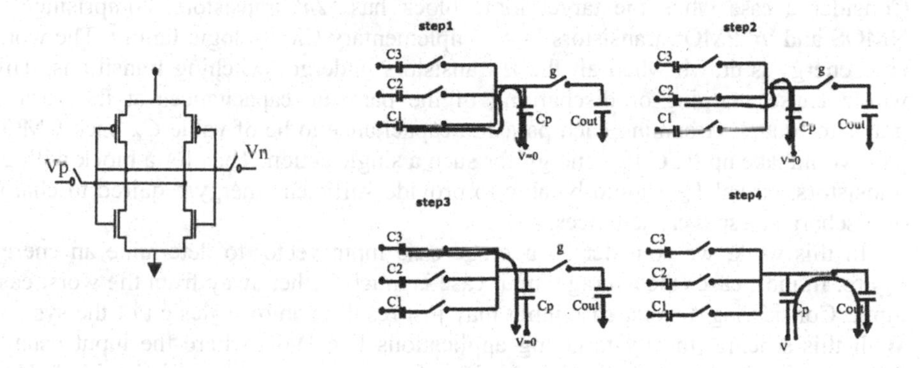

Fig. 3. Voltage Comparator

Fig. 4. Stepwise operation of the charge pump

The design of the voltage comparator (Fig 3) implemented in this scheme is similar to a SRAM cell with one input fixed at a reference voltage level (*Vn*) and the other input (*Vp*) connected to the "virtual ground capacitor". The voltage comparator is like a differential amplifier. Its output settles to a '0' or '1' based on whether *Vp<Vn* or *Vp>Vn*. This output is used to control the gates that connect the ground capacitor to charge pump.

The charge collected in the various input capacitors *C1*, *C2* and *C3* of the charge pump is then transferred in steps to the intermediate capacitor (C_P) as shown in Steps 1, 2 and 3 of Fig 4. The input capacitors are connected to the intermediate capacitor through transmission gates controlled by special step wave signals. Depending on the sizes of the input capacitor and C_P, at the end of each step, certain amount of charge is transferred to C_P. Once the capacitor C_P accumulates all the charge, it is disconnected from the inputs and connected to the output capacitor through '*g*'. During the entire duration of the first three steps (when the capacitor C_P is connected to various inputs), the lower plate of C_P is at ground potential thereby causing charge to flow from input into the charge pump. During Step 4, the clock signal, and thereby the lower plate of C_P is raised high to a voltage *Vp*, forcing the voltage at the top plate of the capacitor C_P to increase by a value *Vp*. When C_P is connected to *Cout*, charge sharing takes place between the two capacitors and energy is dumped from the charge pump into the output capacitor. This process is repeated cyclically until the energy on the output capacitor equals the energy required by the "target logic block" to which this capacitor is to be connected as virtual supply.

Similar to the design of input stages, the charging of the output capacitor is also performed adiabatically to reduce the energy dissipation while dumping the charge from capacitor C_P to *Cout*. This is done by increasing the voltage at the lower plate of capacitor C_P (i.e. *Vp*) in small incremental steps to allow adiabatic charge transfer. This would reduce the energy dissipation as square of ΔV while the same amount of charge settles finally on the output capacitor.

3.3 Determining Output Capacitor Value – A Design Issue

Consider a case when the target logic block has '*2n*' transistors, comprising '*n*' NMOS and '*n*' PMOS transistors in a complementary CMOS logic family. The worst case energy is drawn when all the *n* transistors undergo switching transitions. This would cause charging or discharging of the parasitic capacitances at the various transistor nodes. Assuming each parasitic capacitance to be of value C_x, each CMOS pair would take up $0.5C_xV_{dd}^2$ energy for such a single action. Thus for a block with *2n* transistors, virtual V_{dd} should be able to provide sufficient energy required to charge or discharge these *n* capacitances.

In this work, we consider an average case input vector to determine an energy figure. In most cases the average input case is much farther away from the worst case input. Considering a worst case input may just result in an over-design of the system. With this scheme (mostly targeting applications like DSP) where the input usually follows a standard pattern, the assumption of an average case input is justified. Thus based on the average input case, the output capacitor value is determined such that it can deliver this energy to the logic block without its logic level dropping from V_{dd} to a voltage below the acceptable "logic high".

3.4 Energy Analysis of Charge Pump Circuit

Since the principal focus of the work is to recycle the charge otherwise wasted in order to lower the overall energy consumption of the system, it is essential to maintain the overhead cost in terms of the energy required to drive the charge pump and the associated control circuitry as low as possible. The two sources of input energy are the energy stored in the input capacitor and that provided by the clock signals. The only source of output energy is the energy stored in the output capacitor of the charge pump that is eventually used as the virtual supply. Dissipation occurs in various segments of the charge pump circuit in terms of resistive loss in the switches and charging/discharging of the switch gate capacitance.

The input energy to charge pump does not add to implementation cost as it is obtained from the dumped charge, which would have otherwise been dumped to ground. However, additional energy that is provided by the various control and clock signals, that are essential for the circuit operation, has to be considered in the energy analysis.

Consider a target logic block that needs energy E_0 to undergo certain number of switching transitions while performing a specific functionality. Without the availability of any internal charge recovery scheme this entire energy is drawn from the external clean V_{dd} source. In this case the energy cost would be E_0. With the availability of a scheme such as the one described in this work, the overall cost reduces.

To estimate the reduction in cost, consider again the case of a system in which we have the provision to generate virtual V_{dd}. Let E_{in} denote the energy obtained from virtual ground terminals that act as input to a charge pump. This E_{in} is the result of collection of ground-bound charge after certain switching activity in a "source block" and hence can be considered as a source of "free energy". At any instant of time, the energy drawn from this input source is given by

$$\Delta E_{in} = E_{in}(0) - E_{in}(T) \tag{3}$$

where, $E_{in}(0)$ is the energy available at the input at time $t=0$ and $E_{in}(T)$ is the energy remaining on the input after a duration T. If T refers to the time period of one cycle of charge pump operation, then ΔE_{in} is the magnitude of energy drawn in one iteration of charge pump. Similarly, the energy delivered to output of the charge pump is one cycle is given by

$$\Delta E_{out} = E_{out}(0) - E_{out}(T) \tag{4}$$

where $E_{out}(0)$ and $E_{out}(T)$ refers to the charge present on the output capacitor at time $t=0$ and at the end of one iteration respectively.

Let E_{clk} denote the total energy provided by the clock source during one cycle of charge pump operation in transferring the charge from intermediate capacitor (C_P) to output capacitor. The energy that is dissipated in the charge pump in the form of resistive loss in transmission gates and charging/discharging of the various switches can all be combined as E_{diss}. Thus during every cycle of charge pump operation, energy is drawn from the input capacitor and clock source; part of which is dissipated in the operation of the circuit and the rest is distributed as "usable energy" that is available in the output capacitor and "unusable energy" that is stored in the intermediate capacitor that cannot be tapped directly for usage. Based on the Law of Conservation of Energy, the various energy figures can be arranged as

$$\Delta E_{in} + E_{clk} = \Delta E_{out} + \Delta E_P + E_{diss} \tag{5}$$

where $\Delta E_P = E_P(0) - E_P(T)$ is the resultant "unusable energy" that is stored in the capacitor C_P.

Equation 5 just indicates the relationship between the various energy-factors, and the magnitude of each term varies depending on the input and output logic block and the activity that the blocks undergo. To source a logic block that requires E_0 units of energy, ΔE_{out} needs to be equal to E_0. From (5) the amount of external energy that is required in the form of E_{clk}, to generate $\Delta E_{out} = E_0$, can be computed. The amount of saving with the above scheme, in terms of the total energy consumption is given by E_0-E_{clk}. The percentage of energy saved, which is also an indirect estimate of the efficiency of charge pump circuit is given by

$$\left[\frac{E_0 - E_{clk}}{E_0}\right] * 100 \tag{6}$$

4 Simulation Results and Performance Analysis

The proposed scheme is most suitable for large systems which can be divided into many sub-blocks which function sequentially. This scheme can be implemented even more efficiently if the application can tolerate some inaccuracy. This is because the voltage levels of virtual ground terminals are based on the charge flow through logic blocks on an average basis. If the actual ground-bound charge flow is significantly lower than the average, the following virtual V_{dd} terminals may not achieve desirable voltage levels designed to tolerate reasonable noise margins. Hence noise margins drop resulting in more inaccurate results potentially. In these cases, no backup and voltage monitors need be designed. DSP applications where the least significant bits do not carry much information can tolerate these inaccuracies in the least significant bits. They are also easily pipelined. Three DSP applications, namely FIR, FFT and DCT were chosen because they have architectures which satisfy both the conditions (of sequential computation blocks and inaccuracy tolerance). Also these systems are mostly found in areas where power reduction is usually a major issue, for example in cellular phones. All the three systems were implemented in SPICE (in Cadence design setup) in TSMC0.18μ technology (with supply voltage of 1.8V) and tested with the adiabatic charge pump circuits.

To estimate the energy savings of the proposed scheme, all the three architectures were implemented with and without the charge recovery scheme discussed in this paper, and the resulting total energy savings were computed. The circuits were first divided into smaller blocks like multipliers, adders, or combination stages of both. Certain blocks were chosen for virtual ground (source blocks) and certain for virtual V_{dd} (target blocks).. The source and target blocks were chosen such that they had significant computational delay. An adiabatically controlled charge pump was included in each of these systems. During the period of computation between the source and target blocks, the adiabatic charge pump recycles the charge and generates sufficient virtual V_{dd} for the target block. Table 1 shows a set of energy cost values, i.e., the energy drawn from the power supply pin, for different systems with and without the CPs (charge pumps). Two random sets of inputs (sets 1 and 2) were given to these systems and their energy consumption figures are given in Table 1(a). In the simulations considered, the additional circuitry comprising of the charge pump and control circuitry has less than 50 transistors which form less than 1% of the total transistor count of the circuits considered, as shown in Table 1(b). The energy savings can be increased further by spatial implementation of multiple charge pumps. Though this would result in an increase in transistor count, the additional area overhead would still be a small fraction of the entire device area.

Table. 1(a) Energy costs with and without the charge pump (CP) circuit in various circuit blocks with two random sample inputs (set1 and set2). **(b)** Area comparison as a measure of transistor count with and without the charge pump and control circuitry.

Circuit under considerati--on		*Energy cost w/o CP (pJ)*	*External energy cost with CP (pJ)*	*Total energy savings (%) (incl. control circuit dissipation)*
FIR	set1	2.656	2.06	11.12
	set2	4.617	3.72	12.23
DCT	set1	1.7036	1.232	8.52
	set2	3.77	3.033	10.88
FFT	set1	2.239	1.65	11.67
	set2	6.125	4.975	13.44
64 bit ALU	set1	0.962	0.62	1.56
	set2	1.438	1.009	7.1

(a)

Circuit under conside-ration	*# of transist-ors w/o CP*	*# of transis-tors with CP*	*% increase in area*
FIR	24224	24248	0.09
DCT	20032	20056	0.11
FFT	11552	11576	0.2
64 bit ALU	4360	4384	0.55

(b)

Fig 5 depicts the percentage savings achieved by using this scheme in different circuits. For most of the systems about 15% energy savings are observed. This figure can be further improved by refining the charge pump circuits to draw optimally minimum energy. In Fig 6 we notice that this scheme is most beneficial i.e. has higher savings when used to generate supply for large logic blocks or high activity. This is because, in small logic blocks with less signal activity, where the energy required is very low, the energy dissipation in the control circuitry which is a constant quantity would be more dominant and reduce the efficiency of the scheme. On the other hand, in large circuits with high signal activity, the energy spent in control circuit forms a negligible fraction of the entire energy that is saved.

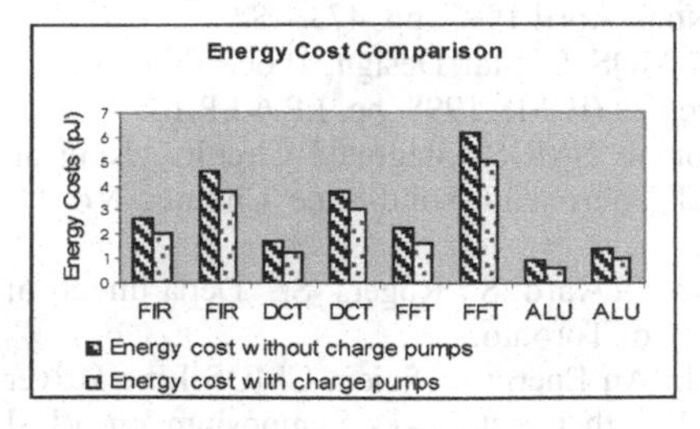

Fig. 5. Energy Cost Comparison

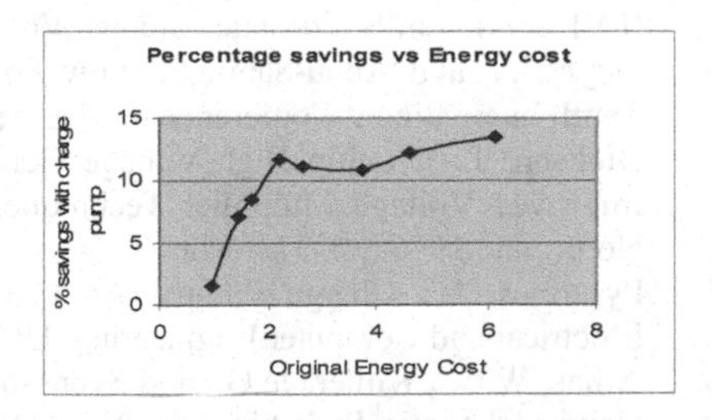

Fig. 6. Energy Savings

5 Conclusions

The typical CMOS logic is designed to draw energy from the power supply at V_{dd} and to dump it into the ground terminal. The adiabatic CMOS design style reduces the energy converted into heat in the transistors of the logic by maintaining as low as possible drain to source voltages. The recovery phase recovers the energy from the logic into an oscillating power supply. The main shortcoming of adiabatic CMOS is the need to slow down the logic to accommodate the adiabatic charge transfer. Additionally, a way of integrating high-frequency RLC power supplies into a reasonable sized silicon area has not been found. The proposed charge recovery mechanism cuts the V_{dd} to ground path by introducing virtual ground nodes to collect the dumped charge which is then recycled into virtual V_{dd} nodes to supply other logic blocks (that are activated later in time, and hence can tolerate the intervening charge pump latency). The charge pumps incorporate adiabatic charge transfer in order to save energy (a charge pump operating normally would result in net energy loss). The proposed charge recycling places adiabatic blocks in the paths that can tolerate its latency, hence this scheme does not result in any performance loss. The pipelined datapath units and algorithms are ideal candidates for the proposed scheme. This schema also incorporates a limited degree of voltage scaling naturally since all the virtual ground and V_{dd} logic blocks operate with a reduced voltage swing. We incorporated this design methodology in several DSP filters such as FIR and DCT/IDCT. The resulting energy savings are of the order of 15%. Future work includes other spatially and temporally multiplexed charge pumps to increase the deployability of the scheme. We are also considering CAD algorithms to identify the logic blocks that can benefit from the proposed scheme both at the logic synthesis and at netlist levels.

References

1. Arslan, T., Horrocks, D.H. and Erdogan, A.T.: Overview and Design Directions for Low-Power Circuits and Architectures for Digital Signal Processing. IEE Colloquium on Low Power Analogue and Digital VLSI: ASICS, June 1995. pp 6/1–6/5.
2. Chandrakasan, A.P., Sheng, S. and Brodersen, R.W.: Low-Power CMOS Digital Design. IEEE Journal of Solid-State Circuits. Vol.27, No.4, April 1992. pp. 473–484.
3. Guyot, A. and Abou-Samra,S.: Low Power CMOS Digital Design. Proceedings of the Tenth International Conference on Microelectronics (ICM), 1998. pp. I.P.6-I.P.13.
4. Dickson, J.: On-chip High-Voltage Generation in NMOS Integrated Circuits Using an Improved Voltage Multiplier Technique. IEEE Journal of Solid-State Circuits, Vol.11, No.6, June 1976. pp. 374–378.
5. Pylarinos, L.: Charge Pumps: An Overview. Edward S. Rogers Sr. Department of Electrical and Computer Engineering, University of Toronto.
6. Athas, W.C. , Koller, Je.G., and Svensson, L.J.: An Energy-Efficient CMOS Line Driver Using Adiabatic Switching. In Proc. Of the Fourth Great Lakes Symposium on VLSI Design, March 1994. pp. 159–164.
7. Svensson, L.J. and Koller, Je.G.: Driving a Capacitive Load Without Dissipating fCV^2. IEEE Symposium on Low Power Electronics, 1994. pp. 100–101.

Reduction of the Energy Consumption in Adiabatic Gates by Optimal Transistor Sizing

Jürgen Fischer[1], Ettore Amirante[1], Francesco Randazzo[2], Giuseppe Iannaccone[2], and Doris Schmitt-Landsiedel[1]

[1] Institute for Technical Electronics, Technical University Munich
Theresienstrasse 90, D-80290 Munich, Germany
juergen.fischer@ei.tum.de
http://www.lte.ei.tum.de

[2] Dipartimento di Ingegneria dell'Informazione,
Università degli Studi di Pisa
Via Diotisalvi 2, I-56122 Pisa, Italy

Abstract. Positive Feedback Adiabatic Logic (PFAL) with minimal dimensioned transistors can save energy compared to static CMOS up to an operating frequency $f = 200$MHz. In this work the impact of transistor sizing is discussed, and design rules are analytically derived and confirmed by simulations. The increase of the p-channel transistor width can significantly reduce the resistance of the charging path decreasing the energy dissipation of the PFAL inverter by a factor of 2. In more complex gates a further design rule for the sizing of the n-channel transistors is proposed. Simulations of a PFAL 1-bit full adder show that the energy consumption can be reduced by additional 10% and energy savings can be achieved beyond f = 1GHz in a 0.13μm CMOS technology. The results are validated through the use of the design centering tool 'WiCkeD' [1].

1 Introduction

In modern technologies with high leakage currents and millions of transistors per chip energy consumption has become a major concern. Static CMOS circuits have a fundamental limit of $E_{charge} = \frac{1}{2}CV_{DD}^2$ for charging a load capacitance C. The typical way to minimize the energy consumption is to lower the power supply V_{DD}. The adiabatic circuits can break this fundamental limit by avoiding voltage steps during the charging and by recovering part of the energy to an oscillating power supply. A comparison among different adiabatic logic families can be found in [2]. Logic families based on cross-coupled transistors like Efficient Charge Recovery Logic (ECRL) [3], 2N-2N2P [4] and Positive Feedback Adiabatic Logic (PFAL) [5] show lower energy consumption than other adiabatic families. The largest energy saving can be achieved with PFAL because the resistance of the charging path is minimized. A standard-cell library for this family was proposed in [6].

In this paper, PFAL is used to determine the dependence of the energy consumption on transistor sizing. After a short description of PFAL, the different

J.J. Chico and E. Macii (Eds.): PATMOS 2003, LNCS 2799, pp. 309–318, 2003.
© Springer-Verlag Berlin Heidelberg 2003

sources of dissipation in adiabatic logic gates are presented. Considering a simple PFAL inverter, the influence of the transistor dimensions on the energy consumption is derived and a basic design rule for transistor sizing is proposed, which allows to halve the energy consumption compared to previously used minimum sized devices. For more complex gates, this simple rule is validated through the simulation of a 1-bit full adder. To achieve higher operating frequencies, second order effects have to be considered. A further design rule is presented, which allows the 1-bit full adder to save an additional 10% of energy at f = 500MHz. The proposed design rules are validated through the design centering tool 'WiCkeD' [1]. With optimal transistor dimensions operating frequencies higher than f = 1GHz can be achieved.

2 Different Sources of Energy Dissipation

Figure 1a shows the general schematic of a PFAL gate. Two cross coupled inverters build the inner latch and two function blocks F and /F between the supply clock V_{PWR} and the output nodes realize the dual rail encoded logic function. Both logic blocks consist only of n-channel MOSFETs and are driven by the input signals In and /In. Figure 1b shows the timing of a PFAL gate. The input signals are in the hold phase while the gate evaluates the new logic value. During the adiabatic charging of the output load capacitance C the resistance R of the charging path determines the energy dissipation, according to the well-known formula [7] of the adiabatic loss:

$$E_{adiab} = \frac{RC}{T_{charge}} C\hat{V}^2 , \tag{1}$$

where $\hat{V}$ is the magnitude of the supply voltage and T_{charge} the charging time.

In a real circuit, leakage currents flow even when the transistors are cut off, providing a second source of energy dissipation:

$$E_{leak} = \hat{V} \cdot I_{off} \cdot T_{hold} \tag{2}$$

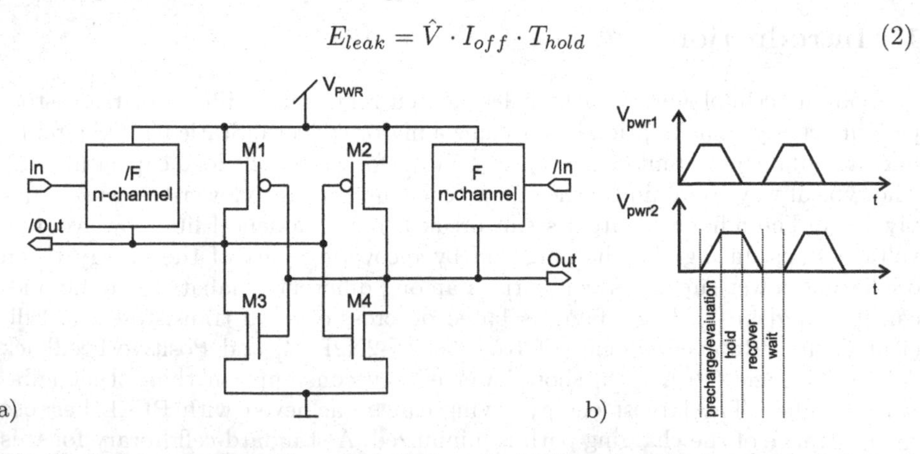

Fig. 1. a) General schematic and b) timing of a PFAL gate. The input signals In and /In are driven by the power supply V_{pwr1} which is a quarter period in advance with respect to V_{pwr2}.

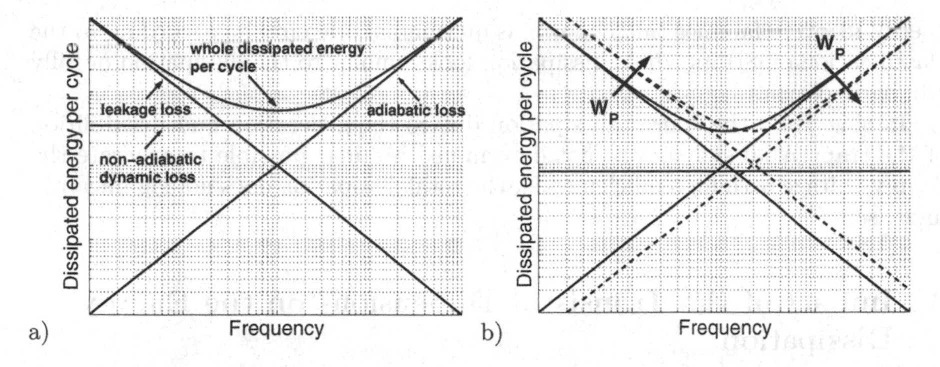

Fig. 2. a) Typical energy dissipation of quasi-adiabatic circuits in former technologies. Beside the leakage loss at low frequencies and the adiabatic loss at high frequencies, the non-adiabatic dynamic losses also affect the energy consumption. b) In modern technologies, the reduced threshold voltage lowers the non-adiabatic losses and therefore only the leakage and the adiabatic losses have to be considered. Both mainly depend on the p-channel transistor width (see figure 1a). The arrows show the expected dependence for increasing width.

where I_{off} is the average off-current and T_{hold} the hold time. In PFAL the hold time $T_{hold} = T_{charge}$. The off-current in static CMOS is comparable to the one in adiabatic logic. However, adiabatic gates dissipate less energy, as the off-current only flows during the hold time, which is typically a quarter period (see figure 7).

Additional energy dissipation is caused by several coupling effects and by partial recovery of energy. During the recovery phase the p-channel transistor cuts off when the power supply falls below the threshold voltage V_{th} and further energy recovery is inhibited. Both effects are referred as non-adiabatic dynamic losses and do not depend on the operating frequency. As the coupling effects are minimized due to the topology of PFAL, the non-adiabatic dissipation mainly depends on the threshold voltage of the p-channel transistor V_{th}:

$$E_{Vth} = \frac{1}{2} C V_{th}^2 \qquad (3)$$

Figure 2a shows the different contributions to the energy dissipation versus frequency. At high frequencies, the adiabatic loss is the dominant effect, whereas the leakage current determines the dissipation at low frequencies. The non-adiabatic dynamic losses which are typical for partial energy recovery circuits can be observed at medium frequencies. In former technologies with larger threshold voltages, the minimum energy dissipation is determined by these losses, and a plateau over a certain frequency range can be noticed. On the contrary, in modern technologies like the 0.13μm CMOS technology used in this paper the intercept point of the curves for leakage and adiabatic losses is above the plateau caused by the non-adiabatic dynamic loss (see Figure 2b, solid lines). Therefore in modern technologies with reduced threshold voltage V_{th} the non-adiabatic losses can be neglected, and the curve for the whole energy dissipation shows a minimum for a particular frequency. Although PFAL belongs to the partial energy recovery families (also known as quasi-adiabatic circuits according to the classification of [8]) its whole dissipation looks similar to the dissipation of fully adiabatic circuits.

In this work, the energy dissipation due to the generation and distribution of the trapezoidal signals was not determined. As it is possible to generate the supply voltage with high efficiency [9], the total dissipation does not significantly increase.

3 Impact of the Transistor Dimensions on the Energy Dissipation

If a high operating frequency is given and load capacitances and supply voltages cannot be changed, a minimization of the energy dissipation can only be achieved by decreasing the resistance R of the charging path. Because in PFAL this resistance mainly consists of the p-channel transistor on-resistance, wider p-channel MOSFET are expected to decrease the energy consumption as shown in figure 2b (dashed line). On the other hand, wider transistors give rise to larger leakage currents. As a result, the whole energy-versus-frequency characteristic is shifted towards higher frequencies without affecting the magnitude of the minimal energy consumption. In logic families using cross-coupled transistors, the resistance of the charging path can be optimized without significantly increasing the load capacitance driven by the former gate. The resistance of the p-channel devices (see figure 3a) mainly determines the charging path resistance in both the evaluation and the recovery phase. In the evaluation phase, the p-channel transistor and the logic block form a kind of transmission gate. Thus, the charging resistance is decreased compared to other families, which only use the p-channel MOSFET for charging the output node such as ECRL. During the recovery phase, the input signals turn off the n-channel MOSFETs MF1 and MF2 in the logic function block. The energy can be recovered through the p-channel device as long as the supply voltage does not fall below the threshold voltage V_{th}. Hence, by properly sizing the p-channel transistors a decrease of the energy dissipation is achieved without increasing the load capacitance seen

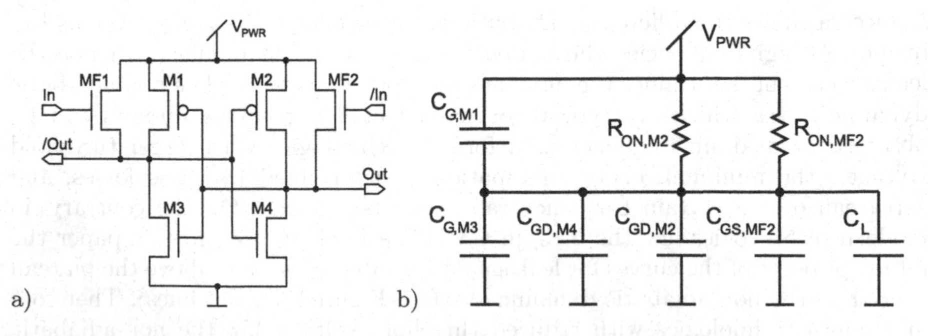

Fig. 3. a) Schematic of a PFAL inverter. b) Equivalent model of a PFAL inverter if the output node Out is charged and discharged.

by the former stage. Through thc sizing of the n-channel transistors M3 and M4 no reduction of the energy consumption can be obtained. Therefore, these transistors are kept minimal in order to avoid larger leakage currents.

In the following, the optimal dimension for the p-channel devices is determined. In adiabatic circuits, conducting transistors work in their linear region because voltage steps across the channel of conducting transistors are minimized. For an estimation of the energy consumption during the charging and discharging of the output node Out, the equivalent model shown in figure 3b is taken into account. Beside the external load capacitance C_L the circuit has to drive its intrinsic load, which consists of gate capacitances C_G, gate-source capacitances C_{GS} and gate-drain capacitances C_{GD}. These capacitances are directly proportional to the width of the corresponding transistors. For simplicity the junction capacitances are neglected.

The on-resistance of the p-channel transistor M2 takes over the main part of the charging and the recovery, so that in this approach the n-channel device MF2 is not considered. In the linear region, the on-resistance is equal to:

$$R_{on}(t) = \frac{1}{\mu_P C_{OX} \frac{W}{L} \cdot [V_{GS}(t) - V_{th}]} = \frac{1}{\mu_P C_{OX} \frac{W}{L} \cdot V_{GSt,avg}} \quad (4)$$

where μ_P is the hole mobility, C_{OX} the oxide capacitance per unit area, L the channel length and $V_{GSt,avg}$ the average gate-overdrive voltage. The average gate-overdrive voltage is assumed to be independent of the transistor widths. Using equation 1 the energy consumption of a gate with the p-channel transistor width W_P amounts to:

$$E_{adiab} = \frac{L \cdot (C_{G,M1} + C_{G,M3} + C_{GD,M2} + C_{GD,M4} + C_{GS,MF2} + C_L)^2}{W_P \, \mu_P \, C_{OX} \, V_{GSt,avg}} \cdot \frac{\hat{V}^2}{T} \quad (5)$$

To summarize the capacitances an effective width W_{eff} is introduced. Because in the linear region $C_{GS} \approx C_{GD} \approx \frac{1}{2} C_G$ the channel widths must be weighted:

$$W_{eff} = \frac{3}{2} W_P + \sum_i \sigma_i W_{N,i} \quad , \text{where} \quad \sigma_i = \begin{cases} \frac{1}{2} & \text{for } C_{GS} \text{ and } C_{GD} \\ 1 & \text{for } C_G \end{cases} \quad (6)$$

and W_N is the width of a n-channel transistor. With this effective width equation 5 can be rewritten as:

$$E_{adiab} = \frac{L}{W_P \, \mu_P \, C_{OX} \, V_{GSt,avg}} \cdot (W_{eff} C'_G + C_L)^2 \cdot \frac{\hat{V}^2}{T} \quad (7)$$

where C'_G represents the gate capacitance per unit width.

The minimal energy dissipation E_{adiab} is found for the following p-channel transistor width W_{opt}:

$$W_{opt} = \frac{2}{3} \sum_i \sigma_i W_{N,i} + \frac{2}{3} \cdot \frac{C_L}{C'_G} \quad (8)$$

This procedure can also be performed with other adiabatic families. For 2N-2N2P a similar result was found, while the optimal width for ECRL is different due to the gate topology.

In the following simulations, a load capacitance $C_L = 10\text{fF}$ is used. For the $0.13\mu\text{m}$ CMOS technology according to the above formula the optimal p-channel transistor width W_{opt} is approximately equal to $4.4\mu\text{m}$ using minimal dimensioned n-channel transistors.

4 Simulation Results

The simulations were performed with SPICE parameters of a $0.13\mu\text{m}$ CMOS technology using the BSIM 3V3.2 model. For the nominal load capacitance the typical value in static CMOS $C_L = 10\text{fF}$ was chosen. The oscillating supply voltage has a magnitude of $\hat{V} = 1.5\text{V}$.

4.1 Results for the PFAL Inverter

A series of simulations was performed for the PFAL inverter varying the p-channel transistor width. Figure 4 shows a 3D-plot of the energy consumption per cycle as a function of the operating frequency and the p-channel transistor width W_P. The dark gray layer represents the theoretical minimum for static CMOS, $E = \frac{1}{2}CV_{DD}^2$, where the capacitance $C = 10\text{fF}$ and leakage currents are not taken into account. A minimum of 0.86fJ for the energy consumption of PFAL can be found at $f = 2\text{MHz}$. For the region between $f = 200\text{kHz}$ and $f = 20\text{MHz}$ the energy dissipated per cycle is less than 2fJ. With minimal dimensioned n-channel transistors and with a p-channel width of $2\mu\text{m}$ the energy dissipation is decreased by a factor of 13.1 compared to the theoretical

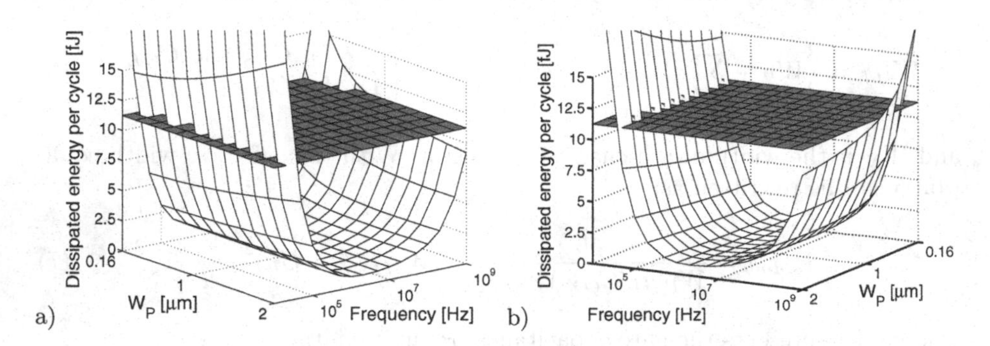

Fig. 4. Energy dissipation versus operating frequency for a PFAL inverter with a load capacitance of 10fF in a $0.13\mu\text{m}$ CMOS technology. Enlarging the p-channel transistor width the energy dissipation is reduced at high frequencies and is increased at low frequencies. The dark gray layer represents the theoretical minimum for static CMOS (11.25fJ) without the effect of leakage currents.

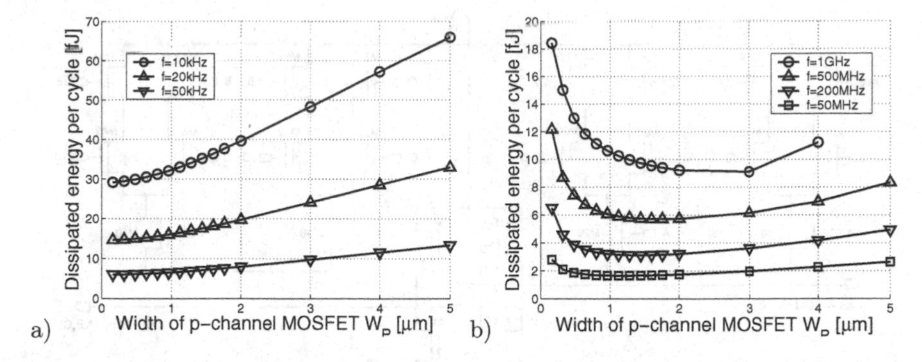

Fig. 5. Energy dissipation versus p-channel MOSFET width for a PFAL inverter in a 0.13μm CMOS technology for different operating frequencies. a) At low frequencies wider p-channel MOSFETs give rise to larger leakage losses. b) At high frequencies a minimum for the energy dissipation between $W_P = 1\mu$m and $W_P = 3\mu$m can be observed. With a p-channel transistor width $W_P = 2\mu$m the energy consumption is reduced by a factor of two.

minimum of static CMOS, which is equal to $E = 11.25$fJ for the used parameters. As expected, at low frequencies the energy consumption increases with increasing p-channel transistor width whereas at high frequencies it decreases. At $f = 1$GHz the dissipation is equal to 18.4fJ with minimal transistors, while with $W_P = 2\mu$m it is reduced to 9.2fJ. Thus, adiabatic circuits consume less energy than the theoretical minimum for static CMOS even at operating frequencies beyond 1GHz.

To determine the minimal energy dissipation in dependence of the transistor width the simulation was extended up to $W_P = 5\mu$m (see figure 5). At low frequencies wider p-channel MOSFETs give rise to larger leakage losses (see figure 5a). At high frequencies a minimum for the energy dissipation between $W_P = 1\mu$m and $W_P = 3\mu$m can be observed as shown in figure 5b. At $f = 1$GHz the gate operates correctly until $W_P = 4\mu$m. With a p-channel transistor width $W_P = 2\mu$m the energy consumption can be reduced by a factor of two in the high frequency range ($f \geq 50$MHz) with respect to the minimal sized implementation. Hence, this is the maximum width used even for the adder presented in the next section.

4.2 Results for a PFAL 1-Bit Full Adder

To investigate the dependence of the energy dissipation on the transistor sizing for more complex gates, a PFAL 1-bit full adder was simulated. The circuit schematic of the sum is shown in figure 6. According to the general schematic (figure 1a) two logic function blocks are implemented using only n-channel transistors. Only two p-channel transistors are needed for the cross-coupled inverters. Above $f = 500$MHz an adder realized with minimal dimensioned transistors dissipates more energy than the equivalent static CMOS implementation (cascaded logic blocks, see [10]). In a first approach, the p-channel transistor width is

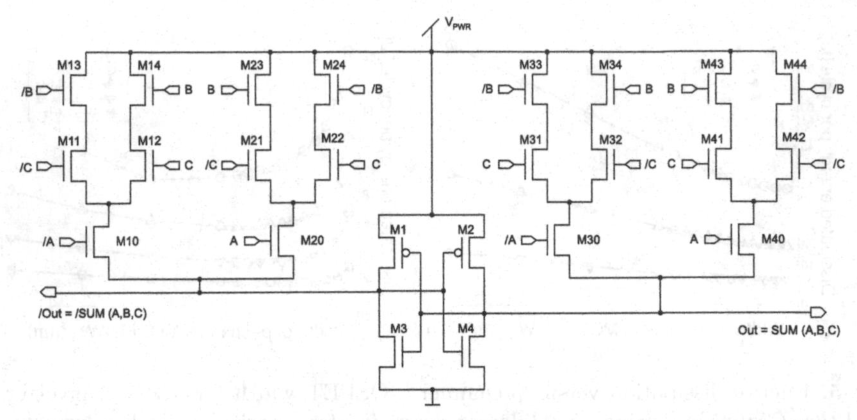

Fig. 6. Schematic of the sum generation circuit in a PFAL 1-bit full adder.

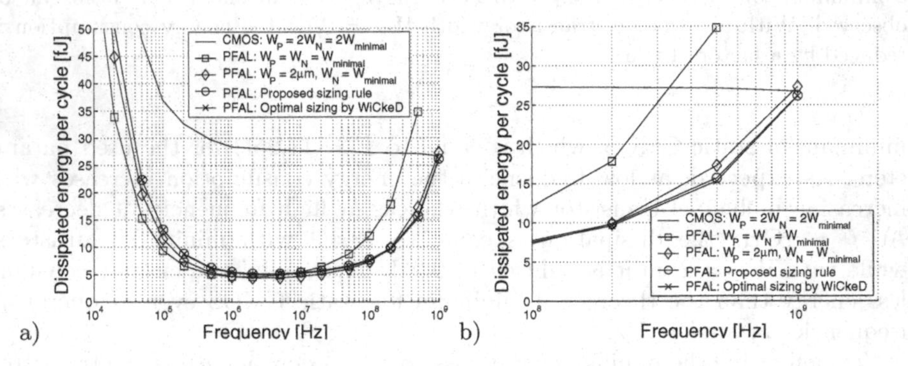

Fig. 7. PFAL 1-bit full adder: Dissipated energy per cycle versus operating frequency for different implementations. a) The whole frequency range is shown. b) Zoom in for f = 100MHz..1GHz.

increased up to $W_P = 2\mu$m while the n-channel transistors are minimal dimensioned. Compared with the realization using only minimal transistor width, the adder with $W_P = 2\mu$m dissipates half of the energy at $f = 500$MHz (Figure 7, diamonds). Through this additional energy savings the dissipation becomes lower than in the conventional static CMOS implementation. The maximal operating frequency in this case is $f \approx 1$GHz.

In a further simulation, the optimal transistor dimensions are determined by means of the design centering tool 'WiCkeD' [1] for the frequency $f = 200$MHz. As expected, the p-channel transistor widths are chosen maximal ($W_P = 2\mu$m). In contrast to the proposed approach, the n-channel devices of the function blocks are not kept minimal. Actually the transistors at the top of the blocks (Figure 6) show the largest average width (see table 1). The transistors in the middle of the function blocks are slightly larger than the ones at the bottom. With this sizing, the RC-delay of the whole function blocks is optimized and the energy dissipation is lowered (Figure 7, cross).

Table 1. Average width of the n-channel transistors in the PFAL gate shown in figure 6 with regard to the position in the function blocks: Optimal values obtained by the design centering tool 'WiCkeD' compared to the values according to the proposed sizing rule, which uses integer multiples of the minimal width.

Position of the n-channel devices in the function blocks	Sizing by WiCkeD	Proposed sizing rule
upper transistors	0.53μm	0.64μm $= 4W_{min}$
middle transistors	0.23μm	0.32μm $= 2W_{min}$
lower transistors	0.20μm	0.16μm $= 1W_{min}$

Regarding layout, it is better to use integer multiples of a standard width to save area and junction capacitances. Therefore, a further sizing rule for the n-channel devices in the function blocks is proposed. The width of the n-channel transistors connected to the output nodes (M10, M20, M30 and M40 in figure 6) are kept minimal ($W_{N,bottom} = W_{min} = 160$nm). The transistors in the middle of the function blocks (M11, M12, M21, M22, M31, M32, M41 and M42) have a width $W_{N,middle} = 2W_{min}$. The transistors at the top of the function blocks (M13, M14, M23, M24, M33, M34, M43 and M44) have $W_{N,top} = 4W_{min}$. For the two p-channel transistors $W_P = 2\mu$m was chosen. The simulation results (Figure 7, circle) show an additional decrease of the energy consumption compared to the proposed basic sizing rule which only considered the p-channel MOSFETs. At $f = 500$MHz the additional reduction of the energy dissipation amounts to 1.6fJ, that is approximately 10%. Compared to optimal sizing obtained by WiCkeD (Figure 7, cross) only a marginal difference is observable at $f = 500$MHz. With this extended sizing rule, energy can be saved with respect to the static CMOS implementation even at $f = 1$GHz.

5 Conclusions

By means of the Positive Feedback Adiabatic Logic (PFAL) it was demonstrated that proper transistor sizing enables adiabatic logic gates to save energy at frequencies beyond 1 GHz in a 0.13μm CMOS technology. Different sources of the energy dissipation in adiabatic circuits were analytically characterized. At high frequencies the energy consumption mainly depends on the resistance of the charging path. Enlarging the width of the p-channel transistors minimizes this resistance without affecting the input capacitances, which represent the load for the former stage. An estimation for the optimal width was presented, which can be used as a starting-point for optimization. Using p-channel transistors with a width $W_P = 2\mu$m, a reduction of the energy dissipation by a factor of 2 was achieved, which enables energy saving even at $f = 1$GHz in simple gates. In more complex gates like a 1-bit full adder, the sizing of the n-channel transistors has to be considered. By means of the design centering tool WiCkeD the optimal sizing for a PFAL 1-bit full adder was determined. In terms of layout a practicable sizing rule was proposed. With this rule the energy dissipation could be decreased by an additional 10% compared to the circuit with minimal dimensioned n-channel transistors. Compared to the static CMOS implementation the PFAL 1-bit full adder has a gain factor of 3.5 at $f = 100\text{MHz}$ and operates up to a frequency of $f = 1\text{GHz}$ with energy savings. This is the highest operating frequency reported for adiabatic logic up to now.

Acknowledgements. The authors would like to thank Prof. Antreich and Dr. Gräb for providing the design centering tool 'WiCkeD'. This work is supported by the German Research Foundation (DFG) under the grant SCHM 1478/1-2.

References

1. Antreich, K., Gräb, H., *et al.*: WiCkeD: Analog Circuit Synthesis Incorporating Mismatch. IEEE Custom Integrated Circuits Conference (CICC), Orlando, Florida, Mai 2000.
2. Amirante, E., Bargagli-Stoffi, A., Fischer, J., Iannaccone, G., Schmitt-Landsiedel, D.: Variations of the Power Dissipation in Adiabatic Logic Gates. Proceedings of the 11th International Workshop on Power And Timing Modeling, Optimization and Simulation, PATMOS'01, Yverdon-les-Bains, Switzerland, September 2001, pp. 9.1.1–9.1.10
3. Moon, Y., Jeong, D.: An Efficient Charge Recovery Logic Circuit. IEEE Journal of Solid-State Circuits, Vol. 31, No. 4, pp. 514–522, 1996
4. Kramer, A., Denker, J. S., Flower, B., Moroney, J.: 2nd order adiabatic computation with 2N-2P and 2N-2N2P logic circuits. Proceedings of the International Symposium on Low Power Design, pp. 191–196, 1995
5. Vetuli, A., Pascoli, S. D., Reyneri, L. M.: Positive feedback in adiabatic logic. Electronics Letters, Vol. 32, No. 20, pp. 1867–1869, 1996
6. Blotti, A., Castellucci, M., Saletti, R.: Designing Carry Look-Ahead Adders with an Adiabatic Logic Standard-Cell Library. Proceedings of the 12th International Workshop on Power And Timing Modeling, Optimization and Simulation, PATMOS'02, Sevilla, Spain, September 2002, pp. 118–127
7. Athas, W.C., Svensson, L., Koller J.G., *et al.*: Low-power digital systems based on adiabatic-switching principles. IEEE Transactions on VLSI System. Vol. 2, Dec. 1994, pp. 398–407
8. Lim, J., Kim, D., Chae, S.: nMOS Reversible Energy Recovery Logic for Ultra-Low-Energy Applications. IEEE Journal of Solid-State Circuits, Vol. 35, No. 6, pp. 865–875, 2000
9. Bargagli-Stoffi, A., Iannaccone, G., Di Pascoli, S., Amirante, E., Schmitt-Landsiedel, D.: Four-phase power clock generator for adiabatic logic circuits. Electronics Letters, 4th July 2002, Vol. 38, No. 14, pp. 689–690
10. Weste, N. H. E., Eshraghian, K.: Principles of CMOS VLSI design: a systems perspective. 2nd edition, Addison-Wesley Publishing Company, 1992

Low-Power Response Time Accelerator with Full Resolution for LCD Panel

Tae-Chan Kim[1], Meejoung Kim[1], Chulwoo Kim[1], Bong-Young Chung[2], and Soo-Won Kim[1]

[1] ASIC Design Lab., Dept. of Electronics Eng., Korea University, Anam-Dong, Sungbuk-Ku, Seoul 136-701, Korea.
{taechan, kmj, ckim, ksw}@asic.korea.ac.kr

[2] System LSI Business, Samsung Electronics Co. Ltd, San #24, Nongseo-Ri, Kiheung-Eup, Yongin-City, Kyunggi-Do, Korea.
{bychung}@samsung.com

Abstract. In recent years, power consumption has become one of the most critical design concerns in designing VLSI systems. The reduction of power consumption in displayers is inevitably required by the emergence of mobiles, fast systems, and et cetera. There are many reported problems associated with the response time of an LCD display, typically termed as afterimage specially when a motion picture video is being played on an LCD panel. To reduce afterimage artifacts, a new 'overdrive' method, named as Response Time Acceleration (RTA) is proposed in this paper. The proposed 'overdrive' method effectively controls truncation and quantization errors present in the conventional ones only with a few incremental gates. Furthermore the proposed 'overdrive' method provides two solutions with a shorten table memory while keeping full resolution and reducing power consumption. Test results show that the power consumption is reduced with the proposed method by 41% compared with conventional methods.

1 Introduction

In general, as the operating speed of displayers is increased, the related amount of power consumption is also increased. Accordingly, power consumption for high-performance and high-speed of displayer devices must be considered in design phase. Active Matrix LCD panels are being rapidly accepted for the PC monitors, TV sets, and mobile gadgets. Because every element of raw material for LCD contributes to the total accumulated response time, it is quite challenging to achieve a switching speed of one-frame-time refreshing rate, for all inter gray levels of video frames. One of approaches for minimizing the switching speed is to introduce new LCD materials for low viscosity and narrow cell spacing such as ferro-electric liquid crystal and precise cell gap control [1]. However, this approach is expensive.

The algorithmic approach is described in this paper. It is quite attractive in the sense that its implementation can be done in parallel with efforts of discovering new materials for LCD. The algorithmic approach is based on the method of

J.J. Chico and E. Macii (Eds.): PATMOS 2003, LNCS 2799, pp. 319–327, 2003.
© Springer-Verlag Berlin Heidelberg 2003

tuning active matrix drivers, so called an 'overdrive' method. It has already been demonstrated its effectiveness in reducing the response time [2] - [9]. The basic principle of the 'overdrive' method is to accelerate the current frame through comparisons with the previous video frames [2]. Fig. 1 illustrates an 'overdrive' method. Without overdriving, (c) is actually fed into a panel driver and cannot reach the desired value in the next frame. However, if an overdrive signal (a) is forced into a panel driver, (b) can be achieved due to overdrive of the accumulated impedance in a panel. A table memory is generally employed in a chip for the frame data comparison. On the other hand, conventional methods are such that either full 8-bits are utilized for full resolution of HDTV and UXGA or n-bits of data are utilized. In the latter case, the implementations are done by discarding LSB (8-n)-bits of data in order to reduce its table memory size as many as $2^{(8-n)} \times 2^{(8-n)}$. This results in noticeable truncation and quantization errors, especially in a large screen LCD panel. This paper presents an algorithm to reduce power consumption with n-bits and to minimize truncation and quantization errors while keeping full resolution with a few incremental gates, nullifying those artifacts for HDTV and UXGA resolution.

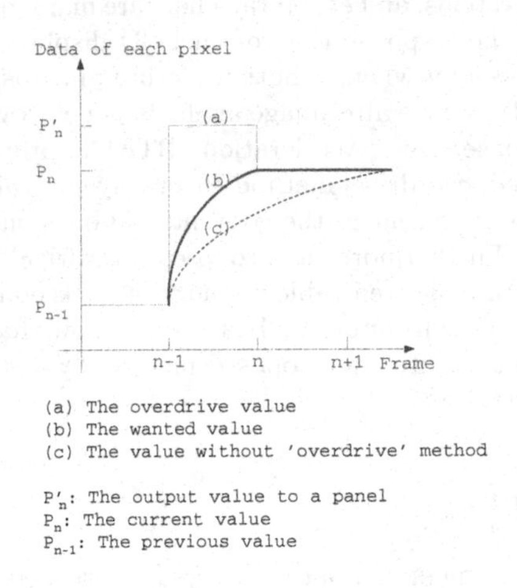

Fig. 1. An overdrive method.

This paper is organized as follows. Section 2 describes architecture of RTA algorithm proposed in this paper. Section 3 describes errors associated with conventional 'overdrive' methods. In Section 4, our new method is detailed. In Section 5, the experimental results are given.

2 Architecture of RTA

The conventional 'overdrive' method is to overshoot or undershoot the driver in order to drive the current frame data value after comparing the previous frame data to current frame data [2]. Fig. 2 illustrates the block diagram of RTA chip.

The RTA chip consists of two line memories, a table memory, and glue logic that controls SDRAM data transfer and provides data to a panel. The external SDRAM and the internal line memories can be accessed with 48 bits/clock. The current frame data can be stored in the external SDRAM, which serves as a buffer for storing a previous frame data, as soon as the frame data is compared with the previous frame data. The table memory is accessed with 36 bits/clock for each RGB channel.

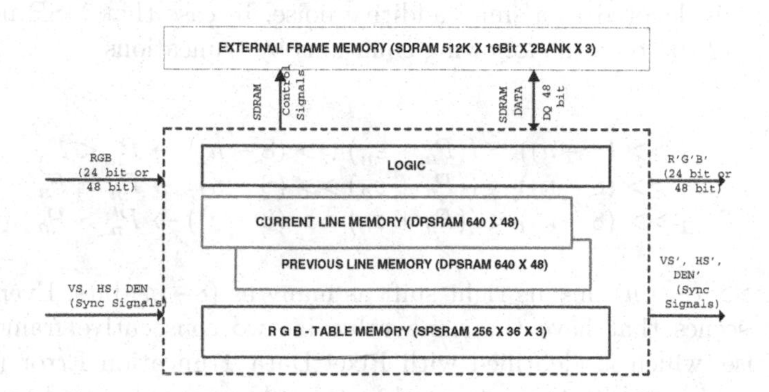

Fig. 2. The block diagram of RTA.

3 Errors of Conventional Methods

3.1 Truncation Error

Conventional methods usually store full 8-bits for full resolution or MSB of 4–6 bits of frame data into the memory in order to save the memory size. However, both methods have disadvantages: the power consumption can be increased if full 8-bits are stored and errors can be easily generated by such truncations of the pixel data if only part of MSB is stored. The algorithm of conventional methods can be represented as follows:

$$\begin{array}{l} P_{n-1} > P_n \rightarrow P'_n < P_n \\ P_{n-1} = P_n \rightarrow P'_n = P_n \\ P_{n-1} < P_n \rightarrow P'_n > P_n \end{array} \qquad (1)$$

where P_{n-1}, P_n, and P'_n are denoted as the pixel values of input video data of the previous frame, current frame, and the output value to a panel, respectively. Note that the value of a previous frame data equals to that of current frame data in case of ideal stationary scenes. Then, the data value of current frame will be driven out. But the addition of noises is inevitable in the real application. Major sources of such noises are identified as non ideal behaviors of Analog-to-Digital

Converter (ADC) or Printed Circuit Board (PCB) that are commercially in use. Therefore Eqs. (1) are modified as the following equations.

$$\begin{aligned} P_{n-1} > P_n + \varepsilon_n \rightarrow P'_n < P_n \\ P_{n-1} = P_n + \varepsilon_n \rightarrow P'_n = P_n \\ P_{n-1} < P_n + \varepsilon_n \rightarrow P'_n > P_n \end{aligned} \qquad (2)$$

where ε_n is denoted as a small additive noise. In case that MSB n-bits are stored, Eqs. (2) become the following equations by truncations.

$$\begin{aligned} (P_{n-1} >> (8-n)) > ((P_n + \varepsilon_n) >> (8-n)) \rightarrow P'_n < P_n \\ (P_{n-1} >> (8-n)) = ((P_n + \varepsilon_n) >> (8-n)) \rightarrow P'_n = P_n \\ (P_{n-1} >> (8-n)) < ((P_n + \varepsilon_n) >> (8-n)) \rightarrow P'_n > P_n \end{aligned} \qquad (3)$$

where '$>> (8-n)$' means right shift as many as $(8-n)$-bits. Even though stationary scenes that have the same value in two consecutive frames come, a small noise, which is identified with Pixel Data Truncation Error (PDTE), should be accelerated. In this paper, when the table size is reduced to $2^n \times 2^n$, the grid of table becomes $2^{(8-n)}$, and therefore the unit of overdriving value also becomes $2^{(8-n)}$. This means that when PDTE is generated, the unit of PDTE becomes $2^{(8-n)}$.

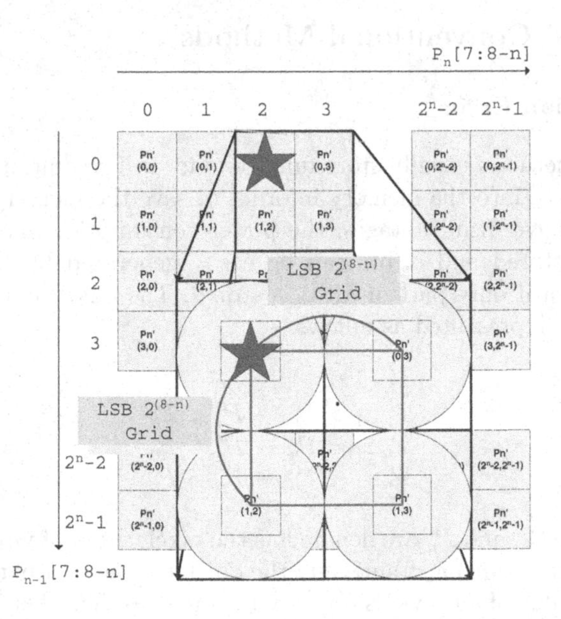

Fig. 3. Bilinear interpolation of the output value to a panel, P'_n, in $2^n \times 2^n$ table.

3.2 Quantization Error

In conventional 'overdrive' methods, the pixel values of input image of both previous and current frames are compared at each RGB channel. And the levels of overshoot and undershoot are classified. The output value to a panel is stored in the table memory and the table size can be determined by the number of bits of pixel data to be stored. Since an image consists of full 8-bits, the size of maximum table can reach to 256×256. Practically, however, the table size might be shortened to $2^n \times 2^n$ rather than 256×256 to reduce the chip cost and the power consumption. Therefore, the range of output value is determined through quantization from 256×256 to $2^n \times 2^n$ as depicted in Fig. 3. LSB $(8-n)$-bits are discarded when representing the pixel values of input image in both previous and current frames. That results in quantization errors which degrades the quality of the image.

4 The Proposed Method

The representation of pixel values of input image in both previous and current frames should use full 8-bits to minimize the effects of an added small noise ε_n and quantization errors even though this requires a larger memory and consumes large power. This paper shows how the previously described PDTE and quantization errors can be reduced with a small addition of table memory. PDTE can be generated by a small noise ε_n that becomes amplified, which affects the panel output. Therefore, Eqs. (2) should be modified as the following.

$$\begin{array}{l} |P_{n-1} - P_n| > Th, P_{n-1} > P_n \rightarrow P'_n < P_n \\ |P_{n-1} - P_n| \leq Th \rightarrow P'_n = P_n \\ |P_{n-1} - P_n| > Th, P_{n-1} < P_n \rightarrow P'_n > P_n \end{array} \quad (4)$$

If an amplified condition is limited by the threshold level Th, PDTE will not be generated when $Th \geq |\varepsilon_n|$. If the pixel values of input image in both previous and current frames use only MSB n-bits and a small noise ε_n becomes smaller than the threshold level Th, a small noise ε_n can not amplify PDTE because the threshold level Th is larger than $2^{(8-n)}$.

For example, if $n = 4$, $P_{n-1} = P_n$, $\varepsilon_n = -1$ and if the bit value can be denoted by the next expression such as $P_{n-1} = 8'b0011\_0000$ and $P_{n-1}[7:4] = 4'b0011, (P_n + \varepsilon_n)[7:4]$ will be $4'b0010$. Where the $8'b$ means the value of full 8-bits and the $[7:4]$ means the value of bits from the 7th to the 4th among full 8-bits. Then the output value to a panel will undershoot and go out to a panel. A table memory of output value to a panel consists of $2^n = 16$ grids and the more PDTE than 16 is generated. If Th is set to 1 (i.e. Th=1) and the relation $|P_{n-1} - (P_n + \varepsilon_n)| \leq Th$ is satisfied, PDTE will not be amplified.

The bilinear interpolation is used to reduce quantization errors. Fig. 3 shows that bilinear interpolation is made of LSB $(8-n)$-bits in four close table data. Note that this table memory can be fully mapped to 256×256. Therefore, if a table memory consists of $2^n \times 2^n$ and a bilinear interpolation is adopted, the quantization errors can be reduced.

Fig. 4 is the block diagram for full acceleration data path with no PDTE and no quantization errors. When the pixel data of both previous and current frames arrive, in our implementation, MSB n-bits go to the table memory and LSB (8-n)-bits go to Ratio Generator block. The coefficients in Fig. 4 are obtained by bilinear interpolation. Data 9-bits in the table memory pass through Table Decoder block. The output level to a panel P'_n and the real level to a panel pP_n are extracted from Table Decoder block. Then the output level to a panel goes through Bilinear Interpolator blocks for Panel Out while the real level to a panel goes to Bilinear Interpolator blocks to choose the data, P'_n or pP_n, in Bilinear Interpolator blocks for SDRAM Out.

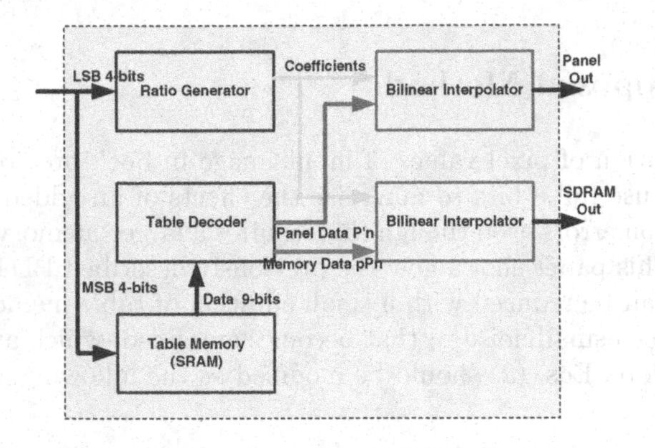

Fig. 4. Datapath in RTA.

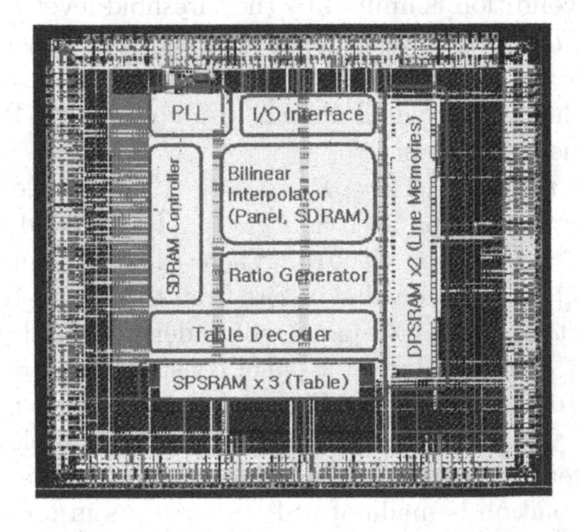

Fig. 5. The layout of RTA chip.

5 Experimental Results

The RTA chip is fabricated in a 0.18 μm 4-layer metal CMOS technology. The chip contains 300,000 gates and consumes 393mW at operating frequency of 135MHz and occupies about 20.2 mm^2. The chip layout is shown in Fig. 5.

Fig. 6. The result picture.

In case of the table size of n=4, PDTE is reduced by threshold level Th according to Eqs. (4). And the quantization errors are reduced by the bilinear interpolation. Fig. 6 demonstrates how effectively the proposed algorithm and its architecture improve the displaying image quality. The picture is moving from right to left with 0.5 m/sec velocity in an LCD panel. Fig. 6(a) shows a blurring boundary for slow response time, while Fig. 6(b) shows a distinct boundary with the proposed algorithm. We observe that the image quality and response time acceleration are dramatically improved. For instance, approximately 10 msec for response time can be achieved with our proposed method. Table 1 and table 2 summarize the measured data.

Table 1. The measured data.

	*Conventional method	Proposed method
PSNR by PDTE (dB)	40.7	45.9
Expression ratio by quantization (%)	7.4	100

* N-bits case in table memory.

6 Conclusions

In this paper, we identified PDTE, the quantization errors, and presented a novel algorithm to overcome such problems. The proposed algorithm shows that PDTE and the quantization errors can be completely reduced by applying the

Table 2. The power consumption of each block in RTA chip.

	*Conventional method	Proposed method
Glue logic (mW)	150	190
SRAM (mW)	390	80
Input/Output (mW)	120	120
PLL (mW)	3	3
Total (mW)	663	393

* Full 8-bits case in table memory.

threshold level and bilinear interpolation to the conventional method that only part of MSB is used. Moreover, with these methods, power consumption can dramatically be reduced while image quality becomes improved. Response time acceleration is improved and afterimage artifact is cleared. The proposed algorithm and architecture can be readily applicable to large sized LCD TVs bigger than 20 inches and PC monitors of 17 inches or bigger.

References

1. Masahito Oh-E and Katsumi Kondo "Respose Mechanism of Nematic Liquid Crystals Using the In-Plane Switching-Mode," *Applied Physics Letters*, vol. 69, no. 5, pp. 623–625, 1996.
2. Baek-Woon Lee, "Reducing Gray-Level Response to One Frame:Dynamic Capacitance Compensation," *SID 01 Digest*, 51.2, 2001.
3. H. Nakamura, J. Crain, and K. Sekiya "Optimized Active-Matrix Drives for Liquid Cystal Displays," *Journal of Applied Physics*, vol. 90, no. 90, pp. 2122–2127, 2001.
4. Hajime Nakamura "A Model of Image Display in the Optimized Overdrive Method for Motion Picture Quality Improvements in Liquid Crystal Devices," *Journal of Applied Physics*, vol. 40, no. 11, pp. 6435–6440, 2001.
5. K. Nakanishi, "Fast Response 15-in. XGA TFT-LCD with Feedforward Driving Technology for Multimedia Application," *SID 01 Digest*, 29.3, 2001.
6. K. Kawabe, "New TFT-LCD Method for Improved Moving Picture Quality," *SID 01 Digest*, 35.4, 2001.
7. Baek-Woon Lee, "Late-News Paper: LCDs: How Fast is Enough?" *SID 01 Digest*, 41.5, 2001.
8. M. Ohta, M. Oh-E, S. Matsuyama, N. Konishi, H. Kagawa, and K. Kondo, "Some LCD modes, for example, the In-Plane Switching modes, are known to have less varying response time between grays," *SID 99 Digest*, 86, 1999.
9. Hajime Nakamura, "Overdrive Method for Reducing Response Times of Liquid Crystal Display," *SID 01 Digest*, 51.1, 2001.
10. M. Cheriet, J. N. Said, and C. Y. Suen, "A Recursive Thresholding Technique for Image Segmentation," *IEEE Trans. Image Processing*, vol. 7, no. 6, pp. 918–921, 1998.
11. E. J. Delp and O. R. Mitchell, "Image Truncation Using Block Truncation Coding," *IEEE Trans. Comm.*, vol. COM-27, pp. 1335–1342, 1979.

12. R. Kohler, “A Segmentation System Based on Thresholding,” *Comput. Graphics Image Process*, vol. 15, pp. 319–338, 1981.
13. N. Otsu, “A Threshold Selection Method from Grey-Level Histograms,” *IEEE Trans. Syst. Man, Cybern*, vol. SMC-8, pp. 62–66, 1978.
14. P. K. Sahoo, S. Soltani, and A. K. C. Wong, “SURVEY: A Survey of Thresholding Techniques,” *Comput. Vis. Graph. Image Process.*, vol. 41, pp. 233–269, 1988.

Memory Compaction and Power Optimization for Wavelet-Based Coders

V. Ferentinos*, M. Milia, G. Lafruit, J. Bormans, and F. Catthoor**

IMEC-DESICS, Kapeldreef 75, B-3001 Leuven-Heverlee, Belgium
{ferentin, miliam, lafruit, bormans, catthoor}@imec.be

Abstract. A methodology for memory optimization in wavelet-based coders is presented. The dynamic memory requirements of the ASAP forward Wavelet Transform (WT) in three different output data grouping modes are studied: (a) independent output blocks with dyadically decreasing sizes; (b) zero-tree blocks and (c) independent equally-sized blocks. We propose an optimal approach of data clustering and calculation scheduling aiming at minimal memory requirements. This goal is reached using an appropriate subdivision of the filter inputs and it is verified with the assistance of an automatic design tool. The importance of the data dependencies between the different functional modules is shown to be dominant.

1 Introduction

In typical multimedia applications large amounts of data have to be stored, transferred and processed, largely dominating the chip area, processing speed and power consumption of system-on-a-chip solutions [1], [2]. Since the market tends to evolve to powerful applications running on very compact and portable systems, special design methodologies for overall system memory optimizations have to be provided. Such optimizations cannot be performed on the system's modules separately: the overall interaction between them has a large impact on the proposed solution. What implemented in this paper takes advantage of an in-house developed memory optimization methodology [1], [2] and accompanying tools [11]. As example we use a simple signal coder, composed of two interacting modules: a data transform (more specifically, the wavelet transform, which is the core of many current/upcoming image and video compression standards [3], [4], [5], [6], [7]) for signal energy compaction and its associated entropy coder for compression (reduction of redundancy). In recent related work [13], [14], [15] memory size and power requirements are optimized at the circuit level or at system level by introducing dedicated architectures. This paper provides a specific high-level optimization with the use of the above referred methodology and tools and imposes the exact implementation that fulfills their constraints.

---

* Also Ph.D. student with K.U.Leuven

** Also professor with E.E. Dept. of K.U.Leuven

J.J. Chico and E. Macii (Eds.): PATMOS 2003, LNCS 2799, pp. 328–337, 2003.
© Springer-Verlag Berlin Heidelberg 2003

In the field of wavelet-based signal compression, different data clustering modes for entropy coding exist. This paper shows that each mode imposes a unique data dependency pattern between the data transform and the entropy coder, leading to very diverse memory optimization results. In particular, we show that two basic features determine the memory requirements of the Wavelet Transform (WT): (i) The schedule (calculation sequence) of executing the transform among its levels and (ii) The grouping/order of the output data depending on the associated entropy coder. Memory complexity results are extracted for different cases of grouping the wavelet data for further entropy coding, according to (a) a so-called ASAP schedule, (b) the MPEG-4 and (c) the JPEG2000 entropy coding approach. A preliminary, theoretical study in [8] suggests large differences in memory requirements between these modes. However, in [8] several assumptions have been made and the memory requirements where based not on the complete execution of the WT but only on a descriptive approach of its data dependencies. In contrast, the current paper is devoted to the practical implementation of the minimal memory requirements (in size and power) for the Wavelet Transform, describing a systematic approach to achieve this and yielding accurate results.

The paper is organized as follows. Section 2 gives a brief explanation of the data dependencies and how the results are calculated along the different levels of the WT. Additionally the three different modes of output grouping are introduced in this section. Section 3 presents the memory optimization used and the memory complexity analysis. Section 4 shows the memory size and power results for the different grouping output modes, as defined above. Section 5 studies the possibility of future extensions of 1D results to 2D. Finally, conclusions are drawn in Section 6.

2 The Data Dependencies of the Wavelet Transform, the ASAP Scheduling, and the Output Modes

The Wavelet Transform (WT) is a repetitive filtering and subsampling procedure, yielding successive signal resolution levels. Figure 1 shows the data processing flow involved in the in-place WT calculations, in which Highpass and Lowpass samples are interleaved. According to the RPA memory efficient algorithm described in [9], for each two newly created Lowpass samples in level *i-1*, a new Lowpass and Highpass sample in level *i* are recursively calculated (through filtering operations). The output data of each level *i* (Lowpass, Highpass) are stored in different arrays (*Li*, *Hi*) respectively. For level0 the array *L0* contains the whole input (and *H0* does not exist). In essence, memory optimization is obtained by performing As Soon As Possible (ASAP) filtering operations "vertically" throughout the levels, instead of calculating the WT "horizontally" level-by-level [10]. This scheduling algorithm is used as a starting point to partition the data involved in the WT calculation and to take maximum advantage of the memory reuse. The data dependency arrows in Figure 1 show the ASAP front up to which the WT is partially calculated, region-by-region. In particular, the ASAP schedule will first perform the calculations along the first data dependency line 28(level0)-12(level1)-4(level2)-0(level3) and then continues by performing filtering operations along the front 36(level0)-16(level1)-6(level2)-1(level3). Between two successive execution front lines, the calculation order follows the light gray arrows starting from the lowest level up to the highest. To resume the calculations after the previous front line, the *2M-1* last Lowpass samples at each level must be stored in a temporary memory. This corresponds to the (gray) memory cells of Figure 1. Highpass samples are also temporarily stored into memory (not shown in the figure) to create data blocks for the entropy coder. This paper shows results of the minimal memory requirement grouping the output (entropy coder input) in three different modes:

- **Mode A**) dyadically decreasing block sizes over the different levels (independent blocks);
- **Mode B**) zero-tree parent-children trees (mode A) with "vertical" synchronization (as used in MPEG-4 VTC [3], [4]);
- **Mode C**) independent equally-sized blocks (as used in JPEG2000 [5], [6], [7]).

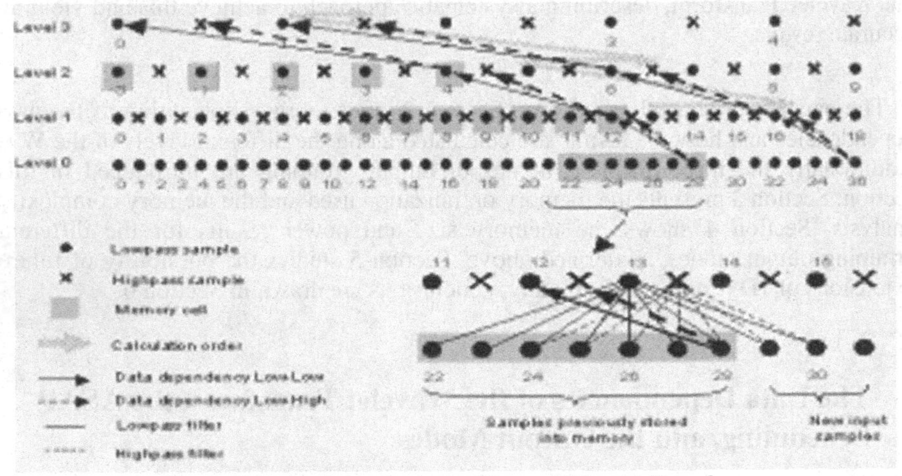

Fig. 1. Data dependencies in the Forward Wavelet Transform

For convenience, a 9/7-tap wavelet analysis filter pair in a 3-level WT is used as example throughout the whole paper. The analysis is mainly shown in 1D, but the results can be extrapolated to 2D for any number *L* of wavelet levels and any wavelet analysis filter pair, up to *2M+1* and *2M-1* taps in respectively the Lowpass and Highpass filter.

3 Memory Optimization

Starting from classical WT C-code, it is possible through successive transformations, supported by our in-house ATOMIUM [11] tool set, to reduce the memory requirements to the minimal amount suggested by Figure 1. During these code transformations each *Li* array is partitioned into *k* smaller arrays (*Lik*). This split is done according to whether parts of the data are reused in the next WT calculation step or not. When the WT between two successive ASAP front lines is performed, one or more instances of each *Lik* array are needed. The size of these *Lik* arrays depends only on the dimensions of the WT filters and the dimension of the block that is used to partition the data of each WT level. However, as it is going to be explored and described in Sections 3.1 and 3.2, the overall memory requirements for *Lik* arrays will be determined not only by the role of each one in the data dependencies, but also by their life duration and ability of being overwritten.

Highpass data containing arrays (*Hi*) are also divided in smaller arrays (*Hik*), although their contents are not participating in the WT calculation. As shown in [12] the front line of the ASAP WT does not always align with the block borders of the output data structure in either of the output mode A, B and C. Therefore, the Highpass output data of each level can be temporarily stored in an intermediate buffer till a whole block or tree output structure is computed for release to the output. The memory requirements and optimizations for the Highpass data are also verified through ATOMIUM.

Although the *Lik* and *Hik* arrays are not aligned among levels, with the appropriate write/read order organization the data can be passed through them (or to the output) without the need of *Li* and *Hi* arrays. Taking advantage of their small size and their reusability, significant optimizations in the memory requirements of the WT can be achieved. All aforementioned array partitioning actions – highly influenced by ATOMIUM's memory optimization capabilities and constraints - reflect themselves inevitably in source code (loop) transformations, therefore considerably increasing the optimized source code control complexity. Nevertheless, the overall gain is worthwhile, as explained in all following sections.

3.1 Array Lifetime

The approach described in the previous paragraphs is illustrated with the 9/7-tap filter, 3-level 1D WT. The input (*L0*) is divided into blocks of 8 (2^3) samples and it is supplied to the WT in block-by-block mode. In this case the Lowpass results have to be stored in 3 different arrays for level0 (calculation of the first level) (*L00*, *L01*, *L02*), in 3 arrays for level1 (*L10*, *L11*, *L12*) and in 2 arrays for level2 (*L21*, *L22*). Similarly, the Highpass have to be stored in 2 arrays for level0 (*H00*, *H01*), in 2 arrays for level1 (*H10*, *H11*) and in 1 array for level2 (*H21*). At each level *i* 3 filtering buffers are used: *filter_in_buf_i*, *filter_out_high_buf_i*, *filter_out_low_buf_i*, where the input, the Highpass result and the Lowpass result are stored respectively. The use of different arrays through levels (*i*) ensures that the same filtering procedure can be applied onto the minimum dimensioned arrays (which are decreasing through levels because of the subsampling).

The Memory Compaction tool of ATOMIUM can extract (formally, based on the C source code) an evolution analysis of the memory usage of the different data structures during the execution of the application. In particular, the tool can analyze what the maximum amount of memory location that is being occupied by each data structure (array) at every point in the application code. This information is visualized in a graphical way (like the graph of Figure 2) that makes it easy to detect memory usage bottlenecks in the code. The vertical axis of this graph corresponds to the flow of the source code that is analyzed. Thus, the maximum usage at a given point in the code can therefore be found through this graph. The horizontal dimension of each structure represents its "active" size (see arrows in figure) at any point in time (which is previously decreased also by the tool to its strict minimum– intra-data optimization). The plot also indicates where write and read accesses of each array occur (indicated by dark and light gray rectangles, respectively). The order of the data structures (arrays) in the graph is the one that corresponds to the ASAP execution order presented in Section 2. With the use of a cursor (the horizontal thin cross line in the middle of the graph) one detects which arrays are alive at each position of the code (which is also mapped to the execution time). Consequently, the arrays that are never alive at the same time (i.e. *filter_in_buf0* and *filter_in_buf1*) can be stored in the same physical location of the memory (inter-data optimization).

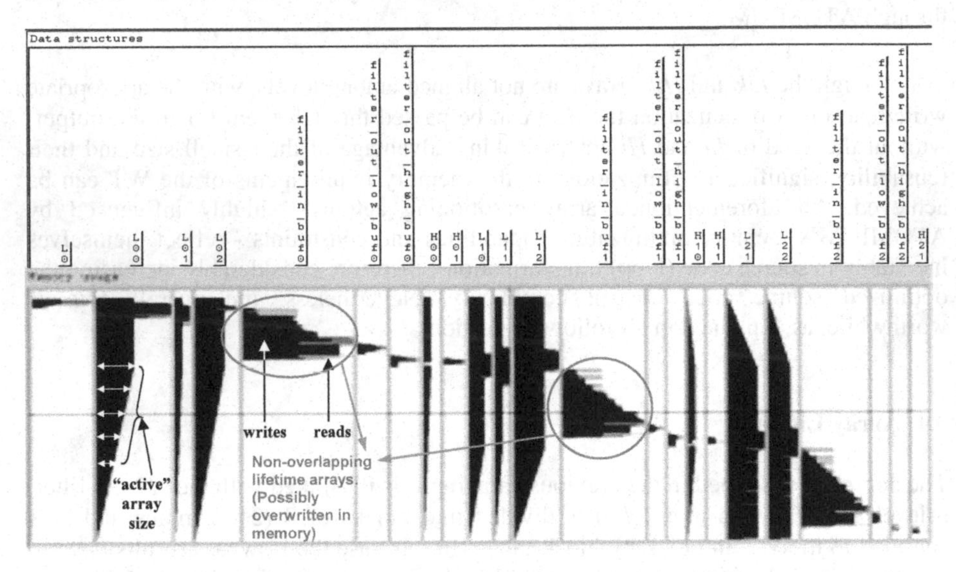

Fig. 2. Results of the memory usage evolution analysis

3.2 Memory Reuse

The finally optimized memory requirements (of the 9/7-taps, 3-level WT ASAP implementation) are shown graphically in Figure 3. The graph represents the memory that is allocated for the execution of the WT and how the arrays are optimally stored in it. Note that the total number is counted in samples and it does not depend on the dimension of the input. The non-simultaneous existence of the array of Figure 2 is exploited to reach the minimal memory requirements: the arrays whose lifetimes do not overlap can be stored in the same place of the memory.

In the implementation of this example also the output is following the ASAP schedule. Each block of Highpass data (with dimension 2^i) at each level i, and each

Lowpass result at the highest level, are released to the output immediately when they are fully calculated (mode A). Thus, the output consists of independent blocks with dyadically decreasing dimensions through levels and follows the skewed front lines of ASAP schedule of Figure 1. In the next session the memory requirements for two more different approaches of grouping the WT results will be extracted and compared to this mode A approach.

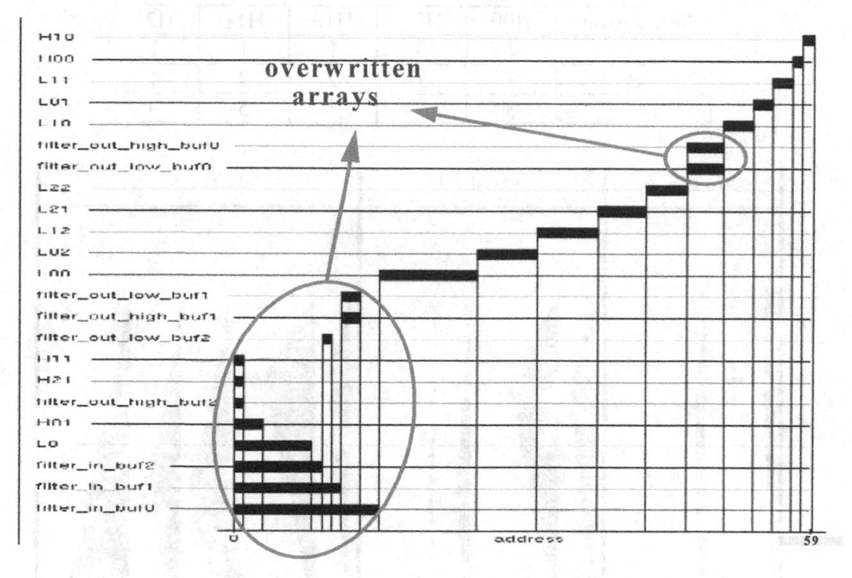

Fig. 3. Final Optimized memory organization

4 Output Formulation Comparisons

In this session two additional approaches of forming the output of the WT are studied, the approach of MPEG-4 VTC which clusters the output in parent-children (zero)trees (mode B), and the approach of JPEG2000 which outputs equally-sized blocks (mode C). It is shown that the Memory Compaction result of an optimal implementation of an MPEG-4 wavelet encoder (VTC – mode B) is always less memory expensive than an optimal implementation of mode C (e.g. JPEG2000).

Memory Size Requirement Results

The basic difference from the memory requirements of the implementation between modes A and B is due to the fact that the Highpass results of the WT in mode B have to be delayed before they can be released to the output (till the appropriate output structure is calculated). Thus, *Hik* arrays have to become bigger in order to temporarily keep these results. In this case it is not only their dimension that is increased, but also their lifetime, as more data has to be alive for a longer period (the time that is needed to a whole structure of output to be calculated). The same effect is what differentiates the mode B and C approaches. More precisely, the exact numbers of Highpass arrays memory requirements (№ of samples) are shown in Table 1 and a graphical representation of their lifetimes in Figure 4.

Table 1. Memory requirements (dimensions in samples) for the Highpass WT result arrays with different output coding modes (A, B, C)

Mode/Array	H00	H01	H10	H11	H21
A	1	3	1	1	1
B	3	9	2	2	1
C	3	9	5	5	9

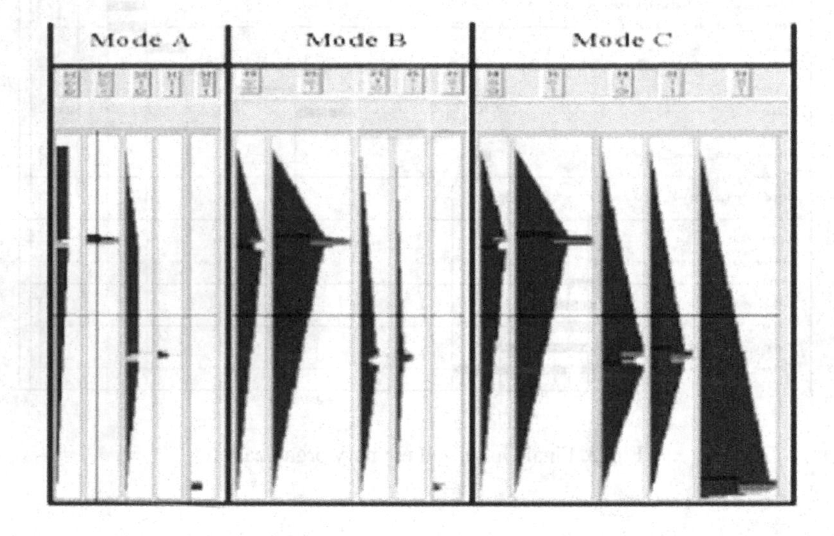

Fig. 4. Lifetime of each optimized Highpass WT result array with different WT output coding modes (A, B, C)

The three different approaches of the WT output data modes are compared in Figure 5 for two cases of WT filters, i.e. 9/7 and 5/3-tap filters. Modes A and B always form the output data in dyadically decreasing blocks such that the highest level block has only one sample. For mode C the data is always grouped in independent blocks of 8, 16 and 32 samples respectively (indicated by Mode C (x), with x = 8, 16, 32).

From Figure 5 the following observations can be made:

- The coder that forms the output data in the ASAP order (mode A) has always the lowest memory complexity.
- The only case where mode C has lower memory requirements than mode B is for M=4 with coding block equal to 8 and 5 level WT (which for JPEG2000 is never commercially implemented [7]).
- In all other cases the optimal implementation of a mode B coder is less memory expensive than an optimal implementation of a mode C coder, whose memory size

increases with the size of the blocks to be coded. Consequently, mode C pays a high penalty because of its constraint of processing equally-sized blocks of B (typically B=32 or 64) wavelet coefficients over all wavelet levels. In fact, the parent-children tree processing of MPEG-4 VTC appears to be a more natural schedule for wavelet data processing (also studied in [8]).

- The memory size differences between modes A, B and C can be as large as a factor 3.
- Finally, in the case of short data dependencies, implied by the wavelet filters (i.e. *M*=2), the memory complexities of modes A and B are equivalent. This is due to the fact that the data dependency pyramid that is formed through levels does not go beyond the borders of a single parent-children tree. As a matter of fact, the front line of the ASAP schedule coincides with the right border of the parent-children tree.

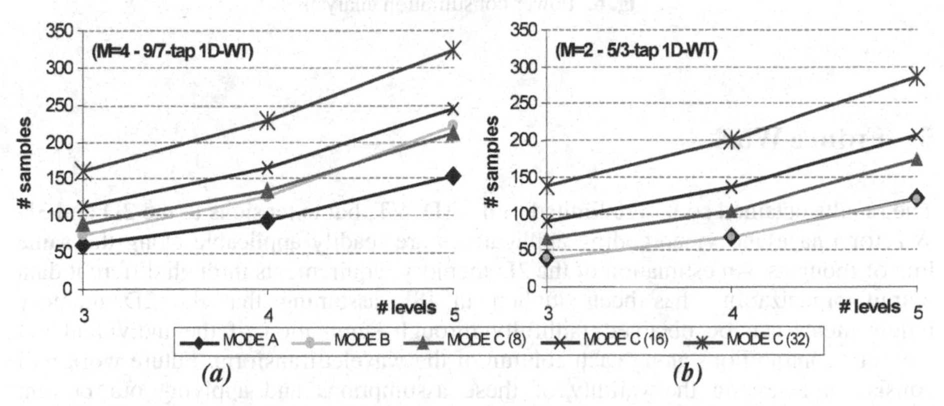

Fig. 5. 1D Memory requirements of ASAP, MPEG-4 VTC and JPEG2000 codecs *(a)* 9/7-taps filter (*M*=4), *(b)* 5/3-taps filter (*M*=2)

Consequently, the output grouping and order of the wavelet coding has a crucial impact on the memory complexity of the encoder. Because of the large variation of the memory requirements, depending on the mode, the upper described results have potentially to be taken as a strong decision parameter w.r.t. the entropy coder format that will be integrated in all the future scalable standards.

Memory Power Consumption Results

The size memory requirements are straightly affecting the power consumption of the memory where the data are stored. A pragmatic assumption for the above results is to use a memory of 100 words for all the cases that require memory size up to this number, a memory of 200 words for the intermediate memory-consuming cases and a memory of 350 words for the remaining. The power consumption values of these memories, according to one of Mietec's embedded SRAM memory modules in 0.35 μm CMOS technology (for specific working frequency), are shown in Figure 6. For the different implementation cases analyzed before (number of WT levels, output grouping case) the power consumption can vary from 14% till 32%. More power optimizations can be obtained through a hierarchical memory structure (e.g. multi-level cache), which analysis is however outside the scope of this paper.

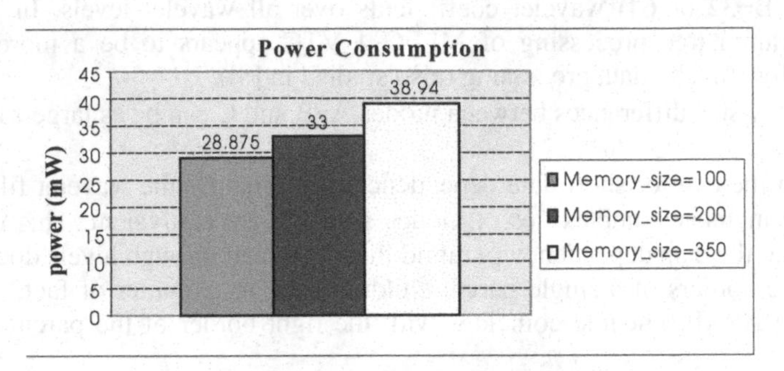

Fig. 6. Power consumption analysis

5 Future Work

The results obtained so far are limited to the 1D WT, but extensions to the 2D and 3D WT for image and video coding applications are readily applicable along the same line of thoughts. An estimation of the 2D memory requirements through different data output organizations has been studied in [8], assuming that the 2D memory requirements can be obtained faithfully through summation of the individual 1D memory contributions along each column of the wavelet transform. Future work will consist in assessing the validity of these assumptions and applying our current methodology to 2D WT for still image wavelet coding and 3D WT for interframe wavelet coding.

6 Conclusion

We have described a systematic approach and accompanying tool flow for reducing the memory requirements in wavelet-based signal coders, taking the inter-dependencies between their constituent modules into account. We observe that the wavelet transform output data organization, imposed by the entropy coder, strongly affects the memory requirements of the encoder chain. As verified in Section 4 the output format with independent, dyadically decreasing block sizes is demonstrated to be the less memory-consuming mode. Each output block is released just after its creation and no temporary buffer is required to store intermediate results. Additionally, the zero-tree blocks grouping mode requires less memory than the independent equally-sized blocks mode.

References

[1] F. Catthoor, S. Wuytack, E. De Greef, F. Balasa, L. Nachtergaele, A. Vandecappelle, " **Custom memory management methodology: exploration of memory organization for embedded multimedia system design,** " ISBN 0-7923-8288-9 Kluwer Acad. Publ, June 1998.

[2] F.Catthoor, K.Danckaert, C.Kulkarni, E.Brockmeyer, P.G.Kjeldsberg, T.Van Achteren, T.Omnes, " **Data access and storage management for embedded programmable processors,** " ISBN 0-7923-7689-7, Kluwer Acad. Publ., Boston, 2002.

[3] —, " **Information technology - Coding of audio-visual objects - Part 2: Visual,**" ISO/IEC JTC 1/SC 29/WG 11 N 3056 Maui, December 1999.

[4] I. Sodagar, H.J. Lee, P. Hatrack, Y.Q. Zhang, " **Scalable Wavelet Coding for Synthetic/ Natural Hybrid Images**," *IEEE Transactions on Circuits and Systems for Video Technology*, Vol. 9, No. 2, pp. 244-254, March 1999.

[5] —, " **JPEG 2000 Image Coding System**," ISO/IEC JTC 1/SC29/WG1, FCD 15444-1.

[6] A. Skodras, C. Christopoulos, T. Ebrahimi, " **The JPEG2000 still image compression standard**," *IEEE Signal Processing Magazine*, Vol. 18, No. 5, pp. 36–58, September 2001.

[7] D. S. Taubman, M. W. Marcellin, " **JPEG2000: Image compression fundamentals, standards and practice**," Kluwer Academic Publishers, 2002.

[8] G. Lafruit, J. Bormans, " **Assessment of MPEG-4 VTC and JPEG2000 Dynamic Memory Requirements**," *International Workshop on System-on-Chip for Real-Time Applications (invited paper)*, pp. 276–286, 2002.

[9] M. Vishwanath, " **The Recursive Pyramid Algorithm for the Discrete Wavelet Transform,** " *IEEE Transactions on Signal Processing*, Vol. 42, No. 3, pp. 673–676, March 1994.

[10] G. Lafruit, L. Nachtergaele, J. Bormans, M. Engels, I. Bolsens, " **Optimal Memory Organization for Scalable Texture Codecs in MPEG-4**," *IEEE Transactions on Circuits and Systems for Video Technology*, Vol. 9, No. 2, pp. 218–243, March 1999.

[11] **http://www.imec.be/atomium**

[12] G. Lafruit, Y. Andreopoulos, B. Masschelein, " **Reduced Memory Requirements in Modified JPEG2000 codec,**" *International Conference on Digital Signal Processing 2002*, CD-ROM T2C.4-pp.1–8, 2002.

[13] C.Y. Chen, Z.L. Yang, T.C. Wang, L.G. Chen, " **A Programmable Parallel VLSI Architecture for 2-D Discrete Wavelet Transform**," *Journal of VLSI Signal Processing 2001*, No. 28, pp. 151–163, Revised April 2000.

[14] T. Simons, A.P. Chandrakasan, " **An Ultra Low Power Adaptive Wavelet Video Encoder with Integrated Memory**," *IEEE Journal of Solid State Circuits*, Vol. 35, No. 4, pp. 218–243, April 2000.

[15] M. Ferretti, D. Rizzo, " **A Parallel Architecture for 2-D Discrete Wavelet Transform with Integer Lifting Scheme**," *Journal of VLSI Signal Processing 2001*, No. 28, pp. 165–185, Revised April 2000.

Design Space Exploration and Trade-Offs in Analog Amplifier Design

Emil Hjalmarson, Robert Hägglund, and Lars Wanhammar

Dept. of Electrical Engineering, Linköping University, SE-581 83 Linköping, Sweden
{emilh, roberth, larsw}@isy.liu.se

Abstract. In this paper, we discuss an optimization-based approach for design space exploration to find limitations and possible trade-offs between performance metrics in analog circuits. The exploration guides the designer when making design decisions. For the design space exploration, which is expensive in terms of computation time, we use an optimization-based device sizing tool that runs concurrent optimization tasks on a network of workstations. The tool enables efficient and accurate exploration of the available design space. As a design example, we investigate three operational transconductance amplifiers, OTAs, implemented in a standard 0.35-μm CMOS process. This example shows that large savings in terms of chip area and power consumption can be made by selecting the most suitable circuit.

1 Introduction

When implementing an analog integrated circuit there are numerous design decisions to be made. One of the most critical parts of the design process is to decide upon what circuit realization to use for the implementation. There typically exist several realizations for most analog functions. Usually, the designer has to select a suitable topology from numerous of possible candidates. Due to the nonlinear and complex behavior of the performance of analog circuits, the time required to manually design several candidates is long. Instead the selection is often based on hand-calculations using simplified models, experience, and rules of thumb. Even when the selection of a circuit topology has been made the sizing of the circuit components, e.g., transistors, still poses a challenging task. Further, in traditional manual design only a fraction of the available design space is examined. Hence, there usually exist other sets of design parameters yielding better performance. It is therefore essential to further explore the available design space in order to make good trade-offs.

Previous work focused on decreasing the time required to compute the possible trade-offs at the expense of the accuracy [1], [2]. Another approach, [3], utilizes the multi-objective formulation of the optimization problem in conjunction with a SPICE type simulator inside the optimization loop. The accuracy of the results is high, but the use of a SPICE like simulator decreases the computational efficiency.

J.J. Chico and E. Macii (Eds.): PATMOS 2003, LNCS 2799, pp. 338–347, 2003.
© Springer-Verlag Berlin Heidelberg 2003

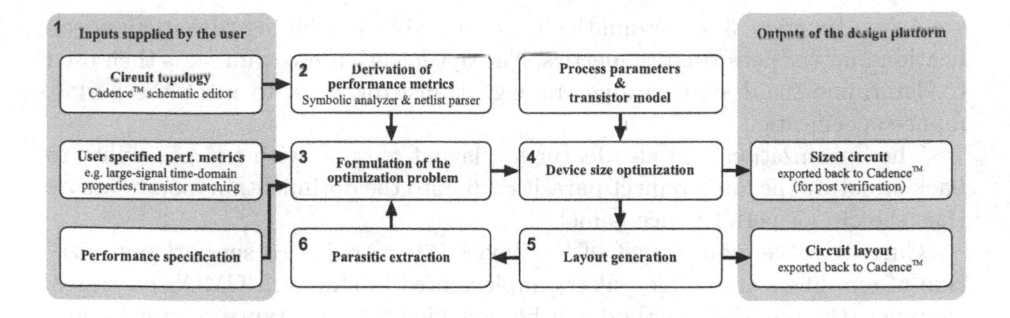

Fig. 1. An overview of the equation-based device sizing tool.

In this paper we present an efficient and accurate method to explore the design space and find suitable design trade-offs for analog circuits. This process involves the use of a device sizing tool [4] for design space exploration.

To illustrate the proposed approach a simple design example involving the selection of an amplifier topology for a sample-and-hold circuit is considered. Three commonly used amplifier topologies are investigated and possible design trade-offs are examined. It is shown that by selecting the best of the evaluated topologies significant savings in terms of chip area and power consumption are possible.

2 The Device Sizing Tool

The device sizing tool has two main objectives; firstly to increase the design efficiency and secondly to obtain reliable and better overall, possibly optimal, circuit performance compared to a manual simulation-based design approach. Increased design efficiency is obtained by using the equation-based optimization approach, where symbolic expressions for the circuit performance are used instead of a circuit simulator inside the optimization loop. A key feature is the automatic generation of these expressions and the formulation of the optimization problem. This yields a short setup time for new circuit topologies which is important when a large number of circuit topologies is to be examined. Reliable and good circuit performance is achieved using high-accuracy device models, such as BSIM3v3, and by examining several sets of device sizes. The tool also enables optimization of large analog circuits by distributing the optimization tasks on several workstations. Furthermore, the memory requirements are kept small by evaluating the small-signal performance metrics using a modified admittance matrix, MNA, instead of fully expanded expressions. An overview of the device sizing tool is shown in Fig. 1. The user input, 1, is a netlist, a performance specification, and possibly user-defined performance metrics. The netlist is analyzed by a netlist parser, 2, which generates a set of circuit dependent constraints and large-signal performance metrics. This parser also generates the input to the symbolic analyzer program used to derive the MNA.

A cost function, 3, is automatically generated by combining the circuit specification and the performance metrics. The optimization program, 4, is then used to determine the design parameters, e.g., transistor sizes, to meet the performance specification.

The optimization tool also features a layout part, 5, with the possibility to back annotate the interconnect parasitics, 6, into the optimization loop to assure that the circuit performance is met.

Currently, the main target of the device sizing tool is to support optimization of amplifiers and analog filters implemented in standard CMOS processes. However, the proposed method can be extended to other types of circuits and technologies.

3 Circuit Specification

As a design example three operational transconductance amplifiers (OTAs) to be used as buffers in an 8-bit sample-and-hold circuit with a 20-MHz sample frequency are evaluated.

Although the tool handles fully differential circuits, we choose to illustrate the concept with single-ended circuits.

The circuit topologies for the three OTAs are shown in Fig. 2a (two-stage), Fig. 2b (folded-cascode), and Fig. 2c (current-mirror).

3.1 Performance Specification

In Table 1 the performance specification for the OTAs is compiled. For sake of simplicity, no distortion or small-signal time-domain properties are included in the cost function, although the tool handles these as well.

Since automatic derivation of expressions for slew rate has not yet been implemented they have been manually specified. For the OTAs, approximate expressions for the slew rate are essentially the supply current (I_d) over the compensation (C_c) and/or load capacitance (C_L) according to

$$SR_{TS} = min\{\frac{I_{d3}}{C_c}, \frac{I_{d5}}{C_L + C_c}\}, SR_{FC} = \frac{I_{d2a}}{C_L + C_c}, \text{and } SR_{CM} = \frac{2I_{d7a}}{C_L + C_c} \quad (1)$$

for the two-stage, folded-cascode, and current-mirror OTA, respectively. Further, the currents through the transistors are computed at their operation points.

The area of the circuits are estimated by

$$Area = L_T \sum W_i + W_R^2 \frac{R_c}{R_{sq}} + \frac{C_c}{C_a} + \frac{4C_p}{C_a^2}\left(2C_p - \sqrt{C_c C_a + 4C_p^2}\right) \quad (2)$$

where W_i is the width of transistor M_i, R_c and C_c are the values of the components in the compensation network. Further, L_T is the distance between the drain and source of the transistors, W_R and R_{sq} are the width of the resistor and the resistivity per square, respectively. C_a and C_p are the capacitance per square and the perimeter capacitance for the capacitor, respectively. The area required for the routing and the substrate contacts are not included.

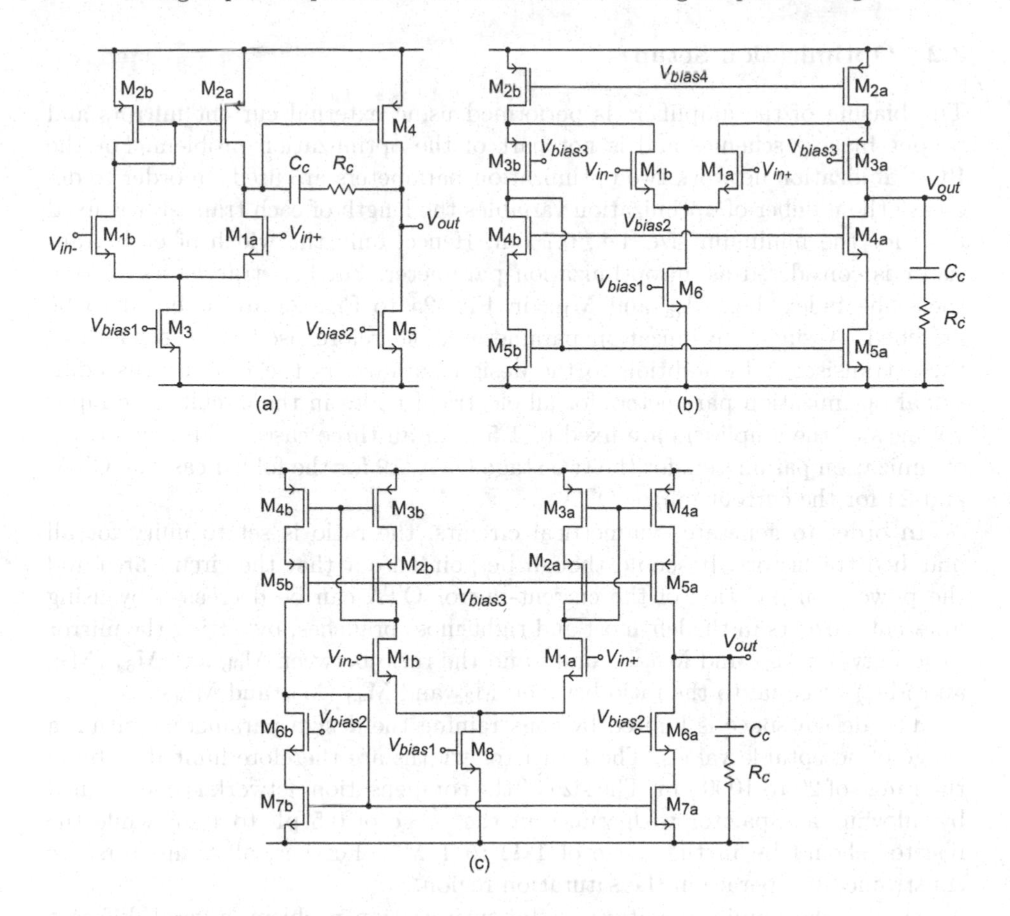

Fig. 2. The schematic of the two-stage (a), folded-cascode (b), and current-mirror (c) OTA.

Table 1. The performance specification of the OTAs.

Performance metrics	Specification	Unit
Load capacitance	5.0	pF
Unity-gain frequency	> 100	MHz
Power supply voltage	3.3	V
PSRR	> 70	dB
CMRR	> 70	dB
Input range	$[1.0, 2.0]$	V
Output range	$[1.0, 2.0]$	V
DC gain	> 70	dB
Slew rate	> 100	V/ms
Phase margin	> 60	$^\circ$
Power consumption	< 5.0	mW
Equivalent output noise @ 1kHz	< 320	$\mu V/\sqrt{Hz}$

3.2 Optimization Setup

The biasing of the amplifiers is performed using external current mirrors and proper biasing schemes and is not part of the optimization problem. For the RC stabilization network two optimization parameters are used. In order to decrease the number of optimization variables the length of each transistor is fixed to twice the minimum size, i.e., 0.7 μm. Hence, only the width of each transistor is considered as an optimization parameter. Further, transistors sharing the same index, i.e., M_{1a} and M_{1b}, in Fig. 2a to Fig. 2c are assumed to be matched. A single optimization parameter is therefore used for the width of these transistors. In addition to the design parameters the tool assigns additional optimization parameters for all electrical nodes in the circuit. The input voltages of the amplifiers are fixed to 1.5 V in all three cases. This leads to 13 optimization parameters for the two-stage OTA, 19 for the folded-cascode OTA, and 24 for the current-mirror OTA.

In order to generate symmetrical circuits, the ratio is set to unity for all matched transistors. It should though be pointed out that the circuit area and the power consumption of the current-mirror OTA can be decreased by using different currents in the left-most and right-most branches, by letting the mirror ratio between M_{3b} and M_{4b} be unity and the ratio between M_{6b} and M_{6a} (M_{7b} and M_{7a}) be equal to the ratio between M_{3a} and M_{4a} (M_{2a} and M_{5a}).

The design space is limited by constraining the design parameters within a range of acceptable values. The transistor widths are therefore limited to be in the range of 20 to 1000 μm. The size of the compensation network is constrained by allowing a capacitor with values in the range of 0.5 pF to 1 nF while the resistor should be in the range of 1 Ω to 1 MΩ. Further, all transistors are constrained to operate in the saturation region.

Due to the nonlinear nature of the optimization problem, several different starting points are required in order to find good solutions. The initial values of the optimization parameters are therefore randomly chosen between their upper and lower bounds. The node voltages are initially selected between the positive and negative power supply voltage.

4 Optimization Results

In the following the optimization results for the OTAs shown in Fig. 2 are presented. The optimization tool is directed to generate circuits that meet the specification in Table 1. The tool can simultaneously sweep several performance metrics in order to fully explore the design space. However, the graphs presented are two-dimensional cuts in this space. These are generated by sweeping the value of a performance metric, e.g., the power supply voltage, while minimizing, e.g., the power consumption.

In the graphs the performance metric being swept is placed on the x-axis while the metric being minimized/maximized is on the y-axis. Each point in the graphs represents the best of several solutions (local minima). For the examined

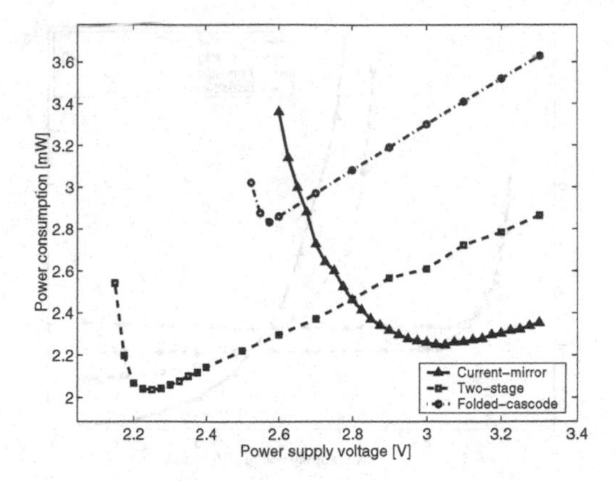

Fig. 3. The power consumption vs. the power supply voltage for the three amplifiers.

topologies there exist several solutions with about the same cost, typically 90% of the solutions, i.e., local minima, are within 10% of the best solution.

The time required to generate the circuit parameters used for compiling the information in each of the graphs below is a few hours on a cluster of workstations.

For the circuits examined the average time required to find one solution is about 3 minutes for the two-stage, 6 minutes the folded-cascode, and 8 minutes for the current-mirror OTA using a single workstation (Intel Pentium^{TM} 4, 2.6 GHz). Hence, the approximate form of the presented curves can be generated in short time with low requirements on computer power.

The curves presented are based on results taken directly from the circuit optimizer, only a limited number of the solutions have been validated in the Spectre^{TM} circuit simulator. These solutions show, however, an excellent agreement with the optimizer results due to the use of the same high accuracy transistor models (BSIM3v3) both in the optimization tool and in the Spectre^{TM} simulator. Furthermore, special care has been taken in order to assure that the presented results are not affected by the setup and formulation of the optimization problem. For example the impact of different relative weights of the performance metrics have been examined. Several different choices of weights have proven to yield the same results as the ones presented here.

4.1 Impact of Scaling the Power Supply Voltage

In Fig. 3 the power consumption for the three amplifiers are shown as a function of the power supply voltage. The voltage has been decreased from 3.3 V until the circuit fails to meet the specification. The circuit area has not been constrained.

The graph shows the trade-offs between the power consumption and the power supply voltage for the circuit topologies. At 3.3 V, the current mirror OTA has the lowest power consumption, however, when lowering the power

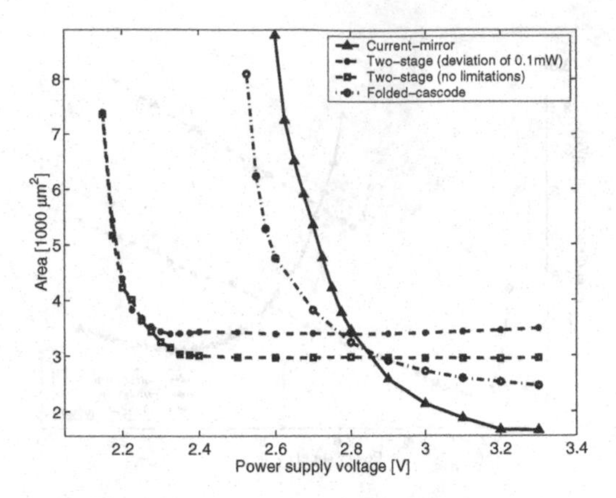

Fig. 4. The circuit area as a function of the power supply voltage for the three different amplifiers. An extra curve for the two-stage OTA where the maximum power consumption is constrained to be within 0.1 mW of the minimum power consumption shown in Fig. 3.

supply voltage below 2.8 V the two-stage amplifier is instead the best as far as the power consumption is concerned.

The graphs also show the range of power supply voltages for which the circuits meet the specification. The lower limit is set by the input and/or output swing of the amplifiers. Further, as the power supply voltage decreases from 3.3 V the power consumption also decreases linearly (constant power supply current) for the two-stage and the folded-cascode OTA. This is due to the slew rate limitation.

The minimum circuit areas required to meet the circuit specification as a function of the power supply voltage are shown in Fig. 4. For the two-stage amplifier two curves are plotted. In the lower-most curve no constraints on power consumption is applied, in the other the maximum power consumption is constrained to be within 0.1 mW from the best solution in Fig. 3 for each power supply voltage. For the folded-cascode and current-mirror OTAs the circuit areas are not affected by limiting the power consumption as discussed in Sec. 4.3. However, there exists a trade-off between these two metrics for the two-stage OTA.

It is clear that the area rapidly increases when the point where the circuit fails to meet the specification is approached. This is due to the increased transistor widths required to meet the input/output swing. When the upper bound of the transistor widths is reached the specification can no longer be met.

4.2 Impact of Scaling the Unity-Gain Frequency

In the sample-and-hold circuit the output from the buffer must settle before the clock arrives. The linear settling-time for the buffer is determined by its first pole.

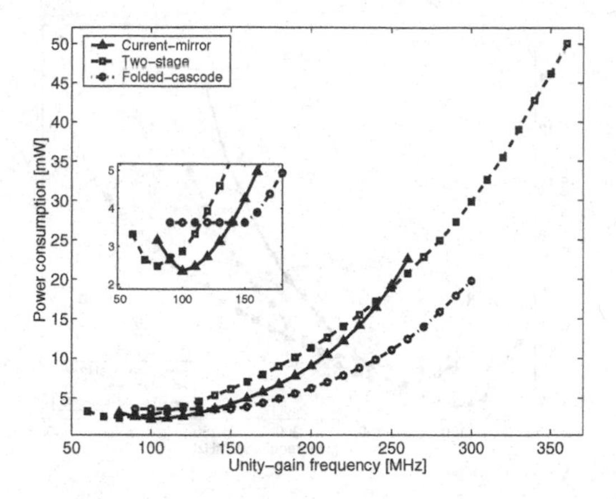

Fig. 5. The power consumption as a function of the unity-gain frequency.

This is approximately the unity-gain frequency of the open-loop amplifier. Hence, the circuit area and power consumption as a function of the unity-gain frequency is of interest. The possible trade-off between these performance metrics are shown in Fig. 5 for a power supply voltage of 3.3 V. The circuit area has not been constrained. The maximum unity-gain frequency is 360 MHz, 300 MHz, and 260 MHz for the two-stage, folded-cascode, and current-mirror OTA, respectively.

For frequencies below 150 MHz the slew rate specification is limiting the minimum power consumption for the folded-cascode amplifier. Further, the graph shows that each circuit topology has a specific range in which the power consumption is smallest. For all topologies, the power consumption increases rapidly with increased unity-gain frequency.

The circuit areas of each amplifier as a function of the unity-gain frequency is shown in Fig. 6. The power consumption is not constrained. The circuit area is increased for larger unity-gain frequencies.

4.3 The Circuit Area vs. Power Consumption Trade-Off

The design parameters that usually are to be minimized as long as the specification is met are the circuit area and the power consumption. These parameters are associated with the cost of the chip. The area vs. power consumption trade-off, at 3.3 V power supply voltage, is shown in Fig. 7 where the circuits are forced to have a specific power consumption and the objective is to minimize the area. Increasing the power consumption do not necessarily lead to increased design margins, in fact, all circuits used in Fig. 9 have minimum margins. Hence, the number of poor circuit realizations is large.

The circuit area of the two-stage amplifier can be made smaller at the expense of increased power consumption. For the folded-cascode and the current-mirror OTA there are no trade-off between these performance metrics. The points (A-D) in Fig. 7 indicate the best choices of circuit realizations for our application.

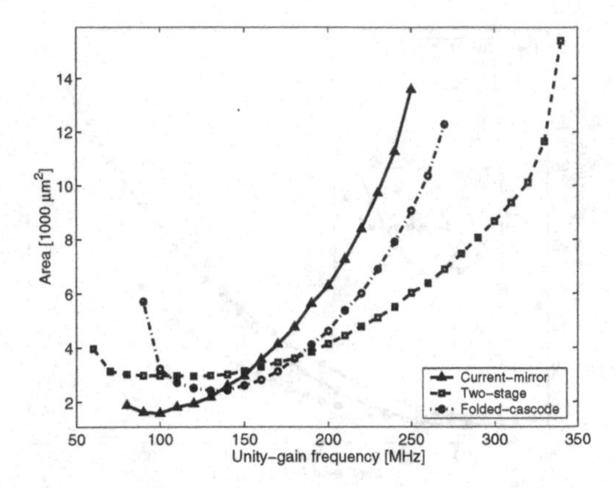

Fig. 6. The circuit area as a function of the unity-gain frequency for the amplifiers.

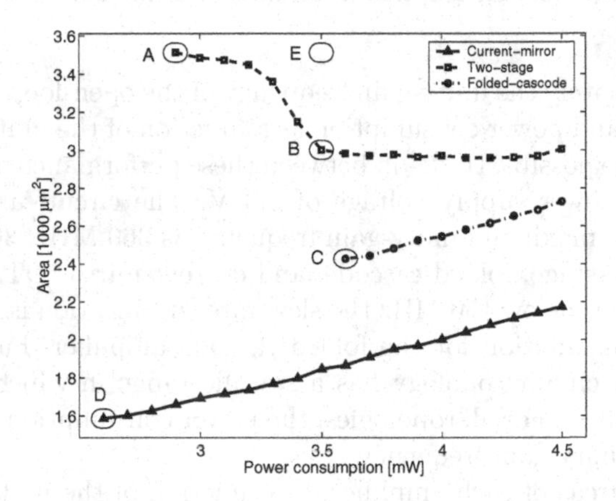

Fig. 7. The circuit area as a function of the power consumption for the amplifiers.

In this case the current-mirror OTA is the best topology with respect taken to both area and power consumption. Selecting circuit D instead of A, B, or C will result in a power reduction of 17%, 31%, and 33%, respectively. For the circuit area the savings are 55%, 47%, and 35% for D compared to A, B, and C.

Notice that the selected circuits are most likely not found using manual design techniques, hence, the savings in area and power consumption can be even larger in that case. Furthermore, since a large number of poor circuit realizations exist, there is a large risk of ending up with an even worse implementation, e.g., the point E, rather than that indicated by the curves in Fig. 7.

4.4 Design Example Summary

The material presented in the design example is dependent on the performance specification. Hence, no general conclusions about the relations between the circuit topologies can be made. However, the described approach can be used to explore the design space in order to visualize trade-offs and limitations within reasonable time for a set of circuit topologies. Thus, increased design efficiency and circuit performance are obtained.

5 Conclusion

We have discussed an approach utilizing an optimization-based device sizing tool for design space exploration of analog amplifiers. The tool is capable of generating sized circuits that meet a set of performance requirements. By varying the performance requirements the design space is explored in order to find limitations and possible trade-offs.

The tool utilizes high accuracy device models which lead to reliable computation of the circuit performance. Hence, the trade-off curves are based on real transistor-level implementations featuring the same accuracy as e.g. HSPICE^{TM}. Furthermore, once the designer has made a design decision a sized circuit meeting the specification is available.

We have also illustrated how the effect of limitations on important design parameters such as power supply voltage, power consumption, and area affect properties of three OTAs. Since the tool only requires the netlist for a new topology it is easy to optimize and compare several possible implementations.

By selecting the best circuit topology for our application, large savings in terms of power consumption and circuit area are made. Although, we have illustrated the proposed approach on small examples, the principle and the optimization tool can be applied to larger circuits.

References

1. G. Van der Plas, J. Vandenbussche, G. Gielen, and W. Sansen, "EsteMate: a tool for automated power and area estimation in analog top-down design and synthesis," *Proc. IEEE Cust. Integ. Circuits Conf.*, pp. 139–142, 1997.
2. R. Harjani, R. A. Rutenbar, and L. R. Carley, "Analog circuit synthesis and exploration in OASYS," *Proc. IEEE Computer-Aided Design Conf.*, pp. 492–495, 1998.
3. B. De Smedt and G. Gielen, "WATSON: Design Space Boundary Exploration and Model Generation for Analog and RF IC Design," *IEEE Trans. Computer-Aided Design*, vol. 22, no. 2, pp. 213–224, Feb. 2003.
4. R. Hägglund, E. Hjalmarson, and L. Wanhammar, "A Design Path for Optimization-Based Analog Circuit Design," *Proc. IEEE Midwest Symp. Circuits Syst.*, pp. I-287–290, Tulsa, OK, Aug. 2002.

Power and Timing Driven Physical Design Automation

Ricardo Reis
Instituto de Informática – Universidade Federal do Rio Grande do Sul
C.P15064 – 91501-970, Porto Alegre, Brazil,
reis@inf.ufrgs.br

Abstract. The paper presents why power and timing can be improved by physical design automation where the cells are generated on the fly. The non-use of cell libraries allows the implementation of any logic function defined at logic synthesis, using simple gates or static CMOS complex gates - SCCG. The use of SCCG reduces the amount of transistor and helps to reduce wire length. It is discussed the main strategies in the automatic layout synthesis, like transistor topology, contacts and vias management, body ties placement, power lines management, routing management and transistor sizing. In both methodologies, standard cell and automatic layout, it is needed accurate pre-characterization tools. The development of efficient physical design automation is the key to find better solution than traditional standard cell methodologies.

1 Introduction

The new fabrication technologies are changing some physical design paradigms. Some electrical effects that were neglected in the past are increasing their importance nowadays and they cannot be neglected anymore. The design of random logic blocks is still now based on the use of cell libraries. This was a good solution for a long time because the timing of a circuit could be calculated by considering only the delay of cells. The standard cell approach was a good solution because the cells of a library were already characterized and it was possible to estimate timing and power by using only cells parameters and the fan-out of them.

Nowadays, as clear stated in the Semiconductor Roadmap [1], the connections are taking the first role in the timing calculation. So, the standard cell approach doesn't allow anymore having performance estimation by only considering the logic cells. This is true because it is needed to take into account the effects of routing wires. The challenge is to have tools to accurate estimate routing effects. Another point to consider is that the cells in a library don't have flexibility for transistor sizing. A cell library normally has cells designed for area, timing or power, but only 1 cell for each objective. As the cell library is designed to support a great fan-out, they are normally oversized than they should be, if it is considered the environment where they are inserted. If many cells are bigger than they should be it means that the final layout will be bigger than it could be. Also, the average wire length will be bigger than it could be. And it is known that nowadays it is very important to reduce wire lengths because it has a great influence in circuit

J.J. Chico and E. Macii (Eds.): PATMOS 2003, LNCS 2799, pp. 348–357, 2003.
© Springer-Verlag Berlin Heidelberg 2003

performance. The option was the development of tools for automatic layout generation, but the problem was the estimation of circuit performance. It is fundamental to construct accurate power and timing estimation tools that takes account of interconnection contribution to circuit performance, for using in the design using deep submicron CMOS technologies. So, pre-characterization is needed in both cases, in the design of standard cell circuits or in the automatic layout generation. But with the synthesis of cells on the fly, during layout synthesis, it is possible to reduce the average wire length and consequently to increase performance. This paper will give more details about physical design automation and the how it can improve layout optimization. The paper also claims that the challenge we have nowadays is the development of efficient tools for layout design automation where the cells are designed to fit well in their specific environment.

2 The Standard Cell Approach

The standard-cell approach is accepted till nowadays as a good solution for the layout synthesis problem, besides that the technology mapping is limited by the cells found in a library and by the sizing possibilities, which are restricted to few options for each logic cell.

The standard-cell approach is currently the approach most used for the layout synthesis of most random logic blocks. The reasons for that are: the cells of a library are pre-characterized and the lack of efficient cell design automation tools. The cell pre-characterization let the designer to evaluate the delays of a circuit with a reasonable accuracy for past fabrication processes, as well a good estimation of power consumption. In the design of circuits using current submicron technologies, the cell pre-characterization is not anymore sufficient to evaluate the delays, because the delay related to connections is becoming more and more important. The challenge is to have efficient physical design tools to perform an automatic full custom physical design, with cells automatically designed to meet the requirements of the environment where they will be inserted. These tools should include automatic circuit pre-characterization.

The increasing development in IC technology brought the transistor gates to submicron dimensions. These advances brought also the need to review layout design methods. Timing and power reduction is an important issue to be considered at all design levels including layout design. So, the goal is to have a set of layout synthesis tools addressing the following:

- **Specific cells**. Each cell should be generated on the fly, considering the specific needs of the circuit environment where it is going to be placed, like cell fan-out.

- **Transistor minimization**. Reduction of the amount of transistors by the use of static CMOS complex gates (SCCG) combinations. The advantages of using SCCG gates are: area, delay and power reduction (when comparing with standard-cell approach).

- **Fan-out management.** The management of the fan-out should be used for tuning of delay, power, routing congestion and wire length reduction.
- **Transistor sizing.** Each transistor should be sized to a good compromise between power and delay, considering the environment where it will be placed.
- **Placement for routing**. The placement algorithms should be guided to reduce the average wire length reduction and to reduce routing congestion.
- **Full over the cell routing – FOTC [2]**. The routing should be done over the cells avoiding the use of routing channels.
- **Routing minimization**. Addressing interconnections routing to reduce the average wire length and to minimize the length of the connections over the critical paths.
- **Performance estimation**. The delay of a circuit should be evaluated with a good accuracy before the generation of it. So, it is fundamental to have timing and power estimation tools that take care of both transistor and connection influence in circuit performance.
- **Area evaluation**. The area and shape of a circuit should be evaluated before the generation of each functional block.
- **Technology independence.** All design rules are input parameters for the physical design tool set. The redesign of a same circuit with a different set of design rules should be done by just changing the technology parameters description file and by running again the layout synthesis tool.

3 Using Static CMOS Complex Gates

One great advantage in the cell generation on the fly is that it is possible to synthesize any logic function by using Static CMOS Complex Gates – SCCG. But it is recommended to avoid the use of more than 3 or 4 NMOS or 3 PMOS serial transistors (see table 1) when using nowadays CMOS technologies.

Table 1. Numbers of SCCGs with limited serial transistors, from [3].

Number of Serial NMOS Transistors	Number of Serial PMOS Transistors				
	1	2	3	4	5
1	1	2	3	4	5
2	2	7	18	42	90
3	3	18	87	396	1677
4	4	42	396	3503	28435
5	5	90	1677	28435	125803

When using SCCGs with a maximum of 3 serial transistors, it is possible to reduce by 30% the number of transistors, when comparing with original ISCAS85 benchmark circuits as shown in [4]. The logic mapping to SCCG using BDDs is shown in [5].

Remark1. The transistor minimization implies in the reduction of the circuit area, thus it also helps to reduce the average wire length and by consequence circuit delay and power consumption.

4 Layout Strategies

The layout synthesis tools should consider several goals as area reduction, wire length reduction, congestion reduction, routability, full over the cell routing and critical path management. The first challenge in automatic layout design is that the number of variables to handle is high and it is difficult to formalize rules to be used by a layout tool. When a designer is constructing a cell by hand, it has many choices for the placement of each cell rectangle. But man ability to visually manage topological solutions for a small set of components can generally find a better solution than can be found by known tools. It can be observed the high density that was found in random logic blocks circuits designed by late seventies (MC 6800, Z8000), where with only one metal layer all routing was implemented over the cells without using exclusive routing channels (Fig.1 shows an example). We don't have yet nowadays tools that can perform similar layout compaction.

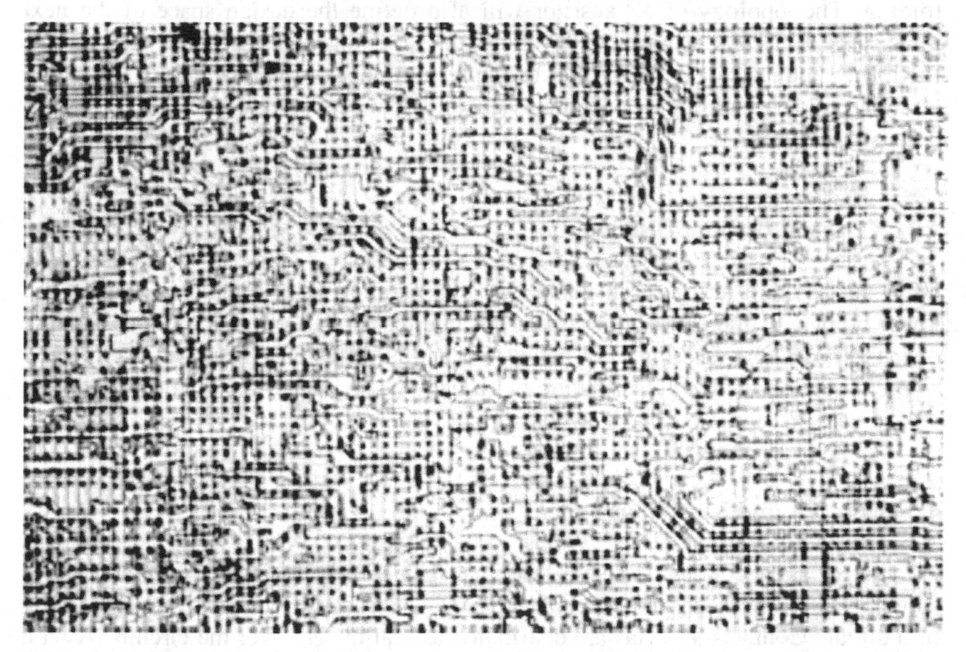

Fig. 1. A random logic block detail taken from the Zilog Z8000, where it can be observed the high layout compaction.

On the other side, man cannot handle large circuits without using CAD tools. So, there is a research space for the development of methodologies and tools that could efficiently handle all the possibilities and choices in the topology definition of each cell. But, it is possible to start by setting up some parameters and by giving flexibility to others. For example, the height of a strip of cells can be fixed and the cell width can change to accommodate cell's transistors.

The first step in the construction of a CMOS layout synthesis tool is to define the layout style, including the definition of the following:

- transistor possible topologies,
- management of routing in all polysilicon and metal layers,
- VCC and Ground distribution,
- clock distribution,
- contacts and vias management.
- body ties management.

4.1 Transistor Topology

The transistors can have several topologies, like vertical transistor, horizontal transistor, transistor with doglegs (right dogleg, left dogleg, etc…) and transistor with folding. The topology of a transistor will also define the design space of the next transistors. To let the management of transistor's topologies it is generally limited the flexibility in transistor topology. For example, all transistors can be vertical ones, but allowing folding and dogleg of the polysilicon wires between a PMOS and a NMOS transistor (see Fig. 2).

4.2 VCC and Ground Distribution

VCC and Ground are generally distributed as parallel wires running at the up and lower borders of the cells, but they can also be distributed as parallel wires running over the transistors (see Fig.2) or between P and N diffusions. They can use metal 1 or metal layer 2 for local distribution but it can be used upper metal layers to do the power lines distribution over the whole circuit.

4.3 Contacts and Vias Management

As routing is becoming strategic, the management of contacts and vias is also strategic. A bad contacts and vias management can drastically decrease the routability of a circuit. Contacts and vias are distributed as real towers over the circuits. A wire in the middle routing layers should not forbid the construction of tower of vias to bring a signal to the upper layers. It is also important to define a good strategy in the placement of body ties.

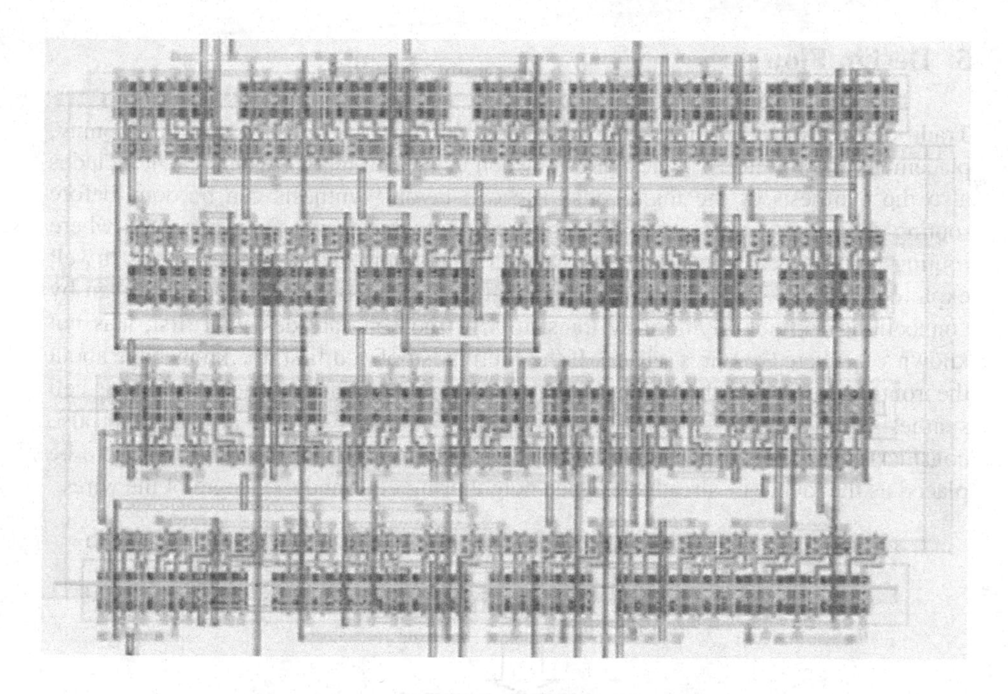

Fig. 2. Sample of a layout obtained automatically with the Parrot Physical Synthesis Tool Set developed at UFRGS, where it can be observed that all transistors are vertical and that the gate polysilicon lines have doglegs between PMOS and NMOS transistors. Metal 2 is used to run power lines over the transistors.

4.4 Routing Management

Routability of a circuit depends a lot in the routing management strategy. One important point is to manage layers transparency. So, in each layer all the connections should have only one direction. When a connection must change its direction it should use a neighboring metal layer to do that, except when it is done a double dogleg to the change of a signal to the neighboring metal track. Neighboring metal layers should have different orthogonal directions.

4.5 Transistor Sizing and Transistor Folding

The layout should be able to handle different transistor sizing. To avoid different diffusion height the solution is to use transistor folding, where a transistor can have several parallel segments. This solution is very useful in the design of buffers. A specific tool should do the transistor sizing.

5 Design Flow

Traditional Physical Design Flow of random logic blocks is in order, partitioning, placement and routing. The design flow when using automatic layout design includes also the synthesis of the logic cells. The cell layout synthesis can be done before routing definition or after routing definition. But, the design flow approach where routing is defined first and cell generation is done later can give better results if well explored (Fig. 3). The reason for that is that the layout is more and more driven by connections and not anymore by transistors. If the cells are designed first, it is not known yet where the wires will be. If the routing is defined first, the knowledge about the routing helps the definition of the placement of contacts and vias in the cell synthesis and also the sizing of transistors. The layout synthesis of a cell can also consider the specific aspects of the circuit environment where the cell is going to be placed as the layout of already designed neighboring cells and the length of the wires.

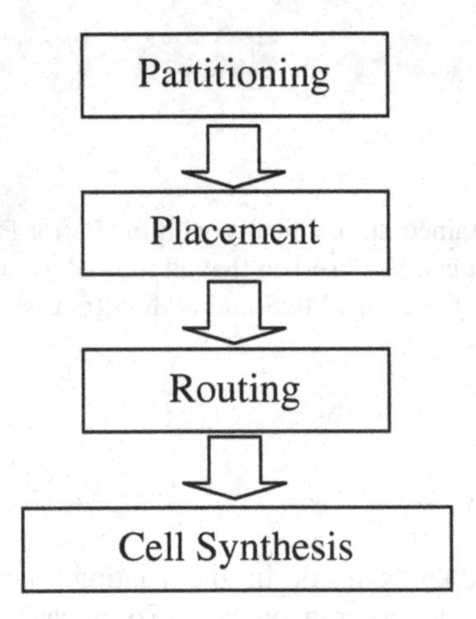

Fig. 3. Basic Physical Design Flow with automatic cell generation

5.1 Partitioning

There are two types of partitioning. The first one is the partitioning used to divide a circuit in parts. For example, to implement a circuit using 2 chips or if it is wanted to divide a random logic block in 2 blocks to well organize the floor planning. In these cases it is important to reduce as much as possible the connection between partitions. The second one is the partitioning used as a first step in the placement process. In this case the partitioning has the goal to define which cells will be kept together and it is not a goal to resume to a minimum the wires between partitions. It is better to have an amount of wired compatible with circuit average wire density. It is not desirable to have a lack of connections between partitions when placing them in the layout.

5.2 Placement

The placement defines the real location of each cell and it is divided in 2 steps. In the first step is used a constructive algorithm to do a first placement. This placement is used as input to the second step, where it is used an interactive algorithm. For the first step there is several algorithms that can provide fast running times with a low quality placement (like the random algorithm) or a good quality placement but with long running times (like the force directed one). The PLIC-PLAC algorithm [6] gives a reasonable good placement with running times similar to the random algorithm.

The choice of the constructive algorithm and the interactive algorithm should be done together. For example, if it is used in the second step a Simulated Annealing (SA) starting in a low temperature, it is important to use in the first step an algorithm that provides a good quality placement as input to the SA. If the SA starts using a high temperature there is no need to use big running times in the first step. In [7], [8], [9] and [10] it is possible to find more data about placement algorithms and experimental results.

The main objectives in the placement step are the reduction of wire length and the reduction of routing congestion. Reduction of congestion is fundamental to improve routability of a circuit. But it is also important to consider critical path in the placement strategy. In [11] it is shown that the quality of placement obtained by existent academic and industrial placement tools are quite far from optimal solution, using the wire length as the quality measure (about 50% to 150% worst than optimal solution).

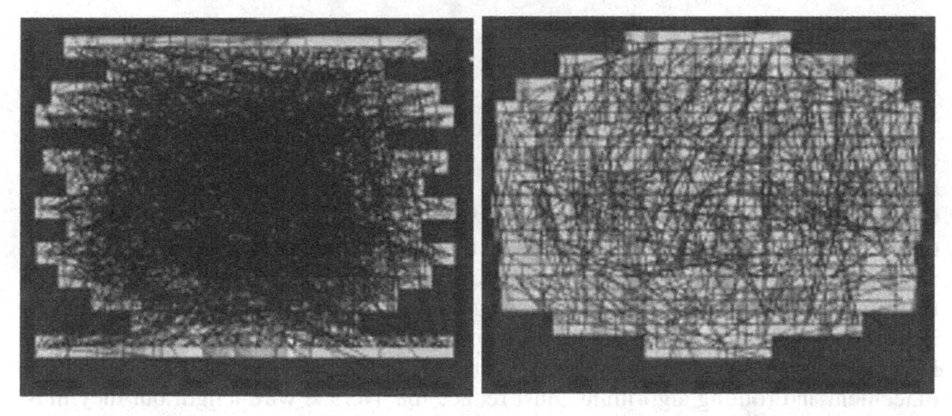

Fig. 4. Example of congestion reduction with the use of a SA algorithm using multiple types of perturbations [7].

5.3 Routing

Routing quality and routability depends a lot on the layout strategies and on the placement quality. So, it is important to define together the cell layout strategies, placement and routing algorithms. The routing is divided in global routing and detailed routing. It is also possible to classify them in general and specific algorithms (like River Routing to do the routing when using exclusive routing channels like in standard cell approach [12]). The good routing algorithms used in standard cell circuits could not be the better algorithms for the routing of circuits designed automatically. In [12] is presented a new algorithm that shows that there is a design space to improve routing algorithms.

6 Automatic Pre-characterization

The main reason that automatic layout generation was not used for a long time was that with standard cell it was possible to have a reasonable estimation of circuit performance. But the increasing role of wires in circuit performance, this performance estimation cannot consider anymore only the logic cells. On the other hand, we need to improve layout strategies where the cells are designed specifically to the location environment where it is going to be placed. In this case the cells must be designed automatically because it would be impossible to design a cell library for each circuit and with a huge number of cells.

For the design of circuits with nanometer scale using standard cells or using automatic layout synthesis it is needed, in both cases, tools for pre-characterization of the circuit. It is fundamental to have accurate timing and power estimations.

7 Conclusions

The paper gives an overview about strategies that should be used in the physical design automation with synthesis of cells on the fly. This automatic cell generation allows the physical implementation of any logic function by also using Static CMOS Complex Gates, SCCG. The logic minimization using SCCG can reduce the number of transistors by around 30 %. This reduction in the number of transistor obviously reduces the circuit area and by consequence the average wire length and it is known that nowadays the wires have an increasing influence in circuit performance. Another important point is that with automatic layout generation the cells could be designed later in the design flow, considering neighboring cells and routing definitions. The placement and routing algorithms must reduce the average wire length but they must also guarantee the routability by the management of routing congestion. As for both standard cell and automatic layout generation, it is needed the development of pre-characterization tools, it is better to use automatic layout generation that can improve power and timing of a circuit mainly by number of transistors reduction and by wire length reduction.

References

1. SIA International Semiconductor Roadmap, Semiconductor Industry Association (2001).
2. Johann, M., Kindel, M., Reis, R. Layout Synthesis Using Transparent Cells and FOTC Routing. 38th IEEE Midwest Symposium on Circuits and Systems, IEEE Circuits and Systems Society, Rio de Janeiro (1995).
3. Detjens, E., Gannot, G., Rudell, R., Sangiovanni-Vinccentelli, A.L. and Wang, A. Technology Mapping in MIS. Proceedings of ICCAD (1987) 116–119.
4. Reis, A.; Robert, M.; Auvergne, D. and Reis, R. (1995) Associating CMOS Transistors with BDD Arcs for Technology Mapping. Electronic Letters, Vol. 31, No 14 (July 1995).
5. Reis, A., Reis, R., Robert, M., Auvergne, D. Library Free Technology Mapping. In: VLSI: Integrated Systems on Silicon, Chapman-Hall (1997) pg. 303–314
6. Hentschke, R.; Reis, R., Pic-Plac: A Novel Constructive Algorithm for Placement, IEEE International Symposium on Circuits and Systems (2003).
7. Hentschke, R., Reis, R. Improving Simulated Annealing Placement by Applying Random and Greedy Mixed Perturbations. In: Chip in Sampa, São Paulo. 16th Symposium on Integrated Circuits and Systems Design. Los Alamitos: IEEE Society Computer Press (2003)
8. Sherwani. Algorithms for Physical Design Automation, Kluwer Kluwer Academic Publishers, Boston (1993).
9. Sarrafzadeh, M., Wang, M., Yang, X., Modern Placement Techniques. Kluwer Academic Publishers, Boston (2003).
10. Cong, J., Shinnerl, R., Multilevel Optimization in VLSICAD. Kluwer Academic Publishers, Boston (2003).
11. Chang, C., Cong, J., Xie M.. Optimality and Scalability Study of Existing Placement Algorithms. ASP-DAC, Los Alamitos: IEEE Society Computer Press (2003).
12. Johann, M., Reis, R., Net by Net Routing with a New Path Search Algorithm. 13th Symposium on Integrated Circuits and Systems Design, Manaus, proceedings, IEEE Computer Society Press (2000) pg. 144–149

Analysis of Energy Consumed by Secure Session Negotiation Protocols in Wireless Networks

Ramesh Karri and Piyush Mishra

Department of Electrical and Computer Engineering
Polytechnic University, Brooklyn, NY, 11201
ramesh@india.poly.edu, pmishr01@utopia.poly.edu

Abstract. Traditionally, researchers have focused on security, complexity, and throughput metrics while designing security protocols for wired networks. Deployment of these security protocols in battery-powered mobile devices has elevated energy consumption into an important design metric. In this paper we study the energy consumption characteristics of secure session negotiation protocols used by IPSec and WTLS protocols. Based on this study we present techniques to optimize energy consumed by secure wireless session negotiation protocols. Proposed techniques achieve 4× to 32× energy savings without compromising the security of wireless sessions. We show that these techniques are platform independent and discuss the impact of platform specific characteristics, such as available system resources, on the presented techniques.

1 Introduction

Confidential communication over public networks demands end-to-end secure connections to ensure data authentication, privacy, and integrity. Security protocols used to provide these services first negotiate a secure session between communicating parties using session negotiation protocol. After successfully establishing a secure session secure data communication is carried out using the agreed upon security associations (SAs), such as data encryption, authentication, and integrity algorithms and their parameters. Mobile devices operating in a wireless environment have limited processing power, bandwidth, and battery life. Absence of clear boundary and other physical constraints in inherently unreliable and discontinuous wireless networks places more stringent security constraints as compared to wired networks.

Security protocols are computation and communication intensive and significantly reduce the operational lifetime of battery-powered mobile devices [1]. Table 1 shows the energy consumed by various components of a secure wireless session on an IBM ThinkPad based mobile test bed using IPSec with Advanced Encryption Standard (AES) encryption, SHA-256 MAC, and a key-refresh rate of 8 KB. A detailed description of the mobile test bed and experimental methodology can be found in section 3.1 and [15]. Security overheads constitute more than 60% of the session energy, of which session negotiation energy can be as high as 98%. Energy overheads

J.J. Chico and E. Macii (Eds.): PATMOS 2003, LNCS 2799, pp. 358–368, 2003.
© Springer-Verlag Berlin Heidelberg 2003

associated with session negotiation increase as transaction becomes shorter (less time duration) or smaller (less data), which is true of most mobile applications.

Table 1. Energy consumed by a mobile client during secure wireless sessions (mJ)

Data Size	32 KB	1 KB
Session negotiation	2936	2936
Encryption + MAC	188	10
Transmission	2112	76
Key refresh	495	-
Total	5731	3022

In this paper we focus on energy consumption characteristics of secure session negotiation protocols used by security protocols for establishing and managing secure sessions for the following reasons:

- Secure session negotiation protocols form core of all security protocols and are independent of the layer at which security protocols operate. Therefore, techniques for optimizing the energy consumed by session negotiation protocols reduce the energy consumed by all security protocols.
- Inherent discontinuities in wireless networks lead to frequent secure session negotiations, thereby resulting in significant energy overheads.
- Lack of well-defined boundaries in wireless networks and other physical constraints imply stronger authentication and authorization, both of which in turn increase the energy consumed by session negotiations.

1.1 Related Research

Woesner et al. studied the power saving features of wireless LAN standards and presented simulation studies for energy-efficient ad-hoc configurations [2]. Ebert et al. presented a packet length dependent power control mechanism for optimal RF power level [3], while Rulnick and Bambos considered autonomous transmitters to determine optimal level of transmit power given a set of quality-of-service constraints and information on the nature of channel and the level of interference at the receiver [4]. Karri and Mishra presented data compression based techniques to reduce the energy consumed by wireless data communications [1]. Kravets and Krishnan presented a transport level protocol to selectively suspend communication and shut down the communication device and studied the tradeoff between power consumption and data delay [5]. Rohl et al. studied the influence of typical parameters on the power saving mechanisms in IEEE 802.11 [6]. Singh and Raghavendra developed a power-efficient multi-access protocol for ad-hoc radio networks and showed that the idea can be easily extended to other access protocols [7]. Zorzi and Rao developed probing schemes for energy-efficient error control protocols and a formal approach to track complex models for power sources including dynamic charge recovery in batteries [8]. Lettieri et al. presented another energy-efficient error control protocol with hybrid combination of an appropriate forward error correction (FEC) code and automatic repeat request (ARQ) protocol that adapts over time for each data stream [9].

2 Secure Session Negotiation

Secure session negotiation entails (1) *mutual authentication* of communicating parties, (2) *parameter negotiation* to agree upon SA primitives, such as data encryption algorithm, authentication algorithm, and key exchange and management protocol and their associated parameters, (3) *key exchange and management* to generate a set of secret keys shared among communicating parties, and (4) *SA establishment* to establish a secure session using the negotiated SAs. We use session negotiation protocols used by network layer security protocol IPSec [10] (of TCP/IP suit) and transport layer security protocol WTLS [11] (of WAP suite) to analyze and reduce the energy consumed by secure wireless session negotiations. Since session negotiation protocols are independent of the layer at which security protocol operates, these techniques are also applicable to security protocols at application layer, session layer, data link layer, and physical layer.

2.1 IPSec Session Negotiation

Internet Key Exchange (IKE) [12] protocol used by IPSec for secure session negotiation has been influenced by Station-to-Station, SKEME, and OAKLEY protocols. Session negotiation protocol supports Diffie-Hellman (D-H) key exchange and management [13] to generate shared secret keys and four authentication mechanisms - pre-shared secret, public key signature, public key encryption, and revised public key encryption [12], [14]. After authenticating the parties the protocol runs public-key based D-H protocol to establish *primary* SA, as shown in Fig. 1 (a). Establishment of *primary* SA can be carried out in two modes - main and aggressive. Aggressive mode collapses the 6 messages of main mode into 3 messages at the expense of constrained negotiation space and unsecured identities. Next, under the protection of *primary* SA session negotiation protocol runs a private-key based key exchange and management protocol to establish and manage multiple IPSec SAs, as shown in Fig. 1 (b). HDR contains session-specific information (such as session ID etc), SA contains a list of cryptographic algorithms, g^x is the D-H parameter, nonces are large random numbers, and IDs are unique identities of communicating parties. Parameters in '[]' are optional.

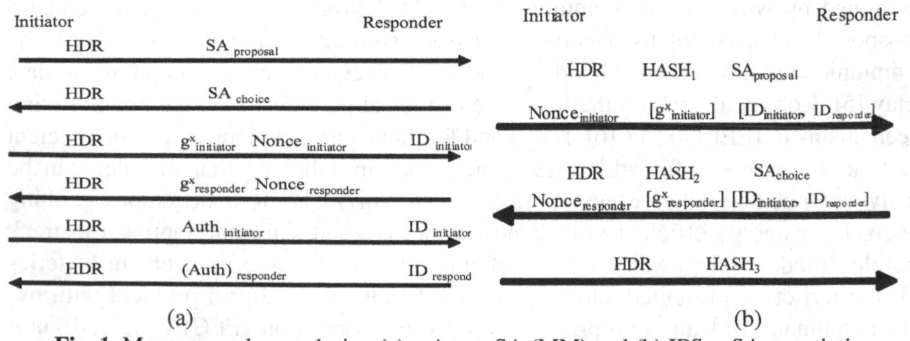

Fig. 1. Message exchange during (a) *primary* SA (MM) and (b) IPSec SA negotiation

2.2 WTLS Session Negotiation

WTLS has been derived by optimizing SSL and TLS protocols for wireless environment. While IPSec provides end-to-end security, WTLS secures only the wireless channel between WAP gateway and mobile client. Wired channel between web server and WAP gateway is secured using SSL protocol. In this study we do not consider the energy consumed by SSL session negotiation protocol since it is similar to the one used by WTLS. Besides, WAP gateways and web servers are assumed to be connected to power sources of much greater capacity. Fig. 2 shows that message flow and cryptographic computations during WTLS secure session negotiation are similar to those during IPSec secure session negotiation. KE denotes key-exchange.

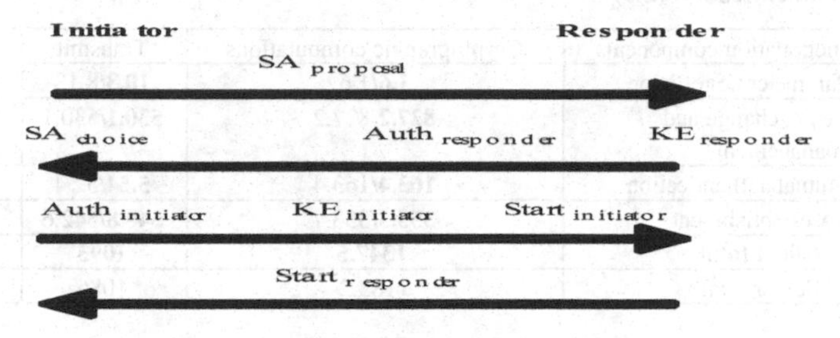

Fig. 2. Session negotiation used by WTLS

3 Energy Consumed by Secure Session Negotiation Protocols

In this section we describe our mobile test bed and experimental methodology for studying energy consumption characteristics of secure session negotiation protocols.

3.1 Mobile Test Bed and Experimental Methodology

Mobile test bed consists of IBM® 600E series ThinkPad™ equipped with 11 Mbps Spectrum24™ WLAN card from Symbol® Tech. Inc. and running Windows 98™ operating system on a 366 MHz Mobile Pentium II™ processor system with 64 MB SDRAM, 64 KB L1 internal CPU cache and 256 KB L2 external CPU cache. The WLAN card can operate in 5 different polling modes, each with unique beacon period and transmit and receive timeouts. We operate the WLAN card in P1 polling mode with "receive timeout" of 200 mS, "transmit timeout" of 500 mS, and average response time of 150 mS since we found that this mode offers the best trade-off between energy consumption and performance. We measure the current drawn by an application executing on the mobile test bed using a Tektronix TCP202 current probe (DC to 50 MHz, Min sensitivity: 10 mA/div, DC accuracy: ±1% with probe calibrator) and a Tektronix TDS 3054 oscilloscope (4 channel, 500 MHz, 5 GS/s). Detailed description of energy profiling experiments and results can be found in [15].

3.2 Energy Consumed by IPSec Session Negotiation

We compute the energy consumed during a session negotiation as energy consumed for authenticating communicating parties (sign and verify certificates) + energy consumed for negotiating parameters (exchange SA primitives) + energy consumed for key exchange and management (generate shared secrets) + energy consumed for SA establishment (generate secret keys for data encryption + generate secret keys for data authentication and express it as a function of size of messages exchanged and cryptographic computations performed.

Table 2. Energy consumed by IPSec secure session negotiation in main mode with pre-shared key based authentication (mJ)

Session negotiation components		Cryptographic computations	Transmit	Receive
Primary	Parameter negotiation	**1.6**/1.6	**10.3**/8.18	**3.72**/4.68
	Key exchange and management	**827.2**/827.2	**530.1**/530.1	**241**/241
	Mutual authentication	**163.4**/163.4	**5.54**/5.54	**2.5**/2.5
IPSec	SA establishment	**355.3**/353.7	**548.8**/542.6	**246.7**/247.5
Client total		**1347.5**	**1095**	**493.9**
Server total		1183.2	1086	495.7

We assume that client is initiator and server is responder; client SA proposal includes AES, RC6, and Twofish encryption algorithms for Encapsulating Security Payload (ESP) protocol and SHA-256 MAC for Authentication Header (AH) protocol; and server selects 128-bit key AES and SHA-256. Table 2 summarizes the energy consumed by various components of IPSec secure session negotiation protocol. Energy values in bold correspond to client. More than 80% of energy consumed during *SA establishment* is due to exchange of large size certificates. Table 3 summarizes the total energy consumed by IPSec session negotiation at the client and the server in main and aggressive modes corresponding to various authentication methods. Aggressive mode consumes approximately the same amount of energy as the main mode since both modes exchange approximately the same amount of information and perform similar cryptographic computations. Further, an IPSec client using *revised public key encryption authentication* based session negotiation consumes 3687 mJ (almost equal to the energy consumed by a client using *public key signature authentication* based session negotiation), of which 89% is consumed by message exchanges and 11% is consumed by public key cryptographic computations.

Table 3. Total energy consumed by IPSec session negotiation protocol in various modes (mJ)

Mode	Authentication method	Client	Server
Main	*Pre-shared key*	2936	2927
	Public key signature	3717	3708
	Public key encryption	3160	3151
	Revised public key encryption	3687	3339
Aggressive	*Pre-shared key*	2935	2927
	Public key signature	3666	3657
	Public key encryption	3160	3151
	Revised public key encryption	3687	3339

3.3 Energy Consumed by WTLS Session Negotiation

Table 4 shows the energy consumed by session negotiation protocol used by WTLS. A WTLS client using D-H key management protocol consumes approximately 2100 mJ, 23% of which is consumed by cryptographic computations and 77% is consumed by message exchanges.

Table 4. Energy consumed by session negotiation protocol used by WTLS (mJ)

	Cryptographic computations	Transmit	Receive
Parameter negotiation	**1.6**	**23**	10
Mutual authentication + Key exchange and management	355	1068	**486**
	355	**1067**	486
SA establishment	**168**	-	-
	168	-	-
Client Total	**522**	**1079**	**485**
Server Total	522	1069	491

4 Reducing the Energy Consumed by Secure Session Negotiation

In this section, based on the analysis of our experimental results, we propose techniques for optimizing the energy consumed by session negotiation protocols and study the impact of these techniques on the energy consumption characteristics of the session negotiation protocols used by IPSec and WTLS security protocols. Message exchanges account for a considerable fraction of the energy consumed by session negotiation (more than 80%). Therefore, reducing the amount of information exchanged during a secure wireless session negotiation can significantly reduce its energy consumption. This can be achieved by reducing either the size or the number, or some combination thereof, of the messages exchanged. Similarly, reducing the number and complexity of cryptographic computations can also reduce the energy consumed by secure session negotiation protocols.

4.1 Protocol Message Compression

Compressing session negotiation messages before transmission and decompressing them after reception may reduce the energy consumed by secure session negotiations. To study the impact of compression we used optimized 'C' implementation of DEFLATE loss-less data compression algorithm [16]. Compression level, history window size, and memory-level are the three important parameters that affect the energy consumed by DEFLATE compression. Table 5 summarizes the energy consumed by DEFLATE while compressing 1KB, 8 KB and 32 KB size benchmarks from Calgary corpus [17]. Column 3 and 4 show that reducing the memory resource while increasing the compression level increases the energy consumed by large (32KB) data block due to the energy-intensive memory fetch operations. On other hand, column 5 and 6 show that for a constant memory level increasing the compression level has a more severe effect on the energy consumed by smaller data blocks since small data size requires more complex operations for achieving high compression ratio. From Table 5, it can be seen that increasing either the compression level or the memory level results in a proportional increase in the energy consumed without a corresponding increase in the compression ratio, while increasing the size of the history window yields a proportional increase in the compression ratio without a corresponding increase in the energy consumed.

Table 5. Energy consumption characteristics of DEFLATE as a function of its paramters (mJ) [1]

Data size		CL = 1 ML = 9	CL = 9 ML = 1	CL = 9 ML = 5	CL = 5 ML = 5
32 KB	Energy	28.32	103.98	52.65	34.58
	CR	3.6800	4.1550	4.6330	4.04720
8 KB	Energy	21.12	16.93	24.03	11.53
	CR	3.3800	3.4700	3.9200	3.08430
1 KB	Energy	18.76	7.89	20.55	7.80
	CR	2.1000	1.9360	2.1760	2.1710

We found that medium compression level (level 5), medium memory level (level 5), and maximum history window size (32 KB) combination achieves a compression ratio close to the best while consuming significantly less energy. We also found that energy consumed by data decompression is very small compared to energy for data compression (~10× less for DEFLATE) since decompression involves fewer and simpler computations. Hence, energy consumed by mobile client can be reduced significantly if server/gateway compresses all messages before transmission. Table 6 shows a 1.7× and 2× reduction in the energy consumed by session negotiation protocols used by IPSec and WTLS respectively by compressing the messages containing certificates and shared secret values.

Table 6. Energy saved by compressing session negotiation protocol messages at the client (mJ)

	IPSec (Main mode, Revised PKE)		WTLS
	primary SA	IPSec SA	
Uncompressed	2536	1151	2088
Compressed	1609	620	1033
Energy saving	1.58×	1.86×	2×

4.2 Choice of Cryptographic Algorithms

Choice of key exchange and management algorithms can also reduce the energy consumed by the session negotiation protocols at the mobile clients. IPSec session negotiation protocol uses D-H key exchange and management for exchanging shared secrets and RSA public-key scheme for mutual authentication [18]. In wired networks,

[1] CR: Compression Ratio, CL: Compression Level, ML: Memory Level

these two operations are isolated to ensure perfect forward secrecy; otherwise anybody who somehow gains access to the RSA private keys can obtain access to the future as well as the past communications. On other hand, in mobile wireless environment such risks are minimal due to the relatively shorter life-span of wireless sessions. For example, both IPSec and WTLS session negotiation protocols derive all secret keys using ephemeral nonces which expire as soon as wireless session terminates. Therefore, energy consumed by session negotiation can be reduced by replacing symmetrical D-H with asymmetrical RSA for both exchanging the shared secret and authentication, as shown in Table 7. Using RSA protocol a server sends its certificate containing public key to client. Client encrypts a small random value (shared secret) using server public key and transmits the result back to server. This reduces the energy consumption by more than 1.5× for both IPSec and WTLS.

Table 7. Energy saved by the choice of key exchange and management protocols at client (mJ)

Key exchange and management protocol	IPSec (Main mode, Revised PKE)		WTLS
	primary SA	IPSec SA	
Diffie-Hellman	2536	1151	208
RSA	1761	550	1319
Energy saving	1.44×	2.1×	1.58×

4.3 Optimized Client Authentication

With increasing mobility of wireless devices across heterogeneous wireless networks frequency of session establishment, and hence authentication, is increasing. Energy consumed by mutual authentication can be reduced by 1.17× and 1.34× respectively for IPSec and WTLS by allowing the client to instead send the URL of its certificate to the server. Table 3 shows that *public key signature authentication* based session negotiation consumes maximum energy since it entails computation of public-key signatures and exchange of large size certificates. *Public key encryption authentication* based session negotiation is comparatively energy-efficient due to the absence of certificate exchanges, even though it requires 2 public-key encryption operations at the client. *Revised public key encryption authentication* based session negotiation is almost as energy-inefficient as the *public key signature authentication* based session negotiation since it also involves exchange of large size certificates. *Pre-shared key authentication* based session negotiation scheme consumes least energy but is unsuitable for large-sized networks and for mobile operations. An optimized IPSec secure session negotiation protocol based on revised public key encryption is most suitable for wireless environment since it allows client to send URL of its certificate for energy-efficiency, is scalable, involves only one public-key encryption at client, does not carry out private-key encryption and transmission of large certificate, and uses ephemeral shared secrets (based on exchanged nonces).

4.4 Security Association Refresh and Energy-Efficient Secret Key Generation

Wireless networks are inherently unreliable and discontinuous resulting in frequent secure session negotiations. Network services may become unavailable due to bad radio coverage, shortage of resources or network roaming. Therefore, session resumption and transaction recovery, together with fast secret key generation are key to energy efficiency of mobile clients. Client energy can be reduced by modifying the session negotiation protocols such that the server looks up the client's certificate from its own source (*session negotiation variant 1*). Embedding the client's shared secret in its certificate and compressing the resulting output can further reduce this energy. When establishing a new session, a client-server pair can exchange new client and server nonces and combine these with previously negotiated security association (*session negotiation variant* 2). Finally, implanting the shared secret in the server and mobile client eliminates the energy consumed by shared secret exchange messages (*session negotiation variant 3*). Table 8 shows the energy consumed by these variants of session negotiation protocols.

We propose an *adaptive session negotiation* protocol that uses *session negotiation variant 1* to establish a new session and *session negotiation variant 2* to refresh SA by exchanging new client and server random numbers if the session lasts beyond a certain number of messages, as determined by security requirements of the session, or if the session is disrupted abruptly due to bad channel conditions or temporary network outage. It is important to note that the proposed *adaptive session negotiation* protocol does not compromise security. Level of security can easily be adjusted by selecting an appropriate security policy, such as the session or key refresh rates, which can be negotiated by communicating parties beforehand. Since *session negotiation variant 3* is inflexible (shared secret is implanted permanently into server and client) and vulnerable to physical and side-channel attacks to recover implanted shared secret, we did not include it in the *adaptive session negotiation* protocol.

Table 8. Energy saved by optimizing session negotiation protocol (mJ)

	Basic		variant 1		variant 2		variant 3	
	WTLS	IPSec	WTLS	IPSec	WTLS	IPSec	WTLS	IPSec
Transmit	1079.6	1623	553.2	561	0.93	1.93	25.2	66.2
Receive	486.4	494	481	164	0.36	1.36	1	2.5
Auth.+ Key-Ex. & Mgmt.	352.5	1407	352.5	1341	-	-	-	-
Key Gen. (Enc.+ Auth.)	168.2	163	168.2	163	168.2	163	168.2	163
Total	2086.7	3687	1554.9	2229	169.49	166	220.6	232
Saving factor			1.3×	1.65×	12.5×	22×	9.5×	16×

5 Impact of System Resources on Session Negotiation Energy

Towards validating the platform-independence of proposed techniques we used another Symbol PPT2800™ Pocket PC based mobile test bed [1]. Choice of a Pocket

PC based mobile test bed was influenced by the rapidly increasing growth of integrated mobile communication and computing devices that combine the functionality of Pocket PC with those of cellular phones. Table 9 shows that proposed techniques are platform independent, though the magnitude of their impact on the energy consumed by secure session negotiation protocols depends upon platform-specific details (such as available system resources). For example, energy savings obtained by the proposed techniques on Pocket PC test bed are higher than those obtained on ThinkPad test bed. This is primarily because all the techniques reduce energy consumed by wireless data communication which forms a significant fraction of the system energy (idle Symbol PPT2800™ Pocket PC draws 60 mA without WLAN card and 350 mA current with WLAN card operating in P1 polling mode). Similarly, matching compression block size to data cache size (8KB for Symbol PPT2800™) provides significantly better tradeoff between energy and compression ratio.

Table 9. Impact of platform-specific characteristics on proposed techniques

Secure wireless session negotiation		IPSec (Main Mode, Revised PKE)	WTLS
Basic	Un-optimized	1521	1062
	Optimized	291	199
	Energy saving	5.23×	5.33×
Re-fresh	Un-optimized	543	1062
	Optimized	51	33
	Energy saving	10.6×	32.2×

References

[1] R. Karri, P. Mishra, “Optimizing the energy consumed by secure wireless session – Wireless Transport Layer Security case study,” Journal of Mobile Networks and Applications, Kluwer Academic Publications, Apr 2003, Vol. 8, No. 2, pp. 177–185.

[2] H. Woesner, J. P. Ebert, M. Schlager, A. Wolisz, “Power saving mechanisms in emerging standards for wireless LANs: The MAC-level perspective,” IEEE Personal Communication Systems, 1998, pp. 40–48.

[3] J. Ebert, B. Stremmel, E. Wiederhold, A. Wolisz, “An energy-efficient power control approach for WLANs,” Journal of Communications and Networks, Sep 2000, Vol. 2, pp. 197–206.

[4] J. M. Rulnick, N. Bambos, “Mobile power management for maximum battery life in wireless communications network,” Proceedings, IEEE ICC, Mar 1996, Vol. 2, pp. 443–450.

[5] R. Kravets, P. Krishnan, “Power management techniques for mobile communication,” Proceedings, ACM/IEEE MOBICOM, Aug 1999.

[6] C. Rohl, H. Woesner, A. Wolisz, “A short look on power saving mechanisms in the wireless LAN standard draft IEEE 802.11,” Advances in Wireless Communications, Kluwer Academic Publishers, 1998.

[7] S. Singh, C. S. Raghavendra, “PAMAS-Power Aware Multi-Access Protocol with signaling for ad-hoc networks,” Computer Communications Review, Jul 1998.

[8] M. Zorzi, R. R. Rao, “Energy constrained error control for wireless channels,” IEEE Personal Communications, Dec 1997, Vol. 4, pp. 27–33.

[9] P. Lettieri, C. Fragouli, M. B. Srivastava, “Low power error control for wireless links,” Proceedings, ACM/IEEE MOBICOM, Aug 1997 pp. 139–150.

[10] "IP Security protocol (IPSec)," http://www.ietf.org/html.charters/ipsec-charter.html.
[11] "Wireless Transport Layer Security Specification," http://www.wapforum.org.
[12] "The Internet Key Exchange Protocol," http://www.ietf.org/rfc/rfc2409.txt.
[13] W. Diffie, M. E. Hellman, "New directions in cryptography," IEEE Transactions on Information Theory, Nov 1976, Vol. IT-22, No. 6, pp. 644–654.
[14] P. C. Cheng, "An architecture for the Internet Key Exchange Protocol," IBM Systems Journal, Vol. 40, No. 3, 2001, pp. 721–746.
[15] R. Karri, P. Mishra, "Modeling energy efficient secure wireless networks using network simulation," Proceedings, IEEE ICC, May 2003, Vol. 1, pp. 61–65.
[16] "DEFLATE Compressed Data Format Specification version 1.3," http://www.kblabs.com/lab/lib/rfcs
[17] T.C. Bell, "Text compression", Prentice Hall, Englewood Cliffs, NJ 1990.
[18] R. L. Rivest, A. Shamir, L. M. Adelman, "A method for obtaining digital signatures and public key cryptosystems," Communications of the ACM, Feb 1978, Vol. 21, No. 2, pp. 120–126.

Remote Power Control of Wireless Network Interfaces

Andrea Acquaviva[1], Tajana Simunic[2], Vinay Deolalikar[2], and Sumit Roy[2]

[1] Istituto di Scienze e Tecnologie dell'Informazione
Universita' di Urbino, Italia
acquaviva@sti.uniurb.it,
[2] Hewlett-Packard Laboratories
1501 Page Mill Road
Palo Alto, CA 94304, USA
{tajana.simunic, vinay.deolalikar, sumit.roy}@hp.com

Abstract. This paper presents a new power management technique aimed at increasing the energy efficiency of client-server multimedia applications running on wireless portable devices. We focus on reducing the energy consumption of the wireless network interface of the client by allowing the remote server to control the power configuration of the network card depending on the workload. In particular, we exploit server knowledge of the workload to perform an energy-efficient traffic reshaping, without compromising on the quality of service. We tested our methodology on the SmartBadge IV wearable device running an MPEG4 streaming video application. Using our technique we measured energy savings of more than 67% compared to no power management being used on the WLAN interface. In addition, we save as much as 50% of energy with respect to the standard 802.11b power management. All of the energy savings are obtained with no performance loss on the video playback.

1 Introduction

Portable devices spend a considerable amount of energy in order to support power hungry peripherals e.g. wireless local area network (WLAN) interfaces, and liquid crystal displays (LCD). A good example of such a device is the SmartBadge IV [21] wearable computer. It consists of a StrongARM-1110 processor and SA-1111 coprocessor, memory, WLAN interface, audio codec and 2.2" LCD. Depending on the network traffic, the WLAN accounts for as much as 63% of the overall system power consumption.

One way to reduce the energy consumption is to use the power management included in the 802.11b standard (802.11b PM) [4]. In the standard, an access point (AP) transmits a beacon every 100 ms, followed by a traffic indication map (TIM). Each client checks the TIM for its turn to send or receive data. When not communicating, the WLAN goes into the doze mode until the next beacon. Unfortunately, the protocol power management is not very effective. First, the

J.J. Chico and E. Macii (Eds.): PATMOS 2003, LNCS 2799, pp. 369–378, 2003.
© Springer-Verlag Berlin Heidelberg 2003

energy efficiency of the 802.11b PM decreases and receiver wait times increase with more mobile hosts, since multiple concurrent attempts at synchronization with the beacon cause media access contention. Second, the response time of the wireless link with 802.11b PM grows because of the delay imposed by sleep periods [22]. These two issues can be resolved by careful scheduling of communication between the server and the client WLAN. Lastly, in a typical wireless network, broadcast traffic can significantly reduce the chances to enter the doze mode. Figure 1 shows the power consumption of a WLAN card with 802.11b PM enabled under light and heavy network broadcast traffic conditions. Clearly, as the amount of broadcast traffic increases, the WLAN spends a large amount of energy listening to it, even if no other application is running on the device. As a result, very little or no energy savings are obtained. One way to solve this problem is to turn off the card. It is important to schedule data transmission carefully, since the overhead of waking up the WLAN from the off state is large.

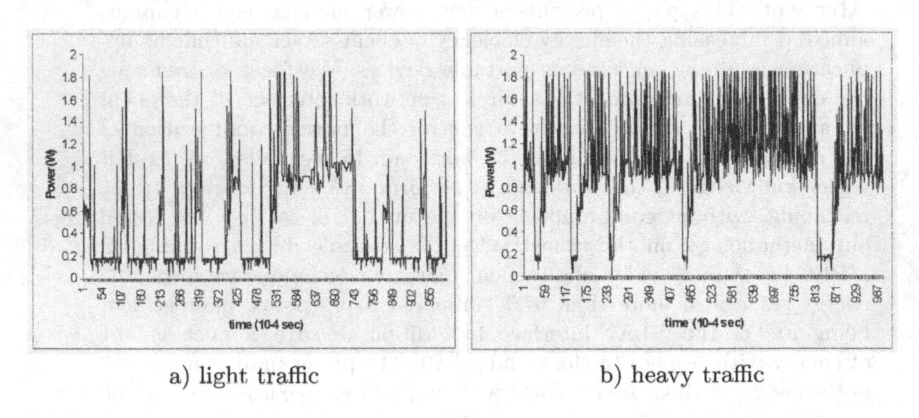

a) light traffic b) heavy traffic

Fig. 1. 802.11 PM under different broadcast traffic conditions

Many current wireless local area networks are organized in a client-server fashion. Multiple WLAN clients connect to wired servers via APs. Servers are great candidates for efficient scheduling of data transmission to clients as they are not power constrained, and know both wired and wireless network conditions.

In this work, we present a server controlled power management strategy. Our technique exploits server knowledge of the workload, traffic conditions and feedback information from the client in order to minimize WLAN power consumption. Our methodology is applicable to a wide variety of applications, ranging from video and audio streaming, to web browsing and e-mail. We define two new entities: a server power manager (server PM) and a client power manager (client PM). Server PM uses the information obtained from the client and the network to control the parameters of 802.11b PM and to perform energy efficient traffic reshaping so that WLAN can be turned off. Client PM communicates through a dedicated low-bandwidth link with the server PM and implements power controls by interfacing with device drivers. It also provides a set of APIs that client programs can use to provide extra information to the server.

In order to illustrate the effectiveness of our approach, we tested our methodology with a streaming video application. By using our approach with this application, we can exploit the server knowledge of stream characteristics. We show that when our methodology is implemented on both the server end, and on the client end, we measure savings of more than 67% in power with respect to leaving the card always on, and more than 50% relative to using default 802.11b PM. Even larger savings are possible for applications that inherently have longer idle periods, such as e-mail or web browsing. Our methodology can also be easily extended to manage other system components (e.g. CPU).

The rest of the paper is organized as follows. Section 2 presents some related work. Section 3 gives an overview of the proposed methodology and describes server and client power managers. Section 4 presents experimental results and Section 5 concludes the paper.

2 Related Work

The wireless network power optimization problem has been addressed at different abstraction layers, starting from physical, to system and application level. Energy efficient channel coding and traffic shaping to exploit battery lifetime of portable devices were proposed in [3]. A physical layer aware scheduling algorithm aimed at efficient management of sleep modes in sensor network nodes is illustrated in [17]. Energy efficiency can be improved at the data link layer by performing adaptive packet length and error control [8]. At the protocol level, there have been attempts to improve the efficiency of the standard 802.11b, and proposals for new protocols [5,6,19]. Packet scheduling strategies can also be used to reduce the energy consumption of transmit power. In [14] authors propose the E^2WFQ scheduling policies based on Dynamic Modulation Scaling. A small price in packet latency is traded for the reduced energy consumption.

Traditional system-level power management techniques are divided into those aimed at shutting down components and policies that dynamically scale down processing voltage and frequency [20,1]. Energy-performance trade-offs based on application needs have been recently addressed [7]. Several authors exploit the energy-QoS trade-off [12,22,11]. A different approach is to perform transcoding and traffic smoothing at the server side by exploiting estimation of energy budget at the clients [16]. A new communication system, consisting of a server, clients and proxies, that reduces the energy consumption of 802.11b compliant portable devices by exploiting a secondary low-power channel is presented in [18]. Since multimedia applications are often most demanding of system resources, a few researchers studied the cooperation between such applications and the OS to save energy [9,2,15,10].

We present a new methodology, where server knowledge of the workload is exploited to control the power configuration of the radio interface. Compared to physical and protocol layer strategies, the power control is performed at the application level, so it does not require hardware modifications. Compared to client-centric approaches, we exploit additional information available at the server, and thus obtain large energy savings without loosing performance. Moreover, with respect to previous application-driven policies, our infrastructure can be used with a wide range of applications, since it exploits very common parameters.

3 Server Controlled Power Management

Our methodology exploits server knowledge of the workload, traffic conditions and feedback information from the client in order to minimize power consumption. The server schedules communication sessions with the client based on the knowledge of both, the wireless and wired networks, e.g favorable channel conditions or channel bandwidth capabilities. When broadcast traffic needs to be monitored, the server can enable the 802.11 PM. Alternatively, it can coordinate the shut down of WLAN and perform on-time wake-up, thus avoiding the performance penalty typically incurred by client-centric approaches. Our application driven infrastructure can be also be used to manage power consumption in stand-alone and ad-hoc applications. The power control strategy can easily be extended to include other system components, such as peripherals or the CPU.

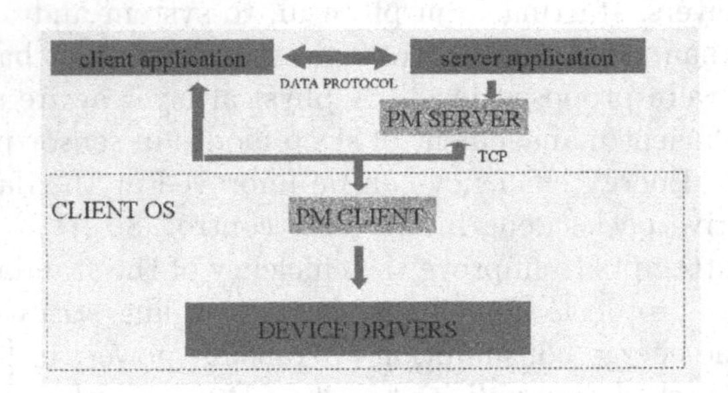

Fig. 2. Server Controlled PM Architecture

In order to exploit the extra information available at the server and the client, the traditional client-centric power manager model has to be extended. Two different power managers are defined: one running on the client (Client PM) and the other on the server (Server PM). The two PMs exchange power control information through a dedicated TCP connection. As shown in Figure 2, the client PM interfaces directly with the drivers on the portable device. It also collects client application dependent information. The server PM interfaces with the server application and the client PM.

Figure 3 shows the communication protocol between the server and client. Control commands are issued by the server PM, interpreted by the client PM, and translated into appropriate device driver calls. Upon request from the server PM, device specific information is fetched by the client PM. Application specific information can also be retrieved by the client PM via API calls.

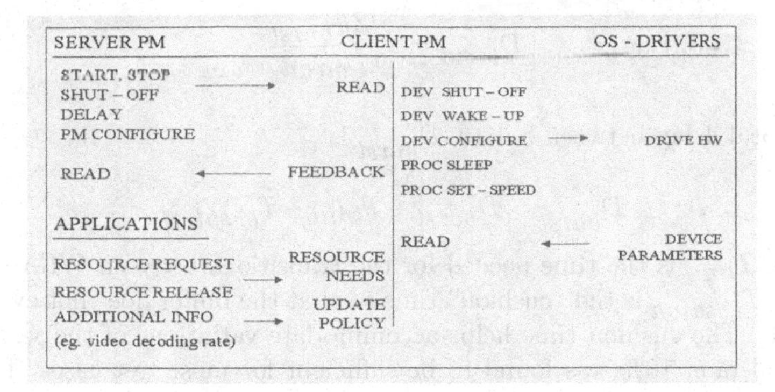

Fig. 3. Communication protocol

3.1 Server Power Manager

The server PM interfaces with the application in order to exploit information available on the host system, in addition to the knowledge of the overall network conditions and the specific feedback information provided by each client. Based on all this knowledge, the server PM controls the power configuration of the WLAN card by communicating with the client PM. The power control commands used include: i) switch-off the WLAN; ii) set the off time; iii) enable the 802.11b PM policy; iv) set 802.11b PM parameters, e.g. the period between wake ups and the timeout before going back to doze mode. The server PM enables:

Client Adaptation. After the application server initializes the server PM, an additional set of APIs becomes available. The initialization routine places the server PM into a state of waiting for incoming client PM requests. Each accessing client PM has a dedicated TCP connection to the server PM. The client PM informs the server PM about the application e.g. input buffer size, the expected value and variance of the service rate, as well as network interface specific information e.g. the WLAN card status and its on/off transition time.

Traffic Adaptation. Server PM monitors both the wired and wireless traffic conditions with minimum overhead. By accounting for the broadcast packet rate, the server decides when to enable the 802.11b PM. For example, in very light traffic conditions, the 802.11b PM might be used instead of a switch-off policy.

Traffic Shaping. The server PM schedules transmissions to the client in bursts, in order to compensate for the client performance and energy overheads during transitions between on and off states. The client WLAN card is switched off once the server has sent a burst of data designed to keep the application busy till the next communication burst. The burst size and delay between bursts are precomputed at the server. The delay should be large enough to almost empty the client input buffer, while the burst size should avoid overflow while keeping the buffer sufficiently filled. This maximizes the off time of the card and reduces the number of transitions between on and off states. The time for the buffer to empty, D_{burst} is the ratio of the total number of packets in the burst, $SIZE_{burst}$, to the average service rate (or buffer depletion rate) at the client, $\lambda_{s,mean}$:

$$D_{burst} = \frac{SIZE_{burst}}{\lambda_{s,mean}} \tag{1}$$

The total delay between bursts, $D^{|}_{burst}$ is:

$$D^{|}_{burst} = D_{burst} - T_{tran} - T_{cushion} \tag{2}$$

Where T_{tran} is the time needed for the transitions between WLAN on and off states, $T_{cushion}$ is the "cushion" time so that the buffer does not ever empty completely. The cushion time helps accommodate variations of the service rate and arrival rate, 10% was found to be sufficient for most test cases. The total energy saved if the client WLAN is turned off is:

$$E_{tot} = P_{on} \cdot T_{on} + P_{off} \cdot T_{off} + P_{tran} \cdot T_{tran} \tag{3}$$

Here P_{tran} is the transition power (hardware dependent). T_{tran} can be computed by multiplying the duration of one transition to the active state, $T_{wake\text{-}up}$, by the total number of transitions, N_{tran}. The number of transitions depends on the size of the burst and the total size of streamed file.

If $\lambda_{a,mean}$ is the average arrival rate at the client input buffer, the total time to send a burst of data to the client will be:

$$T_{on} = \frac{SIZE_{burst}}{\lambda_{a,mean}} \tag{4}$$

Finally, we can obtain the total energy consumed with our methodology:

$$E_{tot} = P_{on} \cdot \frac{SIZE_{burst}}{\lambda_{a,mean}} + P_{off} \cdot \frac{SIZE_{transfer}}{\lambda_{s,mean}} + P_{tran} \cdot T_{tran} \tag{5}$$

In addition to scheduling communication so that the client WLAN can be turned off with no performance overhead, we can also perform the same scheduling while 802.11b PM is enabled. This allows us to save energy during the burst periods, but the overall burst time grows because of the decreased responsiveness and increased contention probability between data and broadcast packets.

3.2 Client Power Manager

The client power manager communicates with the server PM, and also interfaces with the device drivers and client applications. The main client PM tasks are:

Server Interface The client application decides when to set-up the communication between the server PM and the client PM. Once established, the client application provides the information to be forwarded to the server e.g. the buffer size and the depletion rate. The device drivers report the main characteristics of the devices to be managed, e.g. the transition time between on and off states for client WLAN.

Device Interface The client PM calls the appropriate device driver function depending on the command sent by the server PM. Possible actions taken by the client include changing the parameter of the 802.11b PM, switching the WLAN on and off, and reading the interface statistics such as the signal to noise ratio. The client PM can also interact with the CPU by changing its power mode or setting its clock speed.

Application Interface Applications can feedback information that can be exploited by both the server and the client PM. Examples include sending the current backlog level to the server, so that thee server knows exactly how much data to provide in a burst in order to refill the buffer; or providing the buffer size. The application could directly request a WLAN wake-up when its input buffer reaches a minimum value.

Application Driven Infrastructure The client PM can also be used in stand-alone mode (with no server). While in this mode, the applications provide their resource needs to the power manager. The PM then turns on or off devices appropriately. Some of the overhead needed to turn on a device can be masked, as many applications have extra latency due to the initial set-up.

4 Experimental Results

The streaming media server used for this work is a research prototype developed at Hewlett-Packard Laboratories. Real Time Streaming Protocol (RTSP) is used for session initiation and termination between the client and server. The media data units are carried using Real-time Transport Protocol (RTP) over User Datagram Protocol (UDP). Timestamps of the individual video frames are used to determine the deadline for sending data packets sent from the server to the client. The server can exploit client buffering capabilities to reschedule packet transmission times based on a cost function. In our experiments, the server transmits MPEG4 video data to a portable client device via WLAN.

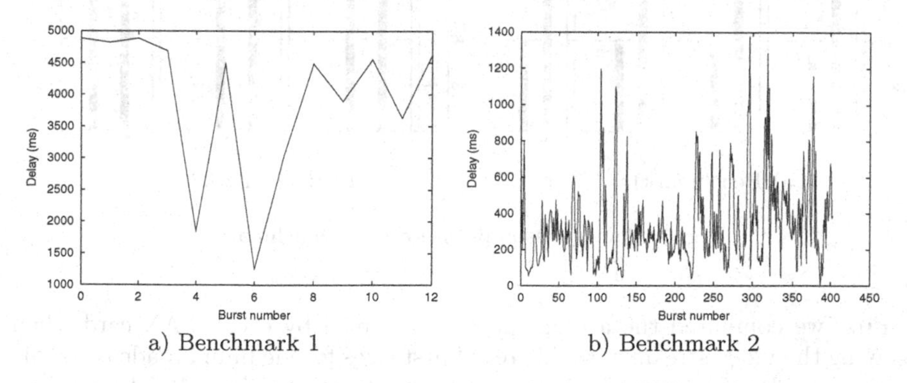

Fig. 4. Computed burst delays for each benchmark

We use two different benchmarks in our tests. Benchmark 1 has 12 bursts and runs at 15 frames/sec, while Benchmark 2 has 402 bursts at 30 frames/sec.

The first benchmark has a large number of single packet frames, while the second benchmark has many large multi-packet frames. The delay between bursts is:

$$D_{burst} = frame\_time \cdot (\frac{n_1}{1} + \frac{n_2}{2} + \frac{n_3}{3} + \ldots + \frac{n_M}{M}) \quad (6)$$

Here *frame_time* is the interval between frames, n_i is the number of frames consisting of i packets and M is the maximum number of packet per frame. Figure 4 shows that the Benchmark 1 has a delay of around 4 seconds, while the Benchmark 2 has a shorter delay because the frames are more complex. Based on the delay computation, the server transmits a burst of data to the client followed by the command to switch off WLAN and the length of time until the next transmission. The client responds by turning off the WLAN and turning it back on at the pre-scheduled time. As a result, no performance penalty due to longer turn on time is incurred.

The approach was tested by measuring power consumption of a SmartBadge IV wearable device, equipped with a CISCO Aironet 350 PCMCIA WLAN card. The off/on transition time was measured at 300 ms, with an average transition power consumption of 0.1 W. The server is connected directly to an AP, and the network is completely isolated in order to perform repeatable experiments. Broadcast traffic is introduced in a controlled manner by using real traces collected from other open networks. Power measurements are performed using a DAQ board accumulating and averaging current and voltage samples (10 ksamples/sec). All measurements include the overhead imposed by our power management protocol.

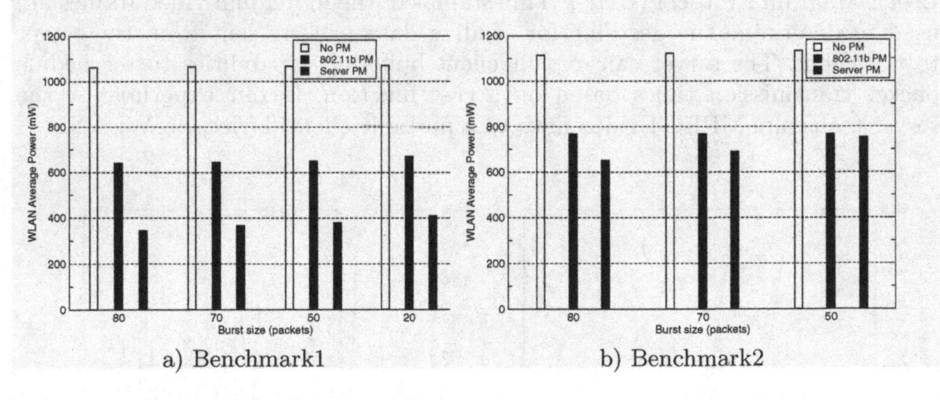

Fig. 5. Power consumption for each benchmark

First we computed the average power consumed by the WLAN card when receiving the video stream with different burst sizes for medium broadcast traffic conditions. The experiment is performed for multiple situations, WLAN with no PM, only 802.11b PM, and with server PM. From Figure 4, for Benchmark 1, the server controlled approach saves 67% of average power compared to no PM, and 50% compared to the default 802.11b PM. The average power savings increase as the burst size increases, since this enables longer times between bursts, thus better compensation for the transition delay between the WLAN on and off states. In all three cases, the video plays back in real time, thus the reported average power savings directly correspond to energy savings. For Benchmark 2, the delays are very short, however, we still save 41% of power with respect to leaving the card always on and 15% with respect to the standard 802.11b power management protocol with bursts of 80 packets. We cannot use a burst size smaller than 50 packets, since the computed delays are too short and the card can never be switched off.

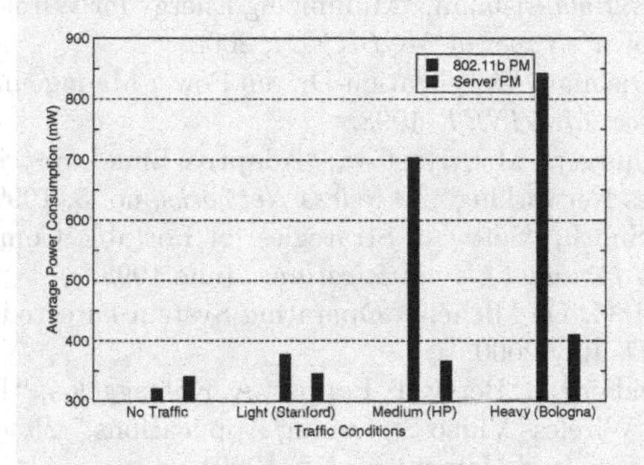

Fig. 6. Power consumption with different traffic levels

Figure 6 presents the average power consumed during streaming video under different broadcast traffic conditions. In the plot we show the difference between using only the default 802.11b PM and server controlled switch off policy for the WLAN card (with no 802.11b PM). The switch off policy is almost insensitive to the traffic conditions, while the 802.11b PM performs better only in light traffic conditions. Clearly, our server controlled power management approach is more efficient than either policy alone as it always selects the best of the two policies.

5 Conclusion

In this work we presented a new methodology to improve the energy efficiency of portable wireless systems that operate in a client-server fashion. Our system enables the server to exploit knowledge of the workload, traffic conditions and feedback information from the client in order to minimize client WLAN power consumption. We tested our methodology on the SmartBadge IV wearable device running a streaming video application. Our server controlled approach saves 67% of energy as compared to leaving the client WLAN card always on, and 50% as compared to the simple 802.11b PM policy.

References

1. A. Acquaviva, L. Benini, B. Riccó, "Software Controlled Processor Speed Setting for Low-Power Streaming Multimedia," *IEEE Trans. on CAD*, Nov. 2001.

2. F. Bellosa, "Endurix: OS-Direct Throttling of Processor Activity for Dynamic Power Management," *Technical Report TR-I4-99-03, University of Erlangen*, June 1999.
3. C. Chiasserini, P. Nuggehalli, V. Srinivasan, "Energy-Efficient Communication Protocols", *Proc. of DAC*, 2002.
4. IEEE LAN/MAN Standards Committee, "Part 11: Wireless LAN MAC and PHY Specifications: Higher-Speed Physical Layer Extension in the 2.4 GHz Band", 1999.
5. C. Jones, K. Sivalingam, P. Agrawal, J. Chen, "A Survey of Energy Efficient Network Protocols for Wireless Networks", *Proc. of DATE*, 1999.
6. R. Krashinsky, H. Balakrishnan, "Minimizing Energy for Wireless Web Access with Bounded Slowdown", *Proc. of MOBICOM*, 2002.
7. R. Kravets, P. Krishnan, "Application-Driven Power Management for Mobile Communication," *Proc. of WINET*, 1998.
8. P. Lettieri, C. Schurgers, M. Srivastava, "Adaptive Link Layer Strategies for Energy Efficient Wireless Networking", *Wireless Networks*, no. 5, 1999.
9. J. Lorch, A. J. Smith, "Software Strategies for Portable Computer Energy Management," *IEEE Personal Communications*, June 1998.
10. Y. Lu, L. Benini, G. De Micheli, "Operating System Directed Power Reduction," *Proc. of ISLPED*, July 2000.
11. C. Luna, Y. Eisenberg, R. Berry, T. Pappas, A. Katsaggelos, "Transmission Energy Minimization in Wireless Video Streaming Applications," *Proc. of Asilomar Conf. on Signals, Systems, and Computers*, Nov. 2001.
12. R. Min, A. Chandrakasan, "A Framework for Energy-Scalable Communication in High-Density Wireless Networks," *Proc. of ISLPED*, 2002.
13. T. Pering, T. Burd, R. Brodersen, "Voltage Scheduling in the lpARM Microprocessor System", *Proc. of ISLPED*, July 2000.
14. V. Raghunathan, S. Ganeriwal, C. Schurgers, M. Srivastava, "E^2WFQ: An Energy Efficient Fair Scheduling Policy for Wireless Systems", *Proc. of ISLPED*, 2002.
15. J. Flinn, M. Satyanarayanan, "Energy-aware adaptation for mobile applications," *Proc. of SOSP*, Dec. 1999.
16. P. Shenoy, P. Radkov, "Proxy-Assisted Power-Friendly Streaming to Mobile Devices," *Proc. of MMNC*, Jan. 2003.
17. E. Shih, P. Bahl, M. Sinclair, "Dynamic Power Management for non-stationary service requests", *Proc. of MOBICOM*, 2002.
18. E. Shih, S. Cho, N. Ickes, R. Min, A. Sinha, A. Wang, A. Chandrakasan, "Physical Layer Driven Protocol and Algorithm Design for Energy Efficient Wireless Sensor Networks", *Proc. of SIGMOBILE*, 2001.
19. K. Sivalingam, J. Chen, P. Agrawal, M. Srivastava, "Design and Analysis of low-power access protocols for wireless and mobile ATM networks", *Wireless Networks*, no.6, 2000.
20. T. Simunic, L. Benini, P. Glynn, G. De Micheli, "Event-driven Power Management," *IEEE Trans. on CAD*, July 2001.
21. M. T. Smith and G. Q. Maguire Jr., "SmartBadge/BadgePad version 4", HP Labs and Royal Institute of Technology (KTH), http://www.it.kth.se/ maguire/badge4.html, date of access: 2003-06-11.
22. E. Takahashi, "Application Aware Scheduling for Power Management on IEEE 802.11" *Proc. of Intl. Performance, Computers, and Communications Conf.*, Feb. 2000.

Architecture-Driven Voltage Scaling for High-Throughput Turbo-Decoders*

Frank Gilbert and Norbert Wehn

Microelectronic System Design Research Group, University of Kaiserslautern
Erwin-Schrödinger-Straße, 67663 Kaiserslautern, Germany
{gilbert,wehn}@eit.uni-kl.de

Abstract. The outstanding forward error correction provided by Turbo-Codes made them part of today's communications standards. Therefore, efficient Turbo-Decoder architectures are important building blocks in communications systems. In this paper we present a scalable, highly parallel architecture for UMTS compliant Turbo decoding and apply architecture-driven voltage scaling to reduce the energy consumption. We will show that this approach adds some additional, more energy-efficient solutions to the design space of Turbo decoding systems. It can save up to 34 % of the decoding energy per datablock, although the supply voltage can not arbitrarily selected. We present throughput, area, and estimated energy results for various degrees of parallelization based on synthesis on a 0.18 μm ASIC-technology library, which is characterized for two different supply voltages: nominal 1.8 V and nominal 1.3 V.

1 Introduction

Wireless communications is without doubt an emerging technology. The transition from 2nd generation to 3rd generation mobile communications systems is accompanied with an increased algorithmic complexity of the communications algorithms. The computational requirements has raised from 100 MIPS for GSM to more than 6300 MIPS for UMTS systems. It is projected in [11], that the requirements for future communications systems will follow this trend. Thus VLSI architectures for advanced communications systems have to accomplish *high throughput signal processing* requirements. *Scalability* is another important architectural feature which allows to adapt the architecture to the actual processing needs. *Low energy consumption* is a further very important issue since many of the devices are handheld devices.

Channel decoding is an important building block in communications systems. It allows to reduce the transmission energy, which is the dominant part in a wireless system. We focus on Turbo-Codes, an iterative channel decoding algorithm. The outstanding forward error correction of Turbo-Codes, made them part of today's communications standards, *e.g.* UMTS [13], CDMA2000, or WCDMA. The high communications performance comes at the expense of high computational complexity. Efficient Turbo-Decoder implementations are an very active research area.

* This work has been partially supported by the *Deutsche Forschungsgemeinschaft (DFG)* under Grant We 2442/1-2

J.J. Chico and E. Macii (Eds.): PATMOS 2003, LNCS 2799, pp. 379–388, 2003.
© Springer-Verlag Berlin Heidelberg 2003

Many papers exist to reduce the energy consumption of Turbo-Decoders on various abstraction levels [10,6,18,7]. But to the best of our knowledge, no publication exists in which voltage scaling was applied to high throughput architectures. Recently in [3] the strong interaction between highly parallel architectures and energy-efficiency was emphasized: maximum parallelism allows to use the minimum clock frequency and supply voltage thus minimizing the energy consumption.

Thus, key for the successful application of voltage scaling is a highly parallel architecture. Although the component decoders in a Turbo-Decoder system have a high degree of inherent parallelism [17], the iterative decoding nature combined with data interleaving hampers a parallelization on system level. However parallelization on system level is a must for high throughput architectures. In this paper, we show how to break open this interleaver bottleneck allowing highly parallel architectures.

The most common metric for architecture evaluation is the architectural efficiency defined as

$$\text{Efficiency} = \frac{\text{Throughput}}{\text{Area} \cdot \frac{\text{Energy}}{\text{Task}}}. \tag{1}$$

Interconnect, communication and feasibility of synthesis based approaches in deep-submicron technologies, however, are key issues for implementation, which limit the design space significantly. State-of-the-art synthesis tools can synthesize up to 100 K gates flat [12]. Thus complex designs must be composed of blocks with this size limitation. Top-level-routing complexity, which leads to time-closure problems [1] has to be reduced by minimizing communication between the building-blocks. Our architecture takes this into account. We compare our architectures with respect to the efficiency metric of Equation 1 and the best solutions for different targeted throughputs are depicted.

The paper is structured as follows: In Section 2 we present the concept of Turbo-Codes, the component decoder algorithm, the windowing technique, and the interleaving process. The Turbo-Decoder system architecture is outlined in Section 3. Our architecture driven voltage scaling approach is proposed in Section 4 and further presents throughput, area, and estimated energy consumption for various degrees of parallelization and two different supply voltages. Section 5 finally concludes the paper.

2 Turbo-Codes

Channel coding in general enables error correction in the receiver side by introducing redundancy (*e.g.* parity information) in the encoder. In Turbo-Codes [2], the original information, denoted as *systematic information* ($\boldsymbol{x}^s$), is transmitted together with the *parity information* ($\boldsymbol{x}^{1p}, \boldsymbol{x}^{2p}_{int}$).

For UMTS [13], the Turbo-Encoder, see Figure 1 (a), consists of two recursive systematic convolutional (RSC) encoders with constraint length $K_c = 4$. Both encoder work on the same block of information bits; for UMTS the blocklength (K) is in the range from 40 to 5114. One RSC encoder can also be interpreted as an 8-state finite state machine (mealy machine). The parity bits depend on the state *transition* of the state machine. In the decoder (receiver side), the knowledge of the state machine's structure can be exploited to calculate the probability of state transitions (based on the sequence of the received information). One RSC encoder works on the block of information in its

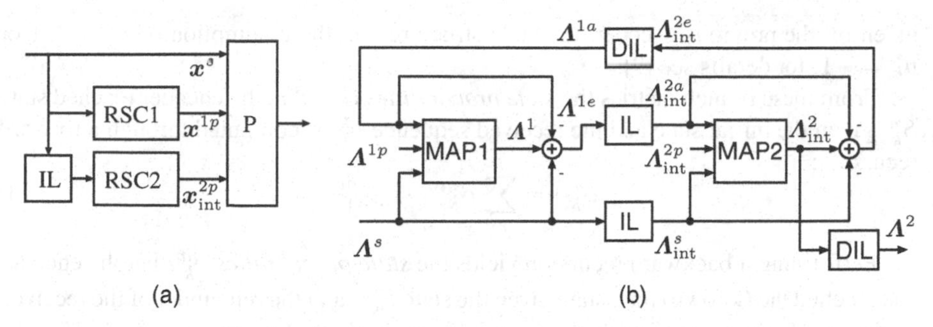

Fig. 1. Turbo-Encoder (a) and Turbo-Decoder (b)

original sequence. The second encoder has the same structure but works on the original data in a different order (an interleaved sequence). In a puncturing unit (P) the code rate R_c can be adjusted.

The Turbo-Decoder receives three sequences of *logarithmic likelihood ratios* (LLRs) Λ_k^s, Λ_k^{1p}, and $\Lambda_{k,\mathrm{int}}^{2p}$ according to $\boldsymbol{x}^s$, $\boldsymbol{x}^{1p}$, $\boldsymbol{x}_{int}^{2p}$. For every RSC encoder a corresponding component decoder exists, see Figure 1 (b). Decoding is an iterative process with the exchange of reliability information. In every iteration each decoder calculates for every received bit a LLR as soft-output (reliability information). The soft-output of each component decoder ($\boldsymbol{\Lambda}$) is modified to reflect only its own confidence ($\boldsymbol{\Lambda}^e$) in the received information bit. The sign of each LLR indicates the received information bit of being sent either as "0" or "1", the absolute values are measures of confidence in the respective 0/1-decision. The *maximum a posteriori* (MAP) decoder has been recognized as the component decoder of choice as it is superior to the *Soft-Output Viterbi Algorithm* (SOVA) in terms of communications performance and implementation scalability [16]. The first MAP-Decoder (MAP1) works on the received bits in the original sequence, the second decoder (MAP2) works on the received bits in the interleaved order. In Figure 1 (b) interleaving and deinterleaving is performed by the blocks IL and DIL respectively. The exchange continues until a stop criterion is fulfilled. The last soft-output is not modified and becomes the soft-output of the Turbo-Decoder ($\boldsymbol{\Lambda}^2$).

2.1 The MAP Algorithm

The reliability information is calculated from the state transition probabilities. For this computation the probabilities of the actual state, its succeeding state and of the structural-caused transitions are required. The MAP algorithm computes the probability for each symbol of $\boldsymbol{u}$ to have been send as $u_k = +1$ or $u_k = -1$. The logarithmic likelihood ratio (LLR) of these probabilities is the soft-output, denoted as:

$$\Lambda_k = \log \frac{\Pr\{u_k = +1 | y_1^K\}}{\Pr\{u_k = -1 | y_1^K\}}. \tag{2}$$

Equation 2 can be expressed using three probabilities, which refer to the encoder states S_k^m, where $k \in \{0 \dots K\}$ and $m, m' \in \{0 \dots 7\}$:

The *branch metrics* $\gamma_{k,k+1}^{m,m'}(u_k)$ is the probability that a transition between S_k^m and $S_{k+1}^{m'}$ has taken place. It is derived from the received signals, the *a priori* information given by the previous decoder, the code structure and the assumption of $u_k = +1$ or $u_k = -1$, for details see [9].

From these branch metrics the *state probabilities* $\alpha_k^{m'}$ that the encoder reached state S_k^m given the initial state and the received sequence y_1^k, is computed through a forward recursion:

$$\alpha_k^{m'} = \sum_m \alpha_{k-1}^m \cdot \gamma_{k-1,k}^{m,m'}. \tag{3}$$

Performing a backward recursion yields the *state probabilities* β_k^m that the encoder has reached the (known) final state given the state $S_k^{m'}$ and the remainder of the received sequence y_{k+1}^K:

$$\beta_k^m = \sum_{m'} \beta_{k+1}^{m'} \cdot \gamma_{k,k+1}^{m,m'}. \tag{4}$$

αs and βs are both called *state metrics*. Equation 2 can be rewritten as:

$$\Lambda_k = \log \frac{\sum_m \sum_{m'} \alpha_k^m \cdot \beta_{k+1}^{m'} \cdot \gamma_{k,k+1}^{m,m'}(u_k = +1)}{\sum_m \sum_{m'} \alpha_k^m \cdot \beta_{k+1}^{m'} \cdot \gamma_{k,k+1}^{m,m'}(u_k = -1)}. \tag{5}$$

The original probability based formulation as presented here involves a lot of multiplications and has thus been ported to the logarithmic domain to become the *Log-MAP Algorithm* [9]: Multiplications turn into additions and additions into maximum selections with an additional correction term. Arithmetic complexity can further be reduced by omitting the correction term (*Max-Log-MAP Algorithm*) which leads to a slight loss in communications performance (about 0.1 - 0.3 dB [8]). Log-MAP and Max-Log-MAP algorithm are common practice in state-of-the-art implementations.

2.2 Windowing

The forward and backward recursion of a MAP-Decoder start at one end of the trellis and stop at the opposite end. However, it is possible to start the recursions on arbitrary positions in the block with approximated initialization values. For this, a recursion on a certain number of proceeding bits (*acquisition*) must be performed to obtain sufficiently accurate estimates of the state metrics. This applies to both forward and backward recursions. Windowing [4,5] exploits this property to divide the data into sub-blocks. Several sub-blocks can thus be decoded sequentially (SMAP) on the same hardware for memory reduction as only the state metrics of one sub-block have to be stored. Moreover, windowing allows to map sub-blocks to individual nodes for parallel processing (PMAP), allowing to trade off hardware for latency. Note, that this is a non-bit-true transformation. But if the number acquisition steps is high enough, the degradation of the communications performance is neglectable.

Figure 2 compares different example windowing schemes. In these examples, the data block of K bits is divided in $\lceil \frac{K}{W} \rceil$ windows (here four), each of them W bits wide. The *y-axis* represents the trellis steps, the *x-axis* represents decoding time (t). In (a) the MAP decoding without windowing is illustrated. The thin line illustrates the progress of the forward recursion, the thick line the progress of the backward recursion *and* the soft-output calculation over the time.

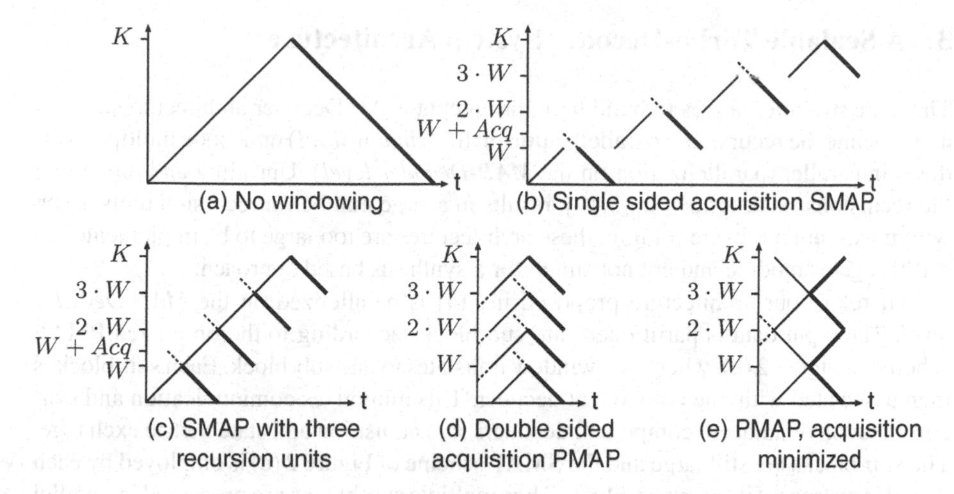

Fig. 2. Windowing schemes for the MAP algorithm

Figure 2 (b) and (c) show windowing schemes of two alternatives for SMAP implementation. Both use single sided acquisition of the backward recursion. The length of the acquisition period is denoted as *Acq*. In (b) a single recursion unit to calculate the state metrics is used for forward recursion, acquisition, and backward recursion sequentially. The SMAP scheme of (c) uses three dedicated recursion units, one for forward recursion, one for acquisition, and one for backward recursion simultaneously. This reduces decoding time at the expense of additional resources.

Figure 2 (d) and (e) illustrate as examples two PMAP windowing schemes. Both use acquisition of the forward *and* backward recursion (double sided acquisition) to allow parallel processing of the windows. In (e) the order of the recursions is rearranged to minimize the acquisition steps.

2.3 Interleaving

Interleaving is scrambling the processing order to break up neighborhood-relations. It is essential for the performance of Turbo-Codes. The interleaver specifies a permutation of the original processing order of the data-sequence. Although in many systems, like UMTS, the interleaving process is specified by an algorithm, the algorithm is not used to do the interleaving but to build an *interleaver table*. The *deinterleaver* reverts the scrambling, setting back the original order. Interleaver and deinterleaver tables contain one-to-one mappings of source addresses to destination addresses.

Due to interleaving, the second MAP-Decoder (MAP2) has to wait for completion of the first decoder (MAP1), *c.f.* Figure 1 (b) and vice versa. Thus, the LLRs produced by the first decoder have to stored in a memory. One LLR of the previous MAP-Decoder has to be read for every new LLR produced. PMAP architectures produce multiple LLRs per time-step. Then *several* LLRs have to be *fetched* from and *stored* to memories *at the same time*. Thus partitioning of the LLR-memory becomes mandatory to provide the concurrent read-access. These resulting concurrent write-accesses to the same single port memory are the real bottleneck in high throughput Turbo decoding.

3 A Scalable Turbo-Decoder System Architecture

There are two alternatives to build high throughput MAP-Decoder architectures: unroll and pipeline the recursions (parallelization on the *window level*) or decode multiple windows in parallel (parallelization on the *MAP-Decoder level*). Unrolling and pipelining the recursions, as described in [5,17], results in a large number of recursion units. Even with maximum hardware folding, these architectures are too large to be implemented in a 100 K gates module and are not suited for a synthesis based approach.

Therefore our architecture proposed in [14] is parallelized on the *MAP-Decoder level*. The input data is partitioned into sub-blocks according to the optimized PMAP scheme in Figure 2 (e), where one window translates to one sub-block. Each sub-block is then associated with one component decoder. This minimizes communication and synchronization of adjacent component decoders for acquisition and state metric exchange. The sub-blocks are still large and the SMAP scheme of Figure 2 (c) is employed by each decoder to process its own sub-block. Thus multiple sub-blocks are processed in parallel each with a dedicated SMAP-unit.

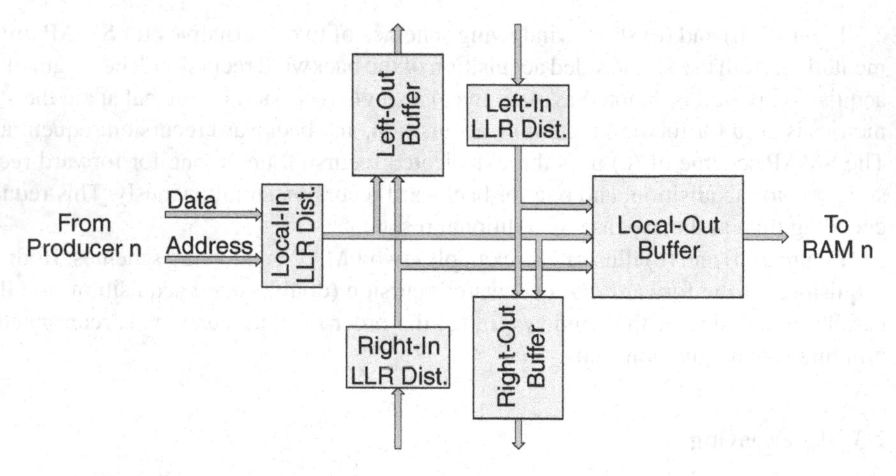

Fig. 3. RIBB-cell as presented in [15]

The inevitable high throughput communication for interleaving is reduced to point-to-point communication of RIBB-cells, as in [15]. There, we presented an optimized concurrent interleaving architecture to solve the conflicts described in Section 2.3. It is a distributed architecture with local entities that resolve local conflicts only, which is called ring-interleaver-bottleneck-breaker (RIBB) architecture. Nodes, each associated with a producer of one LLR/cycle (SMAP-unit), are connected in a ring structure with buffers between them. Each RIBB-cell, see Figure 3, has its own local LLR distributor. It has to decide whether the incoming LLR-value is stored in the local RAM or has to be send left or right. The direction for non-local data is determined based on the shortest path to the target RAM. Two additional LLR distributors are necessary in each cell for the left and right inputs. Data coming in from the left side can either get fed through or stored to the local RAM. The same holds for data from the right side respectively, leading to very simply control.

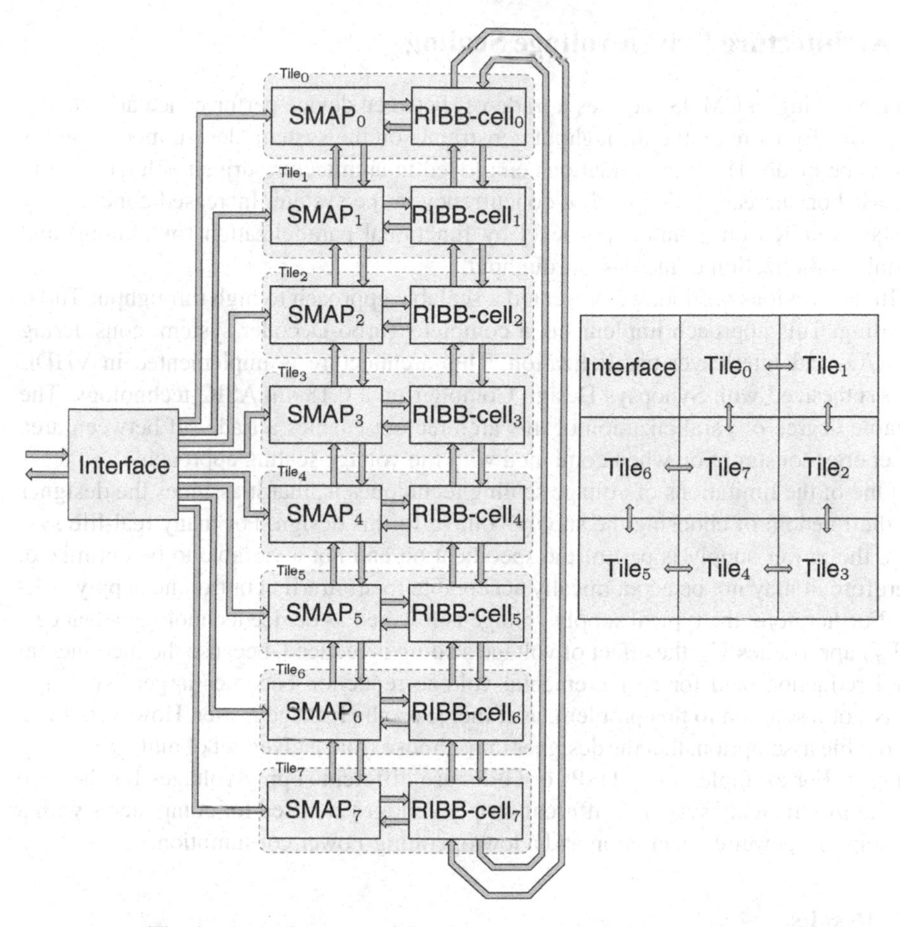

Fig. 4. Block diagram and floorplan for an 8-SMAP Turbo-Decoder

The buffers used inside the RIBB-cell are multiple-data-in single-data-out FIFOs. They are capable of storing up to three LLRs with their target address every clock cylce (two LLRs in the case of the left- or right-out buffers). The data can then accessed sequentially on the buffer's output.

Multiple RIBB-cells are connected in a ring topology: the left-out buffer of one RIBB-cell is connected to the right-in LLR distributor, the right-out buffer is connected to the left-in LLR distributor of the neighboring RIBB-cell. Each RIBB-cell receives data from one port of a SMAP-unit and a corresponding address information *e.g.* from a partition of the interleaver table. This approach allows to integrate one SMAP-unit and the associated interleaving-node (RIBB) within the same 100 K gates module. Furthermore, this architecture allows a regular floorplan, where communication among the SMAP-units is limited to adjacent cells. Wirelength, therefore, is independent of the number of cells. Global communication is limited to the latency-insensitive distribution of input and output data. Figure 4 shows the overall architecture of an 8-SMAP implementation.

4 Architecture Driven Voltage Scaling

Voltage scaling in CMOS requires a trade-off between device performance and energy reduction. To maintain the throughput constraints on the system, design modifications have to be made. The transformations employed in architecture-driven voltage scaling are based on increasing the level of concurrency in the system. Increased concurrency can be exploited on architectural level by functional parallelization (pipelining) and spatial parallelization to increase throughput.

In the previous section, we presented a scalable approach to high throughput Turbo decoding. This approach implements a complete Turbo-Decoder system, considering also I/O- and interleaver parallelization. This architecture is implemented in VHDL and synthesized with Synopsys Design Compiler on a 0.18 μm ASIC-technology. The variable degree of parallelization of this architecture enables a trade-off between area and energy consumption when combined with the voltage scaling approach.

One of the limitations of voltage scaling techniques is, that it assumes the designer has the freedom of choosing the supply voltage for his design. For many real-life systems, the power supply is part of the specification and not a variable to be optimized. Therefore, it may not be economically acceptable to arbitrarily control the supply voltage. Furthermore, the typical supply voltage is lowered as device technology advances. As V_{dd} approaches V_{th} the effect of voltage scaling is reduced, because the incremental speed reduction paid for an incremental voltage reduction becomes larger. Reducing V_{th} is not a solution to this problem, as it increases the leakage power. However, it is a reasonable assumption, that the designer may choose from a given set of multiple supply voltages. For example, many DSPs or GPPs use different supply voltages for the core and the I/O. In many systems, different supply voltages are used for components with a low standby-power consumption and a low operating-power consumption.

4.1 Results

In the scalable Turbo-Decoder system architecture described in Section 3, throughput can be scaled by selecting the number of parallel processing SMAP-units (N_D). Furthermore, as for parallel Turbo decoding architectures the time for data-transfer of the input and output data is not neglectable, the degree of I/O interface parallelization (N_{IO}) can be selected. N_{IO} determines the number of input and output data transferred in each clock cycle. This architecture is implemented in VHDL and synthesized with Synopsys Design Compiler on a 0.18 μm ASIC-technology.

The 0.18 μm ASIC-technology library is characterized for two different supply voltages: nominal 1.8 V and nominal 1.3 V. Timing analysis of the netlist on the 1.8 V library shows a critical path delay of 6 ns (=166 MHz) under worst-case conditions (1.55 V, 125° C). Using the 1.3 V supply voltage increases the critical path delay to 9.5 ns (=105 MHz) under worst-case conditions (1.2 V, 125° C). Table 1 presents area and throughput results together with estimates of the decoding energy of a Log-MAP implementation. The implementation is UMTS compliant, 6 Turbo-iterations are performed on a datablock with a blocklength of 5114 bits. For comparison, the architectural efficiency as defined in Equation 1 has been calculated and was normalized to the implementation with a single SMAP-unit running at 1.8 V.

Table 1. Implementation results of the scalable Turbo-Decoder system architecture (*cf.* Section 3) for different degrees of parallelization (N_D and N_{IO}) and different supply voltages (V_{dd})

Parallel SMAP-Units N_D	1	2	4	2	4	6	6	8
Parallel I/O N_{IO}	1	1	1	1	1	1	2	2
Supply Voltage $V_{dd}[V]$	1.8	1.8	1.8	1.3	1.3	1.3	1.3	1.3
Total Area [mm^2]	3.90	6.12	9.24	6.12	9.24	13.25	13.00	17.25
Energy per Block [μJ]	48.65	54.01	51.68	36.27	34.46	36.55	33.68	36.53
Throughput [Mbit/s]	11.7	23.0	39.0	14.5	24.7	31.9	37.7	45.9
Efficiency (norm.)	1	1.13	1.32	1.06	1.26	1.07	1.40	1.18

Fig. 5. Comparison of area and energy per Mbit/s for $V_{dd} = 1.8\,\text{V}$ and $V_{dd} = 1.3\,\text{V}$. The numbers in the figures indicate the configurations defined by the columns of Table 1.

The results of Table 1 illustrate the trade-off between throughput, area, and decoding energy. For a 10 Mbit/s data service, the designer can choose between an implementation with a single SMAP-unit and 1.8 V supply voltage or an implementation with two SMAP-units and 1.3 V supply voltage. The latter has not only the lower energy per decoded block but is also more efficient regarding Equation 1. For targeted 40 Mbit/s data service, the most efficient implementation is a 4 SMAP-unit solution running at 1.8 V.

Figure 5 compares area and energy per Mbit/s for the two different supply voltages. For a given throughput the area-efficiency is always higher for a supply voltage of 1.8 V, whereas a supply voltage of 1.3 V results in a lower decoding energy per block.

5 Conclusion

In this paper we present a UMTS compliant Turbo decoding architecture which addresses the important design issues: *high throughput signal processing*, *scalability*, and simultaneously *low energy consumption*. Its variable degree of parallelization enables a trade-off between area and energy consumption when combined with the voltage scaling approach. The results show, that this approach can save up to 34 % of the decoding energy per datablock, although the supply voltage can not arbitrarily selected.

Acknowledgement. Part of this work has been sponsered by a cooperation with ST Microelectronics. Our special thanks goes to Friedbert Berens from the Advanced System Technology Group of STM, Geneva.

References

1. L. Benini and G. De Micheli. Networks on Chips: A New SoC Paradigm. *IEEE Computer*, 35(1):70–78, January 2002.
2. C. Berrou, A. Glavieux, and P. Thitimajshima. Near Shannon Limit Error-Correcting Coding and Decoding: Turbo-Codes. In *Proc. 1993 International Conference on Communications (ICC '93)*, pages 1064–1070, Geneva, Switzerland, May 1993.
3. W. R. Davis, N. Zhang, K. Camera, D. Markovic, T. Smilkstein, M. J. Ammer, E. Yeo, S. Augsburger, B. Nikolic, and R. W. Brodersen. A Design Environment for High-Throughput Low-Power Dedicated Signal Processing Systems. *IEEE Journal of Solid-State Circuits*, 37(3):420–431, March 2002.
4. H. Dawid. *Algorithmen und Schaltungsarchitekturen zur Maximum a Posteriori Faltungsdecodierung*. PhD thesis, RWTH Aachen, Shaker Verlag, Aachen, Germany, 1996. In German.
5. H. Dawid, G. Gehnen, and H. Meyr. MAP Channel Decoding: Algorithm and VLSI Architecture. In *VLSI Signal Processing VI*, pages 141–149. IEEE, 1993.
6. D. Garrett, B. Xu, and C. Nicol. Energy efficient Turbo Decoding for 3G Mobile. In *Proc. 2001 International Symposium on Low Power Electronics and Design (ISLPED '01)*, pages 328–333, Huntington Beach, California, USA, August 2001.
7. F. Gilbert, A. Worm, and N. Wehn. Low Power Implementation of a Turbo-Decoder on Programmable Architectures. In *Proc. 2001 Asia South Pacific Design Automation Conference (ASP-DAC '01)*, pages 400–403, Yokohama, Japan, January 2001.
8. F. Kienle, H. Michel, F. Gilbert, and N. Wehn. Efficient MAP-Algorithm Implementation on Programmable Architectures. In *Kleinheubacher Berichte 2003*, volume 46, Miltenberg, Germany, October 2002. to appear.
9. P. Robertson, P. Hoeher, and E. Villebrun. Optimal and Sub-Optimal Maximum a Posteriori Algorithms Suitable for Turbo Decoding. *European Transactions on Telecommunications (ETT)*, 8(2):119–125, March–April 1997.
10. C. Schurgers, M. Engels, and F. Catthoor. Energy Efficient Data Transfer and Storage Organization for a MAP Turbo Decoder Module. In *Proc. 1999 International Symposium on Low Power Electronics and Design (ISLPED '99)*, pages 76–81, San Diego, California, USA, August 1999.
11. R. Subramanian. Shannon vs. Moore: Driving the Evolution of Signal Processing Platforms in Wireless Communications (Invited Talk). In *Proc. 2002 Workshop on Signal Processing Systems (SiPS '02)*, San Diego, California, USA, October 2002.
12. D. Sylvester and K. Keutzer. Rethinking Deep-Submicron Circuit Design. *IEEE Computer*, 32(11):25–33, 1999.
13. Third Generation Partnership Project. 3GPP home page. `www.3gpp.org`.
14. M. J. Thul, F. Gilbert, T. Vogt, G. Kreiselmaier, and N. Wehn. A Scalable System Architecture for High-Throughput Turbo-Decoders. In *Proc. 2002 Workshop on Signal Processing Systems (SiPS '02)*, pages 152–158, San Diego, California, USA, October 2002.
15. M. J. Thul, F. Gilbert, and N. Wehn. Optimized Concurrent Interleaving for High-Throughput Turbo-Decoding. In *Proc. 9th IEEE International Conference on Electronics, Circuits and Systems (ICECS '02)*, pages 1099–1102, Dubrovnik, Croatia, September 2002.
16. J. Vogt, K. Koora, A. Finger, and G. Fettweis. Comparison of Different Turbo Decoder Realizations for IMT-2000. In *Proc. 1999 Global Telecommunications Conference (Globecom '99)*, volume 5, pages 2704–2708, Rio de Janeiro, Brazil, December 1999.
17. A. Worm. *Implementation Issues of Turbo-Decoders*. PhD thesis, Institute of Microelectronic Systems, Department of Electrical Engineering and Information Technology, University of Kaiserslautern, 2001. ISBN 3-925178-72-4.
18. A. Worm, H. Michel, and N. Wehn. Power minimization by optimizing data transfers in Turbo-decoders. In *Kleinheubacher Berichte*, volume 43, pages 343–350, September 1999.

A Fully Digital Numerical-Controlled-Oscillator

S.R. Abdollahi[1], B. Bakkaloglu[2], and S.K. Hosseini [3]

[1]VLSI Circuits and Systems Lab, ECE Department, University of Tehran.
srabdollahi@yahoo.com

[2] Texas Instrument, USA.

[3] Iran Marine Industry (SADRA), Neka, Iran

Abstract. A 3.3V, 0.8 mW programmable Numerical Controlled oscillator Oscillator (NCO) core is designed in 0.6 micron CMOS process and its prototype design is mapped on an Altera MAX9400 CPLD. This architecture is suitable for digital wireless transceivers that use different bands for transmit and receive modes, such as GSM and DECT. Linearity and phase noise of the NCO is analyzed. Thermal drift and power supply level sensitivity is characterized. This architecture can be used for higher frequencies using faster FPGA devices or by implementing it on an advanced deep-submicron process.

1 Introduction

As the complexity, bandwidth and synchronization requirements of modern communications systems increase, use of digital techniques are becoming more versatile and cost effective. An extended performance for digital high-speed VLSI circuit is provided by new sub-micron MOS integrated circuit technologies. A Numerical-Controlled Oscillator (NCO) is the digital counterpart of a voltage-controlled oscillator (VCO) in digital phase locked loops (DPLLs) and is a very important building block in modern communication circuits. There are four types of NCOs. The first type of NCO is the "path-delay oscillator," which is designed by cascading many logic gates to form a circular ring oscillator. The second categories of NCOs are the Schmitt-trigger based current-driven oscillator. In this type of NCO, oscillation frequency setting is achieved by a Schmitt-trigger inverter with a capacitive load. Due to the need for high capacitance, generally external (discrete) capacitors are used.. However, the external capacitance degrades the performance due to DC leakage and ESR (equivalent series resistance). The third category is "current-starved" ring oscillator. This type of NCO has good linearity.

The fourth type is Direct Digital Synthesis NCO (DDS NCO) that is built with a phase accumulator, lookup tables and a D/A converter. The lookup table contains digital samples of Sin(.) or Cos(.) functions, and the D/A converts the memory content to analog values at the output of the NCO. Usually an analog VCO provides the highest linearity. However, it is preferable to adopt a fully digital solution for reasons of cost, size, flexibility and repeatability [1-5]. The analog VCO may be

J.J. Chico and E. Macii (Eds.): PATMOS 2003, LNCS 2799, pp. 389–398, 2003.
© Springer-Verlag Berlin Heidelberg 2003

replaced with a digital NCO to achieve this. However, in order to obtain a linear characteristic, a high order of over sampling is required. This is typically at least 100 times the modulated signal bandwidth. Considering chip rates in the range 1-10 MHz, NCOs that operate at clock rates of 100 MHz-1GHz are then required. While current digital technologies can operate at clock frequencies of several hundred MHz, both the complexity and power consumption increase with the operating frequency. NCO and DDS circuits rely on analog filtering in order to eliminate signal harmonics, and to obtain fine frequency resolution. Another advantage of this design is its ability to operate without a processing clock.

2 Contribution and Paper Organization

Our main contribution in this work is design and implementation of a linear all-digital NCO. The linearity of the oscillator is better than most of the published work [3-5]. We have characterized our design for maximum of seven control bits. The oscillation band is relatively wide compared to the desired free-running oscillation frequency. We can further reduce the nonlinearity by reducing the relative oscillation bandwidth. In this paper, we characterized the linearity of our proposed architecture under temperature and supply variations as well. We simulated our idea in 0.6-micron CMOS process and its power consumption is measured.

The paper is organized as follows: oscillator architecture is discussed in Section 3. Section 4 explains simulation and experimental results. Finally, conclusion and references are presented in Sections 5 and 6.

3 Oscillator Architecture

We have designed and implemented flexible programmable oscillator. NCO contains programmable delay cells, fine-tuning circuits (not included in the existing design) and control unit. NCO control unit receives a NCO control word that contains channel selection bits.

Channel selection input is for selecting the center frequency of oscillation or data rate selection, which is used in GSM or DECT and in Frequency-Hopping-Spread-Spectrum transceivers. Fig. 1 shows a programmable delay cell of NCO. Fig. 2 demonstrates the control signals applied to each delay cell.

Fig. 3 (a) shows details of a NCO programmable delay cell that can be reconfigured by control unit under various channel selection conditions. This block supports increasing or decreasing delay amount of each programmable delay cell. The delay of the main cell is designed to be much larger than the delay of the other controllable delay chains in order to increase the sensitivity and linearity of the NCO as shown in Fig. 4.

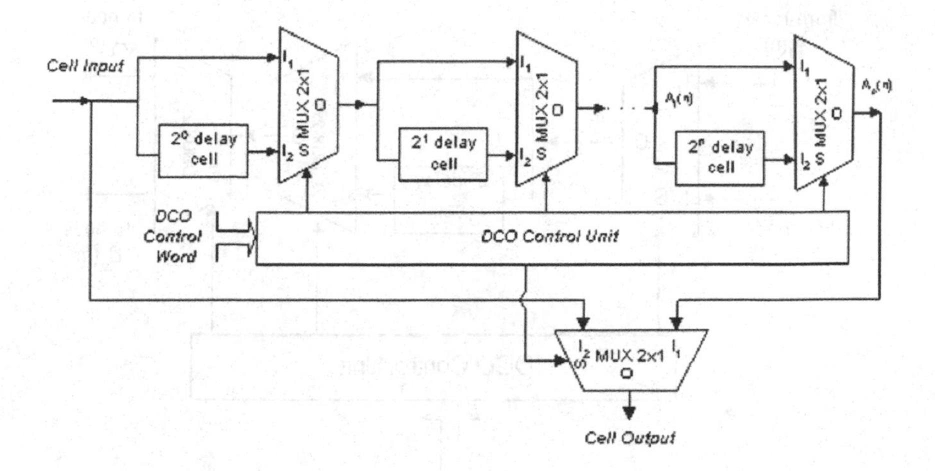

Fig. 1. Details of one programmable delay cell of NCO.

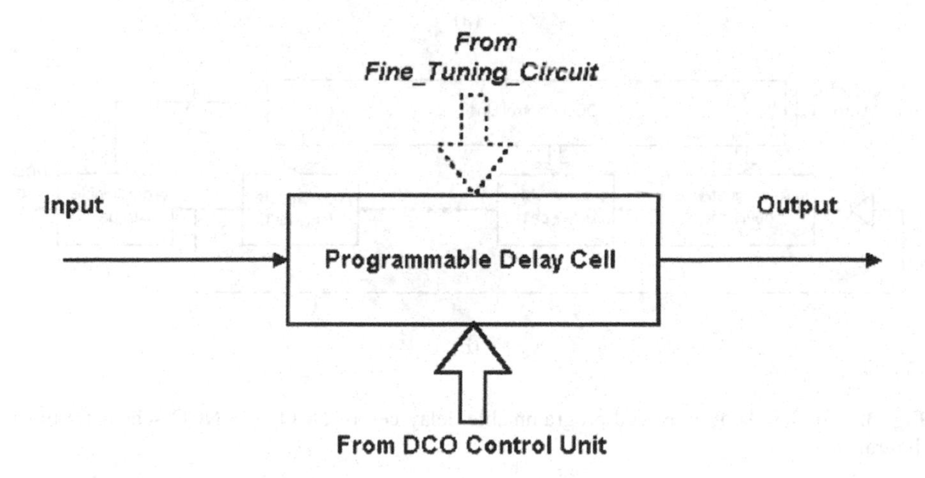

Fig. 2. Programmable delay cell block.

In Fig. 3 node A_I (n) is the output of the (n-1)th delay cell block in programmable delay cell and node A_o(n) is the output of the current delay cell block. Similarly, node B_o(m-1) is the output of the (m-1)th delay cell in the main delay cell block.

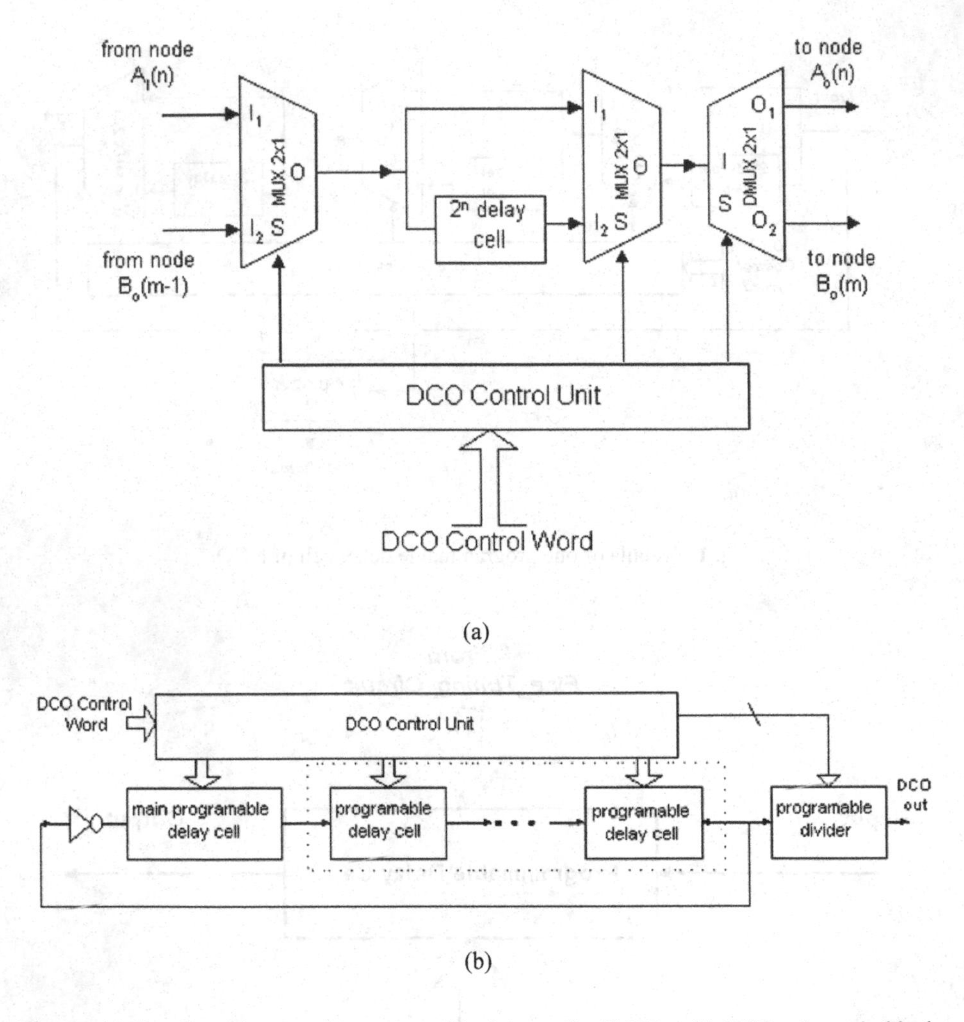

Fig. 3. (a) Details of improved programmable delay cell of NCO. (b) NCO schematic block diagrams.

The sum of three terms, the delay of the inverter gate, Mux2X1, main delay cell and programmable delay cells mainly determine the output frequency. D_{inv}, D_{mdc}, D_{mux} and D_{pdc} represent these delays respectively.

$$D_{PLC} = D_{inv} + D_{mdc} + D_{mux} + D_{pdc} \tag{1}$$

Fig. 4 shows a simulation of the output frequency as a function of the control words at different ratios of $R = D_{mdc}/D_{plc}$ with D_{PLC} fixed. To achieve good linearity and sensitivity, we have to use a large D_{mdc}/D_{PLC} ratio. A large D_{mdc}/D_{PLC} ratio is obtained by using more logic cells for the main cell.

We implemented the proposed NCO circuit at gate level on an Altera MAX9400 CPLD device. By using this method, we can employ the advantages of both "current-starved ring oscillators" and "path delay oscillators". Due to its all-digital structure, equal sized transistors, and compatibility for implementation on FPGAs, this NCO can easily be ported to a submicron CMOS processes. The main advantage of this new NCO is that by using switch tuning, path propagation delay lines and a single inverting gate, it possesses a simple structure, high resolution and good linearity in each band of oscillation. By using more control bits, we are able to have better resolution.

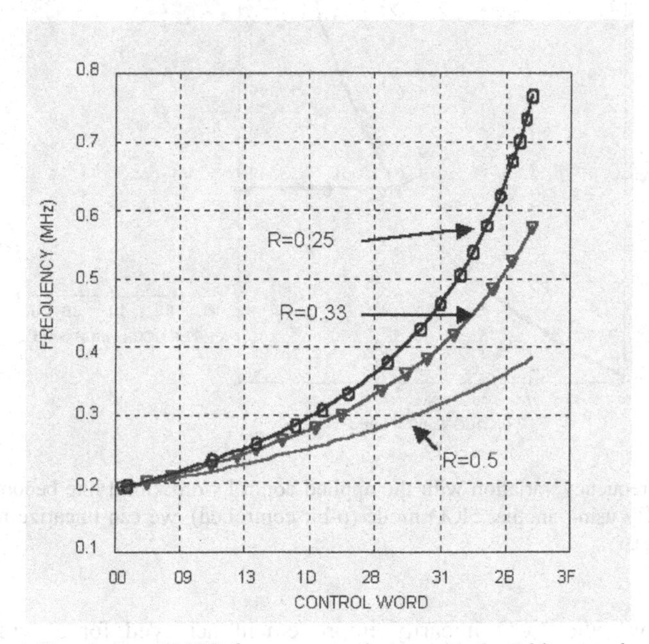

Fig. 4. Simulation of NCO frequency variation with the 7-bit control word.

A programmable delay cell structure is shown in Fig. 1. As shown, Mux2x1 (solid line) is used to create the proper delay line based on the control bit. This tunes the oscillation band and frequency.

By cascading a number of programmable delay cells of Fig. 2, as shown in Fig. 3 (b), we can build multi-band NCOs that can be used for wireless applications. For this purpose, we consider a channel-select bit for selecting the desired channel band for multi-rate transceiver applications.

4 Simulation and Experimental Results

The experimental results indicate, that NCO mapped on FPGA works precisely at ambient temperatures form 0C to 70C. The resolution of 7-bit controlled NCO without fine-tuning circuit is 20Hz per bit, with oscillation band from 4 KHz to 12 KHz at output of the programmable divider. We can extend the oscillation frequency of NCO to several MHz by proper design of programmable delay cells and adjusting number of them. Total static gate count of the NCO is 340 and number of DFF is 20.

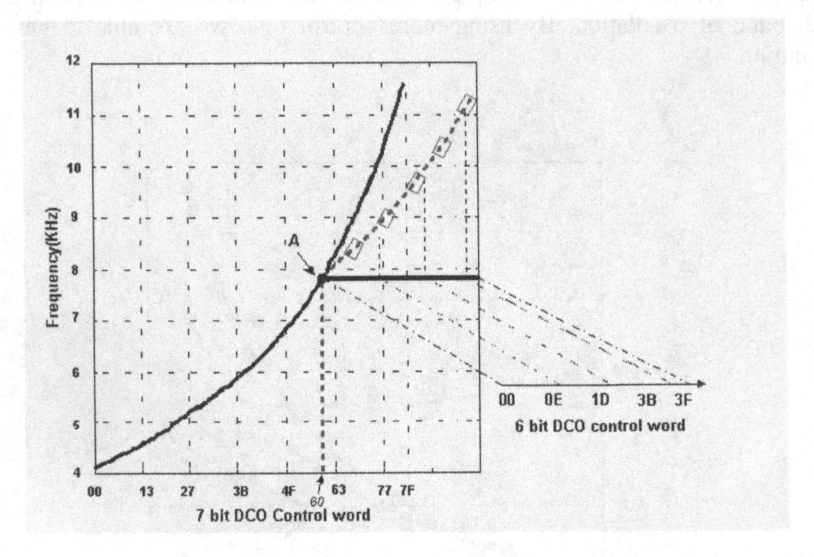

Fig. 5. NCO frequency variation with the applied control word. Solid line becomes nonlinear after point A. By using another NCO mode (6-bit controlled), we can linearize this nonlinear NCO (dash line).

Fig. 5 shows the NCO linearity improvement achieved for a wide range of oscillation frequencies. At point **A** the curve slope starts to be nonlinear, but, at this point a 6-bit NCO is switched in by using a programmable circuit and the NCO control unit shown in Fig. 3 (a). This segmentation method linearizes the high band frequencies as well as the lower oscillation band.

The maximum deviation from center frequency is around 4-7kHz, which is very competitive to existing commercial products.

By using this technique, we have a linear oscillator with the same control slope in each data transmission mode. The control unit segments the control of the NCO between the two MSBs and 6 LSBs.

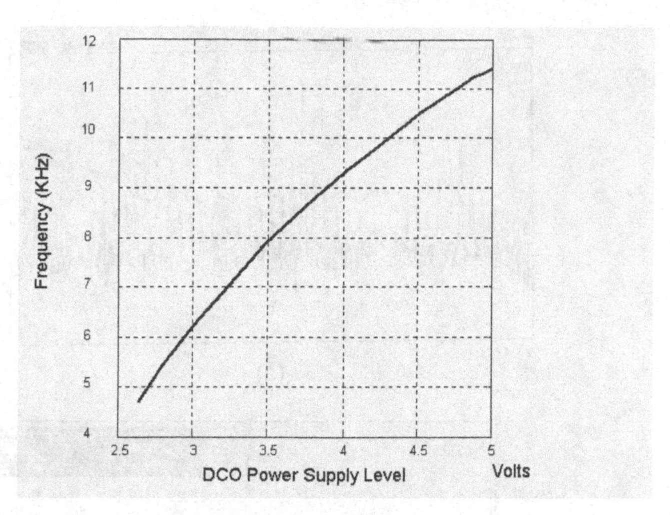

Fig. 6. NCO frequency output per supply level variations.

Fig. 6 shows the frequency deviation of the NCO with the power supply. By using the supply dependency, we can control NCO frequency in the fine-tuning mode.

Fig. 7 represents the NCO frequency variation at a lower power supply voltage. This enables a low power operation and shows lower dependency to thermal variation on the junction temperature.

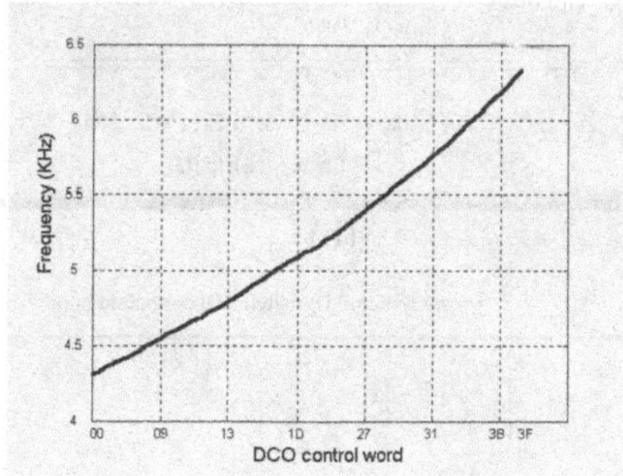

Fig. 7. NCO frequency output variation at VCC=3 volts.

Fig. 8 shows the FFT and PSD of the NCO output in ASIC and prototype design of the NCO with Hanning window. As shown, in prototype design that mapped on MAX 9400 CPLD, the noise level is ~25dB lower than the desired signal. In ASIC Design the noise level is ~ 40dB lower than desired signal. Prototype FFT spectrum is plotted using an HP-56445D Mega zoom logic analyzer and PSD spectrum is plotted by using HP-56445D extracted data in MATLAB software environment, In ASIC design this FFT is simulated in HSPICE software. Because our logic analyzer isn't powerful we can't present better plot. In prototype design NCO, the output waveform is received at the of frequency divider.

Fig. 9 shows the transient waveform of NCO output with "3F" control word.

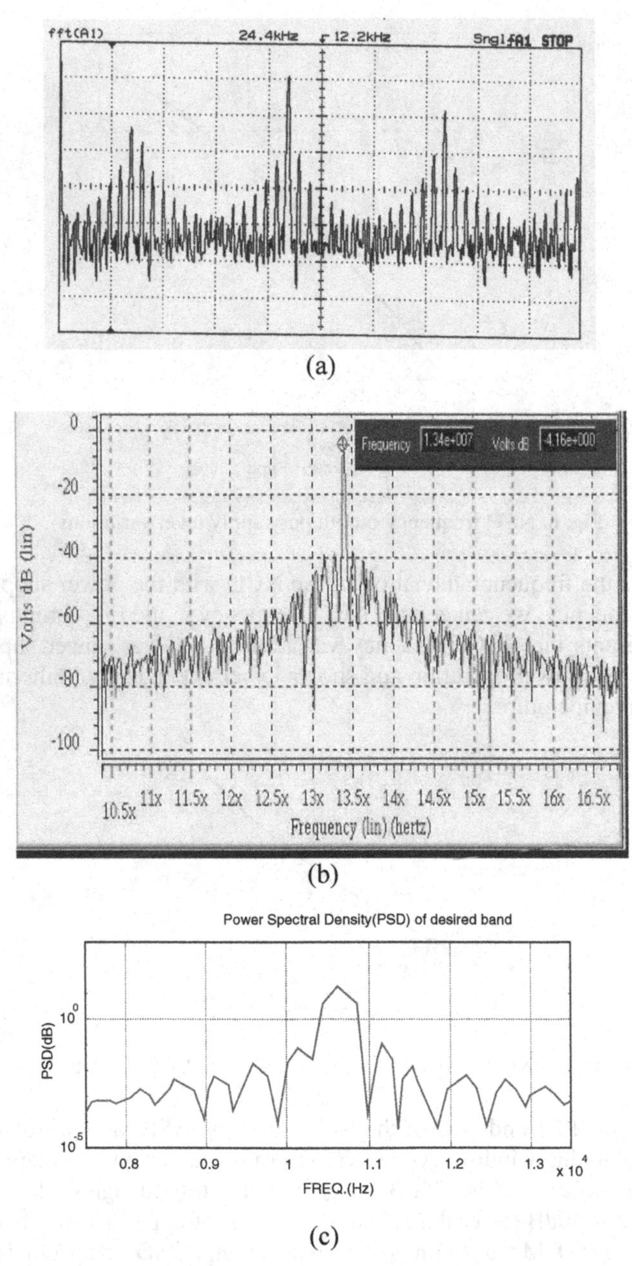

Fig. 8. (a). NCO output FFT with Hanning windows, with sensitivity 10dBV, Offset –45.00 dBV and span 24.41 KHz, with control word "3F". (b) : NCO output FFT of ASIC design of NCO X $=10^6$ (c): PSD of prototype NCO desired band output.

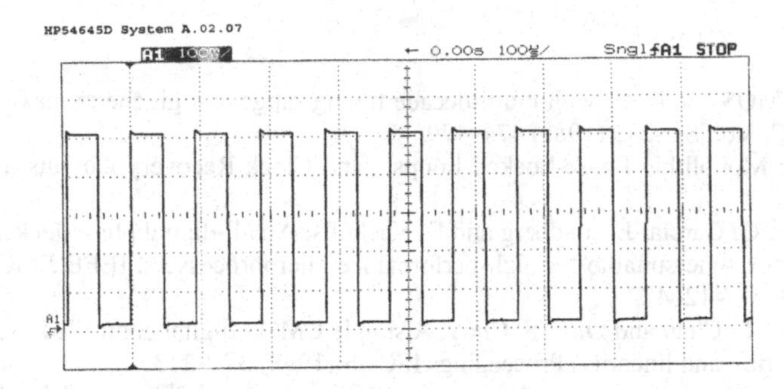

Fig. 9. NCO output signal by using 6-bit control words with control word "3F".

Fig. 10 shows the NCO test board that includes an ALTERA MAX9000 FPGA.

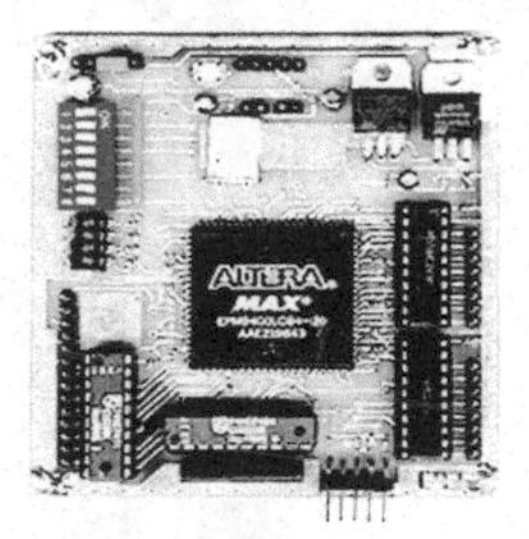

Fig. 10. NCO Emulator Board.

5 Conclusion

Our experimental and simulation results indicate that the all-digital NCO has a proper linear transfer function compared to other mostly nonlinear NCOs [2-5]. By using high-speed and larger FPGAs, we can implement high frequency NCOs with proper resolution. We can use this NCO for multi-band wireless transceivers. Due to its all-digital structure the proposed NCO it is suitable for fabrication in digital CMOS processes. Due to higher transistor speed and higher integration, ASIC implementation of the proposed all-digital structure will have a higher frequency tuning range. In 0.6 micron CMOS process, this NCO dissipates 0.8 mW at 23.81MHz at 3.3V power supply level and its free running frequency was 23.81MHz. By using advanced deep micron CMOS technology, we can increase NCO frequency and decrease its power dissipation.

References

1. M. Banu: MOS oscillator with multi-decade tuning range and gigahertz maximum speed. IEEE JSSC, Vol. SC-23, (1988) 474–479
2. B. Razavi: Monolithic Phase-Locked Loops and Clock Recovery Circuits. IEEE Press, (1996).
3. J. Dunning, G. Garcia, J. Lundberg and E. Nuckolls: An all-digital phase-locked loop with 50-cycle lock time suitable for high-performance microprocessors. IEEE JSSC, VOL. 30, NO. 4, (1995), 412–422
4. C. H. To, C. F. Chan and O. C. S. Choy: A simple CMOS digital controlled oscillator with high resolution and linearity. Proceedings ISCAS (1998) 371–373
5. L. Sabel: On the performance of a low complexity baseband 2nd order delay locked loop incorporating a non-linear NCO. Proceedings of Spread Spectrum Techniques and Applications, Vol. 2 (1998) 469–473

Energy Optimization of High-Performance Circuits†

Hoang Q. Dao, Bart R. Zeydel, and Vojin G. Oklobdzija

ACSEL Lab, Department of Electrical and Computer Engineering,
University of California, Davis, CA 95616
{hqdao, brzeydel, vojin}@ece.ucdavis.edu

As technology scales, energy consumption is becoming an important issue in high-performance VLSI microprocessor designs. Often the critical path of these microprocessors is in the datapath of arithmetic units, resulting in high energy consumption along the datapath. We propose an optimization method that can achieve energy saving versus delay-based optimization at equal performance. It reveals that the source of energy saving lays in the balance of delay and energy consumption among different stages of a circuit. The energy saving is significant, 30%–50%. The results are confirmed with simulation, using Fujitsu's 0.11µm, 1.2V CMOS technology.

1 Introduction

Technology scaling has brought about a real concern and need for energy-efficient designs. It is illustrated in [1], where high-performance VLSI adders are compared in both energy and delay. It has been shown that the energy-delay space for different designs can potentially cross. Therefore, the best energy-delay curve for each design must be used. More importantly, the most efficient design choice is only valid over the range of performance where its energy consumption is the least.

Logical Effort (LE) has provided a method to estimate optimal delay for a given constraint on input and output loading, [2]. It also provides the circuit sizing for optimal delay, from which the energy can be determined. However, this solution is very inefficient in energy, as illustrated later.

Recent work [3,4] in circuit power minimization has focused on a limited portion of the energy-delay space, where a significant saving in energy can be achieved for a certain delay penalty. However, this does not resolve the issue of achieving minimum energy for a given delay nor does it give insight into how these lower energy points can be obtained and the associated costs.

Recent publications [5,6] have shown a method to achieve energy-efficient designs at the micro-architectural level. The optimization of a system is shown to be a function of the energy-delay spaces of functional blocks in the microprocessor. This optimization determines exactly which portion of the energy-delay curve for each

† This work has been supported by SRC Research Grant No. 931.001, Fujitsu Laboratories of America and California MICRO 02-057.

J.J. Chico and E. Macii (Eds.): PATMOS 2003, LNCS 2799, pp. 399–408, 2003.
© Springer-Verlag Berlin Heidelberg 2003

block is of interest, i.e. Energy•Delay$^{\eta}$, where η varies depending on hardware intensity of the functional block.

In this paper, we propose a circuit-level energy optimization method at the same performance as the LE solution. The paper is organized as follows: section 2 briefly explains the gate delay model using logical effort; section 3 describes how to obtain energy-delay space of a system; section 4 discusses the energy optimization method at equal performance and its application on several test circuits; section 5 presents simulation results; and section 6 summarizes the analysis.

2 Delay Modeling Using Logical Effort

The logical effort method [2] describes a convenient model for gate delay. The delay through a gate is determined from its driving capability, its output load and its size. The driving capability of the gate is modeled by logical effort (g); it is unique for each gate input and is independent from loading. Electrical effort (h) is the ratio of output load to gate size, modeling the effect of the output load on delay. Finally, the parasitic (p) models the delay effect of internal parasitic capacitance. Then, the gate delay is:

$$d = g \bullet h + p = f + p, \text{ where } f = g \bullet h = \text{effort delay}$$

The above delay model is normalized to delay unit τ, the slope of the inverter delay versus electrical effort. Intuitively, it is the parasitic-free fanout-1 delay of the inverter. This normalization enables delay comparison over different technologies.

A systematic method has been shown to optimize the delay of a path. An optimal solution exists for single-path circuit where effort delay in each gate equals. For multi-path circuit, the optimal solution is harder to achieve and requires accurate branching factors. However, especially when all paths are balanced, the delay when equal effort delay applies to all gates is close to the optimal solution. Interested readers should refer to [2] for thorough understanding.

The gates used in our analysis are modeled with Fujitsu's 0.11μm, 1.2V CMOS technology. The effective PMOS-to-NMOS ratio of 2 is used. The delay unit τ is 6.2ps. Each gate is characterized for faster input where the critical path is assumingly connected. The results are summarized in table 1.

Table 1. Delay characterization of selected gates

Gates	INV	NAND2	NOR2	AOI	OAI	XOR2
g	1.00	1.25	1.63	1.78	1.66	2.29
p	1.42	1.64	3.41	3.48	3.56	4.37

3 Energy-Delay Space

Shown below is the typical representation of a functional block, including its drivers and loads (Fig. 1). For a given design, the load and driver size are generally pre-estimated. As such, the overall electrical effort of the block is fixed, $H = C_{Load}/C_{Driver}$. For a delay target, gates in the block are typically sized by applying a certain fan-out rule or effort delay value.

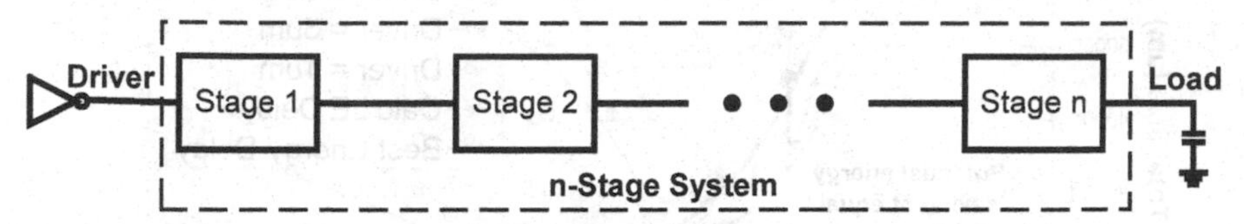

Fig. 1. Functional block structure

The energy-delay space for a fixed input and output can be seen in Fig. 2. Each point on the curve is obtained by fixing a effort delay for all stages. The curve is generated by varying this effort delay value.

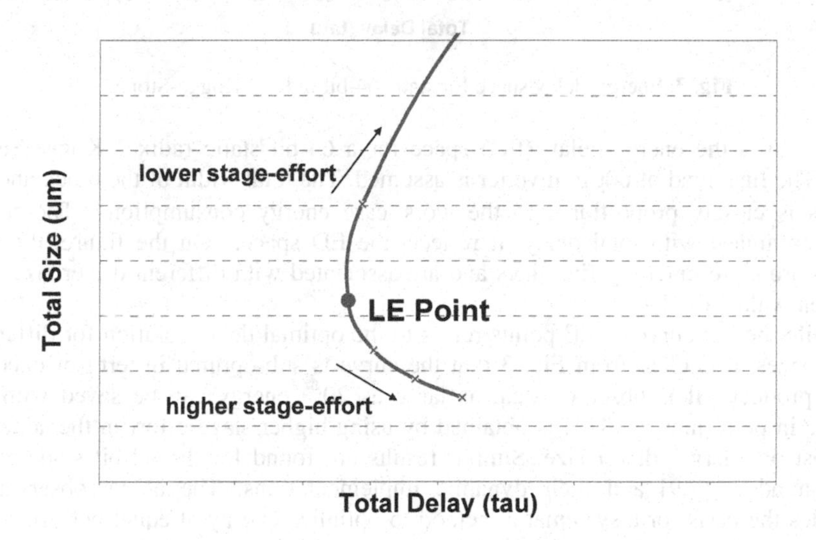

Fig. 2. Energy-Delay Curve for fixed driver and output load

The lower part of the curve has higher stage effort. This results in a decrease of gate sizes and thus a saving in energy occurs. In addition, the delay is increased due to higher percentage delay increase in the block compared to percentage delay decrease in the driver. Conversely, the upper part of the curve has lower stage effort. As a result, the size of the gates increase, and thus an increase in energy is observed. Furthermore, the delay also increases due to higher percentage delay increase in the driver compared to the percentage delay decrease in the block. Hence, there exists a minimal delay point where the percentage delay variations in the block and the driver are balanced. This point typically occurs when the stage efforts of those in the block and the driver are equal – as in LE optimal delay solution. As such, it is called LE point.

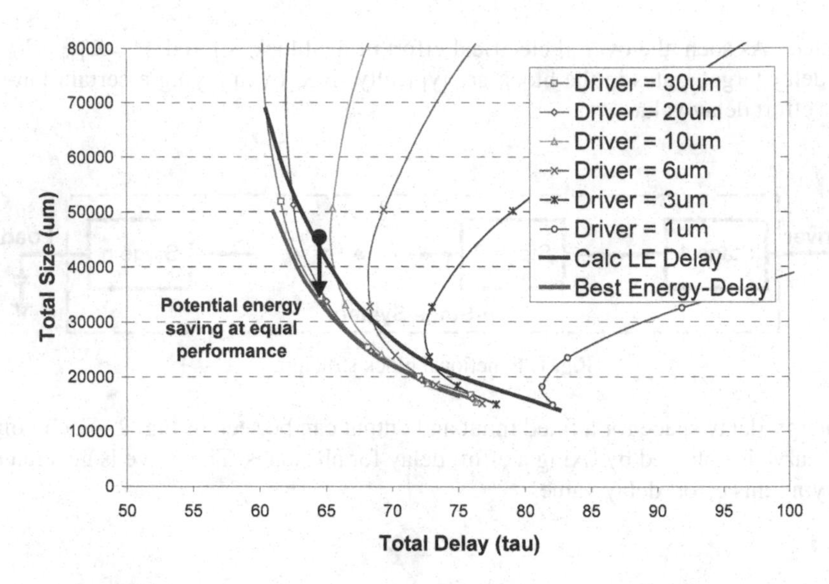

Fig. 3. Energy-delay space for static 64-bit radix-2 Kogge-Stone

Fig. 3 shows the energy-delay (ED) space for a 64-bit static radix-2 Kogge-Stone [7,8]. The final load of 60μm inverter is assumed. The total width of the block and its drivers is closely proportional to the worst-case energy consumption. Therefore, when combined with total delay, it reflects the ED space. On the figure, the ED curves are represented by fine lines and are associated with different driver sizes, or different values of H.

While the ED curve of LE points refers to the optimal delay solution for different driver sizes, it is clear from Fig. 3 that the curve is sub-optimal in term of energy-delay product. It is observed that, as large as 20% energy can be saved with no change in performance. This is obtained by using higher stage effort in the adder at the cost of a larger driver size. Similar results are found for the 64-bit static Han-Carlson adder [7,9] and their dynamic implementations. The above observation provides the basis for a systematic method to optimize energy at equal performance, as shown in the following section.

4 Energy Optimization for Equal Performance

For the sake of simplicity and pending work in progress, we will present the energy optimization for linear systems. Extension to non-linear systems will be shown in our future publication.

4.1 Linear Systems

Definition: An n-stage system is linear when the output load is proportional to the size of all gates in the system and the effort delays of the later output stages. That is,

$$W_{Load} \propto W_{gate,i} \bullet \prod_{j=i}^{n} f_j \,, \quad \text{for } i = 1 \rightarrow n$$

where W_{Load} : output load
$W_{gate,i}$: size of a gate at stage i
f_i : effort delay of stage i

Theorem: A system is linear if and only if loads to any gates are proportional to the effort delays of the later output stages.

Proof: by induction on the stages, from i=n to i=1
* At i=n, $W_n = W_{Load} / f_n \Leftrightarrow W_{Load} = W_n \bullet f_n$

* Assume it is true for all stages i ≥ k+1. That is, $W_{Load} \propto W_{gate,k+1} \bullet \prod_{j=k+1}^{n} f_j$

* For an arbitrary gate in stage k and loaded by m gates in stage (k+1),

$$W_{gate,k} = \left(\sum_{j=1}^{m} \beta_j W_{k+1,j}\right) / f_k, \text{ with } \beta_i = \text{constant}$$

$$\propto \left(\sum_{j=1}^{m} \beta_j\right) \bullet W_{Load} / \left(\prod_{j=k}^{n} f_j\right)$$

$$\Leftrightarrow \quad W_{Load} \propto W_{gate,k} \bullet \prod_{j=k}^{n} f_j$$

Therefore, by induction, the theorem is proved.

Colloquium: In a linear system, the load is proportional to the total size of any stage and the effort delays of the later output stages. That is,

$$W_{Load} \propto \Sigma W_{stage\ i} \bullet \prod_{j=i}^{n} f_j \,, \qquad \text{for } i = 1 \rightarrow n$$

where $\Sigma W_{stage,i}$: total gate size at stage i

It can be shown that when sizing of a system is solely based on its own gate size, the system is linear. When wiring capacitance is included, the system becomes non-linear.

4.2 Energy Optimization for Linear Systems

Assume a linear system consisting of inverter drivers and m-input n-stage multi-path circuit, Figure 1. A fixed load C_{Load} is presented by an equivalent gate size W_L.

One solution for a target delay can be obtained by using equal effort delay for all stages in the system, including its drivers. The total size of a stage is given by

$$\Sigma W_i = k_i \bullet (W_L / f^{n-i+1}), \text{ for } i = 0 \rightarrow n \quad (1)$$

Note that k_i is determined from the system architecture and is constant. It can also include the energy-switching factor. In addition, the driver stage corresponds to i=0. The driver size is determined by

$$W_{dr} = W_0 = k_0 \bullet (W_L / f^{n+1}) \quad (2)$$

The total delay, including driver, is

$$t_{d,Total} = (n+1)\, f + \Sigma p \quad (3)$$

To achieve lower energy consumption at equal performance, effort delay of each stage is varied by a factor $(1+x_i)$

$$f_i = f\,(1+x_i) \qquad \text{with } x_i \in (-1, n), \text{ for } i = 0 \rightarrow n \quad (4)$$

The new total size of the stages, input size and the total delay are

$$\Sigma W_{i,x} = \Sigma W_i / \left(\prod_{j=i}^{n} (1+x_j)\right) \quad (5)$$

$$W_{dr,x} = W_0 / \left(\prod_{j=0}^{n} (1+x_j)\right) \quad (6)$$

$$t_{d,Total,x} = \left(\sum_{i=0}^{n} f_i\right) + \Sigma p = f\left(\sum_{i=0}^{n} (1+x_i)\right) + \Sigma p \quad (7)$$

Combine (3) and (7), under equal performance condition

$$\Delta t_{d,Total} = t_{d,Total,x} - t_{d,Total} = \sum_{i=0}^{n} x_i \equiv 0 \quad (8)$$

(8) is called the equal-delay constraint C.

The total gate size of the system (including drivers)

$$S_W = \sum_{i=0}^{n} \Sigma W_{i,x} = \sum_{i=0}^{n} \frac{\Sigma W_i}{\prod_{j=i}^{n} (1+x_j)} \quad (9)$$

Since the worst-case energy is closely associated with total gate size, minimal energy solution is found by minimizing (9) under constraint (8). It is solved using LaGrange method.

$$F = S_W + \lambda C, \text{ where } \lambda = \text{LaGrange constant} \tag{10}$$

The solution is found by solving (8) and (n+1) partial derivative equations

$$\delta F / \delta x_i = 0 \qquad \text{with } i = 0 \rightarrow n \tag{11}$$

The derivative equations can be simplified to

$$\delta F/\delta x_0 = 0 \Leftrightarrow \Sigma W_0 = \lambda(1+x_0) \prod_{j=0}^{n}(1+x_j) \tag{11.0}$$

$$\delta F/\delta x_1 = 0 \Leftrightarrow \Sigma W_0 + \Sigma W_1 (1+x_0) = \lambda(1+x_1) \prod_{j=0}^{n}(1+x_j) \tag{11.1}$$

(...)

$$\delta F/\delta x_n = 0 \Leftrightarrow \Sigma W_0 + \sum_{i=1}^{n}\{\Sigma W_i \prod_{j=0}^{i-1}(1+x_j)\} = \lambda(1+x_n) \prod_{j=0}^{n}(1+x_j) \tag{11.n}$$

Equations (11.0)–(11.n) can be simplified further by combining adjacent pairs. The results are

$$\Sigma W_0 (x_1 - x_0) = \Sigma W_1 (1+x_0)^2 \tag{12.1}$$

$$[\Sigma W_0 + \Sigma W_1 (1+x_0)] (x_2 - x_1) = \Sigma W_2 (1+x_0)(1+x_1)^2 \tag{12.2}$$

(...)

$$\{\Sigma W_0 + \sum_{i=1}^{n}[\Sigma W_i \prod_{j=0}^{i-1}(1+x_j)]\} (x_n - x_{n-1}) = \Sigma W_n (1+x_{n-1}) \prod_{j=0}^{n-1}(1+x_j) \tag{12.n}$$

Note that equations (12.1–n) are a lot simpler, where factor x_i is determined solely from preceded x_j ($j < i$) and initial sizing. Minimal energy solution is obtained by numerically solving equations (8), (11.0), (12.1–n).

4.3 Application to Single-Path Linear Systems

Assume a 6-gate chain of inverter, nand2's and nor2's, driving a 789μm load. The optimization is done using LE delay model in Table 1. The results are summarized in

Table 2. LE Size corresponds to the sizing using LE method, assuming a stage effort of 4τ. The optimized factor x_i for each gate is numerically found using the above equations. Opt_Size refers to the new energy-optimized gate sizes. The total gate size is reduced from 396μm to 91μm, or 77% saving. In addition, the estimated energy saving is about 26%. The cost is approximately 2x size of the first gate.

Table 2. Energy optimization for 6-gate chain

Gate	LE Size	x factor	(1+x) f	Opt_Size
INV	1.0	-0.804	0.78	1.9
NAND	4.0	-0.650	1.40	1.5
NOR	12.8	-0.477	2.09	2.1
NAND	31.4	-0.256	2.98	4.5
NOR	100.5	0.271	5.08	13.3
NAND	246.7	1.916	11.66	67.7

4.4 Application to Multi-path Linear Systems

The optimization is also applied to two static 64-bit radix-2 adder implementations: Kogge-Stone (KS) and Han-Carlson (HC). Circuit implementation details can be found in [7]. The 3-μm driver and the load of equivalent 60-μm inverter are assumed. Wire capacitance is ignored to guarantee linearity of the systems.

Table 3 summarizes the optimization for the KS adder. The LE stage effort delay is 3.79τ. The total gate size is reduced from 19,421μm to 9,705μm, or 50% saving. The estimate worst-case energy is reduced by 42%. The cost is approximately 4x driver size.

Table 3. Energy optimization for 64-bit static radix-2 Kogge-Stone adder

Level	LE Size	x factor	(1+x) f	Opt_Size
driver	192	-0.809	0.72	749
0	2073	-0.417	2.21	1555
1	1426	-0.259	2.81	613
2	1458	-0.139	3.26	466
3	1893	-0.006	3.77	533
4	1794	0.127	4.27	503
5	1814	0.277	4.84	567
6	1738	0.470	5.57	683
7	7033	0.757	6.66	4036

Similarly, Table 4 summarizes the optimization for the HC adder. The LE stage effort delay is 3.43τ. The total gate size is reduced from 16,067μm to 9,349μm, or 44% saving. The estimate worst-case energy is reduced by 35%. The cost is approximately 3x driver size.

Table 4. Energy optimization for 64-bit static radix-2 Han-Carlson adder

Level	LE Size	x factor	(1+x) f	Opt Size
driver	192	-0.779	0.84	575
0	1532	-0.388	2.32	1135
1	942	-0.241	2.88	412
2	864	-0.138	3.27	288
3	1023	-0.034	3.66	301
4	877	0.067	4.05	249
5	809	0.176	4.46	244
6	704	0.307	4.96	247
7	1890	0.484	5.63	839
8	7774	0.545	5.86	5059

4.5 Discussion

As shown in sections 4.3–4.4, significant total size and energy can be saved at no cost in performance. The exact amounts depend on the distribution of sizes among the stages. The proposed optimization allows the redistribution of delay to balance out the circuit intensity in each stage. More delay is applied to large stages (typically output ones). The result is the ripple effect of gate size reduction in earlier stages. To maintain equal performance, less delay is applied to the remaining stages. The optimal energy is achieved once delay distribution is balanced.

There are a number of side effects from the optimization. First, input size will increase. That introduces extra loading to previous circuit. Second, total delay may vary due to variation of signal rate at each stage. The delay model of logical effort is based on equal signal rates of input and output of a gate. When these rates vary (by varying effort delays of adjacent stages), the delay will change. Preliminary simulation shows that the delay is inversely proportional to effort delay ratio of the output stage to the input stage.

5 Simulation

Results in section 4.3–4.4 are verified with simulation using Fujitsu's 0.11μm, 1.2V CMOS technology. To account for the effect of signal rate variation on total delay, input signal rate is set equal to that of the output. The results are shown in Table 5. Some delay variation is seen. It is mainly due to different signal rate at the input and output of gates, which results from different stage efforts in two adjacent stages. Nonetheless, the significant energy saving is 40%–50%, consistent with previous calculation. As a result, the energy-delay product is reduced by 33%–50%.

Table 5. Simulation results

Design		Delay (ps)	Energy (pJ)	Energy (norm)	EDP (pJ.ps)	EDP (norm)
6-Gate Chain	LE Opt	246	3.79	1.00	931	1.00
	Energy Opt	283	2.19	0.58	621	0.67
64-b KS Adder	LE Opt	412	144.3	1.00	59488	1.00
	Energy Opt	402	72.7	0.50	29211	0.49
64-b HC Adder	LE Opt	419	142.8	1.00	59890	1.00
	Energy Opt	415	83.2	0.58	34503	0.58

6 Conclusion

We have proposed a systematic way to optimize energy at equal performance. The energy saving is significant, 30%–50%. More importantly, the energy optimization is realized from the balance of delay to energy consumption among stages.

References

1. V. G. Oklobdzija, B. Zeydel, H.Q. Dao, S. Mathew, R. Krishnamurthy, "Energy-Delay Estimation for High-Performance Microprocessor VLSI Adders", Proceeding of the 16th Symposium on Computer Arithmetic, June 2003.
2. D. Harris, R.F. Sproull, and I.E. Sutherland, "Logical Effort Designing Fast CMOS Circuits," Morgan Kaufmann Publishers, 1999.
3. V. Stojanovic, D. Markovic, B. Nikolic, M.A. Horowitz, R.W. Brodersen, "Energy-Delay Tradeoffs in Combinational Logic Using Gate Sizing and Supply Voltage Optimization," Proceedings of the 28th European Solid-State Circuits Conference, ESSCIRC'2002, Florence, Italy, September 24–26, 2002. pp. 211–214.
4. R.W. Brodersen, M.A. Horowitz, D. Markovic, B. Nikolic, V. Stojanovic, "Methods for True Power Minimization," International Conference on Computer-Aided Design, ICCAD-2002, Digest of Technical Papers, San Jose, CA, November 10–14, 2002, pp. 35–42.
5. V. Zyuban, and P. Strenski, "Unified Methodology for Resolving Power-Performance Tradeoffs at the Microachitectural and Circuit Levels", IEEE Symposium on Low Power Electronics and Design, 2002.
6. V. V. Zyuban, and P. M. Kogge, "Inherently Lower-Power High-Performance Superscalar Architectures", IEEE Symposium on Low Power Electronics and Desing, 2001.
7. Hoang Dao, Vojin G. Oklobdzija, "Performance Comparison of VLSI Adders Using Logical Effort", 12th International Workshop on Power And Timing Modeling, Optimization and Simulation, Sevilla, SPAIN, September 11–13, 2002.
8. P.M. Kogge, H.S. Stone, "A Parallel Algorithm for the Efficient Solution of a General Class of Recurrence Equations", IEEE Trans. Computers, Vol. C-22, No. 8, 1973, pp.786–793.
9. T. Han, D. A. Carlson, and S. P. Levitan, "VLSI Design of High-Speed Low-Area Addition Circuitry," Proceedings of the IEEE International Conference on Computer Design: VLSI in Computers and Processors, 1987, pp.418–422.
10. V. G. Oklobdzija, "High-Performance System Design: Circuits and Logic", IEEE Press, July 1999.

Instruction Buffering Exploration for Low Energy Embedded Processors*

Tom Vander Aa[1], Murali Jayapala[1], Francisco Barat[1], Geert Deconinck[1], Rudy Lauwereins[2], Henk Corporaal[3], and Francky Catthoor[2]

[1] ESAT/ELECTA, Kasteelpark Arenberg 10, K.U.Leuven, Heverlee, Belgium-3001
{first_name.last_name}@esat.kuleuven.ac.be
[2] IMEC vzw, Kapeldreef 75, Heverlee, Belgium-3001
[3] TU Eindhoven, Electrical Engineering, Den Dolech 2, 5612 AZ Eindhoven, Netherlands

Abstract. For multimedia applications, loop buffering is an efficient mechanism to reduce the power in the instruction memory of embedded processors. Especially software controlled loop buffers are energy efficient. However current compilers do not fully take advantage of the possibilities of such loop buffers. This paper presents an algorithm the explore for an application or a set of applications what is the optimal loop buffer configuration and the optimal way to use this configuration. Results for the MediaBench application suite show an additional 35% reduction (on average) in energy in the instruction memory hierarchy as compared to traditional approaches to the loop buffer without any performance implications.

1 Introduction and Motivation

Low energy is one of the key design goals of the current embedded systems for multimedia applications. Typically the core of such systems are programmable processors. VLIW DSP's in particular are known to be very effective in achieving high performance and sometimes low power for our domain of interest [10]. Examples of such processors are the Trimedia [17] processor from Philips or the 'C6x processors from Texas Instruments [18]. However, power analysis of such processors indicates that a significant amount of power is consumed in the on-chip (instruction) memory hierarchy: 30% of the total power according to [4]. If the appropriate data memory hierarchy mapping techniques are applied first [7, 16], and if all methods to reduce power in the data path are applied [5], we have performed experiments that show this number goes up to 50% if nothing is done here. Hence, reducing this part of the budget is crucial in reducing the overall power consumption of the system.

* This project is partially supported by the Fund for Scientific Research - Flanders (FWO) through projects G.0036.99 and G.0160.02 and the postdoctoral fellowship of G.Deconinck, and by the IWT through MEDEA+ project A502 MESA.

J.J. Chico and E. Macii (Eds.): PATMOS 2003, LNCS 2799, pp. 409–419, 2003.
© Springer-Verlag Berlin Heidelberg 2003

Loop buffering is an effective scheme to reduce energy consumption in the instruction memory hierarchy [15]. In any typical multimedia application, significant amount of execution time is spent in small program segments. By storing them in a small loop buffer (also called L0 buffer) instead of the big instruction cache (IL1), energy can be reduced. However, even the current way of applying this with a single central loop buffer and a standard compiler still leads to a total power contribution of more than 20%.

An important way to reduce this further is by more effectively managing these loop buffers through a dedicated pre-compiler. The pre-compiler should be responsible for mapping the appropriate parts of the application onto these L0 buffers. However, to the best of our knowledge there has been little work on algorithms and tools that use the information available in the application to effectively profit from a software controlled loop buffer. As will be shown, the instruction memory energy consumption can be significantly reduced when the appropriate loop buffer configuration is used in the appropriate way.

An algorithm to explore the loop buffer design space is presented. In that, given a program (e.g. in C), the algorithm finds the most energy efficient loop buffer configuration and also decides what loops should be mapped to the loop buffer. Results show an average reduction in instruction memory energy consumption for typical multimedia applications of 35% as compared to a simple loop buffer mapping method currently applied in existing work.

The rest of this paper is organized as follows: A brief account of the related work is presented in Section 2. In Section 3 the software controlled loop buffer organization is described. In Section 4 the energy model under consideration for our exploration tool is outlined. Section 5 describes the loop buffer exploration algorithm, which is our main contribution. Section 6 presents the simulation results and Section 7 is the conclusion.

2 Related Work

Several loop buffering schemes have been proposed in the past [12,1,2] An overview of the options can be found in [14]. Initially only inner loops without any control constructs could be mapped to the loop buffer. In [9] support is added for control constructs such as if-then-else, subroutine calls and nested loops. Our loop buffer architecture also supports conditional constructs, as well as mapping a set of nested loops. What is not discussed in [9] are the methodologies on how to efficiently use these new features.

Our main reason to use a software controlled loop buffer is to exploit knowledge about the program in the compiler to reduce power. But, as is discussed in [15], software controlled loop buffers also do not need the energy consuming tag memories. Furthermore, they do not suffer any cycle penalty.

The idea to add compiler support has already been proposed in [3]. In that paper, however, a regular cache is used (with tags) and no energy model is used directly in the exploration framework.

The creation of a custom memory hierarchy has already been explored extensively in a data memory context [7,16]. For instruction memory, [8] presents tuning of the loop buffer size for simple loop buffer architectures, only supporting inner loops.

3 Operation of the Low Loop Buffer Organization

Figure 1 illustrates the essentials of the low power loop buffer under consideration. Instructions are fed to the processing unit either from the level 1 instruction cache or through the loop buffer.

Initially the loop buffer is turned off and the program executes via IL1 (Figure 1, left). When a special instruction is encountered marking the beginning of the loop that has to be mapped to the loop buffer, the loop buffer will be turned on. The form of this special instruction is `lbon` <*startaddress*>, <*endaddress*>, where *startaddress* is the address of the first instruction of the loop and *endaddress* that of the last one. These values are stored in the local controller (LC) and will be used during the execution of the loop.

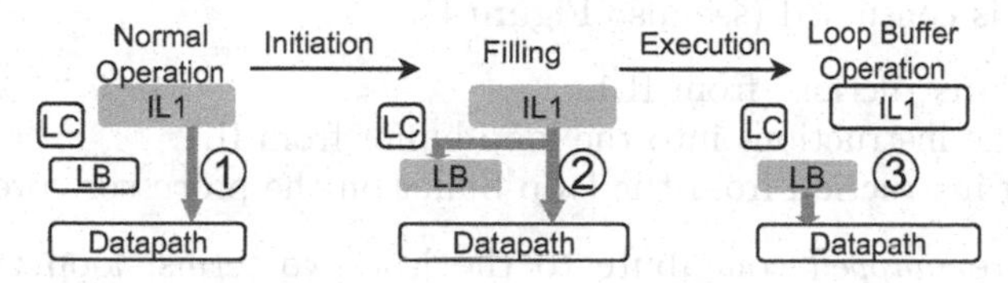

Fig. 1. The processor operates in three phases controlled by the local controller (LC): normal operation, filling of the Loop Buffer and Loop Buffer operation

If a `lbon` is encountered and no *startaddress* is stored in the local controller, or the *startaddress* in the LC is different from the one in the `lbon` instruction, the first iteration will be used to fill the loop buffer. The instruction stream will go from IL1 both to the loop buffer and to the processing unit (Figure 1, middle). After the first iteration the IL1 can be put into low power mode and only the loop buffer will be used (Figure 1, right).

When the loop buffer is used, a local controller will translate the program counter to an index in the loop buffer by using the stored *startaddress.* This mechanism reduces the power by avoiding the expensive tag memory found in normal caches.

When the LC detects the program counter is bigger than *endaddress*, the loop buffer will be turned off.

4 Energy Dissipation Model

A tree like representation for the loops, as shown in Figure 2, is extracted from the source code of the application. From this representation we can identify

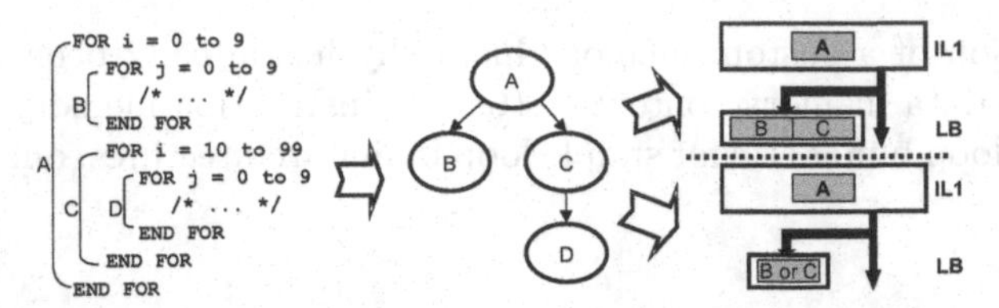

Fig. 2. Loop nest, corresponding tree representation and two possible mappings. A is the parent of B. B is the child of A. B and C are siblings.

different mappings of loops on the loop buffer. The figure shows two possible configurations for a given program, both leading to different loop buffer sizes and different energy consumptions. In the first configuration B and C are mapped to different locations in the loop buffer, leading to a bigger size. In the second, loops B and C are loaded each time the loop starts, leading to more loads from IL1. For a given mapping of loops, the energy E is:

$$E = \begin{array}{ll} \sum_{l \in unmapped}^{l} N_{exec}(l) \times E_{access}(IL1) & (1) \\ + \sum_{l \in mapped}^{l} N_{load}(l) \times E_{access}(IL1) & (2) \\ + \sum_{l \in mapped}^{l} N_{exec}(l) \times E_{access}(lb) & (3) \end{array} \quad (1)$$

The three sums correspond to the three places in the instruction memory where energy is consumed (see also Figure 1):

1. Executing instructions from IL1.
2. Loading the instructions into the loop buffer from IL1.
3. Executing instruction from the loop buffer on the processor core.

Loops that are *mapped* contribute to the last two terms, loop that are not mapped (*unmapped*) to the first term. N is the number of accesses to memory (loop buffer or IL1) due to the loading or execution of the loop. E_{access} is the energy per access of the loop buffer or the instruction level 1 cache. The value of $E_{access}(lb)$ depends on the size of the loop buffer, which on itself can be calculated when you know what loops are currently mapped and if loops are mapped together or reloaded each time the loops is invoked.

The energy values we have used for the memories (E_{access}) are calculated using the Wattch [6] power models. For the IL1 we used a regular cache, for the loop buffer we used a cache without tags. Although the Wattch model has some known limitations, it is still suited for our purpose since we only need good relative energy values.

5 Design Space Exploration

Since the compiler is responsible for inserting the `lbon` instructions, it should decide what is the most energy efficient way to do so. This leads to a design space exploration problem with the following goals:

1. What is the optimal loop buffer size? If the loop buffer is too small, not enough loops fit and there will be too many access to the IL1 cache. If the loop buffer is too big the energy per access of the loop buffer will be too big.

2. What is the best way to map the loops that fit? If we decide to use the loop buffer for a certain loop, we still have several options: Do we map the loop entirely or only parts of the loop? If we load two loops like B and C in Figure 2, do we put them in the same address space or in different address spaces? The former will save us space in the loop buffer, we will have to load the loop each time it starts. The latter case needs a bigger loop buffer, but will save us accesses to IL1 to load the loops (as can be seen from the energy model from the previous section).

5.1 Size of the Design Space

Using a brute force approach to find the optimal mapping of loops to the loop buffer by trying all the configurations is not feasible. The number of combinations you would have to try for the outer-most loops of the program is:

$$\#Sol(OuterLoop) = 2 + \prod_{C \in Child\ Loops} \#Sol(C) \tag{2}$$

We count the two basic solutions (mapping the loop completely or not mapping the loop at all), plus all the combinations of possible solutions of the immediately nested loops (children). For MPEG2 encoding, for example, this would lead to more than 10^{48} combinations. It's clear that we need an effective heuristic to find an optimal mapping.

5.2 Exploration Algorithm

Instead of the brute force approach, we use a recursive algorithm to explore our design space in a more intelligent way (see Algorithm 1). The loops of a program are represented as a tree as already presented in Figure 2. The Algorithm works like this: the procedure `Find_Mappings` will be called with the top loop of the tree as an argument. If the program has multiple top loops, a dummy loop with iteration count of one, can be assumed around the whole program.

Two basic solutions exist for a loop passed to `Find_Mappings`: mapping the loop completely or not mapping the loop at all. Both solutions are kept since they will have a different energy consumption and required loop buffer size. If the loop has children, `Find_Mappings` is called recursively to find the solutions of the child loops. These solutions are then combined in the procedure `Filter_Solution`. Here a heuristic is used to prune the design space: if for a certain loop and loop buffer size several solutions exist, we only keep one solution for each possible size, namely, the most energy efficient solution. Although we do not have a proof, we believe that only this solution might lead to the optimum for the top loop. For several small examples we did an exhaustive search, and for these our assumption was indeed true. Using this heuristic, the number of solutions you have to keep is limited to the number of different loop buffer sizes you will encounter. For a typical application this will be less than 1000 combinations, which is much less than the total number of combinations of Equation 2.

Algorithm 1 Design space exploration to map a given set of loops to a loop buffer. Energy and needed loop buffer size are calculated using Equation 1

Procedure: Find_Mappings
  Input: Loop l
  Output: Set of solutions $S = \{(z_1, E_1), (z_2, E_2), ...(z_n, E_n)\}$ such that for each possible loop buffer size z_i, E_i is the minimal energy for that size}
Begin
  /* 2 base solutions */
  $S \Leftarrow \{($ 0, Compute_Energy(l, unmapped) $)\}$
  $S \Leftarrow S \cup \{($ size(l), Compute_Energy(L, mapped) $)\}$

  /* Find the solutions of the children */
  let $\{c_1, c_2, ..., c_n\} \leftarrow$ children of l
  for $i = 0$ to n **do**
    $Sol_i \leftarrow$ Find_Mappings(c_i)
  end for
  Filter_Solutions($S, Sol_1 \times Sol_2 \times ... \times Sol_n$)
End

Procedure: Filter_Solutions
  Input: $Sol_1 \times Sol_2 \times ... \times Sol_n$, Set of Solutions S
  Output: Updated set of Solutions S
Begin
  /* Combine the solutions of the children */
  /* filtering out the not needed solutions */
  for all $(sol_1, sol_2, ..., sol_n) \in Sol_1 \times Sol_2 \times ... \times Sol_n$ **do**
    let $(z, energy\_current) \Leftarrow$ New solution by combining $s_1, s_2, ..., s_n$
    /* We only keep one (the best) solution per size */
    let $(z, energy\_optimal) \Leftarrow$ Current solution with size z
    if $energy\_optimal$ does not exist yet **then**
      $S \Leftarrow S \cup \{(z, energy\_current)\}$
    else if $energy\_current < energy\_optimal$ **then**
      $S \Leftarrow S \setminus \{(z, energy\_optimal)\} \cup \{(z, energy\_current)\}$
    end if
  end for
End

5.3 Extension: Partial Mapping of Loops

We have extended the algorithm to support partial mapping of loops. If a certain loop body contains basic blocks that are almost never executed (they contain for example exception handling code), it does not make sense to map those to the loop buffer. To add support for this we had to make a minor modification to the algorithm. Instead of only two basic solutions – mapping the loop completely or not mapping the loop – we now also consider all possibilities in between. We start with no basic blocks mapped, and for each new possibility we add the block that is executed the most amongst all of them that are not mapped yet. The solution generated last is precisely the one where the whole loop is mapped.

This scheme can be handled with our loop buffer hardware since the compiler we use applies trace-based code layout [19]. The blocks of the most frequently executed trace are grouped together at the beginning of the procedure. The less executed code is put at the end of the instruction layout. By changing the value of *endaddress* of the `lbon` instruction, you decide what basic blocks are mapped.

6 Results and Discussion

We have used the MediaBench [13] application suite to compare our work, both mapping of only complete and partial loops, to existing implementations of the loop buffer. Table 1 shows the benchmarks we used, and the optimal loop buffer size (in number of instructions) for each benchmark in terms of energy.

Table 1. List of benchmarks used, and optimal loop buffer size

Benchmark	Description	Opt. Size	Benchmark	Description	Opt. Size
ADPCM decode	Audio	45	ADPCM encode	Audio	53
AES	Encryption	88	Blowfish encode	Encryption	51
JPEG decode	Image	128	JPEG encode	Image	59
EPIC	Image	55	g721 decode	Audio	144
g721 encode	Audio	155	ghostscript	Image	85
gsm decode	Audio	81	gsm encode	Audio	81
H.263	Video	182	mesa osdemo	3D graphics	64
MPEG2 decode	Video	57	MPEG2 encode	Video	76
Rasta	Speech Recogn.	76	SHA	Encryption	25
Snake	3D graphics	179	**Overall Optimum**		**99**

6.1 Energy versus Loop Buffer Size

Figure 3 (left) shows for an example application the loop buffer size versus the energy of the optimal mapping on a loop buffer of that size. Since the algorithm we use (the extended one of Section 5.3) only generates solutions for certain loop buffer sizes, the graph is discrete. Figure 3 (right) shows the same information but for all applications tested together. For all sizes between 0 and 255 we took the energy of optimal solutions for that size of all applications. The sum of those values leads for the optimal solution for each size.

The graph also shows you can indeed efficiently explore the energy-size trade-off with the algorithm. For small sizes the total energy is dominated by the IL1 energy. As the loop buffer size increases more loops will be mapped to the small loop buffer and the IL1 energy will go down. With the size, also the energy per access of the loop buffer increases, until at a certain point the most energy will be consumed in the loop buffer itself. Generally, this will happen in the *knee* of the *"% Mapped"*-curve, where it does not pay off anymore to map more loops to the loop buffer.

An interesting feature to notice in the left graph, is that the algorithm produces two solutions with the same loops mapped. One solution needs a size of

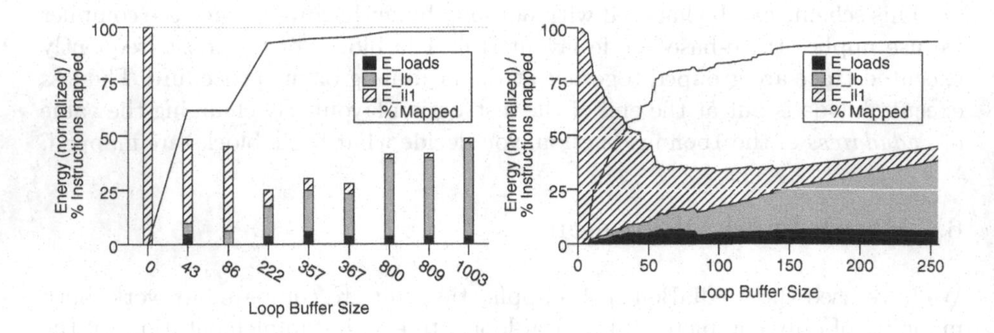

Fig. 3. Instruction memory energy consumption versus the loop buffer size, for one specific benchmark (left) and for all benchmarks together (right)

43, the other a size of 86. This is because the program has two loops of the same size (43), and they can be mapped to different locations in the loop buffer, needing a loop buffer of size 86, or they can be mapped on the same location. For this last solution you are required to load the loop each time you enter it, resulting in more loads from IL1 to the loop buffer. You cannot say beforehand what will be the most energy efficient solution in such case, since this depends on the behavior of the program.

The other graph shows the global energy behavior of all applications together. The minimal energy is consumed for a loop buffer of size 99. So if you would like to build *one* loop buffer for the whole application domain, 99 would be the best choice.

6.2 Comparison of Different Mapping Strategies

Figure 4 shows the normalized energy consumption in the instruction memory of the optimal mapping according to five different mapping strategies:

- **no loop buffer**: All loops are executed from the IL1 memory. Since much previous work has been done on loop buffers and it is well known that multimedia applications contain a lot of loops, making them very suitable for use with a loop buffer, has been done it would not be fair to use this case (no loop buffer) as our reference case. What can be seen, however, is that by simply using a loop buffer, although maybe in a not so clever way, you can already reduce the energy in instruction memory significantly (72% on average).
- **inner loops**: This strategy only maps the most inner loops to the loop buffer. Since this is what is implemented in existing loop buffer architectures, the other energy values are normalized against this strategy.
- **whole loops**: This is the basic version of the algorithm, as described in Section 5.2. The average gain amongst all benchmarks as compared to the previous strategy is 35%, with maxima for some benchmarks of up to 88%. This gain has two main reasons: the first reason for energy reduction is the fact that not only inner loops but also the outer loops can be mapped. If the

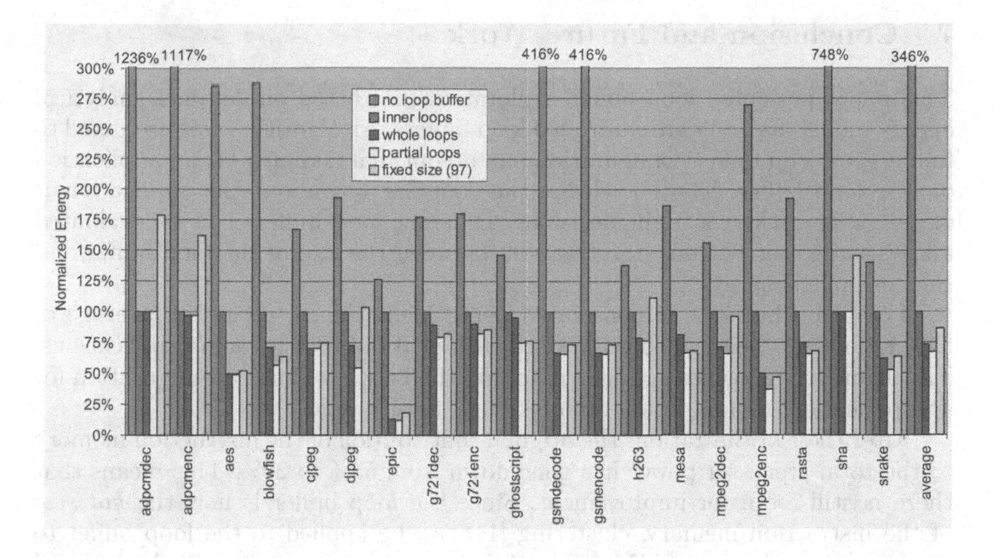

Fig. 4. Energy comparison of different loop buffer mapping strategies applied on the MediaBench suite

inner loops have a too low iteration count, we can also map the surrounding loop(s), because we have support in our loop buffer for nested loops. For this reason simple applications with mostly inner loops, such as ADPCM and SHA, perform already very well with the strategy that only maps inner loops. For complex applications, such as MPEG2 and JPEG, taking into account all the loops and not just the inner one, has indeed a big impact (a gain of a factor 2 or more). The other main reason is the fact that we explore the trade off between mapping two sibling loops separately or together on the loop buffer.

- **partial loops**: This is the extended version of our algorithm, that is able to map loops only partially. For certain benchmarks, such as ghostscript or MPEG2 Decoding, there are loops that contain less frequently executed basic blocks. For these benchmarks not mapping the whole loop gives a significant gain. Averaged over all applications the gain compared to the previous strategy is limited, only 7%.
- **fixed size**: As can been seen in Figure 3, the overall optimal size over all benchmarks is 99 instructions. We have fixed this size and calculated the optimal mapping for all benchmarks. For some benchmarks the optimal mapping (with a free-to-choose loop buffer size, cf. Table 1) is close to 99. For these benchmarks the penalty for not being able to choose the size of the loop buffer is small. For others, however there is a penalty because 99 instructions is not enough or too much for these benchmarks. The average penalty as compared to *partial loops* is 20%. A solution here is to perform additional code transformations to make your application more suitable for a specific loop buffer configuration.

7 Conclusion and Future Work

This paper presented loop buffer exploration tool based on detailed analytical energy models for software controlled loop buffers. An algorithm was presented to find a good loop buffer configuration and an optimal mapping for an application on the loop buffer. The knowledge the compiler has about the application is exploited to make the right decisions. Different loop buffers can be evaluated allowing the user to make the trade-off between the size of the loop buffer and the consumed energy.

The algorithm was demonstrated using MediaBench, giving an average of 65% reduction in energy consumption, with peaks of 88% for some applications. The algorithm also allows designers to find the best loop buffer configuration for an application domain instead of one single application.

After these optimizations the original contribution of the instruction memory to the total processor power has gone down from 53% to 20%. This means that there is still room for improvement. Since the loop buffer is now the *hot spot* of the instruction memory, clustering [11] can be applied to the loop buffer to further reduce the energy. How to optimally use a software controlled, clustered loop buffer will be the subject of future work.

References

1. Tim Anderson and Sanjive Agarwala. Effective hardware-based two-way loop cache for high performance low power processors. In *Proc of ICCD*, September 2000.
2. R. S. Bajwa and et al. Instruction buffering to reduce power in processors for signal processing. *IEEE Transactions on VLSI*, 5(4):417–424, December 1997.
3. Nikolaos Bellas, Ibrahim Hajj, Constantine Polychronopoulos, and George Stamoulis. Architectural and compiler support for energy reduction in the memory hierarchy of high performance microprocessors. In *Proc of ISLPED*, August 1998.
4. L. Benini, D. Bruni, M. Chinosi, C. Silvano, V. Zaccaria, and R. Zafalon. A power modeling and estimation framework for vliw-based embedded systems. In *in Proc. Int. Workshop on Power And Timing Modeling, Optimization and Simulation PATMOS*, September 2001.
5. Luca Benini and Giovanni de Micheli. Sysmtem-level power optimization: Techniques and tools. *ACM TODAES*, 5(2):115–192, April 2000.
6. David Brooks, Vivek Tiwari, and Margaret Martonosi. Wattch: A framework for architectural-level power analysis and optimizations. In *Proc of ISCA*, pages 83–94, June 2000.
7. Francky Catthoor, Koen Danckaert, Chidamber Kulkarni, Erik Brockmeyer, Per Gunnar Kjeldsberg, Tanja Van Achteren, and Thierry Omnes. *Data access and storage management for embedded programmable processors.* Kluwer Academic Publishers, March 2002.
8. S. Cotterell and F. Vahid. Tuning of loop cache architectures to programs in embedded system design. In *Proc of International Symposium on System Synthesis (ISSS)*, October 2002.
9. A. Gordon-Ross, S. Cotterell, and F. Vahid. Exploiting fixed programs in embedded systems: A loop cache example. In *Proc of IEEE Computer Architecture Letters*, Jan 2002.

10. Margarida F. Jacome and Gustavo de Veciana. Design challenges for new application-specific processors. *Special issue on Design of Embedded Systems in IEEE Design & Test of Computers*, April-June 2000.
11. Murali Jayapala, Francisco Barat, Pieter OpDeBeeck, Francky Catthoor, Geert Deconinck, and Henk Corporaal. A low energy clustered instruction memory hierarchy for long instruction word processors. In *Proc of PATMOS*, September 2002.
12. Johnson Kin, Munish Gupta, and William H. Mangione-Smith. Filtering memory references to increase energy efficiency. *IEEE Transactions on Computers*, 49(1):1–15, January 2000.
13. Chunho Lee and et al. Mediabench: A tool for evaluating and synthesizing multimedia and communicatons systems. In *International Symposium on Microarchitecture*, pages 330–335, 1997.
14. Lea Hwang Lee, Bill Moyer, John Arends, and Ann Arbor. Low-cost embedded program loop caching - revisited. Technical report, EECS, University of Michigan, December 1999.
15. Lea Hwang Lee, William Moyer, and John Arends. Instruction fetch energy reduction using loop caches for embedded applications with small tight loops. In *Proc of ISLPED*, August 1999.
16. Preeti Ranjan Panda, Nikil D. Dutt, and Alexandru Nicolau. Memory data organization for improved cache performance in embedded processor applications. *ACM TODAES*, 2(4):384–409, 1997.
17. G.A. Slavenburg, S. Rathnam, and H. Dijkstra. The Trimedia TM-1 PCI VLIW media processor. In *Proceedings Hot Chips VIII Conference*, 1996.
18. Texas Instruments Inc., http://www.ti.com. *TMS320 DSP Family Overview*.
19. Trimaran group, http://www.trimaran.org. *Trimaran: An Infrastructure for Research in Instruction-Level Parallelism*, 1999.

Power-Aware Branch Predictor Update for High-Performance Processors

Amirali Baniasadi

Dept. of Electrical and Computer Engineering, University of Victoria,
3800 Finnerty Road, Victoria, BC, V8P 5C2, Canada
amirali@ece.uvic.ca

Abstract. We introduce Power-Aware Branch Predictor Update (PABU) as a power-efficient branch prediction technique for high performance processors. Our predictor reduces branch prediction energy consumption by eliminating unnecessary branch predictor updates. Our technique relies on information regarding past branch behavior to decide if additional predictor updates result in performance improvements. We avoid updating the predictor for branches where there is already enough information available to correctly predict their outcome. In this work we study energy and performance trade-offs for a subset of SPEC 2k benchmarks. We show that on the average and for an 8-way processor, our technique can reduce branch prediction energy consumption up to 80% compared to a 32k conventional combined branch predictor. This comes with a negligible impact on performance (0.6% max). We show that our technique, on the average, reduces the number of predictor updates by 83%.

1 Introduction

Designing power-efficient processor components has emerged as a first class consideration in high-performance processors.

One of the components that could impact power dissipation highly is the branch predictor. While branch predictors can dissipate as much as 10% of overall processor power [1], their accuracy impacts the overall processor power dissipation. Reportedly, replacing complex power hungry branch predictors with small and simple ones will result in higher overall power dissipation due to an increase in the number of mispredicted instructions executed caused by an increase in the number of mispredicted branches [2]. Accordingly, in this work we focus on low power and accurate predictors.

We propose *Power-Aware Branch Predictor Update* or *PABU* as an energy-efficient extension to combined predictors. While we focus on the combined predictor, our technique could be applied to other branch predictors as well. Combined predictors exploit saturating counters to record the information regarding previous branch behaviors. By

J.J. Chico and E. Macii (Eds.): PATMOS 2003, LNCS 2799, pp. 420–429, 2003.
© Springer-Verlag Berlin Heidelberg 2003

using counters, the predictor can tolerate a branch going an unusual direction one time and keep predicting the usual branch direction. While the predictor increments or dec rements these counters occasionally and based on the branch outcome, quite often, new updates do not change the information stored in the tables. This is the result of the fact that the counters are already saturated. For example, in the case of loops with more than hundreds of iterations, only the first and last few predictor updates provide useful information. The rest of the updates do not change the state of the counters associated with the loop branch instruction. This is also true for other examples of frequent well-behaved branch instructions. PABU aims at identifying and eliminating updates that do not contribute to the predictor accuracy and therefore to performance.

We show that our technique can, on the average, reduce branch predictor energy consumption by 64% over a 32K-entry, conventional combined predictor. Moreover, we show that a processor with PABU is *always* more energy efficient than one without one.

The rest of the paper is organized as follows. In section 2 we explain PABU. In section 3 we present our experimental evaluation. We report update frequency, performance and energy savings. We report relative energy consumption for both the predictor and the entire processor. In section 4 we review related work. Finally, in section 5, we summarize our findings.

2 Power-Aware Predictor Update

Branch prediction is essential to sustaining high performance in modern high-end processors. The combined predictor is one of the most accurate predictors available and is used in many high-performance designs. Combined predictors use three underlying sub-predictors. Two of the sub-predictors produce predictions for branches. They are typically tuned for different branch behaviors. The third sub-predictor is the *selector* and it keeps track of which of the two sub-predictors works best per branch. Typical configurations, use *bi-modal* predictors for one of the sub-predictors and the selector, and a pattern-based predictor like *gshare* for the last sub-predictor. Bi-modal predictors capture temporal direction biases (*e.g.*, mostly taken) in branch behavior and have short learning times. However, they cannot capture complex repeating branch behaviors that do not exhibit a direction bias. Pattern-based predictors like gshare can successfully capture repeating direction patterns and correlated behaviors across different branches. However, these predictors require longer learning times and have larger storage demands. By using a selector predictor, combined predictors offer the best of both underlying sub-predictors: fast learning and prediction for branches that exhibit temporal bias, and slower but accurate prediction for branches with repeatable direction patterns. The sub-predictors use saturating counters to record information. A typical mechanism in such predictors is to increment the associated counter if the branch is taken and decrement it if it is not taken. To reduce the number of bits required, small counters (*e.g.*, 2-bit counters) are used. Once the counter value has reached the maximum (*e.g.*, three), taken branch outcomes will no longer increment the counter. Similarly, once the counter has reached its minimum, not taken branch outcomes will not change the predictor state. In other words, the counter is not decremented past the minimum, nor it is incremented past the maximum. Later, the counters are probed to predict the branch outcome. If the counter value is more than a threshold (*e.g.*, one for 2-bit counters) the branch is predicted taken. If the counter is less or equal the threshold the branch outcome is predicted to be not taken.

We categorize predictor updates to three major groups. The first group includes those updates that do not change the predictor state. Examples are updates caused by taken branches for branches whose associated counter is already saturated to the maximum value. We refer to such updates as *non-effective updates (NEUs)*. The second group of updates are those that change the predictor state but not the immediate outcome. An example of such updates is an update caused by a taken branch whose associated counter is already more than the threshold (*e.g.*, the counter value is two) but less than the maximum value (*e.g.*, three). While such updates do not change the immediate predictor outcome they may impact future decisions. We refer to such updates as *probably effective updates (PEUs)*. The third group of updates are those that change both the predictor state and immediate outcome. An example of such updates is an update causing a counter to pass the threshold. We refer to such updates as *effective updates (EUs)*.

Figure 1 shows the percentage of each kind of the updates for a subset of SPEC'2k benchmarks. Bars from left to right report relative distribution for NEUs, PEUs and EUs. On the average, 95% of the updates appear to be non-effective (NEUs). This reaches a minimum of 90% for *vor* and a maximum of 99.7% for *mes* and *amm*. Note that *mes* and *amm* have the highest branch prediction rates among the studied benchmarks (as we will show later in table 1). On the average, only 3% of the total number of updates are PEUs. This is 2% for the EUs. We conclude from figure 1 that only a very low percentage of updates contribute to performance. On the average, 95% of the updates result in power dissipation without contributing to performance.

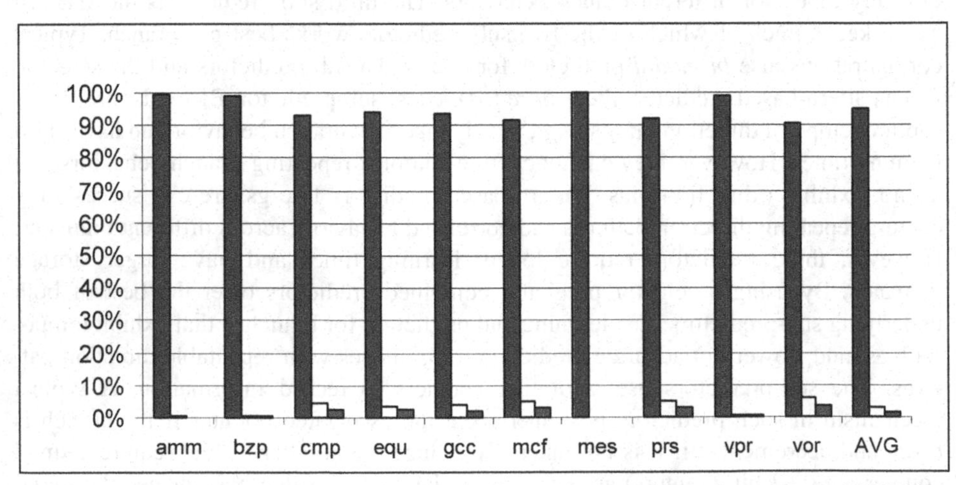

Fig. 1. Bars from left to right show percentage of NEU, PEU and EUs for a subset of SPEC2k benchmarks

In this work we suggest PABU as a dynamic method to identify and eliminate such updates. We introduce PABU as an extension to conventional combined branch predictors. PABU aims at reducing energy consumption while maintaining accuracy virtually intact.

PABU deems a predictor update as an NEU if both of the following conditions are true:

(1) If the update updates the counters associated with a branch, which is already being predicted accurately. These are branches, which there is enough information already available to the predictor to predict their outcome accurately. One simple and power-efficient way to find out if a branch falls in this category is to check if the branch was accurately predicted last time encountered.

(2) If the update updates underlying branch predictors (we use a combined, McFarling predictor) that are already strongly biased (*i.e.*, their counters are saturated). We used this mechanism to produce the "Both Strong"[2] estimation method which marks a branch as high confidence only if the saturating counters for both gshare and bimodal predictors are in a strong state and have the same predicted direction (taken or not-taken). This mechanism also leverages information that is readily available and hence it's power and complexity requirements are negligible.

Intuitively, if the sub-predictors are already biased and the branch outcome is being predicted correctly there is no need for additional predictor updates.

PABU stores the last 8 dynamic accurately predicted branch instructions in a PC-indexed, small 8-entry table which we refer to as the PABU-table. We picked the number of PABU-table entries after testing different alternatives. Our experiments shows that 8 entries provides enough accuracy while imposing very little power overhead. Also in every reservation station we use one bit to record if the underlying predictors are biased and if the branch was accurately predicted last time. This bit is decided at predictor lookup time when the branch predictor is probed in parallel with the PABU-table. We set this bit to one if we find the branch already recorded in the PABU-table and the underlying associated counters biased. Later, and before updating the predictor, we check this bit to decide if updating the predictor is necessary. We remove mispredicted branches from the PABU-table to allow additional updates for such branches.

Provided that sufficient branch locality and biased sub-predictor entries exist, PABU has the potential for reducing branch prediction energy consumption. However, it introduces extra energy overhead and can, in principle, increase overall energy consumption if the necessary behavior is not there. We take into account this overhead in our study and show that for the programs we studied PABU is robust.

3 Methodology and Results

In this section, we present our analysis of the PABU technique. We report update frequency in 3.1. We report performance results in section 3.2. We report energy measurements in section 3.3. While our study shows that power (energy per cycle) results follow the same trend of energy results, we do not report power in the interest of space.

We used programs from the SPEC2000 suite compiled for the MIPS-like architecture used by the Simplescalar v3.0 simulation tool set. We used GNU's gcc compiler (flags: -O2 –funroll-loops –finline-functions). Table 1 reports the branch prediction accuracy per benchmark (this includes all control flow instructions and takes into account not only direction prediction but also target prediction). We simulated one Billion of the instructions after skipping the initialization. We detail the base processor model in table 2.

Table 1. Benchmarks and control flow prediction accuracy (direction and target).

Program	Ab.	BP Acc.	Program	Ab.	BP Acc.
ammp	amm	99%	*mcf*	mcf	92%
bzip	bzp	98%	*mesa*	mes	99%
compress	cmp	92%	*parser*	prs	91%
equake	equ	95%	*vortex*	vor	93%
gcc	gcc	85%	*vpr*	vpr	91%

Table 2. Base processor configuration.

Branch Predictor	32K GShare+32K bi-modal w/ 32K selector
Scheduler	128 entries, RUU-like
Fetch Unit	Up to 8 instr./cycle. Max 2 branches/cycle 64-Entry Fetch Buffer
Load/Store Queue	128 entries, 4 loads or stores per cycle Perfect disambiguation
OOO Core	any 8 instructions / cycle
Func. Unit Latencies	same as MIPS R10000
L1 - Instruction /Data Caches	64K, 4-way SA, 32-byte blocks, 3 cycle hit latency
Unified L2	256K, 4-way SA, 64-byte blocks, 16-cycle hit latency
Main Memory	Infinite, 100 cycles

For our experiments we used a combined predictor with 32K-entries per sub-predictor. We use the 32K-entry predictor after investigating predictors of different sizes studying performance and accuracy. Many metrics for summarizing energy vs. performance trade offs exist. For example, one such metric is the energy, delay product. Since our primary goal was to maintain performance we choose instead on predictors that offer performance within 2% of the best predictor we studied. Figures 2(a) and 2(b) show how predictor size impacts performance and predictor energy consumption relatively to the same processor that uses a 64k-entry predictor. As shown performance-wise most benchmarks show 0% performance slowdown when a 32k predictor is used. Exceptions are gcc (2% slowdown) and prs (0.6% slowdown). The average slowdown stays within 2% even with the 16K-entry predictor. However, worst case slowdown with this predictor is 5% and much worse for predictors of smaller sizes. In figure 2(b) we report predictor energy consumption. Here relative energy consumption is 22%, 25%, 30%, 37%, 50% and 69% for 1k, 2k, 4k, 8k, 16k and 32k predictors when compared to a 64k predictor. Parikh *et. al,* showed that while using smaller predictors saves predictor energy, it also results in higher overall processor energy consumption[1]. Our study shows a similar trend. However we do not report overall energy in the interest of space.

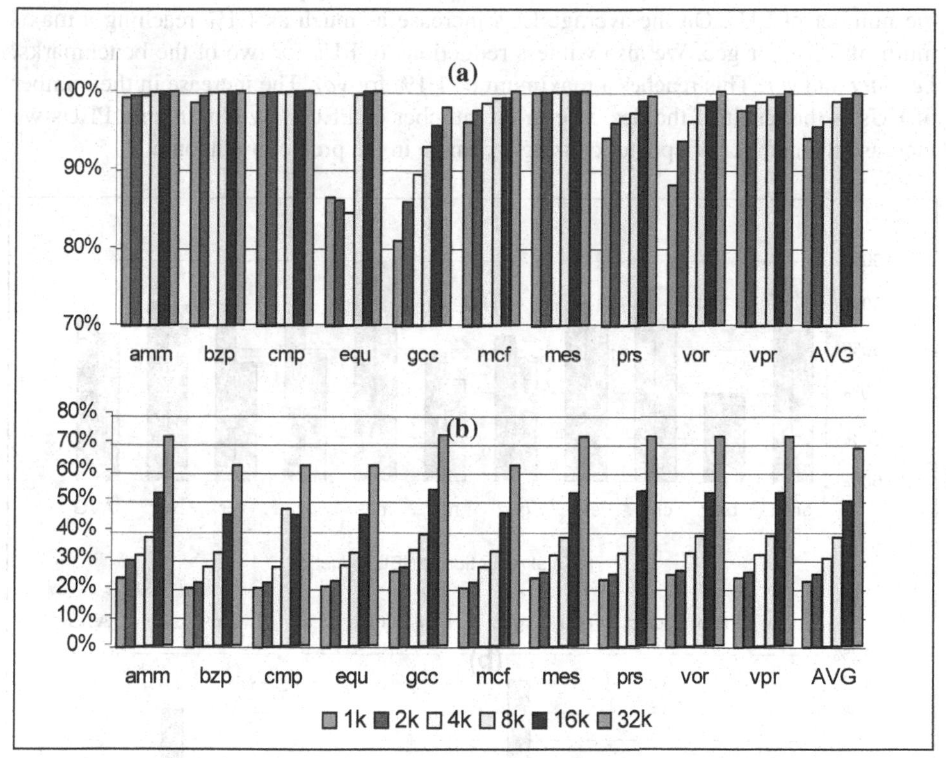

Fig. 2. How different predictor sizes impact (a) performance, and (b) predictor energy consumption. Results are shown relative to a processor that uses a 64K-entry predictor

We used WATTCH [5] for energy estimation.Wattch is a framework for analyzing microprocessor power dissipation at the architectural level. The power estimations are based on a suite of parameterizable power models for different hardware structures and on per-cycle resource usage counts generated through cycle-level simulation. We modeled an aggressive 2GHz superscalar microarchitecture manufactured under a 0.1 micron technology.

3.1 Update Frequency

In figure 3 we report how PABU impacts the total number of predictor updates, the number of EUs, NEUs and PEUs for different benchmarks. In 3(a) bars from left to right report the relative reduction in the total number of updates and the NEUs. On the average this is 83% for the total number of updates. It reaches a maximum of 99.7% for *mes* and a minimum of 69% for *mcf*. On the average we reduce the number of NEUs 87%. This reaches a maximum of 99.6% for *mes* and a minimum of 70% for *bzp*.The reduction in NEUs should be viewed as a success measure for our techniques since our primary goal is to identify and eliminate such updates.

In 3(b) bars from left to right report the relative reduction in the number of PEUs and EUs. On the average we reduce the number of PEUs 3.7% reaching a maximum of 10.2% for *gcc* and a minimum of 0.5% for *bzp*. Interestingly, we witness an increase in the number of EUs. On the average EUs increase as much as 1.1% reaching a maximum of 3.3% for gcc. We also witness reductions in EUs for two of the benchmarks, *i.e.*, *mcf* and *vor*. This reaches a maximum of 1.1% for *vor*. The increase in the number of EUs is the result of the decrease in the number of PEUs. By eliminating PEUs we increase the number of updates causing a change in the predictor outcome.

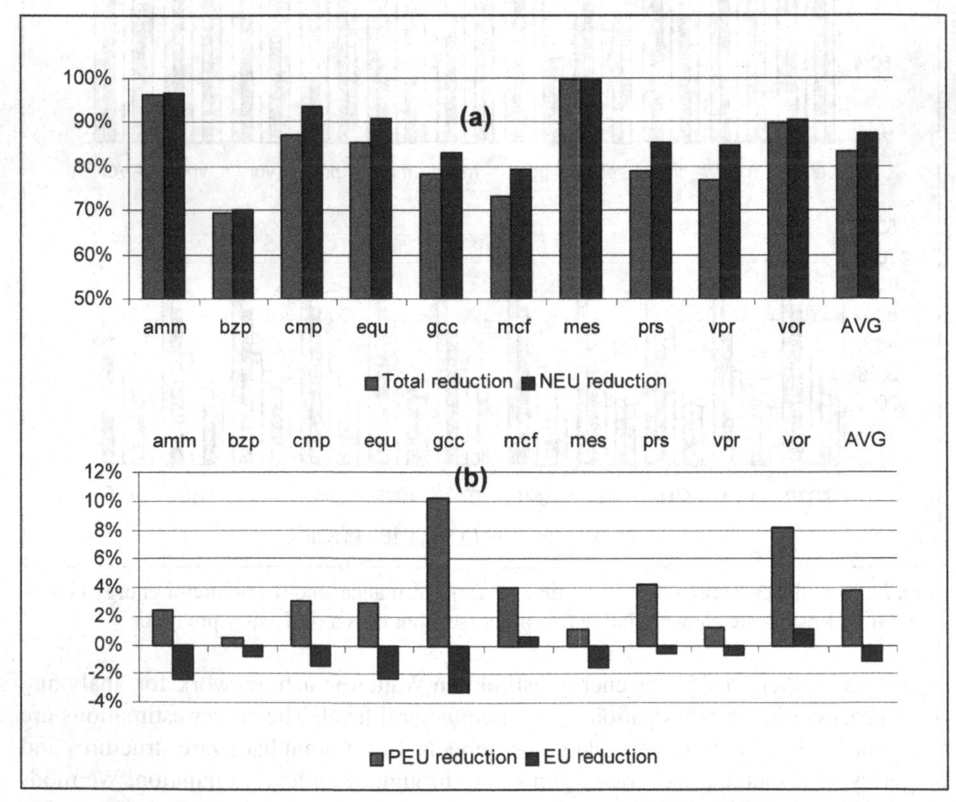

Fig. 3. a) Relative reduction in the total number of updates and NEUs (b) Relative reduction in the number of PEUs and EUs for a PAPU-enhanced predictor compared to conventional combined branch predictors

From figure 3 we conclude that PABU successfully identifies the majority of non-effective predcitor updates. Moreover, PABU successfully avoids eliminating updates that can potentially contribute to performance.

3.2 Performance

PABU can negatively impact accuracy and hence performance. Accordingly, we investigate how PABU impacts performance. To determine whether our technique is indeed worthwhile, we compare it with a conventional processor which uses a predictor that does not eliminate any of the updates. In figure 4 we report performance. Numbers lower than 100% represent slowdowns. On the average, performance slowdown is 0.2%. In the worst case of *mcf*, it is only 0.6%. Performance loss stays zero for 4 of the 10 benchmarks (*i.e.*, *amm*, *bzp*, *mes* and *vpr*).

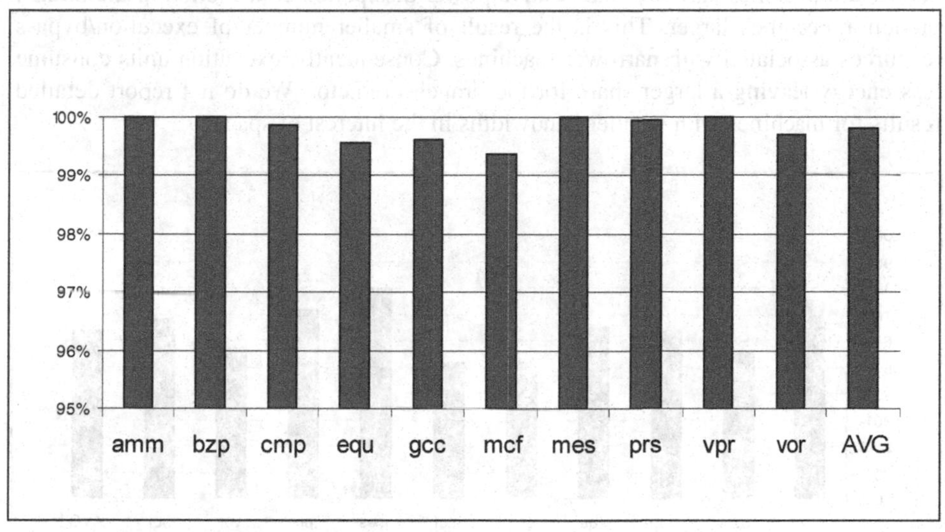

Fig. 4. Performance for PABU-enhanced processor compared to a processor using a conventional combined branch predictor (higher ist better)

3.3 Energy

As a first indicator of energy reduction in figure 3(a) we reported the relative reduction in the number of updates for the PABU-enhanced predictor. In this section we first report predictor energy consumption. While reducing the number of predictor updates reduces predictor's energy consumption, it may potentially increase the total energy consumption. Therefore, we also study overall energy consumption since it is very sensitive to branch prediction accuracy. Note that in our experiments we take into account the energy overhead associated with the PABU-table.

Figure 5(a) reports relative branch predictor energy consumption reduction for PABU-enhanced predictors.On the average, we reduce predictor energy consumption by 64%. This reaches a maximum of 80% for *mes* and a minimum of 51% for *mcf* and *gcc*.In figure 5(b) we report relative total energy consumption reduction. On the average, we reduce total energy consumption by 1.2%. This reaches a maximum of 2.1% for *amm* and a minimum of 0.6% for *prs*.

As reported PABU reduces overall energy consumption across all benchmarks. Moreover, while our overall savings appear to be small in relative terms, they should be weighted against the negligible performance cost. In the best case, *amm*, we save 2.1% of the overall and 75% of the branch prediction energy without any performance cost. For four other benchmarks, *i.e.*, *bzp*, *cmp*, and *mes*, overall energy savings is around 1.5% while performance cost is below 0.1%.

We have also observed that when a similar PABU-enhanced branch predictor is used in machines with lower execution bandwidths (*e.g.*, a 4-way machine), while performance is maintained, the relative overall energy reduction is higher. Our study shows that in a 4-way machine overall energy reduction reaches a maximum of 4.6% which is more than twice of the maximum reached in an 8-way machine. As execution bandwidth becomes narrower the relative power dissipation contribution of the branch predcitor becomes larger. This is the result of smaller number of execution/bypass resources associated with narrower machines. Consequently, execution units consume less energy leaving a larger share for the branch predictor. We do not report detailed results for machines with smaller bandwidths in the interest of space.

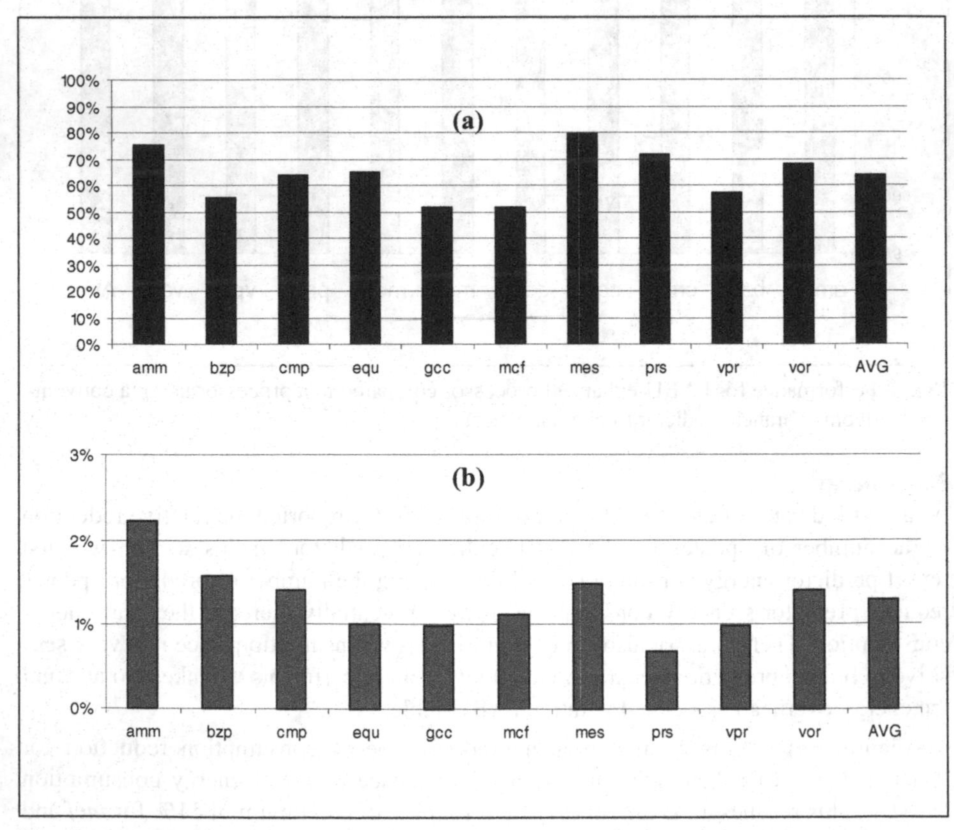

Fig. 5. a) Relative branch prediction energy consumption reduction (b) Relative total energy consumption reduction for a PABU-enhanced processor

4 Related Work

Previous work has introduced *Banking*[1], *Predictor Probe Detection (PPD)*[1] and *Branch Predictor Prediction (BPP)* [6] as methods to reduce branch predictor energy consumption while maintaining performance.

Banking is a natural solution to multi-cycle access times to large on-chip structures such as branch predictors. Our solution could be used on top of banking to achieve higher power savings. PPD aims at reducing the power dissipated during predictor lookups. PPD identifies when a cache line has no conditional branches so that a lookup in the direction predictor can be avoided. Also, it identifies when a cache line has no control-flow instructions at all, so that the BTB lookup can be eliminated. This is done by storing pre-decoded bits in a structure called the *prediction probe detector.* BPP, exploits branch instruction behavior to gate two out of the three sub-predictors. In BPP, a buffer is introduced in the fetch stage. Each BPP entry is tagged by the PC of a recently seen dynamic branch and records the sub-predictors used by the two branches that followed it in the dynamic execution stream.

While both BPP and PPD aim at reducing the number of predictor lookups, PABU aims at reducing the number of branch predictor updates. To the best of our knowledge, PABU is the first study to do so.

5 Conclusion

We presented PABU, a technique for reducing branch predictor energy consumption while maintaining the accuracy advantage of combined branch predictors.We affirmed that it is possible to significantly reduce energy by identifying and eliminating the branch predictor updates that do not contribute to performance. PABU reduces predictor energy consumption up to a maximum of 80% (average of 64%) compared to a conventional combined predictor. This comes with a negligible performance cost. Our study covered a subset of SPEC 2k benchmarks. We have shown that when one considers the overall processor energy consumption, PABU-enhanced processors always dissipate less energy when compared to ones that use the conventional combined predictor. Because of the considerable energy savings and the relatively small cost, PABU is an attractive power-aware enhancement for high-performance processors.

References

1. D. Parikh, K. Skadron, Y. Zhang, M. Barcella, amd M.R. Stan. *Power Issues Related to Branch Prediction. In Proc.* Intl. Symposium on High-Performance Computer Architecture, February 2002.
2. S. Manne, A. Klauser and D. Grunwald. *Pipeline Gating: Speculation Control For Energy Reduction. In Proc.* Intl. Symposium on Computer Architecture, Jun., 1998.
3. D. Grunwald, A. Klusser, S. Manne and A. Plezkun, *Confidence Estimation for Speculation Control,* In Proc. Intl. Symposium on Computer Architecture, Jun., 1998.
4. A. Baniasadi, A. Moshovos, *Instruction Flow-Based Front-end Throttling for Power-Aware High-Performance Processors. In Proc.* ISLPED'01, August 2001
5. D. Brooks, V. Tiwari M. Martonosi "Wattch: A Framework for Architectural-Level Power Analysis and Optimizations", *Proc of the 27th Int'l Symp. on Computer Architecture*, 2000
6. A. Baniasadi, A. Moshovos, *Branch Predictor Predcition a Power-Aware Branch Predcitor for High-Performance Processors. In Proc.* ICCD'02, September 2002

Power Optimization Methodology for Multimedia Applications Implementation on Reconfigurable Platforms[1]

K. Tatas [1] , K. Siozios [1] , D. Soudris [1] , K. Masselos [2] , K. Potamianos [2] , S. Blionas [2] , and A. Thanailakis [1]

[1] VLSI Design and Testing Center, Department of Electrical and Computer Engineering, Democritus University of Thrace, 67100, Xanthi, Greece
{ktatas, ksiop, dsoudris, thanail}@ee.duth.gr

[2] *Intracom SA, Hellenic Telecommunications Industry, Peania, Greece*
{kmas, cpot, sbli}@intracom.gr

Abstract. A methodology for the power-efficient implementation of multimedia kernels based on reconfigurable hardware (FPGA) is introduced. The methodology combines various types of algorithmic transformations and high-level memory hierarchy exploration with register-transfer level design and implementation. An FPGA with an external memory was used for obtaining experimental results which prove the viability of the methodology. Comparisons among implementations with and without this optimization, prove that great power efficiency is achieved.

1 Introduction

Portable devices have become increasingly popular lately. With the tendency being towards constantly increasing functionality, the need for reduced power consumption and high-performance increases exponentially.

Programmable and/or custom hardware processing elements are most commonly used for implementation. Programmable processors have the advantage of flexibility at the expense of performance, while custom processors provide high performance at the expense of flexibility. Recently, reconfigurable hardware has been proposed as a solution [1], since it is located between programmable and custom processors in both performance and flexibility.

It has been proven that the main power components in systems based on programmable processors are the data and instruction memories [2]. The data memory com

[1] This work was partially supported by the project IST-34793-AMDREL which is funded by the E.C.

J.J. Chico and E. Macii (Eds.): PATMOS 2003, LNCS 2799, pp. 430–439, 2003.
© Springer-Verlag Berlin Heidelberg 2003

ponent was reduced by the introduction of an on-chip memory hierarchy, while the instruction memory power component was reduced by inserting a cache memory [3][4]. Similarly, in custom processors, power consumption in data memories is the largest power component [2].

In this paper we intend to shift the paradigm from programmable processors or custom hardware to reconfigurable hardware, while employing a similar strategy to reduce the data memory power consumption. A new design methodology for implementing multimedia applications with reduced power consumption is introduced. This methodology can be considered as an extension of existing approaches [2], [3], [4], targeting custom and programmable processors. Data memory hierarchy power exploration at the high levels of design and register transfer (RT) level implementation are combined in order to achieve the optimum implementation in terms of memory power consumption. An FPGA has been used for the implementation of two common motion estimation algorithms with and without memory power optimization.

The paper is organized as follows: Section 2 describes the target architecture, section 3 the target application and the proposed methodology. Experimental results are provided and analyzed in section 4, and finally the conclusions are presented in section 5.

2 Target Architecture

The architecture we have considered is illustrated in Fig. 1. It consists of an FPGA which implements the required logic and an external (off-chip) memory, which stores the data of the application.

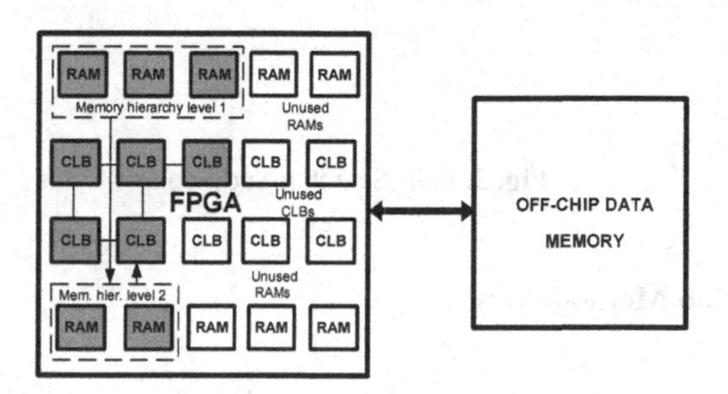

Fig. 1. Target architecture

3 Target Application and Optimization Methodology

We have chosen two popular multimedia kernels for implementation, namely the full-search and hierarchical search motion estimation algorithms [5]. For the sake of completeness, a brief description follows:

3.1 Motion Estimation Algorithms

We consider the following Motion Estimation Algorithms: i) Full Search (FS) [5] and, ii) Hierarchical Search (HS) [6]. Both algorithms are block based, which means that in order to compute the motion vectors between two successive frames, the previous frame and the current frame are divided into a number of blocks, each of which have a fixed number of pixels. The algorithm compares every block from the current frame with the block from previous frames that are placed near the original point. Considering a cost function, the best match between a previous and a current block frame is the one exhibiting the minimum value in the cost function. This results to the motion vectors MV(i,x), MV(i,y), of the i-th block and x- and y-axis, respectively. Full-search searches all locations in the search space and is very computationally intensive, while hierarchical search forms two subsampled images by 2 and 4 respectively for each frame and searches in the subsampled frames first. The two algorithms are depicted in pseudo code in Figs. 2 and 3.

```
for(y=0;y<N/B;y++)
  for(x=0;x<M/B;x++)
  {
    for(i=-p;i<p+1;i++)
      for(j=-p;j<p+1;j++)
      {
        for(m=0;m<B;m++)
        for(n=0;n<B;n++)
        {
           read_from_current_frame;
          Check_Bound_Condition;
          read_from_previous_frame;
         Distance_Criterion();
        }
     }
  }
```

Fig. 2. Full- Search pseudo-code

3.2 Optimization Methodology

In [2] a data memory optimization methodology based on an on-chip memory hierarchy was introduced for custom and programmable processors. The memory hierarchy is reflected in the high-level specification language (C) as code transformations, known as data-reuse transformations. The optimization methodology, is based on the fact that a large, off-chip memory consumes greater power per memory transfer, than a small, on-chip one, therefore the methodology attempts to move the greatest number of transfers from the off-chip memory to on-chip ones, by locating data that are used often. This methodology is modified here for reconfigurable hardware (FPGAs).

The methodology is illustrated in Fig. 4. A high-level description of the application is the input to the first stage, known as the global loop transformation stage. Global loop transformations such as loop merging and tiling are used to increase the regularity of the loop structure of the application. It is most often necessary in multi-kernel algorithms such as hierarchical search in our case, but not essential in the case of full-search which has a very regular nested loop structure as seen in the previous section. The output of this stage is a transformed code with greater regularity.

```
create_current_frame_subsampled_by_two();
create_current_frame_subsampled_by_four();
create_previous_frame_subsampled_by_two();
create_previous_frame_subsampled_by_four();
/* ------------- Level 2 ------------- */
for(i=-p/4;.....) /* ME at fr. subsamp. by 4 */
  for(j=-p/4;.....)
  {
    for(m=0;.....)
     for(n=0;.....)
     {
      read_from_current_frame_4();
      Check_Bound_Condition;
      read_from_previous_frame_4();
      distance_criterion_check();
     }
  }
/* ------------- Level 1 ------------- */
for(i=-1;.....) /* ME at fr. subsamp. by 2 */
  for(j=-1;.....)
  {
   for(m=0;......)
    for(n=0;......)
    {
     read_from_current_frame_2();
     Check_Bound_Condition;
     read_from_previous_frame_2();
     distance_criterion_check();
    }
  }
/* ------------- Level 0 ------------- */
for(i=-1;.....) /* ME at original frame */
  for(j=-1;.....)
  {
   for(m=0;.....)
    for(n=0;.....)
    {
     read_from_current_frame_0();
     Check_Bound_Condition;
     read_from_previous_frame_0();
     distance_criterion_check();
    }
  }
```

Fig. 3. Hierarchical Search motion estimation pseudo-code

The output of the previous stages then undergoes data-reuse transformations that imply a specific on-chip memory hierarchy. The basic concept behind data-reuse transformations is that to place data that are used often in smaller memories, use them as many times as necessary, and then load the memories with new data, effectively increasing the total number of data transfers but moving them to smaller memories.

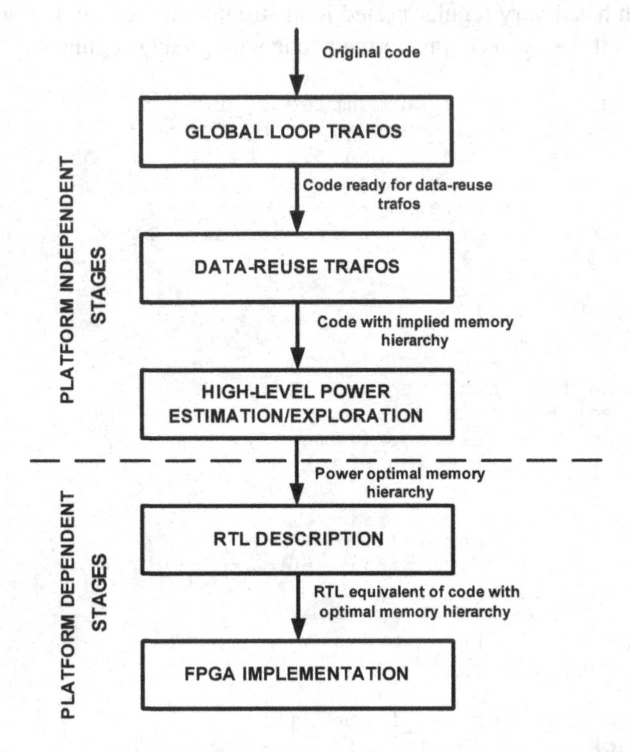

Fig. 4. Power optimization methodology

Then, high level data memory power exploration is performed, by using high-level power estimation models. Comparisons among the various transformations, yield the most efficient memory hierarchy in terms of power. Since, the number of memory transfers required by the application is independent of the specific implementation, the first three steps are platform independent.

The next step is where the platform dependent part of the methodology begins. Since the optimal memory hierarchy has been selected in the previous step, the transformed code must be translated to RT level (VHDL). Obviously the memory hierarchy must be maintained, since this translation is done manually.

The final step is the FPGA implementations using any available FPGA mapping tools, using the RTL description of the previous set as input. The on-chip memory hierarchy implied by the selected transformation, is implemented inside the FPGA.

The main advantage of the methodology is that it uses the high-level, performance independent steps for power exploration, instead of implementing in RTL all the possible memory hierarchies, before evaluating them and selecting the optimal one, which would be impractically time-consuming.

In our, application, among global loop transformations, loop merging was performed in order to merge the subsampling stages with the main hierarchical search motion estimation kernel. No global loop transformations were applied to full-search.

It has been proven [3], [4], [5] that a total of 21 possible data memory hierarchies for our applications exist. These candidate hierarchies are implied from the 21 possible data-reuse transformations, which are applied on the high-level description of the kernels. The power consumption of each memory hierarchy was estimated using Landman's memory model [9]. We have selected the one that exhibited the lowest power consumption. This specific transformation implies two levels of on-chip memory hierarchy: A line of candidate blocks which is loaded from the external memory and a single candidate block is loaded from the line of candidate blocks, used as many times as necessary, and then a new candidate block is loaded until all candidate blocks are exhausted. Then, a new line of candidate blocks is loaded from the external memory. The size of the required on-chip memory in bits is

$$Mem_{on-chip} = [pixel-bits] \times [block\_width \times (block\_width + 2 \times search\_space) + block\_width \times block\_width \quad . \qquad (1)$$

where *pixel-bits* is the number of bits per pixel, i.e. 8 for greyscale frames and 24 for color frames. The *block_width* and *search_space* parameters depend on the frame size. This internal memory hierarchy was implemented in the FPGA, using its available resources. Most recent FPGA devices include blocks of RAM memory, that can be connected in parallel to form larger memories and appropriately to implement the desired memory hierarchy.

4 Experimental Results

For illustration purposes, greyscale frames of 144×176 pixels were used, therefore an external memory of 2×25344×8 bit is necessary in order to contain two such greyscale type of frames. The selected low-power transformation, was described in VHDL and as an FPGA evaluation platform, Xilinx devices were selected. For the on-chip memory hierarchies plus the motion vector storing and the subsampled frames storing in the hierarchical search motion estimation kernel, an appropriate number of Xilinx BlockSselectRAMs were used in their 512×8 configuration [9]. These local memories are connected in parallel with appropriate decoding logic for the addresses to form larger memories. Since the required memory size is not an exact multiple of a BlockSelectRAM size, some waste of resources is unavoidable. The other option would be to implement the memory in registers, but that would result in greater slice occupation, and therefore larger devices would be required for implementation. In fact the optimized version of the hierarchical search algorithm would not fit even in the largest Virtex-II device.

Power consumption for the FPGA was estimated using the Xpower tool [7]. The power consumption of the external memory was estimated using Landman's model [8], which uses the following equation:

$$P = \frac{1}{2} V_{dd}^2 \left(C_{read} \cdot f_{readaccesses} + C_{write} \cdot f_{writeaccesses} \right). \tag{2}$$

where C_{read} and C_{write} denote the capacitances for read and write operation and $f_{readaccesses}$, $f_{writeaccesses}$ are the frequencies of read and write operations, respectively. The values of C_{read} and C_{write} are specified by the model itself, while the access frequencies depend on the chosen application, and they are calculated by the designer. We assumed 0.18 micron technology and 3V supply voltage.

As mentioned above two motion estimation algorithms were implemented with and without data memory hierarchy optimization in a variety of devices, leading to a number of different implementations. Table 1 provides the total power consumption of all implementations in detail, at a frequency of 25 MHz and at the maximum frequency achieved by the optimized version of the algorithm on the same device. The original version of the algorithm in most cases achieves higher clock frequencies and these are shown in the performance comparison later in this section. Both the FPGA and the external memory power consumption are presented. The last column presents the percentage gains of a specific optimized FPGA implementation in comparison to the corresponding non-optimized implementation on the same device. The FPGA power consumption is the sum of its quiescent power and the power dissipated by the logic implemented on it. The quiescent power factor is a constant power consumption that is dissipated by a given device regardless of the amount of logic mapped to it or the operating clock frequency and therefore equals the minimum power dissipation of that device. For example in the case of the original Full-search algorithm and the xc2s15 device implementation, at the frequency of 25 MHz the power dissipation is 180 mW, of which 75 mW is the quiescent power of the device, and the remaining 115 mW are dissipated due to the logic mapped on the device. At the frequency of 54 MHz, the quiescent power remains the same, while the power due to the logic approximately doubles since it is a linear function of frequency, and therefore the total power consumption reaches the value of 239.22 mW. The optimized full-search implementation exhibits 40-60% power gains depending on the device at the frequency of 25 MHz. At the maximum frequency of the optimized implementation the gains are somewhat greater, because the memory power consumption increases more rapidly with the clock frequency than the FPGA power consumption does.

The optimized hierarchical search implementation exhibits smaller power consumption gains (about 25%) than the full-search implementation. There seems to be a direct relation between the computational complexity of the algorithm and the feasible power gains, which has been confirmed by experiments with other similar algorithms at a higher level of abstraction.

Table 2 shows the device utilization for all implementations in number of slices and number of BlockSelectRAMs used. The number of required resources, the number of total available resources and the percentage of utilization is given. For different implementations of the same algorithm, the required resources are the same for all devices, therefore as the device gets larger, the number of total available resources increases, and the utilization percentage drops. In the original version of the applications, only two BlockSelectRAMs are used, for storing the motion vectors. In the case of the transformed versions, additional RAM blocks are required to implement the on-chip memory hierarchy of the respective transformation.

Table 1. Total power consumption comparison of alternative implementations

Application	Device	Clock freq. (MHz)	FPGA power (mW)	External memory power (mW)	Total power (mW)	Power gain (%)
FS (original)	xc2s15	25	160.88	703.76	864.64	
		54.44	239.22	1532.55	1771.77	
	xc2s200	25	275.81	703.76	979.57	
		57.47	359.18	1617.80	1976.98	
	xcv50	25	217.25	703.76	921.01	
		54.8	296.46	1542.64	1839.10	
	xcv1000	25	343.62	703.76	1047.38	
		51.16	421.41	1440.17	1861.58	
	xcv400e	25	586.49	703.76	1290.25	
		66.29	635.11	1866.09	2501.20	
	xc2v10000	25	593.02	703.76	1296.78	
		81	653.97	2280.18	2934.15	
FS (optimized)	xc2s15	25	120.38	211.93	332.31	61.56
		Max (54.44)	160.23	461.51	621.74	64.90
	xc2s200	25	239.70	211.93	451.63	53.89
		Max(57.47)	288.84	487.21	776.05	60.74
	xcv50	25	172.22	211.93	384.13	58.29
		Max(54.8)	215.15	464.80	679.95	63.02
	xcv1000	25	295.2	211.93	507.13	51.58
		Max(51.16)	342.49	433.78	776.27	58.30
	xcv400e	25	555.66	211.93	767.59	40.50
		Max(66.29)	581.52	561.98	1143.50	54.28
	xc2v10000	25	590.98	211.93	802.91	38.08
		Max(81)	649.37	686.55	1335.92	54.46
HS (original)	xcv400e	25	637.61	1093.88	1731.49	
		Max(44.08)	712.95	2238.91	2951.86	
	xc2v10000	25	592.85	1093.88	1686.73	
		Max(78.58)	657.79	3438.51	4096.30	
HS (opt.)	xc2v10000	Max(25.24)	1170.53	110.39	1280.92	24.05

In the original version of the applications, only two BlockSelectRAMs are used, for storing the motion vectors. In the transformed versions, additional RAM blocks are required to implement the on-chip memory hierarchy of the respective transformation. This explains the increase in the number of BlockSelectRAMs between original and optimized algorithms. In the case of Hierarchical Search, additional memory is required to store the subsampled frames. Also, large amounts of memory were required in order to implement the global loop transformations in the case of the Hierarchical Search algorithm, a fact that explains the great difference in required memory size between original and optimized version of this application. Also, the logic becomes more complex in the case of the optimized versions, therefore more slices are needed and the application often does not fit in a small device. That is the reason the number of implementations is smaller in the case of the optimized in comparison to the original applications. The smaller devices are the least costly, and therefore the most cost-efficient implementations would be the ones with the greatest device utilization percentage for each algorithm.

Table 2. Area comparison of alternative implementations

Application	Device	Slice count (% utilization)	BlockRAM count (% utilization)
FS (original)	xc2s15	164/192 (85%)	2/4 (50%)
	xc2s200	164/2352 (6%)	2/14 (14%)
	xcv50	164/768 (21%)	2/8 (25%)
	xcv1000	164/12288(1%)	2/32 (6%)
	xcv400e	164/4800 (3%)	2/40 (5%)
	xc2v10000	164/61440 (0.002%)	2/192 (0.01%)
FS(optimized)	xc2s15	166/192 (86%)	4/4 (100%)
	xc2s200	166/2,352 (7%)	4/14 (21%)
	xcv50	166/768(22%)	4/8 (50%)
	xcv1000	166/12288 (1%)	4/32 (12%)
	xcv400e	166/4800 (3%)	4/40 (10%)
	xc2v10000	166/61440	4/192 (0.02%)
HS (original)	xcv400e	761/768 (99%)	36/40 (90%)
	xc2v10000	761/61440 (1%)	36/192 (18%)
HS (optimized)	xc2v10000	16817/61440 (27%)	161/192 (84%)

Performance measurements can be seen in Table 3. There is a clear relation between the computational complexity of the algorithm and the performance achieved. Full-search is a very computationally demanding algorithm, unlike Hierarchical search, which is less computationally intensive. Also, the data-reuse transformations result to an additional performance penalty, since the total number of memory transfers is increased. Furthermore, the control logic becomes more complex, leading to a reduction in the maximum clock frequency. Since the power optimized version of the hierarchical search algorithm requires almost triple execution time than the non-optimized one, an alternative would be to reduce the clock frequency of the non-optimized implementation, achieving the same performance, while also reducing power consumption. But due to the great static power consumption of the FPGA device (562.5 mW) which is not affected by clock frequency reduction, it was estimated that it would consume approximately 100 mW more than the optimized version.

Implementing the full-search algorithm in small devices leads to low cost and low power consumption, while implementation in large devices leads to superior performance. The hierarchical search algorithm requires a great amount of control logic and far greater on-chip memory in order to store the subsampled images and therefore can only be mapped to large devices.

5 Conclusions

A methodology that combines exhaustive high-level data memory power exploration and RTL design and implementation for FPGA-based multimedia systems was presented. Simulation results indicated significant power gains that support the validity of the proposed methodology.

Table 3. Performance comparison of alternative implementations

Application	Device	# cycles	Clock freq. (MHz)
FS (original)		5702400	25
	xc2s50		Max (87.030)
	xc2s200		Max(80.97)
	xcv50		Max(69.152)
	xcv1000		Max(91)
	xcv400e		Max(95)
	xc2v10000		Max(103)
FS (optimized)		9554688	25
	xc2s50		Max (54.44)
	xc2s200		Max(57.47)
	xcv50		Max(54.8)
	xcv1000		Max(51.16)
	xcv400e		Max(66.29)
	xc2v10000		Max(81)
HS (original)		320544	25
	xcv400e		Max(44.08)
	xc2v10000		Max(78.58)
HS (optimized)	xc2v10000	950994	Max(25.24)

References

1. Katherine Compton and Scott Hauck: Reconfigurable Computing: A Survey of Systems and Software. ACM Computing Surveys, Vol. 34, No. 2 (2002) 171–210.
2. F. Catthoor et al.: Data Acess and Storage Management or Embedded Programmable Processors. Kluwer Academic Publishers, Boston (2002)
3. D. Soudris, N. D. Zervas, A. Argyriou, M. Dasygenis, K. Tatas, C. Goutis, A. Thanailakis: Data-Reuse and Parallel Embedded Architectures for Low-Power, Real-Time Multimedia Applications. IEEE International Workshop on Power and Timing Modeling, Optimization and Simulation (PATMOS), Göttingen, Germany (2000) 243–254
4. N. D. Zervas, K. Masselos, C.E. Goutis: Data-reuse exploration for low-power realization of multimedia applications on embedded cores. Proc. Of 9th Int. Workshop on Power and Timing Modeling, Optimization and Simulation (PATMOS'99) (1999) 71–80
5. V. Bhaskaran and K. Kostantinides: Image and Video Compression Standards. Kluwer Academic Publishers (1998)
6. K. M. Nam, J.-S. Kim, Rae-Hong Park, and Y. S. Shim: A fast hierarchical motion vector estimation algorithm using mean pyramid. IEEE Transactions on Circuits and Systems for Video Technology, vol. 5, no. 4 (1995) 344–351
7. http://support.xilinx.com/support/sw_manuals/xilinx4/manuals.pdf
8. P. Landman: Low-power architectural design methodologies. Doctoral Dissertation, U.C. Berkeley (1994)
9. http://direct.xilinx.com/bvdocs/publications/ds003.pdf

High-Level Algorithmic Complexity Analysis for the Implementation of a Motion-JPEG2000 Encoder

Massimo Ravasi[1], Marco Mattavelli[1], Paul Schumacher[2], and Robert Turney[2]

[1] Swiss Federal Institute of Technology
CH-1015 Lausanne, Switzerland
{Massimo.Ravasi, Marco.Mattavelli}@epfl.ch

[2] Xilinx Research Labs
Longmont, CO, USA
{Paul.Schumacher, Robert.Turney}@xilinx.com

Abstract. The increasing complexity of processing algorithms has lead to the need of more and more intensive specification and validation by means of software implementations. As the complexity grows, the intuitive understanding of the specific processing needs becomes harder. Hence, the architectural implementation choices or the choices between different possible software/hardware partitioning become extremely difficult tasks. Automatic tools for complexity analysis at high abstraction level are nowadays a fundamental need. This paper describes a new automatic tool for high-level algorithmic complexity analysis, the Software Instrumentation Tool (SIT), and presents the results concerning the complexity analysis and design space exploration for the implementation of a JPEG2000 encoder using a hardware/software co-design methodology on a Xilinx Virtex-II™ platform FPGA. The analysis and design process for the implementation of a video surveillance application example is described.

1 Introduction

The evolution of digital silicon technology enables the implementation of signal processing algorithms that have reached extremely high levels of complexity. This fact, among others, has two relevant consequences for the system designer. The first is that processing algorithms cannot be specified in ways other than developing a reference software description. The second important consequence is that the understanding of the algorithms and the evaluation of their complexity have to be derived from such software description. As consequence of the greatly increased complexity, the generic intuitive understanding of the underlying processing becomes a less and less reliable design approach. Considering that, in many cases, the complexity of the processing is also heavily input-data dependent, the system designer faces a very difficult task when beginning the design of a system architecture aiming at efficiently implementing the processing at hand.

This difficulty is evident when considering for instance the case of hardware/software co-design for System-on-Chip integration. A typical design flow for

J.J. Chico and E. Macii (Eds.): PATMOS 2003, LNCS 2799, pp. 440-450, 2003.
© Springer-Verlag Berlin Heidelberg 2003

this implementation case is shown in Fig. 1. All the relevant information must be extracted from the software description; indeed, the analysis of the complexity of single functions does not give any information without the knowledge of the interconnection, occurrence and actual use of all functions composing the algorithm. Some other traditional styles of design, such as complexity analysis based on "*pencil and paper*" or *worst-case* applied to some portions of the algorithm, not only become more and more impractical for the required effort, but can also results in very inaccurate results for not taking into account the correct dependency of the complexity on the input data.

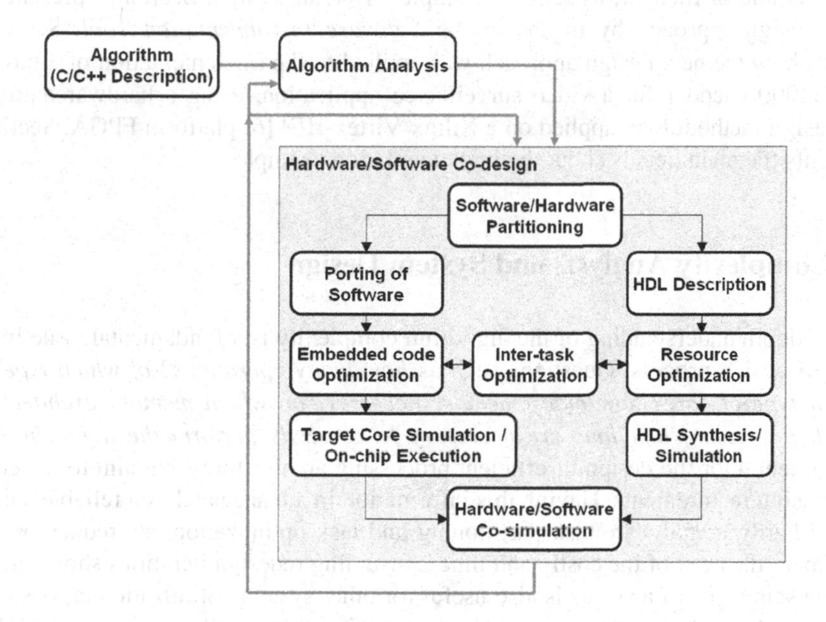

Fig. 1. Typical simplified design flow of a hardware/software embedded system

It can be noticed that for hardware/software co-design [1], [2], a large variety of tools is available at all levels. Conversely no suitable automatic tools are available to assist the fundamental task partitioning stage or to gather detailed and reliable information on the computational complexity of the algorithm for optimizing the implementation, starting from the generic software description.

All these considerations, although relevant for most of signal processing implementation problems, become fundamental for video, still image, audio and multimedia coding, where the latest generation of compression standards (i.e. MPEG-4 [3] and JPEG-2000 [4]) reaches a very high level of complexity that is also extremely sensitive to the encoder optimization choices and strongly data-dependent.

This paper presents a new approach to complex system design by means of a tool-assisted high-level algorithmic complexity analysis based only on the pure software description of an algorithm. This analysis is carried out by means of the *Software*

Instrumentation Tool (SIT) [5], an automatic tool allowing the extraction of relevant information about the complexity of the algorithm under study. The dependencies on the underlying architecture and the compilation process used to verify and analyze the algorithm are not taken into account, because only the software description of the algorithm is relevant for the analysis and not how it is compiled and run on an architecture chosen for verification purposes only. Furthermore, the analysis is performed in real working conditions on real input data, to take into account the input-data dependency of the performance and the complexity of signal processing algorithms.

Section 2 gives a brief overview of current state of the art methods in complexity analysis and of their drawbacks for complex systems design. Section 3 presents the new design approach by means of the *Software Instrumentation Tool*. Section 4 shows how the new design approach was applied to the implementation of a motion-JPEG2000 encoder for a video surveillance application, using a hardware/software co-design methodology applied on a Xilinx Virtex-II™ [6] platform FPGA. Section 5 presents the main details about the implementation example.

2 Complexity Analysis and System Design

An in-depth understanding of the algorithm complexity is a fundamental issue in any system design process. Questions such as *how many operations*? *of which type*? *on which type of data*? *how many memory accesses*? *on which memory architecture*? *which processing functions are necessary to correctly perform the algorithm*? are fundamental for the design of efficient processing architectures that aim to match the processing requirements. Having this information in advance and as a reliable support to the hardware/software task partitioning and task optimization can reduce or even eliminate the need of the costly and time-consuming redesign iterations shown in Fig. 1. The same type of analysis is also useful for other system optimization tasks such as data-transfer and power consumption minimization that require several methodological steps starting from a generic algorithm specification [7].

2.1 State-of-the-Art Approaches to Complexity Analysis

Depending on the specific goals of the desired complexity analysis to be performed, very different approaches and tools can be chosen [8]. All these approaches are perfectly suited for their specific applications, but they present serious drawbacks when applied to the design of complex systems. The most common approaches can be classified into the following categories:

- *Profilers*, modifying the program to make it produce run-time data [8]. Profilers can basically provide two types of results: number of calls of a given section of a program and execution time of that section. The results provided by profilers strictly depend on the architecture on which the code is executed and on the compilation optimizations. These results cannot therefore be easily used for complexity analysis concerning the implementation of the same algorithm onto another architecture and they could even yield misleading complexity evaluations.

– *Static methods*. For these methods state of the art solutions rely on annotation at high-level programming language so as to determine lower and upper bounds of resource consumption [9]. The main drawback of these techniques is that the real processing complexity of many multimedia algorithms heavily depends on the input data, while static analysis depends only on the algorithm. These approaches are better suited for real-time control applications for which strict *worst-case* analysis is required. Moreover restricted programming styles such as absence of dynamic data structures, recursion and bounded loops are required.
– *Hardware Description Languages* and *Hardware/Software Co-Design tools*, allowing describing (at different abstraction levels), synthesizing and simulating hardware or heterogeneous hardware/software systems [1], [2]. Through synthesis and simulation, these approaches allow gathering very reliable results about the implementation complexity and performance of the described algorithm. However, the analysis can only be performed at the end of the design cycle, *after* all architectural choices have already been taken. If it is realized that the a priori architectural choices are not appropriate for the desired performance constraints, a costly redesign of the system is necessary.

3 The *Software Instrumentation Tool* (*SIT*)

It is assumed that a software implementation of an algorithm is available and that it can be run in realistic input data conditions. The goal is to measure the complexity of the algorithm, whose performance can be data-dominated. In other words, the interest is not only about the measure of complexity of the algorithm itself, but also about its dependencies on specific input-data. Moreover, the software implementation is a high-level description of the algorithm whose complexity has to be measured independently of the underlying architecture on which the software is run for verification purposes. This approach is fully in line with methodological approaches, aiming at optimizing data-transfer and memory bandwidths at a high-level description of the algorithm [7].

The implementation of the Software Instrumentation Tool (SIT) [5] is based on the concept of the instrumentation of all the operations that take place during the execution of the software program. Instrumenting code by overloading C++ operators has been already proposed in literature, but it has always been considered an approach presenting severe practical and functional limitations [8]. Major drawbacks were considered the applicability only to C++ program, the impossibility to instrument pointers and other data types such as structures and unions, resulting into not accurate analysis of data-transfer oriented operations and to an extensive manual rewriting of the original code. All known functional and practical limitations of the operator overloading approach have been overcome with SIT. The current version of SIT is able to instrument a C program by translating it into a corresponding C++ program by means of an automatic tool: both programs have the same behavior but, by substituting C simple types with C++ classes and by substituting all C operators with C++ overloaded operators, standard C operations can be intercepted during the execution and counted. The great advantage of this approach is that no manual code rewriting is necessary. Moreover, SIT allows associating an appropriate and customizable memory model to the algorithm, in order to complete the complexity analysis with data-transfer analysis.

The results gathered with SIT are presented on a per-context basis, which can be chosen to correspond to the function call tree or to be extended to the single compound statements for a more detailed analysis. The results of the computational complexity analysis are in terms of executed operations within a context node and are collected on the two axis *operations* and *data-types*. The *operations* axis is an extension of the C operator set (+, +=, etc.) as well as the *data-types* axis is an extension of the C data type set (`int`, `float`, `struct`, etc.). The results of the data-transfer analysis depend on the simulated memory model, which may include the simulation of cache hierarchies.

In order to validate the SIT methodology, a real-world design example was used. The chosen design was a JPEG2000 encoding system for video surveillance applications.

4 Hardware/Software Co-design of a Motion-JPEG2000 Encoder

JPEG2000 standard [4] includes a specification for the encoding and storage of motion sequences [10]. Whereas well-known video standards such as MPEG-2 and MPEG-4 [3] use inter-frame dependencies and motion compensation, motion-JPEG2000 involves encoding each frame independently.

Whereas the standard specifies the bitstream to ensure interoperability between encoder and decoder systems, it leaves the actual implementation open. This section presents the implementation of a JPEG2000 encoder system capable of handling video data rates, created using a hardware/software co-design methodology on a platform field programmable gate array (FPGA). A cohesive and programmable hardware/software co-design is created.

The targeted video surveillance application involves a low-grade video coding system coding. The frame size is 640 × 480 × 24 bits and the rate is 15 frames/sec. High compression of the video data is expected as quality is not of high importance.

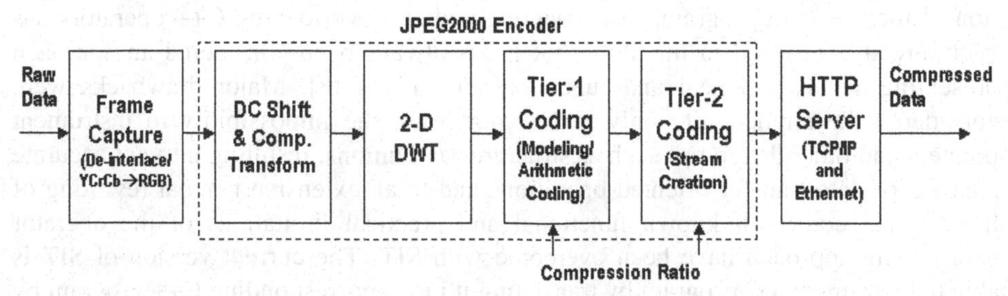

Fig. 2. Block diagram of motion-JPEG2000 system

Fig. 2 shows the block diagram of the designed JPEG2000 encoder system. After the frames are captured and the data is DC shifted, the user can decide to perform a component transformation on the three components for each pixel as specified by the standard [4]. A 2-D DWT is then performed on each tile within each frame. It was decided to have each frame be a single tile in order to eliminate the tiling effects. That is, the DWT is performed on the entire frame, and the number of decomposition levels performed is decided by the user (for the designed system, up to five levels is supported). The 5/3 DWT kernel was chosen for the implementation. The sub-bands of the DWT results are then divided into code-blocks and the *Tier-1* coder operates on each code-block independently. The code-block size for this system was chosen to be 64x64. *Tier-2* coding then involves adding the appropriate headers, compiling the compressed data into packets, and delivering the data as a complete codestream.

4.1 Complexity Analysis

For initial system definition, the JPEG2000 encoder system was defined in ANSI C. This gives an excellent starting point for the eventual co-design, provides a reference and test bed for verification, and allows numerous modes and parameters available from the standard to be tested. For an initial complexity assessment, this software was analyzed with SIT. This software JPEG2000 encoder implements only lossless coding, while the targeted video surveillance application is based on lossy coding, aiming to an average compression ratio of 20:1. For these reasons, the results for the *Tier-1* and *Tier-2* blocks were respectively scaled by 1/3 and 1/6.5 [11], while the performance of *DC Shift*, *Component Transformation* and *2-D DWT* is unaffected by the coding type.

The results of the complexity analysis with SIT, concerning the encoding of one frame, are summarized in Table 1, which shows how the computational complexity and data-transfers are distributed over the main processing blocks of Fig. 2. With *SITview*, the graphical visualization tool of SIT, the results for the computational complexity (operation counts) were mapped onto the instruction set of the targeted Xilinx MicroBlaze™ 32-bit RISC soft processor by means of a set of weights representing the latencies of the operations [12]. Therefore the results in the "*Operations*" columns of Table 1 are an estimate of the clock cycles required by the targeted core to perform the processing of each block, without taking into account the data-transfers.

Table 1. Computational complexity and data-transfers of the main blocks of the encoder

	Operations		Data-Transfers			
	Tot	Tot [%]	R	R [%]	W	W [%]
DC Shift & Comp. Transf.	1.72e7	8 %	3.69e6	7 %	3.69e6	23 %
2-D DWT	8.33e7	37 %	1.95e7	37 %	9.77e6	62 %
Tier-1	1.22e8	55 %	2.96e7	56 %	2.12e6	13 %
Tier-2	2.64e5	< 1 %	5.49e5	1 %	1.78e5	1 %
All Blocks	*2.23e8*		*5.33e7*		*1.58e7*	

As with the results of the computational complexity analysis, the results of the data-transfer analysis were mapped onto the MicroBlaze instruction set, taking into account the latencies of the *load* and *store* instructions of the core [12]. By summing the results of this mapping with the results of the computational complexity analysis and multiplying the obtained total by the desired frame rate of 15 frames per second, an estimate of the required processing power was obtained in term of clock frequencies, as shown in Table 2. This estimate represents the clock frequencies at which the MicroBlaze core should run in order to perform the processing of each block of Fig. 2.

Table 2. Estimation of the minimum clock rate of the MicroBlaze core required to perform the processing of each block at the desired frame rate of 15 frames per second

DC Shift & Comp. Transf.	*2-D DWT*	*Tier-1*	*Tier-2*
479 MHz	2126 MHz	2786 MHz	26 MHz

Considering that the maximum clock frequency for the MicroBlaze core is approximately 125 MHz for a Xilinx Virtex-II™ FPGA, the results in Table 2 clearly show that only the *Tier-2* coder can be implemented in software, while for all the other blocks a hardware accelerator is necessary.

4.2 Hardware Software Co-design

The target platform for implementing the system is the Xilinx MicroBlaze Multimedia Demonstration Board designed around a Xilinx Virtex-II XC2V2000 FPGA, which is the heart of the user-defined video processing engine. The FPGA is supported with five independent banks of 512K × 36-bit ZBT RAM with byte write capability. These memories may be used as microprocessor code/data storage or as video frame buffers. The microprocessor supported is a soft 32-bit RISC processor (MicroBlaze™) which can utilize IBM Power PC™ peripheral busses and IP. MicroBlaze is a pre-synthesized soft core implemented in the FPGA fabric and can be included in the FPGA design source as a black box. The MicroBlaze design supports full 32-bit operands, 32-bit data paths and 32-bit registers to provide high performance.

According to the results of the complexity analysis, the three blocks targeted for hardware acceleration were the *DC shift & Component Transform*, the *2-D DWT* and the *Tier-1* coder. The MicroBlaze core was targeted for the software implementation of the *Tier-2 coder*, as well as for mastering the whole processing by handling the interrupt requests from the hardware blocks and scheduling the different tasks. Furthermore, the data-transfer results of SIT Memory Simulation (Table 1) clearly show that *DC Shift & Component Transform*, *2-D DWT* and *Tier-1* are the most I/O intensive tasks, accounting for about all the read and write operations. Mapping only the I/O operations onto the MicroBlaze instruction set, similarly to what previously done for obtaining the global results of Table 2, yields an estimate of the impact of the I/O operations only on the clock frequencies required by the MicroBlaze to sustain the desired frame rate, as shown in Table 3.

Since the maximum clock frequency of the MicroBlaze core is about 125 MHz, the results in Table 3 clearly show that the modules for the *DC Shift & Component Transform*, *2-D DWT* and *Tier-1* tasks, targeted for hardware implementation, must access their respective input and output data independently of the MicroBlaze core. It is therefore necessary to provide hardware-controlled memory access so that each hardware block has direct access to the off-chip memory and the processor is not involved in these transactions. For this reason, it was decided to dedicate to the aforementioned hardware modules three of the five independent banks of ZBT SRAM. A Multi-Memory/Multi-Port ZBT Interface was created in hardware (Fig 3, Fig. 4), in charge of round-robin interfacing, frame after frame, the three hardware accelerators to the three dedicated ZBT SRAM banks; thanks to the round-robin interfacing scheme and to the fact that the ZBT SRAMs can be accessed independently, the three hardware accelerators work independently and concurrently on their respective data, without charging the MicroBlaze core of any extra data-transfer load.

Table 3. Impact of I/O operations on the estimated performance of the MicroBlaze core

	DC Shift & Comp. Transf.	*2-D DWT*	*Tier-1*	*Tier-2*
Read	111 MHz	584 MHz	888 MHz	16 MHz
Write	111 MHz	293 MHz	64 MHz	5 MHz
Total	221 MHz	877 MHz	952 MHz	22 MHz

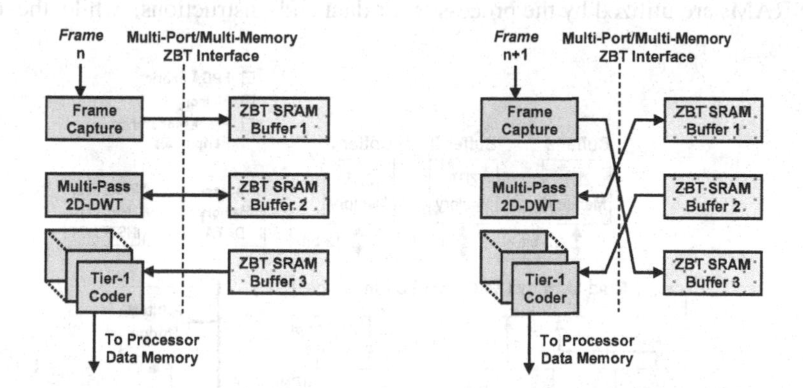

Fig 3. Round-Robin connections between the hardware accelerators and the dedicated ZBT RAMs for two successive frames, as managed by the Multi-Port/Multi-Memory ZBT Interface

The *DC Shift & Component Transform* tasks only rely on a single pixel value for their computations. For this reason it was decided to integrate them into the frame capture streaming data path, resulting in a straightforward implementation and in reducing the overall bandwidth of the corresponding hardware module, since the memory accesses for the temporary intermediate results are eliminated.

The second block that was targeted for hardware acceleration was the *2-D DWT* function. Since the DWT is being performed on the entire frame as a single tile, it becomes unreasonable to store an entire tile on chip. The DWT was implemented using a line-based design so that it can accept a stream of input data, while the output results are written over the already-read values off-chip; line buffers can easily be implemented using the many on-chip 18Kbit BlockRAMs that Virtex-II provides, thus reducing the bandwidth toward the external ZBT SRAM by avoiding storing the temporary data between horizontal and vertical filtering.

The last block that was hardware accelerated was the *Tier-1* coder. This coding involves bit/context modeling and arithmetic coding. A hardware core was designed to accept a single code-block containing up to 4096 words (as specified by the standard), perform the modeling and arithmetic coding on that code-block, and store the compressed byte stream. Three Tier-1 coders were implemented to operate in parallel and guarantee the required processing power for the desired frame rate. As with the other hardware modules, the bandwidth from the ZBT SRAM buffer was optimized by re-scheduling the operations on temporary data in order to reduce of the corresponding I/O accesses.

5 System Implementation

Fig. 4 shows the motion-JPEG2000 encoder system as implemented on the multimedia board using a Xilinx MicroBlaze soft processor. Note that two of the five off-chip ZBT SRAMs are utilized by the processor for data and instructions, while the other

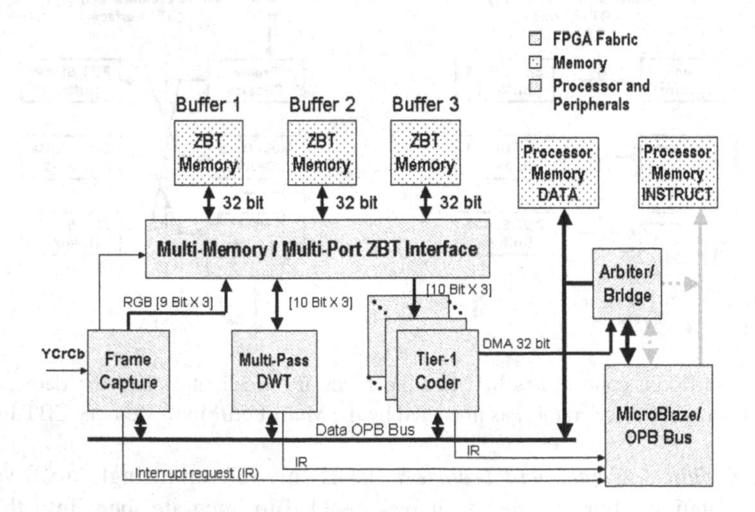

Fig. 4. Block diagram of the motion-JPEG2000 system on multimedia board

three are used for storage of intermediate frame data. First, the frame capture block grabs the YCrCb data from the NTSC camera, performs de-interlacing and conversion to RGB, and also performs the DC shift and component transformation for the JPEG2000 encoder. This data is stored in one of the ZBT buffers. Secondly, the *2-D DWT* block performs the required transformation on all three components for a user-specified number of decomposition levels. Lastly, the DWT coefficients are read and processed by three *Tier-1* coders in parallel, with each coder operating on a 64x64 code-block. The output byte streams of each of these code-blocks is transferred to the processor data memory, where the compiling of the codestream (i.e., *Tier-2* coding) is performed by the MicroBlaze processor. The three tasks: frame capture, DWT, and coding are all given a frame time to complete their job. This gives an overall latency of the system of three frames.

6 Conclusions

This paper presented a new approach to hardware/software co-design based on high-level algorithmic complexity analysis and showed how it was applied to a real design case of a motion-JPEG2000 encoder for a video surveillance application. The target architecture was based on an FPGA with an embedded RISC core, providing an excellent platform for a hardware/software co-design. A software representation of a JPEG2000 encoder was analyzed by means of an automatic tool, the Software Instrumentation Tool, in order to extract relevant information about the algorithmic complexity of the encoder, such as the number of operations and data-transfers. The results of the complexity analysis were applied to the design of the encoder, allowing a fast evaluation of the system implementation requirements as well as of the hardware/software partitioning constraints.

References

1. CoWare Corporation: CoWare N2C Design System. Available on the Internet at www.CoWare.com
2. Synopsys Corporation: Designing Complex Digital Communications for Systems on a Chip. Available on the Interned at www.synopsys.com/products/dsp/digital_br.html
3. Ebrahimi, T. et al.: Dynamic Coding of Visual Information, Technical Description. JTC1/SC2/WG11/M0320, MPEG-4, International Organization for Standardization ISO/IEC (Oct. 1995)
4. Boliek, M., Christopoulos, C., Majani, E. (Editors): JPEG 2000 Part 1 Final Publication Draft. ISO/IEC JTC1/SC29/WG1 N2678 (July 2002)
5. Ravasi, M., Mattavelli, M.: High-Level Algorithmic Complexity Evaluation for System Design. Journal of Systems Architecture, Vol. 48/13–15. Elsevier Science B.V., (May 2003) 403–427
6. Xilinx: Virtex-II Platform FPGA Handbook v1.4. (November 2002)

7. Nachtergaele, L., Moolenaar, D., Vanhoof, B., Catthoor, F., De Man, H.: System-Level Power Optimization of Video Codecs on Embedded Cores: A Systematic Approach. Journal of VLSI Signal Processing, 18 (1008) 89–109
8. Kuhn, P.: Algorithms, Complexity Analysis and VLSI Architectures for MPEG-4 Motion Estimation. Kluwer Academic Publisher (1999) 61–81
9. Li , Y. S., Malik, S.: Performance Analysis of Embedded Software Using Implicit Path Enumeration. IEEE Transactions on Computer-Aided Design of Integrated Circuits and Systems, vol. 16 (December 1997) 1477–1487
10. Fukuhara, T., Singer, D.: Motion JPEG2000 Version 2, MJP2 Derived from ISO Media File Format. ISO/IEC JTC1/SC29/WG1 N2718 (October 2002)
11. Taubman, D.: Software Architectures for JPEG2000. International Conference on Digital Signal Processing, vol. 1 (2002) 197–200
12. Xilinx: MicroBlaze Software Reference Guide, v2.2 (April 2002)

Metric Definition for Circuit Speed Optimization

X. Michel, A. Verle, N. Azémard, P. Maurine, and D. Auvergne

LIRMM, UMR CNRS/Université de Montpellier II, (C5506),
161 rue Ada, 34392 Montpellier, France
{xmichel, verle, azemard, pmaurine, auvergne}@lirmm.fr

Abstract. Designing high performance circuits requires the definition of trade-off between speed, power and area. Based on a design oriented modeling of the delay, this work presents a method for defining metrics for load criticality analysis of different nodes. The purpose of this work is to define indicators for the selection of optimization alternatives. The validation of these indicators is obtained through comparison to the critical loads determined from Spice simulations. The application to various benchmarks shows that, without enumeration, an initial path delay improvement can be obtained at reduced area/power cost by just applying this metric to identify the critical nodes that are the best candidates for speed optimization.

1 Introduction

Combining transistor sizing to buffer insertion supplies one of the most efficient alternative to trade speed for power. Transistor sizing can be performed using circuit simulators and critical path analysis tools to modify iteratively the transistor width in order to satisfy the constraints. In addition to transistor sizing [1], more general speed-up techniques involve buffer insertion [2] and logic transformation [3]. Most synthesis systems use buffer trees to speed-up critical paths. If these techniques may be found efficient for speeding-up combinational paths they may have different impacts in the resulting power dissipation or area. Gate sizing is area (power) expensive and, due to the capacitive loading effects, may slow down adjacent paths. This imposes complex and iterative timing verifications. Buffer insertion preserves path interaction but is only efficient for relatively highly loaded nodes. Intermediate solution can be obtained through inverter insertion but the conservation of the signal logic implies a new transformation at the logical synthesis step. To manage these alternatives it is necessary to evaluate and compare the performances of any decomposition or mapping alternative. That implies a correct interpretation of the timing library format.

Without using any robust indicator, selecting between all these different techniques for the various gates of a library is NP complex. Iterative attempts are processing time explosive. A reasonable selection of speed-up techniques must be based on the

J.J. Chico and E. Macii (Eds.): PATMOS 2003, LNCS 2799, pp. 451–460, 2003.
© Springer-Verlag Berlin Heidelberg 2003

determination of the critical nodes, and the characterization of the gate sensitivity to sizing or buffering alternatives.

Different techniques have been used to control large fanouts in circuits, using separated or combined transistor sizing and buffer insertion. The problem of continuous sizing is often solved using nonlinear programming techniques to get the solution [1]. A discrete or library-specific sizing, in which only a limited number of drives is available for each gate, is a NP-complete [4] combinational optimization problem that is usually solved using heuristic algorithms [5].

Most of these techniques use a simplified Elmore delay model. They are based on successive iterations between the different optimization alternatives applied to all the gates of the specified path. Sutherland [6] minimizes the delay along a single path by imposing equal effort to each stage. It can easily be shown that this solution is exact to minimize the delay for ideal array of gates (without any parasitic or branching capacitance), but does not result in a delay and area-optimal solution for realistic structures. However no indicator has been given to identify the critical nodes or to select the most efficient technique for speeding up the corresponding gate.

In order to help designers in identifying critical gates and selecting speed optimization alternatives, we define in this paper a metric for buffer insertion. This metric is to be used as an efficient indicator for characterizing the logic gates in terms of sensitivity to sizing and buffering techniques. This metric is based on an extension of the logical effort model [6], obtained by considering the explicit sensitivity of the delay, to the logic structure of the gate and to the input transition time.

We first present the delay model, and discuss the buffer insertion conditions. Next in section 3, we explicitly define the limit of load to be considered for fanout optimization. The validation of these limits is given in section 4 and the conclusion in section 5.

2 Delay Model

2.1 Transistor Level Modeling

As a reasonable first order delay model we use, as a basic delay unit, the CMOS generalized Mead [7] step response:

$$\begin{aligned} t_{HLstep} &= \frac{C_L \cdot \Delta V}{I_N} = \tau \cdot \frac{C_L}{C_N} \\ t_{LHstep} &= \frac{C_L \cdot \Delta V}{I_P} = \tau \cdot R \cdot \frac{C_L}{C_P} \end{aligned} \tag{1}$$

where τ is a time unit that characterizes the process and can be calculated or directly determined on specific electrical simulations. C_L, C_N and C_P represent, respectively, the output load and the N and P transistor gate capacitance. Defining τ for the falling

edge, R represents, for identical load and drive capacitance, the ratio of the current value available in N and P transistors.
Following [6] the extension to gates is obtained by reducing each gate to an equivalent inverter. For that we consider the worst case situation. The current capability of the N (P) parallel array of transistors is evaluated as the maximum current of an inverter with identically sized transistors. The array of N (P) series-connected transistors is modeled as an input voltage controlled current generator with a current capability reduced by a factor $DW_{HL,LH}$. These reduction factors (DW) are defined as the ratio of the current available in an inverter to that of an array of transistors. This results in a more general expression of (1) given by

$$t_{HLstep} = \tau \cdot S_{HL} \cdot \frac{C_L}{C_{IN}}$$
$$t_{LHstep} = \tau \cdot S_{LH} \cdot \frac{C_L}{C_{IN}} \qquad (2)$$

where $C_{IN} = C_N(1+k)$ is the gate input capacitance, and k represents the P, N transistor configuration ratio ($k=W_P/W_N$). For simplicity here, the S factors include all the current capability difference between the pull up and pull down equivalent transistors, they are configuration ratio dependent

$$S_{HL} = \frac{(1+k)}{2} \cdot DW_{HL}$$
$$S_{LH} = \frac{(1+k) \cdot R}{2k} \cdot DW_{LH} \qquad (2a)$$

where the DW represents for each edge the ratio of current available in an inverter and a gate of identical size.

$$DW_{HL,LH} = \frac{I_{N,P}(Inv)}{I_{N,P}(Gate)} \cdot \frac{W_{N,P}(Inv)}{W_{N,P}(Gate)} \qquad (2b)$$

These reduction factors represent the explicit form of the logical effort given in [6]. Some typical values of these coefficients are given in Table 1. They have been determined on a 0.25µm process for gates designed with a configuration ratio k=1. Detailed evaluation of these coefficients has been given in [8].

2.2 Delay Computation

Real delay computation must consider finite value of the gate input transition time. As developed in [9] we introduce the effect of the input-to-output coupling and the input slope effects in the model as

$$t_{HL}(i) = \frac{v_{TN}}{2} \tau_{INLH}(i-1) + (1 + \frac{2C_M}{C_M + C_L}) t_{HLstep}(i)$$
$$t_{LH}(i) = \frac{v_{TP}}{2} \tau_{INHL}(i-1) + (1 + \frac{2C_M}{C_M + C_L}) t_{LHstep}(i) \qquad (3)$$

where $\tau_{INHL,LH}$ is the duration time of the input signal, taken to be twice the value of the step response of the controlling gate. C_M is the coupling capacitance between the input and output nodes, that can be evaluated as one half the input capacitance of the P(N) transistor for input rising (falling) edge, respectively or directly calibrated from SPICE simulation. Indexes (i), (i-1) refers to the location of the cell in the array.

As shown in (3), we have considered that the input transition time is short enough, to assume that the output switching of the gate still occurs under a constant maximum value of the current. This assumption justifies the use of the step responses for evaluating the delay.

3 Fanout Optimization

3.1 Fanout Optimization Alternatives

We address here the problem of defining an indicator for selecting between various optimization alternatives. Considering a critical path, defined as a path violating the timing constraint, we propose to identify what are the critical nodes and then to select the best optimization technique. Another alternative, prior to any budget distribution on the path, is that the identification of the critical nodes will permit a local optimization for improving the constraint distribution algorithm. At this step three well-identified actions can be considered.

The first solution is transistor sizing. Increasing the drive of the gate controlling the critical node decreases its load to drive ratio (C_L/C_{IN} = electrical effort of [6]) of this gate, as shown in (1). If this solution improves the delay of that gate, this affects the speed of its preceding gate and due to input slope effect reduces the initial benefit. More over, depending on the number of gate inputs, the speed improvement may be obtained at the expense of a large area increase.

The second and third solutions are to insert one inverter or a buffer between the gate and its output load. The use of one inverter imposes an upstream logic reconfiguration to conserve the parity of the signal. This can be of interest if a heavily loaded gate is replaced by a fastest one (with a lower value of S). The insertion of two inverters (buffer) improves the speed of the design for heavy loading condition. A significant interest in inserting one or two inverters is to speed up the gate switching without modifying its input capacitance, conserving the load of its fan-in gate. Our goal is to determine the best speed up design alternative in order to suppress iterative attempts. One possible solution is to characterize each gate by a critical loading level.

3.2 Metric Definition

In order to evaluate the loading level of each gate (i) along a path, let us use the fanout factor $F_0 = C_{Li}/C_{Ini}$ in which C_{Li} and C_{Ini} represent respectively the total output load and the input capacitance of the gate (i).

We characterize each family of gate by a specific fanout factor, F_{0lim}, to be used as a threshold indicator of the gate sensitivity to sizing or buffer insertion as well as a threshold for selecting an optimization alternative.
We consider simple Nand, Nor gates and inverters with identically sized N and P transistors (k = 1). Different values of the configuration ratio can be considered, as well, updating the value of the reduction factors given in Table 1.

Table 1. Example of reduction factor value for the simple gates of a 0.25μm CMOS library.

	Reduction factor	
Gates, 0.25μm,	S_{HL}	S_{LH}
Inverter k = 1	1	2.3
Inverter k = 2	1.5	1.73
Inverter k = 3	2	1.53
Nand2 k = 1	1.55	2.3
Nand3 k = 1	2.05	2.3
Nor2 k = 1	1	4.3
Nor3 k = 1	1	6.3

In Fig.1 we illustrate the three configurations we have to consider: (a) a gate with an output load C_L, (b) and (c) the same gate after one or two inverter insertion.

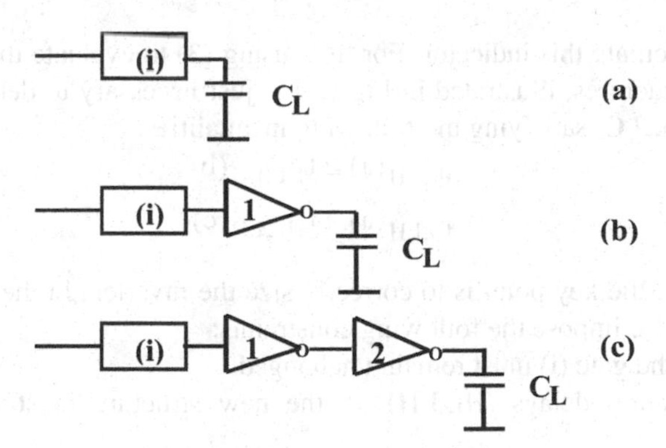

Fig. 1. Array of gate and inverters to be considered for determining the limit of load for buffer insertion.

It clearly appears that for each gate under consideration the implementation with minimum delay depends on the value of the output load and on the sizing of the different inverters. As expected from (3), the insertion of a correctly sized inverter distributes the initial load and speed-up the gate at the expense of an extra propagation delay. The limit value of the output load at which the gain in speed begins to be greater than the propagation delay overhead, or equivalently at which the solution (b) or (c) is faster than the initial implementation, is a good indicator of the criticality of the gate loading. This is illustrated in the Fig.2, which represents the simulated delay sensitivity to the load of a 2 input Nor gate. As shown the insertion of one (curve 1) or two inverters (curve 2) has a significant influence on the load sensitivity of the structure. The intersection between the delay variation (i) and (1) or (2) defines the minimum value of the fanout factor at which the insertion of inverters will improve the speed of the structure.

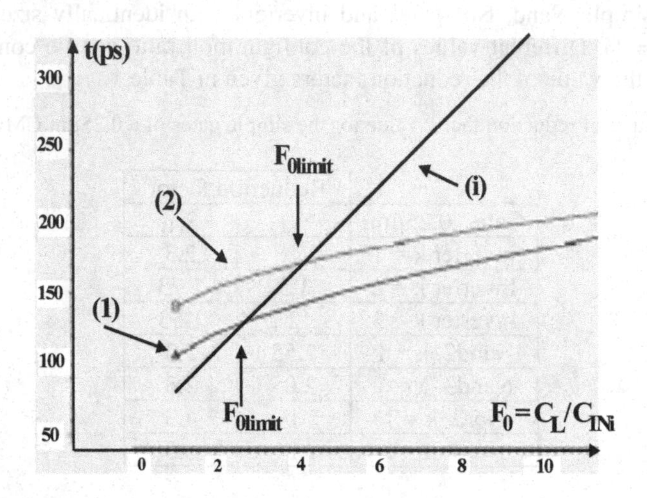

Fig. 2. Sensibility of the delay to the load, for the structures represented in Fig.1, the gate (i) is a 2 input Nor with an input capacitance of 10.4fF.

Let us now calculate this indicator. For that, using (3) to evaluate the input-to-output delay of the structures, illustrated in Fig.1, it is just necessary to determine the value of the output load C_L satisfying the following inequalities:

$$\begin{aligned} t_{HL,LH}(a) &\geq t_{HL,LH}(b) \\ t_{HL,LH}(a) &\geq t_{HL,LH}(c) \end{aligned} \tag{4}$$

Developing (4), the key point is to correctly size the inverters in the configurations b and c. For that we impose the following constraints:

- the size of the gate (i) must remain unchanged,
- the propagation delays (HL,LH) on the new structure must be identical and minimized,
- the sizing solution must give the lowest area alternative that we evaluate as the sum of the transistor widths (for simplicity, we equivalently consider the sum of the corresponding input capacitances).

Then it results that the sizing solution depends on the initial gate (i) to be considered.

A – One Inverter Insertion

If the initial gate is an inverter it can be shown from (1) and (3) that the equality of the fall and rise delay on an array of two inverters is obtained when they have identical configuration ratio value (ki= k1= k in Fig.1b). We also impose, the same condition for Nand and Nor gates. Then the input capacitance of the inserted inverter is obtained from

$$C_{IN1} = \sqrt{\frac{S_{LH1} - S_{HL1}}{S_{LHi} - S_{HLi}}} \sqrt{C_L \cdot C_{IN_i}} \tag{5}$$

Here, to minimize the total area, the configuration ratio of each element is chosen equal to 1, as long as the argument of the square root of (5) remains positive.

B – Two Inverter Insertion

We want to impose the same constraint of symmetry on the delay of the array, while minimizing the total area. With an odd number of stages (configuration (c)) the unique solution is to impose, as in the preceding part, the equality of delays on two of the three stages and to impose the symmetry of edges on the remaining one. In the configuration (c) the area efficient solution consists in imposing the equality of delay between the two first stages, (i) and (1), and to balance the edges of (3) by selecting $k_3 = R$. As given in (1), this is the condition, on the inverter configuration ratio, for balancing the input and falling edges.

This time two parameters must be determined, C_{IN1} and C_{IN2}. The first equation is obtained by applying (5) to the stage (i) and (1). This gives

$$C_{IN1} = \sqrt{\frac{S_{LH1} - S_{HL1}}{S_{LHi} - S_{HLi}}} \sqrt{C_{IN2} \cdot C_{IN_i}} = A \cdot \sqrt{C_{IN2} \cdot C_{IN_i}} \tag{6}$$

The resulting delay expression only depends on C_{IN2}. Equating to zero the derivative of the delay with respect to C_{IN2} results in:

$$C_{IN2} = \left[\frac{(1+R)}{\frac{1}{2}\left(S_{HLi} \cdot A + \frac{S_{HL1}}{A} \right)} \right]^{2/3} \cdot C_{IN_i}^{1/3} \cdot C_L^{2/3} \tag{7}$$

and

$$C_{IN1} = A \cdot \left[\frac{(1+R)}{\frac{1}{2}\left(S_{HLi} \cdot A + \frac{S_{HL1}}{A} \right)} \right]^{1/3} \cdot C_{IN_i}^{2/3} \cdot C_L^{1/3} \tag{8}$$

We have implemented these values of the inverter input capacitance in the delay equations of the configuration (b) and (c) of Fig.1, solving for the value of the output load satisfying (4). The resulting value of the fanout factor characterizes the loading level of the gate at which any insertion of inverter or buffer will be more efficient than any gate sizing.

4 Experimental Results

In Table 2 we compare the resulting minimum value of the fanout factor F_{0lim} satisfying (4) to the value obtained from HSPICE (0.25µm, level49) simulation of the same structure. Defining by Cpar the output parasitic capacitance of each cell, we use the relative parameter p = Cpar/C_{IN} and consider two parasitic loading conditions (p = 0, 1) for each node. As shown, we obtain a good agreement between the calculated and the simulated values. Note that if the fanout limit, obtained for buffer insertion at a node driven by an inverter or a Nand gate corresponds to a quite large value of the load, this limit is lower for Nor gates. This is an indication that this category of gate is not able to drive a load greater than one or two times its input capacitance. Such nodes are privileged nodes for inverter or buffer insertion. Considering, for example, the limit of a three input Nor gate for inserting one inverter, we can easily conclude that any logic synthesis based on three input OR will be faster than using NOR gates.

Table 2. Comparison between simulated and calculated values of the limit fanout factor.

Gate		F_{Olimit} One inverter p=0	One inverter p=1	Two inverters p=0	Two inverters p=1
INV k = 1	Simul.	3.44	4.46	5.7	7.5
	Calcul.	3.36	4.44	5.2	7.2
INV k = 2	Simul.	6.05	7.97	9	12.5
	Calcul.	5.82	7.6	8	11
INV k = 3	Simul.	5.1	7	7.3	10.3
	Calcul.	4.9	6.8	7.3	9.9
NAND2	Simul.	4.1	5.2	6.1	8.6
	Calcul.	4.3	5.6	5.6	7.8
NAND3	Simul.	4.6	5.7	6.6	9.4
	Calcul.	4.5	6	6.1	8.4
NOR2	Simul.	1.7	2.35	2.9	4.1
	Calcul.	1.7	2.4	3	4.1
NOR3	Simul.	1.1	1.46	1.8	2.5
	Calcul.	1.14	1.48	2.4	3.2

Let us consider a critical path subject to a delay constraint. Depending on the feasibility of the constraint to be satisfied, two alternatives can be applied:

- gate sizing on the initial path,
- gate sizing after buffer insertion on the critical nodes of this path.

The first solution has been widely investigated. An elegant solution has been given in [6]. However it may result in gate over sizing. The second solution, we proposed here, is based on the use of the preceding metric. It can be implemented in the following protocol.

For any node of the critical path, compare the fanout factor value at this node to the F_{0limit} defined for the gate controlling this node.

- If $F_0 < F_{0limit}$ the more efficient design alternative to speed up the considered gate is sizing. This can be directly managed during the delay constraint distribution on the path (see [6] for example).

- If $F_0 > F_{0limit}$ a more efficient path implementation may be obtained by inserting one or two inverters. Then the delay constraint distribution on this path will result in a lower area (power) implementation.

This protocol has been used to optimized several benchmarks. The delay constraint imposed on each critical path has been satisfied using the Synopsis optimization tool (Amps). The results are summarized in Table 3 where we compare the area (given by the sum of the transistor width) of the implementation of different benchmarks, with and without buffer inserted on the critical nodes. For the initial implementation all the transistors are at the minimum width. This defines both :

-the "reasonable" maximum value of the delay

-and the corresponding minimum area of the circuit implementation.

Table 3. Efficiency of the buffer insertion on the area of implementation of various circuits.

		Before buffer insertion			After buffer insertion
	Delay Constraint t_c (ns)	Initial delay t_0 (ns)	Initial area ΣW (μm)	Area for the constraint t_c (μm)	Area for the constraint t_c (μm)
BIXAV1 22 gates	2.02	3.5	96	277	169
BIXAV1 22 gates	2.5	3.5	96	152	122
XAV1 9 gates	1.2	2.3	42	79	59
XAV1 9 gates	1	2.3	42	104	83
XAV1 9 gates	0.9	2.3	42	146	110
GXAV1 13 gates	1.56	2.7	54	104	72
GXAV1 13 gates	1.2	2.7	54	175	111
FPD 13 gates	0.8	1.8	13	71	41

As shown, the controlled insertion of buffer results in an appreciable area saving. This area saving may be particularly important for configurations in which the size of the loaded gate can not be varied so much, such as at branching nodes or at the input of a logic array.

5 Conclusion

In this paper we have proposed a new definition of metrics for qualifying the criticality loading level of gates on combinatorial paths. Indicators for buffer insertion have been defined as a function of the physical design parameters.

If this indicator has been developed to select between gate sizing and buffer insertion design alternatives, it could also be apply at others steps of the physical synthesis flow. As an example, used as a metric to classify the gates of a library with respect to their load sensitivity, this indicator could help to choose the best physical implantation of a given Boolean function while designing a library.
Buffering highly loaded nodes, prior to any delay constraint distribution, appears as a very interesting solution for achieving a delay goal at a lower area/power cost.

References

[1] J. M. Shyu, A. Sangiovanni-Vincentelli, J. Fishburn, A. Dunlop, "Optimization-based transistor sizing" IEEE J. Solid State Circuits, vol.23, n°2, pp.400–409, 1988.
[2] S.R. Vemuru, A.R. Thorbjornsen, A.A. Tuszynski, " CMOS tapered buffer", IEEE J. Solid State Circuits, vol.26, n°9, pp.1265–1269, 1991.
[3] P.G. Paulin, F. J. Poirot, "Logic decomposition algorithm for the timing optimization of multilevel logic", Proc. ICCD 89, pp.329–333.
[4] P. K. Chan, "Algorithms for library-specific sizing of combinational logic" in Proc. ACM/IEEE Design Automation Conf., 1990, pp.353–356.
[5] J. Fishburn, A. Dunlop, "TILOS: a posynomial programming approach to transistor sizing" in Proc. Design Automation Conf. 1985,pp.326–328.
[6] I. Sutherland, B. Sproull, D. Harris, "Logical Effort: Designing Fast CMOS Circuits", Morgan Kaufmann Publishers, INC., San Francisco, California, 1999.
[7] C. Mead, M. Rem, "Minimum propagation delays in VLSI", ", IEEE J. Solid State Circuits, vol.SC17, n°4, pp.773–775, 1982.
[8] P. Maurine, M. Rezzoug, N. Azémard, D. Auvergne "Transition time modeling in deep submicron CMOS" IEEE trans. on Computer Aided Design, Vol.21, n°11, pp.1352–1363, nov. 2002.
[9] K. O. Jeppson, "Modeling the influence of the transistor gain ratio and the input-to-output coupling capacitance on the CMOS inverter delay", IEEE J. Solid State Circuits, vol.29, pp.646–654, 1994.

Optical versus Electrical Interconnections for Clock Distribution Networks in New VLSI Technologies

G. Tosik[1, 2], F. Gaffiot[1], Z. Lisik[2], I. O'Connor[1], and F. Tissafi-Drissi[1]

[1] LEOM-Ecole Centrale de LYON
36, avenue Guy de Collongue, 69134 Ecully-CEDEX- France
{grzegorz.tosik, frederic.gaffiot, ian.oconnor}@ec-lyon.fr
[2] ZPP – Technical University of Lodz
ul. Stefanowskiego 18/22, 90924 Lodz Poland
grzegorz.tosik@wp.pl lisikzby@ck-sg.p.lodz.pl

Abstract. Interconnects constitute a severe impediment to the future progress of VLSI integrated systems technology. The use of optics is considered to be an alternative solution that could overcome the limitations of metallic interconnects. Clock distribution networks could particularly benefit from this technology. In this paper , we present a quantitative analysis of the performance gains of using optical interconnect for clock distribution in terms of power and frequency. We present detailed comparative simulations of the features of optical and electrical H-tree clock networks.

1 Introduction

Due to continually shrinking feature sizes, higher clock frequencies, and the simultaneous growth in complexity, the role of interconnect as a dominant factor in determining circuit performance is growing in importance. The 2001 ITRS roadmap [1] shows that by 2010, high performance integrated circuits will count up to $2x10^9$ transistors per chip and work with clock frequencies up to 10 GHz. Coping with electrical interconnects under these conditions will be a formidable task. This is particularly true for clock distribution networks (CDN) where line widths are of the order of a few μm, total length can be of the order of kilometres, and required power consumption represents a large portion of total power. In a modern VLSI circuit with power dissipation of 100W [2], the clock tree uses at least 30% of this power and may even reach 50% [2-4]. Due to natural limits in thermal management, this is a real barrier to further progress. One proposed alternative to overcome these limitations is the use of global optical clock distribution networks, with conversion to the local electrical clocks. This alternative can be acceptable only if it demonstrates significantly improve performance over the all-electrical solution.
The state of the art in clock distribution networks for modern IC's is briefly described in Sec.2. Since optical interconnects used in CDN are new solutions, these are described in more detail in Sec.3. In order to carry out a comparison between optical and electrical main H-trees of clock networks, simulations of its features have been performed in Sec.4, and are discussed in Sec.5.

J.J. Chico and E. Macii (Eds.): PATMOS 2003, LNCS 2799, pp. 461–470, 2003.
© Springer-Verlag Berlin Heidelberg 2003

2 Electrical Clock Distribution Network

Semiconductor technologies operate at increasingly higher speeds and system performance has become limited not by the delays of the individual logic elements but by the ability to synchronise the flow of data signals. A clock network distributes the clock signal from the clock generator, to the clock inputs of the synchronising components. This must be done while maintaining the integrity of the signal and minimising (or at least upper bounding) such clock parameters as clock skew, clock phase delay and sensitivity to parametric variations of the clock skew. Additionally these objectives must be attained while minimising the use of system resources such as power and area.

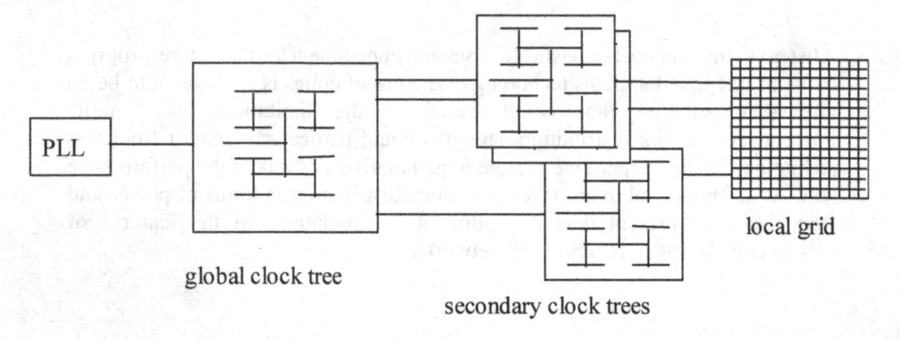

Fig. 1. An example of a clock distribution network.

The clock distribution network of a modern microprocessor uses a significant fraction of the total chip power and has substantial impact on the overall performance of the system. Techniques for synchronous VLSI often utilise a tree and grid-like structure with several levels of hierarchy [5], as shown in Fig.1. This represents the most efficient solution in terms of skew covering the symmetrical topology on the upper metalisation levels. In this work only the main clock tree routed on the upper metal layers is taken into account. To investigate its feature the SPICE model has been worked out. In the model, the branches of the clock tree are replaced

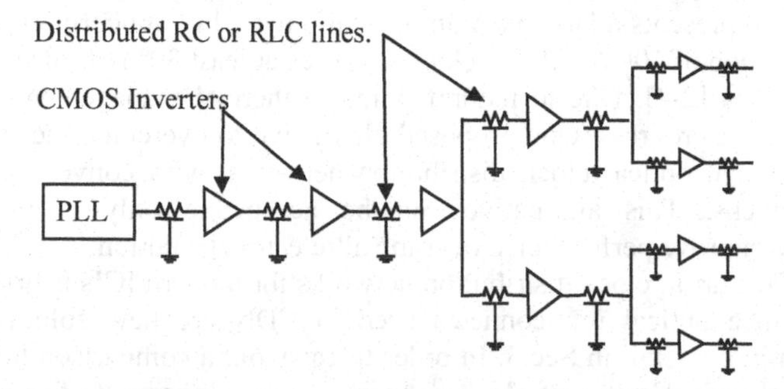

Fig. 2. Equivalent circuit of modeled H-tree.

by RC or RLC transmission lines[6-8] coupled by buffers designed as CMOS inverters (see Fig.2). Delay time is traditionally minimised by splitting long global interconnect into subsections, each with its own buffer. The optimal number and size of the buffers is calculated using the conventional repeater insertion method [9].

Due to the bandwidth limitation of upper level interconnects the global clock signal cannot be distributed at GHz frequency across the chip. To guarantee the high speed clock signal at the latches frequency multiplication circuits placed between the global and local clock network will be needed. This leads to higher power consumption and growth in circuit complexity. We have limited our analysis to non-multiplying architectures.

3 Optical Clock Distribution Network

To overcome the need for clock multiplication circuits and thermal limitations of the electrical system we propose the replacement of the conventional global clock network by an optical H-tree. In the proposed system shown in Fig.3 a low-power vertical cavity surface emitting laser (VCSEL) is used as an off-chip photonic source [10]. The VCSEL is coupled to the symmetrical passive waveguide structure and provides the clock signal to *n* optical receivers. At the receivers, the high speed optical signal is converted to an electrical signal and subsequently distributed by the local electrical networks.

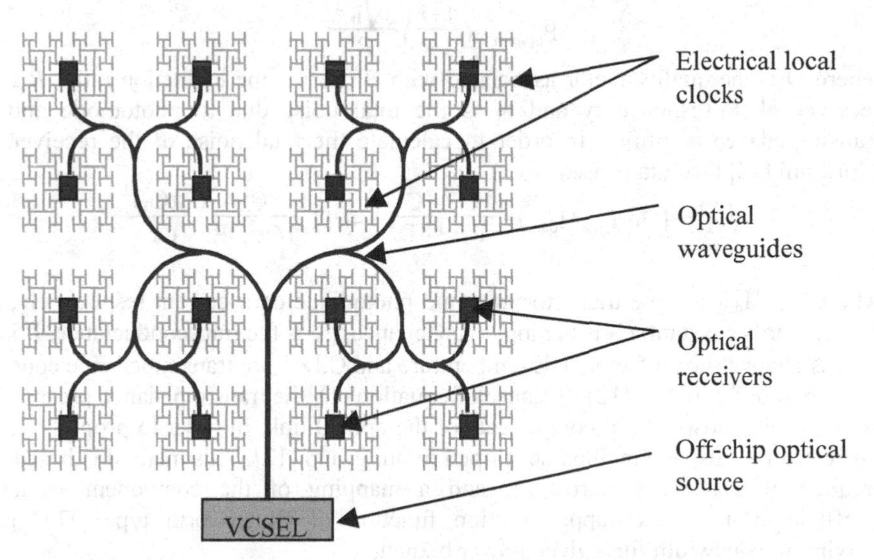

Fig. 3. General approach to optical global clock distribution network.

To form the planar optical waveguide tree we assume the use of Si as the core and SiO_2 as the cladding materials, as shown in Fig.4. We chose Si/SiO_2 structures because they are compatible with conventional silicon technology and transparent for 1.3-1.55μm wavelength. Additionally, such waveguides with high relative refractive index difference $\Delta \approx (n_1^2 - n_2^2)/2n_1^2$ between the core ($n_1 \approx 3.5$ for Si) and claddings ($n_2 \approx 1.5$ for SiO_2) allow the realisation of a compact optical circuit, with bend radius of the order of a few μm [11]. To avoid modal dispersion, improve coupling efficiency and reduce loss, single mode conditions are applied to the waveguide dimensions.

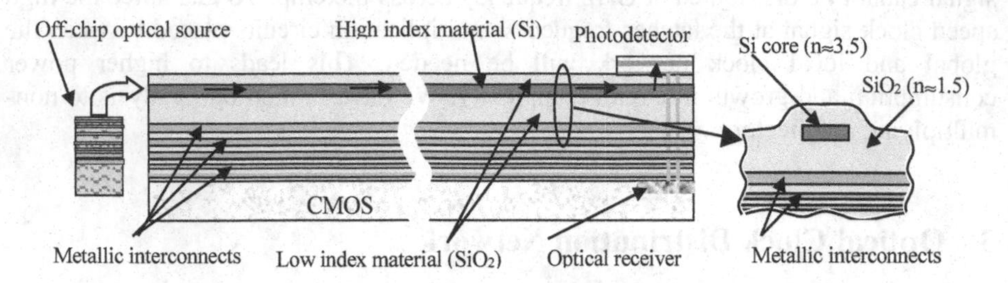

Fig. 4. Cross section of hybridated interconnection structure

The main criterion in evaluating the performance of digital transmission systems is the resulting bit error rate (BER), which may be defined as the rate of error occurrences. Typical BER required by Gigabit Ethernet and by Fibre Channel is 10^{-12} or better. For our on-chip interconnect network, we assume that a BER of 10^{-15} is acceptable. The minimal optical power P_{optmin} required by the receiver to operate at a given error probability can be written as:

$$P_{optmin} = \left(\frac{1+r}{1-r}\right)\frac{Q\sqrt{i_N^2}}{R} \tag{1}$$

where Q is the quality factor associated with BER, r is the extinction ratio, R is the receiver photo-responsivity and i_N is the total noise due to photodiode and the transimpedance amplifier. In order to calculate the total noise of the receiver, the Morikuni [12] formula is used

$$\sqrt{i_N^2} = \left[\, 2q(I_{gate}+I_{dark}) + \frac{4kT}{R_f}\,\right]\frac{C}{4D} + 4kT\Gamma\frac{C^2}{16\pi^2 DE}\frac{(2\pi C_T)^2}{g_m} \tag{2}$$

where I_{gate}, I_{dark} are the transistor gate and photodiode dark current respectively, k is Boltzmann's constant, C_T is the input capacitance, g_m is the transconductance, Γ is the excess channel-noise factor, T is temperature and C,D,E are transimpedance constants described by Morikuni [12]. Because of equation (1), the transimpedance preamplifier is one of the most critical components of the optical link. In order to properly design this circuit we use the method developed in our group [13]. This method is based on a frequential analysis of structures and a mapping of the component values to coefficients in a filter approximation function of Butterworth type. This gives maximum bandwidth for a given power budget.

The optical system performance depends on the minimum optical power required by the receiver and on the efficiencies of passive optical components used in the system. The total loss in the optical system shown in Fig.5 is the sum of the losses (in decibels) of all optical components.

$$L_{total} = L_{CV} + L_W + L_B + L_Y + L_{CR} \tag{3}$$

where L_{CV} is the coupling loss between the photonic source and optical waveguide, L_W is the rectangular waveguide transmission loss, L_B is the bending loss, L_Y is the Yspliter loss and L_{CR} is the coupling loss from the waveguide to the optical receiver. To provide an unambiguous comparison in terms of dissipated power between the optical and electrical on-chip clock distribution networks it is necessary to incorporate all of these components.

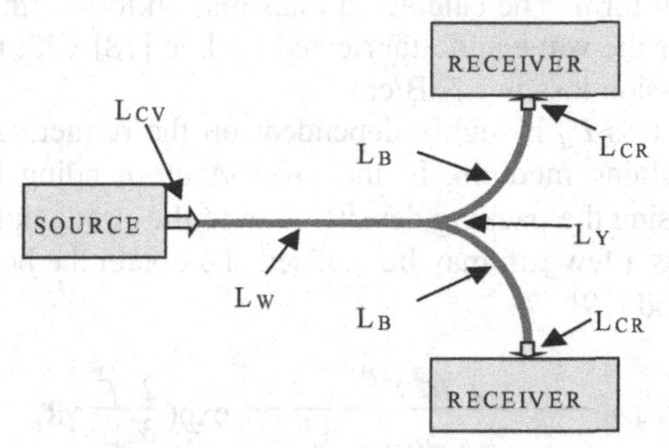

Fig. 5. Loss sources in optical system.

In the present technology there are several methods to couple the beam emitted from the laser into optical waveguide. There are end-fire coupling, end-butt coupling, mirrors and lensing coupling. In the proposed system we assumed 50% coupling efficiency L_{CV} from the VCSEL to a single mode waveguide.

Transmission loss L_W describes the attenuation rate of the optical power, as light travels in the waveguide.. Due to small waveguide dimensions and large index change at the core/cladding interface in the Si/SiO_2 waveguide the side-wall scattering is the dominant source of losses. The basic scattering theory in the optical waveguides due to side-wall roughness was presented by Marcuse [14].

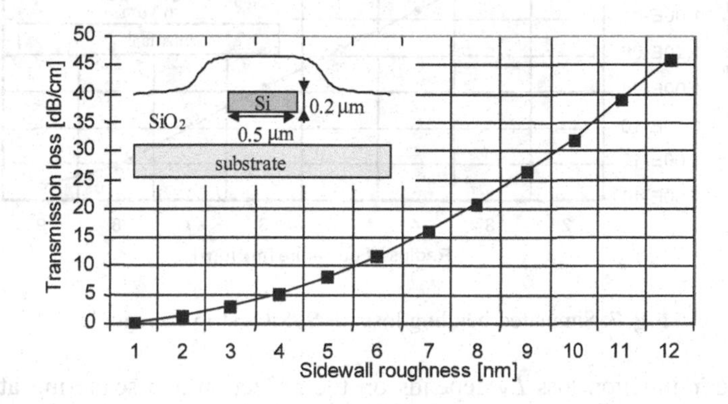

Fig. 6. Calculated transmission loss of the strip waveguide for varying sidewall roughness.

Based on Marcuse's theoretical work, Payne and Lacey formulated a simpler scattering formula [15] by direct calculations of the far field pattern. Using this formula the loss coefficient α can be written as:

$$\alpha = \frac{\sigma^2}{\sqrt{2}\, k_0 d^4 n_1} g f_e \tag{4}$$

where σ is the standard deviation of roughness, k_0 is the free-space wavenumber, d is waveguide thickness and g,f_e are functions of various parameters defined by Payne. To apply the Payne formula to a real dielectric waveguide we use the Effective Index Method [16,17], which reduces a 3-dimensional waveguide problem into a 2-dimensional equivalent form. The calculated transmission loss of the strip waveguide is plotted in Fig.6. For the waveguide fabricated by Lee [18] with roughness of 2nm the calculated transmission loss is 1.3dB/cm.

The bending loss L_B is highly dependent on the refractive index difference between core and cladding medium. In the lower-Δ the bending loss is very high, which prevents increasing the packing density. Due to the strong optical confinement, bend radius as small as a few μm may be realised. To obtain the bending loss α_B we use the Marcuse method [19]

$$\alpha_B = \frac{2\,\gamma\kappa^2 e^{2\gamma d}}{(n_1^2 - n_2^2)/k_0^2 \beta (2d + \frac{1}{\gamma} + \frac{1}{\theta})} \exp(\frac{2}{3}\frac{\gamma^2}{\beta^2}\gamma R_c) \tag{5}$$

where γ, κ and θ are the waveguide eigenvalues, β is the propagation constant and R_c is the radius of curvature. As can be seen from Fig.7, the bending losses associated with a single mode strip waveguide are negligible if the radius of curvature is bigger then 2μm.

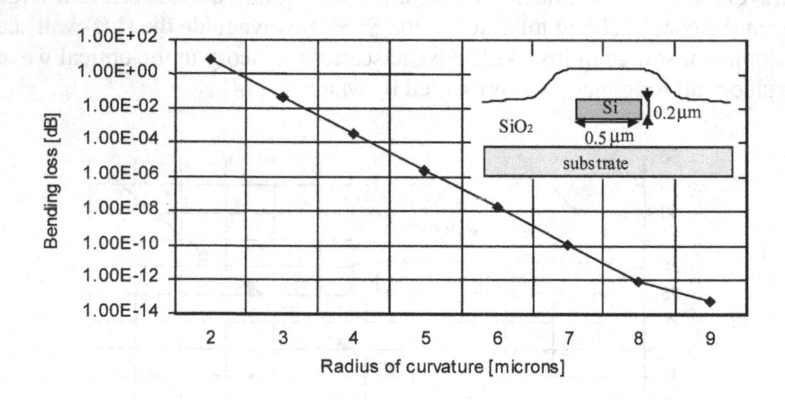

Fig. 7. Simulated bending loss for Si/SiO_2 strip waveguide

The Y-junction loss L_Y depends on the reflection and scattering attenuation into the propagation path and surrounding medium. Different Y-branch structures have been analyzed by several methods [20,21]. For high index difference waveguides the losses for the Y-branch are significantly smaller then for the low-Δ structures and the simulated losses are less then 0.2dB per split [22].

Using currently available materials and methods it is possible to achieve an almost 100% coupling efficiency from waveguide to optical receiver. In the proposed system the coupling efficiency L_{CR} from the waveguide to the optical receiver is assumed to be 87%[23].

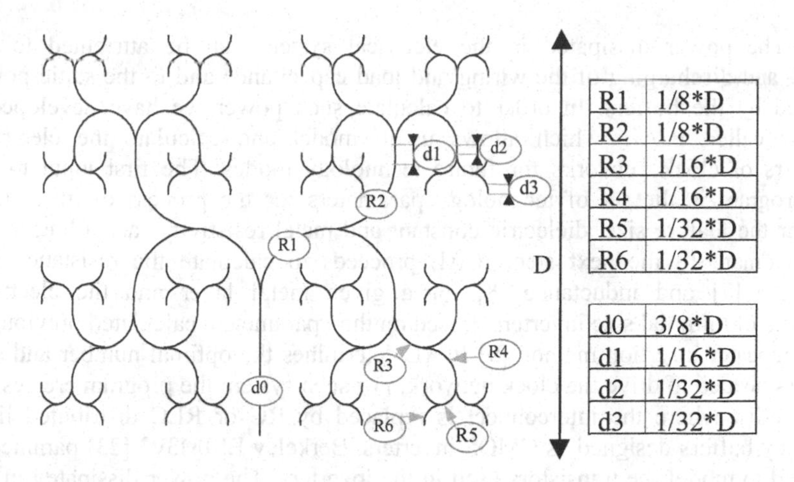

Fig. 8. Optical H-tree network with 64 output nodes. R1..R6 are the radius of curvatures and d0..d3 are the lengths of straight lines linked to the chip width D.

As an example Fig.8. shows a global optical H-tree optimised for 64 output nodes. To achieve minimal optical losses introduced by the H-tree, the radius of curvatures are designed to be as large as possible. This ensures the smallest distance from source to receivers, which leads to the smallest transmission loss and guarantees a minimal pure bending loss at each curvature. For 20mm die width, the smallest radius of curvature is 625μm, which leads to negligible pure bending loss.

4 Comparison of Optical and Electrical Systems

To provide an unambiguous comparison in terms of dissipated power between the optical and electrical clock distribution networks it is necessary to estimate the electrical power dissipated in both systems. In this paper we present the power dissipated in both systems at 70nm technology generation. Power dissipated by electrical and optical CDN have been calculated based on the system performance summarised in Table.1.

Table 1. Optical and electrical systems performances.

Optical system performance		Electrical system performance	
Wavelength λ [μm]	1.55	Technology [μm]	0.07
Waveguide core index (Si)	3.475	Vdd [V]	0.9
Waveguide cladding index (SiO_2)	1.444	Tox [nm]	1.6
Waveguide thickness [μm]	0.2	Chip size [mm^2]	400
Waveguide width [μm]	0.5	Global wire width [μm]	1
Transmission loss [dB/cm]	1.3	Intermediate wire width [μm]	0.5
Loss per Y-junction [dB]	0.2	Local wire width [μm]	0.09
Input coupling coefficient [%]	50	Metal resistivity [μΩ-cm]	2.2
Bit error rate (BER)	10^{-15}	Dielectric constant	3
Photodiode capacitance [fF]	100	Optimal segment length [mm]	1.7
Photodiode responsivity [A/W]	0.95	Optimal buffer size [μm]	90

The power dissipated in the electrical system can be attributed to the charging and discharging of the wiring and load capacitance and to the static power dissipated by the buffers. In order to calculate such power we have developed a simulator called ICAL, which allows us to model and calculate the electrical parameters of clock networks for future technology modes. The first input to the ICAL program is the set of technology parameters for the process of interest, in particular the feature size, dielectric constant and metal resistivity according to the ITRS roadmap. In the next step, ICAL proceeds to calculate the resistance [6], capacitance [7] and inductance [8] for a given metal layer and the electrical parameters of minimal size inverters. Based on the parameters calculated previously, and the repeater insertion method [9] ICAL determines the optimal number and size of buffers needed to drive the clock network. For such system the program creates the SPICE netlist where the interconnect is replaced by RC or RLC distributed lines coupled by buffers designed as CMOS inverters. Berkeley BSIM3v3 [23] parameters were used to model the transistors used in the inverters. The power dissipated in the system is extracted from transistor-level simulations.

Table 2. Optical power budget for 20mm die width at 3GHz frequency.

Number of nodes in H-tree	4	8	16	32	64	128
Transmission loss in straight lines [dB]	1.3	1.3	1.3	1.3	1.3	1.3
Transmission loss in curved lines [dB]	1	1.31	1.53	1.66	1.78	1.85
Y-dividers [dB]	6	9	12	15	18	21
Loss in Y-couplers [dB]	0.4	0.6	0.8	1	1.2	1.4
Output coupling loss [dB]	0.6	0.6	0.6	0.6	0.6	0.6
Input coupling loss [dB]	3	3	3	3	3	3
Total optical system loss [dB]	12.3	15.81	19.23	22.56	25.88	29.15
Min. power at the receiver [dBm]	-22.3	-22.3	-22.3	-22.3	-22.3	-22.3
VCSEL optical output power [mW]	0.1	0.25	0.5	1.1	2.30	4.85

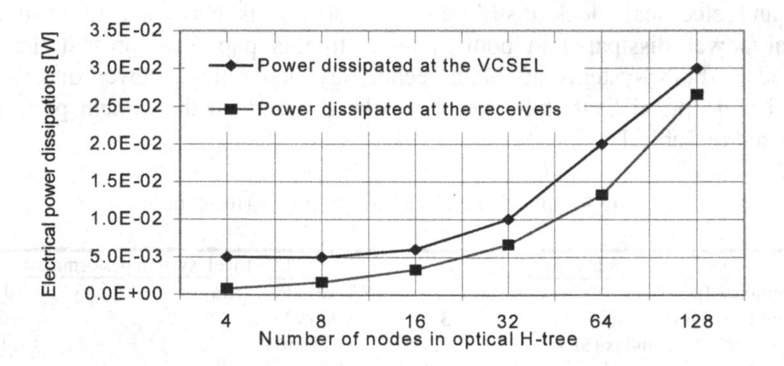

Fig. 9. Electrical power dissipated by the VCEL and receivers.

In the optical clock distribution there are two main sources of electrical power dissipation. First is the power dissipated by the optical receivers and second is the energy needed by the optical source to provide the required optical output power. Based on given BER and system performances (summarised in Table.1). the receiver circuits have been designed and its power dissipation for various frequencies have been extracted from transistor-level simulations. In order to determine the optical power emitted by the VCSEL it is necessary to estimate the power required by each receiver to provide a given error probability and to estimate the losses incurred throughout passive optical components. For a BER of 10^{-15} the minimal power required by the receiver is -22.3dBm (at 3GHz). Losses incurred by passive components for various nodes in the H-tree are summarised in Table.2.

Since the VCSEL optical output power is determined (Table.2) we can use the laser light-current characteristics given by Amann [7] to find the energy needed by the off-chip optical source (VCSEL). Figure.9. shows the electrical power dissipated by the VCSEL and by the receivers for various H-tree nodes and 3GHz frequency.

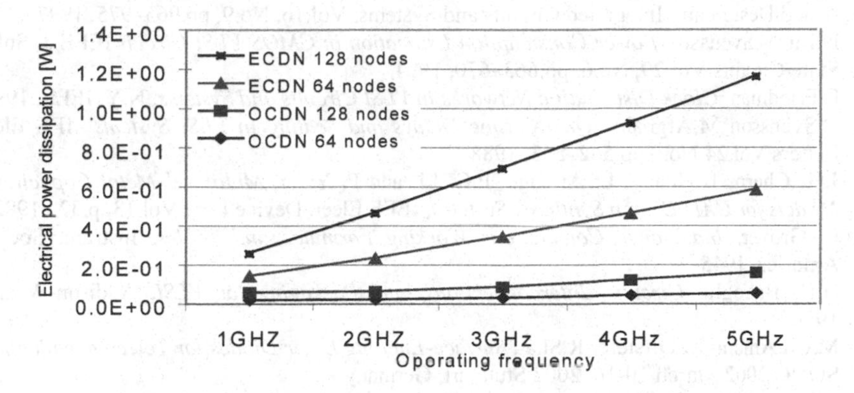

Fig. 10. Electrical power dissipated by electrical and optical CDN for varying frequency.

The unambiguous comparison in terms of dissipated power between the optical and electrical global clock distribution networks is shown in Fig.10. It can be seen that the power dissipated by the electrical system is highly dependent on the operating frequency, while in the optical system remains almost the same. Difference between the power dissipated in both systems is higher if we increase the frequency and number of nodes in H-trees.

5 Conclusion

Integrated optics is a possible alternative to overcome limitations due to metallic interconnections. The aim of this work is to provide an unambiguous comparison in terms of dissipated power between electrical and optical clock distribution networks. The power consumption in the main balanced H-tree of a classical CDN have been calculated. In the case of electrical distribution the power consumption takes into account the power dissipated in the buffers and in the wires of an optimised balanced H-tree. In the case of optical CDN, the calculations take into account the power dissipated in the optoelectronic conversion circuits and in the optical path.

First results are presented for the 70nm technology mode. They show that for a classical 64 output nodes H-tree at 5GHz frequency the power consumption in the optical CDN should be 5 times lower then in an electrical network. The proposed solution allows distribution of high local frequency signals across the chip, with significantly smaller power dissipation than the electrical system. Although only the power consumption is compared we expect lower crosstalk and clock skew in optical system. These parameters will be investigated in future work.

References

1. International Technology Roadmap of Semiconductor 2001 Edition. http ://public.itrs.net/
2. Kavech Azar *The History of Power Dissipation*, Electronics Cooling, vol 6, No.1, 2000.
3. A.Vittal M.Marek-Sadowska *Low Power Buffered Clock Tree Design*, IEEE Trans. Comp. Aided Design and Integrated Circuits and Systems, Vol.16, No.9, pp.965–975. 1997.
4. D.Liu C.Svensson *Power Consumption Estimation in CMOS VLSI Circuit* IEEE J. Solid-State Circuits Vol.29, No.6, pp.663–670, 1994.
5. E.Friedman *Clock Distribution Networks in VLSI Circuits and Systems*, N.Y. IEEE, 1995.
6. C.Svensson M.Afghahi *On RC Line Delays and Scaling in VLSI Systems* IEE Elect. Letters Vol.24 No.9 pp.562–563, 1988.
7. J.H. Chern, J. Huang, L. Arledge, P.C. Li, and P. Yang, *Multilevel Metal Capacitance Models for CAD Design Synthesis Systems*, IEEE Elect. Device Lett. Vol.13, p.32, 1992.
8. F. Grover, *Inductance Calculations Working Formulas and Tables*, Instrum. Soc. of America, 1945.
9. H.B. Bakoglu, *Circuits, Interconnections, and Packaging for VLSI*, Addison-Wesley, 1990.
10. M.Ch.Amann M.Ortsiefer R.Shau *Surface-Emitting Laser Diodes for Telecommunications* SODC 2002 March 10-16, 2002 Stuttgart, Germany.
11. A.Sakai G.Hara T.Baba *Propagation Characteristics of Ultrahigh-D Optical Waveguides on Silicon-on-Isulator Substrate* Jpn. J. Applied Physics Vol.40 No.4 pp.383–385 2001.
12. J.J Morikuni A.Dharchoudhury Y.Leblebici S.M.Kang *Improvements to the Standard Theory for Photoreceivers Noise*, J. Lightwave Techn. Vol.12, No.4 pp.1174–1184, 1994
13. I.O'Connor F.Mieyeville F.Tissafi-Drissi F.Gaffiot Exploration Parametrique d'Amplificateurs de Transimpedance CMOS a Bande Passante Maximisee pp.73-76TAISA 2002.
14. D. Marcuse *Mode Conversion Caused by Surface Imperfections of a Dielectric Slab Waveguide* Bell System Technical J. Vol.48 pp.3187–3214, 1969.
15. F.P.Payne J.P.R.Lacey *A Theoretical Analysis of Scattering Loss from Planar Optical Waveguide*, Optical and Quantum Electronics Vol.26 pp.977–986 1994.
16. H.Nishihara M.Haruna T.Suhara *Electro-Optics Handbook* McGraw"Hill 2000.
17. C.M.Kim B.G.Jung C.W.Lee *Analysis of Dielectric Rectangular Waveguide by Modified Effective-Index Method* Electron. Letters Vol.22, No.6 pp.296–298, 1986.
18. K.K.Lee D.R.Lim L.C.Kimerling *Fabrication of Ultralow-Loss Si/SiO$_2$ Waveguides by Roughness Reduction. Optical Letters* Vol.26 No.23 2001.
19. D..Marcuse Light Transmission Optics New: York Van Nostrand Reinhold, 1973.
20. F.S. Chu P.L.Liu Low-Loss Coherent-Coupling Y-Branches Optics Lett. Vol.16 No.5 pp.309–311 1992
21. M.Rangaraj M.Minakata S.Kawakami Low Loss Integrated Optical Y-Branch J. Lightwave Technol. Vol.7 No.5 pp.753–758 1989.
22. A.Sakai T.Fukazawa. T.Baba Low Loss Ultra-Small Branches in a Silicon Photonic Wire Waveguide IEICE Trans. Electron. Vol.E85-C, No.4. 2002.
23. S.M.Schultz E.N.Glytsis T.K.Gaylord Design, Fabrication, and Performance of Preferential-Order Volume Grating Waveguide Couplers Appl. Opt. Vol.39 p.1223, 2000.
24. http://www-device.eecs.berkeley.edu

An Asynchronous Viterbi Decoder for Low-Power Applications

B. Javadi[1], M. Naderi[1], H. Pedram[1], A. Afzali-Kusha[2], and M.K. Akbari[1]

[1] Department of Computer Eng. and Information Technology,
Amirkabir University of Technology, Hafez Ave., Tehran, Iran
{javadi, naderi, pedram, akbari}@ce.aut.ac.ir

[2] IC Design Lab., Department of Electrical and Computer Eng., University of Tehran,
Kargar Ave., Tehran, Iran
afzali@ut.ac.ir

Abstract. This paper presents a robust and low-power Viterbi Decoder designed based on asynchronous architecture. The design is based upon Quasi Delay Insensitive (QDI) timing model which leads to a robust functionality for the decoder. To lower the power consumption of the decoder further, an optimization technique to reduce the power dissipation is applied to add-compare-select (ACS) unit of the decoder. The simulation results shows a 20% reduction in the power consumption for the asynchronous design compared to the synchronous design in 0.35μm CMOS technology with a power supply of 2.5V. The throughput for the circuit is 50 MS/s.

1 Introduction

Many of digital transmission and storage systems use Forward Error Correction (FEC) techniques for correcting errors occurring during the transmission, storage, or retrieval of data. Viterbi Algorithm [480,2] belongs to a large class of FEC known as convolution codes which are used in a wide range of applications such as wireless communications, digital mobile telephony, digital TV broadcast, CD-ROMs, and magnetic disks. The basic idea behind the Viterbi decoder is to maximize the correlation between received vector and the table of possible codewords while sequentially performing the opposite operation of the encoder. The Viterbi decoder operates by finding the maximum likelihood of the decoding sequence.

The quality of a Viterbi decoder design is mainly measured by three criteria: coding gain, throughput, and power dissipation [3]. High coding gain results in low data transfer error probability while high throughput is necessary for high-speed applications. The design of Viterbi decoders with high coding gain and throughput is made challenging by the need for a low power circuit implementation. This requirement mainly stems from the fact that the Viterbi decoders are often placed in communication systems running on batteries where the power reduction is a must. Several attempts for reducing the power dissipation in the Viterbi decoder have been

J.J. Chico and E. Macii (Eds.): PATMOS 2003, LNCS 2799, pp. 471–480, 2003.
© Springer-Verlag Berlin Heidelberg 2003

reported in the literature [4,5,6,7,8,9]. These research activities may be categorized in two following groups. The first group is based on changing the decoder architecture to reduce the power while the second group utilizes different circuit implementation techniques to minimize the power dissipation. Among the methods in the first group, one can mention the alteration of the architecture for Add Compare Select (ACS) unit [4,5], changing the memory management method [6], and altering the trace-back circuit [7]. Examples in the second group include SPL (Single ended Pass transistor Logic) implementation [8], and the self-timed circuit implementation [9]. Except for the Viterbi decoder based on self-timed architecture, other designs are synchronous.

In this paper, we present an asynchronous design for a Viterbi decoder based on Quasi Delay Insensitive (QDI) timing model. In this model, no constrain on the delay of the circuit element exists except for the isochoric forks [10]. Another advantage of QDI for FEC applications is its robust functionality for a wide range of parameter values. This paper is organized as follows. In Section 2, the conventional Viterbi decoder is described while the proposed design of the decoder, asynchronous synthesis and optimization techniques used to reduce the power dissipation are presented in Section 3. Finally, Section 4 contains the results and the conclusion of the work.

2 Viterbi Algorithm

A brief description of the Viterbi algorithm is given in this section where a detailed one can be found in [2]. A small system is chosen to illustrate the Viterbi encoding and decoding process. The algorithm used by the Viterbi Decoder belongs to a class of algorithms known as convolution codes. The rate of the convolution coder is defined as the number of input bits to the output bits.

2.1 Encoder Operation

A typical encoder is depicted in Fig. 1 where the rate of encoding is 1/2, i.e. the system encodes 1 input bit to 2 output bits.

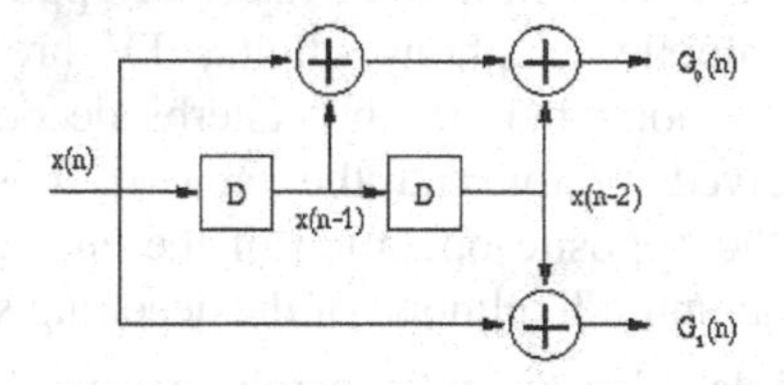

Fig. 1. Block diagram of the Viterbi Encoder.

In order to encode bit $x(n)$ from the input stream, this encoder creates two bits, namely $G_0(n)$ and $G_1(n)$ using the last 3 input bits, i.e. $x(n\text{-}1)$, $x(n\text{-}2)$, and $x(n)$. The encoding bits are produced from the following equations.

$$G_0(n) = x(n) + x(n-1) + x(n-2) \bmod 2 \quad (1)$$

$$G_1(n) = x(n) + x(n-2) \bmod 2 \quad (2)$$

The number of bits used for encoding one bit is called *constraint length* (k) (in this case $k = 3$), and the equations that describe the encoding are called *generator polynomials*.

2.2 Decoder Operation

At the destination, the decoder utilizes the *trellis diagram* to decode the received stream by finding the sequence with the maximum likelihood [1,2]. Fig. 2 shows the trellis diagram of the encoder in Fig. 1. Each node in the trellis diagram denotes one of the four potential pairs ($x(n\text{-}1)$, $x(n\text{-}2)$) of the last two decoded bits. The trellis can be seen as a flow-control diagram where each node represents a state and transitions happen depending on the input stream. From any node we can make a transition to one of two other nodes corresponding to receiving a 0 or a 1 as bit $x(n)$ at the input. The way the trellis diagram is constructed depends on the constraint length but not on the generator polynomials. The two numbers shown on every transition in the figure are the results of the above two generator polynomials. Each *stage*, which is associated with one decoded bit and two encoded bits, comprised of 2^{k-1} nodes.

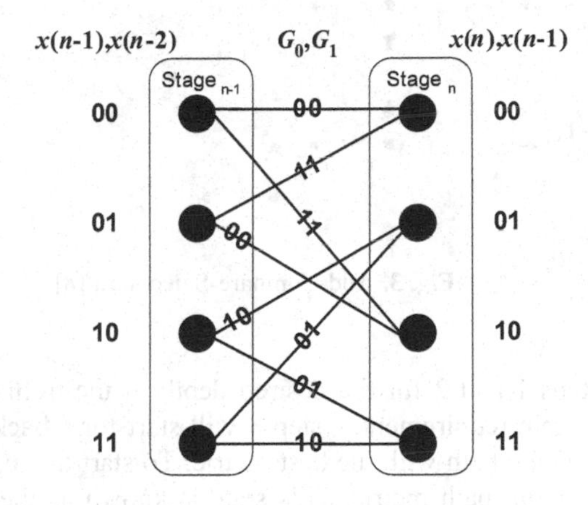

Fig. 2. Two stages of the trellis diagram.

There will be three distinct steps for the implementation of the decoding algorithm. The first step is called *branch metric* calculation where the received data symbols are compared to the ideal outputs of the encoder from the transmitter. The second step, called *path metric* computation where the path metrics of a stage is calculated by adding the branch metrics associated with the received symbol to the path metrics from the previous stage of the trellis diagram. The final step is *trace-back* process, when the survivor path, the path with the highest likelihood, and the output data are identified.

In the first step, the branch metric for each symbol of the input sequence is generated by calculating the Hamming distance between the input symbol and the expected symbol for each connection of the trellis diagram. The Hamming distance is the number of bits which two codewords are different. It can be realized by XORs and an adder. In the second step, the path metric is calculated. The calculations performed in this step, constitute a major part of the arithmetic operations performed in a Viterbi decoder. In this step, the decoder makes use of *Add-Compare-Select* (ACS) module as shown in Fig. 3. An ACS unit has 4 inputs, namely, two branch metrics (BM_1 and BM_2) and two path metrics (PM_{IN1} and PM_{IN2}), and two outputs, namely, the new path metric (PM_{OUT}) and the *Decision Bit*. The decision bit is the most important information generated by the ACS. It indicates that which sum between an input path metric and a branch metric generates the smallest result and was selected as output path metric or *local winner*. All decision bits generated at one stage are saved in a trace-back memory and later used for the trace-back operation.

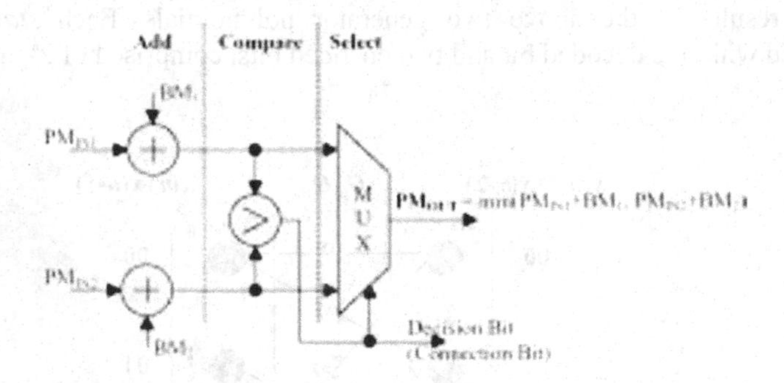

Fig. 3. Add-Compare-Select unit [8].

After repeating steps 1 and 2 for the desired depth of the trellis (determined by the memory and code gain requirements), step 3 will start for a back-trace path and find the *maximum likelihood* path with the best metric. To start this operating we need the state with the minimum path metric. This state is known as the *global winner*. The trace-back operation starts from the global winner state and trace the memory based on the local winner in the active state of each stage. The block diagram of the Viterbi decoder is shown in Fig. 4.

The global winner detection is not needed in each stage of the metric computation. It could be done based on the depth of the memory and the trace-back unit. Because the path metrics are ascending integer numbers they may overflow and in such situation the global winner detection may find a false winner. The false global winner will cause a malfunctioning of the trace-back unit leading to an incorrect output. The essential operation that should be added to the standard global winner detection is the *overflow prevention*. At the end of the trace-back step, the survivor path will be determined and the output data is reconstructed according to the trellis diagram.

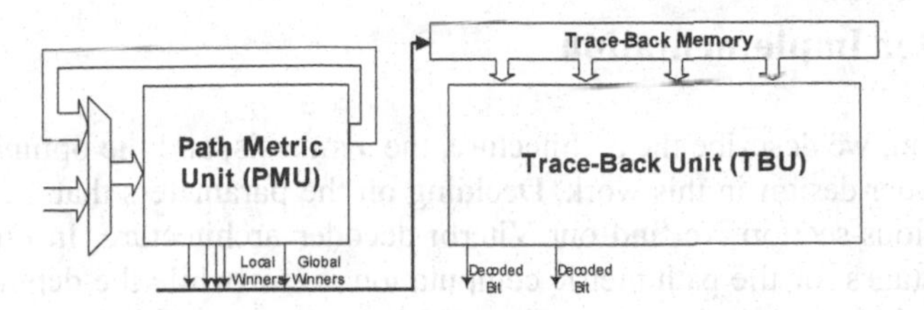

Fig. 4. Block Diagram of the Viterbi Decoder.

2.3 Decoder Structure

Five stages of the trellis diagram in the Viterbi decoder is shown Fig. 5 where the input bits are received at the top and the decoder gives the output decoded bits at the bottom. Each block in this figure is an ACS unit. The dashed lines correspond to two-way communications, i.e., the left state sends the path metric to the right state, while the right state sends a signal to the left state during the backtracking process.

Instead of a direct implementation of the diagram shown in Fig. 5, a more cost effective solution could be to implement with a smaller number of stages and reuse them in every cycle. The output metric of the last stage can be stored in a register to be used by the first stage in the next cycle. The comparison results of the states can also be held in the resisters that are used by the back-trace logic.

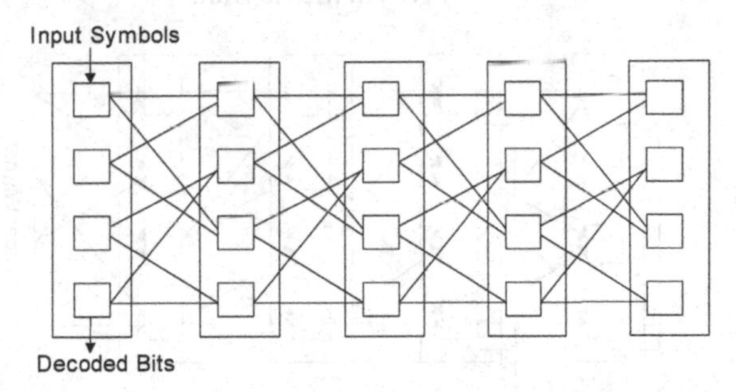

Fig. 5. Stages of the Viterbi decoder (k=3). Each block is an ACS unit.

The path metric unit could be implemented only in one stage. To obtain higher throughputs, one could utilize more stages to compute the metrics for many input symbols in each cycle of circuit operation. The number of bits for the metrics is an important parameter where a longer metric could lead to a higher decoding performance at a higher implementation cost. The last parameters are the depth of the trace-back unit and the number of decoded bits in each trace-back step. These parameters, which determine the Viterbi decoder, architecture will change the speed and the decoding performance of the circuit.

3 Decoder Implementation

In this section, we describe the architecture, the synthesis, and the optimization of the Viterbi decoder design in this work. Deciding on the parameters that are explained at end of previous section we find our Viterbi decoder architecture. In our design, the number of stages for the path metric computation is two while the depth of the trace-back section is four and two bits are decoded in each stage. Fig. 6 and fig. 7 contains the detailed designs of the Path Metric Unit and the Trace-Back Unit of fig. 4. The path metrics are 8-bit wide while the branch metrics are 2-bit wide.

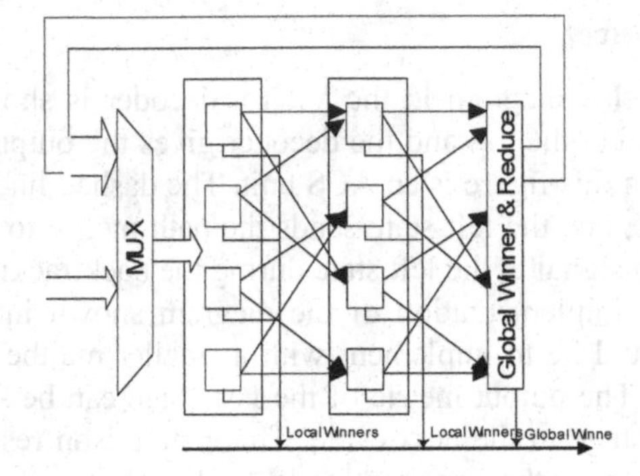

Fig. 6. Path Metric Unit.

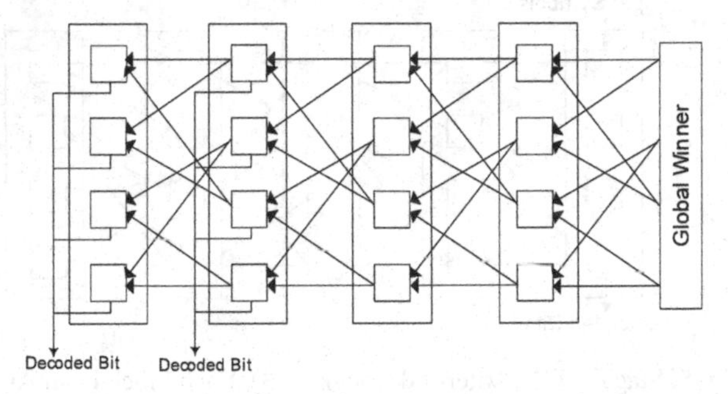

Fig. 7. Trace Back Unit.

3.1 Asynchronous Synthesis

For the implementation of the decoder, we have used the asynchronous design style. In general, the asynchronous implementation has a greater potential for low power applications, although the synthesis tools are less availability for this design style. In our design, an asynchronous ripple carry adder is utilized in the ACS unit. Since the numbers of bits for the branch metric and the path metric are different, the adder only adds the required bits. The asynchronous adder consumes less power compared to the synchronous counterpart due to the fact that the metrics are small numbers, in most cases, and, hence, a large part of the circuit is not active. Another advantage of using this style for the decoder is that in the trace-back only one unit within each stage is required to be active, which lead to lower power dissipation in the asynchronous implementation.

Asynchronous design mythologies are classified based on the timing model being used and may be categorized to bounded-delay and delay insensitive models [12]. These two main categories have many sub-models. The bounded-delay models have timing consideration similar to the synchronous design techniques with this difference that the clock is not used. The delay insensitive models use handshaking for the synchronization of the data transfer. The implementation of pure delay-insensitive circuits is very complex and has a significant overhead. To reduce the complexity and the overhead, one needs to tolerate some timing constrains. One of the most powerful with minimum timing constrain models is Quasi Delay Insensitive (QDI) method that was introduced by A. Martin [10].

We have used QDI model and Martin synthesis method in design of this circuit. Martin synthesis method converts a description of circuit in CSP (Communicating Sequential Process) directly to CMOS instead of converting it to a gate-level description as many synchronous synthesis tools do. For this research, some of the synthesis steps were synthesized using the tools developed by the asynchronous design group of [13] while the rest were synthesized manually.

Here, the design was based on the dual-rail encoding for data transfer and the four-phase handshaking protocol for the communications between the circuit elements. Many of the functional blocks are based on PCFB (Pre-Charge Full Buffer) template [14]. With these blocks, there is no need for buffers because they have an internal buffer.

3.2 ACS Optimization

As mentioned before, a large part of the power is consumed in the ACS units and, therefore, the optimization of these units can reduce the total power consumption of the circuit. Each ACS consists of two add and one compare operations. Considering a block of two ACS neighbors in butterfly architecture [15], each block consists of four 8-bit to 2-bit adder and two 8-bit comparators/selectors where the path metric has 8 bits while the branch metric has 2 bits. In [4], a re-arrangement of the butterfly operations is proposed to reduce the number of operations. This architecture is shown in fig. 8 where in this figure, BM and PM are Branch Metric and Path Metric, respectively Sa and Sb indicate inputs and S0, and S1 shows two possible path for 0 and 1 as output. his re-arrangement reduces the circuit to two 2-bit subtractors, one 8-bit subtractor, two 8 to 3-bit comparators/selectors, and two 8 to 2-bit adders, hence, a considerable reduction in the power. Except for the comparators, all the components are implemented using QDI method.

The asynchronous implementation of the comparators using QDI model leads to large area and power consumption. This is due to the fact that they are active almost all the times and all the input bits are used in this operation. This gives rise to a rather large power and area overhead for the normal dual-rail coding in QDI design, and, therefore, the two comparators are implemented using Micro-Pipeline method [16], which has a lower overhead. The Micro-Pipeline has a normal function covered by a delay-insensitive asynchronous control circuit.

4 Results and Discussion

We used a C++ program to simulate the behavior of the design and to measure the coding gain (decoding performance) of the decoder. After designing the circuit, the data flow was simulated using Verilog. The Martin method was used to synthesize the decoder to switch level description [10]. The description was converted to CMOS circuit implementation using a 0.35μm technology. Each block was simulated independently using SPICE to obtain its timing behavior and power consumption. Finally, the entire circuit was simulated in SPICE.

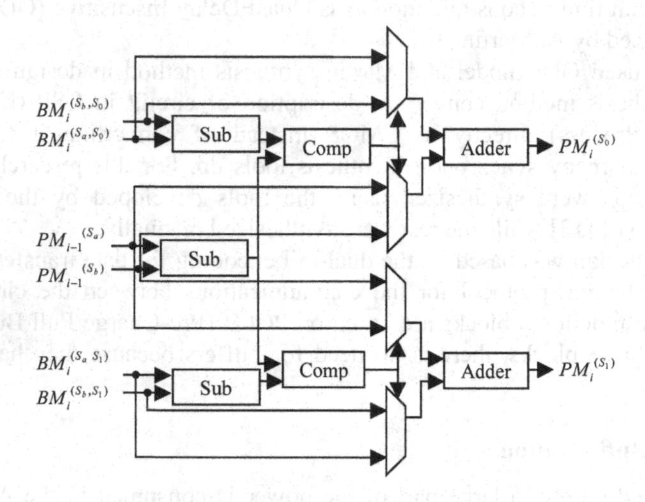

Fig. 8. Re-arranged butterfly ACS [5].

Table 1 shows the estimated power consumption for the circuit by adding all the blocks power and the power consumption from simulation. The results show that the power consumption is half of the worst-case estimation. So, at the average half of the circuit blocks are inactive during normal functionality of the circuit.

Table 1. SPICE simulation results

Simulation	Power (mW)	
	V_{dd}=3.3V	V_{dd}=2.5V
Worst-Case Estimation (Block Simulation)	0.285	0.175
Total Simulation	0.166	0.085

As the table 1 shows about half of the power is consumed in ACS units. Optimizing the power of ACS units can reduce the total power consumption of the circuit. The reduction in the power consumption achieved for the rearranging the ACS structure in the proposed model compared to the same change in the conventional butterfly ACS is given in Table 2.

The simulation results show that the asynchronous design has more potential for low-power applications. This is due to the in the asynchronous implementation, the blocks are only active when needed and so the activity factor is lower for asynchronous circuit.
Also, in [4] by adding a glitch reduction logic, a reduction of 7% in the power dissipation was obtained while the latency was not changed at a cost of 3% increase in the area compared to the ACS unit introduced in Fig. 8. Since the QDI timing model is hazard-free model [10], there is no need to the glitch reduction logic circuit.

Table 2. Re-arranged butterfly ACS architecture results

Architecture	Technology	V_{dd}	Optimization
Synchronous butterfly ACS [5]	0.8 μm	5 V	30.2%
Asynchronous butterfly ACS [This work]	0.35 μm	2.5V	42%

Table 3 shows a comparison of the Viterbi decoders of this work and some other low-power Viterbi decoders reported in the literature. The power consumption in the case of the optimized decoder with a power supply of 2.5V is lowest among the decoders of Table 3. This circuit could operate up to 50 MS/s.

Table 3. Comparison of Viterbi decoder designs.

Design	Technology	V_{dd} (V)	Power (mW)
Synchronous Reference [9]	0.35μm	n/a	203
Systolic Array [7]	0.5 μm	3.3	380
SPL [8]	0.35 μm	2.5	88
Self-Timed [9]	0.35 μm	n/a	9.2
This work	0.35 μm	3.3	166
This work	0.35 μm	2.5	85
This work-Optimized ACS	0.35 μm	3.3	109
This work-Optimized ACS	0.35 μm	2.5	62

5 Summary and Conclusion

The design and implementation of a low-power asynchronous Viterbi decoder was presented. The design was based upon Quasi Delay Insensitive (QDI) timing model which can be used for robust circuits. It was shown the ACS unit consumes about half of the power of the decoder and, hence, to reduce the power, an optimization technique was applied to this unit. The simulation results of the optimized asynchronous decoder showed a 20% reduction in the power consumption compared to the synchronous design in 0.35μm CMOS technology with a power supply of 2.5V. Therefore, the asynchronous Viterbi decoder could be a good candidate for low-power applications.

Acknowledgement. The authors wish to thank Dr. Babak Sadeghian for his useful remarks and discussions and Dr. Mehdi Sedighi for his helps about SPICE simulations and power estimation methods.

References

1. J. Viterbi. Error bounds for convolutional codes and an asymptotically optimum decoding algorithm, in *IEEE Trans. Information Theory*, vol. 13, pp. 260–269, April 1967.
2. G. Davis and Jr. Forney. The Viterbi algorithm, *Proceedings of IEEE*, vol. 61, no. 3, pp. 268–278, Mar. 1973.
3. Xun Liu and M. C. Papaefthymiou. Design of a High-Throughput Low-Power IS95 Viterbi Decoder, in *Proc. of Design Automation Conf. (DAC)*, pp. 263–268, June 2002.
4. J.H. Ryu, S.C. Kim, J.D. Cho, and H.W. Park, and Y. H. Chang. A New Lower Power Viterbi Decoder Architecture With Glitch Reduction, in *Proc. of the First IEEE Asia-Pacific Conf. on ASICs,* pp. 83–86, Aug. 1999.
5. Chi-Ying Tsui, R.S.K. Cheng, and C. Ling. Low power ACS unit design for the Viterbi decoder, in *Proc. of the IEEE Symp. on Circuits and Systems*, pp. 137–140, 1999.
6. K. Hu, M.D. Caldwell. A Viterbi Decoder Memory Management System Using Forward Traceback and All- Path Traceback, in *Proc. of Int. Conf. on Consumer Electronics*, pp. 68–69, 1999.
7. D. Ahmadian, M.N. Azarmanesh, and Kh. Hadidi. VLSI Design of a Viterbi Decoder using Systolic Array for Trace-Back operation in 0.5u µm CMOS, in *Proc. of 10th Iranian Conf. on Elec. Eng.*, Tabriz, Iran, pp. 524–532, May 2002.
8. I. Bogdan, M. Mumunteanu, P.A. Ivey, N.L. Seed, and N. Powell. Power Reduction Techniques for a Viterbi Decoder Implementation, *ESPLD 2000 (European Low Power Initiative for Electronic System Design) Third International Workshop*, Rapallo, Italy, ISBN 90-5326-036-6, pp 28–48, July 2000
9. P.A. Riocereux, E.M. Brackenbury, M. Cumpstey, and S.B. Fruber. A Low-Power Self-Timed Viterbi Decoder, in *Proc. 7th International Symp. on Asynchronous Circuits and Systems,* pp. 15–24, 2001.
10. A. J. Martin. Programming in VLSI: From Communicating Processes to Delay-Insensitive Circuits, in *UT Year of Programming Institute on Concurrent Programming,* C. A. R. Hoare, Ed. MA: Addison-Wesley, pp. 1-64, 1989
11. S. Ranpara and Ha Dong Sam. A low-power Viterbi decoder design for wireless communications applications, in *Proc. 12th Ann. IEEE Int. AISC/SOC Conf.,* pp. 377-381, Sep. 1999.
12. Scott Hauck. Asynchronous Design Methodologies, *Proceedings of the IEEE*, vol. 83, no. 1, pp. 69-93, Jan. 1995.
13. http://ce.aut.ac.ir/async.
14. A.J. Martin. Asynchronous Data paths and the Design of an Asynchronous Adder, *Formal Methods in System Design*, vol. 1, no. 1, pp. 119-137, July 1992.
15. I. Kang and A. N. Willson. A 0.24mW, 14.4 kbps, r=1/2, k=9 Viterbi Decoder, in *the Proc. of IEEE CICC*, pp. 603-606, 1997.
16. I. E. Sutherland. Micropipelines, *Communications of the ACM*, vol. 32, no. 6, pp. 720-738, Jun. 1989.

Analysis of the Contribution of Interconnect Effects in Energy Dissipation of VLSI Circuits

Eugeni Isern[1], Miquel Roca[1], and Francesc Moll[2]

[1] Electronic Technology Group, Physics Department, Universitat Illes Balears
Cra. Valldemossa km.7.5. 07071-Palma Mallorca, Illes Balears. Spain.
eugeni.isern@uib.es, miquel.roca@uib.es,
http://www.uib.es//depart/dfs/GTE

[2] Electronic Engineering Department, Universitat Politècnica de Catalunya,
C/ Gran Capità s/n. Barcelona, Spain.
moll@eel.upc.es
http://pmos.upc.es/blues

Abstract. The study of interconnects in VLSI circuit is becoming a crucial issue. A model for energy calculation is developed. An study of the different contributions to energy dissipation through electrical simulation for several interconnect structures and different driver strengths is also performed. The main goal is to compare the different contributions related to the aspects mentioned above. The importance of the interconnect contribution in the energy dissipation of the chip is analyzed in a 0.18μm CMOS technology showing its relevance even for not very long lines.

1 Introduction

Technological trends in microelectronic circuits show an increase in the density of transistors per chip, in the number of metallization layers, in the clock frequency, and in the power dissipation per chip, while the level of integration is also increasing. That is, the minimum channel length in a MOS transistor is decreasing for each generation (a 70nm length could be achieved in today technologies). These aspects are translated in an increase in the subthreshold leakage current, operation speed, size and complexity of VLSI circuits as well.
This trends causes that interconnect lines are closer together, and an important capacitance between lines appears [1]. In the usual analysis of switching power consumption this capacitance is ignored, or at least, included implicitly in a total capacitance to ground [2]. However, the effect of this coupling capacitance is very different from the capacitance to ground because it depends on the switching activity of the drivers. The effect of the coupling capacitance, besides signal integrity problems [3], [4], as glitches [5] or crosstalk delay [6], is an increasing in dynamic power consumption. This aspect is not considered in usual dynamic power consumption estimation tools and must be very important as is mentioned in [7], where it is estimated that

J.J. Chico and E. Macii (Eds.): PATMOS 2003, LNCS 2799, pp. 481–490, 2003.
© Springer-Verlag Berlin Heidelberg 2003

simultaneous transitions to opposite values of two bus lines in a 0.25 μm technology imply four times more energy dissipation than in the case of same sign transitions. Evidently this increase in energy dissipation is due to crosstalk problems.

The increase in power dissipation due to interconnect structure presents two different contributions, the first one is the spurious signal produced because of the coupling capacitance which causes an extra dissipation in the coupled lines driver resistance. The second contribution is due to the propagation through logic of this spurious signal, in the following nodes of the coupled signal.

In this paper the first contribution will be quantified, showing how much more power is consumed by the presence of capacitance between lines. This aspect was mentioned in [8]. Several encoding schemes have recently appeared trying to minimize the total number of transitions and to avoid opposite transitions in neighboring lines [7,9,10]. These papers base their results on the energy drawn from the power supply, considering a resistive model for the driver. In this work an analysis of the energy consumption considering transistor devices including formally shortcircuit, subthreshold and input-output capacitance current contribution is developed.

A set of different line structures is analyzed through electromagnetic and electrical simulation in order to obtain the lines parasitic parameters and also through electrical simulation to calculate the energy dissipated. An evaluation of the different contributions to the total energy dissipated will be done, allowing to introduce some rules depending on the structure, very useful in energy estimation tools to optimize the algorithms used in these tools.

The paper is organized as follows. Section 2 presents the model for a set of two coupled lines with CMOS inverters as drivers. In section 3 the simulation procedure is introduced and a description of the different cases analyzed is detailed. Section 4 summarizes the results obtained in the simulation analysis, while the main conclusions of the work are commented in section 5.

2 Model for Energy Calculation of Two Coupled Lines

In order to calculate the dynamic current consumption of two coupled lines, let us consider the circuit depicted in Fig. 1. Two interconnect lines are conducted respectively by two CMOS inverters. Each line is described as a capacitance to ground modeling the ground parasitic capacitance and a coupling capacitance between them, modeling capacitive crosstalk.

For the electrical structure of Fig. 1, the relation between the different currents is:

$$i_{p1} - i_{n1} = (C_1 + C_{12} + C_{C1})\frac{dV_1}{dt} - C_{C1}\frac{dV_{i1}}{dt} - C_{12}\frac{dV_2}{dt} \tag{1}$$

$$i_{p2} - i_{n2} = (C_2 + C_{12} + C_{C2})\frac{dV_2}{dt} - C_{C2}\frac{dV_{i2}}{dt} - C_{12}\frac{dV_1}{dt} \tag{2}$$

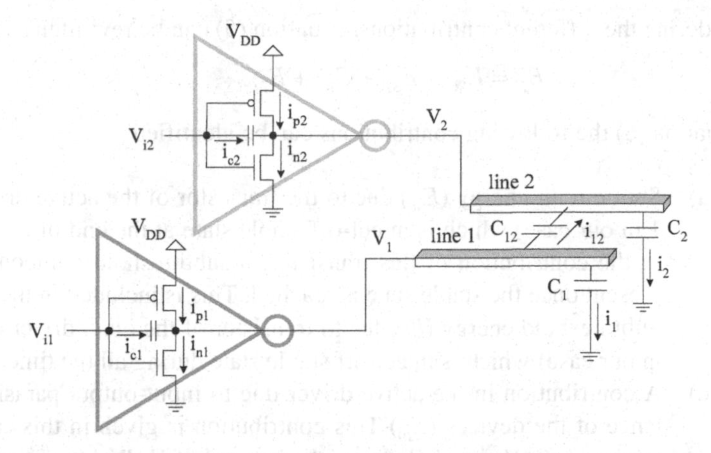

Fig. 1. Electrical circuit modeling the structure of two coupled lines with inverter drivers.

The energy dissipated in the transistors during a certain time interval can be obtained by integrating the power, that is voltage drop in each transistor times the current:

$$E_T = V_{DD}\int i_{p1}dt + V_{DD}\int i_{p2}dt - \int V_1\left(i_{p1} - i_{n1}\right)dt - \int V_2\left(i_{p2} - i_{n2}\right)dt \quad (3)$$

Let us first analyze the case of a positive transition in V_{i1}, while the driver input of the other node (V_{i2}) is kept at a constant value, V_{DD} in this case. Once the steady state is reached, the final output of line 2 is the same as in the beginning, hence $\Delta V_2 = 0$. In this case V_{i2} is constant and therefore its derivative and increment are null. As the driver 1 is making a positive transition, the current through transistor P_1 is mainly short-circuit contribution. In the same way, as the driver 2 have a constant value V_{DD} at its input, the current through transistor P_2 is only subthreshold contribution. The remaining terms can be calculated as voltage differences using equations (1) and (2). Therefore, the total dissipated energy can be written as follows:

$$\begin{aligned} E_T = {} & V_{DD}\int i_{p1,SC}dt + V_{DD}\int i_{p2,ST}dt - \left(C_1 + C_{12} + C_{C1}\right)\frac{1}{2}\Delta\left(V_1^2\right) + \\ & + C_{C1}\int V_1 \frac{dV_{i1}}{dt}dt - \left(C_2 + C_{12} + C_{C2}\right)\frac{1}{2}\Delta\left(V_2^2\right) + C_{12}\Delta\left(V_1 V_2\right) \end{aligned} \quad (4)$$

Taking a sufficiently long integration period, the output voltages will reach stable values, and the resulting voltage differences will be $\Delta V_1 = -V_{DD}$, $\Delta(V_{12}) = -V_{DD}^2$, $\Delta(V_1V_2)=0$ and $\Delta V_2 = \Delta(V_2^2)=0$. Inserting these values in equation (4) leads to the following expression:

$$\begin{aligned} E_T = {} & V_{DD}\int i_{p1,SC}dt + V_{DD}\int i_{p2,ST}dt + C_{C1}\int V_1 \frac{dV_{i1}}{dt}dt + \\ & + \frac{1}{2}\left(C_1 + C_{12} + C_{C1}\right)V_{DD}^2 \end{aligned} \quad (5)$$

Considering the different contributions, equation (5) can be rewritten as follows:

$$E_T = E_{SC} + E_{ST} + E_{IO} + E_{LINES} \tag{6}$$

In equation (6) the following contributions can be identified:

a) Shortcircuit energy (E_{SC}) due to the transistor of the active driver (driver 1 in our case) which is in cut-off stable state at the end of the transition. In the contribution of this transistor, a subthreshold component is also present once the stable state is reached. This is included in the same i_{p1}.
b) Subthreshold energy (E_{ST})due to transistor of the quiet driver (driver two in our case) which is in cut-off stable state during all the time.
c) A contribution in the active driver due to input-output parasitic capacitance of the devices (E_{IO}) This contribution is given in this case by two terms, $1/2C_{CI}V_{DD}^2$ and C_{CI} times the integral of $V_1 dV_{i1}/dt$.
d) A contribution due to interconnects (E_{LINES}), in this case this contribution is equal to $1/2(C_1+C_{12})V_{DD}^2$.

Until now, the case of a positive transition at the input of driver 1, while input of driver 2 is at V_{DD} have been considered. Evidently other cases must be considered, as positive transition in V_{i1} while V_{i2} is at *GND*, negative transition in V_{i1} while V_{i2} is at *GND* and at V_{DD}. The methodology to analyze this consists of calculate the energy from equations (1) and (2), leaving explicitly in the integral the current of the transistor which is in subthreshold region (cut-off of quiet driver) and the current of the transistor which contribute with a shorcircuit current (transistor of the active driver which is in cut-off state at the final stable state). The first transistor will depend on the constant voltage applied at the input of the quiet driver, while the second transistor will depend on the signal transition sign at the input of the active driver. With this consideration, and taking into account the value of voltage differences in the nodes of the circuit, equation (6) give us the total dissipated energy during the transition. In this equation the different contributions are:

a) E_{ST} and E_{SC} corresponding to the cut-off transistor in the quiet driver (subthreshold current) and the cut-off transistor of the active driver at its final state (shortcircuit current).
b) $E_{io} = \frac{3}{2} C_{C1} V_{DD}^2 + C_{C1} \int V_1 \frac{dV_{i1}}{dt} dt$ for positive transitions and

   $E_{io} = \frac{1}{2} C_{C1} V_{DD}^2 + C_{C1} \int V_1 \frac{dV_{i1}}{dt} dt$ for negative transitions.
c) E_{LINES} is given by $1/2(C_1+C_{12})V_{DD}^2$ in all cases.

Contribution a) and b) depends on the semiconductor devices while the c) contribution depends on the interconnect structure. Evidently, the same results would be obtained if the quiet driver was driver 1 and the active driver was driver 2, only changing line and device parameters.

A different problem is done if there are simultaneous transitions in both lines. In that case a more complicated mathematical treatment, and two new parameters play an important role, the sign of the transitions (opposite or equal sign) and specially the relative delay between transitions. This problem will not be analyzed in this work.

3 Simulation Analysis

In this section the simulation procedure used to analyze the contribution of the coupling on the total power consumption is presented. Simulations were conducted on the same circuit used to derive the analytical expressions in the previous section (Fig. 1). A 0.18 μm CMOS technology with one poly and 6 metallization levels is considered. HSPICE simulations have been done using a level 49 model for the MOSFET devices, and a supply voltage of V_{DD} = 1.8 V. The energy dissipation values have been estimated by numerically integrating the simulated transient current through the transistors and multiplying by the voltage drop.
To derive the parameters that characterize the behavior of the coupled conductors (C_1, C_2 and C_{12}) an electromagnetic field solver integrated in HSPICE have been used. It must be noted that C_1 and C_2 includes in addition to the substrate coupling capacitance an estimated fanout capacitance of C_g = 5 fF (equivalent to 6 minimum size gates). From the description of the 2-dimensional geometry the field solver provides the capacitance matrix of the coupled structure, which can be used as input to run an electrical simulation. This approach provides important advantages in front of using other field solver tools, as for example RAPHAEL [11], since the results can be directly introduced in the electrical simulator, making possible the consideration of accurate semiconductor device models and advanced transmission line models. Fig. 2 shows an schema of the simulation procedure.
Four different structures have been considered in order to analyze the effect of different line geometries in the power dissipation contribution of coupling (Fig. 3).

For the coupling configuration labeled structA, different cases have been considered regarding to the size of the driver transistors:

- Equal and unbalanced drivers (equally sized transistors L = 0.18μm, W = 1μm).
- Equal and balanced drivers (L= 0.18μm, W_n = 1μm, $W_p = 2.5W_n$).
- Different and balanced drivers ((L= 0.18μm, W_{n1} = 1μm, $W_{p1} = 2.5W_{n1}$, W_{n2} = 5 W_{n1}, W_{p2} = 5 W_{p1}).

For the other coupling structures, only the case of equal drivers with equally sized transistors is reported.

For all the considered cases, three energy values have been obtained:

- Total energy including crosstalk effects ($C_{12} \neq 0$), by HSPICE simulation.
- Total energy excluding crosstalk effects ($C_{12} = 0$), by HSPICE simulation.
- Line contribution to the total energy, (E_{LINES} in Eq.(6)), analytically calculated.

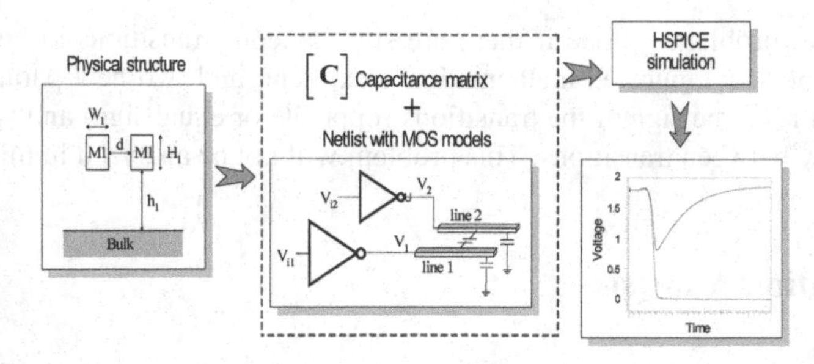

Fig. 2. Simulation procedure.

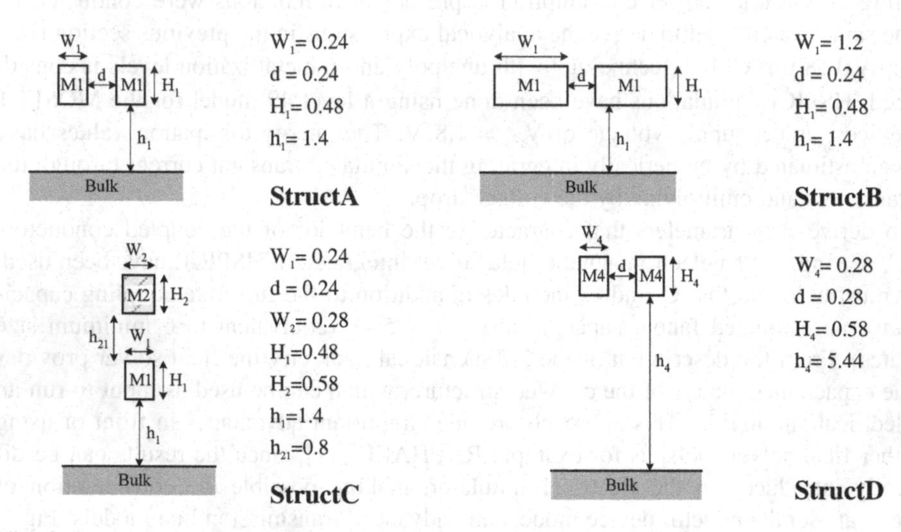

Fig. 3. Structures used to analyze the influence of the coupling geometry in the energy consumption (dimensions in μm).

4 Results

In this section we will present the results obtained in our analysis. We will focus in the impact of two different contributions to total energy dissipation. First the comparison between device and lines contributions. This result is obtained by comparing HSPICE estimation of the total energy with the value computed through the contribution E_{LINES} obtained from Eq.(6). The difference between these two values corresponds to the device contribution. Secondly, to calculate the contribution of crosstalk coupling in the total energy dissipation we compare the same HSPICE simulation with a new simulation performed with $C_{12} = 0$. The difference between these two computed energies corresponds to the crosstalk contribution to the total power dissipation.

Let us first focus on geometry structA of Fig. 3, that is, two equal lines of minimum width and minimum separation in the first level of metallization. The results obtained are depicted in Fig. 4. for different line lengths. It can be seen as the device contribution represents 50% of the total energy dissipation for a very short line (1 μm), and decreases rapidly when line length increases. For example, for a line of 250μm length the contribution of devices is only around 10% of total energy consumption. This result shows that in energy dissipation analysis of VLSI circuits, the effect of lines never can be neglected, while the effect of devices could be neglected for lines longer than a certain value. This result can be used in energy dissipation estimation tools to obtain accurate results and improve time analysis.

In Fig 4. the contribution of crosstalk is also analyzed. The importance of this contribution for long lines is highlighted in this figure. It can be seen as for 100μm length lines crosstalk contribution is around 40% of total energy dissipation. Only for short lines this contribution can be neglected. This implies that energy estimation tools can not disregard crosstalk contribution (coupling parasitic capacitance) because it represents a very important percentage of total energy dissipation.

Absolute values of total energy consumption are provided in Fig 5. for both the cases were crosstalk are considered or neglected ($C_{12} = 0$).

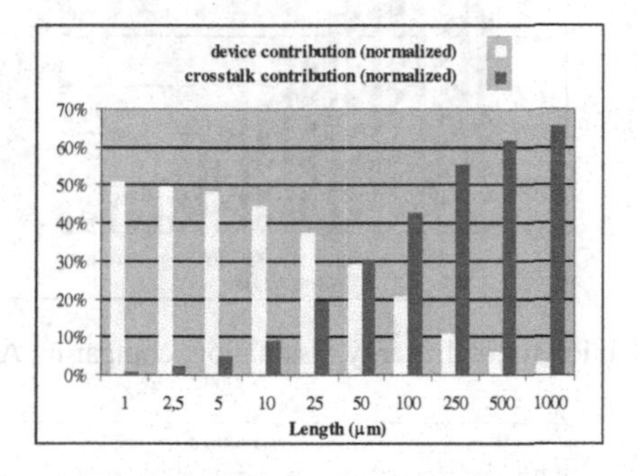

Fig. 4. Device and crosstalk contributions to total energy consumption for structure A.

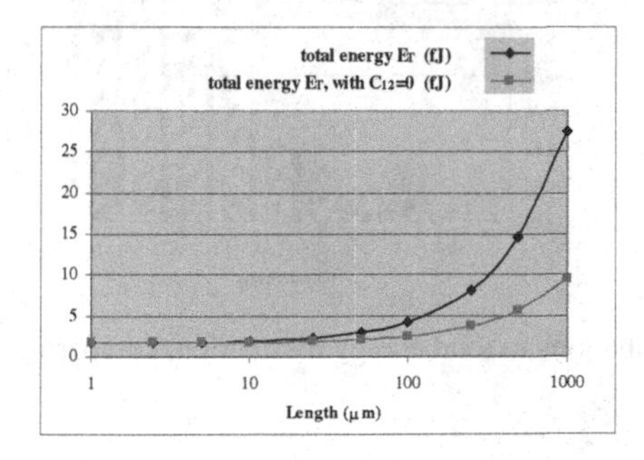

Fig. 5. Total energy consumption for structure A, considering and neglecting crosstalk.

Fig. 6. shows a comparison of device contribution to energy dissipation for the four structures of Fig. 3 (A, B, C and D). Same trends in all structures can be observed. The device contribution is around 50% (equal contribution of devices and lines) for short lines (1μm length), and decreases for longer lines. For 100μm length lines the device contribution is around 20% of total energy dissipation, except for structure C, where this contribution is near 30%. It must be noted that for structure C, the device contribution is more important than for the other structures, due to the screening effect of metal 1 to the upper metal layer, resulting in a small value of metal 2 ground capacitance, thus reducing the total capacitance involved in this case.

Fig. 7. shows the importance of crosstalk contribution to total energy dissipation for the analyzed structures. Same trends in all the structures are obtained. For short lines the effect of crosstalk is negligible, while for long ones it is very important. For lines up to 100μm the crosstalk contribution represents more than 40% of total energy dissipation, except for structure C. Structure C presents less crosstalk importance (20% of total energy dissipation), due to the above mentioned reasons.

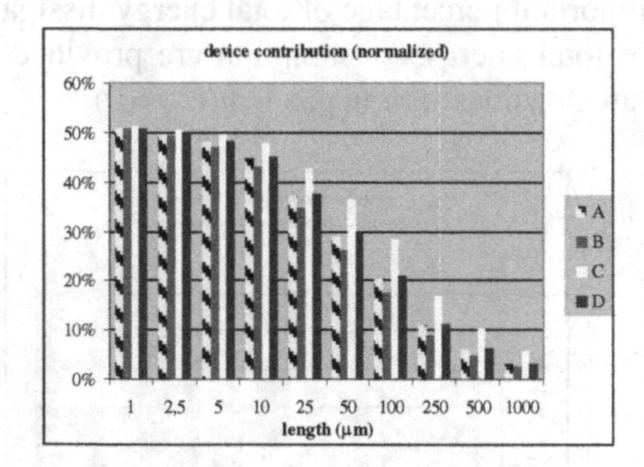

Fig. 6. Device contribution to total energy dissipation, comparing A, B, C and D structures.

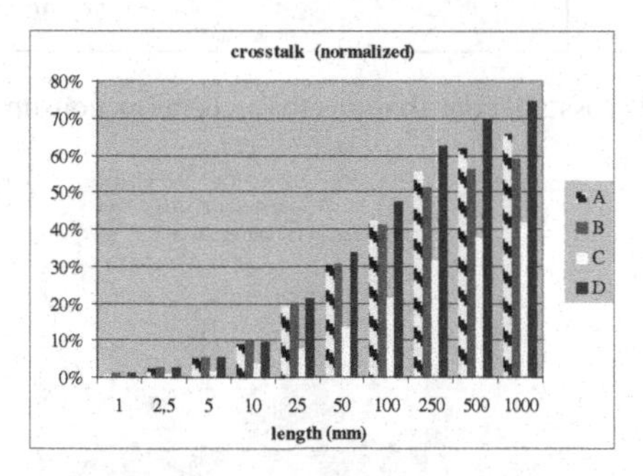

Fig. 7. Crosstalk contributions to total energy dissipation, comparing A, B, C and D structures.

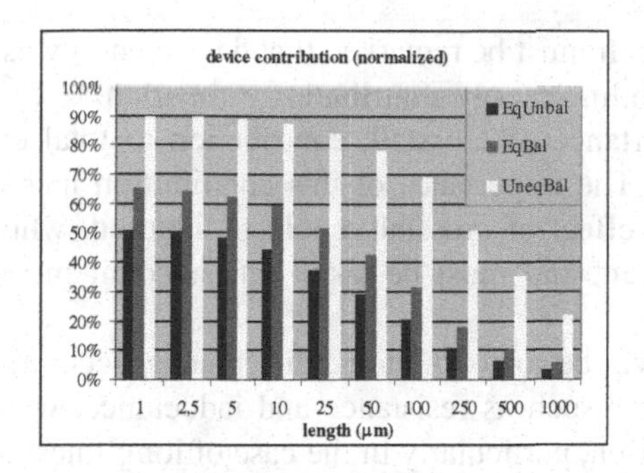

Fig. 8. Driver dependence in energy dissipation of lines.

In Fig 7. it is also seen that crosstalk effect is more important in structures A and D. This fact correlates with the well known result that crosstalk noise increases when the ratio between coupling and ground capacitance increases. This is true also for extra energy dissipation. Comparing structures A and B, coupling capacitance is almost the same, while ground capacitance is bigger in B structure due to the wider lines considered, therefore the ratio between coupling and ground capacitance decreases. The same happens comparing with D structure, but in this case it is not so intuitive due to the different values of width, high and distance between lines.

Fig 8. analyzes the effect of driver sizes. It is observed that when length lines increases, evidently the relative device contribution to total energy consumption decreases. Considering the different driver sizes, it can be seen as the importance of device contribution is more relevant when the size of drivers increases. Comparing cases unequal balanced and equal balanced (the first have transistors with channel width five times channel width of second case), the result is that in unequal balanced case the importance of device contribution is more important than for the equal balanced case (for l=100μm, the values change from 70% to 30% respectively).

5 Conclusions

VLSI circuits are more complex with a very high density of devices and interconnects. This fact and the increase in clock frequencies implies an increase in energy dissipation, therefore accurate energy consumption estimation of VLSI circuits is mandatory in today's technologies. A very simple model to estimate energy dissipation in VLSI circuits is presented. The model distinguish different contributions to the total energy dissipation, which can be included in two sets, device and interconnect contribution.

An analysis of the importance of different contributions to the total energy dissipation for different structures and driver sizes have been performed through circuit simulation. The line contribution to the total energy dissipation is very important, the parasitic parameters of the lines never can be neglected in energy estimation tools, if accuracy is the goal. For long lines the contribution of devices to the energy dissipation is much smaller. It must be remarked that device energy estimation is much more time consuming than line energy contribution estimation.

Finally the importance of crosstalk contribution to total energy dissipation have been demonstrated. The importance of this contribution increases with length lines. For short lines, the effect of crosstalk could be ignored, while for long lines it becomes very significant and must be taken into account in energy estimation algorithms.

Future research will be focused on two different aspects: (a) analyzing the effect of other line parameters, such as resistance and inductance, which could contribute to the power consumption, particularly in the case of long lines; and (b) handling multiple transitions to evaluate how the relative delay influences the energy dissipation.

Acknowledgement. This work has been partially supported by the Spanish Ministry of Science and Technology through projects TIC2001/2337 and TIC2002/1238.

References

1. Werner C., Göttsche R., Wörner A. and Ramacher U. "Crosstalk Noise in Future Digital CMOS Circuits". IEEE Design Automation Test Europe DATE 2001.
2. Chandrakasan A. P. and Brodersen R. W. "Low Power Digital CMOS Design". Kluwer Academic Publishers, 1995.
3. Caignet F., Delmas S. and Sicard E. "The Challenge of Signal Integrity in Deep-Submicrometer CMOS Technology". Proc. of the IEEE, Vol. 89, No 4, 556–573, 2001
4. Choi S. H., Paul B. C. and Roy K. "Dynamic Noise Analysis with Capacitive and Inductive Coupling". Proceedings of the 15th International Conference on VLSI Design, 2002.
5. Roca M., Moll F. and Rubio A. "Electric design rules for avoiding crosstalk in microelectronic circuits". PATMOS 94, 76–83, October 1994.
6. Moll F., Roca M., Rubio E. And Sicard E. "Analysis and measurement of crosstalk induced delay errors in integrated circuits", Chapter 12 of Signal Propagation of Interconnects, edited by Grabinsky H. and Nordholz P. 139–147, Kluwer Academic Publishers, 1998.
7. Sotiradis P. P. and Chandrasakan A. "Bus Energy Minimization by Transition Pattern Coding (TPC) in Deep Submicron Technologies", ICCAD, November 2000.
8. Kim K. W., Jung S. O., Narayanan U., Liu C. L. and Kang S. M. "Noise-Aware Power Optimization for On-Chip Interconnect", in Proc ISLPED 00, 2000.
9. Kim K. W., Baek K. H., Shanbhag N., Liu C. L. and Kang S. M. "Coupling-driven signal encoding scheme for low-power interface design", ICCAD, November 2000.
10. Machiarulo L., Macii E. and Poncino M. "Wire placement for crosstalk energy minimization in address buses". Design Automation and Test in Europe DATE 2002.
11. *Raphael Reference Manual*, Avant! Corporation, December 1998.

A New Hybrid CBL-CMOS Cell for Optimum Noise/Power Application

Raúl Jiménez[1], Pilar Parra[2], Pedro Sanmartín[2], and Antonio J. Acosta[2]

[1] Dpto. DIESIA, Universidad de Huelva, Cra. Huelva-La Rábida, s/n 21071-Huelva, Spain
[2] Instituto de Microelectrónica de Sevilla/Universidad de Sevilla, Avda. Reina Mercedes s/n, 41012-Sevilla, Spain
{naharro,parra,sanmart,acojim}@imse.cnm.es

Abstract. The design of a new configurable hybrid current-mode/static CBL-CMOS cell is presented. This cell can be used in order to obtain the optimum partitioning between conventional and low-noise logic in the digital part of a mixed-signal circuit, resulting in a optimum power/noise solution. This new cell has been compared with the original logic families obtaining acceptable results with low hardware cost. A combinational multiplier has been designed as a demonstrator example of the utility of the proposed cells.

1 Introduction

The advances in integration technology allows the implementation of complex mixed signal circuits in a same wafer. So, the influence of switching noise becomes as one of the main problems in this kind of circuits. Traditionally, this problem has been treated using layout techniques [1,2] oriented to avoid the influence of switching noise instead to reduce the noise generation.

Recently, some efforts from the digital view point are done in low noise applications. Among these approaches, we can find some references about the influence of timing schemes [2,3], techniques to reduce the noise generated by an specific logic family [4,5], new logic families oriented to low noise applications [6-8], etc. Generally, a good solution to this problem will consist to combine several digital solutions in order to reduce the switching noise.

The usage of low-noise current-mode logic families has as main drawback the static power consumption. This undesired property limits the use of these low-noise families in those zones near to the analog part where low-noise requirements become critical, and the rest of the digital part must be integrated using a more conventional logic family as it is shown in fig. 1. With this solution, a trade-off between power consumption and switching noise can be found. However, this kind of solution implies as further new question where should the conventional logic and the current-mode family be placed in the digital part.

This work has been sponsored by the Spanish MCYT TIC2000-1350 MODEL and TIC2001-2283 VERDI Projects

J.J. Chico and E. Macii (Eds.): PATMOS 2003, LNCS 2799, pp. 491–500, 2003.
© Springer-Verlag Berlin Heidelberg 2003

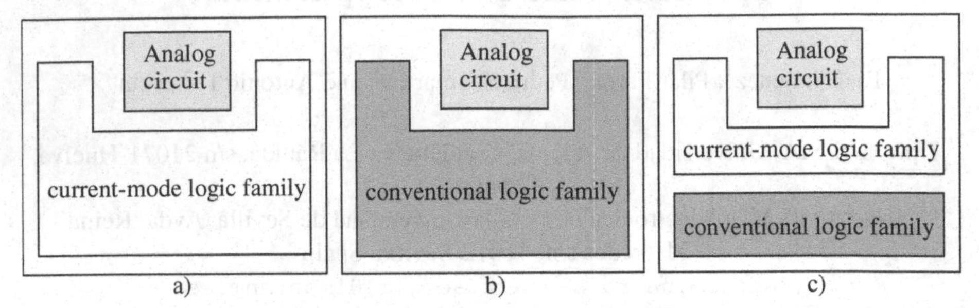

Fig. 1. Scheme of a mixed-signal layout circuit considering a) a power expensive solution, b) a noise expensive solution and c) a trade-off noise-power solution.

To find the optimum partitioning of the digital part into current-mode and conventional logic, several attempts of the digital design must be integrated, resulting in a very expensive solution in terms of design time. The solution proposed in this paper uses a configurable logic family to implement the digital part mainly in those zones near the boundary of the analog part. This cell could be configured both as a conventional CMOS as well as a low-noise current mode family. So, for a unique single design, the determination of the optimum partitioning would consist of varying the configuration of the different cells.

The communication is divided as following. First, the design of this new configurable cell will be considered. After that, electrical simulations will be performed; in these simulations, we will consider a comparison with the original logic families, as well as a 4x4 multiplier as an example application. Finally, we will expose the conclusions obtained with this work.

2 Hybrid CBL-CMOS Cells

The proposed cell must have the low-power characteristic of a conventional logic family and the low-noise feature of a current-mode family when neccesary and in a configurable way. These characteristics forced are the almost zero static power consumption of a conventional family and the approximately constant supply current of a low noise one.

The chosen families incorporated in the hybrid cell are the static CMOS and CBL [7,8] families. A generic CBL cell is shown in fig. 2a. Its operation is based on a NMOS gate, and the stability of supply current is due to it flows through NMOS tree when the tree is ON, or through transistor T when the tree is OFF. In order to obtain a higher stability, the PMOS transistor and NMOS tree must be very well matched with transistor T. This necessity is one of the main design aspects in CBL family. On the other hand, a well known generic static CMOS cell is shown in fig. 2b. Its operation is based on having ON only a tree, either NMOS or PMOS, while the other one is OFF, resulting in a dominant dynamic power solution.

Among the reasons to choose these families, we can find a wide use of static CMOS family, and a very similar logic values of both families. In fact, the high level in both cases is the supply voltage Vdd, and the low level of CBL family can be considered as

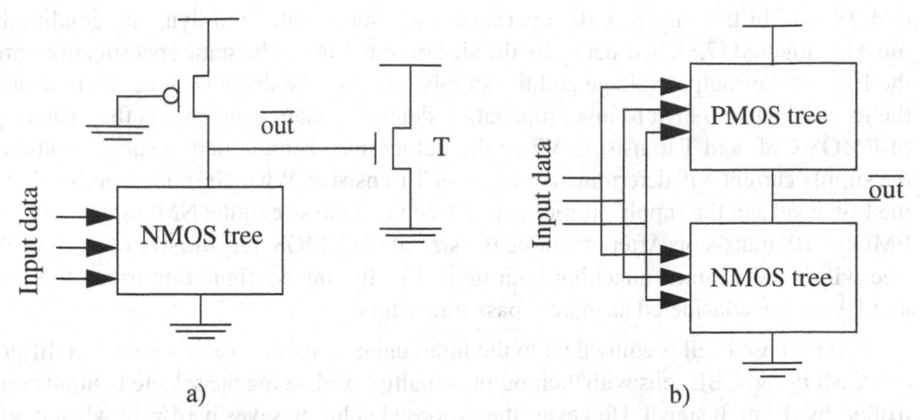

Fig. 2. Scheme of a) a generic CBL cell and b) a static CMOS cell.

a degraded low level of CMOS family. So, no interface between CMOS and CBL cells will be necessary.

The scheme of the proposed CBL-CMOS hybrid cell is shown in fig. 3. In this scheme, the NMOS tree will be shared for both configurations. The operation is determined by the signal m_b, loading the two configuration transistors N1 and P1. When the signal m_b is high, the cell is configured as a static CMOS cell. In this case, the CBL PMOS transistor is OFF, and hence, the high level is supplied to the output by the PMOS tree. Also, the transistor T is OFF due to transistor N1, and hence, there does not exist current flow through it. So the static power consumption is negligible and a low power solution is achieved. When the signal m_b is low, the cell is configured as a CBL cell. In this case, the CBL PMOS transistor is ON, disabling the PMOS tree. Also, the transistor T is connected to the output node in order to permit the flow of supply current when the NMOS tree is OFF. So, the flow of supply current is approximately constant, yielding a low noise solution.

The following step in the design process consists in obtaining the optimum size of different transistors. The operation way with more restrictions is the CBL one, so first we impose the conditions to the correct operation of CBL, and after, the conditions of

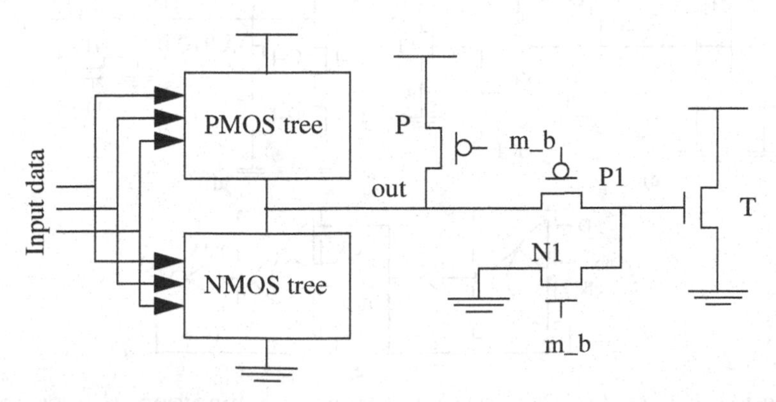

Fig. 3. Scheme of the proposed CBL-CMOS hybrid cell.

CMOS cell. In the case of CBL operation way, some static and dynamic conditions must be imposed [7,8], in order to fix the size of transistors. The static specifications are the low level of output voltage and the supply current. The dynamic specifications are the low to high and high to low propagation delays. A main constraint is the matching of PMOS CBL and T transistor. When the output node is high, tied to supply voltage, the supply current will determinate the size of T transistor. When the output node is low, the low level and the supply current will determinate the size of the NMOS tree and the PMOS CBL transistor. After obtaining the size of the NMOS tree, the size of the PMOS tree will be determined matching both trees. Finally, the configuration transistors N1 and P1 will be considered as merely pass transistors.

The proposed cell is equivalent to the interconnection of two equivalent, but different, CMOS and CBL cells with their outputs multiplexed, being the selected output controlled by the m_b signal. However, the proposed solution saves hardware when comparing to the multiplexed solution, because the NMOS tree is shared by the two cells and the multiplexor is reduced to only two pass transistors.

The schematics of hybrid CBL-CMOS cells used for comparison are shown in fig. 4. The XNOR gate has been implemented using the AND2-NOR2 gate, and the com-

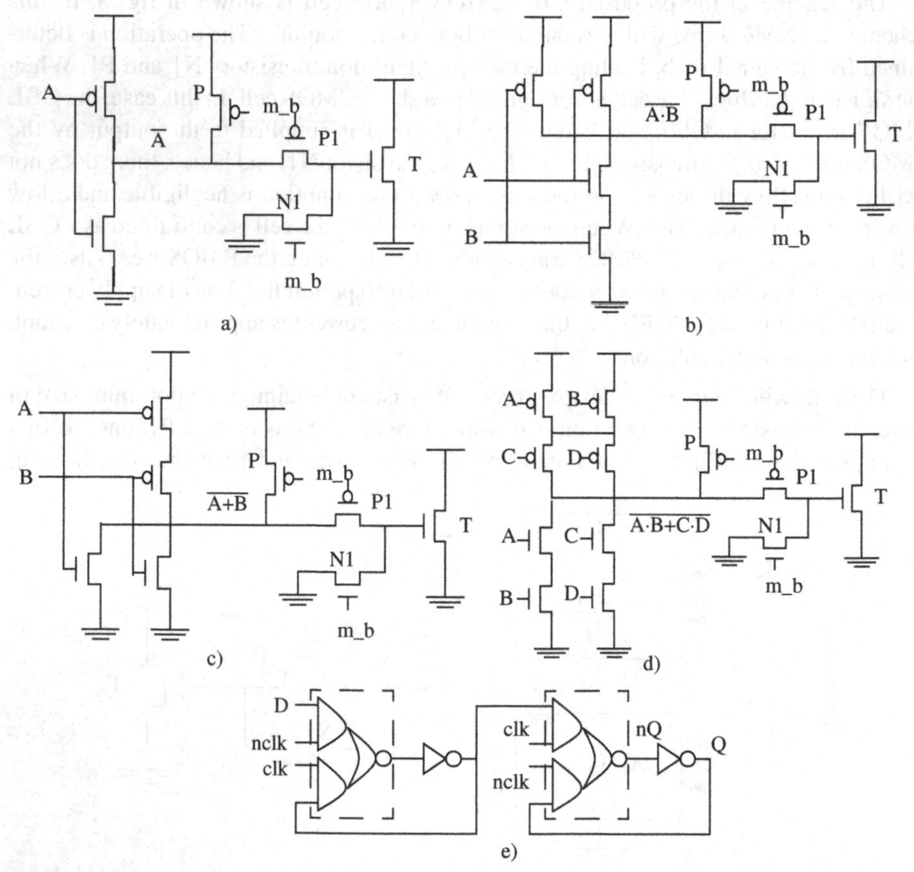

Fig. 4. Schematics of a) inverter, b) NAND2, c) NOR2, d) AND2-NOR2 gates and e) D flip-flop implemented as new hybrid CBL-CMOS cells.

plemented inputs (C=$\overline{A}$ and D=$\overline{B}$) have been taken from input inverters. The chosen structure of D flip-flop is due to use only standard cells, more concretely inverters and AND2-NOR2 gates. Besides, this structure is generated by an automatic synthesis tools of low noise CSL gates [9]. The size of different transistors are: 1.5μm/0.3μm for NMOS tree, 4.5μm/0.3μm for PMOS tree of CMOS cell, 0.7μm/0.3μm for PMOS CBL transistor, 0.6μm/0.75μm for T transistor and 1.5μm/0.3μm and 4.5μm/0.3μm for NMOS and PMOS configuration transistors respectively. The size of transistors of static CMOS and CBL cells is the same than in the corresponding hybrid CBL-CMOS cell. A comparison between the hardware resources of the proposed and reference cells in terms of transistor-count is shown in table 1

Table 1. Hardware resources for the pure cells, the classical multiplexed solution (a CBL and a CMOS cells loading a 2:1 multiplexer) and the proposed hybrid CBL-CMOS cell. Hardware reduction when the hybrid CBL-CMOS and the multiplexed solutions are compared.

	CBL	CMOS	Cells multiplexed	Hybrid CBL-CMOS	Hardware reduction
Inverter	3	2	11	6	46 %
NAND2	4	4	14	8	43 %
NOR2	4	4	14	8	43 %
AND2-NOR2	6	8	20	12	40 %
D-Flip-flop	18	20	44	36	19 %

3 Simulation Results

A negative consequence of merging the CBL and CMOS functionality in the same cell could be a decrease in the fan-out of CBL, when it is directly connected to a cell configured as CMOS. Also, a high influence of load capacity is among the characteristics of CBL cells. To arise these limitations evident and to check the performances of the proposed cell, a comparison among hybrid CBL-CMOS and the original CBL and CMOS cells has been done, in order to obtain the variations of the different parameters. The comparison was done over a basic inverter, NAND, NOR, XNOR gates and a D-type edge-triggered flip-flop, shown in fig. 4. The simulations have been performed within a 0.35 μm technology using Eldo simulator, from Mentor Graphics.

The simulations have been performed under the scheme shown in fig. 5, where one inverter loads all input signals, and the cell loads one inverter. The cell is connected to supply through a inductor of 1nH, modeling the inductance of a chip wiring. The switching noise will be measured both as maximum peak of supply current and as maximum variation of this current (the voltage drop across the inductor). The input patterns considered are such that output signals change in every change of input signals. In fig. 5 the simulation waveforms of the CBL-CMOS inverter for both configurations (first part in CBL and the second part in CMOS) are shown. As important keynotes, it can be

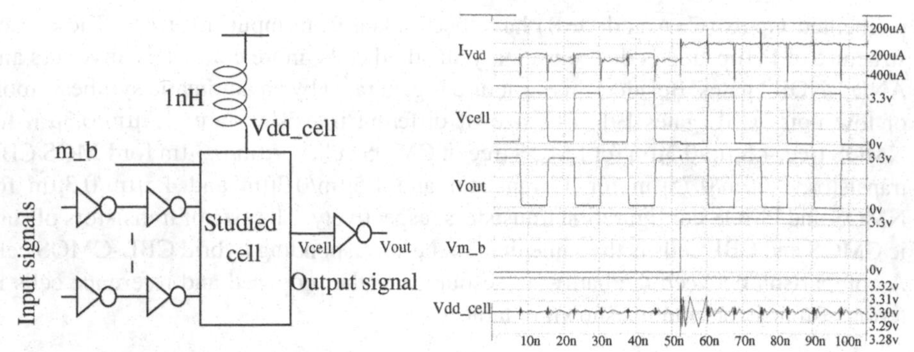

Fig. 5. Simulation environment and waveforms of the hybrid CBL-CMOS inverter cell.

seen the differences in the voltage of the low level: in the case of CBL mode is about 0.16 v. When the NMOS tree is more complex, this value can increase, but the logic function remains correct. Also, we can see the difference of switching noise, both in supply current and in voltage drop across the inductance. The lowest noise is observed in the operation as CBL mode. The highest generation of noise is obtained when switching the configuration, that is, in the change of signal m_b, but this situation is not considered because the operation mode will be normally fixed to one specific level.

In table 2, we show the average propagation delay and power-delay product for input patterns switching every 10 ns. As expected, there exists a higher deviation when considering the CBL mode, due to its higher dependency with output capacitive load.

In the case of propagation delay, the CMOS configuration behaves better than CBL one. In the case of rising transition in outputs, both configurations show similar values. In the CBL mode, when the PMOS tree is ON the equivalent PMOS CBL transistor must be stronger and hence, the low-to-high propagation delay decreases, making lower the low-to-high propagation delay in the proposed hybrid CBL-CMOS cell.

In the case of power-delay product, the influence of capacity load is also significative. In fig. 6, we can see the variation of PDP vs period of input signal for the D flip-flop cell. It is noticed that variation of hybrid cell with period is the same than in the original CBL and CMOS cells. In the case of CBL configuration, the hybrid cell shows worse behavior due to the influence of capacity, generating more power and delay. In CMOS configuration, the deviation is lower indicating lower influence of capacity.

In the case of switching noise, the behavior depends on the complexity of gate. In the case of CMOS configuration and with increased complexity, the noise figure is similar to that of the original cell, because of as for more complex gates, the effect of added capacity is lower. However, an improvement in noise behavior is observed in the hybrid CBL-CMOS cell when configured as CMOS when compared to the original CMOS cell. This situation is originated by the increase in delay of the hybrid cell and hence, the operation is more distributed in time and it generates less noise.

4 A 4x4 Multiplier as Demonstrator Example

To assess the feasibility of the proposed cell from the noise analysis point of view, a combinational 4x4 bit multiplier has been simulated. The chosen structure is shown in

Table 2. Cells comparison for input patterns switching every 10 ns. tp: propagation delay; PDP: power-delay-product; Noise_I: peak in supply current; Noise_V: peak in inductance voltage

Parameter	Inverter					
	Hybrid CBL mode	pure CBL	ratio	Hybrid CMOS mode	pure CMOS	ratio
tp (ns)	0.146	0.083	57%	0.103	0.085	83%
PDP (fJ)	94.2	47.6	51%	3.37	1.56	46%
Noise_I (mA)	0.29	0.12	41%	0.51	0.65	127%
Noise_V (mV)	0.007	0.003	43%	0.027	0.019	70%
Parameter	AND2-NOR2 gate (configured as XNOR2 gate)					
	Hybrid CBL mode	pure CBL	ratio	Hybrid CMOS mode	pure CMOS	ratio
tp (ns)	0.162	0.120	74%	0.147	0.122	83%
PDP (fJ)	104.1	60.8	58%	13.8	10.5	76%
Noise_I (mA)	0.58	0.08	14%	0.74	0.78	105%
Noise_V (mV)	0.017	0.003	18%	0.027	0.021	78%
Parameter	D-type edge-triggered flip-flop					
	Hybrid CBL mode	pure CBL	ratio	Hybrid CMOS mode	pure CMOS	ratio
tp (ns)	0.424	0.237	56%	0.335	0.305	91%
PDP (fJ)	1040	526.1	51%	75.1	61.8	82%
Noise_I (mA)	0.73	0.10	14%	1.12	0.92	82%
Noise_V (mV)	0.013	0.003	23%	0.071	0.013	18%

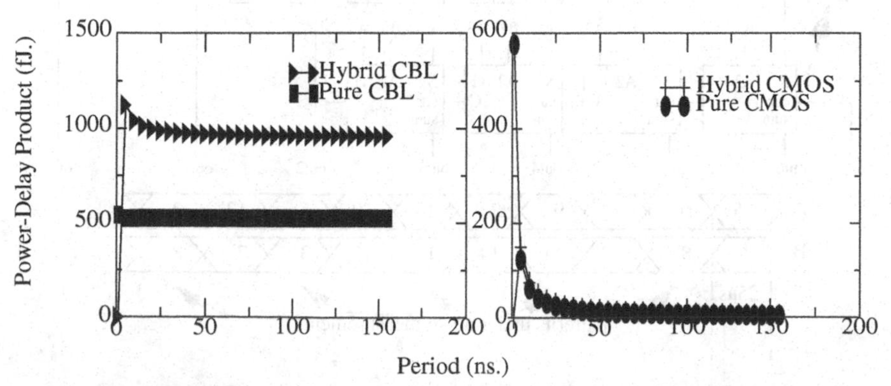

Fig. 6. PDP versus period of input signals in the case of D flip-flop.

fig. 7, and it is a full-adder based array with AND gates and inverters. In order to observe the variation of noise depending on the configuration of each cell, the rows of the multiplier can be separately programed through the configuration signal of input inverters (m_bi), of AND gates (m_ba), of first, second and third rows of full-adders (m_bs1, m_bs2 and m_bs3).

The noise study has been done by observing the data variation obtained for the most and least noisy operations for both configurations. The noisiest operation is for input combination A=0000 and B=1000 for both configurations and both I and V noise figures. The input combination generating less noise are 1001x1011, 0101x0011, 1101x0001 and 0010x1000 (A x B). In order to obtain the worst case operation, the input pattern selected changes all input bits to obtain the pattern shown in fig. 7.

The variation of peak in supply current is about 50% higher in hybrid as CMOS than CBL. In the case of the voltage of inductor, the variation is about 10% greater in CMOS than hybrid cells, and about 90% higher in hybrid than CBL. This increment is due to the excess of hardware in the multiplier.

In fig. 8, we can see the measurements of different configurations, remarking those generating more and less noise and all rows as CMOS and CBL. It is worthy of noticing that, considering switching noise, there exist optimum configurations different than pure CBL configuration of the whole multiplier. This situation leads to an important result from the noise optimization viewpoint, and it is due to the synchronization in operations of different cells. The optimum configurations to both I and V noise measurement are different, depending on the higher influence of the current difference or dI/dt.

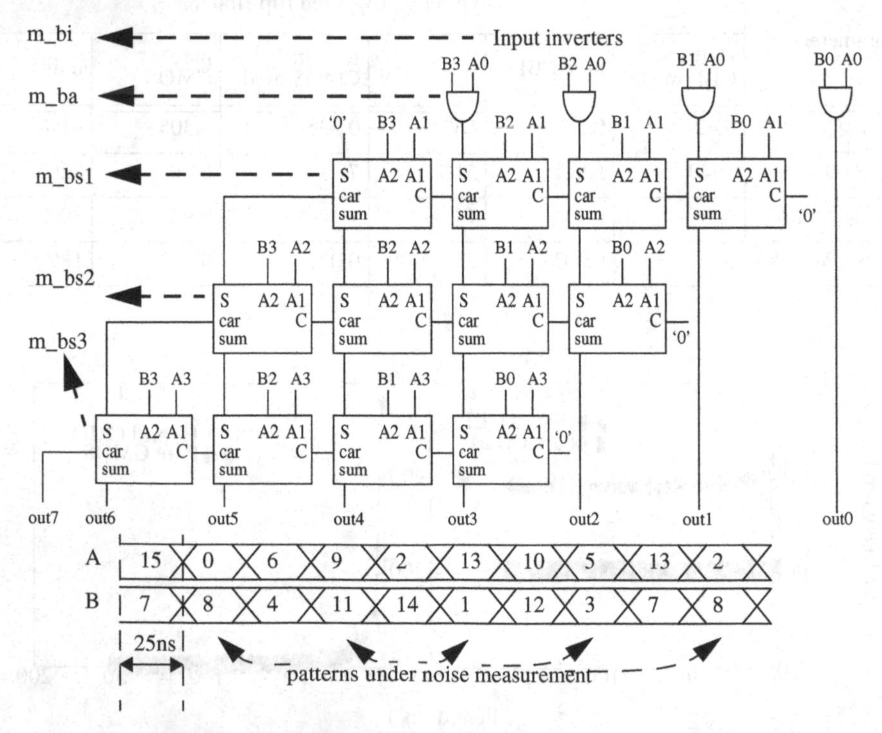

Fig. 7. Scheme and input patterns of multiplier simulated.

In the measurements of peak of current, power consumption and PDP, four well-differenced groups are determined by the configuration rows of full-adders, with better behavior when the configuration as CMOS is dominant. For propagation delay, a zone with lower almost constant delay when most of full-adders are configured as CMOS is observed. In the other zone, there exists a higher variation of delay. The fig. 9 shows some measurements of switching noise and PDP varying the input pattern.

5 Conclusions

In this work, a hybrid CBL-CMOS cell has been designed in order to obtain the optimum configuration of a circuit that generates as less noise as possible. The parameters of these hybrid cells are influenced by the extra capacity of the excess circuitry: two configuration transistors and, the PMOS tree (when it operates as a CBL cell) and the PMOS CBL transistor and the current source (when it operates as a CMOS cell). The variation is more appreciable in CBL configuration because these cells are more sensitive to the load capacity. The proposed cell saves hardware if we compare it with a classical multiplexed solution.

No interface between the two different configurations is necessary because the logic levels of both families are similar. The high level is the same in both configurations and the low level of CBL configuration can be consider as a degraded low level of CMOS configuration.

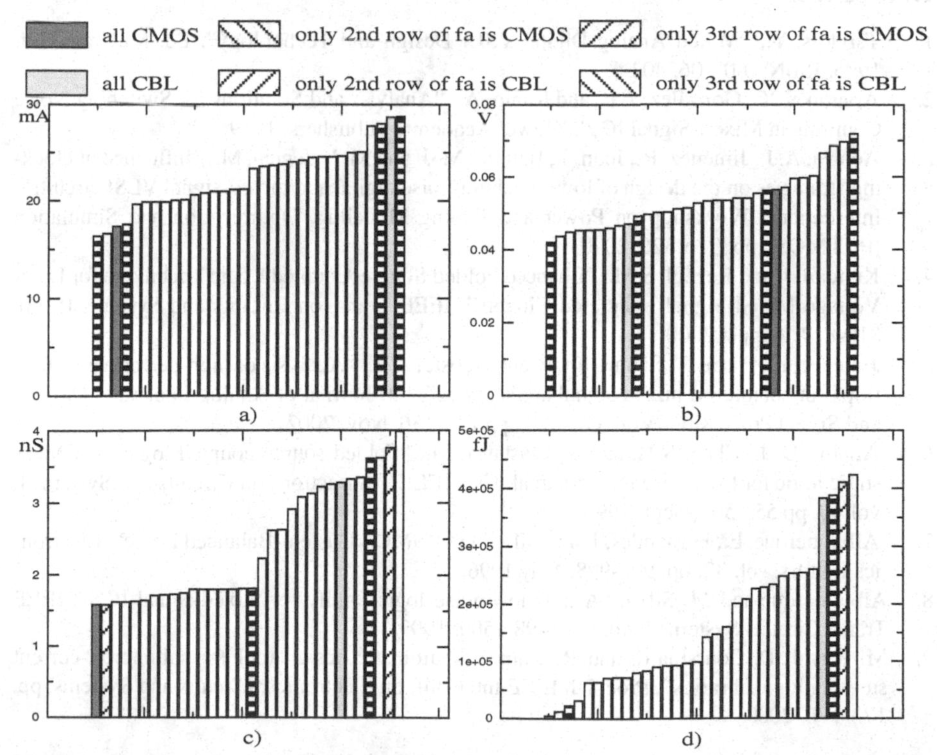

Fig. 8. a) Switching noise measured as peak of supply current; b) variation of inductor voltage; c) average propagation delay; d) power-delay product for different configurations of the multiplier.

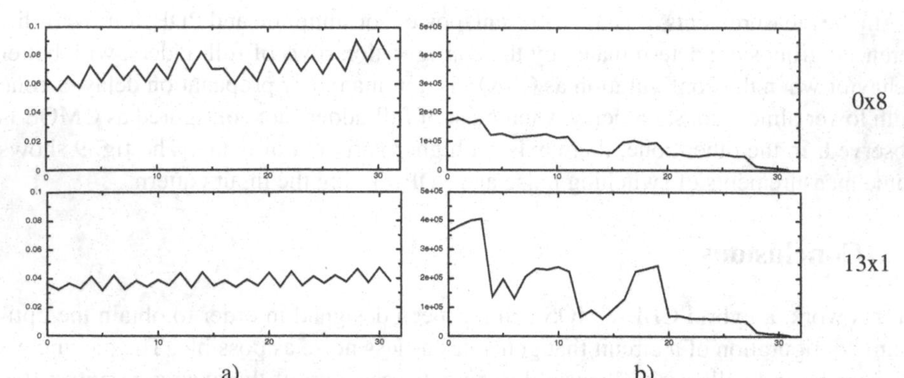

Fig. 9. a) Switching noise as peak of supply current (mA) and b) power-delay-product (fJ) for different configurations of the multiplier and operations. The X axis is the decimal value of the configuration signals in the multiplier, i.e., the 18 value is for (m_bi,m_ba,m_bs1,m_bs2,m_bs3) = (1,0,0,1,0). Thus, 0 and 31 represent the multiplier operating as CBL and CMOS, respectively)

As an application and demonstration example, we have chosen a 4x4 bit combinational multiplier. In this circuit we have obtained that the optimum configuration in noise generation is different than all cells configured as CBL, due to the desynchronization of different cell configurations for a non trivial configuration of multiplier rows.

References

1. Tsividis, Y., "Mixed Analog-Digital VSLI Design and Technology". Ed. McGraw-Hill, 1995. ISBN: 0-07-065402-6.
2. Aragonès, X., González, J. L. and Rubio, A., "Analysis and Solutions for Switching Noise Coupling in Mixed-Signal ICs", Kluwer Academic Publishers, 1999.
3. Acosta, A. J., Jiménez, R., Juan, J., Bellido, M. J. and M. Valencia, M., "Influence of clocking strategies on the design of low switching-noise digital and mixed-signal VLSI circuits", International Workshop on Power and Timing Modeling, Optimization and Simulation (PATMOS), pp. 316-326, 2000.
4. Kundan, J. and Rezaul, S.M.: "Ehanced Folded Source-Coupled Logic Technique for Low-Voltage Mixed-Signal Integrated Circuits", IEEE Trans. on Circuits and Systems II, pp. 810-817, August 2000.
5. Jiménez, R., Parra, P., Sanmartín, P. and Acosta, A. J., "Analysis of high-performance flip-flops for submicron mixed-signal applications", Int. Journal of Analog Integrated Circuits and Signal Processing. Vol. 33, No. 2, pp.145-156, Nov. 2002.
6. Allstot, D. J., Chee, S-H. and Shrivastawa, M.: "Folded source-coupled logic vs. CMOS static logic for low-noise mixed-signal ICs". IEEE Transactions on Circuits and Systems I, vol 40, pp 553-563, Sept. 1993.
7. Albuquerque, E., Fernandes, J. and Silva, M.: "NMOS Current-Balanced Logic". Electronics Letters, vol. 32, pp. 997-998, May 1996.
8. Albuquerque and M. Silva, "A new low-noise logic family for mixed-signal IC's", IEEE Trans. Circuits Systems I, 46, pp. 1498-1500, 1999.
9. M. Kayal, D. Coursinard, and R. Kanan, "Automatic design tool for submicron current steering logic libraries", Proc. 7th IEEE Int. Conf. on Electronics, Circuits and Systems, pp. 173-176, 2000.

Computational Delay Models to Estimate the Delay of Floating Cubes in CMOS Circuits

D. Guerrero[3,4], G. Wilke[1], J. L. Güntzel[2], M. J. Bellido[3,4], J. Juan Chico[3,4], P. Ruiz-de-Clavijo[3,4], and A. Millan[3,4]

[1] Universidade Federal do Rio Grande do Sul
Instituto de Informática
Porto Alegre - RS (Brazil)
Tel.: +55 (51) 3316-6159 - Fax: +55 (51) 3316-7308
http://www.inf.ufrgs.br/
wilke@inf.ufrgs.br

[2] Universidade Federal de Pelotas
Departamento de Matemática, Estatística e Computação
Pelotas - RS (Brazil)
Tel.: +55 (53) 275-7000 - Fax: +55 (53) 275-9023
http://www.ufpel.tche.br/
guntzel@ufpel.tche.br

[3] Instituto de Microelectrónica de Sevilla - Centro Nacional de Microelectrónica
Sevilla (Spain)
Tel.: +34 955056666 - Fax: +34 955056686
http://www.imse.cnm.es

[4] Departamento de Tecnología Electrónica - Universidad de Sevilla
Sevilla (Spain)
Tel.: +34 954556160 - Fax: +34 954552764
http://www.dte.us.es
{guerre, bellido, jjchico, paulino, amillan}@dte.us.es

Abstract. The verification of the timing requirements of large VLSI circuits is generally performed by using simulation or timing analysis on each combinational block of the circuit. A key factor in timing analysis is the election of the delay model type. Pin-to-pin delay models are usually employed, but their application is limited in timing analysis when dealing with floating mode or complex gates. This paper does not introduce a delay model but a delay model type called Transistor Path Delay Model (TPDM). This new type of delay model is specially useful for timing analysis in floating mode, since it is not required to know the whole input sequence to apply it, and can manage complex CMOS gates. An algorithm to get upper bounds on the stabilization time of each gate output using TPDM is also introduced.

1 Introduction

One of the most important tasks in the design process of VLSI circuits is the verification of the system. Timing verification may be performed by electric-level

J.J. Chico and E. Macii (Eds.): PATMOS 2003, LNCS 2799, pp. 501–510, 2003.
© Springer-Verlag Berlin Heidelberg 2003

simulation, but it demands huge execution times. An alternative is timing simulation, that is faster because it uses less accurate delay models, although still requires exercising all possible input vector sequences.
Designers can also rely on the input-independent approach for estimating the critical delay of VLSI circuits. This approach represents each combinational block of the circuit as a direct acyclic graph (DAG) [1], where nodes represent gates and edges represent connections.
The most simple solution relies on disregarding logic behaviour of gates and assuming the delay of the longest path as the critical delay of the combinational block. Hence, the critical delay problem of a combinational block is reduced to finding its longest path, which can be solved in linear time by the well-known topological sort algorithm. Such approach is referred to as static or topological timing analysis (TTA).
However, there may not exist any input pattern that exercises the longest path in the circuit, or conversely, it may never transmit any signal transition and hence, the critical delay may be smaller than the delay of the topologically longest path. Paths that never transmit a signal transition are called false paths [2] or unsensitizable paths.
Unlike TTA, functional timing analysis (FTA) takes into account the logic behaviour of gates so it is more accurate.
This paper introduce a new type of delay model targeting FTA. We begin with a reviw of timing analysis related terminology. In section 3 we will see the aplication of a pin-to-pin delay model in timing analysis. In section 4 we will see how TPDM can solve lacks of pin-to-pin delay models. In section 5 we will generalizate TPDM to deal with complex gates. Finally we will introduce algorithms to employ TPDM in cube simulation.

2 Floating Delay: Delay of a Cube

Timing analysis by pairs of vectors is computationally expensive and can be too optimistic, since it assumes that primary inputs change simultaneously while memory elements may present different propagation times that can lead to misalignment at the inputs.
Another approach is to get a safe upper bound for all the possible vector sequences ending in the same vector V. Such a bound is called the delay of the floating vector V. If we calculate the delay of all the possible floating vectors, the maximum of those delays will be an upper bound on the delay of the circuit. The delay thus obtained is referred to as the floating delay of the circuit.
To calculate the delay of a floating vector V, every node is assumed to be at an unknown state before instant 0, and the primary inputs are assumed to be stable with value V after instant 0. An upper bound on the instant when each node becomes stable is then systematically calculated.
Let be I the set of primary inputs of a logic circuit C, an input vector of C can be defined as a function $V : I \to \{0,1\}$ whose domain is I. Every W subset of

V is called a cube, that is, W is a function whose domain is a subset of I and such that for all i in $Dom(W)$, $W(i) = V(i)$.
Let W be a cube, let $vectors(W)$ be the set $\{V \in input\ vectors\ of\ C / W \subseteq V\}$, the delay of cube W is an upper bound on the set $\{floating\_delay(V)/V \in vectors(W)\}$. To get such an upper bound, every node is assumed to be in an unknown state before instant 0, and every input $i \in Dom(W)$ is assumed to be stable with value $W(i)$ after instant 0. An upper bound on the instant when each node becomes stable is then systematically calculated. Let $M = \{W_1, .., W_n\}$ be a finite non empty set of cubes such that any possible input vector is contained in $vectors(W_j)$ for at least an $W_j \in M$, then $max\{delay(W)/W \in M\}$ is an upper bound on the delays of all the floating vectors, so it is also and upper bound on the delay of the circuit.

3 Application of Pin-to-Pin Delay Models in Cube Simulation

Suppose a gate G that receives a single transition in input a at instant t, that is, all the inputs have been and will be always stable except input a, that changes only in instant t, and the output has always been stable before instant t. Under such conditions, the delay of pin a is the time elapsed from t to the transition at the output of G. In simple gates this only makes sense when all the inputs but a are in non-controlling values. A pin can have different delays for raising transitions and falling transitions. This delay have been modeled in [3], [4] and [5].
Sometimes it is possible to use a pin-to-pin delay model to get an upper bound on the instant when a gate output will become stable. This happens when we get an upper bound on the instant when one of the inputs becomes stable and we know that its final value is the controlling value of the gate. For example, suppose that the nand gate in fig. 1 is part of a circuit. If during the computation

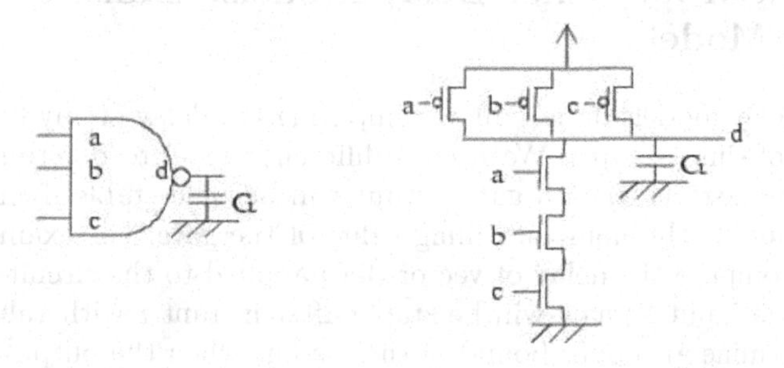

Fig. 1. A 3 input nand gate

of the delay of a cube we find that input b will be stable at instant t (or before) and that its final value will be the controlling value of the gate (i.e. 0), then we know then that the final value of output d is 1. However we do not know the instant when it becomes stable. To be pessimistic we should suppose that:

- The output capacitance C_L is utterly discharged at instant t, so no PMOS transistor will be active before instant t ($a = b = c = 1$ before instant t).
- Only the PMOS transistor of input b will charge C_L after instant t ($a = c = 1$ after instant t).

Hence a pessimistic vector sequence for this gate would be that shown in fig. 2. Note that tp_b is the pin-to-pin delay of input b. Then an upper bound on the

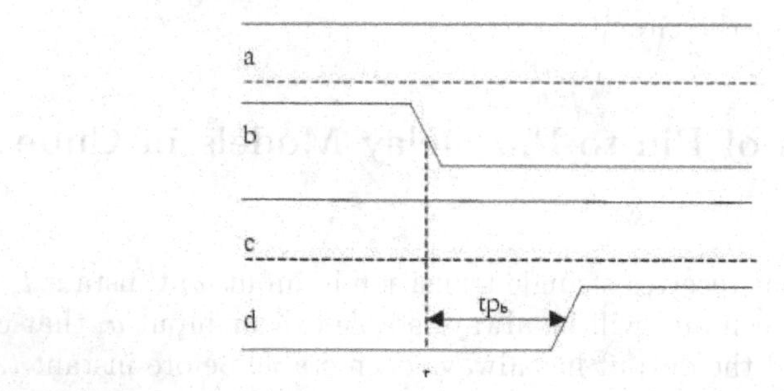

Fig. 2. Pessimistic vector sequence for a 3 nand gate

instant when d becomes stable is $t + tp_b$. If we also find that input c becomes stable at instant t' or before and its final value is 0, then another upper bound on the instant when d becomes stable would be $t' + tp_c$, where tp_c is the pin-to-pin delay of input c. Of course to be as accurate as possible we should always take the lowest upper bound.

4 The Need for Other Delay Models: Transistor Path Delay Model

Pin-to-pin delay models do not allow computing the delay of any floating vector of a circuit of simple gates. We need a different model to determine an upper bound on the instant when a gate output will become stable if the final value of all its inputs is the non-controlling value of the gate. For example, suppose we have to compute the delay of vector (0,0) applied to the circuit of fig. 3. We know that the input signals will be stable after instant t with value 0, and we have to determine an upper bound of the instant when the output will become stable with its final value (i.e. 1). To be pessimistic, we can assume that:

- C_L and any internal gate node in the path from V_{dd} to the output is utterly discharged before instant t ($b = 1$ before instant t).
- Every input receives a falling transition at instant t, so no PMOS transistor will be in saturation till the end of those transitions.

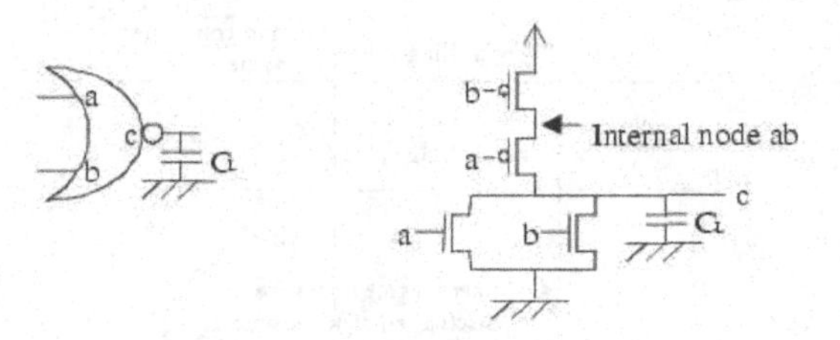

Fig. 3. A 2 input nor gate

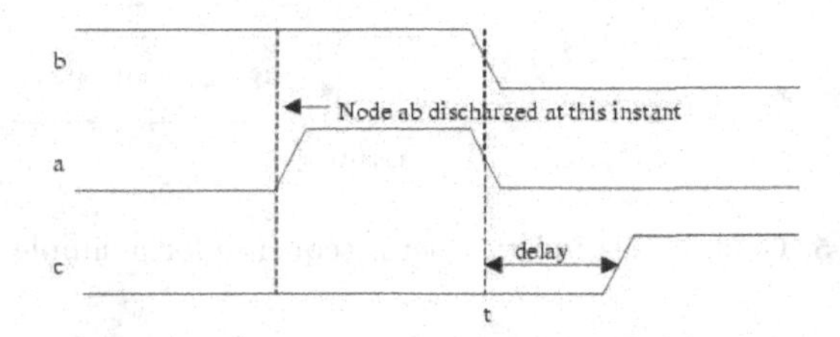

Fig. 4. Pessimistic vector sequence for a nor gate

So a pessimistic but possible vector sequence would be the showed in fig. 4. In this vector sequence we can not use a pin-to-pin delay model since none of the PMOS transistors is in saturation just before instant t (that is, no input is at non controlling value just before instant t). We present a new type of delay model called Transistor Path that solves this by modeling the behaviour of the gate when all its inputs change simultaneously to non-controlling value. In general, if we have a simple gate of inputs $i_1,..,i_n$ (where i_n is the input whose PMOS transistor is connected to V_{dd} if it is a NOR gate, or the input whose NMOS transistor is connected to ground if it is a NAND gate) that are set to non-controlling value respectively at instants $t_1,..,t_n$ (or before), we can get an upper bound on the instant when the output turns stable by simulating the vector sequence shown in fig. 5. To simplify we can assume that the transition time of all the input transitions above is the same, but it must be an upper bound of all the transition times. The effect of multiple input switches in simple gates have been studied in [7] and [8] modeling the delay as a function of the skew betwen input transitions. As we can see, the characterization process can be simplified by modeling only the behaviour of the gate for the most pesimistic skew.

5 Generalization of Transistor Path Delay Model for Circuits Containing Complex Gates

In complex gates the concept of controlling or non-controlling value does not make sense so we need a more general delay model. The final logic value of a complex gate is known when a transistor path from V_{dd} or from GND to the output is activated. For example suppose that in the complex gate of fig. 6 we

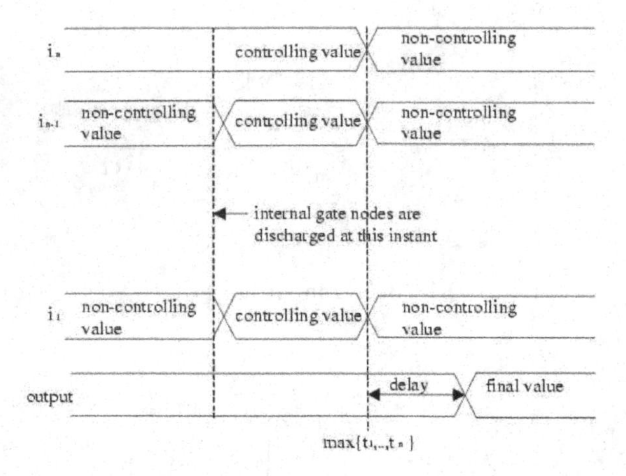

Fig. 5. Generic pessimistic vector sequence for a simple gate

know that input c is stable with value 0 after instant t_1 and that input b is stable with value 0 after instant t_2. There will be a path of conducting PMOS transistors from V_{dd} to the output after instant $max\{t_1, t_2\}$ so we know that the final logic state of the output will be 1. We know that input signals b and c will

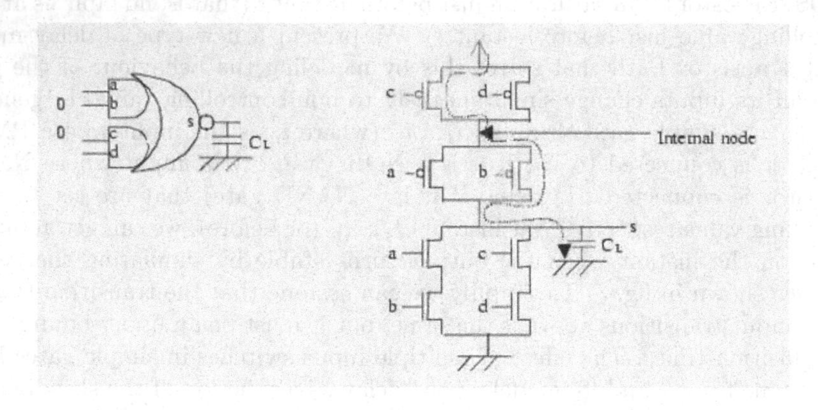

Fig. 6. A conducting transistor path in a complex gate

be stable after instant $max\{t_1, t_2\}$ with value 0, and we have to find an upper bound on the instant when the output will become stable with its final value (i.e. 1). To be pessimistic and to simplify the upper bound computation, we can assume that:

- C_L and any internal gate node in the active path from V_{dd} to the output is utterly discharged before instant t ($c = d = 1$ before $max\{t_1, t_2\}$).

- There will be a falling transition at every input corresponding to a PMOS transistor in the path at instant $max\{t_1, t_2\}$, hence no pmos transistor in the path will be conducting before those transitions.
- Only the pmos transistors in the path will charge C_L ($a = d = 1$ after instant $max\{t_1, t_2\}$)

So a pessimistic vector sequence would be the shown in fig. 7. We have simulated

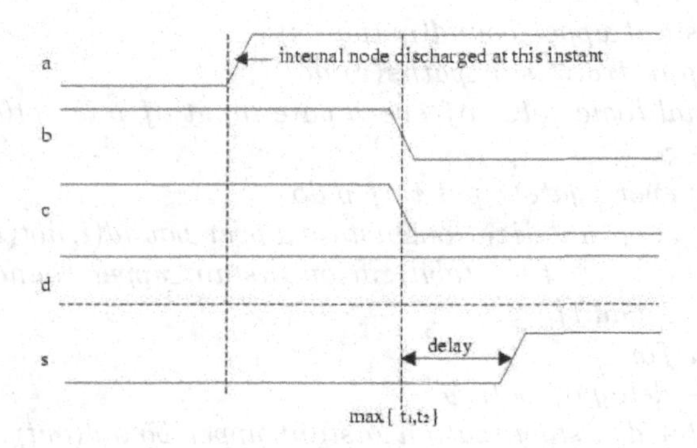

Fig. 7. Pessimistic vector sequence for a complex gate

this vector sequence with the electric simulator SPECTRE using 0.35 μm CMOS technology. When inputs b and c change simultaneously we have a delay of 0.219 ns. If b changes 0.1 nanoseconds before c we have a delay of 0.210 ns. If we change c 0.1 ns before b we have a delay of only 0.154 ns, because the internal node loads before the last input transition. In order to get an upper bound on the gate delay we need to model the behaviour of the gate when all the transistors of the path are activated simultaneously. The activation instant of a transistor path P is the maximum among the activation instants of its transistors. If t_1,..,t_n are respectively upper bounds on the activation instants of those transistors, then $max\{t_1, .., t_n\}$ is an upper bound on the activation instant of P. If at instant t (or before) a path with delay d is activated and in instant t' (or before) a path of the same gate with delay d' is activated, then $t + d$ and $t' + d'$ are upper bounds of the instant when the gate output becomes stable so we should take $min\{t + d, t' + d'\}$ as the upper bound.

6 Application of TPDM to Estimate the Delay of Floating Cubes

For every gate type we must keep two transistor path sets: one for the set of paths from V_{dd} to the output and another for the set of paths from GND to the output. For every transistor path we must codify the set of gate inputs corresponding to transistors in that path and the set of delay parameters corresponding to that path. The set of gate inputs of the path can be implemented with an array of bits of dimension n, where n is the number of gate inputs. Let $transistor\_paths(0)$ be the set of transistor paths of a gate that go from GND to the output and let $transistor\_paths(1)$ be the set of transistor paths that go from V_{dd} to the output, if the input values of the gate set the final logic state of the gate output to v, to get an upper bound of the instant when the gate output becomes stable we can follow this algorithm:

```
stabilization_instant_upper_bound(output, v) ← ∞
for every path p in transistor_paths(v) do
        if the final logic value of every gate input of p is not(v) then
                t ← ∞
                for every gate input i of p do
                        if stabilization_instant_upper_bound(i, not(v)) > t then
                                t ← stabilization_instant_upper_bound(i, not(v))
                        end if
                end for
                d ← delay of path p
                if t + d < stabilization_instant_upper_bound(output, v) then
                        stabilization_instant_upper_bound(output, v) ← t + d
                end if
        end if
end for
```

In the algorithm, $stabilization\_instant\_upper\_bound(s, x)$ is the lowest known upper bound on the instant when signal s becomes stable when its final logic value is x. For example suppose the gate in fig. 6. The set of transistors of each path from V_{dd} to the output could be codified using a bit vector for each path as shown in table 1. The vector component corresponding to a input will be set to 1 if and only if the path of this vector has a transistor whose gate is connected to that input. During the computation of the delay of a cube we have that input a is stable with value 0 after instant t_a , input b is stable with value 0 after instant t_b and input c is stable with value 0 after instant t_c. Since the first and third path shown in table 1 will be active we know that the final state of the gate output will be 1 so we must compute $stabilization\_instant\_upper\_bound(output, 1)$ using the algorithm above. We computed the delays for the known conducting paths and obtained a delay d for the first path and a delay d' for the third path. Valid upper bounds are then $max\{t_a, t_c\} + d$ and $max\{t_b, t_c\} + d'$. At the end of the algorithm, $stabilization\_instant\_upper\_bound(output, 1)$ will be equal to the lowest known upper bound.

If the final logic state of the gate output is not set, we must calculate $stabilization\_instant\_upper\_bound(output, 0)$ and $stabilization\_instant\_upper\_bound(output, 1)$,
but we must use a different algorithm. This is because, from all the transistor paths that can be activated, we do not know which one will actually be acti-

Table 1. *transisto_paths*(1) for the complex gate of fig. 6

	path 0	path 1	path 2	path 3
transistors connected to input a	1	1	0	0
transistors connected to input b	0	0	1	1
transistors connected to input c	1	0	1	0
transistors connected to input d	0	1	0	1
delay parameters (depend on the delay model)	...	...	...	...

vated. To be pessimistic we must take the greater stabilization instant upper bound determined by a path that can be activated. The algorithm to get the upper bound when the final value of the gate output is unknown is the following:

```
stabilization_instant_upper_bound(output, v) ← ∞
for every path p in transistor_paths(v) do
    if the final logic value of every gate input of p is not v or is unknown then
        t ← ∞
        for every gate input i of p do
            if stabilization_instant_upper_bound(i, not(v)) > t then
                t ← stabilization_instant_upper_bound(i, not(v))
            end if
        end for
        d ← delay of path p
        if t + d > stabilization_instant_upper_bound(output, v) then
            stabilization_instant_upper_bound(output, v) ← t + d
        end if
    end if
end for
```

For example suppose the gate in fig. 6. The set of transistors of each path from GND to the output could be codified using a bit vector for each path as shown in table 2. During the computation of the delay of a cube we have that input d is stable with value 1 after instant t_d and input a is stable with value 0 after instant t_a. Since there are no known active paths in table 1 or 2 the final logic state of the gate output is unknown. We must compute *stabilization_instant_upper_bound*(*output*, 0) and *stabilization_instant_upper_bound*(*output*,1), using the algorithm above. To get *stabilization_instan_upper_bound*(*output*, 1) we computed the delays for the possible conducting paths in table 1 and obtain a delay d for the first path and a delay d' for the third path. Possible upper bounds then are $max\{t_a, t_{cf}\} + d$ and $max\{t_{bf}, t_{cf}\} + d'$, where t_cf=*stabilization_instant_upperbound*(c, 0) and t_{bf}=*stabilization_instant_upper_bound*(b,0).
At the end of the algorithm *stabilization_instant_upper_bound*(*output*, 1) will be equal to the biggest upper bound. To get *stabilization_instant_upper_bound*(*output*, 0) we computed the delay for the only possible conducting paths in table 2 and obtain a delay d'' for the second path. The only possible upper bound then is

$max\{t_{cr}, t_d\} + d''$, where $t_{cr} = stabilization\_instant\_upper\_bound(c, 1)$. At the end of the algorithm $stabilization\_instant\_upper\_bound(output, 0)$ will be equal to this upper bound.

Table 2. $transistor\_paths(1)$ for the complex gate of fig. 6

	path 0	path 1
transistors connected to input a	1	0
transistors connected to input b	1	0
transistors connected to input c	0	1
transistors connected to input d	0	1
delay parameters (depend on the delay model)	...	...

We can use both algorithms to systematically find and upper bound on the instant when each node of a circuit becomes stable for a given input cube.

7 Conclusions

There are several approaches to the timing analysis problem. We have introduced a new type of delay model specially suitable for timing analysis under the floating mode. Transistor Path Delay Model (TPDM) is defined for simple and complex gates. It solves several lacks of pin-to-pin models. We have also presented an algorithm to find an upper bound on the stabilization time of gate outputs using TPDM.

References

1. GÜNTZEL, J. L. Functional Timing Analysis of VLSI Circuits Containing Complex Gates, Porto Alegre: PPGC da UFRGS, 2000
2. CHEN, H.-C.; DU, D. Path Sensitization in Critical Path Problem, IEEE Transactions on CAD of Integrated Circuits and Systems, Los Alamitos, California, v.12, n.2, p. 196-207, February 1993
3. J. J. Chico, Propagation Delay Degradation in Logic CMOS Gates, Tesis doctoral, Universidad de Sevilla, 2000
4. J. M. Daga, D. Auvergne: A Comprehensive Delay Macro Modeling for Submicrometer CMOS Logics, IEEE Journal of Solid-State Circuits. Vol. 34, No. 1, January 1999
5. M. R. Casu, G. Masera, G. Piccinini, M. Ruo Roch, M. Zamboni: A high accuracy-low complexity model for CMOS delays. Actas IEEE International Symposium on Circuits and Systems (ISCAS) 2000, pp. I-455-458
6. L. Chen, S. K. Gupta, M. A. Breuer, A New Gate Delay Model for Simultaneous Switching and its Applications, Design Automation Conference, p. 289-294
7. V. Chandramouli, K. A. Sakallah, Modelling the Effects of Temporal Proximity of Input Transitions on Gate Propagation Delay and Transition Time. 33rd Design Automation Conference, 1996

A Practical ASIC Methodology for Flexible Clock Tree Synthesis with Routing Blockages

Dongsheng Wang, Peter Suaris, and Nan-chi Chou

Mentor Graphics Corporation
8005 S.W. Boeckman Road, Wilsonville, Oregon 97070, USA
{dongsheng_wang, peter_suaris, nanchi_chou}@mentor.com

Abstract. In this paper, we propose a practical ASIC methodology for flexible clock tree synthesis (CTS). The allowed flexibility for clock network leads us to be able to synthesize some complex clock networks which may contain clock driver, sequential components, buffers, inverters, gated components. Macro blockages are also allowed to be presented in clock routing area to make CTS more practical. With multiple timing constraints applied, our CTS method first introduces node clustering and buffering to construct an initial clock tree in a bottom-up fashion, pursuing the minimum clock skew with macro blockages eluded. Then tree node pulling up and buffer insertion may be used to further improve clock tree performances. Experiments of CTS program using this methodology show that our CTS method works very well for some complex clock networks with timing closure achieved.

1 Introduction

In the deep submicron era, an efficient and high performance clock tree synthesis (CTS) is crucial to achieve timing closure for ASIC design flow. The task of CTS is to create a clock tree for given clock networks to meet all the constraints specified. In synchronous circuits, each clock network is composed of a source instance that sends clock signal, some sink instances, such as flip flops, latches and so on, on each of which there is a clock sink pin receiving the clock signal, and some internal instances which may be buffers, inverters or gated components. The constraints include clock propagation delay, skew, transition time and so on. Propagation delay is defined as the delay from clock source instance to the synchronized clock pins. Clock skew is defined as the maximum difference in propagation delays of a clock. Transition time is defined as delay from an instance node to its immediate successive instance node of the clock tree. Real designs may have one or more clock domains. Each Domain may contain one or more clock definitions. Each clock definition represents a clock network. All the clock pins belonging to same domain must be synchronized. For simplicity, CTS discussed in this paper will focus on single clock network. A clock tree is composed of three types of tree nodes. Leaf nodes represent all the clock sink

J.J. Chico and E. Macii (Eds.): PATMOS 2003, LNCS 2799, pp. 511–519, 2003.
© Springer-Verlag Berlin Heidelberg 2003

instances staying at the bottom level of the clock tree. Internal nodes represent all the buffers, inverters and gated components staying at internal levels. Root node represents the clock source instance staying at very top of the clock tree.
There are some clock net routing approaches have been proposed. H-tree algorithm, a symmetric clock distribution tree, was proposed in [1] for clock tree construction, which tried to balance skew of clock tree using the symmetry of H wire structure. Followed is a clock routing scheme using MMM (Method of Means and Medians) approach [2] to generalize the H-tree approach.
Zero skew routing approach was widely researched for the balanced clock tree. The first zero skew routing approach was proposed in [3] which was a good base of later researched on clock skew issues. Thereafter, some zero-skew routing improvement approaches were proposed [2], [6], [7], [9], [10]. The DME (Deferred-Merge Embedding) algorithm in [4] used two phases to minimize wire length with zero skew for clock tree. First a bottom-up phase constructs a tree of merging segments which represent loci of possible placements of internal nodes in a zero-skew tree (ZST) T. Then a top-down embedding phase determines exact locations for the internal nodes in T. A bottom-up zero-skew clock tree construction algorithm in [5] reduced the total wire length by 15% with time complexity O(n*n*log n). A clustering based greedy DME algorithm [6] improved the time complexity to O(n log n) without any increase in total wire length. Furthermore, a linear time bucket algorithm was proposed in [7] which was much simpler and more efficient than clustering-based algorithm [6]. Recently, an integrated zero- skew clock tree construction algorithm was proposed in [8] which performed simultaneous routing, wire sizing and buffer insertion to minimize propagation delay as well as to reduce wire length and to use less buffers.
Bounded-skew clock routing approach, a more practical way to build clock tree, was proposed to control the skew with minimum wire length cost. Research in [9] raised the bounded skew clock and Steiner routing problem. They studied the minimum cost bounded-skew routing tree under the linear delay model and proposed a extended MDE algorithm introducing the concept of merging region instead of merging segment. Then the BME (Boundary Merging and Embedding) approach was proposed in [10] to study the minimum-cost bounded-skew routing tree problem under Elmore delay model. By extending traditional clock routing algorithm, researchers in [11] proposed a more practical bounded-skew clock routing algorithm in the presence of obstacles.
Researchers in [12] provided some good trade-offs for clock routing: trade-offs between skew and wire length, between skew and propagation delay and between congestion and wire length.
There are two problems associated with the above researches. One is that they focus on clock net routing without clock buffering which may not be able to be used as a complete clock tree synthesis approach because the complexity of real clock network designs can not be properly synthesized without clock buffering. Some researches has touched clock buffering issue[13]. But it can synthesize simple clock networks with sequential components only. Another problem is that they assume that clock network contains clock driver and sequential components only which is not true in most real

clock network designs. In fact, in addition to clock driver and sequential components, clock network may contain buffers, inverters, gated components and there may be some macro routing blockages in the routing area of the clock network. These blockages may enforce some detouring during clock net embedding, which may mess up the timing performances of the synthesized clock network.
In this paper, we propose a practical methodology for flexible clock tree synthesis. The flexibility allows us to solve the above two problems. We shall show how the methodology solve the problems in two phases: initial clock tree construction with consideration of gated components and clock tree performance improvement.

2 Initial Clock Tree Construction

2.1 Gated Clock Network

Handling gated component for CTS is one of features of the methodology. Gated clock network propagates clock signal from clock driver to all clock signal receivers through some combinational logic gates. Most clock signal receivers are clock pins of sequential components. Some may be other input pins such as primary output IO pins, some input pins of macro blocks, and so on. The combinational logic gate is defined as gated component of the clock network. The gated components specify some necessary logic needed for the clock network and they can not be removed and the original logic of the whole clock network must remain unchanged during clock tree synthesis. Figure 1 shows a gated clock net.
The gated components bring two difficult issues to CTS. First, in clustering process, CTS tries to cluster tree nodes (clock pins, buffers, inverters, etc.) that are physically very close to each other. If gated components exist in a clock network, different nodes rooted by different gated components can not be clustered into same cluster even if the nodes are physically close to each other. Second, if some inverters exist in the original clock network, for the purpose of smaller skew of clock tree, CTS prefers to remove these inverters before starting clock tree synthesis. This removal may have clock phase inversion become headachy problem.

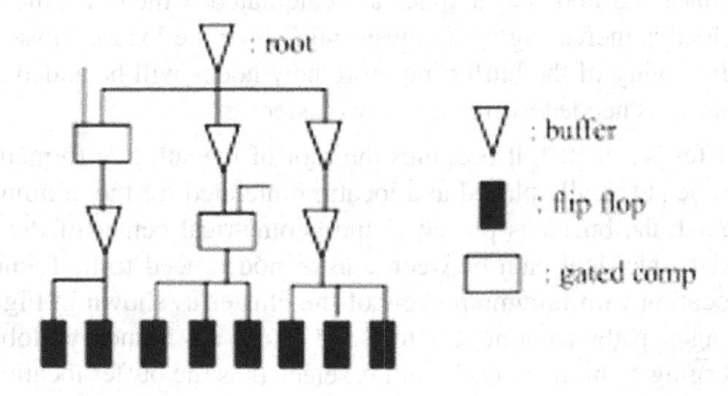

Fig. 1. Gated clock net

2.2 Clustering

Our CTS approach introduces node clustering and buffering to construct an initial clock tree in a bottom-up fashion. A clock tree node represents an instance of the clock tree. There are three kinds of nodes, leaf nodes that represent flip flops, latches and IO ports, root node that represents the clock root instance, and internal nodes that represent buffers, inverters and gated components. Each node except for root node has one and only one parent node. Each node except for leaf nodes has at least one child node. Root node has no parent node and leaf nodes have no child nodes.
For given clock network, starting at the clock root node, trace down to all the leaf nodes, remove all the buffers and inverters, and create a internal tree node for each gated component. Let all the tree nodes stay at bottom level of the tree. Then, cluster all the clock tree nodes level by level in a bottom up fashion. Each cluster groups some nodes. The cluster size is determined by buffer drivability and the total load contributed by all the nodes in the cluster. Each cluster is rooted by a newly inserted buffer or inverter which become a new internal tree node staying at one level higher than the nodes of the cluster. The clustering process stops when only one unclustered node is left. Thus, the initial clock tree construction is completed.

2.3 Buffering

Clock buffering is associated with clustering. For an initial cluster, some new nodes may be added to it bases on buffer selection. A good buffer or inverter should be selected to drive a proper number of nodes with consideration of the buffer's drivability, the physical distances among the nodes to be clustered, the clock network structure, and so on. The maximum number of nodes that the selected buffer can drive is estimated based of the buffer's drivability and the allowed maximum transition time specified by user. If no blockages exist within the clock tree area, the distance between nodes can be the Manhattan distance. If there are some blockages in presence, physical paths should be found among the nodes by using a global router. The length of each physical path is calculated instead of Manhattan distance. Also the wire capacitance and node capacitance are calculated, which become the load of the buffer. The load is increasing while clustering. Once the load is close or equal to the maximum drivability of the buffer, no more new nodes will be added into the cluster and a new buffer is needed to create a new cluster.

Once the buffer is selected, it becomes the root of the sub-tree formed by the cluster and needs to be physically placed at a location intended for the minimum skew. If no blockages exist, the buffer is placed at the geometrical center of the cluster. If any blockage exists, physical path between cluster nodes need to be found to determine the buffer location with minimum skew of the cluster as shown in Figure 2. Because of the blockages, paths from node 1 to 3 and 2 to 4 are found by global router and a balanced merging point along each path is selected as the buffer location.

3 Clock Tree Improvement

Once an initial clock tree is built, we improve it in three stages, that is, node pulling up to speed the slow paths, buffer insertion to fix the transition time violations and buffer insertion to slow down the fast paths.

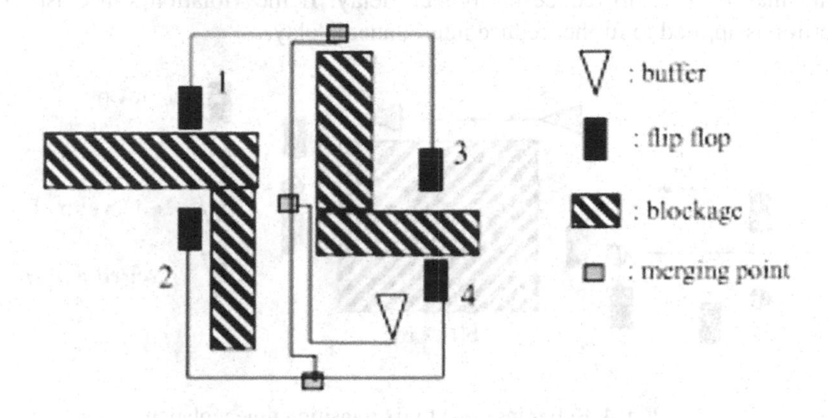

Fig. 2. Buffering with blockages

Some slow paths of the clock network are picked up and their leaf instance nodes may be pulled up to directly connect to higher tree levels in order to speed up the delay from clock root to these nodes. The valid pulling up shown in Figure 3 is an example. The difficulty here is that some gated components, if exist, on the clock tree may restrict some node pulling ups because any pulling up can not let any node go beyond any gated components at higher tree level which is shown as the invalid pulling up in Figure 3. Invalid pulling ups will cause logic errors. By pulling up, not only the maximum propagation delay is reduced, but skew of the tree is reduced as well without any new buffers inserted. Compared to buffer insertion method as discussed below, node pulling up can benefit the clock tree by using fewer buffers.

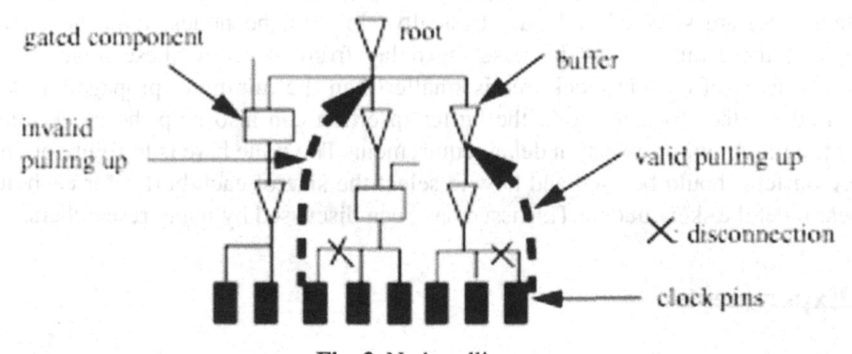

Fig. 3. Node pulling up

Given a clock tree, second stage considers a tree branch to be the connection between a tree node to its immediate successive child node. As defined in section 1, the transition time is the branch delay, including node internal delay and interconnect delay. The transition time violation of a branch means that the delay of this branch exceed user-defined value. If the parent node represents a buffer or inverter, buffer sizing may be used to reduce the branch delay. If the violation still exists, buffer insertion is applied to further reduce interconnect delay.

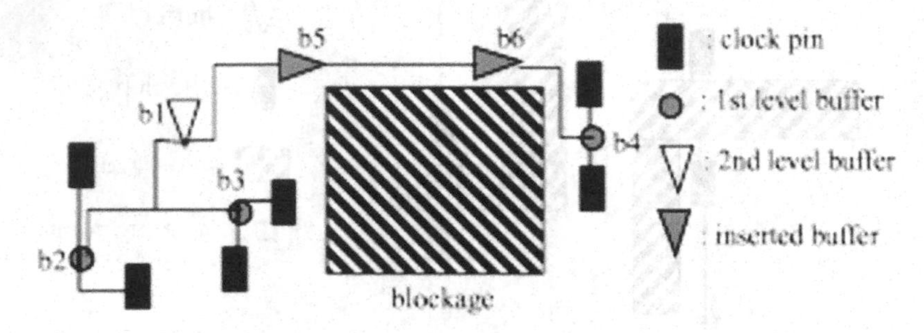

Fig. 4. Buffer insertion to fix transition time violation

Figure 4 indicates the buffer insertion process. Suppose interconnect delay from buffer b1 to b4 causes transition time violation of the branch. Because of the blockage, a physical path from b1 to b4 is found as shown. The drivability of candidate buffers and the estimation of wire load of the path determine what kind of buffer should be selected and how many selected buffers should be inserted along the path to remove the violation. Then the selected buffers should be evenly placed along the path to achieve better timing results. In Figure 4, two buffers, b5 and b6, are selected and evenly placed on the path from buffer b1 to b4.

When the first two stages are finished, if the clock tree still does not meet the maximum skew requirement, a skew minimization stage is involved. In this stage, the clock tree is traced again to find those nodes with smallest delays and then a buffer is inserted at input of each of these nodes to increase their delays. As shown in Figure 5, buffers with proper sizes (different buffers with different sizes may be applied to same nodes) are selected and put physically close to the nodes. It is the internal delays of these buffers that increase the delay from driver to these nodes. If the smallest delay of current clock tree is smaller than the minimum propagation delay specified for the clock network, the buffer insertion can also help the clock tree to meet the minimum propagation delay requirement. The issue here is to figure out how many buffers should be used and how to select the size of each buffer for each node to best meet the skew budget.This issue has been discussed by many researchers.

4 Experiments

The CTS methodology described above has been implemented by programming on Sun unix workstation. Six industrial ASIC design benchmarks with six clock networks shown in Table 1 are used to test our CTS method. The major characteristics of each clock network in the table include the number of gated components connected in the clock network, the number of clock pins representing the clock network size, the number of blockages (blks in table) at clock routing area and the timing constraints specified in nanoseconds by users, including the maximum skew, minimum and maximum propagation delay. The six clock networks selected can be considered the representatives of clock networks with various situations mentioned in this paper that may cause difficulties in CTS. Clock clk1 is a regular small test case. Clock clk2 is a small one and clock clk3 is a large one, both of which are with a lot of gated components which may cause difficulty in delay balance and solving clock phase inversion problems. Although clock clk4 is small, it has one IO port needing to be synchronized with the clock pins, and the IO port is physically locates far away from clock pins of the net. This situation makes CTS hard to achieve good skew. A lot of gated components are attached to large clock network clk5 and the blockages are irregularly distributed within the network area, which may cause difficulty in clock buffering. Clock clk6 is the most difficult case with some very large blockages distributed within the clock network routing area and they overlap with each others.

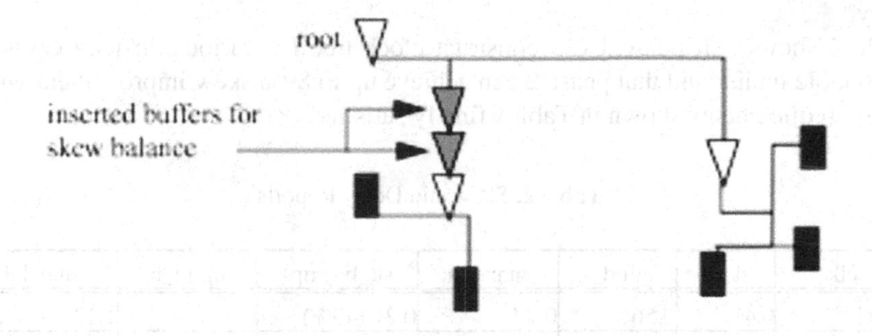

Fig. 5. Buffer insertion to reduce skew

Table 1. Clock Network Specifications

clks	blks	gates	clk pins	max skew	min delay	max delay
clk1	0	2	139	0.3	0	2.0
clk2	3	112	6,026	0.5	2.0	4.0
clk3	22	49	14,298	0.5	0	5.0
clk4	112	3	1,424	0.5	0	5.0
clk5	23	430	36,546	0.3	0	5.0
clk6	12	8	62,186	1.0	0	7.0

Table 2 reports the skew, minimum and maximum propagation delays of each clock network obtained by phase 1, initial clock tree construction phase, represented in sk_a, mind_a and maxd_a in nanoseconds, respectively. The skews and delays improved by phase 2, clock tree improvement phase, presented in sk_b, mind_b and maxd_b in nanoseconds, respectively, are also given in the table for the comparisons. The numbers in the parentheses in column sk_b indicate the percentages of the skew improvement achieved by phase 2 over phase 1. As shown, the two phases of our CTS method work together to construct clock tree to meet the timing requirements. A good example is clock clk1 as shown in Table 2. By completing phase 1, skew 0.41 in column sk_a is relatively large and does not meet the maximum skew requirement 0.3. After phase 2, the skew is reduced to 0.21 with a 49% improvement compared to phase 1. And the minimum and maximum propagation delay still remain within the range specified in Table 1. Another example is clk4. Both skew and maximum delay are reduced quite bit by phase 2. In the case of clk2, phase 1 produced a clock tree with smallest propagation delay smaller than the minimum propagation delay requirement. Phase 2 increased the minimum delay with skew reduced and the maximum delay staying within the range. For the most difficult case, clk6, phase2 successfully reduced skew by 50% with no increase in its maximum propagation delay.
Table 2 shows that phase 1 can construct clock tree for various difficult cases with reasonable timing and that phase 2 can achieve up to 89% skew improvement with all timing requirements shown in Table1 finally satisfied.

Table 2. Skew and Delay Reports

clks	sk_a	mind_a	maxd_a	sk_b (imp)	mind_b	maxd_b
clk1	0.41	0.56	0.97	0.21 (49%)	0.9	1.12
clk2	1.88	0.68	2.56	0.45 (76%)	2.12	2.57
clk3	2.02	2.18	4.20	0.23 (89%)	4.1	4.33
clk4	1.96	4.05	6.01	0.5 (74%)	3.05	3.56
clk5	1.30	1.84	3.14	0.30 (77%)	4.14	4.44
clk6	1.66	3.27	4.93	0.83 (50%)	4.10	4.93

5 Conclusions

A practical ASIC methodology for flexible clock tree synthesis with routing blockages in presence has been proposed in this paper. The flexibility of the method has given us a solution to handle some difficult CTS issues such as gated clock networks, clock phase inversion, macro blockages, very large design size, and so on. The test cases used in this paper have reasonably reflected the difficulty of these issues. Experiments have shown that the two-phase methodology is effective, efficient and practical. The clock tree construction phase can construct an initial tree for complex clock networks with reasonable timing performance and the clock tree improvement phase provides very powerful approach in receiving timing closure of the clock tree. Up to 89% skew improvement have been achieved by this phase for the designs used in this paper. These features make our CTS method very practical and promising in achieving chip-level timing closure in real ASIC industrial world.

References

1. H. Bakoglu, J.T.Walker and J.D.Meindl, "A Symmetric clock-distribution tree and optimized high-speed interconnections for reduced clock skew in ULSI and WSI circuits", ICCD'86, pp. 118–122, 1986.
2. M.A.B.Jackson, A.Srinivasan and E.S.Kuh, "Clock Routing for High-Performance ICs", 27th DAC, pp. 573–579, 1990.
3. R.S.Tsay, "Exact Zero Skew", ICCAD'91, pp. 336–339, 1991.
4. T.H. Chao, Y.C.Hsu and J.M.Ho, "Zero Skew Clock Net Routing", 29th DAC, pp. 518–513, 1992.
5. Y.M.Li and M.A.Jabri, "A Zero-Skew Clock Routing Scheme for VLSI Circuits", ICCAD'92, pp. 458–463, 1992.
6. M. Edahiro, "A Clustering-Based Optimization Algorithm in Zero-Skew Routings", 30th DAC, pp. 612–616, 1993.
7. M. Edahiro, "An Efficient Zero-Skew Routing Algorithm", 31st DAC, pp. 375– 380, 1994.
8. I. Liu, T. Chou, A. Aziz and D.F. Wong, "Zero-Skew Clock Tree Construction by Simultaneous Routing, Wire Sizing and Buffer Insertion", ISPD 2000, pp. 33–38, 2000.
9. D.J.Huang, A.B.Kahng and C.A.Tsao, "On the Bounded-Skew Clock and Steiner Routing Problems", 32nd DAC, pp. 508–513, 1995.
10. J.Cong, A.B.Kahng, C.K.Koh and C.W.A.Tsao, "Bounded-Skew Clock and Steiner Routing Under Elmore Delay", ICCAD'95, pp. 66–71, 1995.
11. A.B.Kahng and C.A. Tsao, "More Practical Bounded-Skew Clock Routing", 34th DAC, pp. 594–599, 1997.
12. J. D.Cho and M.Sarrafzadeh, "A Buffer Distribution Algorithm for High- Speed Clock Routing", 30th DAC, pp/537–543, 1993
13. M.Toyonaga, K.Kurokawa, T. Yasui and A. Takahashi, "A Practical Clock Tree Synthesis for Semi-Synchronous Circuits", ISPD 20000, pp. 159–164, 2000.

Frequent Value Cache for Low-Power Asynchronous Dual-Rail Bus

Byung-Soo Choi and Dong-Ik Lee

Concurrent Systems Laboratory
Department of Information and Communications
K-JIST(Kwangju Institute of Science and Technology)
1 Oryong-dong Puk-gu Gwangju, 500-712, South Korea
{bschoi,dilee}@kjist.ac.kr

Abstract. We study a power reduction method for the asynchronous dual-rail bus. A preliminary analysis of data communication patterns between a processor and a memory module reveals that many communications deliver a set of data items repeatedly. To exploit such communication characteristics, a *frequent value cache*(FVC) method is proposed that delivers not always data itself but sometimes an index of data item of FVC. Because of the lower switching activity, FVC reduces the power consumption of the asynchronous dual-rail bus. Simulation results illustrate that FVC reduces the power consumption of the normal asynchronous dual-rail bus by 25% and 30% at maximum for integer and floating-point benchmarks, respectively.

1 Introduction

Because of the steady increase of the number of components in a chip, SOC design methods have been studied intensively. To succeed in the market, the time-to-market and the reliability of a SOC are very important. To help the design efforts for a short design time and reliability of SOCs, asynchronous design methods [1] have been studied recently. On the other hand, most SOCs are the parts of portable systems that require a low power consumption. Therefore, asynchronous design methods have been investigated for the purpose of a short design time, reliability, and low power consumption of SOCs.

Because of a large portion of power consumption of a SOC, many researches have been interested in a bus structure to reduce the power consumption [2], [3], [4], [5], [6], [7], [8]. There are two kinds of buses as an address and a data buses. Since the address patterns usually have spatial and temporal localities, many methods have already decreased the power consumption of the address bus [9]. However, since the data patterns have no such locality, many researches have been doing with different approaches, and it is also the target of the present research.

Data buses are categorized as a shared and a dedicated buses. For a shared bus, several methods have already been proposed to reduce the power consumption [10]. On the other hand, a shared bus is a generalized form of a dedicated

J.J. Chico and E. Macii (Eds.): PATMOS 2003, LNCS 2799, pp. 520–529, 2003.
© Springer-Verlag Berlin Heidelberg 2003

bus so that the low power method of a dedicated bus is more general, and it is the research point of this study.

To reduce the power consumption of a bus, two methods have been proposed as the reduction of switching activity [5], [7], [8], [9], [11], [12] and the number of a certain signal level [13]. The switching activity reduction method is generally applied to reduce the power consumption of the internal bus of a chip, whereas the certain signal level reduction method is generally exploited to reduce the power consumption of the external I/O bus of a chip. In the present paper, we concentrate on the internal bus structure and hence study a new low power method by reducing the switching activity of a bus.

On the other hand, for a reliable asynchronous bus structure in SOC designs, the dual-rail data encoding method [14] has been intensively investigated. Unfortunately, the dual-rail data encoding method causes much power because it requires two transitions of one of two physical lines for each one-bit data. Therefore, power reduction methods for the dual-rail data encoding method should be taken into account. One-of-four data encoding method reduces the power consumption of the dual-rail encoding method by the reduction of switching activities [15]. Meanwhile, the data pattern analysis illustrates that many data items are repeatedly transmitted in accordance with the result in [16]. As a result, the dual-rail and one-of-four data encoding methods waste the power when the data bus transmits the same data items repeatedly. To reduce such waste of power, a different method should be investigated.

In the paper, we propose a *frequent value cache*(FVC) with several update algorithms. The basic idea is to use a buffer to store the frequently transmitted data items and to send an index of the buffer when the data item is already stored in the buffer. Therefore, for the repeatedly transmitted data items, only the index lines work and the other lines are not changed. As a result, it can reduce the switching activity. Simulation result reveals that FVC can reduce the switching activity by 75% at maximum and finally the power consumption by 25% and 30% at maximum for integer and floating-point benchmarks, respectively, compared with the normal asynchronous dual-rail bus.

This paper is organized as follows. Section 2 analyzes data patterns over the data bus. The proposed FVC are explained in Section 3. The effect of FVC for the switching activity and the power consumption is illustrated in Section 4. This paper is concluded in Section 5 with a disscusion of the advantage of FVC.

2 Data Analysis

To avoid the use of an ambiguous configuration of a SOC model, we analyze a simple but fundamental bus model used between a general processor and a memory module. A SimpleScalar processor [17] is used for the processor. For the benchmark programs, SPEC [18] integer and floating-point programs are used. To investigate the data patterns over the data bus, load and store operations are investigated.

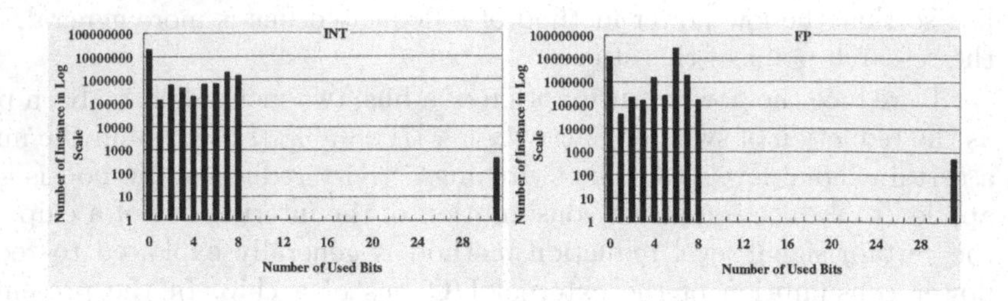

Fig. 1. Minimum Bit Width of Data Items over Bus

To investigate how much bandwidth is wasted, we measure the bandwidth requirement by calculating the minimum bit width for each data item. For an internal bus in a general processor, a 32-bit bus is generally used. The minimum bit width of a data item is calculated by the first bit position of 1 from the MSB. Figure 1 shows the statistics of the minimum bit width for the target bus configuration. From the statistics, several conclusions are derived as follows: 1) Most data items require 0 bit. For both integer and floating-point benchmarks, many data items can be represented with 0 bit; 2) Most data items can be represented with 8 bits because most load and store operations deliver 8-bit operands; and 3) Few data items require 32 bits.

The above analyses illustrate that most data items do not require the supported maximum bandwidth, so that the physically supported maximum bandwidth is wasted for most data communications. Specifically, a pattern that most data items require only 8 bits leads to a conclusion that many data items occur in a repeated manner. From such conclusions, we expect that the power waste can be reduced by reducing the repeated delivery of the same data item.

3 Frequent Value Cache

To reduce the power consumption of a bus, we proposes a new communication method that exploits the feature of repeatedly transmitted data item. The basic idea is to reduce the power waste due to the repeated transmissions of the same data item by using a buffer. The proposed buffer stores data items and sends an index for a data item when the data item to be sent is already stored in the buffer. Since the index requires fewer number of bits than the data itself, the wasted bandwidth or the switching activity can be decreased, and finally the power consumption can be decreased. To do that we propose a *frequent value cache*(FVC) and several replacement algorithms.

3.1 Frequent Value Cache

Figure 2 describes FVC very briefly that stores data items of each communication. The normal *sender* and *receiver* deliver a data item with a normal fashion,

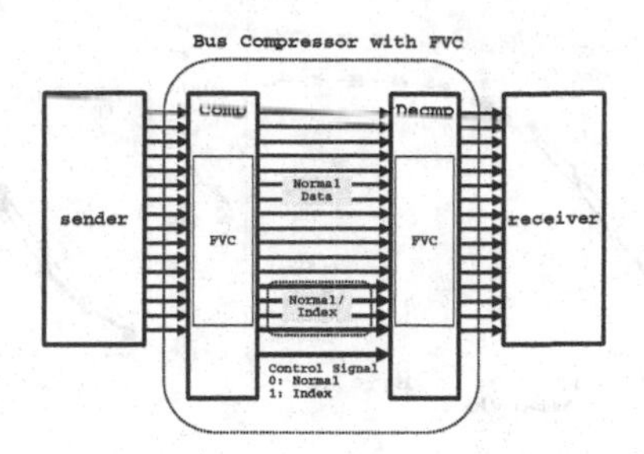

Fig. 2. Frequent Value Cache

while the *Comp* and *Decmp* deliver a data item by a data itself or an index of *FVC* depending on the hit of *FVC*. When a data itself is transferred, all bus lines are used; however, when an index of the data item is transferred, only the index lines are used. Thus, the index lines are used for both an index and a data item. To distinguish whether a transmitted information represents an index or a normal data item, a *control* signal is used. By using such hardware components, *FVC* sends an index when *FVC* hits or a normal data item itself when *FVC* miss. More specifically, the *Comp* and the *Decmp* work as follows.

- **Comp(Compressor):** When a target data item is stored in *FVC*, an index of the corresponding entry is transferred over only the index lines, and *FVC* is updated with the information of the data item. When a target data item is not stored in *FVC*, the target data item itself is transferred over the all bus lines, and *FVC* replaces an entry into the target data item. Depending on the hit of *FVC*, the *control* line changes the signal level as low for a miss and high for a hit case.
- **Decmp(Decompressor):** When the *control* signal is high, the received information from the index lines is used to lookup *FVC* to get a stored data item, and the retrieved data item is transferred to the final receiver component. At the same time, *FVC* changes the information of the corresponding entry with the same update algorithm used in *Comp*. When the *control* signal is low, the received information from the all bus lines is considered as a normal data item, and it is trasferred to the receiver. At the same time, *FVC* is updated using the same update alogirithm used in *Comp*.

For the correct communication, two *FVC*s in *Comp* and *Decmp* should contain the same data entries.

3.2 Update Algorithms

The performance of FVC depends on the hit ratio, and hence the update algorithm is important. In the paper, we propose three update algorithms as follows.

- **FIFO:** The oldest entry is replaced. It requires the simplest implmentation hardware. When FVC hits, no update is occured.

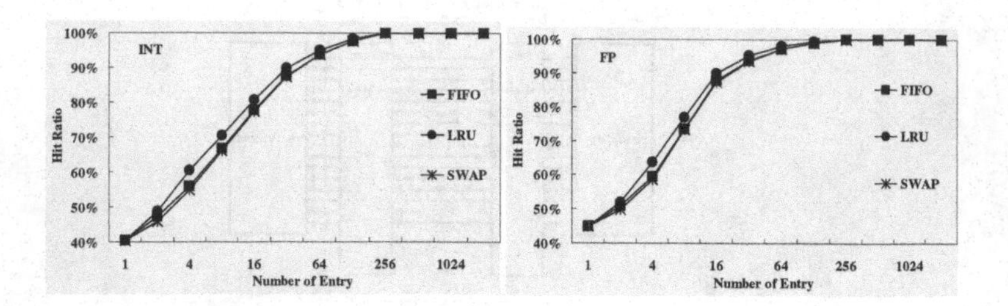

Fig. 3. Hit Ratio of FVC

Table 1. Dual-Rail Data Encoding Method

Logical Bit	Line_0	Line_1
0	1	0
1	0	1
Space	0	0

- **LRU:** The least recently used entry is replaced. It requires the most complex implementation hardware since it must contain the reference time of each entry. When FVC hits, the entry is changed into the most recently used entry.
- **SWAP:** The oldest entry is replaced. It is similar to the FIFO algorithm, but when FVC hits, the corresponding entry is changed as one step recently used entry. The hardware complexity is medium between the FIFO and LRU algorithms.

4 Analysis

We have investigated three measures as hit ratio, switching activity reduction ratio, and power consumption reduction ratio. The hit ratio is the most important one since it decides the switching activity reduction ratio that finally determines the power consumption reduction ratio.

Hit Ratio. The hit ratio of FVC is shown in Figure 3. The figure reveals the following conclusions: 1) Even only one entry of FVC can detect 40% of the repeatedly transmitted data items; 2) Over 256 entries can represent most data items; and 3) Three update algorithms show little difference.

Switching Activity Reduction Ratio. From the high hit ratio of the FVC, we need to know how much switching activity can be reduced. In the paper, only the change of signal levels between consecutive data items are measured to calculate the switching activity ratio of a bus.

The data encoding method affects the switching activity ratio of the asynchronous bus. As explained in the previous section, the dual-rail encoding method is generally used for the asynchronous bus for the design of a reliable SOC. Hence, we investigate the switching activity ratio when the dual-rail encoding method is used. For the dual-rail encoding method, two physical lines are used to represent one-bit data as shown in Table 1. To represent a logical data 1, the *Line_1* is high and the *Line_0* is low; to represent a logical data 0, the signal level of two physical lines are changed alternatively. However, both lines cannot have the same signal level 1 at the same time. To distinguish consecutive data items, a space state must be inserted between valid data items. From the switching activity point, an one-bit data line causes two transitions of a physical line because an one-bit data begins at the space state(*Line_0*=0, *Line_1*=0), changes into a valid state(01 or 10), and ends at the space state(00). Based on such encoding methods, the switching activity of the normal bus and the proposed FVC model are different as follows.

- The normal dual-rail bus utilizes all 32-bit logical bits, hence the switching activity is 32×2.
- FVC delivers an index for a hit case and a normal data item for a miss case. In addition, the control signal changes for every communications, hence it changes two times for each communcation. Therefore, the switching activity when FVC is used is calculated by Equation 1.

$$P_{hit} \times \{1 + \log(\#entry)\} \times 2 + (1 - P_{hit}) \times (1 + 32) \times 2 \quad (1)$$

Based on the above analyses, the switching activity reduction ratio of FVC over the normal dual-rail bus model is calculated by Equation 2.

$$\frac{P_{hit} \times \{1+\log(\#entry)\} \times 2 + (1-P_{hit}) \times (1+32) \times 2}{32 \times 2} \quad (2)$$

Figure 4 illustrates the switching activity reduction ratio of FVC over the normal dual-rail bus model. The figure illustrated that FVC reduces the switching activity by 75% at maximum. However, the switching activity reduction ratio is decreased after the maximum point because of increase of the number of index bits. Therefore, the proposed FVC can decrease the switching activity ratio drastically compared with the normal dual-rail bus model.

Power Consumption Reduction Ratio. Until the previous section, we have analyzed the potentials of the FVC method in the point of switching activity reduction ratio. However, the total power consumption should include the power consumption of the FVC tables although the power consumption ratio of the table would be below 5% as explained in [19]. In addition, the power consumption of the bus itself should be considered as well, as analyzed in [20]. Therefore, in this subsection we investigate the total power consumption when the power consumption of FVC tables and the bus lines are considered.

To measure the power consumption of FVC table and bus lines, we assume that 0.25 micron technology is used, and the length of the bus line is 10 mm,

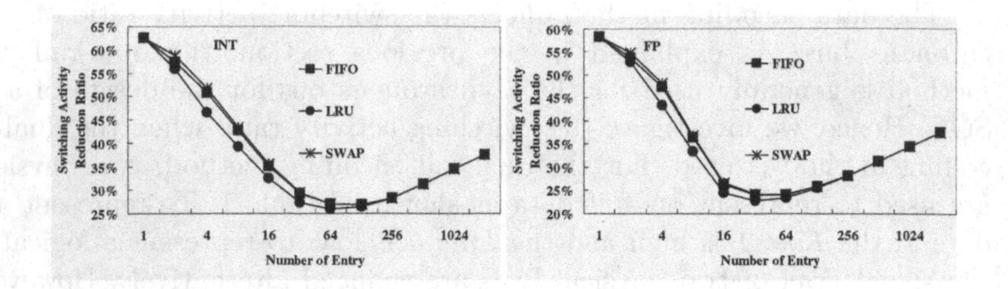

Fig. 4. Switching Activity Reduction Ratio

Table 2. Power Consumption of FVC Table(nJ)

# entry	power	# entry	power	# entry	power	# entry	power	# entry	power
2	0.0498	4	0.0561	8	0.0745	16	0.0985	32	0.1523
64	0.3193	128	0.5648	256	0.9246	512	1.9488	1024	3.3807

which follows the 2001 ITRS [21]. Power consumptions of the normal model and the FVC model are as follows:

- **Normal Model:** The power consumption is only caused by the dual-rail bus for logical 32-bit bus lines. Based on the 0.25 micron technology, 10 mm bus lines consume about 0.4 nJ, which is measured by the NanoSim power analysis tool of Synopsys CAD environment.
- **FVC Model:** The power consumption is caused by two parts as the FVC table and bus lines. To measure the power consumption the FVC table, we use the CACTI tool [22]. Since all entries should be checked at the same time, we assume that the FVC table is a fully associative memory. Table 2 shows the power consumption with the number of entries. On the other hand, the power consumption of the bus line depends on the reduction of switching activity. Therefore, we can formulate the power consumption of FVC model as Equation 3. Specifically, the power consumption of the FVC table is multiplied by two because FVC model requires two FVC tables for a sender and a receiver.

$$Table\_Power \times 2 + Bus\_Power \times Switching\_Activity\_Reduction\_Ratio \tag{3}$$

Finally, we can derive a power consumption reduction ratio of the FVC model over the normal model as shown in Equation 4.

$$\frac{Table\_Power \times 2 + (0.4nJ) \times Switching\_Activity\_Reduction\_Ratio}{0.4nJ} \tag{4}$$

In Equation 4, the *Table_Power* is measured in Table 2, and the *Switching_Activity_Reduction_Ratio* is based on Figure 4. Figure 5 shows the power consumption reduction ratio when the FVC model is used. From the figure,

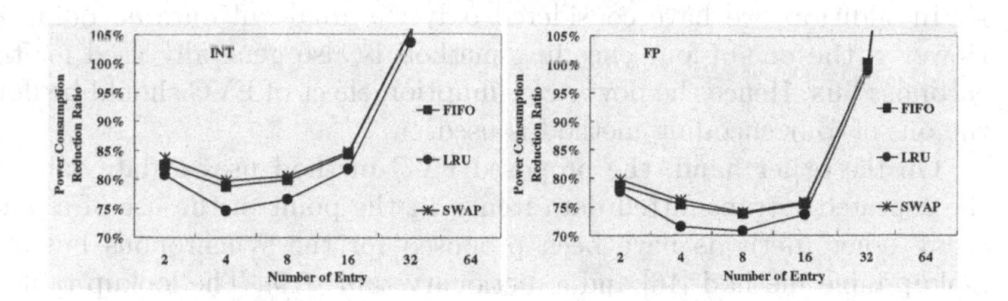

Fig. 5. Power Consumption Reduction Ratio

we can conclude the power consumption reduction effect of the FVC model as follows:

- FVC reduces the total power consumption by about 25% and 30% at maximum for integer and floating-point benchmarks, respectively. Because of the power consumption of two FVC tables, the power consumption reduction ratio is not so high compared with the switching activity reduction ratio.
- Only small number of entries can justify the power consumption reduction. In the figure, we can know that when 4 and 8 entries are used for integer and floating-point benchmarks respectively, the power consumption reduction ratio is the maximum. However, large number of entries increases the power consumption of FVC table itself, and hence reduces the advantage of the FVC method. Therefore, a certain number of FVC entries should be chosen to maximize the reduction of power consumption.

5 Conclusion and Future Work

The reduction of the power consumption for the asynchronous dual-rail data bus have been investigated by using a buffer to store the repeatedly transmitted data items and by reducing the switching activity of the bus when the target data item is already stored in the buffer. To reduce the power consumption caused by the repeatedly transmitted data items, we have proposed FVC that utilizes not always all bus lines but sometimes the fewer bus lines for an index. In addition, several update algorithms have been investigated, which show a similar performance. Simulation results have revealed that FVC reduces the power consumption of the normal asynchronous dual-rail bus by 25% and 30% at maximum for integer and floating-point benchmarks, respectively.

Meanwhile, we have not discussed the power consumption reduction of the detection logic for the dual-rail bus. When an index is transmitted, a detection logic needs to check the validity of the index lines, not all bus lines. Therefore, the power consumption of the detection logic will be decreased when FVC is utilized.

In addition, we have considered only the dual-rail data encoding method. However, the one-of-four encoding method is also generally used for the asynchronous bus. Hence the power consumption effect of FVC should be done when the one-of-four encoding method is used.

On the other hand, the proposed FVC method uses a data table to store the repeatedly transmitted data items. In the point of the use of a data table, many other methods have been proposed for the synchronous bus such as a lookup table method [16] and a dictionary table [19]. The lookup table method uses a different communication method by using an one-hot coding method; the dictionary table stores not all data bits, but the higher bits of data item to get higher hit ratio of the table. Although, those methods have been proposed for the synchronous bus model, they can be utilized for the asynchronous bus. Therefore, a future research should be performed when the different lookup table structure and data format for an entry are utilized for the asynchronous bus.

As well, we have not discussed the effect of the bus width and the layout the index lines of FVC model. When the bus width is small, the effect of FVC will be decreased as analyzed in [23]. Meanwhile, the crosstalk effect between bus lines could be considered as an alternative choice of the layout of index lines in FVC. In the paper, we have assumed that the index lines are located in the lower bit position. However, we can locate the index lines evenly distributed locations. In that case, the crosstalk effect of index lines can be decreased. Therefore, future research will be gone to find the effect of the bus width and the layout of index lines.

Acknowledgments. The authors gratefully acknowledge the reviewer's helpful comments in preparing the final manuscript. This work was supported in part by the KAIST/K-JIST IT-21 Initiative in BK21 of Ministry of Education, in part by the Korea Science and Engineering Foundation (KOSEF) through the Ultrafast Fiber-Optic Networks Research Center at Kwangju Institute of Science and Technology, and in part by the University Research Program supported by Ministry of Information and Communication in republic of Korea, and in part by the IC Design Education Center(IDEC).

References

1. Hauck, Scott: Asynchronous Design Methodologies: An Overview, Proc. of the IEEE, Vol.83, No.1, (1995), 69–93
2. Rung-Bin Lin and Chi-Ming Tsai: Theoretical Analysis of Bus-Invert Coding, IEEE Trans. on VLSI, Vol.10, No.6, (2002), 929–935
3. Mircea R. Stan and Wayne P. Burleson: Low-Power Encoding for Global Communication in CMOS VLSI, IEEE Trans. on VLSI, Vol.5, Issue 4, (1997), 444–455
4. Enrico Macii, Massoud Pedram, and Fabio Somenzi: High-Level Power Modeling, Estimation, and Optimization, IEEE Trans. on VLSI, Vol. 17, No. 11, (1998), 1061–1079
5. Lang, T., Musoll, E., and Cortadella, J.: Extension of the Working-Zone-Encoding Method to Reduce the Energy on the Microprocessor Data Bus, Proc. of the Conf. on Computer Design, (1998), 414–419

6. Pamprasad, S., Shanbhag, N.R., and Hajj, I.N.: Coding for Low-Power Address and Data Busses: A Source-Coding Framework and Applications, Proc. of the Conf. on VLSI Design, (1998), 18–23
7. Youngsoo Shin, Soo-Ik Chae, and Kiyoung Choi: Partial Bus-Invert Coding for Power Optimization of System Level Bus, Proc. of the Symp. on Low Power Electronics and Design, (1998) 127–129
8. Youngsoo Shin and Kiyoung Choi: Narrow Bus Encoding for Low Power Systems, Proc. of the Asia and South Pacific Design Automation Conf., (2002), 217–220
9. Musoll, E., Lang, T., and Cortadella, J.: Working-Zone Encoding for Reducing the Energy in Microprocessor Address Buses, IEEE Trans. on VLSI, Vol.6, No.4, (1998), 568–572
10. Kapadia, H., Benini, L., and De Micheli, G.: Reducing Switching Activity on Datapath Buses with Control-Signal Gating, IEEE Journal of Solid-State Circuits, Vol. 34, No.3, (1999), 405–414
11. Mircea R. Stan and Wayne P. Burleson: Bus-Invert Coding for Lower-Power I/O, IEEE Trans. on VLSI, Vol.3, No.1, (1995), 49–58
12. Madhu, M., Murty, V.S., and Kamakoti, V.: Dynamic Coding Technique for Low-Power Data Bus, Proc. of the IEEE Computer Society Annual Symp. on VLSI, (2003), 252–253
13. Rung-Bin Lin and Jinq-Chang Chen: Lower Power CMOS Off-Chip Drivers with Slew-Rate Difference, Proc. of the Asia and South Pacific Design Automation Conf., (1999), 169–172
14. T. Verhoeff: Delay-Insensitive Codes: An Overview, Distributed Computing, Vol. 3, (1988) 1–8
15. Bainbridge, W.J. and Furber, S.B.: Delay Insensitive System-on-Chip Interconnect using 1-of-4 Data Encoding, Proc. of the Symp. on Asynchronous Circuits and Systems, (2001), 118–126
16. Benjamin Bishop and Anil Bahuman: A Low-Energy Adaptive Bus Coding Scheme, Proc. of the IEEE Workshop of VLSI, (2001), 118–122
17. D. Burger and T. Austin: The SimpleScalar Tool Set, Version 2.0, Technical Report, CS-TR-97-1342, University of Wisconsin Madison, (1997)
18. SPEC CPU Benchmarks: http://www.specbench.org/osg/cpu95
19. Lv, T., Henkel, J., Lekatsas, H., and Wolf, W.: An Adaptive Dictionary Encoding Scheme for SOC Data Buses, Proc. of the Design, Automation and Test in Europe Conf. and Exhibition, (2002), 1059–1064
20. P. Sotiriadis and A. Chandrakasan: Low-Power Bus Coding Techniques Considering Inter-Wire Capacitances, Proc. of the IEEE Conf. on Custom Integrated Circuits, (2002), 507–510
21. The Semiconductor Industry Association: The International Technology Roadmap for Semiconductor, (2001)
22. Premkishore Shivakumar, Normal P. Jouppi: CACTI 3.0: An Integrated Cache Timing, Power, and Area Model, HP Labs Technical Reports, WRL-2001-2, (2001)
23. Zangi, U., Ginosar, R.: A Low Power Video Processor, Proc. of the Symp. on Low Power Electronics and Design, (1998) 136–138

Reducing Static Energy of Cache Memories via Prediction-Table-Less Way Prediction

Akihito Sakanaka[1] and Toshinori Sato[1,2]

[1] Department of Artificial Intelligence
Kyushu Institute of Technology
[2] Precursory Research for Embryonic Science and Technology
Japan Science and Technology Corporation
toshinori.sato@computer.org

Abstract. Power consumption is becoming one of the most important constraints for microprocessor design in nanometer-scale technologies. Especially, as the transistor supply voltage and threshold voltage are scaled down, leakage energy consumption is increased even when the transistor is not switching. This paper proposes a simple technique to reduce the static energy. The key idea of our approach is to allow the ways within a cache to be accessed at different speeds. We combine variable threshold voltage circuits with way prediction technique to activate only the way which will be referred, and propose a simple prediction mechanism which eliminates history tables. Experimental results on 32-way set-associative caches demonstrate that any severe increase in clock cycles to execute application programs is not observed and significant static energy reduction can be achieved, resulting in the improvement of energy-delay product.

1 Introduction

Power consumption is becoming one of the most important concerns for microprocessor designers in nanometer-scale technologies. Until recently, the primary source of energy consumption in digital CMOS circuits has been the dynamic power that is caused by dynamic switching of load capacitors. The trend of the reduction in transistor size reduces capacitance, resulting in less dynamic power consumption. Microprocessor designers have relied on scaling down the supply voltage, resulting in further dynamic power reduction[9,10,14]. In addition, many architectural technologies to reduce power have been proposed by reducing the number of switching activities[8,12,16]. To maintain performance scaling, however, threshold voltage must also be scaled down with supply voltage. Unfortunately, this increases leakage current exponentially. The International Technology Roadmap for Semiconductors (ITRS) predicts an increase in leakage current by a factor of two per generation[19]. Borkar estimates a factor of 5 increases in leakage energy in every generation[1].

Many of techniques[3,11,13,18] proposed to address this problem have focused on cache memory that is a major energy consumer of the entire system because

J.J. Chico and E. Macii (Eds.): PATMOS 2003, LNCS 2799, pp. 530–539, 2003.
© Springer-Verlag Berlin Heidelberg 2003

leakage energy is a function of the number of transistors. For example, the Alpha 21264 and the StrongARM processors use 30% and 60% of the die area for cache memories[16]. Current efforts at static energy reduction have focused on dynamically resizing active area of caches[3,11,13,18]. These architectural techniques require additional circuits to control cache activities. The control circuitry including history tables consumes power, and thus these dynamic approaches are not suitable for embedded processors due to the area and power overhead and design complexity. In order to solve the problem, we propose a simple technique for leakage energy reduction.

The organization of the rest of this paper is as follows: Section 2 discusses the motivation of our work. Section 3 presents our concept to reduce leakage energy in caches, and explains a prediction-table-less way prediction mechanism. Section 4 presents experimental results and discussion on the effectiveness of our approach. Finally, Section 5 concludes the paper.

2 Motivations

2.1 CMOS Power Consumption

Power consumption in a CMOS digital circuit is governed by the equation:

$$P = P_{active} + P_{off} \tag{1}$$

where P_{active} is the active power and P_{off} the leakage power. The active power P_{active} and gate delay t_{pd} are given by

$$P_{active} \propto f C_{load} V_{dd}^2 \tag{2}$$

$$t_{pd} \propto \frac{V_{dd}}{(V_{dd} - V_t)^{\alpha}} \tag{3}$$

where f is the clock frequency, C_{load} the load capacitance, V_{dd} the supply voltage, and V_t the threshold voltage of the device. α is a factor dependent upon the carrier velocity saturation and is approximately 1.3–1.5 in advanced MOSFETs[7]. Based on Eq.(2), it can easily be found that a power-supply reduction is the most effective way to lower power consumption. However, Eq.(3) tells us that reductions in the supply voltage increase gate delay, resulting in a slower clock frequency, and thus diminishing the computing performance of the microprocessor. In order to maintain high transistor switching speeds, it is required that the threshold voltage is proportionally scaled down with the supply voltage.

On the other hand, the leakage power can be given by

$$P_{off} = I_{off} V_{dd} \tag{4}$$

where I_{off} is the leakage current. The subthreshold leakage current I_{off} is dominated by threshold voltage V_t in the following equation:

$$I_{off} \propto 10^{-\frac{V_t}{S}} \tag{5}$$

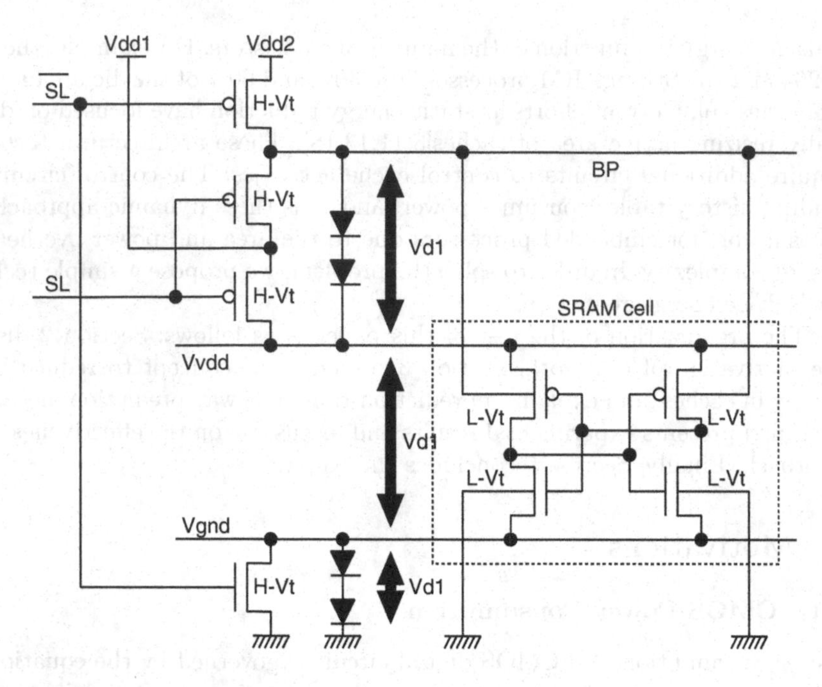

Fig. 1. ABC-MOS Circuit

where S is the subthreshold swing parameter and is around 85mV/decade[19]. Thus, lower threshold voltage leads to increased subthreshold leakage current and increased static power. Maintaining high transistor switching speeds via low threshold voltage gives rise to a significant amount of leakage power consumption.

2.2 Variable Threshold Voltage CMOS Circuits

There are several circuits proposed to reduce leakage current by dynamically raising the threshold voltage, for example by modulating backgate bias voltage[15, 17]. Figure 1 shows ABC-MOS circuit[17]. The four transistors with high threshold voltage (denoted as `H-Vt`) become the switch that cut off leakage current. During normal operation, when sleep signal `SL` is deasserted, the virtual source line, `Vvdd`, and the back gate bias line, `BP`, are set to appropriate power supply, `Vdd1`. During sleep mode, when `SL` is asserted, P-well is biased using the alternative power supply, `Vdd2`, at a higher voltage level. Diodes perform backgate bias effect and thus threshold voltage of all transistors is increased. This reduces leakage current, which flows from `Vdd2` to the ground. As shown in Table 1, Hanson et al.[5,6] report that leakage current in sleep mode is reduced by a factor of 160 compared with that in normal mode. VT-CMOS[15] also controls the backgate bias to reduce leakage current in sleep mode by rising up threshold voltage.

Table 1. Impact of V_t on Leakage Current

Mode	I_{off} (nA)
Sleep	12
Normal	1941

2.3 Related Works

Current efforts at static energy reduction have focused on dynamically resizing active area of caches[3,11,13,18]. These architectural techniques employ circuit techniques. VT+ADR cache[3] and SA cache[11] use VT-CMOS[15] and ABC-MOS[17], respectively. Decay cache[13] and DRI cache[18] use gated-V_{dd}[18], which shuts off the supply voltage to SRAM cells to reduce leakage current. The circuit technique has a disadvantage. Gated-V_{dd} loses the state within the memory cell in the sleep mode. Thus, additional cache messes might occur, resulting in an additional dynamic power consumption. In contrast, VT-CMOS and ABC-MOS can retain stored data in the sleep mode. However, these architectural techniques require additional circuits to control cache activities. The control circuitry, especially large cache-structured history table, consumes power, however, most studies did not consider its effect. In summary, these dynamic approaches are not suitable for embedded processors due to the area and power overhead and design complexity.

3 Prediction-Table-Less Way Prediction

In this paper, we propose a simple technique for leakage energy reduction. We combine the variable threshold voltage circuit with way prediction to limit the number of ways in normal mode. Way prediction is a technique for set-associative cache, which reduces conflict misses and maintains the hit speed of direct-mapped cache. It predicts the way within the set of the next cache access. A hit prediction means short access latency, since only a single way is speculatively referred. However, a miss results in checking the remaining ways in subsequent cycles and thus in long access latency. A dynamic power reduction technique based on way prediction has been proposed[8], and it employs a way prediction policy based on the most recently used (MRU) algorithm.

In this study, we use way prediction for leakage energy reduction. Based on every prediction, only one way, which will be referred in the next cache access, is activated. Remaining ways are set to sleep mode. In other words, only one way consumes large static energy and the other ways do negligible one. We name the cache *non-uniform-access-latency (NUAL) cache*. NUAL cache based on way prediction resembles VT+ADR cache based on address prediction[3] and SA cache based on block prediction[11]. However, while they require large history table to predict the region which will be accessed next, NUAL cache eliminates the history table as explained next.

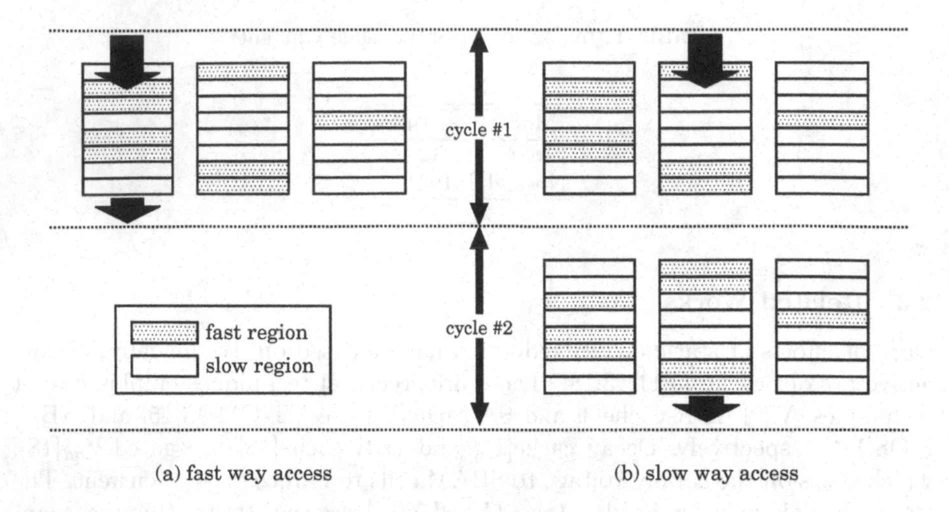

Fig. 2. Non-uniform Access-Latency Cache

We use least recently used (LRU) bits that is already implemented in every set-associative cache. The LRU bits are utilized to determine which way is replaced for the next cache refill process, and keep the order of way accesses. According to the LRU bits, least recently used ways go to sleep mode. The remaining ways stay in normal mode. As you can easily find, no history table is required for NUAL cache. While we use a 3-way set-associative cache for explanation, our proposal is applicable to any set- and full-associative caches. In this explanation, in every set, two least recently used ways go to sleep mode, and only one way is in normal mode. We call the way in normal mode *the primary fast way* and those in sleep mode *slow ways*. When a cache access refers the primary fast way, the datum is provided in 1 cycle as shown in Fig.2(a). The primary ways are in gray. In contrast, it refers a slow way, activating the slow way requires 1 cycle and the datum is provided in the following cycle, as shown in Fig.2(b).

We name the NUAL cache presented in Fig.2 *fine-grained* NUAL cache, because activating or inactivating is each set basis. This requires a backgate bias control circuit per way in every set. To reduce the number of the backgate bias control circuits, it is possible to make the activate policy way basis. This requires an additional N-bit LRU register for N-way set-associative cache and only N backgate bias control circuits. We name this type of NUAL cache *coarse-grained* NUAL cache. The behavior of coarse-grained NUAL cache is explained in Fig.3. When a cache access refers the primary fast way, the datum is provided in 1 cycle as shown in Fig.3(a). There is no difference from fined-grained NUAL cache. In contrast, it refers a slow way, activating the slow way requires 1 cycle and the datum is provided in the following cycle, as shown in Fig.3(b). In this case, all ways regardless of sets are activated, different from fine-grained cache. As you can find, in coarse-grained NUAL cache, it is impossible to optimally predict the

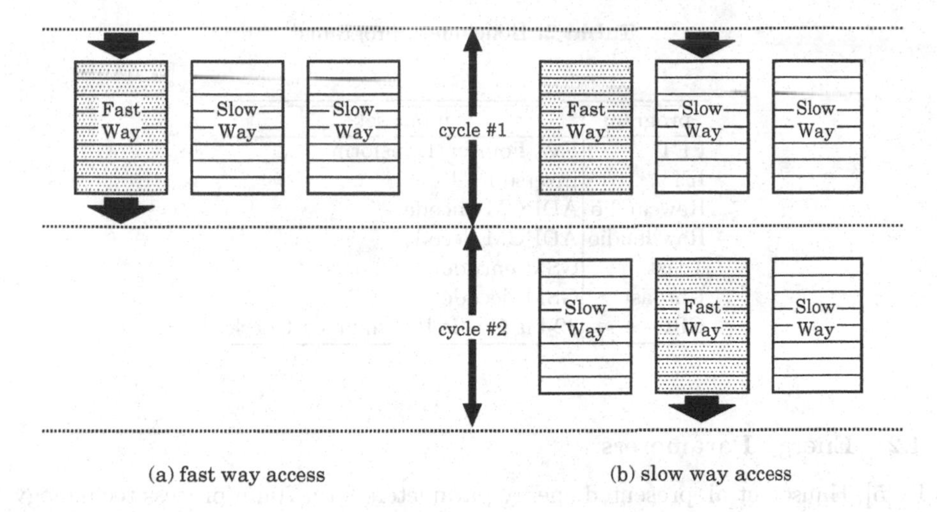

Fig. 3. Coarse-Grained NUAL Cache

next way for every set, and thus average access latency may be longer than that of fine-grained NUAL cache. In the next section, we will evaluate this.

4 Experimental Results

4.1 Simulation Methodology

We implemented our simulator using the SimpleScalar/ARM tool set[2]. We use ARM instruction set architecture in this study. The processor model evaluated is based on Intel XScale processors[9]. 4KB, 4B block, 32-way set-associative L1 instruction and data caches are used. They have a load latency of 1 cycle and a miss latency of 32 cycles. The non-uniform-access-latency 32-way set-associative L1 instruction and data caches consist of one primary fast way and remaining 31 slow ways. It has the same load latency of 1 cycle when the requested data is placed in the primary way and otherwise has a latency of 2 cycles, 1 cycle of which is required to activate the sleeping the way[5,6,11] and another cycle of which is for reading the way. We also evaluate caches whose load latency is always 2 cycles for comparison. The replacement policy is based on LRU. No L2 cache is used. A memory operation that follows a store whose data address is unknown cannot be executed.

The MiBench[4] is used for this study. It is developed for use in the context of embedded, multimedia, and communications applications. It contains image processing, communications, and DSP applications. We use original input files provided by University of Michigan. Table 2 lists the benchmarks we used. All programs are compiled by the GNU GCC with the optimization options specified by University of Michigan. Each program is executed to completion.

Table 2. Benchmark programs

program	input set
FFT	Fast Fourier Transform
IFFT	Inverse FFT
Rawcaudio	ADPCM encode
Rawdaudio	ADPCM decode
Toast	GSM encode
Untoast	GSM decode
CRC	32-bit Cyclic Redundancy Check

4.2 Energy Parameters

In [5], Hanson et al. presented energy parameters for a 70nm process technology. there is no consensus for ABC-MOS to be effective for 70nm process technologies yet. Fortunately for us, Hanson et al. re-evaluated their energy parameter using HSPICE instead of CACTI and found the parameters are more closely aligned with 100nm technologies[6]. Thus, we can use data from Hanson et al.'s[5], and we assume a 100nm process technology, a 2.5GHz clock frequency, a 0.75V supply voltage, and a 110C operating temperature. To compute total energy reduction, we compute the leakage energy using the numbers shown in Table 1 and the number of clock cycles to execute each program. However, since Hanson et al.[5, 6] reports the energy required to awake every way is only 50 fJ per bit, we ignore the energy in our evaluation.

4.3 Results

Figures 4 and 5 shows way hit ratio in the cases of coarse- and fine-grained NUAL caches. The hit means that a cache access refers the primary fast way. The left bar is for the instruction cache, and the right is for the data cache. In the case of the coarse-grained NUAL cache, instruction and data caches suffer approximately 20% and 50% miss ratios, respectively. Every miss requires additional latency to activate the way. This might occur severe performance loss. In contrast, both fine-grained NUAL instruction and data caches achieve 90% of the way hit ratio for most programs.

Figure 6 shows relative processor performance. For each group of 4 bars, the first one (see from left to right) indicates the performance which has the conventional set-associative cache, whose load latency is always 1 cycle (denoted by `Fast`). The middle two bars indicate those of the coarse- and fine-grained NUAL caches (denoted by `Coarse` and `Fine`), respectively, and the right indicates that of the slow conventional cache, which has a load latency of 2 cycles (denoted by `Slow`). Every result is normalized by that of the `Fast` model. We can find the model with the `Fine` NUAL cache has no considerable difference from that with the conventional fast and energy-hungry `Fast` cache. In addition, it has significant performance gain over that with slow and energy-efficient `Slow` cache. In

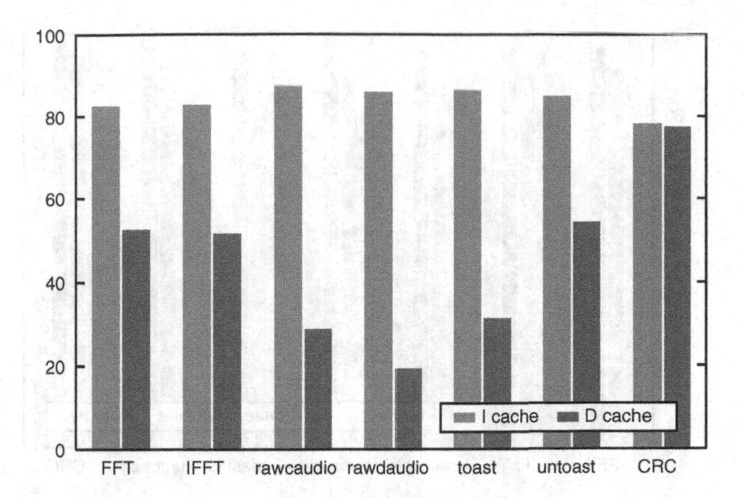

Fig. 4. %Way Hit Ratio (coarse-grained)

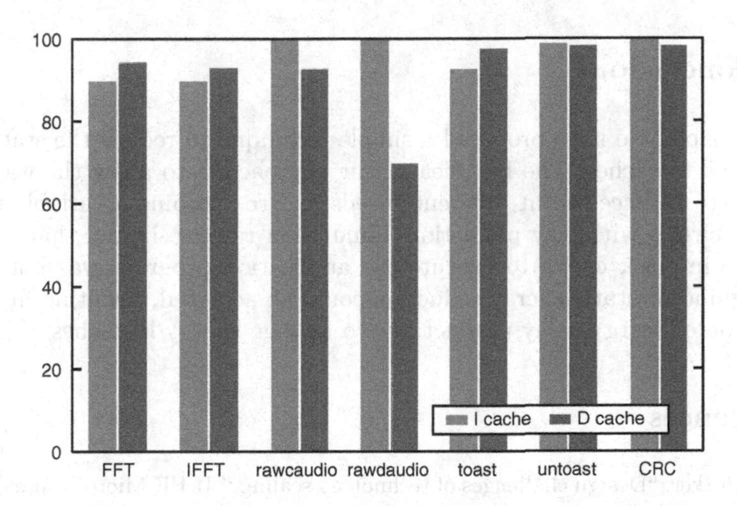

Fig. 5. %Way Hit Ratio (fine-grained)

contrast, the `Coarse` NUAL cache suffers approximately 10% performance loss, as we expected due to relatively large miss ratio (see Fig. 4).

The static energy consumed in NUAL cache is reduced by approximately the factor of 30, since only 1 of 32 ways is activated and we do not consider the energy required to awake every way. In other words, we only consider leakage energy. Based on the discussion in Section 4.2, in the case of the processor model with the fine-grained NUAL cache, we calculate the leakage energy and find over 96% reduction of energy-delay product due to leakage energy for all benchmark programs.

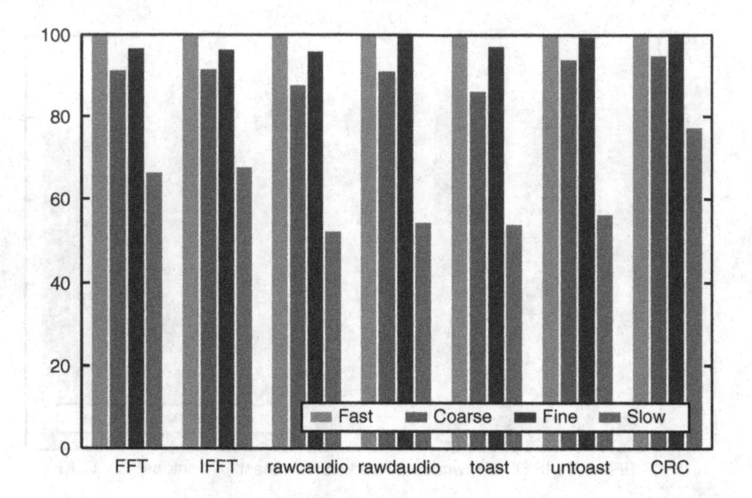

Fig. 6. %Processor Performance

5 Conclusions

In this paper, we have proposed a simple technique to reduce the static energy consumed in caches. The key idea of our approach is to allow the ways within a cache to be accessed at different speeds and to combine a variable threshold voltage circuit with way prediction. Simulation results showed that any severe increase in clock cycles to execute the application program was not observed and significant static energy reduction could be achieved, resulting in over 96% reduction of energy-delay product due to leakage energy in caches.

References

1. S. Borker, "Design challenges of technology scaling," IEEE Micro, volume 19, number 4, 1999.
2. D. Burger and T. M. Austin, "The SimpleScalar tool set, version 2.0," ACM SIGARCH Computer Architecture News, volume 25, number 3, 1997.
3. R. Fujioka, K. Katayama, R. Kobayashi, H. Ando, and T. Shimada, "A preactivating mechanism for a VT-CMOS cache using address prediction," International Symposium on Low Power Electronics and Design, 2002.
4. M. R. Guthaus, J. S. Ringenberg, D. Ernst, T. M. Austin, T. Mudge, R. B. Brown, "MiBench: A free, commercially representative embedded benchmark suite," Workshop on Workload Characterization, 2001.
5. H. Hanson, M. S. Hrishikesh, V. Agarwal, S. W. Keckler, and B. Burger, "Static energy reduction techniques for microprocessor caches," International Conference on Computer Design, 2001.
6. H. Hanson, M. S. Hrishikesh, V. Agarwal, S. W. Keckler, and B. Burger, "Static energy reduction techniques for microprocessor caches," IEEE Transactions on VLSI Systems, to appear.

7. T. Hiramoto and M. Takamiya, "Low power and low voltage MOSFETs with variable threshold voltage controlled by back-bias," IEICE Transactions on Electronics, volume E83-C, number 2, 2000.
8. K. Inoue, T. Ishihara, and K. Murakami, "Way-predicting set-associative cache for high performance and low energy consumption," International Symposium on Low Power Electronics and Design, 1999.
9. Intel Corporation, "Intel XScale technology," `http://developer.intel.com/design/intelxscale/`, 2002.
10. T. Ishihara and K. Asada, "A system level memory power optimization technique using multiple supply and threshold voltages," Asia and South Pacific Design Automation Conference, 2001.
11. T. Ishihara and K. Asada, "An architectural level energy reduction technique for deep-submicron cache memories," Asia and South Pacific Design Automation Conference, 2002.
12. A. Iyer and D. Marculescu, "Power aware microarchitecture resource scaling," Design, Automation and Test in Europe Conference and Exhibition, 2001.
13. S. Kaxiras, Z. Hu, G. Narlikar, and R. McLellan, "Cache-line decay: a mechanism to reduce cache leakage power," Workshop on Power Aware Computer Systems, 2000.
14. A. Klaiber, "The technology behind Crusoe processors," Transmeta Corporation, White Paper, 2000.
15. T. Kuroda, T. Fujita, S. Mita, T. Nagamatsu, S. Yoshioka, F. Sano, M. Norishima, M. Murota, M. Kato, M. Kinugasa, M. Kakumu, and T. Sakurai, "A 0.9V, 150MHz, 10mW, 4mm^2, 2-D discrete cosine transform core processor with variable-threshold-voltage scheme," International Solid State Circuit Conference, 1996.
16. S. Manne, A. Klauser, and D. Grunwald, "Pipeline gating: speculation control for energy reduction," International Symposium on Computer Architecture, 1998.
17. K. Nii, H. Makino, Y. Tujihashi, C. Morishima, and Y. Hayakawa, "A low power SRAM using auto-backgate-controlled MT-CMOS," International Symposium on Low Power Electronics and Design, 1998.
18. M. Powell, S. H. Yang, B. Falsafi, K. Roy, and T. N. Vijaykumar, "Gated-Vdd: a circuit technique to reduce leakage in deep-submicron cache memories," International Symposium on Low Power Electronics and Design, 2000.
19. D. Sylvester and H. Kaul, "Power-driven challenges in nanometer design," IEEE Design & Test of Computers, volume 18, number 6, 2001.

A Bottom-Up Approach to On-Chip Signal Integrity

Andrea Acquaviva and Alessandro Bogliolo

Information Science and Technology Institute (STI) – University of Urbino
61029 Urbino, Italy
acquaviva@sti.uniurb.it
alessandro.bogliolo@uniurb.it

Abstract. We present a new approach to accurately evaluate signal integrity in digital integrated circuits while working at the logic level. Our approach makes use of fitting models to represent the key properties of drivers, interconnects and receivers and the effects of all noise sources (supply noise, timing uncertainty, crosstalk). Such models are then combined to evaluate the correctness of each bit sent across the line. The overall result is a parameterized bit-level model of a noisy on-chip communication channel. The model can be used at the logic level to evaluate the transmission-error probability for an arbitrary bit stream, sent at an arbitrary bit rate, under arbitrary noise source assumptions.

1 Introduction

Signal integrity is a primary concern for designers of deep sub micron integrated circuits [1]. Shrinking technologies and reducing noise margins expose digital circuits to the effects of several noise sources (voltage drops causing common-mode and differential supply noise, cross-talk, inter-symbol interference, clock skew and jitter) that may lead to logical errors. The integrity of digital signals is particularly critical on long interconnects, because of: the large distance between driver and receiver (possibly leading to different effective supply voltages and misaligned clock signals), the large parasitic parameters of the line (causing a signal degradation along the line), the large coupling capacitance (responsible of cross-talk) [2],[3].
Traditional techniques for dealing with noise sources make use of conservative noise margins that have two main limitations: first, they lead to over-conservative designs, second, they are not directly related to bit-level error probabilities.
On the other hand, long interconnects are a bottleneck for high-performance low-power integrated circuits, since they are the main responsible of propagation delay and power consumption. Hence, on-chip interconnects need to be pushed to their limits. In this context, accurate and efficient models of communication channels and noise sources are required to enable a thorough design space exploration.

In this paper we propose a parameterized black-box model of noisy on-chip interconnects that enables accurate signal integrity estimation at the logic level. In particular: i) we use *noise sensitive areas* (NSAs) to represent the noise margins that should be respected at the inputs of the receiver in order to guarantee the correct sampling of the

J.J. Chico and E. Macii (Eds.): PATMOS 2003, LNCS 2799, pp. 540–549, 2003.
© Springer-Verlag Berlin Heidelberg 2003

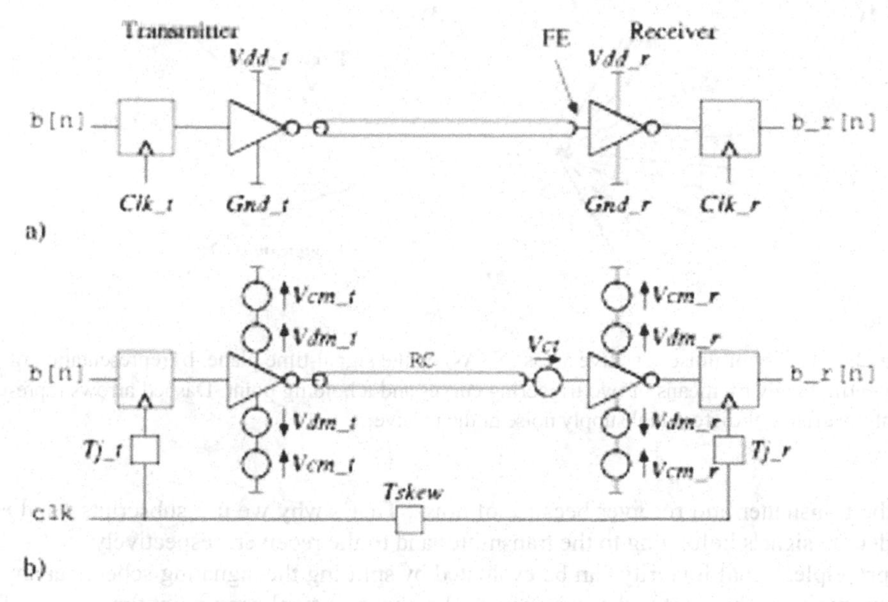

Fig. 1. a) Reference signaling scheme. b) Circuit-level model of noise sources.

received bit stream [4]; ii) we model the effect of noise sources on the NSA; iii) we use the average signal slope at the far end of the line to represent the performance of the driver for a given line; iv) we model the effect of noise sources on the maximum signal slope; v) we combine noisy signal slopes and parametric NSA to obtain bit error probabilities.

The rest of the paper is organized as follows. In Section 2 we provide the overall picture of the proposed approach. In Section 3 we introduce parameterized NSA and we describe the characterization approach. In Section 4 we characterize the maximum signal slope at the receiver. In Section 5 we combine signal slopes and NSAs to obtain logic-level error probabilities from noise source distributions, for given bit rates and bit streams. In Section 6 we validate the approach by comparing logic-level and electrical-level simulation results and we exemplify the application of the proposed approach.

2 The Proposed Approach

We refer to the voltage-mode signaling scheme shown in Figure 1.a. The interconnect is driven by a transmitter (e.g., a CMOS buffer) that takes the input stream from a local flip flop. The receiver is composed of an amplifier (e.g., a CMOS buffer) and a flip flop that samples the received bit stream. Although transmitter and receiver have the same nominal clock signal and supply voltage, the actual signals may be different

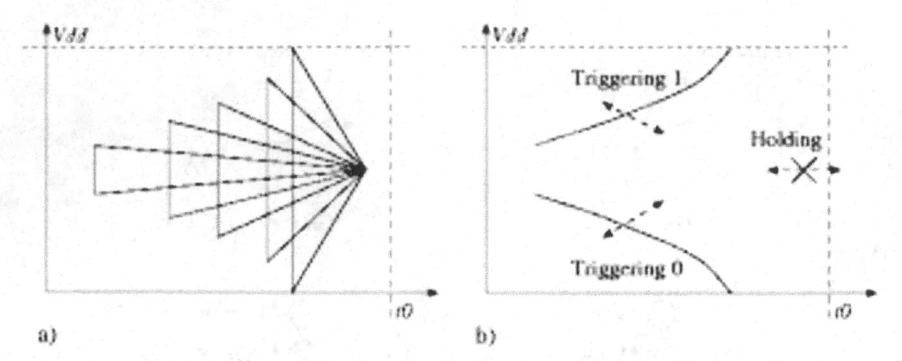

Fig. 2. a) Family of noise sensitive areas (NSAs) on the signal-time plane. b) representation of the entire family by means of two triggering curves and a holding point. Dashed arrows represent the effects of differential supply noise at the receiver.

at the transmitter and receiver because of noise. That's why we use subscripts *t* and *r* to denote signals belonging to the transmitter and to the receiver, respectively.

In principle, signal integrity can be evaluated by splitting the signaling scheme at any point into a *driving* part and a *driven* part. We chose as a splitting point the far end of the line (i.e., the input of the receiver, denoted by FE in Figure 1.a). The driving half of the scheme is composed of the transmitter and of the interconnect, while the driven half of the scheme is composed of the receiver and of the output flip flop. The driving subcircuit provides a signal waveform (hereafter called Vin), while the driven subcircuit imposes necessary conditions to the shape of Vin in order to guarantee correct sampling of the received bit. Signal integrity may be evaluated in terms of error probability per bit by comparing the signal provided by the driving subcircuit with the requirements imposed by the driven subcircuit. If the requirements are strictly necessary, any violation gives rise to a logic error.

We represent receiver's requirements by means of noise sensitive areas (NSAs) [4] and driver's performance in terms of signal slope S. Noise sources may affect both NSA and S.

The advantage of choosing the far end of the line as a splitting point is two-fold: the driving and driven subcircuits are affected by disjoint noise sources, and most noise sources can be implicitly taken into account when combining NSA with signal slopes, as discussed in Section 5.

We construct and characterize black-box parameterized models of both NSA and S taking into account all noise sources. In particular, the size and shape of the NSA is automatically determined based on differential supply noise at the receiver, while signal slope is parameterized in terms of differential supply noise at the transmitter. All other noise sources (common-mode supply noise, clock skew and jitter, crosstalk) affect the relative position of the NSA and of the signal edge of slope S. Hence, they can be implicitly accounted for when combining signal slope and NSA.

3 Parameterized Noise-Sensitive Areas

For a given implementation of the receiver, the input waveform should meet specific requirements in order to guarantee correct sampling of the received bit. Such requirements can be characterized by means of electrical simulations and represented as a region in the signal-time plane that is forbidden to the signal waveform. If the signal waveform crosses the forbidden region, the wrong symbol is sampled by the output flip flop, leading to a bit-level error. Noise sensitive areas (NSAs) provide an informative representation of the forbidden region [4].
In general, the NSA depends both on the nature of the receiver and on the effect of noise sources. We call *inherent NSA* the forbidden region obtained without taking noise sources into account. In presence of noise, the actual NSA becomes larger than the inherent one. The only noise source that may affect the shape of the NSA is differential supply noise at the receiver. Common-mode supply noise, clock skew and jitter cause only vertical and horizontal shifts of the NSA on the signal-time graph.

Consider a receiver consisting of a CMOS inverter followed by a master-slave edge-triggered D flip-flop. Figure 2.a shows a schematic representation of a family of inherent NSAs for the receiver. The inherent NSAs are not unique since the larger the swing of the input signal the lower the transition time of the receiver. Hence, we can obtain a continuous of NSAs by changing the swing of the input signal.
Each NSA has a triangular shape. The left-most vertices represent triggering conditions, while the right-most point represents holding conditions determined by the logic threshold and performance of the input inverter. While holding conditions can be represented by a single point common to all NSAs of the family, triggering conditions must be represented by curves in the voltage-time plane, as shown in Figure 2.b. In practice, the family of inherent NSAs for our case-study receiver is represented by 3 elements: two curves representing positive and negative triggering conditions and a point representing the holding condition.

For correct sampling, at least one of the NSAs must be respected. In fact, each NSA represents sufficient conditions for correct sampling. A logic error may occur only if the signal waveform enters all the NSAs associated with the receiver. In particular, if either triggering or holding conditions are not met, at next clock edge the flip flop maintains its current state, determined by the last received bit rather than by the current one. Hence, the actual bit-level error probability also depends on the input stream. This is an important feature of the proposed approach, since it allows us to evaluate the effects of bit-level encodings on signal integrity.

Since differential supply noise affects the performance of the receiver, different NSA families will be obtained for each noise level. Positive (negative) differential supply noise increases (decreases) the effective supply voltage, thus improving (reducing) the performance of the receiver without changing its logic threshold. This causes the curves representing triggering conditions to shrink and the point representing holding conditions to shift as represented in Figure 2.b by means of dashed arrows.

To provide a simple and practical parametric model of the NSA we use quadratic fitting models to approximate the triggering curves obtained for a fixed value of differential supply noise. Then we construct and characterize a fitting model for each coefficient of the quadratic curve, representing noise dependence. The overall parametric NSA is completely described by the following equations, where V_{n1} denotes the differential supply noise at the receiver.

$$V_{trig0}(t,V_{n1}) = \frac{1}{c_{0,0}(V_{n1}) + c_{0,1}(V_{n1})t + c_{0,2}(V_{n1})t^2} \tag{1}$$

$$V_{trig1}(t,V_{n1}) = \frac{1}{c_{1,0}(V_{n1}) + c_{1,1}(V_{n1})t + c_{1,2}(V_{n1})t^2} \tag{2}$$

$$c_{i,j}(V_{n1}) = a_0^{(i,j)} + a_1^{(i,j)} V_{n1} \quad \forall \begin{cases} i \in \{0,1\} \\ j \in \{0,1,2\} \end{cases} \tag{3}$$

$$V_{hold} = V_{LT} \tag{4}$$

$$t_{hold}(V_{n1}) = b_0 + b_1 V_{n1} \tag{5}$$

For a given receiver, 14 fitting coefficients (namely, $a_h^{(i,j)}$ and b_h for h,i in {0,1} and j in {0,1,2}) need to be determined. This is done by means of least square fitting against the results of Spice simulations.

NSAs are receiver-specific, so that parametric NSAs need to be characterized for each candidate receiver in a library. However, characterization is performed only once and for all.

4 Parameterized Signal Slope

In voltage-mode signaling across a long interconnect, the rising and falling edges of the signal waveform at the far end of the line are well approximated by linear ramps with constant slope, since the input of the transmitter changes much faster than its output, so that most of the output transition is sustained by a constant input signal.
The signal slope S at the far end of the line tells us how fast the signal can switch from properly recognized voltage levels, so that it can be combined with the NSA of the receiver in order to estimate the maximum achievable bit rate [4],[5].
In general, S depends on the driver and on the line and it can be easily characterized by means of transient Spice simulations of the given circuit. If the driver has a symmetric characteristic, rising and falling edges have the same slope.
Since the current drawn by the MOS transistor depends on the supply voltage, S is affected by the differential supply noise at the transmitter. To model such dependence we use the following linear model (where V_{n2} represents the differential supply noise at the transmitter):

$$S(V_{n2}) = d_0 + d_1 V_{n2} \tag{6}$$

requiring the characterization of 2 fitting parameters.

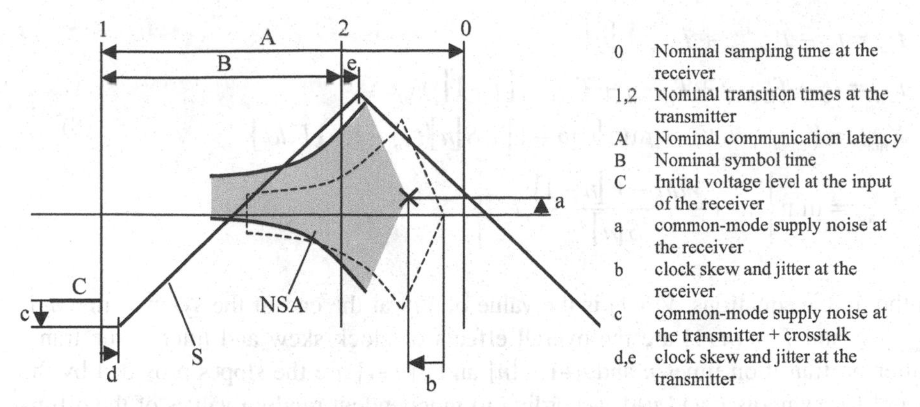

Fig. 3. Schematic representation of the noise sensitive area (NSA), of the signal slope (S) and of the effects of all noise sources. This timing diagram would lead to the correct sampling of symbol 1.

5 Bit-Level Error Probability

Figure 3 shows how to combine NSA and S to evaluate the integrity of a received bit taking into account all possible noise sources. With respect to Figure 3, we assume the current bit (namely, b[n]) to be transmitted in the time interval between time instants 1 and 2, and to be sampled at the receiver at time instant 0. Depending on the length of the line, on the propagation delay of the receiver and on the bit rate, the nominal sampling instant at the receiver may fall outside the symbol time of the transmitted bit (as in the example of Figure 3). The distance between time instants 1 and 0 is the *latency* of the communication channel, while the distance between time 1 and 2 is the *symbol time.* The shape of the NSA and the slope of Vin are provided by their pre-characterized models, parameterized in terms of differential supply noise at the receiver and transmitter. All other noise sources induce vertical and horizontal shifts that are represented in Figure 3.

While NSA is directly provided by the model, the local behavior of Vin has to be constructed based on the model of slope S. Moreover, if the sampling time of b[n] falls outside the corresponding symbol time at the transmitter, than the waveform of Vin has to be constructed for both symbols b[n] and b[n+1] to evaluate the correctness of the n-th received symbol. For instance, when b[n]=1 and b[n+1]=0, as in the case of Figure 3, the waveform of Vin can be expressed as follows:

$$Vin(t)=\begin{cases} V[n-1]+V_{cmsn\_t}[n]+V_{ct}[n]+S[n](t-t_{1n}) & t_{1n}\le t<T_{max} \\ V_{max} & T_{max}\le t<t_{2n} \\ V_{max}-S[n+1](t-t_{2n}) & t\ge t_{2n} \end{cases} \quad (7)$$

where:

$$
\begin{aligned}
t_{1n} &= t_0 - T_{latency} + T_{noise\_t}[n] \\
t_{2n} &= t_0 - T_{latency} + T_{symbol} + T_{noise\_t}[n+1] \\
V_{max} &= V_{cmsn\_t} + V_{ct} + \min\{V[n-1] + S[n](t_{2n} - t_{1n}), Vdd\} \\
T_{max} &= \min\left\{t_{1n} + \frac{Vdd - V[n-1]}{S[n]}, t_{2n}\right\}
\end{aligned}
\tag{8}
$$

In the above equations, $V[n\text{-}1]$ is the value of *Vin* at the end of the symbol time *n*-1, $T_{noise\_t}[n]$ and $T_{noise\_t}[n+1]$ are the overall effects of clock skew and jitter at the transmitter on transition times *n* and *n*+1. $S[n]$ and $S[n+1]$ are the slopes provided by the model for symbols *n* and *n*+1, according to independent random values of the differential supply noise at the transmitter.

If Vin is above the upper triggering curve of NSA for at least a time instant, and it is above the holding point, then the received bit is b_r[n]=1. If Vin is below the lower triggering curve of NSA for at least a time instant, and it is below the holding point, then the received bit is b_r[n]=0. In all other cases b_r[n]=b_r[n-1]. Notice that the comparison between NSAs and signal waveforms can be performed numerically by leveraging the convexity of the triggering curves and the linearity of the signal edge.

The pseudo-code of an algorithm that determines the value of received bit b_r[n] is shown in Figure 4.

6 Experimental Results

To test the effectiveness and accuracy of the proposed technique we implemented the algorithm of Figure 4 in C and we characterized the NSA and signal slope for voltage-level receiver and transmitter implemented in 0.18μm technology with 2V power supply. For the interconnect we used a five-stage RC model with R=50Ω and C=60fF at each stage.

All voltage noise sources were assumed to be independent and uniformly distributed between -0.2V and +0.2V. Similarly, clock jitters at transmitter and receiver were modeled as independent random variable uniformly distributed between -100ps and +100ps.

To validate the model and the error-estimation approach we simulated the transmission of the same bit stream using both our C model and Spice. Random values of all noise sources were generated off-line and coherently injected in both simulations. In particular, arrays of noise values (representing the value of each noise source for all bits in the stream) were directly provided to our C model and used as outlined in the previous section. Noise injection in Spice was implemented by means of the modified circuit of Figure 1.b: all voltage noise sources were implemented as independent voltage sources, while clock jitter and skew were implemented by providing noisy clock signals. The waveforms of both clock signals and voltage noise generators were specified as piece-wise linear functions (PWL) automatically generated by means of a C routine taking in input an array of noise values per symbol and providing the corresponding PWL description. The results provided by our model were

```
V[0]=Vinitial
b[0]=Binitial
generate Vdmsn_t[1], Tnoise_t[1]
for each transmitted bit b[n] n>0
  generate Vdmsn_r[n]
  construct NSA(Vdmsn_r[n],t0)
  generate Vcmsn_r
  generate Tskew_r, Tjitter_r
  NSA(t) = NSA(t-Tskew_r-Tjitter_r)+Vcmsn_r
  generate Vdmsn_t[n+1]
  compute S[n]   based on Vdmsn_t[n],b[n-1],b[n]
  compute S[n+1] based on Vdmsn_t[n+1],b[n],b[n+1]
  generate Vcmsn_t[n], Vct[n]
  generate Tnoise_t[n+1]
  t1n = -Tlatency+Tnoise_t[n]
  t2n = -Tlatency+Tsymbol+Tnoise_t[n+1]
  construct Vin  based on V[n-1],S[n],S[n+1],t1n,t2n
  if Vin is above NSA
    b_r[n] = 1
  else if Vin is below NSA
    b_r[n] = 0
  else
    b_r[n] = b[n-1]
  end if
  V[n] = Vin(t2n)
end for
```

Fig. 4. Algorithm for determining the value of each received bit.

always coherent with Spice simulations. On the other hand, our model provided a speedup of more than three orders of magnitude if compared to Spice simulations. Comparison was made using the same time resolution for Spice and for our approach.

To exemplify the application of the proposed approach, we used the model for determining the achievable bit rate of a given channel. We assumed the same noise conditions described above, and we simulated the propagation of a bit stream of 10000 symbols randomly generated according to a given signal probability p1. For each value of p1, different bit rates were simulated, assuming channel latency equal to the symbol time (this is a common assumption in many synchronous signal schemes). Figure 5 reports the estimated error probability as a function of the bit rate, for different values of p1. For a symbol time of 400ps, providing error probability above 20% with signal probability 0.5, further experiments were performed by varying the channel latency (i.e., by varying the distance in time between the transmission of a symbol and its sampling at the receiver). Interestingly, some of the errors were eliminated by using a latency longer than the symbol time. In particular, the error probability reduced from 21% to 15% for a channel latency of 450ps.

Finally, we remark that the error probabilities of Figure 5 depend on the signal probability. In other words, they depend on the bit stream. In fact, the error probabil-

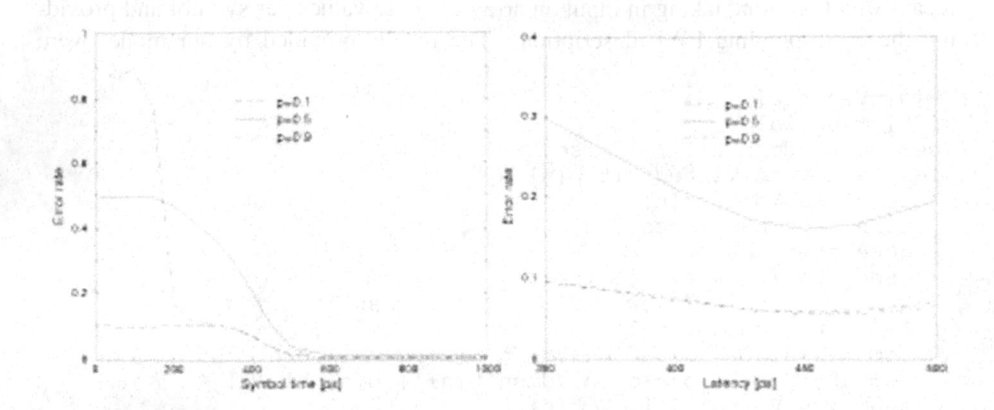

Fig. 5. Error probability for different signal probabilities, as a function of (a) the symbol time, and (b) the latency of the channel.

ity is maximum when p=0.5, because of the larger number of signal transitions. Notice that, for symbol times below 200ps, the line never switches. Hence, the error probability is equal to the probability of the incoming bit to be different from the initialization value of the line (0). The dependence of the error probability on the bit stream demonstrates the sensitivity of the proposed approach to bit-level signal statistics, making it suitable for exploring bit-level encodings and their effects on signal integrity.

7 Conclusions

In this paper we have proposed a parameterized model of on-chip communication channels that provides a logic-level representation of the main effects of noise sources on the signal integrity of digital signals. The model is implementation-specific and needs to be pre-characterized for each transmitter and receiver by means of least-square fitting on electrical-level simulation results.

We have developed an algorithm that exploits the parameterized models to evaluate the effects of all noise sources on the error probability of a bit stream sent across the channel.

We validated the approach by comparing estimated error probabilities with the results of electrical-level simulations. Our approach provided bit-by-bit 100% accuracy with a speed up of three orders of magnitude with respect to Spice simulations.

Finally, we exemplified the application of the proposed technique to evaluate the maximum-bit-rate of a given communication channel and the effects of signal statistics (i.e., of bit-level encodings) on the error probability.

References

1. Caignet F., Delmas-Bandia, S., Sicard, E.: The challenge of Signal Integrity in deep-submicrometer CMOS technology. In Proceedings of the IEEE, vol. 89, no. 4, 556–573 (2001)
2. Dally, W.J., Poulton, J.W.: Digital Systems Engineering. Cambridge University Press (1998).
3. Hall, S.H., Hall, G.W., McCall, J.A.: High-Speed Digital System Design, A Handbook of Interconnect Theory and Design Practices. John Wiley & Sons (2000)
4. Bogliolo, A., Olivo, P.: Dealing with noise margins in high-speed voltage-mode signaling. In Proc. of IEEE Workshop on Signal Propagation on Interconnects (2002).
5. Bogliolo, A.: Encodings for High-Performance Energy-Efficient Signaling. In Proc. of IEEE/ACM Int.l Symposium on Low-Power Electronics and Design (2001).

Advanced Cell Modeling Techniques Based on Polynomial Expressions

Wen-Tsong Shiue and Weetit Wanalertlak

School of Electrical Engineering and Computer Science
Oregon State University, Owen 220, Corvallis, OR 97331
shiue@ece.orst.edu

Abstract. This paper presents an advance cell modeling technique based on polynomial expressions for physical and logic design. Till now, there lacks a modeling method to handle multiple operating points such as input slew, output capacitance, voltage, process, temperature, etc. at a time for the logic synthesis and physical design. Our novel curve-fitting algorithm for cell modeling can process a table with thousands of data points in a modeling equation compared to piecewise equations used in the existing methods [1]-[4]. The number of data storage and CPU run time are significantly reduced without compromising the accuracy of the data. This ensures that the novel curve-fitting algorithm can exactly recover the original data points in a short period of time. In addition, the table-partitioning algorithm is developed to reduce the number of storage and flatten the peak situation occurring at some of the unsampled data. Furthermore, in order to filter the "pass" points with 99% of all data points, the algorithm is developed to select the best order for input variables, further causes to shorten CPU run time again. Our benchmark shows that the new approach has significantly better accuracy and less storage for both the sampling data and unsampling data.

1 Introduction

Nanometer complexity forces designers to perform an increasing number of analyses at various voltage and temperature combinations which, in turn, drives library developers to characterize multiple operating points and publish numerous versions of libraries. The Synopsys' Scalable Polynomial Delay Model (SPDM) [1][2] provides an excellent alternative technique that promises to limit the data explosion of multi-operating point analysis. Furthermore, SPDM preserves accuracy while reducing CPU execution times for static timing analyses. However, there still lacks an accurate curve-fitting algorithm to handle multiple operating points such as input slew, output capacity, process, voltage, and temperature at a time in the SPDM format.

Some of papers present the modeling methods. Feng and Chang [1][4] present a systematic approach of cell delay modeling. Murugavel et. al. [7][8] presents statistical techniques for power estimation in digital circuits based on sequential and recursive least square error. Some of curve-fitting algorithms are mentioned in [5]. A detailed least squares regression analysis in terms of linear algebra is presented in [6]. There still lacks an efficient modeling approach based on curve-fitting algorithms to

J.J. Chico and E. Macii (Eds.): PATMOS 2003, LNCS 2799, pp. 550–558, 2003.
© Springer-Verlag Berlin Heidelberg 2003

handle multiple operating points at a time for sampling data (recovering the original data points) and unsampling data (predicting the unknown data points).

This paper presents a novel methodology to deal with a large set of sampling points by utilizing curve-fitting methods to transform characterization data into computationally efficient polynomial equations while maintaining accuracy. In addition, we develop table-partitioning algorithm to select the best order for the input variables such that the fluctuation is flattened for the unknown data points. Our benchmarks show that these models produce optimization solution in terms of less memory storage, fast CPU ruin time, higher accuracy for the cell modeled in a polynomial expression compared to conventional nonlinear delay model (NDLM) and K-factor techniques with the added benefit of only producing a single multi-PVT library. This technique is also developed to support the scalable polynomial delay model (SPDM) format used in Synopsys' tools, PrimeTime and Library Compiler [3] but unlimited. This implies that the design time is significantly shortened if this method is incorporated into the design stage of the logic synthesis and physical design.

The rest of the paper is organized as follows: Section 2 describes our approach versus SPDM. Section 3 describes our modeling solution. Section 4 describes tool implementation and benchmark results. Section 5 concludes the paper.

2 Our Approach vs. SPDM

A new delay model, the Scalable Polynomial Delay Model [2] (SPDM), had been introduced by Synopsys to enable semiconductor vendors to develop and distribute multi-term, multi-order equations as a function of voltage, temperature, load capacitance, and input slew for the most requested modeling capabilities for Deep Sub-Micron technology. Polynomials play a key role in many numerical computations. The fundamental theory is based on the Taylor Expansion, where an analytical function can be expressed as a finite series of polynomials. These polynomials are specified in the expanded decomposed form. Synopsys' Liberty syntax [3] allows SPDM to handle up to six dimensions with a single equation. Assuming m, n, p, q, r, and s are the orders of the each of the six dimensions (x1, x2, x3, x4, x5, and x6). The number of coefficients (or called storage) will be (m+1)(n+1)(p+1)(q+1)(r+1)(s+1). The polynomial equation for the function f is expressed as

$$f(x_1,x_2,x_3,x_4,x_5,x_6) = \Sigma_{(f=0:s)}\Sigma_{(e=0:r)}\Sigma_{(d=0:q)}\Sigma_{(c=0:p)}\Sigma_{(b=0:n)}\Sigma_{(a=0:m)}A_{abcdef}\,x_1^a x_2^b x_3^c x_4^d x_5^e x_6^f$$

With this approach, memory utilization is significantly improved since relatively few polynomial coefficients are stored as compared to the look-up table values, and only one library is required instead of a complete library for each operating condition. SPDM uses multi-dimension polynomials to model cell delay and other electrical parameters. It simplifies modeling by utilizing standard form equations across different input slew ranges, output capacitance, temperature, and voltage. The SPDM format [2] is shown below for the table of cell rise and rise transition. The coefficients and the orders for the input variables are derived from their own curve-fitting algorithm.

However, this SPDM format did not provide the curve-fitting algorithms for the designers. Therefore, the designers have to develop their own modeling solutions to meet the SPDM format. This degrades the design time for the designers. Nevertheless, our modeling solutions come to play for this issue and the resulting better scalable equation syntax allows optimization for computational efficiency. Additionally, by enabling a direct solve solution for functions of voltage and temperature; gate-level tool capabilities can be expanded to accurately perform multi-operating point analysis within a single session with a single SPDM library.

```
timing() {
   cell_rise(eq1) {
   orders(1,1,2,1);
   coefs("1,2,3,4,5,6,7,8,9,10,1,2,3,4,5,6,7,8,9,20,1,2,3,4");
  /* 1+2a+3b+4ab+5c+6ac+7ab+8ac+9c2+10ac2+1bc2+2abc2+3d+4ad+5bd
  6abd+7cd+8acd+9bcd+20abcd+21c2d+22ac2d+23bc2d+24abc2d */}
   rise_transition(eq2) {
     domain(D1) {
     variable_1_range(0,6);
     orders(1);
      coefs("1,2");  /* 1+2a */ }
      domain(D2) {
     variable_1_range(7,10);
      orders(2);
      coefs("5,6,7");  /* 5+6a+7a2 */ }}}
```

3 Our Modeling Solution

The look-up table (LUT) is formed where all the operating points derived from HSpice are collected. Let's imagine that there are 8 tables consisting of the cell rise, cell fall, rise transition, fall transition, and etc. for each 2-input cell such as AND, OR, NAND, etc. Each table may contain thousands of data points. During the synthesis, the data points will be extracted from these tables for each arc of the cell for the static timing analysis to get the delay for each cell located at the critical path of the circuits. This consumes a lot of time for the data searching from those tables and the storages for these data points are huge. This does not work for the efficient design in logic synthesis.

Our modeling solution is presented to process one or more tables (if they have a similar slope) into a polynomial equation. Here the coefficients for the equation are quite less compared with the data points in look-up table. Additionally, this equation can be written as a SPDM format incorporating with the Synopsys' PrimtTime and Library Compiler [3] for logic synthesis. Note that our curve-fitting algorithm can process different output data such as delay, power, and output slew. The modeled power equation can also be written as scalable polynomial power model (SPPM) format to support the Synopsys' tools. Furthermore, our modeling solution can be used to support any kind of tools without the limitation.

3.1 Examples

We take an example to illustrate our modeling technique based on the polynomial equation. In Figure 1, this table is formed, as a liberty's format [2] and input data points are located in the shaded areas. One of the input variables is S by means of input slew and the other is C by means of output capacitance. The data located in the un-shaded area are output data points such as delay. For instance, while the input slew

(S) is 0.01 and the output capacitance (C) is 0.014, then the delay is 0.131. When the input slew (S) is 0.193 and the output capacitance (C) is 0.359, then the delay is 0.415.

S\C	*0.014*	*0.080*	*0.146*	*0.359*	*0.677*	*1.341*
0.010	**0.131**	**0.181**	**0.209**	**0.267**	**0.327**	**0.415**
0.102	**0.200**	**0.254**	**0.284**	**0.346**	**0.408**	**0.500**
0.193	**0.265**	**0.320**	**0.351**	**0.415**	**0.479**	**0.572**
0.486	**0.463**	**0.522**	**0.554**	**0.622**	**0.689**	**0.786**
0.926	**0.753**	**0.815**	**0.848**	**0.918**	**0.988**	**1.089**
1.841	**1.348**	**1.411**	**1.445**	**1.516**	**1.589**	**1.695**

Fig. 1. Actual data points with Liberty's format [slides of SPDM from Synopsys].

The Best Order Selection. Figure 2 shows the results after incorporating our modeling solution. We explore the different orders for input variables such as S and C. The first column shows the order for S and the second column shows the order for C. Compared with the estimated data points and actual data points based on the criteria of the least square error (LSE) and least square relative error (LSRE) as above, where the z(i) means the actual data points and Z(i) means the estimate data points.

S	C	LSE	LSRE	1.E-06	1.E-04	1.E-03	1.E-02
5	5	4.86E-09	2.58E-09	*100%*	*100%*	*100%*	*100%*
5	6	3.10E-09	1.65E-09	*100%*	*100%*	*100%*	*100%*
5	7	5.93E-10	3.23E-10	*100%*	*100%*	*100%*	*100%*
6	5	5.25E-09	2.77E-09	*100%*	*100%*	*100%*	*100%*
6	6	5.03E-09	2.66E-09	*100%*	*100%*	*100%*	*100%*
6	7	1.53E-09	8.08E-10	*100%*	*100%*	*100%*	*100%*
7	5	7.58E-09	4.00E-09	*100%*	*100%*	*100%*	*100%*
7	6	1.84E-09	9.80E-10	*100%*	*100%*	*100%*	*100%*
7	7	1.37E-09	7.23E-10	*100%*	*100%*	*100%*	*100%*
4	5	3.61E-03	8.43E-03	17%	36%	61%	*100%*
4	6	3.61E-03	8.43E-03	17%	36%	61%	*100%*
4	7	3.61E-03	8.43E-03	17%	36%	61%	*100%*
			:				
2	1	1.89E-01	5.14E-01	0	0	0	14%
3	1	1.87E-01	4.99E-01	0	0	0	11%
4	1	1.86E-01	4.95E-01	0	0	0	8%
5	1	1.86E-01	4.95E-01	0	0	0	8%
6	1	1.86E-01	4.95E-01	0	0	0	8%
7	1	1.86E-01	4.95E-01	0	0	0	8%

Fig. 2. The best order selection for the input variables such that the 100% of data points pass.

$$R_{lse} = \sqrt{\frac{\sum_{i=1}^{n}[z^{(i)} - Z^{(i)}]^2}{\frac{1}{n}\sum_{i=1}^{n}\left|z^{(i)}\right|^2}} \; ; \qquad R_{lrse} = \sqrt{\sum_{i=1}^{n}\frac{[z^{(i)} - Z^{(i)}]^2}{\left|z^{(i)}\right|^2}}$$

These criterions of LSE and LSRE help to determine the best order for the variables such that the estimated data points are well matched with the actual data points. For instance, the order of 55 has better LSE and LSRE compared to the order of 66. This means the order 55 is more accurate than the order 66. In addition, we also compare the relative error rate for each data point for the output variable. Here the error rate is defined as 1e-6, 1e-4, 1e-3, and 1e-2. For instance, 100% points are matched above the row of order 77 for the error rate 1e-6. At this decision, we choose the best order of 55 because it has less terms and saves the number of storages. For the error rate of 1e-2, we choose the best order 45 instead. The modeling equation for order 55 is shown below. Here the estimated data points are the same as all of the sampling data points in Figure 3.

$$Delay_{(55)} = 0.1085+1.123C-4.966C^2+12.77C^3-14.25C^4+5.3C^5+0.7506S+2.124SC-12.37SC^2+35.38SC^3-42.01SC^4+16.20SC^5-0.1234S^2-12.44S^2C+84.33S^2C^2-251.9S^2C^3+304S^2C^4-118S^2C^5-657S^3+30.72S^3C-221.3S^3C^2+673.7S^3C^3-818.7S^3C^4+318.8S^3C^5+-0.1829S^4-28.53S^4C+210.8S^4C^2-646.9S^4C^3+788.6S^4C^4-307.4S^4C^5-0.06784S^5+8.261S^5C-61.71S^5C^2+190.1S^5C^3-232.1S^5C^4+90.55S^5C^5$$

Note that the number of terms is 36 (i.e. 36 coefficients) for the order of 55 compared to the number of data points, 36*3=108, shown in the look-up table. This case saves 66% in storage.

Figure 3 shows our estimated data points (LSE: 0.2, 89% of points matching with the error rate of 0.1) compared to Synopsys' estimated data points (LSE: 1.3, 31% of points matching with the error rate of 0.1) with the order selection of 11 for the input variables. The benchmark shows that our curve-fitting algorithm (blue solid lines in Figure 4) is much better than theirs (red dash lines in Figure 4). Here the green solid lines are the accurate data points). Here the Synopsys' modeling equation is Delay(1,1)= 0.1972+0.9014S+0.1093C+0.0092CS, and our modeling equation is Delay(1,1) = 0.1742+ 0.667S+0.2084C+0.0213CS. Note that the number of coefficients is 4 for the order 11 compared to the entire 108 data points in the table. Here it saves 96.3% in storage! The more order used, the more accuracy for the sampled data points. The less order used, the more storage savings. There is a tradeoff between the accuracy and the number of storage. This flexibility helps designers to make the decision by themselves.

Actual data points	Synopsys' estimate data points (order 11)	Our estimate data points (order 11)	Our estimate data points (order 55)
1.31E-01	2.08E-01	1.84E-01	1.31E-01
2.00E-01	2.91E-01	2.45E-01	2.00E-01
2.65E-01	3.73E-01	3.06E-01	2.65E-01
4.63E-01	6.37E-01	5.01E-01	4.63E-01
7.53E-01	1.03E+00	7.95E-01	7.53E-01
1.35E+00	1.86E+00	1.41E+00	1.35E+00
1.81E-01	2.15E-01	1.98E-01	1.81E-01
2.54E-01	2.98E-01	2.59E-01	2.54E-01
3.20E-01	3.80E-01	3.20E-01	3.20E-01
5.22E-01	6.44E-01	5.16E-01	5.22E-01
8.15E-01	1.04E+00	8.10E-01	8.15E-01
1.41E+00	1.87E+00	1.42E+00	1.41E+00

Fig. 3. Comparison of our estimated data points and Synopsys' estimated data points with different orders.

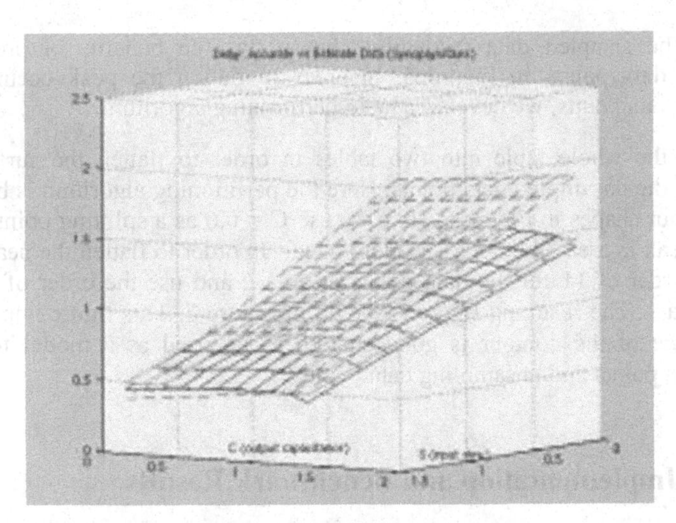

Fig. 4. Plots of Figure 3.

The Fluctuation Situation Occurring at Some of the Unsampled Data Points. In this section, we discuss the fluctuation occurring for the order 55 at some of the unsampled data points even though the order 55 recovers all the sampled data points (100% matching at the error rate of 1e-6). Compared to order 34 in Figure 2, it shows 89% matching at the error rate of 1e-2. This implies that order 34 may not be the best

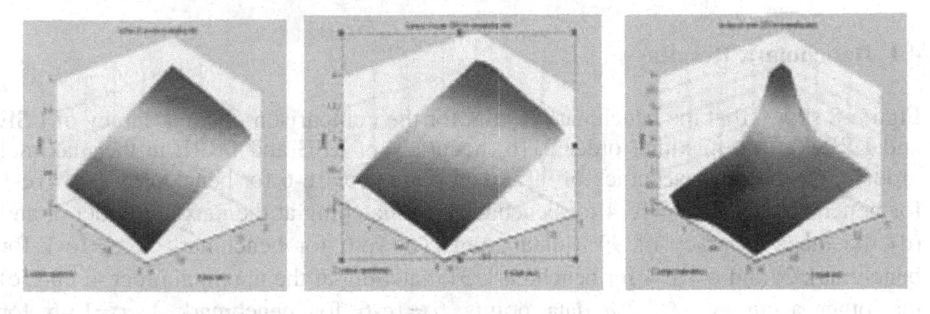

Fig. 5. Prediction of the unsampled data points.

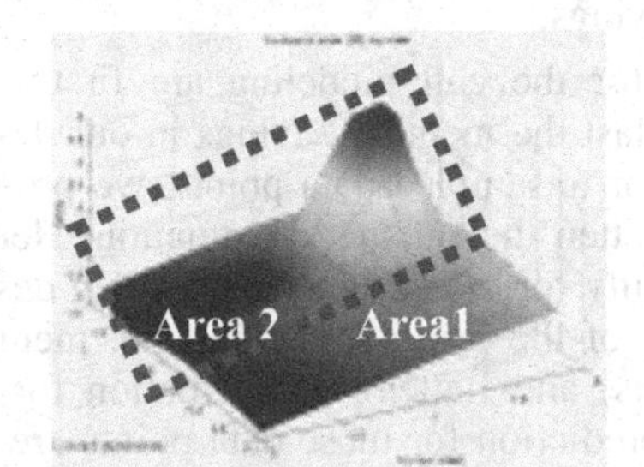

Fig. 6. Table-partitioning for the fluctuation area.

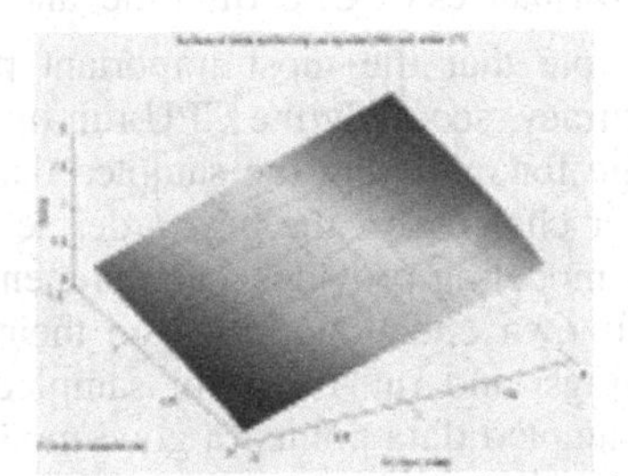

Fig. 7. The fluctuation area is flattened.

order for the sampled data recovering, but it has the benefits of predicting the unsampled data points in Figure 5. In order to flatten the peak occurring at the unsampled data points, we develop a table-partitioning algorithm.

We split the whole table into two tables in order to flatten the surface for the prediction of the unsampled data points. Here the partitioning algorithm is based on the surface contour shapes in Figure 6. Here we use C = 0.6 as a splitting point. There is a cave and a peak in area 2 but area 1 is more linear. In order to flatten the peak and cave, we use the order of 11 for the data points in area 2 and use the order of 55 for data points in area 1. The resulting figure is shown in Figure 7. This figure implies that the curved surface of the contour is good enough to be used as a model for both the sampling data points and unsampling data points.

4 Tool Implementation and Benchmark Results

We develop a tool based on our methodology and algorithms for the optimization-based cell modeling techniques based on polynomial expressions. This tool is implemented using commercial Matlab version 6r13 on a Sun Ultra Ax-i2. The benchmark results show that the accuracy for the data recovered from an optimized polynomial equation, which is derived from a large table, has a very fast CPU time and a large number of memory savings.

4.1 Benchmark Results

Figure 8 shows that the benchmark results for the comparison of (a) accuracy of LSE and LSRE at the maximal order 2, (b) accuracy of LSE and LSRE at the maximal order 3, (c) the other accuracy of 99% data points (σ=1e-6 for benchmark 1, σ=1e-1 for benchmark 2, and σ=1e-4 for benchmark 3) matching at the maximal order 3, and (d) the other accuracy of 99% data points (σ=1e-6 for benchmark 1, σ=1e-1 for benchmark 2, and σ=1e-4 for benchmark 3) matching at the maximal order 4, and (e) the other accuracy of 99% data points (σ=1e-6 for benchmark 1, σ=1e-6 for benchmark 2, and σ=1e-4 for benchmark 3) matching at the maximal order 5, for the performances of CPU run time, and memory savings.

Note that the most important parameters for the cell modeling are firstly the accuracy, secondly the CPU run time, and the last the memory savings in our design stage for modeling the sampled data points. For unsampled data points, we provide other choices for the best order selection to flatten the fluctuation situation. Hence, our modeling provides a comprehensive flexibility for the designers. They can design their own equations based on their preference of the priority of run time, memory savings, and run time (for sampled data points) and flatten the fluctuation for the unsampled data points (a good model for the prediction for those data points are not sampled).

5 Conclusion

This paper presents a comprehensive modeling solutions based on polynomial expressions handling multiple data points at a time with less storage, fast CPU run time, and accuracy for recovering the sampling data and predicting the unsampling data. Our modeling solution can be used to the design stage for the logic synthesis, physical design, and analog circuit design without compromising time-consumed SPICE simulation such that the accuracy is maintained and the memory storage is saved. This optimization-based cell modeling techniques have been implemented as a tool named OSU-CMS (Oregon State University – Cell Modeling Solution in Figure 9). It has been used for thousands of benchmarks and it generates the solution in seconds for thousands of data points with multiple input and output variables in look-up table. The recovering data points are totally matching with the sampling data points.

Searching the best order for the maximal order 2						
Bench mark	Data points in LUT	Best Order	Memory Saving	CPU Run time(sec)	LSE	LSRE
1	294	22221(lse)/ 22222(lsre)	45% (lse)/ 17.3% (lsre)	2.96	2.63E-02	5.62E-02
2	108	22	91.67%	0.1	9.05E-02	2.60E-01
3	75	22	88%	0.06	3.89	23.8

(a) maximal order 2.

Searching the best order for the maximal order 3						
Bench mark	Data points in LUT	Best Order	Memory Saving	CPU Run time(sec)	LSE	LSRE
1	294	33132	12%	48.5	1.22E-11	1.65E-11
2	108	33	85%	0.11	4.55E-02	1.21E-01
3	75	32(lse)/33(lsre)	84%(lse)/78%(lsre)	0.12	3.00E+00	1.80E-01

(b) maximal order 3.

Searching the best order for the 99% of data points matching (the maximal order is 3)						
Bench mark	Data points in LUT	Best Order	Memory Saving	CPU Run time(sec)	LSE	LSRE
1	**294 (σ=1E-6)**	**23121**	**51%**	**7.18**	**5.85E-11**	**6.97E-11**
2	108 (σ=1E-1)	No results				
3	75 (σ=1E-4)					

(c) maximal order 3 with 99% matching.

Searching the best order for the 99% of data points matching (the maximal order is 4)						
Bench mark	Data points in LUT	Best Order	Memory Saving	CPU Run time(sec)	LSE	LSRE
1	294 (σ=1E-6)	24111	59%	17.21	7.23E-11	9.16E-11
2	108 (σ=1E-1)	43	81.5%	0.13	4.34E-02	1.15E-01
3	**75 (σ=1E-4)**	**44**	**67%**	**0.15**	**1.08E-07**	**3.34E-06**

(d) maximal order 4 with 99% matching.

Searching the best order for the 99% of data points matching (the maximal order is 5)						
Bench mark	Data points in LUT	Best Order	Memory Saving	CPU Run time(sec)	LSE	LSRE
1	294 (σ=1E-6)	24111	59%	112.9	7.23E-11	9.16E-11
2	**108 (σ=1E-6)**	**55**	**66.7%**	**0.33**	**4.86E-09**	**2.58E-09**
3	75 (σ=1E-4)	44	66.7%	0.21	1.08E-07	3.24E-06

(e) maximal order 5 with 99% matching.

Fig. 8. Benchmark results for the performances of different accuracy criterion, CPU run time, and memory savings.

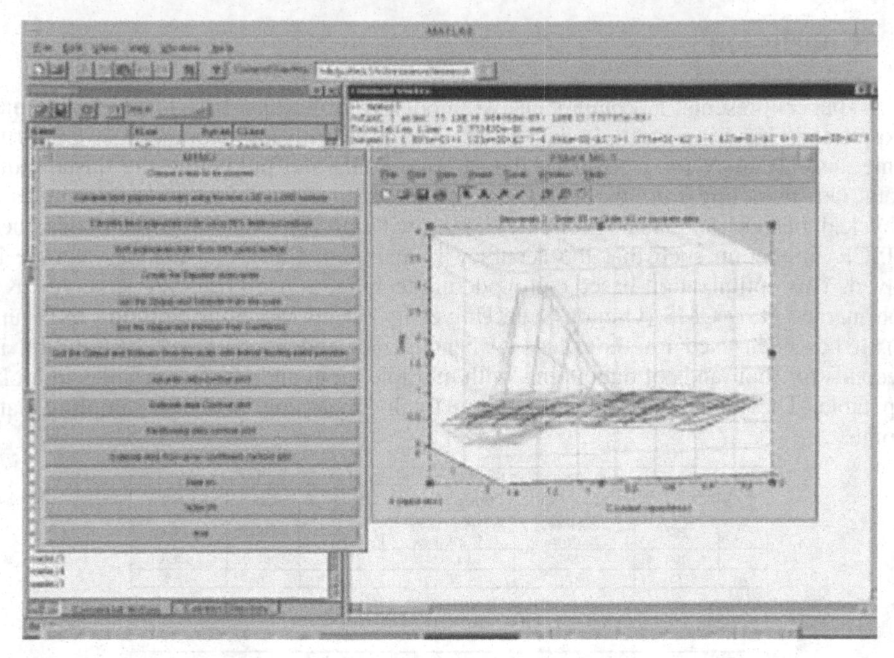

Fig. 9. OSU-CMS alpha version 0.99 (Tool development for our modeling solution).

References

[1] F. Wang and Shir-Shen Chang, "Scalable Delay Model for Logic and Physical Synthesis", The 16th IFIP World Computer Congress, ICDA, August 2000. on VLSI Systems, Vol. 7, No. 4, Dec. 1999.

[2] Shir-Shen Chang, "Liberty Technology Rollout 2000.11, Synopsys Liberty Extensions for Physical Design," 2000.

[3] Reference Manual, "Library Compiler User Guide," vol. 2, 2000.11, Synopsys.

[4] Shin-Shen Chang, "Liberty Updates: Physical Extension & Polynomial Approach," April 3, 2001, Synopsys.

[5] W. Press, S. Teukolsky, W. Vetterling, and B. Flannery, "Numerical Recipes," Cambridge University Press, Nov 1989.

[6] Enders A. Robinson, "Least Squares Regression Analysis In Term of Linear Algebra," Goose Pond Press, 1981.

[7] A. Murugavel, et.al., "Least-Square Estimation of Average Power in Digital CMOS Circuits," *IEEE Transactions on VLSI System*, Vol. 10., No.1, Feb 2002.

[8] A. K. Murugavel et. al., "Average power in digital CMOS circuits using least square estimation," Int. Conf. VLSI Design, pp. 215–220, 2001.

RTL-Based Signal Statistics Calculation Facilitates Low Power Design Approaches

Paul Fugger

Infineon Technologies Design Center Graz,
Babenbergerstr. 10, 8020 Graz, Austria
Paul.Fugger@infineon.com

Abstract. Design approaches, which take into account signal statistics, show good results with respect to power optimization. Since HDL coding and verification by RTL simulation is the preferred flow in the design of SoCs, this level of design abstraction comprises a high potential for low-power digital design. In this paper an approach is presented, how calculation of RTL signal statistics can be accomplished and which types of on-line statistics can be calculated efficiently. With two small examples it is furthermore shown, how the knowledge about signal statistics can be exploited for low-power design approaches.

1 Introduction

Digital CMOS power dissipation results from three major sources:

$$P_{total} = P_{switching} + P_{short\text{-}circuit} + P_{leakage}. \tag{1}$$

$P_{switching}$ represents the switching component of the power consumption,

$$P_{switching} = \tfrac{1}{2} \cdot \alpha \cdot C_L \cdot V_{dd}^2 \cdot f_{clk}, \tag{2}$$

where the switching activity α is the average number of power-consuming transitions per clock period, C_L is the load capacitance, V_{dd} is the supply voltage and f_{clk} is the clock frequency. $P_{short\text{-}circuit}$ is due to the direct-path short circuit current, $I_{short\text{-}circuit}$, which arises, when both the NMOS and PMOS transistors are simultaneously active, conducting current directly from supply to ground. $P_{leakage}$ is dependent on the leakage current, $I_{leakage}$, which can arise from substrate injection and subthreshold effects, primarily determined by the fabrication technology [6].

Due to the fact that $P_{switching}$ is the main contributor to power consumption in CMOS, low-power design mainly focuses on reducing the switching activity. This can be achieved by simply powering down the circuit or parts of it. Clock gating or optimized circuit architectures can also be used to minimize the number of transitions. Moreover, it is important that the effects of signal statistics on power consumption is taken into account, since signals in a digital circuit typically do not switch randomly. The influence of signal statistics on power consumption can be illustrated by considering a 2 input NOR gate. Let P_a and P_b be the probabilities that the inputs a and b are **1**. In this case the probability that the output z is a **1** is given by

J.J. Chico and E. Macii (Eds.): PATMOS 2003, LNCS 2799, pp. 559–568, 2003.
© Springer-Verlag Berlin Heidelberg 2003

$$P_z = (1 - P_a) \cdot (1 - P_b). \tag{3}$$

Hence, the switching activity is

$$\alpha = P_{z=0} \cdot P_{z=1} = (1 - (1 - P_a) \cdot (1 - P_b)) \cdot (1 - P_a) \cdot (1 - P_b). \tag{4}$$

This simple example clearly shows that switching activity (and thus switching power) strongly relies on signal statistics.

Many power-optimization techniques exist which can be applied statically, like equal partitioning of memory (usage of many small memory blocks instead of one large) or clock gating, which is already implemented in a commercial available tool (Synopsys PowerCompiler [12]). These methods have in common, that they only regard the (static) structure of a design, but do not take into account the (dynamic) data behavior. However, most recently a lot of methods have been presented which primarily focus on the statistical behavior of the processed data, like bus encoding techniques [1, 3], data cache compressing [2, 7], dynamic memory mapping [4, 8]. All these methods have in common that they massively apply profiling techniques on typical applications. These techniques are usually performed upon the system level. However, since coding on RTL is the preferred flow in the design of today's SoCs, this level of design abstraction comprises a high potential for data analysis tasks. Due to the fact that every digital designer is used to handle verification by RTL simulation, RTL offers this potential by nature.

All digital simulators offer possibilities to add C functions that provide procedural access to information within an HDL simulation run. These C routines can, for example, be simulation models of certain building blocks or even interact with the simulator to give information on module structure or event ordering, and allow reading and writing of HDL data. A user-written application can apply these functions to traverse the hierarchy of an design, get information about and set the values of Verilog/VHDL objects in the design, get information about a simulation run, and control it (to some extent).

For Verilog, the PLI (Programming Language Interface) is used and standardized (see e.g., [5]). For VHDL, there exists no unique API (application program interface) like the PLI for Verilog. Instead of that, one can use simulator-specific interfaces, like the FLI (Foreign Language Interface) of Mentor's ModelSim [9]. Clearly, this simulator-dependent handling of "foreign" models by VHDL is a huge drawback in comparison with Verilog, since no C-based routine can run on different simulators without changes.

2 An RTL Simulation-Based Data-Analysis Tool

The tool offers an easy-to-use application for interactive data analysis within ModelSim (the choice of ModelSim has simply been motivated by the familiarity with its usage. However, also other simulators should offer the same programming opportunities). ModelSim's FLI library was used within the C code. Due to the choice of C as programming language the tool is very easy to extend with further features.

2.1 Command Structure of the Tool

The tool is currently command-line driven and adds a set of new commands to the simulator. The command structure is:

```
analyze <subcommand> [<arguments>],
```

where `<subcommand>` specifies one of the following subcommands:

```
add <signal_list>[/-all] [-r[ecursive]]
      [-clock <clock_signal> [-rising/-falling]
                                   [-period time]]

remove <signal_list>[/-all] [-r[ecursive]]

clock <clock_signal> [-rising/-falling]

report <signal_list>[/-all] [-r[ecursive]]

on

off
```

These commands give the user the possibility to:

- Define signals to be analyzed and define the related clocks.
- Run the simulation and thus data analysis.
- Generate the signal statistics and print out the results.

The steps, which are necessary for an analysis run, can be divided into the following classes:

1. ***Specify signals*** to be analyzed ***and their related clocks***. If no clock is available in the design, a "virtual" clock can be added, which is simply specified by its clock period and generated in the PLI routine. Furthermore it is possible to use more than one clock signal per simulation. Thus signals from different clock domains can be analyzed in one single run. The sensitive edge of each clock can be specified.
2. ***Sensitize the PLI routine*** (= turn analysis `on`). From now on every active clock tick (sensitive edge) will cause the PLI routine to calculate statistical information.
3. ***Run the simulation*** for an appropriate time and
4. ***Report/write out signal statistics***.

The following example shows the commands, which are applied during a typical simulation run (analyze-tool commands are written bold and red):

```
run 1us;                         # overruns the reset-phase
analyze add data_i -clock clk    # specifies 'clk' as clock signal
                                 # and adds the signal 'data_i'
analyze on                       # sensitizes the PLI routine
run 100us                        # runs the simulation for 100us
analyze report data_i            # calculates and shows statistics
analyze off                      # desensitizes the PLI routine
```

```
run -all                         # continue simulation without
                                 # statistical calculation
```

For runtime reasons only the basic "running" statistical elements (such as number of clock pulses, transitions, sum and sum of squares) are calculated and updated during the simulation. The main statistical information is calculated offline during the report generation.

2.2 Calculation of Running Statistics

To show the need for "running" statistics, the typical way how the standard deviation σ is calculated is exemplified in Eqs. (5) and (6). This "static" calculation allows the calculation of σ only, if μ is already known for a set of sample data:

$$\mu = \frac{1}{N}\sum_{i=0}^{N-1} x_i , \tag{5}$$

$$\sigma = \sqrt{\frac{1}{N-1}\cdot\sum_{i=0}^{N-1}(x_i - \mu)^2} . \tag{6}$$

This means that μ has to be calculated first, and thereafter σ can be calculated. This method is adequate for many applications. However, it has two limitations:

- If the mean is much larger than the standard deviation, Eq. (6) involves subtracting two numbers that are very close in value. This can result in excessive round-off errors in the calculations.
- It is often desirable to recalculate the output result as new samples are acquired and added to the signal. This type of calculation is called ***running statistics*** [11]. Eqs. (5) and (6) can be used for running statistics, but they require that all of the samples must be involved in each new calculation. Obviously, this is a very inefficient use of computational power and memory.

A solution to these problems, which yields an effective equation for σ with the same result as Eq. (6), but with less round-off error and greater computational efficiency, can be found in the following formula:

$$\sigma = \sqrt{\frac{1}{N-1}\cdot\left[sum(x^2) - \frac{(sum(x))^2}{N}\right]} . \tag{7}$$

While moving through the signal, a running tally must be kept of three parameters:

- N, the number of processed samples (i.e., the number of clock pulses),
- $sum(x)$, the sum of these samples,
- $sum(x^2)$, the sum of the squares of the samples.

After any number of samples has been processed, the mean and standard deviation can efficiently be calculated using only the current value of these three parameters.
Clearly, the above-described principle can also be applied to more complex statistical values, like the correlation coefficient. The correlation coefficient describes the *strength of an association between variables*. An association between variables means

that the value of one variable can be predicted, to some extent, by the value of the other. The correlation coefficient is calculated by

$$\rho(X,Y)=\frac{\left[\sum_{i=0}^{N-1}x_i y_i-\frac{1}{N}\cdot\left(\sum_{i=0}^{N-1}x_i\right)\left(\sum_{i=0}^{N-1}y_i\right)\right]}{\sqrt{\left[\sum_{i=0}^{N-1}x_i^2-\frac{1}{N}\cdot\left(\sum_{i=0}^{N-1}x_i\right)^2\right]\cdot\left[\sum_{i=0}^{N-1}y_i^2-\frac{1}{N}\cdot\left(\sum_{i=0}^{N-1}y_i\right)^2\right]}}. \tag{8}$$

Eq.(8) can be expressed in terms of running statistics in the following manner:

$$\rho(X,Y)=\frac{sum(x\cdot y)-\frac{sum(x)\cdot sum(y)}{N}}{\sqrt{\left[sum(x^2)-\frac{(sum(x))^2}{N}\right]\cdot\left[sum(y^2)-\frac{(sum(y))^2}{N}\right]}}. \tag{9}$$

The autocorrelation is the correlation coefficient of one signal with a delayed version of itself. This means that in Eq. (8) the variable y_i is substituted by x_{i-k}, where k = 1, …, N-1. The autocorrelation gives useful information about the switching activity of a data stream: highly correlated data show a small amount of switching activity, whereas uncorrelated data switch randomly.

Table 1 shows an example data sheet, that reports running mean, standard deviation and autocorrelation as new sample values are taken into account.

Table 1. "Static" vs. "running" statistics.

Sample		Static Statistics			Running Statistics						
Num	x_i	μ	σ	ρ	N	Σx_i	Σx^2	Σxy	μ_x	σ_x	ρ
0	0				1	0	0	0	0,0000	–	–
1	10				2	10	100	0	5,0000	7,0711	–
2	20				3	30	500	200	10,0000	10,0000	0,8660
3	27				4	57	1229	740	14,2500	11,7863	0,9231
4	31				5	88	2190	1577	17,6000	12,6610	0,9416
5	31				6	119	3151	2538	19,8333	12,5764	0,9399
6	27				7	146	3880	3375	20,8571	11,7959	0,9202
⋮	⋮				⋮	⋮	⋮	⋮		⋮	⋮
121	-27				122	69	60343	56339	0,5656	22,3244	0,9393
122	-31				123	38	61304	57176	0,3089	22,4142	0,9401
123	-32				124	6	62328	58168	0,0484	22,5106	0,9411
124	-29				125	-23	63169	59096	-0,1840	22,5697	0,9419
125	-23				126	-46	63698	59763	-0,3651	22,5710	0,9422
126	-14				127	-60	63894	60085	-0,4724	22,5138	0,9418
127	-3	-0,4922	22,4261	0,9409	128	-63	63903	60127	-0,4922	22,4261	0,9409

2.3 Word-Level vs. Bit-Level Statistics

One additional issue worth mentioning is the distinction between word-level and bit-level statistics. Whereas the first gives information about the dynamic behavior of data words as a whole, the second provides this information on a bit-oriented basis. To see the difference, again the autocorrelation can be taken as an example. The autocorrelation on word level is expressed by one single value, which embodies the self-similarity (correlation) of a signal. Highly correlated data, like human speech, show typical autocorrelation values of 0.8 to 0.95, whereas random (uncorrelated) data have an autocorrelation value near 0. However, it was shown in [10] that a bit-wise view on the correlation (respectively the switching activity, since these two values are strongly related to each other, but the latter is much easier to calculate) yields results according to Fig. 1.

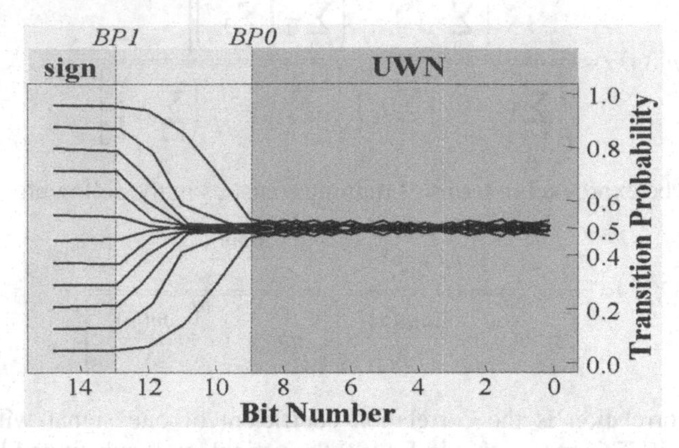

Fig. 1. The dual bit type model.

The *DBT (Dual Bit Type) model* splits data into random activity (uncorrelated) LSBs and correlated MSBs. The LSBs with bit numbers smaller than breakpoint *BP0* are treated as uniform white noise (UWN), MSBs with bit numbers larger than breakpoint *BP1* are treated as autocorrelated. A small intermediate region separates the two constant-activity regions.

2.4 Statistical Reports

As stated above, during the simulation run only the basic running statistics are calculated. After the simulation period the statistical results are reported. The example in Table 2 is used to show the currently available signal properties.

Table 2. Example for reported signal statistics.

Word-level statistics	Bit-level statistics
Signal 'filter_data_i':	[7]: Transitions : 6648
Related Clock : 'clk'	Trans./Clockpulse: 0.0532
Bit Width : 8	Mean : 0.0266
Maximum : 127	[6]: Transactions : 6740
Minimum : -128	Trans./Clockpulse: 0.0539
Transitions : 64607	Mean : 0.0270
Clockpulses : 125001	[5]: ...
Trans./Clockpulse : 0.5169	...
Duty Cycle : 1	[2]: ...
Mean : 0.3623	[1]: Transitions : 37317
Standard Deviation : 12.6848	Trans./Clockpulse: 0.2985
Autocorrelation : 0.967712	Mean : 0.1493
RMS = : 17.63519	[0]: Transitions : 43855
BP0 = : 3.665027	Trans./Clockpulse: 0.3508
BP1 = : 5.263661	Mean : 0.1754

The ***main word-level statistics*** are:

- `Maximum` and `minimum`: If these values do not span the whole range defined by the bitwidth, they can indicate that the range might be to large (e.g., in the VHDL source the definition `integer` was used without specifying a range).
- `Transitions`, `clockpulses`, `transitions/clockpulse`: These values simply represent the absolute activity, the number of samples and the switching activity α (on word level) as needed in Eq. (2).
- `Duty cycle`: This value represents the smallest detected periodic duty cycle in the simulation interval. Signals that receive new data on every *n*:th clock cycle will have a duty cycle of *n*. This means that cyclically (!) only every *n*:th clock pulse is an active one. Note that this differs from an average word-level activity of $1/n$! Such signals can typically be efficiently clock-gated in order to lower the power consumption.
- `Mean`: The (arithmetic) mean informs about where the point of balance is in a signal. For the electronics case this can be seen as the DC (Direct Current) value.
- `Standard deviation`: The standard deviation can be used as a measure of how far the signal fluctuates from the mean.
- `Autocorrelation`: The autocorrelation describes the strength of the self-similarity of a signal. It is obtained by substituting y_i by x_{i-1} in Eq. (8).
- `RMS` (Root Mean Square): As the standard deviation only measures the AC portion of a signal, the RMS value takes both the AC and DC components into account.
- `Crest factor`: The crest factor is also known as the peak-to-RMS ratio.
- `BP0` and `BP1`: According to the DBT model data are split into uncorrelated LSBs and correlated MSBs. The location of the breakpoints that are shown in Fig. 1 can be estimated by [10]:

$$BP1 = log_2(|\mu| + 3\sigma), \tag{10}$$

$$BP0 = log_2\sigma + log_2(\sqrt{1-\rho^2} + |\rho|/8). \tag{11}$$

The ***main bit-level statistics*** are:

- `Transitions` and `transitions/clockpulse`: These values represent the absolute activity, and the (bit-level) switching activity α as needed in Eq. (2).
- `Mean`: On bit level this value expresses the static probability, if the value is more often '0' or '1' (statistically). This value is also referred to as signal probability.

As already mentioned, an extension to the calculated statistics can be easily made, since the calculation is done in a C routine. The only facts that matter here are the running statistics issues, since they directly take influence on the runtime of the analysis (and thus of the whole simulation). However, running statistics calculation is kept as short and small as possible to guarantee smallest possible runtime overhead due to the PLI.

3 Experimental Results

The experiments carried out had two objectives. First, to get a feeling about the performance loss (= increase of runtime) of the RTL simulation. Second, to analyze the usability of the tool. This means, that the question should be answered, if a designer can get benefits out of the tool usage w.r.t. power optimization.

3.1 Runtime Analysis

The increase in runtime was measured on three different designs. The results are shown in Table 3. A maximal runtime increase of ~150% has been noted. However, this increase has not been observed on the largest design, but on the design with the highest clock/data frequency ratio. Since the clock is the trigger for sampling (and thus for calculating the basic running statistics), this signal plays the most important role in the increase of the runtime.

A second remark has to be made on the choice of the selected RTL signals. In the test designs the worst case scenario has been investigated: All signals of the VHDL test designs have been added to the analysis list and sampled with the highest frequency clock. In a typical "real-life" application this wouldn't be the case because of two main reasons: First, signals like resets or configuration signals from primary inputs are typically static signals and thus of no interest for a dynamic statistical analysis. Second, typically not all signals are clocked with the highest frequency, i.e., a separation into clock domains is useful and can help to reduce the runtime overhead. As stated above, multiple clock signals are supported by the tool.

Table 3. Performance comparison for test designs.

Design no.	RTL signals	#flipflops (gate level)	Clock freq.	Data freq.	Runtime w/o PLI	Runtime with PLI	Runtime increase
1	9	64	16 MHz	16 MHz	1.35s	2.63s	94.8%
2	30	206	128 MHz	32 MHz	3.57s	8.94s	150.4%
3	372	~4000	36 MHz	36 MHz	21.4s	48.9s	128.5%

3.2 Facilitating Low-Power Design

One of the intentions of the tool is to give the RTL designer some ideas where to improve a design in terms of switching activity and thus power consumption.

Signal analysis on word level – low activity signals and periodicity:

Important information is obtained from the observation, whether a signal is changed periodically or not. Fig. 2 shows two different multi-bit signals with the same number of changes in the simulation interval. Clearly, the word-level switching activity is the same for both signals. However, the first one, `signal1`, is changed periodically with a minimal duty cycle of **4**, whereas the second one, `signal2`, is changed acyclically. The tool reports a minimal duty cycle of **1**.

A low-power design approach for the first case is to move the whole design to another clock domain with lower frequency or to apply clock gating (e.g., with

Synopsys PowerCompiler [12]) with a cyclic enable signal. Such a design approach, for example, would be appropriate for test design #2 (decimation filter) from Table 3, since the built-in duty-cycle calculator detects a minimal duty cycle of 4.

If acyclic signals with low activity like `signal2` are detected, one solution to reduce the power consumption is to generate an additional signal by XORing the register outputs with the inputs and use this signal as a clock-gating enable condition for the low-activity register bank. Test design #3 (video processing unit) shows a word-level switching activity of about 40% for several signals. Thus this low-power approch can be utilized.

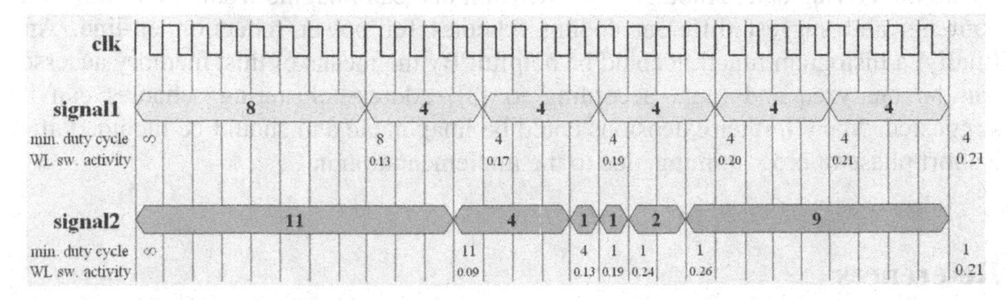

Fig. 2. Periodic and acyclic data with same word-level switching activity.

Signal analysis on bit level – dual bit type model:

An examination of signals on bit level can result in the observation that data behave according to the DBT model (like shown in Fig. 3). A possible low-power solution for such register banks is that LSBs and MSBs are separately clock-gated with individual (XORed) enable conditions, one for the LSBs and one for the MSBs. In test design #3, for example, analysis shows that only the 3 LSBs (out of 8 bit) have a switching activity of about 40%, the MSBs toggle with less than 10%.

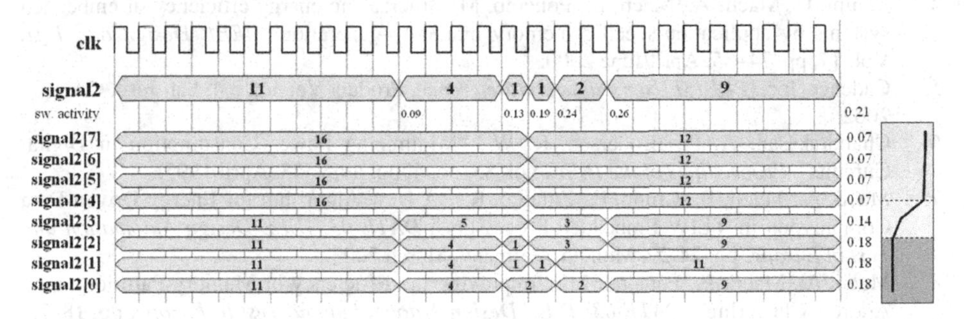

Fig. 3. Word-level and bit-level switching activity.

4 Conclusion and Outlook

An easy-to-use tool was presented, which performs statistical data analysis on RTL. The tool is implemented in C and embedded in the digital logic simulator via the API of the simulator. It extends the simulator command-set to define signals to be analyzed together with their related clocks, to run data analysis and to generate signal statistics reports. One of the intentions of the tool is to give the designer a simple possibility of analyzing data in a well-known environment and also facilitate (basic) low-power design approaches.

The possibilities for future extensions of the tool with respect to low-power design-assistance are manifold. Some of them are planned and under construction. The first is to hide the numerous data and make the tool working in a graphical way. Especially a visualization of the DBT-related information could help to increase the understanding of the underlying data. Another tool extension one can imagine would be to trace bus contents and suggest different coding schemes for power reduction on-line. And finally, a histogram function could be helpful: By the means of this, memory accesses can be analyzed and (e.g., according to [8]) address clustering schemes can be suggested. Many further extensions could be imaginable and should be included after a short phase of programming, due to the implementation in C.

References

1. Aghaghiri, Y., Fallah, F., Pedram, M., "Irredundant Address Bus Encoding for Low Power", *ISLPED-01: ACM/IEEE International Symposium on Low Power Electronics and Design*, pp. 182–187, Huntington Beach, CA, August 2001.
2. Benini, L., Bruni, D., Macii, A., Macii, E., "Hardware-Assisted Data Compression for Energy Minimization in Systems with Embedded Processors", *DATE-02: IEEE Design Automation and Test in Europe*, pp. 449–452, Paris, France, March 2002.
3. Benini, L., De Micheli, G., Macii, A., Macii, E., Poncino, M., "Reducing Power Consumption of Dedicated Processors Through Instruction Set Encoding", *GLS-VLSI-98: IEEE/ACM Great Lakes Symposium on VLSI*, Lafayette, IN, February 1998.
4. Benini, L., Macii, A., Macii, E., Poncino, M., "Increasing energy efficiency of embedded systems by application-specific memory hierarchy generation", *IEEE Design and Test*, Vol. 17, pp. 74–85, April/June 2000.
5. Cadence, Inc., *PLI 1.0 User Guide and Reference*, Product Version 3.4, San Jose, CA, 2002.
6. Chandrakasan, A. P., Brodersen, R. W., "Minimizing Power Consumption in Digital Curcuits", *Proceedings of the IEEE*, vol. 83, no. 4, pp. 498–523, April 1995.
7. Macii, A., Macii, E., Crudo, F., Zafalon, R., "A New Algorithm for Energy Driven Data Compression in VLIW Embedded Processors", *DATE-03: IEEE Design Automation and Test in Europe*, pp. 24-29, Munich, Germany, March 2003.
8. Macii, A., Macii, E., Poncino, M., "Improving the Efficiency of Memory Partitioning by Address Clustering", *DATE-03: IEEE Design Automation and Test in Europe*, pp. 18–23, Munich, Germany, March 2003.
9. Model Technology Inc., *Modelsim Foreign Language Interface*, Version 5.6b, Portland, OR, 2002.
10. Landman, P., Rabaey, J. M., "Architectural Power Analysis: The Dual Bit Type Method", *IEEE Transactions on Very Large Scale Integration (VLSI) Systems*, Vol. 3, No. 2, June 1995, pp. 173–187.
11. Smith, S. W., *The Scientist and Engineer's Guide to Digital Signal Processing Second Edition*, California Technical Publishing, San Diego, CA, 1999
12. Synopsys Inc., *Power Compiler Reference Manual v.2001.08*, Mountain View, CA, 2001.

Data Dependences Critical Path Evaluation at C/C++ System Level Description

Anatoly Prihozhy, Marco Mattavelli, and Daniel Mlynek

Signal Processing Laboratory 3, Signal Processing Institute,
Swiss Federal Institute of Technology, Lausanne
LTS3/ELB-Ecublens, Lausanne, CH-1015, Switzerland
prihozhy@yahoo.com, marco.mattavelli@epfl.ch, dmlynek@txc.com
http://lsiwww.epfl.ch/

Abstract. This paper presents a model metrics, techniques and methodology for evaluating the critical path on the Data Flow Execution Graph (DFEG) of a system on chip specified as a C program. The paper describes an efficient dynamic critical path evaluation technique generating no execution graph explicitly. The technique includes two key stages: (1) the instrumentation of the C code and the mapping into a C++ code version, (2) the execution of the C++ code and actual evaluation of the critical path. The model metrics and techniques aim at the estimation of the bounding speed and parallelization potential of complex designs specified at system level.

1 Introduction

Nowadays, processing and compression algorithms, communication protocols and multimedia systems have reached an extremely high level of sophistication. Architectural implementation choices become extremely difficult tasks, leading to the need of more and more intensive specification and validation by means of C/C++ software implementations. At the highest possible algorithmic level, the evaluation of complexity and parallelization potential of the algorithm have to be derived from such software descriptions under real input conditions [2, 4, 8] in order to be able to take meaningful and efficient partitioning decisions. Depending on the specific goal of the complexity analysis, different approaches and tools have been developed in [4, 6, 12, 14]: (1) profilers, modifying the program to make it produces run-time data, (2) compilers, applying result-equivalent code replacements, (3) static methods, getting information from the source code, (4) descriptions by means of hardware description languages, (5) hardware specific tools, providing computational information to some extent. Nowadays, such approaches and tools are unsuited for implementation exploration at the C/C++ system-level description because either they apply at a stage where architectural decision is already been taken, either they provide results that depend too much on the simulation platform used or do not take into account the complexity and parallelization potential features that depend only on the algorithm itself and on the input data. Conversely, the SIT Software Instrumentation Tool [14, 13] is an automatic general instrumentation environment able to extract and measure data flow and complexity of algorithms/systems specified in C/C++. SIT measures the

J.J. Chico and E. Macii (Eds.): PATMOS 2003, LNCS 2799, pp. 569–579, 2003.
© Springer-Verlag Berlin Heidelberg 2003

processing complexity in terms of executed operators, data types and bandwidths of the data exchanges, while still taking into account real input data conditions.

The problem of identifying one of the longest paths in a circuit is called a critical path problem [7]. The length of the critical path plays a key role in setting the clock cycle time and improving the chip performance. In message passing and shared-memory parallel programs [1], communication and synchronization events result in multiple paths through a program's execution. The critical path of the program is simply defined as the longest time-weighted sequence of events from the start of the program to its termination. The critical path profiling is a metric explicitly developed for parallel programs [1] and proved to be useful. The critical path profile is a list of procedures and the time each procedure contributed to the length of the critical path. Critical path profiling is an effective metric for tuning parallel programs and is especially useful during the early stages of tuning a parallel program when load imbalance is a significant bottleneck. In [7] the parallel algorithms for critical path problem at the gate level are studied, while [1] describes a runtime non-trace-based algorithm to compute the critical path profile of the execution of message passing and shared-memory parallel programs. The idea described in [16] is to insert parallelism analysis code into the sequential simulation program. When the modified sequential program is executed, the time complexity of the parallel simulation based on the Chandy-Misra protocol is computed. The critical path analysis also gives an effective basis for the scheduling of computations. The work presented in [3] proposes a task scheduling algorithm that allocates tasks with corrected critical path length. The technique described in [5] schedules task graphs, analyzing dynamically the schedule critical paths. The techniques based on the minimization of critical path length estimated as the maximal clique weight of the sequential and parallel operator graphs [9, 10] constitute an efficient approach to the generation of concurrent schedules.

Summarizing the previous results, it can be concluded that the majority of already developed tools aim at the critical path profiling for tuning existing parallel programs (Fig. 1a). In this paper, the objective is to propose a critical path model metric in order to be able to find out in which degree a given algorithm described in C satisfies the parallel implementation conditions (Fig.1b). The paper is organized as follows. Section 2 defines the critical path model metrics on data dependences. Section 3 describes techniques for evaluating the critical path length. Section 4 presents the mapping of the C-code into a C++-code version. Transformations for reducing the critical path length are discussed in Section 5. Experimental results on parallelization potential on two real world analysis cases are presented in Section 6.

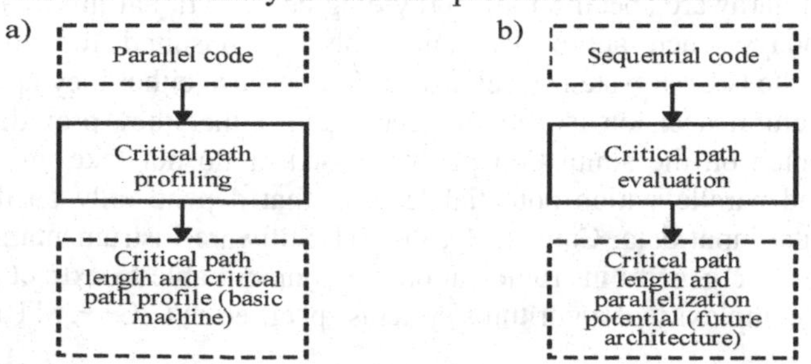

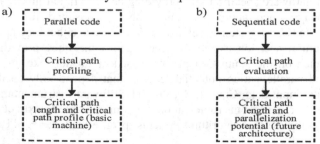

Fig. 1. Critical path profiling (a) of parallel code on event graphs versus critical path evaluation (b) of sequential code on data dependences graphs.

2 Definition of Critical Path on Data Dependences

The principles constituting the basis for the critical path model metric definition on a sequential C-code are the following:

- The critical path is defined on the C-code's execution data flow without taking into account the true control flow.
- The critical path length and the system parallelization potential are defined in terms of C language basic operations (including *read* and *write* operations) complexity. The parameters of the machine executing the instrumented C-code during evaluating the critical path are not taken into account.
- In the definition of the critical path, the Data Flow Execution Graph results from the partial computation of the C-code using true input data. Therefore, such Data Flow Execution Graph is used for the critical path definition instead of the traditional Data Flow Graph.

The DFEG is represented as a finite non-cyclic directed weighted graph constructed on the two types of node. The first type includes name-, address-, and scalar value-nodes. The second type includes operator-nodes. The name- and address-nodes are represented as [i] and the value-nodes are represented as [i]. The operator-nodes are denoted using the usual C-language notation: =, [], ++, --, +, *, %, ==, /=,<, >, +=, /=, *read* (*r*), *write* (*w*) and others. The graph nodes may be connected by two types of arcs: the data dependence arc denoted → and the conditional dependence arc denoted ⋯→. The data dependence arc connects input names, addresses and values with an operator and connects an operator with its output value or address. The conditional dependence arc connects a conditional value with an operator covered by a conditional instruction. A graph node without incoming arcs is called an initial node and a graph node without outgoing arcs is called a final node. A DFEG fragment for *if c then d*=2;* C-code is shown in Fig. 2a. It contains four value-nodes, one operator-node, three data and one conditional dependence arcs.

The DFEG is weighted with node complexities. All the complexities are accumulated at the operator-nodes and each C-operator is represented by a fragment in DFEG as shown in Fig. 2b. *Read* and *write* operators are associated with incoming and outgoing arcs of the operator-node respectively. The critical path on the DFEG is defined as a sequence of the graph nodes with the maximal sum of weights connecting an initial node with a final node. The C-code complexity together with the critical path length describes its parallelization potential. The complexity of the fragment shown in Fig. 2b equals 4 basic operators. The internal critical path length on the graph fragment equals 3 because two *read* operations are executed in parallel.

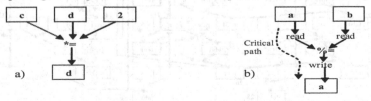

Fig. 2. (a) Example DFEG fragment for *if c then d*=2;* C-code. (b) Evaluation of the complexity and critical path of *a%=b;* C-code.

Similarly, assuming each basic operator complexity be equal to 1, Table 1 represents the C-language operator complexities and internal critical path lengths. When the basic operator complexities are different, the table can be easily modified to map the critical path length on any target architecture.

Table 1. Complexity and critical path length of C language operators.

Operation	Operator	Complexity	Critical path
Assignment	=	1	1
Reference	&	2	2
Dereference	*	2	2
Arithmetic	+, -, *, /, %	3	2
Arithmetic-assignment	+=, -=, *=, /=, %=	4	3
Indexing	[]	3	2
Increment (decrement)	++, --	3	3
Unary minus and others	-	2	2

```
void main() {
    unsigned long n=21414;
    int m[5], k=0;
    for(int i=0; i<5; i++) m[i]=0;
    while(n) {m[n%10]=1; n/=10;}
    for(int j=0; j<5; j++) if(m[j]) k++;
}
```

Fig. 3. C-code for counting 1, 2, 3, 4 and 5 digits in integer *n*.

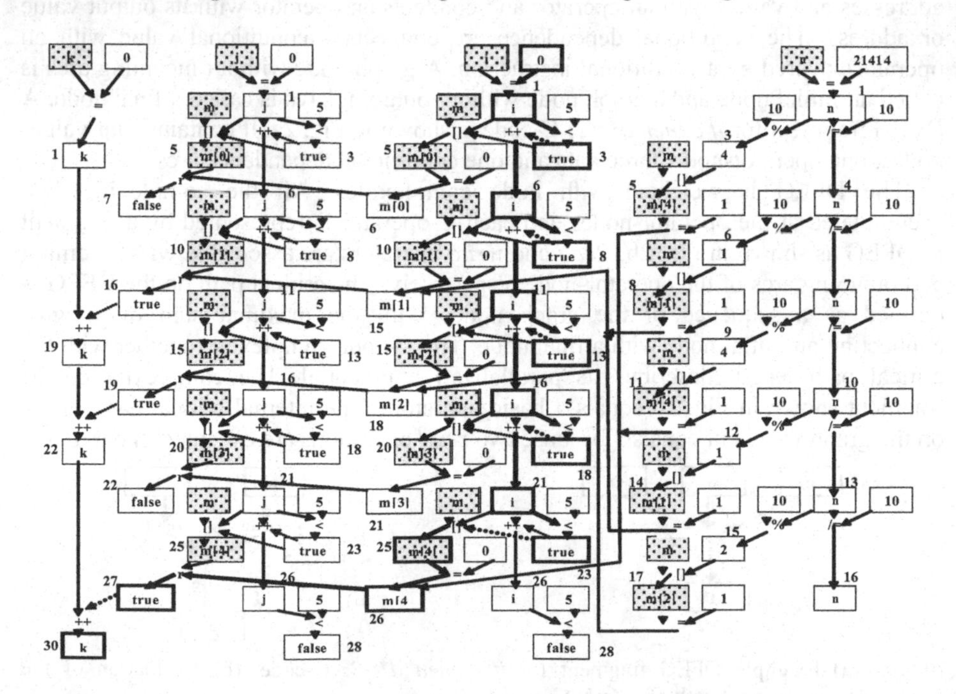

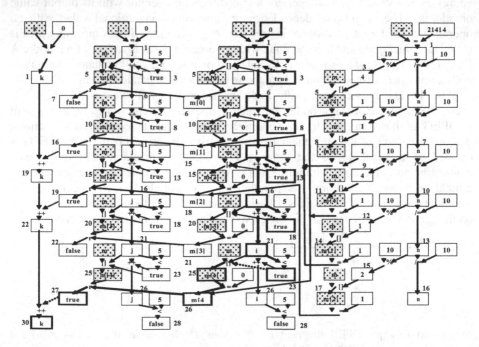

Fig. 4. The Data Flow Execution Graph (DFEG) for the C-code shown in Fig. 3.

Table 2. Evaluation of the complexity of graph shown in Fig. 4.

Operator	Operator complexity	Number of operators	Total complexity
=	1	14	14
[]	3	15	45
<	3	12	36
++	3	13	39
/=	4	5	20
%	3	5	15
read	1	5	5
			$\Sigma = 174$

An example C-code for counting digits in an integer is presented in Fig. 3. The corresponding DFEG is shown in Fig. 4. The array components are treated as separate scalar elements. The results evaluating the algorithm complexity are reported in Table 2. The overall complexity is 174 basic operators.

3 Techniques for Critical Path Evaluation

The approach described in this paper to evaluate the critical path length consists in the preliminary generation of the DFEG by means of performing partial computations on the C-code's DFG (Fig. 5a) under certain meaningful input data sets. Given the complexity and internal critical path length of each operator-node in the DFEG, we can evaluate the external critical path for each address-, value- and operator-node in DFEG using the following simple recursive technique:

1. If *val* is an initial name-, address- or value-node then its external critical path length $cpl(val)=0$.
2. If *val* is a value- or address-node and $op_1, \dots, op_r$ are operator-predecessors of *val* (Fig. 6b), then its critical path length $cpl(val) = max(cpl(op_1), \dots, cpl(op_r))$.
3. If *op* is an operator-node and $val_1, \dots, val_k$ are value-address-predecessors of *op* (Fig. 6a), then the operator critical path length is $cpl(op) = cplint(op) + max(cpl(val_1), \dots, cpl(val_k))$, where $cplint(op)$ is the *op* operator's internal critical path length.

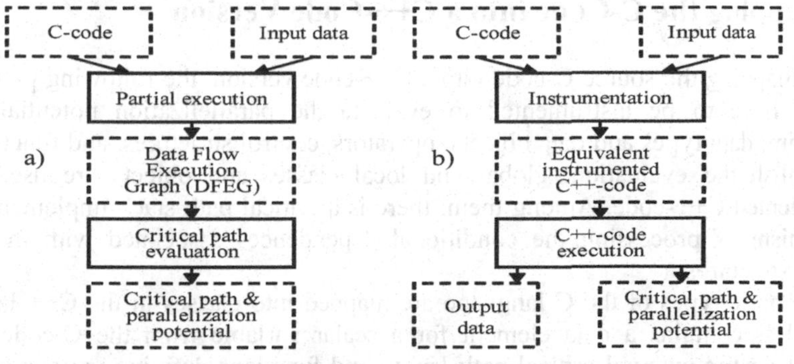

Fig. 5. (a) Critical path evaluation by means of explicitly generation of DFEG; (b) Dynamic evaluation of the critical path by means of instrumenting and executing the C-code.

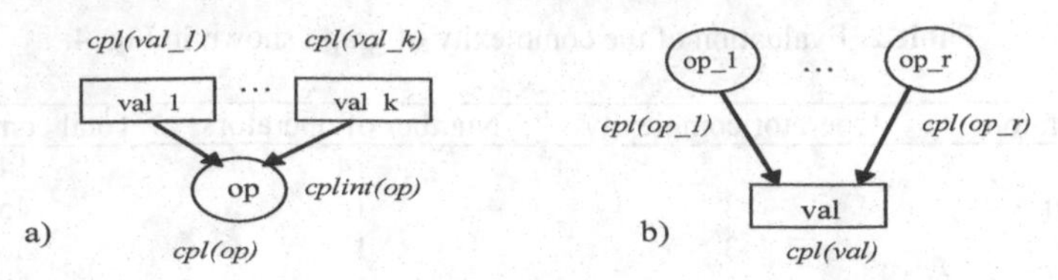

Fig. 6. (a) The graph fragment for evaluating the critical path for an operator; (b) The graph fragment for evaluating the critical path for a value (address).

Since the number of nodes in the DFEG is equal to the number of operation calls during the program's execution, explicitly building the graph is not practical for long running programs. One way to overcome this limitation is to develop a technique that does no require building the graph. Such a technique is based on the flow shown in Fig. 5b. First, the C-code is instrumented and is transformed into an equivalent C++-code, in terms of the operators applied to the input data. Second, the C++-code is executed under the given input data, computing output data and evaluating the critical path and parallelization potential of the algorithm. In the C++-code, an additional variable *cpl* is bound with each actual scalar variable of the C-code (Fig. 7). The execution of a C-code operation also results in computing a new value of the associated additional variable in the C++-code.

In Fig. 4, address- and value-nodes are weighted with external critical path lengths. The critical path of the whole DFEG is shown in bold. The critical path length equals 30. The DFEG total complexity equals 174. The bounding possible acceleration describing the parallelization potential of the C-code is equal to 5.8.

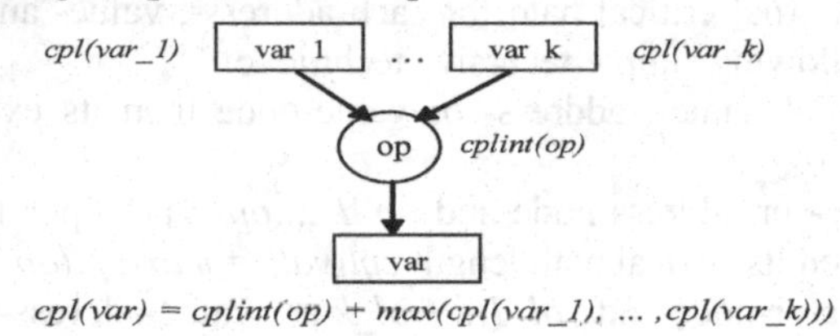

Fig. 7. The graph fragment for dynamic evaluation of the critical path.

4 Mapping the C-Code into a C++-Code Version

While mapping the source C-code into a C++-code version, the following parts of the C-code have to be instrumented to evaluate the parallelization potential of the algorithm: data types and data objects, operators, control structures, and functions. To accomplish the evaluation, global and local classes and objects are used in the instrumented C++-code. Among them, there is a critical path stack implementing the mechanism of processing the conditional dependences associated with the nested control structures.

The basic types of the C language are mapped into classes in the C++ language. Each class contains a data element for a scalar variable from the C-code, a data element for the external critical path length and functions defining various operators on the data. Declarations of pointers to the basic types in the C-code are replaced with appropriate classes in the C++-code. An array of elements of a basic type in the C-code is mapped to an array of objects of the corresponding instrumented type. Other composite types of the C language are instrumented in the similar way. All the operators on addresses and values that will be performed during the C-code execution are instrumented during transition from the C-code to the C++-code. The operators on the C-types are overloaded by member functions of the instrumenting C++-classes. The C-functions are instrumented in such a way as to create transparency in transmission of the dependences from the external environment to the function body and from the function body to the external environment. The C-code complexity and critical path can be evaluated for any part of the C-code by means of control elements.

5 Critical Path Reduction by Means of C-Code Transformation

The true control structures of the C-code do not permit the direct implementation of the bounding acceleration [11]. The transformation methodology is a mechanism searching for appropriate architectural implementations. It allows the reduction of the execution time (control steps and clock cycles) at the same constraints on resources and approaches to the bonding acceleration. Two types of transformations are investigated in the context of architectural synthesis. The first type transformations aim at reducing the critical path. The transformations of the second type aim at breaking the true control structures in order to increase the effectiveness of behavioral synthesis and scheduling techniques. The transformation methodology allows the implementation of the parallelization potential in architectures through C-code transformation. The transformations approach DFG to DFEG.

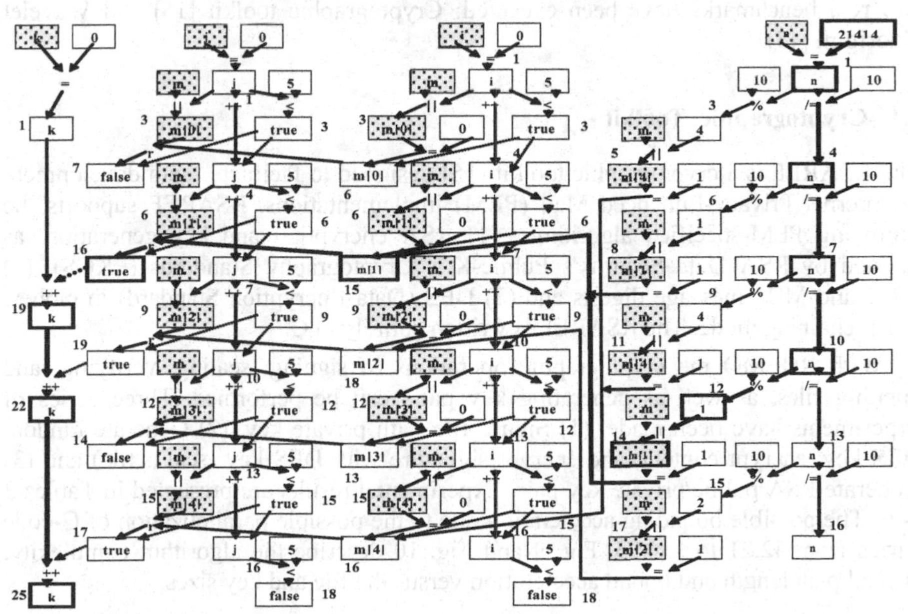

Fig. 8. The DFEG for the C-code shown in Fig. 3 and transformed by means of extracting computations from conditional statements.

The equivalent transformation of the source C-code is a way of reducing the critical path length and increasing the parallelization potential. The control and data flow transformation rules allowing the critical path reduction are defined as follows: (1) restructuring, split and transformation of statements; (2) extraction of computations from control structures; (3) algebraic transformation of arithmetic, logic and other type of expressions; (4) merge of expressions and statements; (5) unfolding loops and others. In order to be able to apply other transformation rules to the source C-code, the loop statements should be preliminary transformed by moving the iteration scheme into the loop body. Although the extraction of operators from control structures and other transformations can imply introducing additional variables and computations, it is an efficient way of accelerating the computations and performing the computations in advance and in parallel.

The procedure of increasing the parallelization potential of C-code is an iterative process. First the C-code is transformed and rebuild. Then it is instrumented and mapped to a C++-code version. After the execution of the C++-code and estimation of the critical path and bound acceleration, the source C-code can be transformed again to perform the next iteration. Fig. 8 illustrates the usefulness of transformation rules on the C-code presented in Fig. 3. The critical path length is reduced from 30 to 25. The bounding possible acceleration potential increases from 5.8 to 6.96.

6 Results

Several experiments on the critical path and parallelization potential evaluation for two real benchmarks have been executed: Cryptographic toolkit [15] and Wavelet algorithm [13].

6.1 Cryptographic Toolkit

The RSAREF is a cryptographic toolkit [15] designed to facilitate rapid development of Internet Privacy-Enhanced Mail (PEM) implementations. RSAREF supports the following PEM-specified algorithms: (1) RSA encryption and key generation, as defined by RSA Data Security's Public-Key Cryptography Standards (PKCS), (2) MD2 and MD5 message digests and (3) DES (Data Encryption Standard) in cipher-block chaining mode. The RSAREF is written entirely in C.

With RDEMO the cryptographic operations of signing, sealing, verifying, and opening files, as well as generating key pairs can be performed. Three series of experiments have been made: (1) Sign a file with private key, (2) Generate random DES key, encrypt content, and encrypt signature with DES key (seal a file) and (3) Generate RSA public/private key pair. Experimental results are presented in Tables 3 to 6. The possible bounding acceleration due to the possible parallelization of C-code varies from 42.21 to 136.93. Fig. 9 and Fig. 10 describe the algorithm complexity, critical path length and bound acceleration versus the file and key sizes.

Table 3. Experimental results for RSAREF (sign a file).

Content size (Bytes)	Algorithm complexity	Critical path	Acceleration
281	21'804'816	502'000	43.44
621	21'826'260	509'660	42.83
971	21'846'076	517'582	42.21

Table 4. Experimental results for RSAREF (seal with sign).

Content size (Bytes)	Algorithm complexity	Critical path	Acceleration
281	26'781'944	502'685	53.28
621	29'553'488	510'345	57.91
971	32'455'414	518'267	62.62

Table 5. Experimental results for RSAREF (seal without sign).

Content size (Bytes)	Algorithm complexity	Critical path	Acceleration
281	4'978'890	77'491	64.25
621	7'728'936	77'491	99.74
971	10'611'022	77'491	136.93

Table 6. Experimental results for RSAREF (keypair generation).

Key size (Bits)	Algorithm complexity	Critical path	Acceleration
508	0.6E9	13.8E6	43.89
767	3.0E9	90.0E6	33.41
1024	21.9E9	806.4E6	27.12

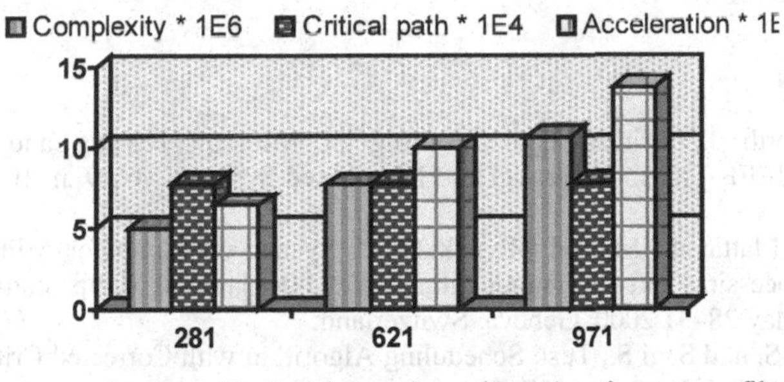

Fig. 9. Algorithm complexity, critical path length, and acceleration versus file size for seal.

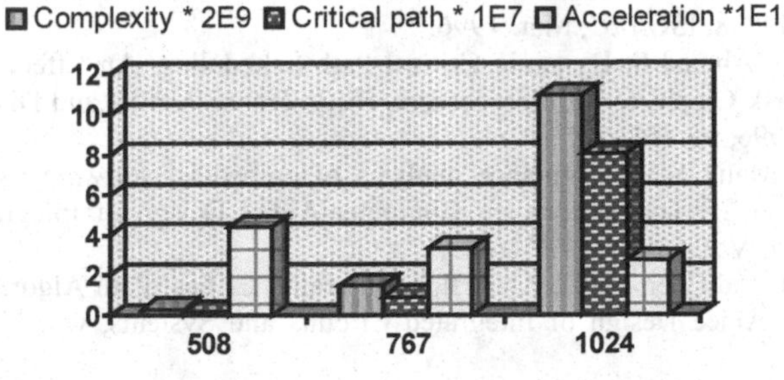

Fig. 10. Algorithm complexity, critical path length, and acceleration versus key size for key pair generation.

6.2 Wavelet Algorithm

More impressive results have been obtained for the two-dimensional Wavelet codec implementations proposed in [13]. The possible bounding acceleration varies from 30'932 to 1'112'331. The data flow computations to be incorporated in the C-code implementation constitute 37.9% and the control flow computations constitute 62.1%. After the equivalent transformation of WAVELET C-code and modifying its DFG, the critical path length has been reduced by 16.7%.

7 Conclusion

The parallelization potential evaluation tool together with the transformation techniques complements the behavioral synthesis tools. Analyzing the results obtained by the tool, the most promising algorithm can be selected among many alternatives. Moreover, the equivalent transformation of algorithm such as for the bounding of possible acceleration of the future parallel architecture can be performed. The critical path length influences significantly the scheduling results at any constraints on resources. The schedule cannot be faster than the critical path length. The reduction of the critical path guarantees more powerful scheduling results and implies the improvement in the trade off "complexity–delay" in behavioral synthesis [11].

Acknowledgements. The authors are grateful to Massimo Ravasi and Clerc Christophe for very useful discussions on the topic under consideration and their help in creating the tool.

References

1. Hollingsworth J., Critical Path Profiling of Message Passing and Shared-Memory Programs, IEEE Trans. on Parallel and Distributed Systems, vol. 9, n. 10, 1998, pp. 1029–1040.
2. Juarez E., Mattavelli M. and Mlynek D., A System-on-a-chip for MPEG-4 Multimedia Stream Processing and Communication, IEEE International Symposium on Circuits and Systems, May 28–31 2000, Geneva, Switzerland.
3. Kobayashi S. and Sagi S., Task Scheduling Algorithm with Corrected Critical Path Length, ISS, Vol. J81-D-I, No.2, pp. 187–194.
4. Kuhn P., Instrumentation tools and methods for MPEG-4 VM: Review and a new proposal, Tech. Rep. M0838, ISO/IEC, Mar. 1996.
5. Kwong Y.-K., Ahmad I., Dynamic Critical-Path Scheduling: An Effective Technique for Allocating Task Graphs to Multiprocessors, IEEE Trans. Parallel and Distributed Systems, vol. 7, n. 5, 1996, pp. 506–521.
6. Li Y.S. and Malik S., Performance analysis of embedded software using implicit path enumeration, IEEE Transactions on Computer-Aided Design of Integrated Circuits and Systems, 1997, Vol. 16, pp. 1477–1487,.
7. Liu L., Du D. and Chen H.-C., An Efficient Parallel Critical Path Algorithm, IEEE Trans. on Computer Aided Design of Integrated Circuits and Systems, vol. 13, n. 7, 1994, pp. 909–919.
8. Mattavelli M. and Brunetton S., Implementing real-time video decoding on multimedia processors by complexity prediction techniques, IEEE Transactions on Consumer Electronics, vol. 44, pp. 760–767, Aug. 1998.

9. Prihozhy A., Mlynek D., Solomennik M. and Mattavelli M., Techniques for Optimization of Net Algorithms, PARELEC 2002 – Parallel Computing in Electrical Engineering, IEEE CS Press, 2002, pp. 211–216.
10. Prihozhy A., Net Scheduling in High-Level Synthesis, IEEE Design & Test of Computers, 1996 spring, pp. 24–33.
11. Prihozhy A., High-Level Synthesis through Transforming VHDL Models, in Book "System-on-Chip Methodologies and Design Languages", Kluwer Academic Publishers, 2001, pp.135–146.
12. Pushner P. and Koza C., Calculating the maximum execution time of real-time programs, Journal of Real- Time Systems, vol. 1, pp. 160–176, Sept. 1989.
13. Ravasi M., Mattavelli M. High-level algorithmic complexity evaluation for system design, to appear on the International Journal on System Architectures, 2003.
14. Ravasi M., Mattavelli M., Schumacher P., Turney R., High-Level Algorithmic Complexity Analysis for the Implementation of a Motion-JPEG2000 Encoder, submitted to PATMOS'2003
15. RSA Data Security, Inc. PKCS #1: RSA Encryption Standard. Version 1.4, June 1991.
16. Wong Y.-C., Hwang S.-Y. and Lin Y., A Parallelism Analyzer for Conservative Parallel Simulation, IEEE Trans. on Parallel and Distributed Systems, vol. 6, n. 6, 1995, pp. 628–638.

A Hardware/Software Partitioning and Scheduling Approach for Embedded Systems with Low-Power and High Performance Requirements

Javier Resano, Daniel Mozos, Elena Pérez, Hortensia Mecha, and Julio Septién

Dept. de Arquitectura de Computadores, Facultad de Informática, UCM, Madrid
{javier1, mozos, eperez, horten, septien}@dacya.ucm.es

Abstract. Hardware/software (hw/sw) partitioning largely affects the system cost, performance, and power consumption. Most of the previous hw/sw partitioning approaches are focused on either optimising the hw area, or the performance. Thus, they ignore the influence of the partitioning process on the energy consumption. However, during this process the designer still has the maximum flexibility, hence, it is clearly the best moment to analyse the energy consumption. We have developed a new hw/sw partitioning and scheduling tool that reduces the energy consumption of an embedded system while meeting high performance constraints. We have applied it to two current multimedia applications saving up to 30% of the system energy without reducing the performance.

1 Introduction

Low-power has become one of the major design concerns. First of all, the designer must guarantee that his design does not exceed the power constraints of the target platform, since it will generate heating problems. Moreover, due to the proliferation of portable, battery-dependent devices, low-energy consumption has become one of the key features for the success of a design.

The current trend for portable embedded systems is to create heterogeneous systems, with one or more low-power processors, some additional hardware (hw) logic (ASICs and/or FPGAs), and some memory hierarchy. Current technologies allow creating the whole system in a single chip (SoC).

One of the most important steps to carry out in order to implement an application over such a system is to partition the application functionality among the different processing elements. This process drastically influences both the energy consumption and performance of the system. Figure 1 presents a simple example where the partitioning process can lead to energy savings. If the designer selects the fastest solution (sch1), the execution time is 139 time-units and the energy 21 energy-units. However, if the deadline for the application is 150, the designer can try to find a slower solution that meets this constraint while consuming less energy. In this case sch2 would be selected since its execution time is less than the deadline and its energy consumption is 16. Thus, the energy consumption decreases 25%.

J.J. Chico and E. Macii (Eds.): PATMOS 2003, LNCS 2799, pp. 580–589, 2003.
© Springer-Verlag Berlin Heidelberg 2003

	PE1		PE2	
	T	E	T	E
Node1	78	13	88	8
Node2	61	8	73	9

Fig. 1. Partitioning example. Two nodes must be partitioned between two Processing Elements (PE). T means time. E means energy. Sch1 and Sch2 are two selected solutions.

Since our partitioning tool is still under construction, currently we just support a software (sw) processor, an FPGA, a system bus and one or several memory blocks. However, partitioning an application to such a system is still a NP-complete problem. Moreover, there are several existing prototype platforms as well as commercial platforms that follow this scheme providing a sw processor and some reconfigurable hw resources e.g. Garp [1], Morphosys [2] and the Virtex II-Pro XC2VP4 and VP7 [3].

The system bus and the memory blocks require a careful study, since both elements can significantly affect the system performance and energy consumption, especially because both hw and sw performance are improving much faster than communication channels and memories do. In order to estimate accurately the impact of the memories and buses in the system performance and energy consumption their physical features must be taken into account. Ideally the vendor should provide either estimators or at least time and power models, but unfortunately, this is not always the case, then, time and power models are needed, some examples of existing useful models are [4] for USB, and PCI buses (just timing considerations), and [5,6] for memories.

However, even after accurately estimating all the tasks, communications and memory accesses, computing the overall execution time it is not trivial, since it involves a scheduling that must take into account data and control dependencies as well as the accesses to the shared resources. Thus, we have developed a tool that schedules the tasks and the accesses to the system bus, and the shared memories during the partitioning process. This scheduling is the only way to accurately evaluate a solution, since otherwise, it is impossible to determine the impact of the communications or the delays introduced due to the conflicts on the accesses to shared resources (In [7] this problem is explained in detail). In addition, this scheduling prevents the need for arbitration logic in the bus controllers. Since the scheduler is integrated in a partitioning tool that must evaluate a great amount of different partitions one of our major concerns was to achieve near-optimal scheduling without increasing significantly the execution time of the partitioning tool.

The rest of the paper is structured as follows: section 2 presents an overview of the related work; section 3 explains in detail the format of the initial specification for our partitioning tool; section 4 describes the cost function that steers the design space exploration; sections 5, 6, and 7 explain how the energy, execution-time and hardware area are estimated for a given partitioning. Section 8 presents the experimental results and finally section 9 remarks some conclusions as well as future work to be done.

2 Related Work

Hardware/software partitioning is a very well known problem. Several partitioning tools have been proposed in literature (e.g. [8, 9]). Most of these previous approaches accomplish the partitioning problem at a high abstraction level, adding the platform low-level details and scheduling the tasks on the processing elements (PEs) in a subsequent step called co-synthesis. Moreover, even during co-synthesis often the communications between different PEs are neglected, thus, these communications are included in a following step called communication synthesis. After these three steps the resultant solution is co-simulated, and likely, the results will not be the expected, so the process will have to start again with another solution. The main problem of this approach is that some of the features neglected during partitioning are critical for the system performance. Thus, it is almost impossible to found near-optimal solutions when communications are neglected during the partitioning process. Another lack of most of the existing approaches is that they just consider either hardware area or execution time minimization. However, as mentioned in the introduction, currently minimizing the energy consumption is often one of the more important designer concerns.

Recently several scheduling and/or partitioning approaches for multiprocessors have been presented. They attempt to minimize the system consumption either applying Dynamic Voltage Scheduling (DVS) or applying different supply voltages to each processor; some of the more relevant are [10, 11,12]. DVS techniques schedule the voltage supplied to each processor during its execution. This is a powerful way to achieve power savings, since in CMOS technologies the power consumption decreases quadratically with the power supply. However, currently there is not support for DVS in most of the commercial processors, and to the best of our knowledge, there is not support at all for DVS in FPGAs platforms. Hence, nowadays, this is not a feasible approach for hw/sw co-design.

[13] is the first hw/sw partitioning tool for low-power that we have found, it starts from a full sw implementation in a microprocessor (μP), and reduces the energy consumption migrating part of the functionality to hw, the energy savings are achieved turning off the μP (in addition clock gating is applied in the hw partition). This approach does not perform a full partitioning design exploration. Moreover, it expects some data for the designer, like the number of ALUs, multipliers, shifters, etc., based on some previous designer experience, so the results of the partitioning will highly depend on the designer capabilities. PAP [14] is a recent partitioning tool that attempts to minimize the hardware area while meeting the timing and power constraints, thus they do not minimize the overall energy consumption but take care that infeasible solutions (those that consume more power than the allowed by the platform) will not be selected. Finally, in [15] a scheduling technique for dynamically reconfigurable FPGAs with support for partial reconfiguration is presented. The scheduling process attempts to minimize the energy consumption optimising the number of partial reconfigurations. However, this scheduling is carried out after the partitioning process, hence, most of the flexibility is lost since the partition has been previously fixed. According to this paper, currently, FPGAs dynamic reconfiguration

is extremely power inefficient, since in their experiments up to 50% of the FPGA energy consumption was due to these reconfigurations.

Although there is substantial work spent in partitioning and scheduling for low-power, we believe that our approach is the first one that accomplishes a deep design space exploration of the partitioning and scheduling process for hardware/software low-power embedded systems, attempting to meet the real-time timing constraints while minimising the overall system energy consumption, and including the system bus, and memories in the performance and energy consumption estimations.

3 Initial Specification

The initial specification is described as a Directed Acyclic Graph (DAG), where each node represents a computational task, or an access to the shared memory, and the edges correspond to dependencies among the nodes. Three different dependencies are considered, namely: communication, internal, and temporal dependencies. A communication dependency edge (CDE) either connects two nodes of PEs, or corresponds to a memory access; therefore, it represents a data transfer that must be carried out using the system bus. An internal dependency edge (IDE) connects two nodes allocated in the same PE, thus, it represents a data transfer, but in this case there is no access to the system-bus. A temporal dependency edge (TDE) represents a dependency between two nodes in the same PE that has been imposed by the scheduler. Each node of the graph must be characterized by its execution, power and area estimations for every possible platform. Each CDE is tagged with the amount of data to be transferred, and the execution time and energy consumption estimations. These estimations must include both the access to the system bus, and when needed, the access to the shared memory.

4 Cost Function

The cost function of a codesign system typically includes different elements like the hw area, the execution time, the energy consumption, or the amount of communications. One of the more difficult issues when designing a partitioning system is how to mix all these completely different magnitudes into a cost function that should be able to lead the design space exploration in a near-optimal fashion.

In literature several codesign approaches can be found where cost functions are built like the following:

$$F = c_a * \sum_{i=0}^{n} Area_i + c_t * \sum_{i=0}^{n} Time_i + c_e * \sum_{i=0}^{n} Energy_i \tag{1}$$

Thus, for a given partition, each node of the DAG is characterizes with a number for every magnitude considered (three in this example). The cost function is then easily computed adding these numbers and multiplying them by some coefficients. Often, the user must fix these coefficients, thus, he has to identify the equivalence between a second, a Joule, and a mm^2. There is not an evident criteria about how to fix these coefficients, therefore these heterogeneous cost functions often lead to inefficient design-space explorations. In order to avoid this problem, our partitioning tool is led by a straightforward cost function that can be identified either with the energy consumption, the hw area or the execution time. Thus, the tool supports three different design-space explorations; the first one attempts to find the solution that consumes less energy and meets three restrictions, namely, maximum execution time, maximum hardware area and maximum power consumption restrictions. The first restriction guarantees that the application meets its real-time deadline; the second guarantees that there are enough hw resources to implement the hw partition; and the third restriction prevents the heating problems. If the system is not battery-dependent, the cost function can be identified either with the execution time, or with the area. When the execution time is selected as cost function, the tool attempts to find the fastest solution that meets the given area and power restrictions, otherwise, when area is selected, the tool will try to find the solution with less hw area that meets the execution time and power restrictions. It is up to the designer to decide which one is the goal of the design-space exploration. Table 1 shows all the possibilities.

Table 1. Cost functions and restrictions that can steer the design space exploration

Available Cost Functions	Available Restrictions
Energy	Time
Time	Energy
Area	Area
	Power

5 Energy Consumption Estimations

First of all, each node and each edge of the DAG must be characterized with its energy consumption for every possible processing element. These estimations must be carried out using the tools provided by the vendors if possible; otherwise generic power models must be applied. In addition to the energy consumption due to the nodes execution (including those nodes that represent the accesses to the shared memory) and communications, we assume that the PEs also consume energy when they are idle. If the PE is a processor, the power consumption in the idle state is commonly provided in the data sheet. The energy can be computed multiplying the power by the idle time. The same approach is used for the memory blocks. If the PE is implemented in the FPGA and clock gating is applied to it, the power that consumes when is idle will be just the device quiescent power. Otherwise, if clock gating is not implemented the logic dissipates more power apart from the quiescent power, since the clock signal continues switching. This case is estimated considering the power consumption of the circuit when the toggle rate of the inputs is set to 0, thus we assume that when the circuit is idle all the inputs are fixed, if this is not correct, a proper toggle rate should be estimated profiling the system.

Besides the energy considerations, the partitioning tool must check if a given partition meets the power dissipation constraints of the platform. To this end, the average power consumption of each node and each communication is included in the DAG.

6 Execution Time Estimations

The execution time estimator, receive as input a given partitioning where the execution time of each node and each access to the system-bus have been previously estimated (we assume cycle accurate estimations). Nodes representing accesses to the shared memory have always a 0 time-units execution time assigned, since the latency of accessing the shared memory is considered as part of the communication delay. With this input the estimator schedules the execution of every node as well as all the accesses to the system bus. This scheduling is a NP-complete problem, however the estimation must be done as fast as possible since it has to be computed for every explored partition. Thus, we have developed a fast heuristic, based on list scheduling techniques, which provides a near-optimal scheduling with a low computational complexity *($O(N^2)$)*. Fig. 2 depicts the scheduling pseudo-code.

> ***A)*** *Assign a weight to each node.*
> ***B)*** *Choose the execution order for the SW nodes.*
> ***C)*** *Recalculate the weights taking into account the new dependencies.*
> ***D)*** *Schedule those nodes that are not waiting for a communication.*
> ***E)*** *While there is a communication waiting for execution do:*
>   ***E1)*** *Choose one communication and schedule it.*
>   ***E2)*** *Schedule those nodes that are not waiting for a communication*

Fig. 2. Scheduling heuristic pseudo-code

Step A: The weights are used to steer the scheduling process trying to minimize the global execution time. The weight of a node is the maximum time-distance from that node to the end of the execution in the initial graph. This distance is computed carrying out an ALAP scheduling that takes into account all the dependencies. Thus, those nodes, which are in the DAG critical path, have higher weights.

Step B and C: The initial DAG allows parallel execution between their nodes, but those nodes assigned to sw must be executed sequentially. The sw execution order is decided sorting the nodes by their weights. To impose this order new TDE dependencies are added to the initial DAG. It is easy to prove that this sw execution order does not allow the new dependencies to create cycles in the graph. Since these new dependencies can significantly affect the system performance, a new weight is assigned to each node. These weights are computed in the same way that in step A, but considering the new dependencies.

***Steps D and E*:** An enhanced list-scheduling heuristic that attempts to minimize the global execution time has been developed for the scheduling process. This heuristic decides when each node and each communication is executed, assigning to them a $\mathbf{t_{start}}$ and a $\mathbf{t_{end}}$ times. The motivation of the heuristic is to detect the system bus access conflicts and the delays created by them.

The scheduling starts assigning $\mathbf{t_{start}} = 0$, and $\mathbf{t_{end}}=\mathbf{t_{ex}}$ to the first node, where $\mathbf{t_{ex}}$ is its execution time in the partition where it has been assigned. Then the algorithm continues scheduling the successors of the first node. A greedy policy is followed to schedule nodes while there is no need for hw/sw communications. When a scheduled node requests a hw/sw communication with another node this request is stored in a list. Once all the nodes that do not need a hw/sw communication have been scheduled, one of the requested communications is selected and scheduled.

There are two selecting criteria (E1):

- If at a given time **t** the system bus is not carrying out any communication and there is just one previous request, the communication channel is assigned to this request, and the bus is tagged as busy until this communication ends.
- Otherwise, if there is more than one request, the one with the greatest weight will be selected. The weight of a communication is computed as the weight of the destination node plus the time needed to execute the communication.

Once the selected communication has been scheduled the graph is examined (E2) and all the nodes that can start their execution without waiting for another HW/SW communication are also scheduled. The loop continues until all the communications are scheduled.

7 Area Estimation

We apply the following equation to estimate the area needed to implement the nodes assigned to hw in the FPGA:

$$Area = \sum_{i=0}^{N-1} A_i + A_{driver} + A_{control} + A_{storage} \tag{2}$$

$\mathbf{A_i}$ is the area of the node i. $\mathbf{A_{driver}}$ is the area needed to implement the communication driver. A_i and A_{driver} are estimated from a core library. When a new core is added to the library its area is estimated using a synthesis tool. $\mathbf{A_{control}}$ is the area needed for the control logic that schedules the communications. In this approach the scheduling control is assumed by a state machine, so the area requested is estimated as a function of the number of communications. $\mathbf{A_{storage}}$ is the area needed for storing the data to transfer until a communication is executed. This storage space is computed during the communication scheduling. During this process a record keeps the maximum storage space required.

8 Results and Analysis

All the estimators has been integrated into a partitioning tool based on genetic algorithms (GA) [16]. This tool creates a random initial population of valid solutions. A solution is valid if meets the given area, time and power constraints. Invalid solutions are rejected to save computational time, as well as to prevent the algorithm from converging to a non-valid area. During the design space exploration solutions evolve by reproducing themselves, generating new offspring of solutions. The crossover and the mutation operators carry out the reproduction process. Population is kept constant deleting the solution surplus. The 80% of the survivors are selected choosing the best solutions, and the 20% remaining is randomly selected in order to prevent a premature convergence. The designer can establish the population and the crossover and mutation probabilities. In addition, he can also select the cost function (between time, energy, and area) and fix the area, time and power restrictions.

The partitioning tool allows the designer to select between two different scheduling modes, the first implements our heuristic while the second carries out a full search of the design space applying a branch&bound (b&b) algorithm, hence this mode guarantees that always the best schedule is found. As a first experiment, in order to validate our heuristic, we have run the partitioning tool in these two different modes for a set of 100 randomly generated DAGs. These DAGs were created using the TGFF tool [17], and their sizes are limited to any number between 10 and 20 nodes (for greater sizes it is not feasible to apply the b&b algorithm). The results obtained show that the b&b algorithm finds slightly better schedulings (on average 10% less execution time), but at the price of increasing 800 times the computational time needed to carry out the partitioning process (which it is reasonable since it performs a full search of the design space). These results confirm that our scheduling heuristic finds near-optimum schedulings with an almost negligible overhead. In this experiment the average time needed to schedule one of the DAGs with our heuristic was less than 2.5 μs using a Pentium II running at 350 MHz.

In our second experiment we attempt to compare the results obtained when using the energy and the execution time as cost function. To this end, we have analyzed two current multimedia applications, namely a JPEG decoder and a pattern recognition application that compute the Hough Transform of a matrix of pixels in order to find simple geometric patterns. The Hough Transform is commonly applied in robotics and astronomical data analysis.

It is very simple to reduce the energy consumption when it is also possible to reduce the performance. Therefore, in this experiment we check whether it is possible to reduce the energy consumption while keeping almost the highest performance.

Hence, we have run first the partitioning tool using the execution time as cost function to find the fastest solution. Then, we have rerun it using the energy instead of the time as cost function, but this time we have imposed that the solutions must be at most 10% slower than the fastest solution found in the previous step. Therefore, the tool is going to found the solution that consumes less energy while keeping almost the highest performance. For this experiment we have estimated the energy, execution time, area and power consumption of the application using the XILINX Foundation 5.i tool for the FPGA and the system bus, an ARM processor simulator for the sw processor and a 128 MB MICRON SRAM memory datasheet for the shared memory.

Each application has been partitioned to a platform composed by a XILINX Virtex FPGA, an ARM processor running at 233 MHz, a 128 MB memory block and a system bus with 16 bit width and clocked at 33 MHz. The measurements were repeated 5 times for 5 different FPGA sizes. The results are shown in table 2. It is remarkable that we can decrease up to 30% the energy consumption (on average 17%), whereas the execution time remains almost the same (it increases less than 3% on average).

Table 2. Results for the Pattern Recogniton Application (a) and the JPEG decoder (b). T1, and E1 are the execution time and the energy consumption for the fastest solution, whereas T2 and E2 correspond to the solution found using the energy as cost function.

a) Pat. Rec.	T1	T2	Time %	E1	E2	Energy %
FPGA1	8412	8493	+ 4%	188324	164147	- 13%
FPGA2	9581	9586	+ 0%	222294	182407	- 20%
FPGA3	9940	9945	+ 0%	226020	186133	- 18%
FPGA4	10841	11027	+ 2%	247887	216630	- 13%
FPGA5	11069	11519	+ 4%	256264	227436	- 11%
Average			+ 2%			- 15%

b) JPEG	T1	T2	Time %	E1	E2	Energy %
FPGA1	927	950	+ 2%	15354	15194	- 7%
FPGA2	1480	1490	+ 1%	16400	14800	- 10%
FPGA3	2329	2385	+ 2%	14265	10947	- 23%
FPGA4	2801	2898	+ 3%	13200	9800	- 26%
FPGA5	3848	4183	+ 8%	8938	6160	- 30%
Average			+ 3%			- 19%

9 Conclusions and Future Work

We have presented the first (to the best of our knowledge) hw/sw partitioning tool that can steer the design space exploration of the partitioning process to minimize the energy, the execution time or the area. In addition this is one of the few tools that accomplishes a full scheduling during the partitioning process including the accesses to the system bus and shared memories. This scheduling is the only way to accurately estimate the goodness of a given partition.

We believe that this tool can be especially useful to decrease the energy consumption of a given application while meeting hard real-time constraints. Thus, we have applied our tool to two current multimedia applications, saving up to the 30% of the energy consumption, whereas the performance remains almost constant. Moreover, it must be remarked that is unimportant that the performance slightly decreases as long as the timing constraints are met.

Although our tool fulfills the requirements to partition an application to several existing platforms, several extensions are needed to apply it to platforms with multiple processors and more complex interconnection networks.

Acknowledgements. This work has been partially supported by Spanish Government research grant TIC2002-00160.

References

1. J. R. Hauser and J. Wawrzynek, "Garp: A mips processor with a reconfigurable coprocessor," in IEEE Workshop on FPGAs for Custom Computing Machines, pp. 24–33, 1997
2. H. Singh et al, "MorphoSyS: An Integrated Reconfigurable System for Data-Parallel and Computation-Intensive Applications", IEEE Trans. on Computers, pp. 465–481, Vol. 49, No. 5, 2000.
3. http://www.xilinx.com
4. M. Gasteier, M. Munich, M. Glesner. "Generation of Interconnect Topologies for Comuni cation Synthesis", DATE'98, pp. 36–42. 1998.
5. K. Itoh et al., "Trends in Low-Power Ram Circuits Technologies", Proc. IEEE, 83(4):524–543, Apr. 1995.
6. M. Kamble and K. Ghose, "Analytical Energy Disipation Models for Low Power Caches", Proc. Int'l Sym. Low Power Electronics and Design, p. 143, Aug. 1997.
7. J. Resano et al, "Analyzing Communication Overheads during Hardware/Software Partitioning", ESCODES'02, pp. 16–21, 2002.
8. R.P. Dick and N.K. Jha, "CORDS: Hardware-Software Co-Synthesis of Reconfigurable Real-Time Distributed Embedded Systems", ICCAD'98, pp. 62–67, 1998.
9. J. Noguera, R.M. Badía, "A HW/SW partitioning algorithm for dynamically reconfigurable architectures", DATE'01, pp. 729–734, 2001.
10. P. Yang et al., "Energy-Aware Runtime Scheduling for Embedded-Multiprocessors SOCs", IEEE Journal on Design&Test of Computers, pp. 46–58, 2001.
11. G. Qu et al., "Power Minimization using System-Level Partitioning of Applications with Quality of Services Requirements", Proc of Int. conf. on CAD. pp. 343–346, 1999.
12. I. Hong et al., "Power Optimization of Variable-Voltage Core-Based System", IEEE Trans. on CAD of Integrated Circuits and Systems, vol. 18, no 12, pp. 1702–1714, 1999.
13. J. Henkel, "A low power hardware/software partitioning approach for core-based embedded systems", DAC'99, pp. 122–127, 1999.
14. R. Mahapatra and P. Vijay, "PAP: Power Aware Partitioning for Reconfigurable System", To be published in Proc. of HPCA Workshop 2003, feb. 2003.
15. L. Shang et al., "Hw/Sw Co-synthesis of Low Power Real-Time Distributed Embedded Systems with Dynamically Reconfigurable FPGAs", ASP-DAC'02, pp. 345–360, 2002.
16. J. Holland. "Adaptation in natural and artificial systems", MIT Press, 1992.
17. R.P. Dick et al, "TGFF: Task Graphs for Free", Int'l Workshop HW/SW Codesign, pp. 97–101, 1998

Consideration of Control System and Memory Contributions in Practical Resource-Constrained Scheduling for Low Power

Chee Lee and Wen-Tsong Shiue

School of Electrical Engineering and Computer Science
Oregon State University, Owen 220, Corvallis, OR 97331
shiue@ece.orst.edu

Abstract. This paper presents a fast and efficient heuristic resource-constrained scheduling algorithm that works to reduce power consumption by maximally utilizing the resources operating at multiple voltages while at the same time maintaining a low latency. Unlike other resource-constrained scheduling algorithms, our algorithm also takes into account the contribution from the control system or circuitry as well as the memory required. By doing so, we are able to gain more insight into the items affecting power, area, and timing as well as the tradeoffs. Scheduling with the control system and memory in mind, we see that there is a direct relationship between the number of resources, the number of control cycles, and the memory needed, which affect the overall power, area, and timing. Experimental results show that taking these extra factors into account we are able to achieve an average power savings that is better than the power savings from other resource-constrained scheduling algorithms.

1 Introduction

Although there are many different low power scheduling schemes [1-4], this paper discusses resource-constrained scheduling since it directly considers resource limitations making it more suitable to mobile devices. The goal of resource-constrained scheduling is to schedule an algorithm or application represented by a data flow graph (DFG) given a resource limitation such that the solution is low power and low latency.

In [5] and [6], the resource-constrained scheduling algorithm uses multiple voltages (5.0V, 3.3V, and 2.5V) and considers items such as switching and level shifters to achieve a low power and low latency scheduling of a DFG. While these algorithms do a good job of reducing the power consumption, they fail to consider the contributions from the memory and control system, which play a vital part in the total power consumption.

Therefore, this paper presents a novel resource-constrained scheduling scheme that considers the control and memory contributions by utilizing multiple voltages and register assignment. Unlike [5] and [6], which focus primarily on reducing power usage, our resource-constrained scheduling scheme works to find a balance between

J.J. Chico and E. Macii (Eds.): PATMOS 2003, LNCS 2799, pp. 590–598, 2003.
© Springer-Verlag Berlin Heidelberg 2003

the latency and power usage, thus, addressing the control and memory contributions. Latency is kept low by reducing the emphasis on using resources that operate at multiple voltages. These resources are still used, but they are used sparingly. Power usage is kept low by considering the data dependence and the maximum number of registers needed. By exploiting data dependencies, the maximum number of registers needed can be reduced, which, in turn, helps to reduce the overall power usage. By balancing the latency and power and considering the control and memory contributions, our resource-constrained scheduling scheme ends up performing, on average, better than the resource-constrained scheduling schemes in [5] and [6].

2 Preliminaries

This section defines terms, equations, etc. that are used in the typical resource-constrained scheduling scheme, our resource-constrained scheduling scheme, and our simulations.

2.1 Resource-Constrained Scheduling Definitions

The input to a resource-constrained scheduling scheme is a data flow graph (DFG) and a resource constraint. A resource constraint is a restriction or a limit imposed on the number of each type of resource such as an adder, multiplier, or shifter. A DFG is a directed acyclic graph whose nodes represent operations and edges represent dependencies between the operations. Each node also has information linked to it in order to help determine its priority. In addition, our resource-constrained scheduling scheme also takes a conflict flow graph (CFG) as an input. A CFG is similar to a DFG except that the edges represent conflicts between nodes. The CFG helps our algorithm address register concerns.

2.2 Delay and Power Models

The delay and power of the different functional units (adders, multipliers, and registers) operating at various voltages have been obtained from simulations using Mentor Graphic's *Design Architect* and *Accusim* in a Sun/Solaris unix environment. *Design Architect* is a schematic tool that allows the user to create circuits by placing components (resistors, capacitors, transistors, etc.) and wiring them together. *Accusim* is a circuit simulation tool similar to spice except that it operates through a graphical user interface (GUI).

We constructed and simulated a 32-bit carry-ripple adder, a 32-bit carry-ripple multiplier, and a 32-bit register for three technologies (AMI 1.2, AMI 0.5, and TSMC 0.35) operating at 5V, 3.3V, 2.4V, 2.2V, 1.8V, 1.5V, 1.2V, and 1.0V. There are different adder and multiplier architectures. However, we chose the ones we did for simplicity and as a starting point. As our library grows, we will include different adder and multiplier architectures as well as other functional units. Only three technologies were used since we were only able to gain access to these technologies. Simulations were done at the specified voltages in order to address the national technology roadmap of semiconductor summarized in Table 1 from [7].

Table 1. National Technology Roadmap of Semiconductor

Year	1997	1999	2001	2003	2006	2009	2012
L (nm)	250	180	150	130	100	70	50
Vdd (V)	2.2	1.8	1.5	1.2	1	0.8	0.6

The delay is a measure of the time it takes for the output to see a change in the input. It is calculated based on the following equation from [8]: $(T_{PLH} + T_{PHL}) / 2$. T_{PLH} and T_{PHL} correspond to a low to high transition at the input and a high to low transition at the input, respectively. The power is determined using a power meter circuit similar to the one in [9].

Table 2 shows the delay and power characteristics for the adder, multiplier, and register for the AMI 0.5 technology and the different operating voltages. Only AMI 0.5 results are shown for the sake of brevity and since this is the technology we chose to use in our simulations.

Table 2. Delay and Power Characteristics for a 32-bit Carry-Ripple Adder, Carry-Ripple Multipler, and Register.

AMI 0.5 um	adder32		mult32		reg32	
Voltages (V)	Delay (ns)	Power (uW)	Delay (ns)	Power (uW)	Delay (ns)	Power (uW)
5	19.03	9,946.00	52.57	30,290.00	5.20	8,473.50
3.3	31.16	4,246.00	86.10	12,930.96	7.21	3,618.20
2.4	38.38	2,213.40	106.04	6,740.79	10.38	1,909.70
2.2	43.94	1,846.70	121.42	5,624.02	11.85	794.04
1.8	64.67	1,081.86	178.70	3,294.73	16.86	545.39
1.5	104.82	419.17	289.65	1,276.56	27.20	368.18
1.2	263.88	129.28	729.18	393.72	66.71	75.53
1	1,117.20	16.37	3,087.15	49.86	80.05	12.76

Notice that, in general, the 32-bit register is approximately four times faster than the 32-bit adder and that the 32-bit adder is approximately three times faster than the 32-bit multiplier. Also, the register consumes the least amount of power followed by the adder and then the multiplier. Another important relationship to notice is that as the supply voltage decreases, the delay increases, and the power decreases.

The power characteristics for level shifters, which are needed to transfer data between resources operating at different voltages is derived from [10]. Table 3 summarizes the power characteristics. Delay characteristics are ignored because they are significantly smaller than the other functional unit delays.

Table 3. Power (uW) Characteristics for 32-bit Level Shifters

Vx to Vy	1.8	2.2	3.3	5
1.8	0	96	146	220
2.2	70	0	160	320
3.3	124	90	0	356
5	184	220	260	0

Overall, our library allows us to provide more meaningful benchmark results since we have more accurate models of typical resources.

2.3 Simulation Environment

All our resource-constrained scheduling simulations are run on a Pentium 4, 1.7 GHz laptop with 512 MB of RAM running Microsoft Windows XP. Each of the resource-constrained scheduling schemes have been coded in Microsoft Visual C++ 6.0. For our resource-constrained scheduling simulations, we chose the delay and power characteristics for AMI 0.5 and the operating voltages of 5V, 3.3V, 2.2V, and 1.8V. This was an arbitrary choice. We could have easily used different characteristics. Implementing the resource-constrained scheduling schemes in C++ allows us to easily change the delay and power characteristics should the need arise. Also, we have assumed that a 32-bit shifter has the characteristics as the 32-bit adder we built.

3 Control and Memory Contributions

Control and memory systems are a vital part of any application. Therefore, it is important to be aware of the contributions they make. In applications targeted for mobile devices, it is even more important due to the stricter power, area, and timing requirements. Below are the contributions that come from the control and memory systems.

3.1 Control System

The number of control cycles needed to accomplish a task is directly related to the size of the control system or state machine needed to control that task. Other resource-constrained scheduling algorithms such as those in [5] and [6] emphasize the importance of the number of control cycles and its relation to the power used by the datapath. However, they fail to take into consideration the effect that control cycles have on the power usage of the control system. We will illustrate that the control system should not be ignored, as its power contributions are significant.

There are many different codes to select from when deciding how to encode a state machine. However, we focus specifically on one-hot encoding. One-hot is attractive because it is simple and easy to implement. However, a state machine using one-hot encoding can grow quite large resulting in a large power consumption. This is illustrated in Figure 1 where the number of states or control cycles range from 1 to 40.

Figure 1 clearly shows that for the range of 1 to 40 states or cycles, one-hot encoding can increase in power usage quite quickly. This occurs because a flip-flop must be used for each additional state. Hence, it is important to minimize the control cycles, which is directly related to latency, as much as possible. We assume the control system operates at 5.0V only.

3.2 Memory

A single 32-bit register operating at 5V consumes approximately 8,474 uW of power and has a delay of 5.2 ns. So as more and more memory is needed, the power and delay will increase. Although the power and delay of a single 32-bit register is small, they make up a large part of a circuit making the number of registers an important part in reducing latency and power. Figure 2 illustrates the contribution of registers in a circuit. We consider registers operating only at a single voltage since it is not very feasible to have registers operating at multiple voltages in a single device. The level shifter overhead would be larger than the benefit obtained from operating registers at multiple voltages.

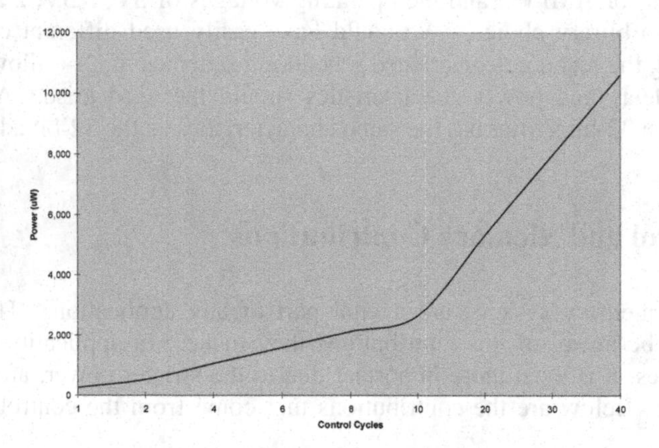

Fig. 1. One-Hot Encoding Power Consumption

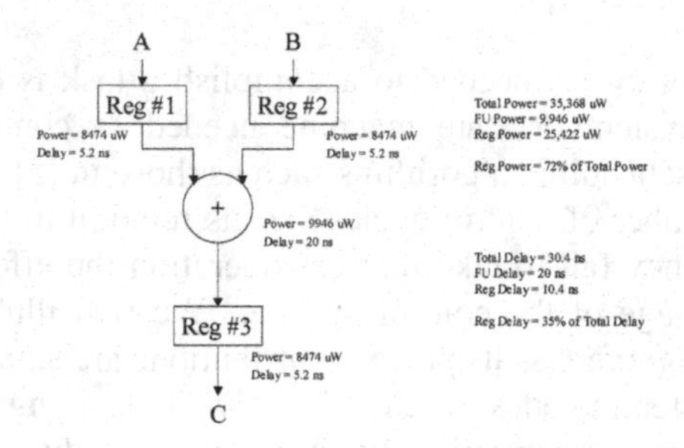

Fig. 2. Register Contributions to Power and Delay

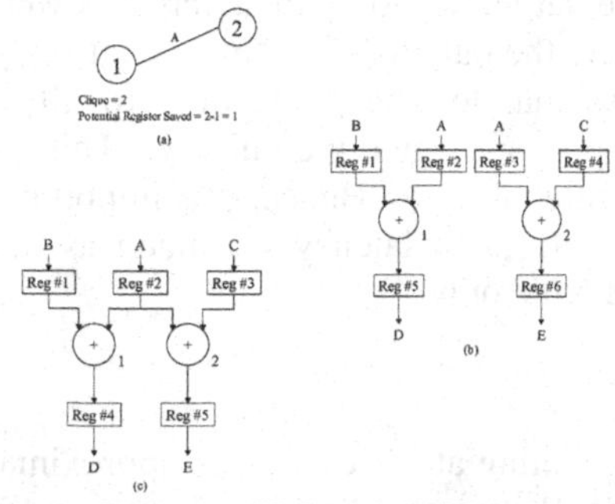

Fig. 3. Register savings from a CFG. (a) CFG. (b) DFG before utilizing CFG. (c) DFG after utilizing CFG.

Even for a simple circuit with only one adder operating at 5V, the registers contribute approximately 72% of the total power and approximately 35% of the total delay. This can change drastically if the registers have a longer lifetime than one cycle—the registers have to hold the data for more than one cycle. Hence, it is important to minimize the number of registers needed in order to efficiently reduce the power consumption. The delay plays less of a vital role. However, reducing the number of registers will also help reduce the latency and, thus, the size of the control system as well.

4 Resource-Constrained Scheduling

We employ a simple technique to address the register problem. Recognizing the data dependency in the DFG, we use a CFG to help determine where to schedule nodes in order to reduce the number of registers needed. We note that if we schedule all the nodes in the CFG that share a conflict in the same cycle, then we can reduce the number of registers needed by the total number of conflicts or clique minus one.

In Figure 3a, we see that node 1 and node 2 share an edge or have a conflict. If node 1 has an input of A and B and node 2 has an input of A and C, the CFG tells us that if we are able to schedule node 1 and node 2 in the same clock cycle, then we would be able to reduce the number of registers by one. The reason this is possible is because node 1 and node 2 can now use one register to store value A instead of two registers (see Figure 3b and 3c).

1) *Determine the ASAP, ALAP, depth and mobility of each node in the DFG.*
2) *Construct the ready set.*
3) *For the current cycle:*
   a. *For all nodes with a mobility of zero or less,*
      i. *Place all nodes with the same mobility into a single group and prioritize the nodes in each group according to highest depth first.*
      ii. *Prioritize the groups according to the lowest mobility first.*
      iii. *Schedule the prioritized nodes starting with the fastest available resource first and moving to the slowest resource.*
      iv. *Continue scheduling until all nodes with mobility less than zero have been scheduled or all resources have been used. If there are resources remaining, go to Step 3b. Otherwise go to Step 4.*
   b. *For all remaining nodes with a mobility greater than zero,*
      i. *Calculate the number of conflicts with each other.*
      ii. *Place all nodes with the same number of conflicts into a single group and prioritize the nodes in each group according to highest depth first.*
      iii. *Prioritize the groups according to the highest number of conflicts first.*
      iv. *Schedule the prioritized nodes starting with the fastest available resource first and moving to the slowest resource.*
      v. *Continue scheduling until all nodes with a conflict greater than zero have been scheduled or all resources have been used. If there are resources remaining, go to Step 3c. Otherwise go to Step 4.*
   c. *For all remaining nodes with no conflicts,*
      i. *Place all nodes with the same mobility into a single group and prioritize the nodes in each group according to the highest depth first.*
      ii. *Prioritize the groups according to the lowest mobility first.*
      iii. *Schedule the prioritized nodes starting with the fastest available resource first and moving to the slowest resource.*
      iv. *Continue scheduling until all nodes have been scheduled or all resources have been used. Then go to Step 4.*
4) *Recalculate the mobility of each node, go to the next cycle, and repeat starting with step 2.*

To reduce the size of the control system, our algorithm uses a priority function that includes the depth and mobility of a node in order to reduce the latency, which then reduces the control system. Our heuristic, list-based resource-constrained scheduling algorithm tries to balance the conflicting requirements of reducing the control system (latency) and utilizing resources operating at multiple voltages with the need to reduce the number of registers needed at each cycle and operates as above.

Step 3a in algorithm addresses the control system (latency) concerns. Step 3b deals with the number of registers needed at each cycle. And step 3c focuses on the functional units. Resources operating at multiple voltages are utilized throughout each step.

5 Benchmarks

This section presents the results obtained from optimizing a 2nd order lattice filter, a 5th order elliptic filter, an 8-point fast fourier transform (FFT), and the classic discrete cosine transform (DCT) with our resource-constrained scheduling scheme and the resource-constrained scheduling scheme from [5] and [6]. Four optimizations were performed for each algorithm and then compared to the non-optimized algorithm—the one with no resource constraints and all functional units operating at 5V. The results of each benchmark for the two resource-constrained scheduling schemes are shown in Table 4 and 5. Section 5.1 describes the optimizations for each benchmark. Section 5.2 discusses the results.

5.1 Optimizations

For the lattice filter, Opt 1 has four adders and four multipliers all operating at 5V. Opt 2 has four adders—one at 5V, one at 3.3V, one at 2.2V, and one at 1.8V—and four multipliers—one at 5V, one at 3.3V, one at 2.2V, and one at 1.8V. Opt 3 has two adders—one at 5V and one at 3.3V—and two multipliers—one at 5V and one at 3.3V. And Opt 4 has one adder and one multiplier both at 3.3V.

For the elliptic wave filter, Opt 1 has four adders and two multipliers all operating at 5V. Opt 2 has four adders—one at 5V, one at 3.3V, one at 2.2V, and one at 1.8V—and two multipliers—one at 5V and one at 3.3V. Opt 3 has two adders—one at 5V and one at 3.3V—and one multiplier at 5V. And Opt 4 has one adder and one multiplier both at 3.3V.

For the fast fourier transform (FFT), Opt 1 has eight adders and four multipliers all operating at 5V. Opt 2 has eight adders—two at 5V, two at 3.3V, two at 2.2V, and two at 1.8V—and four multipliers—one at 5V, one at 3.3V, one at 2.2V, and one at 1.8V. Opt 3 has four adders—one at 5V, one at 3.3V, one at 2.2V, and one at 1.8V—and two multipliers—one at 5V and one at 3.3V. And Opt 4 has two adders—one at 5V and one at 3.3V—and one multiplier at 3.3V.

Finally, for the fast discrete cosine transform (DCT), Opt 1 has eight adders, nine multipliers, and six shifters (assumed to have the same delay and power characteristics as adders) all operating at 5V. Opt 2 has eight adders—two at 5V, two at 3.3V, two at 2.2V, and two at 1.8V—nine multipliers—three at 5V, two at 3.3V, two at 2.2V, and two at 1.8V—and six shifters—two at 5V, two at 3.3V, one at 2.2V, and one at 1.8V. Opt 3 has four adders—one at 5V, one at 3.3V, one at 2.2V, and one at 1.8V—five multipliers—two at 5V, one at 3.3V, one at 2.2V, and one at 1.8V—and three shifters—one at 5V, one at 3.3V, and one at 2.2V. And Opt 4 has two adders—one at 5V and one at 3.3V—three multipliers—one at 5V, one at 3.3V, and one at 2.2V—and two shifters—one at 5V and one at 3.3V.

5.2 Results

Table 4 and 5 summarize the performance of our resource-constrained scheduling scheme with the resource-constrained scheduling scheme from [5] and [6] for all the simulations. All though we collected information about the functional units, level shifters, etc., we show only the power reduction, control and memory contribution, total power, register power, and latency for the sake of brevity.

From the results, we see that our scheduling scheme, on average, performs better in reducing power and latency than the resource-constrained scheduling scheme using just multiple voltages. Our scheme has an average power reduction of 6.0% while the multiple voltages scheme has an average power reduction of 4.0%. Our scheduling scheme latency is also better with an average of 27 cycles compared to 30 cycles. Our scheme achieves a scheduling solution that optimally balances the latency and power allowing it to reduce the overall control and memory contributions ("C&M Contribution (%)" in Table 4 and 5)—57% in our case compared to 59% from [5] and [6]—and perform better than resource-constrained scheduling schemes from [5] and [6]. Looking at each of the simulations separately, this is also generally true. On average, our scheduling scheme provides solutions that are better than the solutions provided by [5] and [6].

Table 4. Performance of Resource-Constrained Scheduling Scheme from [5] and [6]

	Lattice Filter					Elliptic Wave Filter					FFT					Fast DCT				
	Opt 1	Opt 2	Opt 3	Opt 4	AVG	Opt 1	Opt 2	Opt 3	Opt 4	AVG	Opt 1	Opt 2	Opt 3	Opt 4	AVG	Opt 1	Opt 2	Opt 3	Opt 4	AVG
Power Reduction (%)	0.00%	5.18%	5.59%	17.24%	7.00%	0.00%	11.84%	2.12%	4.13%	4.52%	0.00%	12.42%	1.76%	-12.42%	0.44%	0.00%	15.08%	5.25%	-3.78%	4.14%
C&M Contribution (%)	48.30%	55.48%	55.69%	73.44%	58.23%	53.72%	58.27%	55.80%	65.74%	58.38%	45.68%	58.02%	60.37%	56.33%	55.10%	49.90%	57.98%	62.61%	63.18%	58.42%
Total Power (uW)	442,856	415,493	413,673	362,076	408,525	1,077,114	938,781	1,043,484	1,021,806	1,020,296	1,544,189	1,336,906	1,501,644	1,720,583	1,525,830	1,396,019	1,171,520	1,308,745	1,434,877	1,327,790
Reg Power (uW)	211,838	228,022	228,022	263,102	232,746	576,198	543,744	578,825	666,526	591,323	703,301	771,766	903,317	964,708	835,773	694,827	675,296	815,617	903,317	772,264
Latency (# cycles)	15	21	20	40	24	20	30	29	57	34	15	35	30	39	30	13	36	38	33	30

Table 5. Performance of Our Resource-Constrained Scheduling Scheme

	Lattice Filter					Elliptic Wave Filter					FFT					Fast DCT				
	Opt 1	Opt 2	Opt 3	Opt 4	AVG	Opt 1	Opt 2	Opt 3	Opt 4	AVG	Opt 1	Opt 2	Opt 3	Opt 4	AVG	Opt 1	Opt 2	Opt 3	Opt 4	AVG
Power Reduction (%)	0.00%	8.25%	8.25%	20.22%	9.18%	0.00%	10.12%	3.26%	5.57%	4.74%	0.00%	12.19%	4.92%	-7.67%	2.36%	0.00%	14.44%	5.80%	-0.53%	4.93%
C&M Contribution (%)	48.30%	54.90%	54.90%	72.77%	57.72%	53.72%	55.46%	54.82%	65.48%	57.37%	45.68%	55.85%	58.50%	54.31%	53.58%	49.90%	54.84%	60.15%	61.19%	56.52%
Total Power (uW)	442,856	406,329	406,329	353,293	402,202	1,077,114	968,077	1,042,014	1,017,071	1,026,069	1,544,189	1,356,008	1,468,209	1,662,660	1,507,766	1,396,019	1,194,502	1,315,068	1,403,400	1,327,247
Reg Power (uW)	211,838	220,311	220,311	254,205	226,666	576,198	533,831	567,725	660,933	584,672	703,301	754,142	855,824	898,191	802,864	694,827	652,460	788,036	855,824	747,786
Latency (# cycles)	15	23	23	40	25	20	26	29	55	33	15	27	28	39	27	13	22	28	29	23

6 Conclusion

From our simulations, we show that, when performing resource-constrained scheduling, it is important not only to consider the functional units or the number of resources, but it is also important to consider the control and memory as well. We also illustrate the relationships and trade-offs between the number of available resources, the latency, the control system, and the memory system and how they affect power. One of the many important relationships and trade-offs to notice is that as the power is reduced by directly targeting the number of resources or functional units, the control and memory systems begin to contribute more and more to the total power. They become the limiting factor in obtaining a small, low power, low latency solution. Therefore, in order to find an optimal solution, these relationships and trade-offs need to be delicately balanced. Our resource-constrained scheduling scheme that utilizes multiple voltages and register assignment is capable of finding the most optimal solution by balancing the latency and power and considering the control and memory.

References

[1] "Report: Worldwide Handheld Shipments Decline in 2Q; Pocket PC to Surpass Palm PDAs in 2004", MobileVillage, September 12, 2002, http://www.mobilevillage.com/news/2002.09.12/ PDAmarket.htm (October 9, 2002).

[2] H.S. Yun and J. Kim Yun, "Power-aware modulo scheduling for high-performance vliw processors," Proc. of the ISPLED 2001, pp. 40–45.

[3] Rabaey D. Singh and others, "Power conscious cad tools and methodologies: A perspective," Proceedings of the IEEE 83, April 4, 1995, pp. 570–594.

[4] R.S. Martin and Knight, "Optimizing power in asic behavioral synthesis," IEEE Design & Test of Computers 13, pp. 58–70.

[5] Wen-Tsong Shiue and Chaitali Chakrabarti, "Low-Power Scheduling with Resources Operating at Multiple Voltages," IEEE Transactions on Circuits and Systems II: Analog and Digital Signal Processing, vol. 47, no. 6, June 2000, pp. 536–543.

[6] Ali Manzak and Chaitali Chakrabarti, "A Low Power Scheduling Scheme with Resources Operating at Multiple Voltages," IEEE Transactions on Very Large Scale Integration (VLSI) Systems, vol. 10, no. 1, February 2002.

[7] Semiconductor Industry Association (SIA), "The National Technology Roadmap for Semiconductors: Technology Needs," 1997, http://www.macs.ece.mcgill.ca/~roberts/ ROBERTS/COURSES/SUPPORT/SIAroadmap97.pdf (October 12, 2002).

[8] Sedra and Smith, "Microelectronic Circuits, 4th Edition," Oxford University Press, Oxford, New York, 1998.

[9] Sung-Mo Kang and Yusuf Leblebici, "CMOS Digital Integrated Circuits Analysis and Design, Third Edition," McGraw-Hill, New York, 2003.

[10] J. Chang and M. Pedram, "Energy Minimization Using Multiple Supply Voltages," IEEE Transactions on Very Large Scale Integration (VLSI) Systems, vol. 5, December 1997.

Low-Power Cache with Successive Tag Comparison Algorithm

Tae-Chan Kim[1], Chulwoo Kim[1], Bong-Young Chung[2], and Soo-Won Kim[1]

[1] ASIC Design Lab., Dept. of Electronics Eng., Korea University, Anam-Dong, Sungbuk-Ku, Seoul 136-701, Korea.
{taechan, ckim, ksw}@asic.korea.ac.kr

[2] System LSI Business, Samsung Electronics Co. Ltd, San #24, Nongseo-Ri, Kiheung-Eup, Yongin-City, Kyunggi-Do, Korea.
{bychung}@samsung.com

Abstract. In recent years, power consumption has become one of the most critical design concerns in designing VLSI systems. The reduction of power consumption is inevitably required by the emergence of highly efficient and fast systems, which include CPU (Central Processor Unit), MCU (Micro Controller Unit), cache, and et cetera. This paper introduces a new low-power cache controller with successive tag comparison algorithm. Using these methods, the power consumption of a cache can be reduced. Simulation results show that the power consumption of a cache using the proposed method is reduced by 42% compared with conventional methods.

1 Introduction

The current electronic systems that are controlled by a Micro Controller Unit (MCU) or a Micro Processor Unit (MPU) continuously evolve as the operational rate and performance of the processors improve. The structure of a processor is designed to be suitable for 8-bit, 16-bit, 32-bit, 64-bit, and more than 64-bit according to the width of data bus. Also, the technologies related to the processor structure track the trend toward an increased number of data bits, and lend an improvement in the performance of the electronic systems [1], [2]. In addition, as the operating speed of the processor and the number of bits on the data bus are increased, the power consumption has also been increased. Accordingly, power consumption for high-speed processors and other devices, especially mobile gadgets, must be reduced and low-power technologies have therefore become popular [3], [4].

In general, the memories operate at lower-operating-speed than that of processors. Data from an external memory to a processor should be supplied according to the relatively fast-operating-speed of the processor. Since the access speed of memories is relatively low, cache memories are typically employed to compensate for the relatively low-operating-speed of the external memories. The operating speed of cache memories tends to also increase with that of the processors. Thus, as the amount of power consumption in a cache is increased,

J.J. Chico and E. Macii (Eds.): PATMOS 2003, LNCS 2799, pp. 599–606, 2003.
© Springer-Verlag Berlin Heidelberg 2003

distribution of power consumption of a cache memory becomes an important factor [5] - [14].

The remainder of this paper is organized as follows. Section 2 describes a cache structure. In Section 3, our proposed model, successive tag comparison, is explained. In Section 4, we present our experimental results. Finally, we conclude the paper in Section 5.

2 Conventional Cache Structure

Fig. 1 illustrates the power consumption distribution in a 32-bit MCU, a cache, cache SRAM, and SRAM, respectively. Cache in MCU, SRAM in cache, cell array in SRAM, and cache data and directory SRAM in cache SRAM dominate the power consumption of each hierarchical level, respectively.

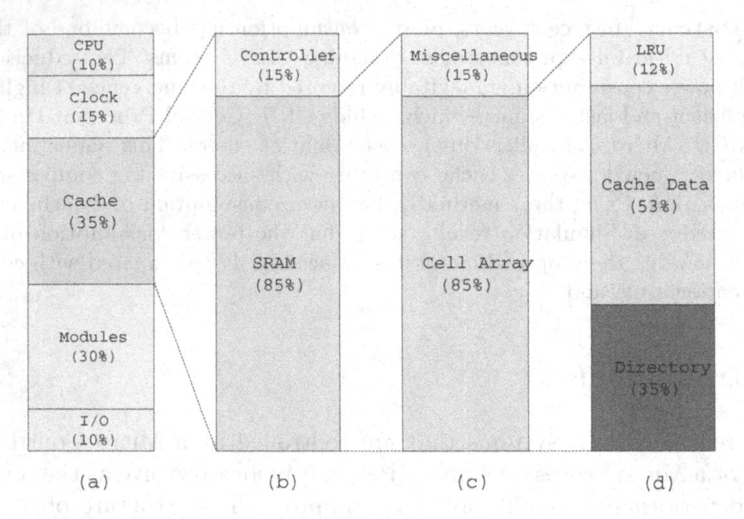

(a) A 32-bit MCU with 8K-byte SRAM.
(b) A cache with 8K-byte SRAM.
(c) SRAM itself and miscellaneous in cache with 8K-byte SRAM.
(d) SRAM in cache.

Fig. 1. Power consumption distribution.

Fig. 2 is a block diagram illustrating the construction of a cache memory for explaining a conventional set-associative cache. Directory SRAM in Fig. 2 saves memory address that indicates cache data SRAM location. The power consumption of a cache can be expressed by the following equation.

$$P_{Cache} = P_{Controller} + \alpha \times P_{SRAM} \quad (1)$$

where P_{Cache}, $P_{Controller}$, P_{SRAM}, and α are the power consumptions of a cache, cache controller, SRAM, and SRAM access activity, respectively. Power consumption of cache controller is smaller than that of cache SRAM as described in

Fig. 1. High SRAM access activity as well as SRAM itself causes large power consumption of a cache. Therefore, this paper suggests that SRAM access activity should be reduced.

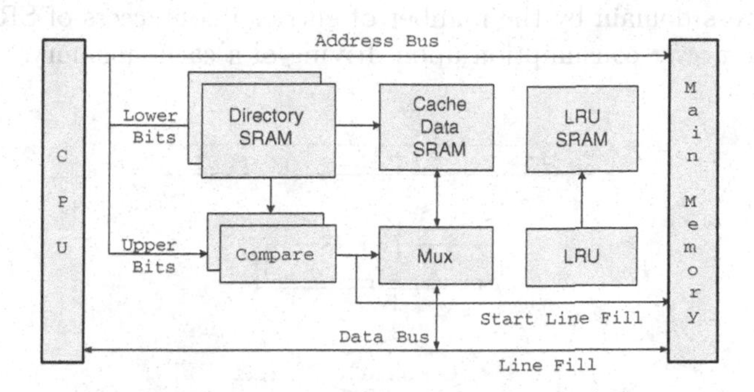

Fig. 2. Conventional set-associative cache structure.

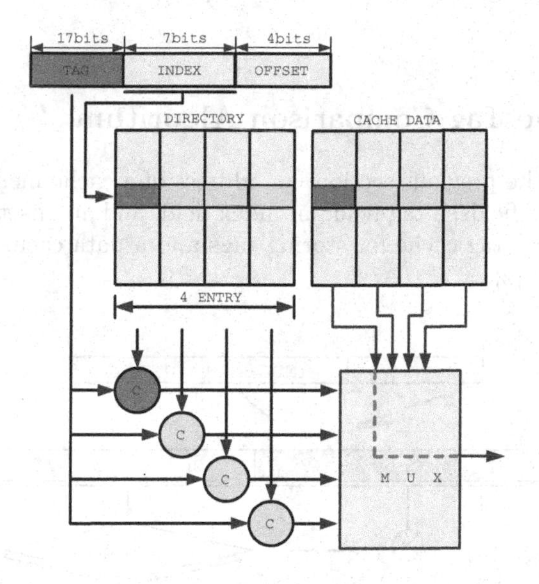

Fig. 3. Conventional set-associative cache operation.

Fig. 3 illustrates an address allocated to a cache memory from a processor, which is divided into three fields, i.e., a tag field, an index field, and an offset field. Typically, a cache memory includes a tag cache to store tags, a data cache to save data or commands, and comparators to compare the tags in the address allocated to a cache memory from the processor with the tags stored in the tag cache. In the conventional approach, during the tag comparing operation, the comparators simultaneously compare all the bits of the tag allocated to a cache memory from the processor with all the bits of the tags stored in the tag cache. Fig. 4 shows conventional tag comparison algorithm. An amount of current is drawn corresponding to a value obtained by multiplying the number of bits in the tag address domain by the number of entries upon access of SRAM, which increases the power consumption upon driving of a cache memory.

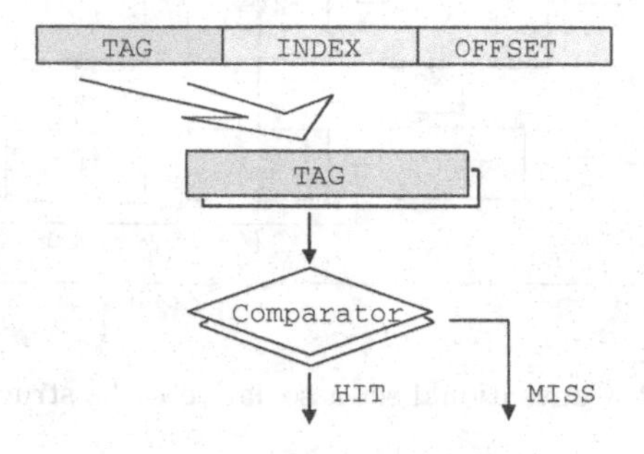

Fig. 4. Conventional tag comparison algorithm.

3 Successive Tag Comparison Algorithm

As explained in the previous section, an address of a cache memory is generally divided into three fields, a tag field, an index field, and an offset field. The cache memory includes a tag cache for storing tags and a data cache for storing data or commands [1], [2].

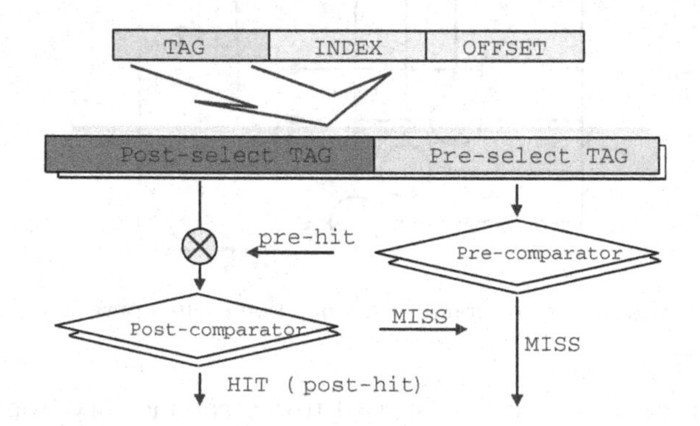

Fig. 5. Successive tag comparison algorithm.

In the proposed algorithm described in Fig. 5, the tag is further divided into at least two fields, a pre-select tag field and a post-select tag field, respectively. Moreover, the cache memory includes two types of comparators: pre-comparators and post-comparators. The pre-comparators compare pre-select address bits stored in the pre-select tag field with address bits corresponding to the pre-select tag field within the processor so as to generate a first hit/miss signal (pre-hit) for an entry that has been hit. Then the post-comparators compare post-select address bits stored in the post-select tag field with address bits corresponding to the post-select tag field within the processor in entries selected by the first hit/miss signal.

The transfer circuit may optionally and selectively limit the transfer of the post-select tags, based on the result of the first phase. Namely, if the pre-comparators determine whether a miss has occurred, there is no need to perform the full comparisons at post-comparators, thereby limiting the amount of current drawn for the comparisons. Through the mechanism of the proposed algorithm, access activity is reduced, and the total power consumption of the cache is decreased.

The ratio of decrease in power consumption can be expressed in the following equation.

$$\frac{N_{PSB} \times N_W + (N_{TADDB} - N_{PSB}) \times N_{PW}}{N_{TADDB} \times N_W} \tag{2}$$

where $N_{PSB}, N_W, N_{PW},$ and N_{TADDB} are the number of pre-select bits, the number of ways, the number of post-ways, and the number of tag address bits, respectively. Here, the tag address bit number, N_{TADDB}, is the number of bits in a tag address field stored in directory SRAM. Post-select TAG cache and post-comparators perform as many as the number of post-ways, N_{PW}. This number can become a value between 0 and N_W, and most of time becomes 1. Therefore, if an entry has been selected perfectly by the pre-select bits, Eq. (2) can be modified as the following.

$$\frac{N_{PSB} \times N_W + (N_{TADDB} - N_{PSB})}{N_{TADDB} \times N_W} \tag{3}$$

It can be seen that the number of entries, the number of pre-select bits, and the number of tag address bits affect the decrease ratio of power consumption.

That is, the proposed algorithm dramatically reduces the power consumption as the number of entries increases and as the number of pre-select bits becomes smaller than the number of bits in the tag address field. If the increased power consumption by the added circuitry is smaller than the decreased power consumption by the proposed algorithm, the power savings by the proposed algorithm is more effective. In fact, the percentage of increased power consumption by the additional controller circuitry is 3% of the total power consumption. A net power consumption is dramatically reduced in this example.

An asynchronous method can be adopted in order to solve the increased latency problem between pre-select TAG cache and post-comparators because of two-phase operation in the proposed algorithm. Fig. 6 illustrates how TAG cache and comparators between pre and post are asynchronously operated. The clock of post-select TAG cache and post-comparators is delayed to get the setup time for next latches or flip-flops. Hence the proposed algorithm can be achieved within 1-clock period and 220MHz operation speed under $0.18\mu m$ CMOS process at 1.8V.

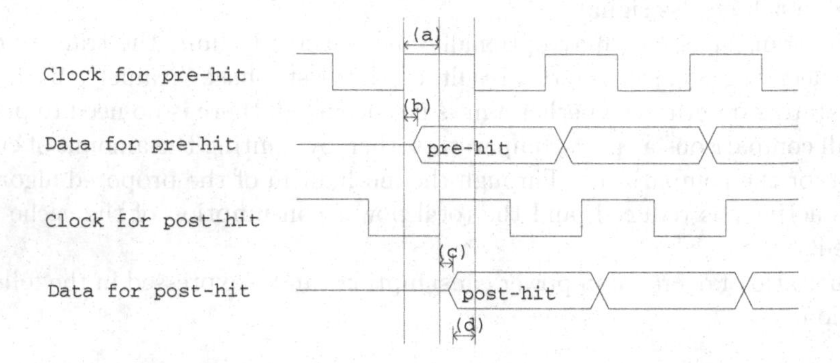

(a) : Clock skew between pre-hit and post-hit.
(b) : Pre-select SRAM access time and pre-comparator process time.
(c) : Post-select SRAM access time and post-comparator process time.
(d) : Setup time for next latches or flip-flops.

Fig. 6. Asynchronous method for increased latency problem.

4 Experimental Results

We use a 32-bit MCU to verify the effect of the proposed algorithm. The experimented cache is a 4-way set-associative cache with 8K-byte SRAM. Benchmark program is a kind of real application program for the MCU. Design parameter is a $0.18\mu m$ CMOS process at 1.8V, 200MHz, and 25^oC. In Table 1, the number of gates in a typical structure is compared with the number of gates in the proposed structure. A design based on the proposed structure requires additional 164 gates.

Table 1. The gates comparison between conventional and proposed structures

The number of glue gates in conventional structure (except SRAM)	1,909
The number of glue gates in proposed structure (except SRAM)	2,073
The number of SRAM gates in conventional structure	140,051
The number of SRAM gates in proposed structure	146,686

Fig. 7 illustrates the experimental results for reflecting the contents of a cache memory structure of the proposed algorithm with an actual design of the cache memory. The proposed algorithm reduces the power consumption of directory SRAM depending on the number of bits in a pre-select tag address field provided the proportion of selection of an entry by the pre-select bits is 100%. It has been shown that if a total of 17 bits are in a tag address field and 7 bits are in a pre-select tag address field, the total power consumption ratio of a tag cache is 58% as shown in Fig. 7.

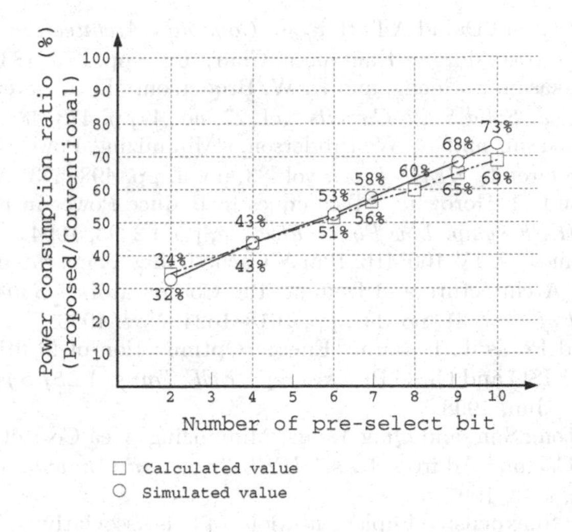

Fig. 7. The experimental results of successive tag comparison algorithm.

In addition, it can be understood from Fig. 7 that the number of bits in the pre-select tag address affects the amount of power dissipation in the cache memory. Particularly, if an application program portion of a processor selectively discriminates respective entries with a small number of pre-select bits, an allocation of a small number of pre-select bits can reduce a large amount of power consumption.

5 Conclusions

In this paper, we described a successive tag comparison algorithm. The proposed algorithm significantly reduces power consumption compared to the conventional algorithm even though the latency increases. The solution with latency within 1-clock period and 220MHz operation speed, however, was proposed with an asynchronous method in this paper. Simulation results show that the power consumption of a cache is reduced by 42% compared with conventional methods. Accordingly, it can be expected that a set of combination cache including many entries such as a 64-way set-associative cache will save more power than a 4-way set-associative cache. As can be seen from the foregoing, using the proposed algorithm, the successive tag comparison process allows for implementation of a low-power cache memory, thereby decreasing overall power consumption in many applications such as MCU, MPU, a lot of entry caches, and especially mobile gadgets which do not need to operate at high speed.

References

1. Michael J. Flynn, *Computer Architecture*, Jones&Bartlett, Charpter 5, pp. 265–344, 1995.
2. John L Hennessy and Dabid A Pattersom, *Computer Architecture : A Quantitative Approach*, New York:Morgan Kaufmann, Charpter 5, pp. 373–483, 1996.
3. A. P. Chandrakasan, S. Sheng, and R. W. Broderson, "Low-Power CMOS Digital Design," *IEEE J. Solid State Circuits*, vol. 27, no. 4, pp. 473–483, Apr. 1992.
4. A. P. Chandrakasan and R. W. Broderson, "Minimizing Power Consumption in Digital CMOS circuit," *IEEE Proc.*, vol. 83, no. 4, pp. 498–523, Apr. 1995.
5. B. Amrutur and M. Horowitz, "Techniques to Reduce Power in Fast Wide Memories," *Proc. IEEE symp. Low Power Electron*, pp. 92–93, 1994.
6. Hiroyuki Mizuno, "A 1V 100MHz 10mW Cache Using a Separated Bit-Line Memory Hierarchy Architecture and Domino Tag Comparators," *Solid-State Circuits, IEEE Journal of* , vol. 31, no. 11 , pp. 1618–1624, Nov. 1996.
7. Uming Ko and Poras T. Balsara, "Energy Optimization of Multilevel Cache Architecture for RISC and CISC Processors," *IEEE Trans. VLSI Systems*, vol. 6, no. 2, pp. 299–308, Jun. 1998.
8. Hong Wang, Tong Sun, and Qing Yang, "Minimizing Area Cost of On-Chip Cache Memories by Caching Address Tags," *IEEE Trans. on Computers*, vol. 46, no. 11, pp. 1187-1201, Nov. 1997.
9. R. E. Kessler, "Inexpensive Implementations of Set-Associativity," *Proc. 16th Int. Symp. on Computer Architecture*, pp. 131–139, Jun. 1989.
10. L. Liu,"Cache Design with Partial Address Matching," *Proc. 27th Int. Symp. on Microarchitecture*, pp. 128–136, Nov./Dec. 1994.
11. N. P. Jouppi and S. J. E. Wilson, "Tradeoffs in Two-Level On-Chip Caching," *Proc. 21st Annu. Int'l. Symp. Comput. Architect.*, pp. 34–45, Apr. 1994.
12. M. D. Hill, "A Case for Direct-Mapped Caches," *Computer*, pp. 25–40, Dec. 1988.
13. Changkyu Kim, Doug Burger, and Stephen W. Keckler, "An Adaptive Cache Structure for Future High-Performance Systems", *IBM Austin Center for Advanced Studies 3rd Annual Austin CAS Conference*, 15, Feb. 2002.
14. M. Miranda, F. Catthoor, M. Janssen, H. De Man, IMEC, and Belgium, "Efficient Hardware Address Generation in Distributed Memory Architectures", *ISSS Int. Symp. System Synthesis*, 6–8, Nov. 1996.
15. *ARM7T RISC Processor Manual*, ARM corp., 1997.
16. M. Farrens, G. Tyson, and A.R. Pleszkun, "A Study of Single-Chip Processor/Cache Organizations for Large Number of Transistors," *Proc. 21st Annu. Int'l. Symp. Comput. Architect.*, pp. 338–347, 1994.
17. J. R. Goodman, "Using Cache Memory to Reduce Processor-Memory Traffic," *Proc. 10th Annu. Int'l. Symp. Comput. Architect.*, pp. 124–132, Apr. 1983.
18. C. Dubnicki and T. LeBlanc, "Adjustable Block Size Coherence Caches," *Proc. 19th Annu. Int'l. Symp. Comput. Architect.*, pp. 170–180, May. 1992.
19. N.P. Jouppil, "Cache Write Policies and Performance," *Proc. 20st Annu. Int'l. Symp. Comput. Architect.*, pp. 191–201, May. 1993.

FPGA Architecture Design and Toolset for Logic Implementation[1]

K. Tatas[1], K. Siozios[1], N. Vasiliadis[2] D.J. Soudris[1],
S. Nikolaidis[2], S. Siskos[2], and A. Thanailakis[1]

[1] VLSI Design and Testing Center, Department of Electrical and Computer Engineering, Democritus University of Thrace, 67100, Xanthi, Greece
{ktatas, ksiop, dsoudris}@ee.duth.gr

[2] Electronics and Computers Div., Department of Physics, Aristotle University of Thessaloniki, 54006 Thessaloniki, Greece
nivas@skiathos.physics.auth.gr,
{snikolaid, siskos}@physics.auth.gr

Abstract. In this paper, the design of an embedded FPGA architecture (i.e. configurable logic blocks) is presented and a complete tool-supported design flow starting from architecture level (i.e. RT-level) and ending with the derivation of the reconfiguration bitstream for the FPGA programming is introduced. The proposed design flow consists of new and modified and extended academic tools. In particular, new tools were developed in order to complement certain critical steps in the implementation flow, since existing academic tools do not combine for a cohesive and complete flow. The remaining design steps are implemented by modified existing academic tools. The FPGA architecture and the tool development is an interactive task, depending on what architectures can be supported by the tools. Using this design support tool set, we designed and simulated in 0.18 TSMC technology an FPGA architecture. More specifically, the detailed design characteristics of the Configurable Logic Block Architecture as well as the interconnect network are determined. Finally, experimental results in terms of energy consumption and delay are given.

1 Introduction

FPGAs have recently benefited from process advances to become significant alternatives to ASICs. The advantage of bypassing fabrication and testing time, effectively reducing the design cycle, has made them increasingly popular and attractive for most applications, since they have become dense and performance-efficient enough to implement a wide range of applications. In today's era of Systems-on-Chip (SoC), embedded FPGA architectures have emerged [1].

Another important feature that has made FPGAs particularly attractive is a logic mapping and implementation flow that is similar to the ASIC implementation flow.

[1] This work was partially supported by the project IST-34793-AMDREL which is funded by the E.C.

J.J. Chico and E. Macii (Eds.): PATMOS 2003, LNCS 2799, pp. 607–616, 2003.
© Springer-Verlag Berlin Heidelberg 2003

Commercial tools exist that implement logic described in RT-level (VHDL or Verilog) in a FPGA device [2], [3]. Furthermore, there has been an effort from academia (e.g. the University of Toronto[4] and UCLA[5]) to build certain tools that cover certain steps of the design implementation flow. These tools are used mostly for purposes of exploration and even when combined, they do not compose a concrete tool chain from RT-level to FPGA bitstream generation.

The goal of the proposed work is twofold: Firstly, the development of a design framework for supporting the design of embedded FPGAs and secondly, the hardware design of an efficient FPGA architecture. This flow is comprised mostly of suitably extended, open-source academic tools. Additional tools were developed in order to allow other tools in the tool chain to communicate with each other, since their inputs and outputs are normally incompatible and generate the configuration bitstream. Furthermore, the design toolset is linux-based. Taking into account the characteristics of the design tools, for instance the smallest structure is the Basic Logic Element (BLE) and exhaustive exploration results, an energy and delay-efficient architecture of Configurable Logic Block and interconnect network are designed.

This paper is organized as follows: Section 2 presents the proposed tool chain and the FPGA architecture options the tools support. Section 3 describes the proposed FPGA CLB and interconnect network architecture in detail. Finally, conclusions and future work are discussed in section 5.

2 Proposed Tool Chain for Logic Implementation Flow

The implementation of logic on a given FPGA device has an established design flow, which is quite similar to the ASIC implementation flow.

2.1 General Logic Implementation Flow

The general logic implementation flow on a FPGA device is illustrated in Fig. 1. For commercial devices, the manufacturer provides the appropriate tools for mapping logic to the device.

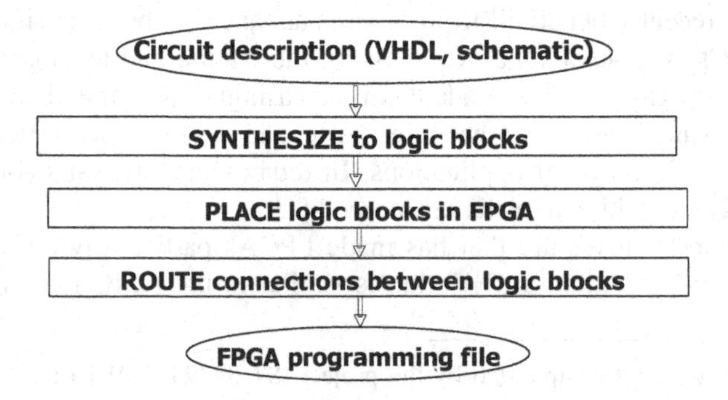

Fig. 1. Logic implementation flow on FPGAs

2.2 Proposed Design Flow

The proposed logic implementation tool chain can be seen in Fig. 2. It is comprised of academic tools (some of them suitably modified), a novel tool (defiler), and a commercial tool (Leonardo). Some of academic tools required certain modifications. Additionally, they were all compiled to the Linux operating system, since the provided binary code was for the Solaris operating system. This was done so the entire toolset can operate under the same operating system. Linux was selected because it is portable an open-source operating system. Therefore, its source code is available, which makes it more appealing in case of possible required modifications to its libraries/packages.

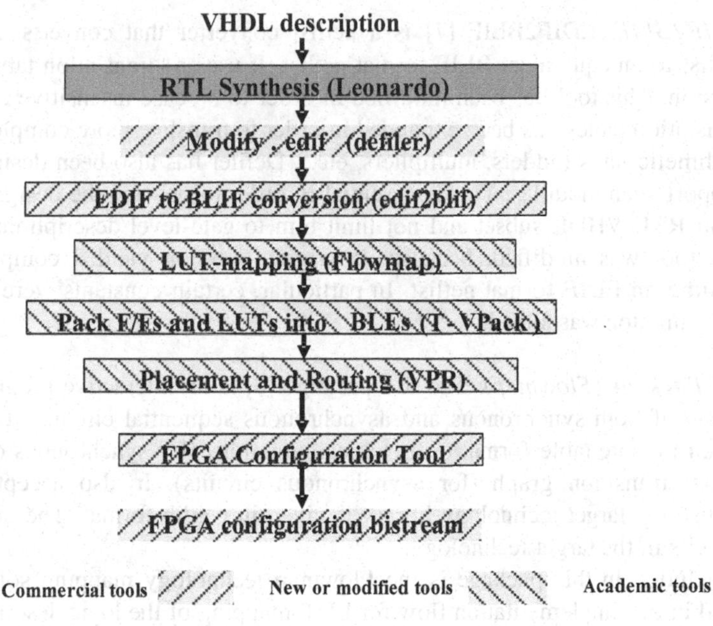

Fig. 2. Proposed tool chain and logic implementation flow

Leonardo. Leonardo [6] is a fairly popular commercial RTL synthesis tool for both FPGA and ASIC technologies by Exemplar. Besides synthesis from a VHDL or Verilog description to gate level netlist, Leonardo also performs LUT-mapping for a target FPGA technology.

In our flow, Leonardo is used only for producing a technology independent netlist. No particular technology is selected and synthesis is done using Leonardo's "primitives" and "operators" libraries, which contains basic elements (i.e. gates and Flip-Flops) and arithmetic units. Therefore the input is a VHDL description of the circuit to be implemented on the novel FPGA, and the output is a technology independent netlist in EDIF format.

Defiler. Defiler is an EDIF to EDIF converter. Its purpose is to convert the EDIF file produced by Leonardo into an EDIF file compatible with the next tool in the tool chain (EDIF2BLIF). Obviously the two netlists must be equivalent. It essentially translates the cells from the Leonardo libraries ("primitives" and "operators") to equivalent functionality cells in the edif2blif library, changing naming conventions wherever necessary, and substituting the references to Leonardo libraries with references to EDIF2BLIF libraries. It was developed from scratch specifically to fill this gap in the tool flow. This tool can be modified accordingly to support any expansions in the library of edif2blif as long as there are corresponding cells in the libraries of Leonardo.

Academic Tools – Modifications.

EDIF2BLIF. EDIF2BLIF [7] is a netlist converter that converts an EDIF format netlist to an equivalent BLIF format netlist. It uses a "translation table" for this conversion. This tool has been modified in order to be case insensitive and its library of translation tables has been expanded in order to translate more complex elements like arithmetic units (adders, multipliers, etc.). Defiler has also been designed in order to support such modules. This is essential in order to provide the designer with a sufficient RTL VHDL subset and not limit him to gate-level descriptions. The parser of this tool was modified, because its original version yielded compile errors when reading an EDIF format netlist. In particular, certain constants were modified and a new function was added.

SIS Package (Flowmap). The SIS package [8] is an interactive program for the synthesis of both synchronous and asynchronous sequential circuits. The input can be given in state table format or as logical equations (for synchronous circuits), or as a signal transition graph (for asynchronous circuits). It also accepts BLIF format netlists. A target technology library is given in genlib format. The output is a netlist of cells in the target technology.

Included in this package is the Flowmap technology mapping software, which is used in our implementation flow for LUT-mapping of the logic described in the BLIF format netlist which is the output of the EDIF2BLIF tool. Therefore, this tool receives the BLIF format gate-level netlist output of the previous tool as input, and outputs a BLIF format netlist of LUTs and FFs.

Powermodel: A power model with an activity estimator have been integrated with VPR in order to estimate the power consumption of an FPGA. The tool estimates power consumption after placement and routing is complete, it does not attempt to guite the placement and routing processes for more power-efficient results. This tool was modified, because the way the BLIF format netlist was read was erroneous, resulting to compile-time error. Therefore, it was modified to resolve this problem.

T-Vpack. T- Vpack [4] reads in a BLIF format netlist of a circuit that has been technology-mapped to LUTs and flip-flops, packs the LUTs and flip-flops into the desired FPGA logic block, and outputs a netlist in VPR's netlist format. VPACK can target a logic block consisting of one LUT of K inputs and one D-FF, as shown in Fig. 3, as this is a common FPGA logic element. Clusters of this block, known as Basic Logic Element (BLE), are also allowed. The modifications that are done during the integration of the Powermodel tool make T-Vpack non-functional. Specifically, the BLIF format netlist input files were read erroneously. Therefore, manual modifications were necessary to the function that reads the input .blif files.

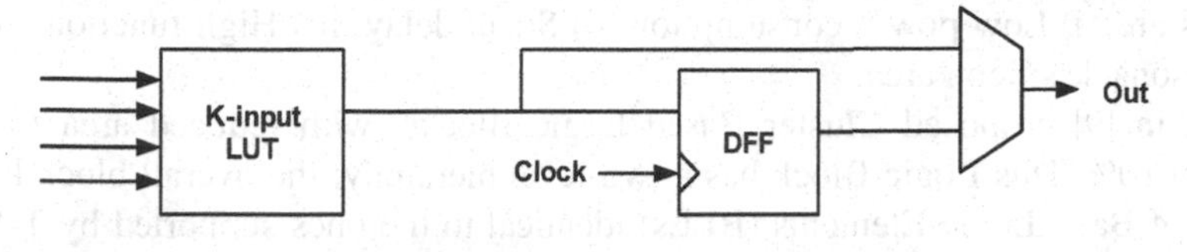

Fig. 3. T-Vpack BLE

VPR. The VPR (Versatile Place and Route) tool [4] is a FPGA placement and routing tool developed at the University of Toronto. The inputs to VPR consist of a technology mapped netlist and a text file describing the FPGA architecture. VPR can place the circuit, or a pre-existing placement can be read in. VPR can then perform either a global route or a combined global/detailed route of the placement. VPR's output consists of the placement and routing, as well as statistics useful in assessing the utility of an FPGA architecture, such as routed wirelength, track count, and maximum net length.

Bitstream generation tool. The bitstream generation tool receives the placement and routing information from VPR, the BLIF format netlist, and the T-Vpack output and generates the FPGA configuration file. This file allows the configuration of the entire FPGA, in other words the entire global interconnect network, the local interconnect network, all LUTs and D-FFs are configured accordingly. This effectively means that the size of the produced configuration bitstream is independent of the logic implemented, but it is directly proportional to the size of the embedded FPGA, in other words its total available resources. Also, the tool for loading the configuration bitstream to the FPGA through a computer has been developed.

Graphic User Interface. The developed Graphic User Interface (GUI), integrates all the above tools, giving the designer a great deal of flexibility, while still hiding unwanted details from the user. At this point it is still under development. This graphic environment will be able to run from any computer, even if not physically accessible from the computer where the design tools are installed. This will be feasible through web pages. The BLIF format netlists and the FPGA architecture file will be uploaded and the design parameters selected. After the data is processed by the server where the tools will be installed, the results will be displayed to the user's computer, in graphic form (placement and routing) and in file format for download.

3 FPGA Architecture

3.1 Configurable Logic Block (CLB) Architecture

The choice of the CLB Architecture is crucial to the CLB granularity, performance, density and power consumption. The CLB structure impacts the interconnect architecture and therefore the characteristics of the whole FPGA. Desirable CLB characteristics are: i) Low power consumption, ii) Small delay, iii) High functionality, and iv) Reasonable silicon area.

Betz in [9] proposed Cluster Based Logic Blocks, with reduced area from 5% down to 10%. This Logic Block has a two-level hierarchy; the overall block is a collection of Basic Logic Elements (BLEs) identical to the ones supported by T-Vpack, mentioned in the previous section. In addition with the reduced area, local interconnect between BLEs results to better routing flexibility.

Fig.4 shows the structure of the two-level hierarchy Logic Cluster. Fig.4a illustrates the structure of the BLE which is formed by a Look-Up Table, a D-FF and a two in one multiplexer. Then these BLEs, connected together, with the use of $I+N$ to one multiplexers as in Fig.4b, forms the Logic Cluster.

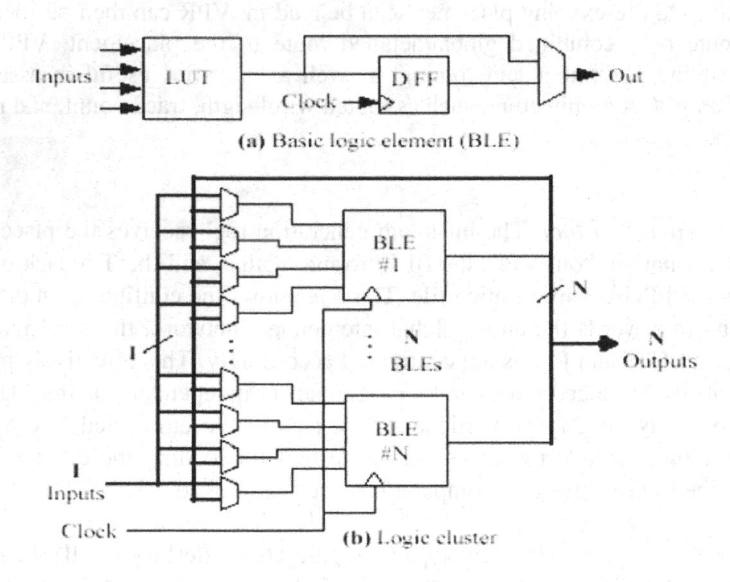

Fig. 4. Structure of Basic Logic Element (BLE) and Logic Cluster

LUT Inputs. A LUT with more inputs can implement more logic and hence fewer logic blocks are needed to implement a circuit. However, LUT complexity grows exponentially with the number of inputs. Ahmed in [10] indicates that a LUT with four inputs results in a more efficient area-delay product. But the main point for a low power approach is the one in [11] that proves that a LUT with four inputs, results in an FPGA with the minimum power consumption.

The D-FF. The D-FF can play an important role in the performance, density and power consumption of the BLE. So a careful design approach is essential. We use a double edge triggered FF to achieve a decrease in power consumption. Finally, a gated clock like the one in Fig. 5, should be used to prevent the unwanted FF transitions and therefore unwanted power consumption. Our simulation results with HSPICE and TSMC 0.18u process, indicates that a save in power consumption up to 60% can be obtained when the FF is idle.

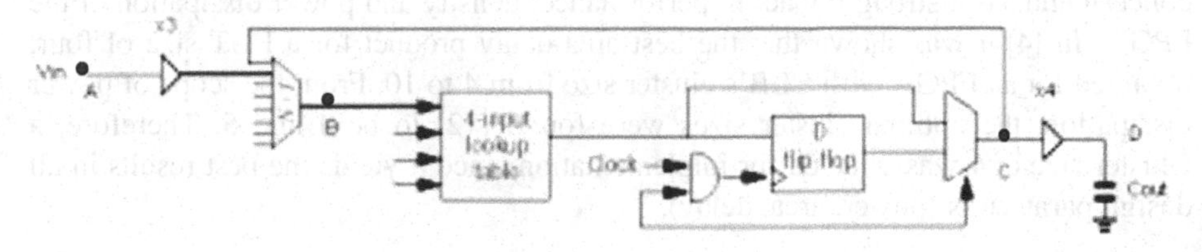

Fig. 5. Gated clock BLE

Table 1 gives the output capacitance and the delays in picoseconds from various points in the BLE architecture which can be seen in Fig.5. The total energy dissipated is 4,356×10^{-13} J.

Table 1. Output capacitance and delay of CLB

Cout(pF)	A to B(ps)	B to C(ps)	C to D(ps)	Total(ps)
0,1	377	340	281	998
0,5	377	340	1127	1844
0,75	377	340	1644	2361
1	377	340	2163	2880
3	377	340	5537	6254

Local Interconnect. From the various modes of Local Interconnection within the CLB, the fully connected one is selected. A fully connected CLB is one which every input of the CLB and every output of each BLE can connect to every input of all BLEs. This structure, with the addition of logical equivalence between the inputs and outputs of the CLB results in a much more flexible CLB, which requires fewer than the full number of inputs (N×K) to achieve high logic utilization.

CLB Inputs. The number of CLB Inputs (*I*) should be a function of *K* and the number of Cluster size (*N*). The larger the number of inputs, the larger and slower the multiplexers feeding the LUTs inputs and more configurable switches will be needed to connect to the logic block. Ahmed [10] shows that to achieve high logic utilization up to 98% for a fully connected CLB the required number of CLB inputs is:

$$I = \frac{k}{2} \times (N+1)\ . \tag{1}$$

Cluster Size. The Cluster Size, and thus the number of BLEs in a CLB is of great concern and has a strong impact in performance, density and power dissipation of the FPGA. In [4] it was shown that the best area-delay product for a LUT size of four, obtained for an FPGA with CLB's cluster size from 4 to 10. From the scope of power dissipation, the optimal cluster sizes were found [12] to be 4 and 5. Therefore, a Cluster size of 4 was selected for implementation since it yields the best results in all design parameters (power, area, delay).

CLB Interface. Another critical issue for the FPGA performance is the CLB I/O pin positioning around the CLB perimeter. It has been shown in [13] that the full perimeter pin positioning versus top/bottom pin positioning results in better performance.

Further exploration was attempted to specify the exact best positioning. Our experiment flow used some of the MCNC benchmarks circuits [14]. We tried the same architectures with different pins positioning. Fig.6 shows a Rounded Pins Positioning in which I/Os pins are distributed evenly around the CLB perimeter and Biased Pins Positioning in which Outputs indicated only at the right of the CLB. Our exploration showed that Rounded Pins Positioning results in a more routable FPGA and therefore better area, speed and power efficiency.

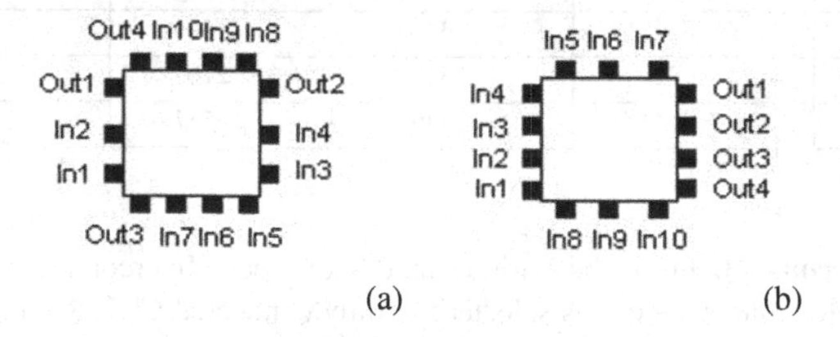

Fig. 6. Rounded (a) and Bias (b) Pins Positioning

Finally, because of the fully connected CLB with logical equivalence I/Os, there is no need to examine connections of I/O pins with specific BLE I/Os. Every connection is equivalent.

3.2 Interconnect Network Architecture

The Interconnect Network Architecture is possibly the most important parameter in the design of an FPGA. It dominates area and dissipates more power than the CLB. In addition, a good CLB could not work efficiently without appropriate interconnection. An interconnect network with low routability will result in an FPGA with small flexibility. On the other hand, high routability beyond necessity could result to large area, delay and power dissipation. We focused on three critical primitives of the interconnect network:

a) Segment Length
b) Connection Box Topology
c) Switch Box Topology

Segment Length. It has been found [11] that power dissipation increases with increase on the segment length. Also, [13] indicates that combination of 83% length four segments with pass transistor switches and 17% length eight segments with buffered switches yields the best delay product while combination 50% length four segments with pass transistor switches and 50% length eight segments with buffered switches gives the most area efficient FPGA. We must again point out that those experimental results were obtained for very large benchmarks. A more fine-grain approach might result to segment length two which is the best for power efficiency and area-delay product.

Connection Box Topology. Two are the major concerns regarding the Connection Box. The first is the type of the connections. The second is the fraction of tracks at which every I/O CLB pin can connect. These fractions are known as $F_{C,Input}$ and $F_{C,Output}$ for the inputs and the outputs respectively.

Since one output pin can drive at the same time several tracks those connections should be made with pass transistors. On the other hand different tracks could be connected to an input pin but only one can drive at a time this pin. So those connections should be made via multiplexers.

Betz [13] proved that a reduction in the FPGA area of 2% to 5% can be obtained, depending on the exact architecture, if $F_{C,Input}$ =0.5W and $F_{C,Output}$ =0.25W, where W is the number of tracks in the channel. The need for smaller $F_{C,Input}$ than $F_{C,Output}$ is obvious since the pass transistors connections require more transistors than the multiplexer (if we take the SRAM bits into account). Performing exploration, we adopted the same design parameters.

Switch Box Topology. Among the different Switch Box topologies, the disjoint one is found to be the least flexible. However, the Disjoint Switch Box requires only one switch connection for each wire that bypasses the Switch Box while other topologies require four. Because of that, this Switch Box results to more power efficient FPGAs when the segment length is greater than one [11]. Since, power consumption is considered critical, this switch box topology was selected.

4 Conclusions

A novel FPGA architecture was presented, with a complete tool chain that allows the mapping of logic described in a subset of RTL VHDL on this architecture. The tool chain is comprised of one commercial tool and a number of modified open-source and newly developed software tools.

References

1. Katherine Compton and Scott Hauck: Reconfigurable Computing: A Survey of Systems and Software. ACM Computing Surveys, Vol. 34, No. 2 (2002) 171–210
2. http://direct.xilinx.com/bvdocs/publications/ds003.pdf
3. http://www.altera.com/products/software/sfw-index.jsp
4. Vaughn Betz and Jonathan Rose: VPR: A New Packing, Placement and Routing Tool for FPGA Research. in Proc. of FPL '97, London, UK (1997) 213–222.
5. UCLA
6. http://www.mentor.com/leonardospectrum/datasheet.pdf
7. http://www.eecg.toronto.edu/~jayar/software/edif2blif/edif2blif.html
8. M. Sentovich, K. J. Singh, L. Lavagno, et al.: SIS: A System for Sequential Circuit Synthesis. UCB/ERL M92/41 (1992)
9. V. Betz and J. Rose: Cluster-Based Logic Blocks for FPGAs: Area-Efficiency vs. Input Sharing and Size. IEEE Custom Integrated Circuits Conference, Santa Clara, CA (1997) 551–554
10. Elias Ahmed, Jonathan Rose: The effect of LUT and cluster size on deep-submicron FPGA performance and density. FPGA 2000 (2000) 3–12
11. Kara KW Poon: Power Estimation for Field-Programmable Gate Arrays. Master of Applied Science Dissertation, University of British Columbia (2002)
12. V. Betz and J. Rose: Effect of the Prefabricated Routing Track Distribution on FPGA Area-Efficiency. IEEE Transactions on VLSI Systems (1998) 445–456
13. V. Betz and J. Rose: FPGA Routing Architecture: Segmentation and Buffering to Optimize Speed and Density. ACM/SIGDA International Symposium on Field Programmable Gate Arrays, Monterey, CA (1999) 59–68
14. MCNC benchmarks

Bit-Level Allocation for Low Power in Behavioural High-Level Synthesis*

María C. Molina, Rafael Ruiz Sautua[1], José M. Mendías, and Román Hermida

Dpto. Arquitectura de Computadores y Automática,
Universidad Complutense de Madrid,
c/ Juan del Rosal 8, 28040 Madrid, Spain,
{cmolinap, mendias, rhermida}@dacya.ucm.es [1] rsautua@acm.org

Abstract. An allocation algorithm at the bit-level specially suited for data dominated applications is presented. In addition to classical low power methods, it implements novel design strategies to reduce power consumption. These new features consist of the successive transformation of specification operations until a circuit implementation with minimum power consumption in functional and storage units is obtained. Thus, some of the specification operations are executed over a set of narrower functional units, linked by some glue logic to propagate partial results and carry signals as necessary. Due to the operation transformations performed, the types of the functional units may be different from those of the operations executed over them. Experimental results show a substantial power consumption reduction in the implementations obtained, compared to other low power algorithms. In addition, circuit areas are dramatically smaller.

1 Introduction

Historically power consumption has not been a crucial constraint in the semiconductor industry. Designers tried mainly to reduce both the area and the latency of circuits. Nowadays circuit area has been reduced thanks to high levels of integration, and performance has been improved with higher clock frequencies. Thus, circuit power consumption has augmented, becoming a parameter as important as area or delay.

The power dissipated by a circuit may be handled from every different abstraction level of the design, obtaining the greater reductions at the higher levels [1]. The total power dissipated in a CMOS circuit sums the static dissipation caused by leakage, to the dynamic dissipation caused mainly by charging and discharging load capacitance of gates. In High-Level Synthesis (HLS) most research has been focused on reducing the dynamic dissipation by decreasing transition activity in Functional Units (FUs), registers, multiplexers and buses [2-10]. A very effective method to reduce this transition activity is by increasing the correlation of input data, which in turn may be performed by changing operation scheduling [3-4], operation binding [6-8], loop pipelining [7], loop interchange, operand reordering [8], and operand sharing. In [9]

* Supported by Spanish Government Grant CICYT TIC-2002/750

J.J. Chico and E. Macii (Eds.): PATMOS 2003, LNCS 2799, pp. 617–627, 2003.
© Springer-Verlag Berlin Heidelberg 2003

authors present a low power method to select hardware (HW) resources. The binding algorithm presented in [10] divides some FUs into two parts, letting the computation of just one of these parts when the range of the input operands allows the execution of the operation over it. This design strategy, the execution of operations over a set of linked FUs, has also been used in HLS scheduling and allocation algorithms with the main purpose of minimizing the datapath area by increasing FUs reuse [11-13].

In this paper, we propose a bit-level allocation algorithm aiming to reduce the datapath power consumption. The algorithm takes into account the next features:

- In general the reduction of the number and width of the datapath FUs, registers, and routing resources results in lower power consumption. This reduction is achieved by the execution of some of the specification operations over a set of linked FUs, whose types may be different from the operation type. Register area is also reduced if some variables are stored in a set of several narrower registers during the same cycle, and also if several variables are stored in the same wider register during the same cycle.
- The application of the commutative property to find an appropriate input operand order may result in some switching activity reduction. This occurs when operations with shared operands are executed over the same FU, and no other operation is executed over this HW resource in between.
- In most datapaths, FUs that do not perform useful operations during some cycles may be found. The power consumption of these idle FUs could be reduced by latching their operands.
- In the circuits obtained by HLS we often find FUs that execute narrower operations. If some of these FUs are fragmented into several narrower ones, some of these new FUs will remain idle during some cycles. Thus, power consumption can be reduced by latching their operands. Due to FUs fragmentation some specification operations are executed over a set of linked FUs.

2 Proposed Algorithm

The proposed algorithm performs jointly the resource selection and binding in order to minimize the power consumption of the circuits obtained. Its current version treats in a special way both multiplications and *additive* operations (operations which may be transformed into additions), and uses a classical FU selection and binding algorithm (not shown in this paper) to deal with the remaining operations.

It takes as inputs both a circuit specification and a schedule proposed by any HLS tool. The output comprises a complete multiple precision datapath and a controller. The datapath is composed of a set of multipliers, a set of adders, a set of other *non-additive* FUs, some glue logic to link partial results and carry chains, a set of storage units, a set of multiplexers, and a set of latches. In order to minimize the number and width of the FUs, registers, and routing resources, the design strategies included lead to datapaths with the following features:

1. The sum of all multiplier widths is less or equal to the maximum number of bits multiplied simultaneously in a cycle.

$$\sum_{mul \in \text{DataPath}} \text{width}(mul) \leq \overset{\lambda}{\underset{i=1}{\text{Max}}}\left(\sum_{m \in OP_i^M} \text{width}(m)\right) \quad (1)$$

2. The sum of all adder widths is equal to the maximum number of bits added simultaneously in a cycle.

$$\sum_{add \in \text{DataPath}} \text{width}(add) = \overset{\lambda}{\underset{i=1}{\text{Max}}}\left(\sum_{a \in OP_i^A} \text{width}(a)\right) \quad (2)$$

3. The sum of all registers widths is equal to the maximum number of bits stored simultaneously in a cycle.

$$\sum_{reg \in \text{DataPath}} \text{width}(reg) = \overset{\lambda}{\underset{i=1}{\text{Max}}}\left(\sum_{v \in VS_i} \text{width}(v)\right) \quad (3)$$

4. The sum of all multiplexers widths is less or equal to the maximum number of bits transmitted simultaneously in a cycle.

$$\sum_{mux \in \text{DataPath}} \text{width}(mux) \leq \overset{\lambda}{\underset{i=1}{\text{Max}}}\left(\sum_{v \in VT_i} \text{width}(v)\right) \quad (4)$$

where λ is the circuit latency, OP_i^M is the set of multiplications scheduled in cycle i, OP_i^A is the set of additions scheduled in cycle i, VS_i is the set of variables stored in cycle i, and VT_i is the set of variables transmitted in cycle i.

These novel features result in datapaths where the number, type, and width of the HW resources used are in general independent of the circuit specification. The algorithm comprises the following seven phases, where the two most relevant ones (3 and 4) are explained in detail later on.

Phase 1. Commutative Property Application. The algorithm applies the commutative property to reorder the input operands of those specification operations with shared operands. This is performed for all operations with shared operands, independently of their types and execution cycles.

Obviously two operations of different types cannot be allocated to a same FU unable to execute operations of both types. But operations may be transformed to let the execution of operations of different types over a set of common FUs and some glue logic. Some of these transformations are performed in the next phase of the algorithm.

If only two operations sharing operands are scheduled in the same cycle, it becomes impossible to allocate both to the same FU. However, if a third operation (sharing any operand with the previous ones) were scheduled in a different cycle, then the best choice to maintain the number of different choices during the allocation (which implies some power consumption reduction) would be the application of the commutative property to all the three operations.

Phase 2. Common Kernel Extraction of Specification Operations. In order to increase the reuse of the datapath FUs (reducing their number and width in the proposed circuits), and thus lowering the circuit power consumption, the algorithm extracts the common operative kernel of the specification operations.

Additive operations (i.e. those with an additive kernel as subtractions, comparisons, maximum, minimum, absolute value, data limiters, code converters, etc) are transformed into a set of additions and some glue logic.

Multiplications are treated separately. Signed multiplications in the specification are transformed into a set of unsigned multiplications and additions. In [12] may be found some of the additive transformations applied during this phase, and how a two-complement signed multiplication may be transformed into one unsigned multiplication and two additions.

Phase 3. Multiplier Selection and Binding. At this stage the algorithm instances a set of unsigned multipliers, which will be in use to execute multiplications of its same width during *every* execution cycle. Some of the specification multiplications are allocated to the instanced multipliers, and in order to increase the number of multiplications with equal width, some others are transformed into a set of smaller ones and some additions. Hence some of these new multiplications are also allocated to the instanced multipliers. At the end of this phase every unallocated multiplication either present in the original specification or obtained from any of the multiplication transformations, is converted into a set of chained additions, to be allocated afterwards.

Phase 4. Adder Selection and Binding. The algorithm instances a set of adders able to execute all the additions. These additions may come from three different sources:

- the original specification,
- the kernel extraction of specification operations (phase 2),
- the transformation of the unallocated multiplications (phase 3).

In the given schedule if the number of bits added per cycle is different in any of its execution cycles, then some of the instanced adders will be partially unused.

Phase 5. Register Selection and Binding. A set of registers able to store every circuit variable is instanced. In order to minimize the number and width of registers, variables are allocated to them taking into account the next two features:

- a variable may be fragmented and each fragment stored in a different register in the same cycle if the sum of all register widths is greater or equal to the variable width,
- several variables may be stored in the same register in a certain cycle if the sum of all variable widths fits in the register width.

The allocation of variables is performed trying to minimize the number of routing resources. The algorithm selects, if variables lifetimes allow it, the same set of registers to store the right operands of those operations allocated to the same FUs, the same set of registers to store their left operands, and the same set to store their results.

Phase 6. Multiplexer Selection and Binding. At this stage the algorithm instances the set of multiplexers needed to execute every operation over the selected set of FUs, and to store every variable over the selected set of datapath registers.

Phase 7. Latch Selection and Binding. Latches are inserted at the inputs of FUs that remain idle during some cycles. Due to the policy followed to instance FUs, only adders may remain idle during any cycle. And due to the policy followed to allocate operations to FUs, just a reduced set of adders remain idle during several cycles. Therefore the number of latches inserted in the circuit is minimized.

In the next subsections the novel multiplier and adder selection and binding phases are explained in detail. To ease their understanding some concepts are first defined:

- *Multiplication order*: Be $m \geq n$ and $k \geq l$ then $m \times n > k \times l$ ($m \times n$ is bigger than $k \times l$) if ($m>k$) or ($m=k$ and $n>l$).
- *Occurrence* of width w in cycle c: number of operations of width w scheduled in cycle c.
- *Candidate*: set of operations formed by zero or one operation scheduled in every cycle. Many different operation alignments of every candidate formed by operations of different widths are possible. But in order to reduce the algorithm complexity we have only taken into account the alignments with the least significant bits of the operands aligned. Thus, if one operation is executed over a wider FU the most significant bits of the result produced are discarded.
- *Power reduction* of C candidate PR(C):

$$\mathrm{PR}(C) = \operatorname*{Max}_{i=1}^{\lambda}\left(\mathrm{Max}(\mathrm{OpeRight}(C,i), \mathrm{OpeLeft}(C,i))\right) \quad (5)$$

  where
  OpeRight(C, i): number of operations of C candidate which share the right operand with C candidate operation scheduled in cycle i.
  OpeLeft(C, i): number of operations of C candidate which share the left operand with C candidate operation scheduled in cycle i.
- Interconnection saving of C candidate IS(C):

$$\mathrm{IS}(C) = \mathrm{BitsOpe}(C) + \mathrm{BitsRes}(C) \quad (6)$$

  where
  BitsOpe(C): number of bits of C candidate left and right operands that may come from the same sources (lifetimes of variables do not overlap).
  BitsRes(C): number of bits of C candidate results that may be stored in the same register (their lifetimes do not overlap).

2.1 Multiplier Selection and Binding

In this phase the algorithm handles only the yet unallocated multiplications. It consists of a loop that finishes when no remaining unallocated multiplication is left in any cycle. Every iteration of this phase of the algorithm executes the next two steps:

Step 1. Multiplier Instanciation and Allocation. For every different multiplication width $m \times n$, the algorithm instances as many multipliers of that width as the minimum occurrence of width $m \times n$ per cycle. Next, the algorithm allocates operations to them. For every instanced multiplier of width $m \times n$, it calculates the candidates formed by as many multiplications of width $m \times n$ as the circuit latency, and the PR of every candidate. The algorithm allocates to every multiplier the operations of the candidate with greater PR. If several candidates have equal PR, the one with the greatest IS is finally allocated. As a result, every instanced multiplier executes one operation of its same width per cycle, and thus is never idle.

Step 2. Multiplication Fragmentation. This step is reached when it is not possible to instance a new multiplier of the same width as any of the yet unallocated multiplications (which may be in use to execute one operation of its same width per cycle).

A maximum of one multiplication per cycle is fragmented in order to obtain a multiplication fragment of width $m \times n$. It results in the final instance of a multiplier of that width in the next iteration of the loop. In those cycles already having one unscheduled multiplication of the fragment size $m \times n$, no multiplication is further fragmented. The operations to be fragmented are the biggest multiplications scheduled in every cycle (in accordance with the multiplication order defined previously). If there were several multiplications of that size in any cycle, then the algorithm would select the one sharing operands with the biggest number of multiplications scheduled in other cycles and already selected for fragmentation. The fragment size is the minimum of each of the operands widths of the selected operations to be fragmented, that is:

$$m \times n = \mathrm{Min}(l_1, l_2, l_3, \ldots) \times \mathrm{Min}(r_1, r_2, r_3, \ldots) \tag{7}$$

being l_i and r_i the widths of respectively the left and right operands of the biggest unallocated multiplication scheduled in cycle i.

Once both the operations to be fragmented and the desired fragment size have been chosen, the algorithm fragments the selected operations, obtaining at least one multiplication of the fragment size in each case. In the next paragraphs the different ways of having one $k \times l$ fragment from one $m \times n$ multiplication are explained.

There are many ways of fragmenting one unsigned multiplication into a set of narrower multiplications and additions. In particular, the number of different fragmentations of one $m \times n$ multiplication into five multiplications and four additions, obtaining one $k \times l$ fragment at least is $(m-k+1) \times (n-l+1) \times 16$ (being $m \geq k$ and $n \geq l$). The number of different fragmentations grows with the width of the operation to be fragmented, and decreases as the width of the operation fragment augments. In the above formula $(m-k+1) \times (n-l+1)$ is the number of different ways of choosing one $k \times l$ multiplication fragment from one $m \times n$ multiplication. And for each, 16 different fragmentations of the remaining part of the original multiplication may be found.

The calculus of all possible fragmentations requires exponential time, so our algorithm only takes into account a reduced set of them formed by the particular cases of the general method explained above, which fragment minimally the original multiplication. Fig. 1.a shows the set of fragmentations considered. Each of these 8 ways

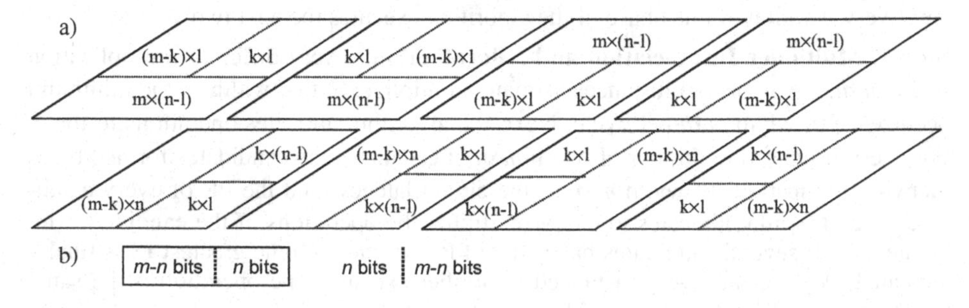

Fig. 1. a) Fragmentations considered by the algorithm to obtain one $k \times l$ fragment from one $m \times n$ multiplication, b) and to obtain one n bits addition fragment from one m bits addition

fragments the original multiplication into a set of 3 multiplications and 2 additions. From the 8 fragmentations explored by our algorithm, the one with the smallest cost in adders is selected. At the end of this phase, the unallocated multiplications are transformed into additions when no unallocated multiplication is left in any cycle

2.2 Adder Selection and Binding

This phase consists of a loop that finishes when every addition has been allocated. Each time the algorithm only considers the yet unallocated additions and their corresponding cycles. The following two steps are executed in every iteration.

Step1. Adder Instanciation and Allocation. For every different addition width the algorithm instances as many adders of that width as the minimum occurrence of that width per cycle (considering only those cycles with unallocated operations). Next, the algorithm allocates operations to them. For every instanced adder, it calculates the candidates formed by as many additions of that width as the number of cycles with unallocated additions, and calculates the PR of every candidate. The algorithm allocates to every adder the operations of the candidate with the greatest PR. If several candidates have equal PR, the one with the greatest IS is finally allocated. Some adders instanced in this step may be idle during some cycles (in fact the ones instanced in the last iterations).

Step 2. Addition Fragmentation. This step of the algorithm is only performed when it is not possible to instance a new adder of the same width as any of the yet unallocated additions (which may be in use to execute one operation of its same width in those cycles with unallocated additions).

A maximum of one addition per cycle (of those with unallocated additions) is fragmented to obtain one addition fragment of a certain width. It results in the final instance of an adder of that width in the next iteration of the loop. In those cycles already having one unscheduled addition of the fragment size, no addition is further fragmented. The operations to be fragmented are the biggest additions scheduled in each cycle. If we find several additions of that biggest size in any cycle, then the algorithm selects the one sharing operands with the biggest number of additions scheduled in other cycles and already selected for fragmentation. The fragment size is the minimum of the widths of the selected operations, i.e.

$$fragment\_width = \mathrm{Min}(w_1, w_2, w_3, \ldots) \tag{8}$$

being w_i the width of the biggest unallocated addition scheduled in cycle i.

Once both the operations to be fragmented and the desired fragment size have been chosen, the algorithm fragments the selected operations to obtain at least one addition of the fragment size in each case. The next paragraph explains the different ways of obtaining one n bits fragment from one m bits addition.

Many ways of fragmenting one addition into a set of narrower additions may be found. In particular, the number of different fragmentations of one m bits addition into three additions, obtaining at least one n bits fragment, is m-n+1 (being $m \geq n$). The number of different fragmentations grows with the width of the operation to be fragmented, and decreases as the width of the operation fragment augments. The calculus

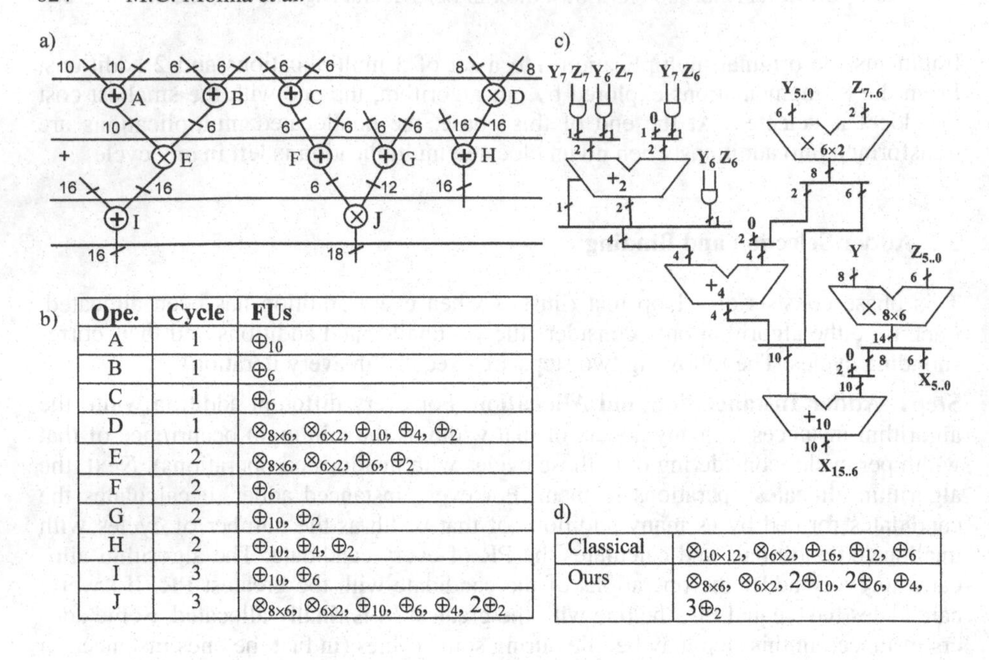

Ope.	Cycle	FUs
A	1	$\oplus_{10}$
B	1	$\oplus_{6}$
C	1	$\oplus_{6}$
D	1	$\otimes_{8\times6}, \otimes_{6\times2}, \oplus_{10}, \oplus_{4}, \oplus_{2}$
E	2	$\otimes_{8\times6}, \otimes_{6\times2}, \oplus_{6}, \oplus_{2}$
F	2	$\oplus_{6}$
G	2	$\oplus_{10}, \oplus_{2}$
H	2	$\oplus_{10}, \oplus_{4}, \oplus_{2}$
I	3	$\oplus_{10}, \oplus_{6}$
J	3	$\otimes_{8\times6}, \otimes_{6\times2}, \oplus_{10}, \oplus_{6}, \oplus_{4}, 2\oplus_{2}$

Classical	$\otimes_{10\times12}, \otimes_{6\times2}, \oplus_{16}, \oplus_{12}, \oplus_{6}$
Ours	$\otimes_{8\times6}, \otimes_{6\times2}, 2\oplus_{10}, 2\oplus_{6}, \oplus_{4}, 3\oplus_{2}$

Fig. 2. a) Specification and schedule, b) sets of FUs selected by our algorithm to execute every operation, c) set of FUs used to execute operation *D*, d) FUs of the datapaths proposed by a classical low power algorithm and ours

of all possible fragmentations requires exponential time, so our algorithm only takes into account a reduced set of them formed by the particular cases of the general method that fragment minimally the original addition. Fig. 1.b shows the 2 fragmentations considered to split the original addition into 2 new additions. The fragmentation that results in the best PR and IS is selected in every case.

Fig. 2 illustrates an example. Fig. 2.a shows a specification formed by 3 multiplications and 7 additions of different widths and a possible schedule of it. It has been synthesized by the proposed algorithm and a low power conscious one, which implements classical methods to reduce power consumption (operand reordering and operand retaining using latches). Fig. 2.d shows the set of FUs of the implementations obtained in both cases. Note that in our implementation the type, number, and width of these resources are independent of those of the specification operations. Fig. 2.b illustrates the set of FUs selected by our algorithm to execute every specification operation, where some additions are executed over a set of adders, and every multiplication is executed over a set of multipliers and adders. Fig. 2.c shows how operation *D* (X=Y×Z) is executed over 2 multipliers and 3 adders. This implementation saves around 33% in power consumption and 40% in area.

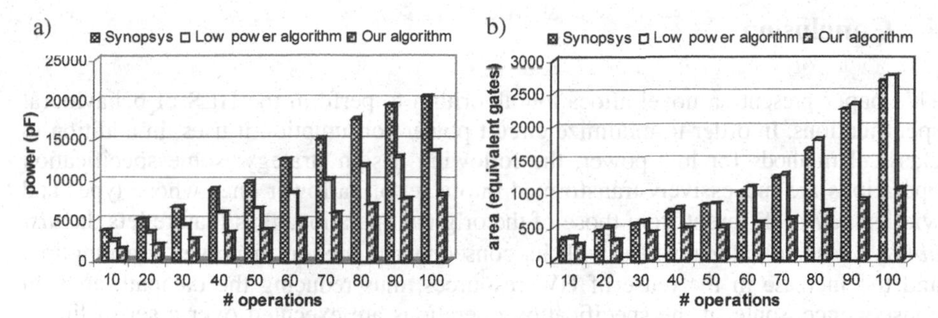

Fig. 3. a) Power consumption estimate of some synthesized circuits, b) circuits area

3 Experimental Results

In order to estimate the power consumption of the circuits synthesized by the proposed and other algorithms, the estimation model presented in [14] has been used. The power consumption estimates of the circuits produced by the proposed algorithm have been compared to those obtained by a low power conscious algorithm, which implements classical methods to reduce power consumption (operand reordering and operand retaining using latches), and to a conventional HLS algorithm (the one implemented in the commercial tool Synopsys Behavioral Compiler).

We have synthesized a wide range of behavioural specifications formed by *additive* operations and multiplications. Specification sizes ranged from 10 to 100 operations (about 40% were multiplications), and latencies varied from 5 to 50 cycles.

Results show that the power consumption estimates made for our circuits are always dramatically smaller than the other two. For the circuits synthesized, the average power consumption reduction achieved by our algorithm is around 45% compared to the implementations obtained by the low power conscious algorithm, and around 60% compared to the commercial tool. Fig 3.a shows the power consumption reductions achieved by the proposed algorithm for some of the examples synthesized.

Additionally, in order to compare the areas of the circuits given by the three algorithms, every implementation obtained has been described in VHDL, and its area (in number of equivalent gates) measured with Synopsys Behavioral Compiler. Results show that the areas of our approach are substantially smaller. It saves around 40% of total area compared to the low power conscious algorithm, and around 45% compared to the commercial tool. Fig 3.b shows the area saved by our algorithm for some of the examples synthesized.

Experimental results show also that the amount of power consumption and area reductions achieved by our approach grows in general with both the number of different operation types and widths in the specification, and the homogeneity of the operation to cycle distribution.

4 Conclusion

This paper presents a novel allocation algorithm to perform the HLS of behavioural specifications. In order to minimize circuit power consumption it uses, in addition to classical methods for low power, the following design strategy: some specification operations are successively transformed into a set of narrower ones whose types and widths may be different from those of the original operation. This feature lets the *disability* of some datapath FUs with its consequent reduction in power consumption, and the increase in the reuse of HW resources, thus reducing the datapath area. In consequence, some of the specification operations are executed over a set of linked FUs whose types and widths may be different from those of the original operations.

Experimental results show that the power consumption estimates of the circuits synthesized with this allocation algorithm are reduced substantially. In addition, the areas of the circuits proposed by our algorithm are also dramatically smaller. The amount of power consumption and area saved by our approach grows in general with both the number of different operation types and widths in the specification and the homogeneity of the operation to cycle distribution.

We are now working on the integration of this allocation algorithm with a bit-level scheduling one which, in addition to the capabilities shown in this paper, schedules fragments of a same operation in different cycles (non-necessarily consecutive ones). It promises additional power consumption and area reductions.

References

1 A. Chandrakasau, and R. Brodersen. "Low Power Digital CMOS Design". Kluwer Academic Publishers, 1995.
2 M. Ercegovac, D. Kirovski, and M. Potkonjak. "Low-power behavioural synthesis optimization using multiple precision arithmetic". In Proceedings of DAC, 1999.
3 J. Liu, P.H. Chou, N. Bagherzadeh, F. Kurdahi. "Power-Aware Scheduling under Timing Constraints for Mission-Critical Embedded systems". In Proceedings of DAC, 2001.
4 Y. Zhang, X.S. Hu, and D.Z. Chen. "Task Scheduling and Voltage Selection for Energy Minimization". In Proceedings of DAC, 2002.
5 L. Benini, A. Bogliolo, and C. De Micheli. "A survey of design techniques for system-level dynamic power management". IEEE Transactions on VLSI systems, June 2000.
6 E. Mussoll, and J. Cortadella. "High-level synthesis techniques for reducing the activity of functional units". In Proceedings of ISSS, 1999, pp. 99–104.
7 D. Kim, and K. Choi. " Power conscious high level synthesis using loop folding". In Proceedings of DAC, 1997, pp. 441–445.
8 D. Shin, and K, Choi, "Lower power high level synthesis by increasing data correlation". In Proceedings of ISLPED, 1997, pp. 441–445.
9 S. Gailhard, O. Sentieys, N. Julien, and E. Martin. "Area/Time/Power Space Exploration in Module Selection for DSP High Level Synthesis". In Proceedings of PATMOS, 1997.
10 J. Choi, J. Jeon, and K. Choi. "Power Minimization of Functional Units by Partially Guarded Computation". In Proceedings of ISLPED, 2000, pp. 131–136.
11 M.C. Molina, J.M. Mendías, and R. Hermida, "Multiple-Precision Circuits Allocation Independent of Data-Objects Length". In Proceedings of DATE, 2002.
12 M.C. Molina, J.M. Mendías, and R. Hermida, "Bit-level Scheduling of Heterogeneous Behavioural Specifications". In Proceedings of ICCAD, 2002.

13 M.C. Molina, J.M. Mendías, and R. Hermida, "High-Level Allocation to Minimize Internal Hardware Wastage". In Proceedings of DATE, 2003.
14 S. Katkoori, and R. Vemuri. "Architectural Power Estimation Based On Behavior Level Profiling". Journal on VLSI Design, Special Issue on Low Power, 1996.

Author Index